Thermochronological Methods: From Palaeotemperature Constraints to Landscape Evolution Models

Geological Society books refereeing procedures

The Society makes every effort to ensure that the scientific and production quality of its books matches that of its journals. Since 1997, all book proposals have been refereed by specialist reviewers as well as by the Society's Books Editorial Committee. If the referees identify weaknesses in the proposal, these must be addressed before the proposal is accepted.

Once the book is accepted, the Society Book Editors ensure that the volume editors follow strict guidelines on refereeing and quality control. We insist that individual papers can only be accepted after satisfactory review by two independent referees. The questions on the review forms are similar to those for *Journal of the Geological Society*. The referees' forms and comments must be available to the Society's Book Editors on request.

Although many of the books result from meetings, the editors are expected to commission papers that were not presented at the meeting to ensure that the book provides a balanced coverage of the subject. Being accepted for presentation at the meeting does not guarantee inclusion in the book.

More information about submitting a proposal and producing a book for the Society can be found on its web site: www.geolsoc.org.uk.

It is recommended that reference to all or part of this book should be made in one of the following ways:

LISKER, F., VENTURA, B. & GLASMACHER, U. A. (eds) 2009. *Thermochronological Methods: From Palaeotemperature Constraints to Landscape Evolution Models.* Geological Society, London, Special Publications, **324**.

GLEADOW, A. J. W., GLEADOW, S. J., BELTON, D. X., KOHN, B. P., KROCHMAL, M. S. & BROWN, R. W. 2009. Coincidence mapping – a key strategy for the automatic counting of fission tracks in natural minerals. *In*: LISKER, F., VENTURA, B. & GLASMACHER, U. A. (eds) *Thermochronological Methods: From Palaeotemperature Constraints to Landscape Evolution Models.* Geological Society, London, Special Publications, **324**, 25–36.

GEOLOGICAL SOCIETY SPECIAL PUBLICATION NO. 324

Thermochronological Methods: From Palaeotemperature Constraints to Landscape Evolution Models

EDITED BY

F. LISKER
Universität Bremen, Germany

B. VENTURA
Universität Bremen, Germany

and

U. A. GLASMACHER
Ruprecht-Karls-Universität Heidelberg, Germany

2009
Published by
The Geological Society
London

THE GEOLOGICAL SOCIETY

The Geological Society of London (GSL) was founded in 1807. It is the oldest national geological society in the world and the largest in Europe. It was incorporated under Royal Charter in 1825 and is Registered Charity 210161.

The Society is the UK national learned and professional society for geology with a worldwide Fellowship (FGS) of over 9000. The Society has the power to confer Chartered status on suitably qualified Fellows, and about 2000 of the Fellowship carry the title (CGeol). Chartered Geologists may also obtain the equivalent European title, European Geologist (EurGeol). One fifth of the Society's fellowship resides outside the UK. To find out more about the Society, log on to www.geolsoc.org.uk.

The Geological Society Publishing House (Bath, UK) produces the Society's international journals and books, and acts as European distributor for selected publications of the American Association of Petroleum Geologists (AAPG), the Indonesian Petroleum Association (IPA), the Geological Society of America (GSA), the Society for Sedimentary Geology (SEPM) and the Geologists' Association (GA). Joint marketing agreements ensure that GSL Fellows may purchase these societies' publications at a discount. The Society's online bookshop (accessible from www.geolsoc.org.uk) offers secure book purchasing with your credit or debit card.

To find out about joining the Society and benefiting from substantial discounts on publications of GSL and other societies worldwide, consult www.geolsoc.org.uk, or contact the Fellowship Department at: The Geological Society, Burlington House, Piccadilly, London W1J 0BG: Tel. +44 (0)20 7434 9944; Fax +44 (0)20 7439 8975; E-mail: enquiries@geolsoc.org.uk.

For information about the Society's meetings, consult *Events* on www.geolsoc.org.uk. To find out more about the Society's Corporate Affiliates Scheme, write to enquiries@geolsoc.org.uk.

Published by The Geological Society from:
The Geological Society Publishing House, Unit 7, Brassmill Enterprise Centre, Brassmill Lane, Bath BA1 3JN, UK

(*Orders*: Tel. +44 (0)1225 445046, Fax +44 (0)1225 442836)
Online bookshop: www.geolsoc.org.uk/bookshop

British Library Cataloguing in Publication Data

A catalogue record for this book is available from the British Library.
ISBN 978-1-86239-285-4

Typeset by Techset Composition Ltd, Salisbury, UK
Printed by CPI Antony Rowe, Chippenham, UK

Distributors

North America
For trade and institutional orders:
The Geological Society, c/o AIDC, 82 Winter Sport Lane, Williston, VT 05495, USA
Orders: Tel. +1 800-972-9892
Fax +1 802-864-7626
E-mail: gsl.orders@aidcvt.com

For individual and corporate orders:
AAPG Bookstore, PO Box 979, Tulsa, OK 74101-0979, USA
Orders: Tel. +1 918-584-2555
Fax +1 918-560-2652
E-mail: bookstore@aapg.org
Website: http://bookstore.aapg.org

India
Affiliated East-West Press Private Ltd, Marketing Division, G-1/16 Ansari Road, Darya Ganj, New Delhi 110 002, India
Orders: Tel. +91 11 2327-9113/2326-4180
Fax +91 11 2326-0538
E-mail: affiliat@vsnl.com

Contents

Preface vii

LISKER, F., VENTURA, B. & GLASMACHER, U. A. Apatite thermochronology in modern geology 1

Part I: New approaches in thermochronology

GLEADOW, A. J. W., GLEADOW, S. J., BELTON, D. X., KOHN, B. P., KROCHMAL, M. S. & BROWN, R. W. Coincidence mapping – a key strategy for the automatic counting of fission tracks in natural minerals 25

HASEBE, N., CARTER, A., HURFORD, A. J. & ARAI, S. The effect of chemical etching on LA–ICP-MS analysis in determining uranium concentration for fission-track chronometry 37

DOBSON, K. J., PERSANO, C. & STUART, F. M. Quantitative constraints on mid- to shallow-crustal processes using the zircon (U–Th)/He thermochronometer 47

CARTER, A. & FOSTER, G. L. Improving constraints on apatite provenance: Nd measurement on fission-track-dated grains 57

MURRELL, G. R., SOBEL, E. R., CARRAPA, B. & ANDRIESSEN, P. Calibration and comparison of etching techniques for apatite fission-track thermochronology 73

TIMAR-GENG, Z., HENK, A. & WETZEL, A. Convective heat transfer in a steeply dipping fault zone and its impact on the interpretation of fission-track data – a modelling study 87

WANG, W. & ZHOU, Z. Reconstruction of palaeotopography from low-temperature thermochronological data 99

Part II: Applied thermochronology – long-term evolution studies

GLOTZBACH, C., SPIEGEL, C., REINECKER, J., RAHN, M. & FRISCH, W. What perturbs isotherms? An assessment using fission-track thermochronology and thermal modelling along the Gotthard transect, Central Alps 111

DUNKL, I., FRISCH, W., KUHLEMANN, J. & BRÜGEL, A. Pebble population dating as an additional tool for provenance studies – examples from the Eastern Alps 125

MALUSÀ, M. G., ZATTIN, M., ANDÒ, S., GARZANTI, E. & VEZZOLI, G. Focused erosion in the Alps constrained by fission-track ages on detrital apatites 141

DEL RÍO, P., BARBERO, L. & STUART, F. M. Exhumation of the Sierra de Cameros (Iberian Range, Spain): constraints from low-temperature thermochronology 153

XU, C., MANSY, J. L., VAN DEN HAUTE, P., GUILLOT, F., ZHOU, Z., CHEN, J. & DE GRAVE, J. Late- and post-Variscan evolution of the Ardennes in France and Belgium: constraints from apatite fission-track data 167

VENTURA, B., LISKER, F. & KOPP, J. Thermal and denudation history of the Lusatian Block (NE Bohemian Massif, Germany) as indicated by apatite fission-track data 181

KOHN, B. P., LORENCAK, M., GLEADOW, A. J. W., KOHLMANN, F., RAZA, A., OSADETZ, K. G. & SORJONEN-WARD, P. A reappraisal of low-temperature thermochronology of the eastern Fennoscandia Shield and radiation-enhanced apatite fission-track annealing 193

KUHLEMANN, J., KRUMREI, I., DANIŠÍK, M. & VAN DER BORG, K. Weathering of granite and granitic regolith in Corsica: short-term ^{10}Be versus long-term thermochronological constraints 217

DE GRAVE, J., BUSLOV, M. M., VAN DEN HAUTE, P., METCALF, J., DEHANDSCHUTTER, B. & MCWILLIAMS, M. O. Multi-method chronometry of the Teletskoye graben and its basement, Siberian Altai Mountains: new insights on its thermo-tectonic evolution 237

DASZINNIES, M. C., JACOBS, J., WARTHO, J.-A. & GRANTHAM, G. H. Post Pan-African thermo-tectonic evolution of the north Mozambican basement and its implication for the Gondwana rifting. Inferences from $^{40}Ar/^{39}Ar$ hornblende, biotite and titanite fission-track dating 261

KOUNOV, A., VIOLA, G., DE WIT, M. & ANDREOLI, M. A. G. Denudation along the Atlantic passive margin: new insights from apatite fission-track analysis on the western coast of South Africa 287

RUIZ, G. M. H., CARLOTTO, V., VAN HEININGEN, P. V. & ANDRIESSEN, P. A. M. Steady-state exhumation pattern in the Central Andes – SE Peru 307

EMMEL, B., JACOBS, J. & DASZINNIES, M. C. Combined titanite and apatite fission-track data from Gjelsvikfjella, East Antarctica – another piece of a concealed intracontinental Permo-Triassic Gondwana rift basin? 317

YAMADA, R., ONGIRAD, H., MATSUDA, T., OMURA, K., TAKEUCHI, A. & IWANO, H. Fission-track analysis of the Atotsugawa Fault (Hida Metamorphic Belt, central Japan): fault-related thermal anomaly and activation history 331

Index 339

Preface

This volume comprises a selection of papers presented on the European Conference on Thermochronology 2006 in Bremen. The conference is part of a series of international and European meetings that were initiated in 1980 as International Workshops on Fission-Track Dating, and have continued since 2000 as conferences on Fission-Track Dating and Thermochronology. The change of the meeting titles testifies to the increasing popularity and importance of this field in various respects: the meetings grew from small workshops with a limited number of participants to large conferences, and the discipline Thermochronology (Thermochronometry) developed from merely fission-track dating to a broad field incorporating a whole spectrum of low-temperature radiometric methods. This evolution of thermochronology from a hobbyhorse of nuclear physicists almost 50 years ago to one of the most innovative methods within modern Earth sciences was dignified in three remarkable and emotional talks during a special session of the Bremen conference. John Garver (Union College, Schenectady) commenced this historic review with an introduction into the early work of Fleischer, Price and Walker who recognized the potential of lattice damages in minerals as age indicators, while Peter Van den Haute (University of Ghent) referred to the first geological applications of fission-track dating and the concept of the partial annealing zone (PAZ). Barry Kohn (University of Melbourne) reviewed the last decade of fission-track and (U–Th–Sm)/He thermochronology with its increasing influence on various geological disciplines. This Special Publication intends to continue from this point as it not only provides a variety of applied studies but shows some exciting progress with respect to improving existing and developing new thermochronological techniques.

Accordingly, the 21 papers of this Special Publication are arranged into two sections following an introductory review chapter by **Lisker *et al.*** This opening contribution provides an overview of the application of thermochronological methods to geological problems.

The first section, *New approaches in thermochronology*, comprises seven papers that present the most recent developments in thermochronological methods, and show the growing interest in alternative thermochronometers and modelling techniques. It starts with a contribution by **Gleadow *et al.*** who present the first fully automated counting system for fission tracks in natural minerals – for long only a chimera in fission-track analysis. **Hasebe *et al.*** promote the use of laser ablation inductively coupled plasma mass spectrometry (ICP-MS) to directly measure uranium concentration for both apatite and zircon fission-track chronometry as an alternative to traditional time-consuming thermal neutron-induced fission tracks. **Dobson *et al.*** present a method for determining zircon (U–Th)/He ages, and show the potential of the zircon He chronometer to constrain thermal histories, and to quantify cooling in different tectonic settings. This new thermometer may fill the 'gap' between $^{40}Ar/^{39}Ar$ and zircon fission-track ages and apatite fission-track data. **Carter & Foster** describe a new approach in detrital thermochronology based on measuring Nd isotopic compositions on single apatites by laser ablation ICP-MS, and show how the combination of $^{143}Nd/^{144}Nd$ ratios with apatite fission-track ages provide better constraints on the source rock. **Murrel *et al.*** verify the influence of etching on the kinetic parameter *D*par, and therefore on the thermal history modelling of apatite fission-track data. **Timar-Geng *et al.*** analyse the influence of hydrothermal fluids circulating in fault zones on fission-track ages by numerical modelling techniques. **Wang & Zhou** propose a simple method to reconstruct past relief using thermochronological data.

The 14 papers of the second section, *Applied thermochronology*, refer to the long-term landscape evolution of various geological settings, to provenance studies and to small-scale tectonic processes. Eight papers describe Phanerozoic geological processes in Europe, with the first three of them being Alpine studies. **Glotzbach *et al.*** investigate the influence of the topography on isotherms on the example of the Gotthard Massif, whereas **Dunkl *et al.*** and **Malusà *et al.*** present provenance studies. Based on fission-track data of pebbles and detrital apatites from modern river sediments, respectively, they refine the denudation history of specific source regions within the Alpine orogen. **del Río *et al.*** report the thermal and denudation history of the Iberian Range in NW Spain. **Xu *et al.*** and **Ventura *et al.*** examine the post-Variscan evolution in Central Europe, with special focus on the Ardennes, and on the northern Bohemian Massif, respectively. **Kohn *et al.*** address the problem of apatite fission-track and (U–Th–Sm)/He cross-over ages on examples of the Fennoscandian Shield and similar settings in southern Canada and Western Australia. Their arguments against substantial influence of α-radiation-enhanced annealing of fission tracks at low temperatures contribute to an ongoing debate. Combining apatite fission-track and terrestrial cosmogenic nuclide data, **Kuhlemann *et al.*** study the dependency of weathering rates of granite and regolith from precipitation and brittle deformation in Corsica.

The remaining six papers present a number of study cases from various geological settings from other continents. **De Grave *et al.*** reconstruct the thermo-tectonic history of the Siberian Altai basement from the Early Palaeozoic to the present using a multi-method chronometric approach. Two multi-thermochronometer-based African studies of **Daszinnies *et al.*** and **Kounov *et al.*** focus on rifting processes and eventually the dispersal of Gondwana in the vicinity of Mozambique and South Africa, and conclude on the evolution of the respective passive margin segments. **Ruiz *et al.*** apply apatite and zircon fission-track analyses on plutonic rocks from the Western Cordillera of Peru to verify the relationship between uplift and denudation within this part of the Andes. **Emmel *et al.*** concentrate on the thermal and denudation history of Dronning Maud Land (East Antarctica), and relate its landscape development to tectonic activity. **Yamada *et al.*** use apatite and zircon fission-track data to detect an ancient thermal anomaly associated with fault displacements within the regional Atotsugawa Fault of central Japan.

The editors are very grateful to the Geological Society of London for making this book possible, in particular to Angharad Hills for her constant support, and to Tamzin Anderson, Helen Floyd-Walker and Phil Leat. We would like to thank the following friends and colleagues and six anonymous referees who kindly reviewed the articles submitted to this book: P. Andriessen, J. Barbarand, L. Barbero, M. Bernet, P. Bierman, A. Blythe, J. Braun, M. Brix, B. Carrapa, A. Carter, C. Cederbom, P. Clift, I. Dunkl, G. Fellin, P. Fitzgerald, R. Galbraith, K. Gallagher, U. A. Glasmacher, A. Gleadow, A. Hartley, A. Henk, A. Hurford, J. Jacobs, R. Jonkheere, P. Kamp, R. Ketcham, P. Koons, F. Lisker, L. Nasdala, J.-O. Näslund, P. O'Sullivan, C. Persano, M. Raab, M. Rahn, L. Ratschbacher, G. Ruiz, C. Schlüchter, E. Sobel, F. Stuart, K. Stüwe, R. Thiede, P. Van den Haute, P. Van der Beek, B. Ventura, G. Viola, G. Wagner, M. Wipf and M. Zattin.

F. LISKER
B. VENTURA &
U. A. GLASMACHER

Apatite thermochronology in modern geology

F. LISKER[1]*, B. VENTURA[1] & U. A. GLASMACHER[2]

[1]*Fachbereich Geowissenschaften, Universität Bremen, PF 330440, 28334 Bremen, Germany*

[2]*Institut für Geowissenschaften, Ruprecht-Karls-Universität Heidelberg, Im Neuenheimer Feld 234, 69120 Heidelberg, Germany*

**Corresponding author (e-mail: flisker@uni-bremen.de)*

Abstract: Fission-track and (U–Th–Sm)/He thermochronology on apatites are radiometric dating methods that refer to thermal histories of rocks within the temperature range of 40°–125 °C. Their introduction into geological research contributed to the development of new concepts to interpreting time-temperature constraints and substantially improved the understanding of cooling processes within the uppermost crust. Present geological applications of apatite thermochronological methods include absolute dating of rocks and tectonic processes, investigation of denudation histories and long-term landscape evolution of various geological settings, and basin analysis.

Thermochronology may be described as the quantitative study of the thermal histories of rocks using temperature-sensitive radiometric dating methods such as $^{40}Ar/^{39}Ar$ and K–Ar, fission track, and (U–Th)/He (Berger & York 1981). Amongst these different methods, apatite fission track (AFT) and apatite (U–Th–Sm)/He (AHe) are now, perhaps, the most widely used thermochronometers as they are the most sensitive to low temperatures (typically between 40 and *c.* 125 °C for durations of heating and cooling in excess of 10^6 years), ideal for investigating the tectonic and climate-driven surficial interactions that take place within the top few (<5 km) kilometres of the Earth's crust. These processes govern landscape evolution, influence climate and generate the natural resources essential to the wellbeing of mankind.

This introductory chapter provides a brief overview of apatite thermochronology and its application to geological studies. We focus on three topics: (1) methodological developments; (2) concepts and strategies for the interpretation of thermochronological data; and (3) applications to various geodynamic settings. For more detailed insights on apatite thermochronology the reader is referred to published reviews by Green *et al.* (1986, 1989*b*), Laslett *et al.* (1987), Duddy *et al.* (1988), Wagner & Van den Haute (1992), R. W. Brown *et al.* (1994), Gallagher *et al.* (1998), Gleadow *et al.* (2002), Ehlers & Farley (2003) and Reiners & Brandon (2006).

Fission-track thermochronology

Basics of the method

Fission-track thermochronology/-chronometry (for differentiation cf. Reiners *et al.* 2005) is based on the analysis of radiation damage trails ('fission tracks') in uranium-bearing, non-conductive minerals and glasses. It is routinely applied on the minerals apatite, zircon and titanite. Fission tracks are produced continuously through geological time as a result of the spontaneous fission of ^{238}U atoms. They are submicroscopic features with an initial width of approximately 10 nm and a length of up to 20 μm (Paul & Fitzgerald 1992) that can be revealed by chemical etching. Crucially, fission tracks are semi-stable features that can self-repair (shorten and eventually disappear) by a process known as annealing at a rate that is a function of both time and temperature. The extent of any track shortening (exposure to elevated temperatures) in a sample can be quantified by examining the distribution of fission-track lengths.

The determination of a fission-track age (a number that relates to the observable track density) depends on the same general equation as any radioactive decay scheme: it requires an estimate of the relative abundance of the parent isotope and of the daughter product. However, unlike most methods of radiometric dating, it measures the effect, rather than the product, of a radioactive decay scheme, that is it refers to the number of ^{238}U atoms and the number of spontaneous fission tracks per unit volume. This fission-track density is obtained by counting the number of spontaneous tracks intersecting a polished internal surface of a mineral grain viewed under high magnification (1000×–1250×) using an optical microscope. Depending on the sample and aims of the study, a typical fission-track sample age consists of a weighted mean of 20–100 single-grain ages. Further details and background information on practical aspects of fission-track age determination are provided by, for example, Fleischer *et al.* (1975), Naeser (1979) and Donelick *et al.* (2005).

From: LISKER, F., VENTURA, B. & GLASMACHER, U. A. (eds) *Thermochronological Methods: From Palaeotemperature Constraints to Landscape Evolution Models*. Geological Society, London, Special Publications, **324**, 1–23.
DOI: 10.1144/SP324.1 0305-8719/09/$15.00

An important dimension to fission-track thermochronology is the semi-stable nature of tracks, whereby annealing can change the significance of a measured age. The observable density of spontaneous tracks in a sample (age) is a function of track length (probability of intersecting the plane of observation). All newly formed tracks in apatite have a length of approximately 16 μm (*c.* 11 μm in zircon). If a sample (e.g. a volcanic apatite) was created at 10 Ma and then resided at low temperatures (<40 °C), the population of tracks reduces in length to a mean value of approximately 15 μm causing an insignificant (not resolvable) reduction in track density and a measured age within an error of 10 Ma. However, if, during its history, the same sample experienced elevated temperatures (but not sufficient to cause total resetting) in its history there will be significant track shortening to a level defined by the maximum heating. This will cause a reduction in observable track density and, therefore, measurable age.

Some milestones in the evolution of the fission-track method

Despite early recognition of fission tracks (Baumhauer 1894; Silk & Barnes 1959), it was not until the early 1960s that their application to geological dating was first proposed (Price & Walker 1963) and subsequently developed (Wagner 1966; Gentner *et al.* 1967; Naeser 1967). Early dating studies were tasked with finding practical ways to etch tracks, measure uranium contents that mirror the dated grain and define the time–temperature stability fields of fission tracks in different uranium-bearing minerals, tektites and glasses. Studies conducted between 1970 and 1983 highlighted the fundamental issues that needed to be resolved in order to enable routine and accurate age determination. Foremost was a lack of consensus on the value of the spontaneous fission decay constant for ^{238}U, to be used in the equation for age calculation. In order to circumvent this and other fundamental problems associated with the measurement of neutron fluence, Hurford & Green (1983) advanced a suggestion made by Fleischer and Hart at a meeting in Austria in 1971 for a comparative approach to AFT dating through the use of a proportionality constant. The resultant 'Zeta' calibration method (Hurford & Green 1983) has become the standard approach to fission-track age determination (Hurford 1990).

Until 1980 most fission-track ages were calculated using a pooled age, based on the ratio of total counts of spontaneous and induced tracks. However, Green (1981) highlighted the need for an alternative method for calculating an age when there is significant (or over-) dispersion within the population of single-grain ages. Green pointed out that where there is evidence for heterogeneity (extra Poissonian variation) within a dataset, detected by statistical tests such as χ^2 (Galbraith 1981), the conventional pooled age, based on the ratio of the number of spontaneous and induced tracks (N_s/N_i), with its Poisson standard error, becomes meaningless. An alternative approach based on the mean of ratios of individual track densities (ρ_s/ρ_i) was used to give a larger estimate of the error to allow for an extra-Poisson component in the dispersion of single-grain ages, but this approach implied a sample should have a single age. In reality, there are a number of different causes of heterogeneity within a dataset (beyond a bad experiment), such as variable responses to partial resetting due to variations in apatite grain composition (see later) and/or a range of provenance ages. Thus, it is important to assess to what the over-dispersion is due rather than make to allowance for it with larger errors.

In 1984 Hurford *et al.* proposed the use of probability density diagrams, a type of continuous histogram that plots each grain-age error as a Gaussian density function, as a way of visualizing a mixed age dataset. However, this type of approach can obscure useful information by inappropriately weighting it with poor information (i.e. an overlap effect associated with broad, imprecise, peaks). To overcome this problem Galbraith developed the radial plot (Galbraith 1988, 1990) (Fig. 1), which is now routinely used across the chronological community. Coupled to this development Galbraith & Laslett (1993) also produced the widely adopted random effects model that gives a central age estimate of the population of grains ages with a relative standard deviation of the population of ages known as the age dispersion (normally expressed as a percentage variation).

Having established protocols to measure fission-track ages and assess data quality and structure, the next major developments were related to understanding the significance of the determined ages. Whilst track-length measurement had been used to detect track annealing more or less since the methods inception, it was not until the mid 1980s that studies demonstrated the utility of such data. Advances included moving away from using semi-tracks (projected track) to length measurement based on surface-parallel confined tracks. Although more numerous, semi-tracks contain less information and have significant sources of bias, particularly towards longer lengths (Laslett *et al.* 1994). A key paper by Gleadow *et al.* (1986*b*) demonstrated the utility of confined track measurement and laid the foundation for data interpretation based on thermal history.

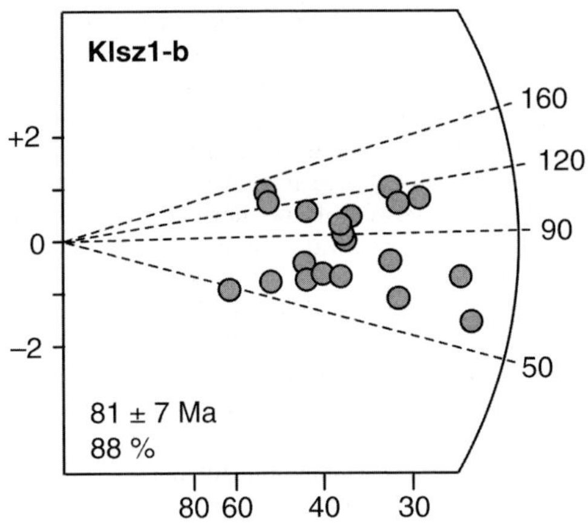

Fig. 1. Radial plots were designed to graphically display single-grain age estimates, taking into account different standard errors (Galbraith 1988, 1990; example from Ventura *et al.* 2009). Single-grain ages (z) with standard error, σ, are plotted (point x, y) according to x (precision) $= 1/\sigma$ and y (standard estimate) $= (z - z_0)/\sigma$, where z_0 is the central age. The error attached to each point is standardized on the y-scale. The value of the age (z) and the 2σ uncertainty can be read off the z-scale by extrapolating lines from point 0, 0 through the plotted age (point x, y) and projecting onto the radial age axis.

Contrary to other radiometric dating methods where the measured age equates to the time a sampled cooled below its closure temperature, the fission-track age records cooling through a temperature interval between total resetting of fission tracks and relative stability, known as the partial annealing zone or PAZ (Wagner 1979). The temperatures relating to the PAZ were defined by systematic investigation of the annealing of fission tracks across laboratory and geological timescales, monitored by changes in confined track-length distribution (Green & Durrani 1977; Gleadow & Duddy 1981; Laslett *et al.* 1982, 1987; Gleadow *et al.* 1986*a*; Green *et al.* 1986, 1989*a*, *b*; Duddy *et al.* 1988; Green 1988; Crowley *et al.* 1991; Carlson *et al.* 1999; Barbarand *et al.* 2003). Quantification of the time–temperature conditions that control the annealing of fission tracks in apatite provide the means to interpret fission track-ages by linking the level of track-length shortening and density (age) reduction to sample thermal history (e.g. Gleadow *et al.* 1986*b*). Typical temperature ranges for the PAZ for heating durations of 10^7 Ma are 60–110 °C for fluorapatite (e.g. Gleadow & Duddy 1981), 170–330 °C for zircon (Zaun & Wagner 1985; Yamada *et al.* 1995) and 265–310 °C for titanite (Coyle & Wagner 1998). Above the upper-temperature values fission tracks undergo total annealing, which removes all traces of fission tracks from the host crystal lattice.

Annealing studies recognized that track fading in apatite is not only governed by temperature, but also by heating duration, chemical composition and crystallographic orientation. Early descriptions of time-dependence on annealing were presented as Arrhenius plots, which generally had a fanning like form. Green *et al.* (1985) noted that fanning plots were probably the result of variable apatite grain composition, such that in effect these fanning plots were no more than a series of overlapping parallel plots. A series of isothermal fission-track annealing studies on apatites (Green *et al.* 1986; Crowley *et al.* 1990; Carlson *et al.* 1999; Barbarand *et al.* 2003) showed that substitution of fluoride and hydroxide ions by chloride ions appears to induce the greatest effects, although other substitutions on the halogen site can also have an influence but have proven much harder to deconvolve from laboratory timescale track-annealing experiments (O'Sullivan & Parrish 1995; Carlson *et al.* 1999). As a consequence of these studies, AFT data interpretation generally takes into account grain composition either by direct measurement of the Cl-content [generally by electron probe micro-analyser (EPMA)] or by assessing grain bulk composition by measuring the solubility of apatite through the *c*-axis parallel length of track etch pits (D_{par}® of Donelick 1991; Burtner *et al.* 1994; cf. also Murrell *et al.* 2009). Green & Durrani (1977) first described the influence of the crystallographic orientation of spontaneous fission tracks on the annealing of those tracks. Tracks orthogonal to the *c*-axis anneal more rapidly than tracks parallel to the *c*-axis (Green 1988). This anisotropy increases with annealing (Green 1981; Laslett *et al.* 1984; Donelick *et al.* 1990; Galbraith *et al.* 1990; Donelick 1991).

The developments outlined above (and many others not cited) have helped establish AFT analysis as a routinely used tool in geological studies, examples of which are published in this Special Publication. Methodological developments continue and it is now possible, following marked improvements in the precision of laser ablation inductively coupled plasma mass spectrometry (LA-ICP-MS) technology, to make direct measurements of uranium content within the counted grains of apatite, providing a fast and precise alternative to neutron irradiation (e.g. Hasebe *et al.* 2004, 2009). A further advance is the arrival of a functioning system for automatically determining the fission-track density and measuring fission-track lengths (Gleadow *et al.* 2009). This will not only increase speed and reliability of data acquisition, but also provides new options, especially with respect to the size (quality) of density and track length datasets.

(U–Th–Sm)/He thermochronology

The last decade has seen the extension of low-temperature studies to include AHe dating. Radioactive decay of the elements uranium and thorium to stable helium represents one of the earliest radiometric schemes available to geologists to investigate the ages of rocks and minerals (Rutherford 1905). Despite initial promise (e.g. Keevil 1943), the (U–Th)/He method was discounted as a dating technique until Lippolt *et al.* (1982) proposed its application as a low-temperature thermochronological tool. Zeitler *et al.* (1987) followed Lippolt *et al.*'s suggestion that, rather than defining a sample's age (implicitly its formation age), (U–Th)/He data provide useful constraints on the sample's thermal history, similar to AFT analysis, and further developed the technique. Fundamental experimental work by Farley and co-workers at Caltech through the 1990s resulted in the development of accurate and precise instrumentation for the extraction and measurement of helium. In practice, an AHe age is obtained by measuring radiogenic ^{4}He trapped in apatite grains by laser or furnace outgassing, and then measuring the relative amounts of uranium and thorium in the sample by solution ICP-MS. A careful selection of inclusion and crack-free idiomorphic apatite crystals with homogeneous U, Th and Sm distribution is essential for this procedure.

AHe thermochronology relies on the accumulation of ^{4}He during the α-disintegration of ^{238}U, ^{235}U, ^{232}Th, their daughter products and ^{147}Sm. The closure temperature (T_C) of mineral grains is dependent on activation energy, a geometry factor for the crystal shape, thermal diffusivity (D_0), the length of the average diffusion pathway from the interior to the surface of the grain and the cooling rate at closure temperature. In addition, ^{4}He diffusion in apatite is impeded by radiation-induced damage to the apatite structure. Therefore, the kinetics is an evolving function of time. The ^{4}He production–diffusion model predicts that the effective ^{4}He closure temperature of apatite will vary with cooling rate and effective U- and Th-concentration, and may differ from the commonly assumed T_C of 75 °C Ma^{-1} by up to ±15 °C (e.g. Farley *et al.* 1996; Wolf *et al.* 1996, 1998; Farley 2000, 2002; Shuster *et al.* 2006; Flowers *et al.* 2009; Shuster & Farley 2009). During radioactive decay alpha (α)-particles are emitted with high kinetic energy and travel significant distances. This poses a complication for the He dating method, as α-particles may be ejected out of the crystal being dated or injected from the surrounding mineral grains.

Farley *et al.* (1996) and Reiners *et al.* (2004) pointed out that implantation or ejection of α-particles may generally obscure (U–Th–Sm)/He dating, whereas He implantation needs to be considered if dealing with low U–Th–Sm phases. Therefore, Farley *et al.* (1996) propose a correction for these effects either numerically by correcting the measured He ages against the specific α retentivity (F_T) or by removal of the outer rim of the crystal (by chemical dissolution or mechanical abrasion) prior to dating. While correcting for α ejection has become a routinely performed practice in recent years, Spiegel *et al.* (2009) demonstrate the potential for possible overcorrection, indicating that abrading the outer rim of a crystal may be favourable. However, mechanical abrasion may lead to erroneous ages when crystals are strongly zoned with respect to uranium and thorium (Farley *et al.* 1996). In this case, a U–Th–Sm zonation-dependent α correction, as proposed by Hourigan *et al.* (2005), needs to be applied.

Usually, an apatite T_C in the range of 70 ± 7 °C (for a monotonic cooling rate of 10 °C Ma^{-1}, a subgrain domain size >60 μm, an activation energy of about 36 kcal mol^{-1} and a log (D_0) of 7.7 ± 0.6 cm^2 s^{-1}), and a He partial retention zone of between 40 and 75 °C, are assumed. The He production–diffusion model relies on homogenous distribution of U, Th and Sm in secular equilibrium, He loss confined to volume diffusion, and spherical diffusion geometry. Meesters & Dunai (2002*a*, *b*) generalized a production–diffusion equation to diffusion domains of various shapes and arbitrary cooling histories. Their set of equations allows α ejection corrected ages to be calculated and accounts for the non-homogeneous distribution of U, Th and Sm.

Preliminary studies on the applicability of (U–Th)/He thermochronology on zircon and titanite suggest closure temperatures about 80 and 130 °C higher than for apatite. These systems have the potential to close the gap between the various mineral diffusion temperatures of the ^{40}Ar/^{39}Ar thermochronological system, and the zircon and titanite fission-track annealing temperatures on one side, and AFT and AHe on the other side (e.g. Farley 2002; Reiners & Brandon 2006; Dobson *et al.* 2009). Moreover, (U–Th)/He analysis of magnetite (T_C: 200 °C/10 Ma) opens new perspectives for dating volcanic rocks (Blackburn *et al.* 2007).

Modelling of low-temperature thermochronological data

Recovery of thermal history information from AFT and AHe datasets is contingent upon a quantitative understanding of how the combination of time and temperature controls fission-track annealing and helium diffusion. Early attempts to use AFT data to track cooling were based on qualitative indicators (e.g. Wagner & Reimer 1972; Gleadow *et al.* 1986*a*, *b*): samples with a single population of long track lengths indicate rapid cooling in contrast to slower and/or stepped cooling paths, which result in shorter and/or more complex length distributions. However, this type of approach limits interpretation to relative rates of cooling and provides no constraints on timing or temperatures. With the advent of annealing studies, algorithms were formulated to describe the time–temperature dependencies of track annealing. The models based on laboratory experiments were extrapolated to geological timescales and verified against well-constrained data from the geological record. The first and, previously, most widely used model (Laslett *et al.* 1987) describes the annealing behaviour of a single type of apatite (Durango). With the realization that annealing is also influenced by grain composition, new experiments extended the annealing database. Carlson *et al.* (1999) published the first multi-kinetic annealing model. In 2007 this model was updated to include the dataset of Barbarand *et al.* (2003) and improved data-fitting techniques. The new multi-kinetic model, now in wide use, is based on 579 experiments and 26 compositionally different types of apatite (Ketcham *et al.* 2007).

Equipped with quantitative descriptions of track annealing it is possible to extract sample thermal histories by using forward or data inversion modelling techniques. Early programs focused on forward modelling data to check annealing models against well-constrained geological examples (e.g. Green *et al.* 1989*b*) and as a guide to the interpretation of real samples (e.g. Willett 1992). However, forward modelling is not a very efficient means of finding solutions for unknown or poorly constrained samples and, since there is no unique solution to a given dataset, such an approach is open to user bias. Data-driven inverse modelling helps to reduce this bias. Most commonly adopted and publicly available AFT modelling programs are Monte Trax (Gallagher 1995, designed for Apple Macintosh), AFTINV (Issler 1996; Willet 1997), AFTSolve (Ketcham *et al.* 2000) and HeFTy (Ketcham 2005, all Windows). These programs differ in modelling approach (Monte Carlo and/or genetic algorithm), annealing models, input parameters and statistical tests to evaluate the level of fit between model results and observed data (cf. Ehlers *et al.* 2005; Ketcham 2005). Some of the modelling programs refer only to AFT data, whilst others derive cooling histories from combined AFT, AHe and VR data (e.g. HeFTy). DECOMP (Meesters & Dunai 2002*a*, *b*) is a popular program to model thermal histories from AHe data. Most recent developments in modelling include a strategy for modelling sample thermal histories jointly (Gallagher *et al.* 2005) and the use of Markov Chain Monte Carlo (MCMC) methods to address the problem of characterizing uncertainties in modelled thermal histories in two and three dimensions (2D and 3D) (e.g. Stephenson *et al.* 2006).

Some concepts in apatite thermochronology

Methodological advancements were accompanied by the development of new strategies for the interpretation of thermochronological data, as the derivation of time–temperature constraints and the conversion of temperature information into geological and geomorphological processes.

Fission-track age types

An important part of fission-track thermochronology relates to using the distribution of measured track lengths in a sample to determine whether the measured age directly records an 'event' or a more complex thermal history. Early on in the AFT methods developmental history Günther Wagner suggested a classification of AFT 'ages' as *event ages*, *cooling ages* and *mixed ages* based on sample thermal history (Wagner 1972).

According to this concept, an *event age* refers to a rock that cools rapidly through the PAZ (e.g. a volcanic rock) and resides at low, near-surface, temperatures thereafter. The AFT age is essentially identical to the age of entrance into the PAZ, and is associated with a narrow distribution of track lengths about a mean value of 15 μm. Slow linear cooling of a sample through the PAZ produces a *cooling age* that is significantly younger than the entrance into the PAZ, with a broader and shorter track-length distribution. Such a pattern is relatively common in old basement terrains that have undergone cooling over very long periods of geological time. *Mixed ages* refer to an at least two-stage cooling history, in which the first generation of tracks resides at relative higher temperatures within the PAZ, prior to final cooling to lower temperatures. The resulting track-length distribution in such cases is typically bimodal with a short peak representing the higher temperature tracks and a second, long peak being added after final cooling. Crucially, both *cooling* and *mixed ages* have no direct significance in terms of the timing of any geological event.

Uplift, exhumation and denudation

With the availability of low-temperature chronological data, the resolution of cooling events improved considerably, and it became possible to analyse increasingly younger and shallower processes. Early AFT studies usually interpreted

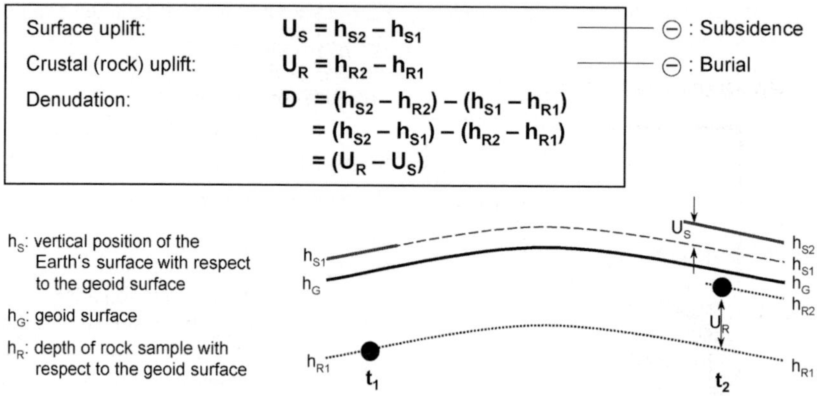

Fig. 2. Sketch illustrating the relationship between surface uplift, crustal (rock) uplift and denudation (see the text for definitions). Abbreviations: h, elevation; R, rock; S, surface; t, time.

cooling patterns in terms of 'uplift and erosion'. However, as this interpretation implicitly assumes that 'uplift' will always be matched by an equivalent amount of 'erosion' (cf. Parrish 1983), a much more precise usage of terminology was required. Terms like denudation, erosion, exhumation and uplift (surface, crustal and rock uplift) were (re-) defined by England & Molnar (1990), and explained further by Summerfield (1991), Summerfield & Brown (1998) and Ring *et al.* (1999) (Fig. 2).

According to these definitions, *surface uplift* refers to changes in the elevation of a surface. It is equal to rock uplift minus exhumation. *Crustal* (rock/bedrock) *uplift* refers to changes in the vertical position of rocks with respect to a fixed reference frame, such as the geoid.

Denudation and *exhumation* both result from the removal of material from a region or point, respectively, on the Earth's surface. Whereas exhumation refers to the unroofing of a point (a single sample or a vertical rock section), denudation applies to an area (cf. Ahnert 2003; see Summerfield & Brown 1998). Reference frame for both is the geometry of the past land surface. Denudation can occur in response to erosion and/or tectonics (e.g. Ring *et al.* 1999). *Erosion* describes the removal of weathered products by geomorphic agents. *Tectonic denudation* typically occurs through processes of extension and normal faulting, and it can result in the rapid removal of large rock volumes. Ductile crustal thinning as a third mode of denudation is not applicable for the uppermost crustal level and related processes.

Thermal histories modelled from thermochronological data, and potentially supplemented by information from vertical profiles, can be converted into an amount of denudation by using a value (measured or inferred) for the local geothermal gradient and taking into account average surface temperature.

Vertical profiles

Much more information than for a single sample alone is available when a suite of samples in a *vertical profile* can be analysed, such as may be obtained by sampling from a deep drill hole or over a significant range of vertical relief. Such vertical arrays are collected across the thermal gradient through which the samples have cooled. A consequence of fission-track annealing/helium diffusion is that AFT and AHe ages gradually decrease from some observed value at the Earth's surface to an apparent value of zero at the depth where no fission tracks/He are retained (Wagner & Reimer 1972).

The decrease in the AFT age with depth, and the associated variation of the length distribution, often deviate from a simple, linear pattern. Gleadow & Fitzgerald (1987) recognized that the exposed base of a fossil (exhumed) PAZ within a vertical sample profile produces a characteristic change in the regression of the age–elevation plot. Both the onset as well as the amount of cooling can be derived from this *break in slope* (Fig. 3). The age of the break in slope approximates the initiation of the last cooling below about 100 °C. Samples from below this break contain only tracks accumulated during and after this cooling stage. In contrast, samples above the break in slope contain two generations of tracks, one from before and one from after the onset of final cooling. The amount of cooling and exhumation can thus be obtained from the elevation of the break in slope within a vertical AFT (or, similarly, an AHe) profile. The actual shape of the age–elevation plot, and the trend of the track-length distributions with elevation, will depend on the geothermal gradient, and on the amount and rate of exhumation (R. W. Brown *et al.* 1994).

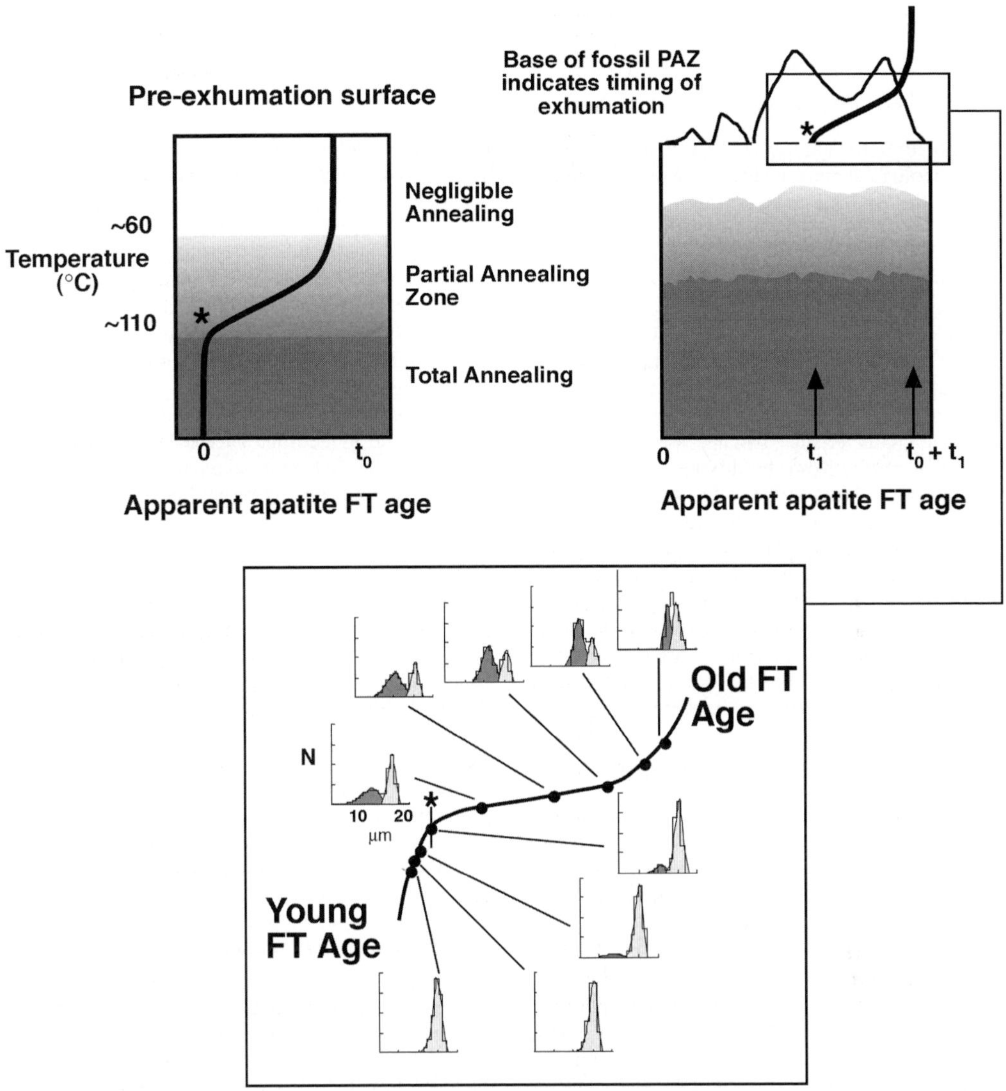

Fig. 3. The concept of an exhumed PAZ (adapted from Fitzgerald *et al.* 1995; Gallagher *et al.* 1998). The left panel illustrates the pre-denudation apatite fission-track age crustal profile, with the initial age as t_0. Denudation at time t_1 exposes different levels of this pre-cooling profile, while the deeper samples begin to retain tracks (central panel). The right panel shows the expected trend in the fission-track data with respect to elevation, that is age increases. The length distribution has two components: tracks formed prior to cooling (dark shading) and those formed after cooling (light shading). The latter are all long, and the composite length distribution depends on the relative proportion of these two components. Only the data below the break in slope (marked by an asterisk) provide the timing for the onset of the cooling/denudation event.

Another advantage of analysing vertical AFT profiles is the possibility of calculating the palaeogeothermal gradient that existed prior to the onset of cooling (Gleadow & Duddy 1981; Bray *et al.* 1992). The palaeogeothermal gradient can be determined by weighted least-squares regression of modelled maximum palaeotemperatures against sample elevation. Palaeogeothermal gradients are not only invaluable for estimating burial depths and amounts of denudation, but also bear important constraints on modes and mechanisms of the related processes (e.g. rifting). A comparison of the palaeogeothermal gradient with a recent one allows the cause of high palaeotemperatures, and the cause of the subsequent cooling to the present temperatures, to be determined.

Thermochronological data patterns

Thermochronological transects (horizontal profiles) across geological units or structures, or regional patterns of the AFT parameters, often allow exhumation trends to be detected or distinct igneous or tectonic events to be verified. Before modelling programs were available, the characteristic relationship between AFT ages, track-length distributions and thermal history was used to qualitatively constrain amount and timing of cooling (denudation) on a regional scale. A plot of AFT age against mean track length of an area that has undergone cooling shows a characteristic 'banana' or 'boomerang' shape that results from the mixing of two cooling components (Green 1986). In such plots, the longest mean track lengths are preserved in samples that experienced rapid cooling without subsequent reheating. Between the two end points of a boomerang there exists a series of transitional bimodal track-length distributions where the abundance of inherited annealed tracks becomes reduced in favour of newly formed tracks while the AFT ages of the samples decrease.

Apatite thermochrononology and topography

Topographic effects and fast rates of cooling. The temptation to collect vertical profiles in active mountain belts where there is considerable relief can introduce a number of interpretative problems for thermochronological data, especially if measured ages are young (<5 Ma). Young AFT (or AHe) ages require rapid rates of cooling and high rates of rock uplift and exhumation, but the calculation of true exhumation rates in such young samples is problematic owing to the combined influences of topographic wavelength and thermal advection. High rates of rock uplift cause perturbation of upper-crust thermal structure as heat is advected at rates that exceed normal heat loss by conduction, driving isotherms closer to the Earth's surface (e.g. modelling studies of Stüwe *et al.* 1994; Mancktelow & Grasemann 1997; cf. also Wang & Zhou 2009). Depending on wavelength and amplitude (height), the underlying thermal structure can undulate with the topography, causing the distance between closure isotherm and surface to vary with sample location. For example, beneath valley floors (lowest elevation samples) the depth to the PAZ will be less than beneath topographic ridges, that is the geothermal gradients will vary with location. Failure to take this effect into account when constructing age–elevation plots to determine exhumation rates can lead to incorrect interpretations. Detailed thermal modelling studies by Braun (2002) show how variable thermal structure can bias the interpretation of low-temperature thermochronological data.

Landscape evolution modelling. Much effort has been devoted recently to understand the coupling between tectonic and surface processes in the formation of recent topography (e.g. Braun 2002; Burbank & Anderson 2001; Burbank 2002; Ehlers & Farley 2003; Braun *et al.* 2006). Quantification of the rate at which landforms adapt to a changing tectonic, heat flow and climate environment (i.e. 'dynamic topography evolution') are performed by combining geomorphological analytical work, low-temperature thermochronological data and 3D thermokinematic modelling. Thermokinematic modelling with the 3D finite-element computer code Pecube (Braun 2003, 2005) predicts time–temperature (t–T) paths for all rock particles that, at the end of the computations, occupy the locations of the nodes at the surface of the finite-element mesh. From the t–T paths, apparent AHe and AFT ages are generated by varying topography, erosion rates, uplift rates and heat flow values. Thus, Pecube allows an overall uplift rate to be created or a block of infinite space, which is bordered by normal faults and/or thrusts, to be defined. When movement is localized at the faults or thrusts, computer-code-generated age data are subsequently compared with the determined real thermochronological age data. As a result, it is now possible to match age data with geomorphological results and cosmogenic nuclide-based age dating in order to test landscape evolution models by processing different timescale resolutions.

Detrital thermochronology

The 'standard' approach to deriving an AFT age is typically based on the measurement of the track density (age) for 20–30 (more if track densities are low) single-grain ages. Detrital thermochronology requires 50–120 grains per sample to enable statistical deconvolution into source-age components. Depending on the objectives of the study, up to 117 grains should be dated if the analyst wants to ensure that no fraction of the dated population comprising more than 0.05 of the total is missed at the 95% confidence level (Vermeesch 2004). Once the desired numbers of grains have been counted, the dataset can be divided into principal age components. Binomfit is a freeware program written by Mark Brandon that calculates ages and uncertainties for mixed distributions of fission-track grain ages. It uses an algorithm based on the decomposition method of Galbraith & Green (1990). Age components are compared with the age patterns of the hinterland and correlated with specific source areas (e.g. Brandon 1996; Bernet & Spiegel 2004; Bernet & Garver 2005).

Detrital thermochronology is routinely used in *provenance analysis*, and for denudation and landscape evolution studies (e.g. Hurford & Carter 1991; Garver *et al.* 1999; Bernet *et al.* 2006). Specific tectonic and geomorphological applications include dating of hinterland uplift to reconstruct early exhumation rates in active orogenic belts, determining sediment-source regions, and reconstructing (palaeo-) drainage systems. A particularly useful interpretative method is to plot the *lag time* (or erosion–transport interval) between sediment deposition age and youngest detrital exhumation age. This provides key information on exhumation history, and conceptually it is possible to use the lag time to monitor the evolution of a mountain belt as it passes from growth stages (decreasing lag times) into topographic and exhumation steady state (constant lag time) into orogenic decay (increasing lag time). Other provenance applications include constraining minimum depositional ages of sediments and the correlation of stratigraphic horizons.

Most sediment is derived from the erosion of pre-existing rocks, and therefore detrital apatite and zircon grains may contain tracks that accumulated in the original source rock. Weathering and physical erosion do not affect the retention and stability of fission tracks (Gleadow & Lovering 1974), but the preservation of provenance-related tracks does depend on the temperature history experienced by eroding source regions and subsequently by the sediment as the basin evolves (Fig. 4). Whereas zircons provide excellent provenance indicators (e.g. Hurford & Carter 1991; Garver & Brandon 1994; Garver *et al.* 1999), the use of apatite for source-rock information is restricted to shallow basins (typically less than *c.* 2 km of burial) and drainage systems (e.g. Corrigan & Crowley 1992; Lonergan & Johnson 1998; Malusà *et al.* 2009). Detrital analyses often combine thermochronological and geochronological data (U–Pb dating of zircon, $^{40}Ar/^{39}Ar$ dating of white mica), heavy mineral assemblages, grain size/shapes (zircon typology), and isotope signatures of detrital minerals (e.g. Dunkl *et al.* 2001; Carter & Foster 2009). The amount of provenance information can be increased by using double or triple dating techniques that involve more than one dating method on the same mineral grain, for example combined U–Pb or $^{40}Ar/^{39}Ar$ and AFT dating (Carter & Moss 1999; Carrapa *et al.* 2009) and/or AHe dating (Rahl *et al.* 2003). A detrital dating technique that utilizes petrographic information is fission-track analysis of single pebbles or pebble populations using conglomerates from foreland basin deposits (Spiegel *et al.* 2001; Dunkl *et al.* 2009).

Applications

Application of low-temperature thermochronology is not confined to the obvious, immediate purpose of dating rock formations, as this is rarely possible due to thermal resetting. Much more powerful

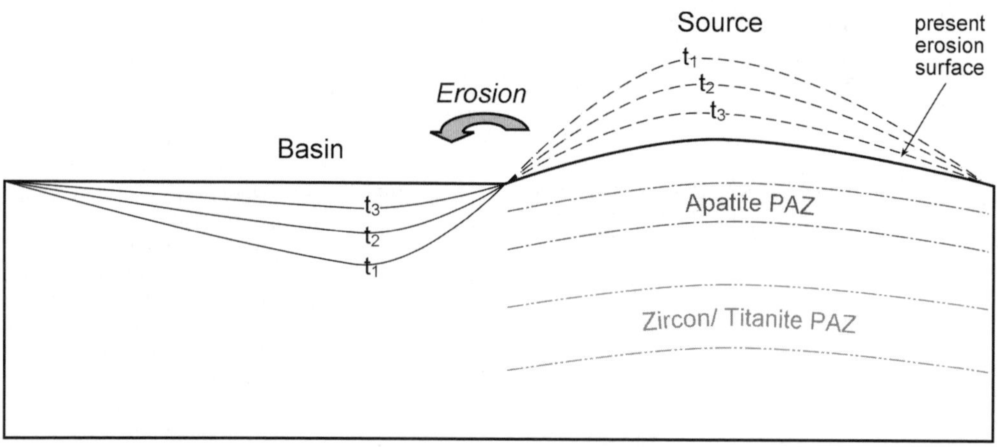

Fig. 4. Cartoon to show the relationship between the progressive exhumation of a source region and deposition of the eroded exhumation record in an adjacent sedimentary basin (modified after Garver *et al.* 1999). The time slices (t_1–t_3) correspond to three progressive and continuous intervals of erosion and deposition. Subsequently, a 'stratigraphy' of apatite fission-track ages develops within the basin, which is the inverse of the source-region exhumation age trend. It should be noted that with progressive burial, sediments may be heated sufficiently to cause annealing of the inherited AFT provenance records. Rarely is such burial-related heating sufficient to reset fission tracks in detrital zircons and titanites. Also, note how topography may cause the upwarping of the lower temperature isotherms.

(and more widely used) are the applications that exploit thermal resetting to reconstruct rock exhumation histories. The main fields of application of AFT and AHe thermochronology include provenance studies, thermal history analysis of sedimentary basins, the evolution of orogenic mountain belts and applications in non-orogenic settings. Recent work in these areas has focused on the coupling between climate and tectonics (e.g. Koons 1989; Willet 1999; Beaumont *et al.* 2000), and thermochronlogical datasets have been used to evaluate the role of climatically driven erosion as a component of exhumation (Blythe & Kleinspehn 1998; Reiners & Brandon 2006).

Fission-track and (U–Th–Sm)/He data are often combined with other sources of data such as thermochronology and geochronological dating to constrain higher temperature histories and/or rock formation age (e.g. $^{40}Ar/^{39}Ar$ and K–Ar: Daszinnies *et al.* 2009; De Grave *et al.* 2009), and terrestrial cosmogenic nuclide dating to compare with more recent erosion rates (e.g. Cockburn *et al.* 2000; Kuhlemann *et al.* 2009). In addition, interpretations increasingly integrate data from stratigraphic archives, geomorphology, structural geology, remote sensing, petrology, fluid inclusion analysis, vitrinite reflectance, clay mineralogy, conodont colour alteration, zircon typology and seismic data. These multi-method approaches strengthen the interpretation of AFT and AHe data, and help to extend geological histories through time.

Absolute dating

Owing to fission-track annealing and He diffusion at relatively low temperatures, geological events can be dated in well-defined settings, when rocks very rapidly passed the PAZ and/or partial retention zone and resided at the surface or at a very shallow-crustal level thereafter. Rapid cooling to low temperatures mainly occurs subsequent to volcanic or hydrothermal activity (Duddy *et al.* 1998), dyke emplacement, faulting and friction or meteorite impacts (e.g. Miller & Wagner 1979). For these cases, thermochronological ages are more or less identical to those obtained by conventional radiometric techniques, and can be used directly as discrete time constraints. Fission-track analysis is also used as a conventional method for dating glasses (e.g. Fleischer & Price 1964; Bigazzi & De Michele 1996), and has been applied to date stone tools and fossils (e.g. Morwood *et al.* 1998). Moreover, both fission-track and (U–Th)/He techniques have been used successfully to date the formation of supergene minerals such as some phosphates or vanadates, hematite, goethite, limonite, manganese oxides and carbonate minerals (e.g. Bender 1973; Lippolt *et al.* 1995; Shuster *et al.* 2005; Boni *et al.* 2007; Copeland *et al.* 2007).

Denudation and long-term landscape evolution studies

Denudation and long-term landscape evolution studies represent the most common and broadest field of applied low-temperature thermochronology. Studies range from compressional to extensional settings and 'stable' cratonic interiors.

Orogenic belts. Orogenic mountain ranges are characterized by substantial relief and immense uplift/denudation rates, resulting in large-scale advective transfer of heat and increased thermal gradient. The obvious correlation between exhumation and cooling predetermined this setting for an early application of thermochronological research, and expanded the scope of the method(s) from a purely 'age determination' approach to a unique thermotectonic tool. Wagner (1968) and Wagner & Reimer (1972) first used AFT data to provide estimates of the time and rates at which rocks approach the surface and cool as a result of 'uplift and erosion'. At present, low-temperature thermochronology is the most efficient method to quantify denudation rates on geological timescales. Changes in erosion rates with time can be constrained using multiple chronometers with different closure temperatures on the same rock sample, or from the distribution of cooling ages from a single system along a vertical transect. Spatial–temporal patterns of thermochronometrically determined erosion rates help to constrain the flow of material through orogenic wedges, orogenic growth and decay cycles, palaeorelief, and relationships with structural, geomorphological or climatic features (cf. Reiners & Brandon 2006). Subsequent to the first apatite thermochronological works in the Alps (e.g. Schaer *et al.* 1975; Wagner *et al.* 1979; Grundmann & Morteani 1985; Hurford 1986; Hurford *et al.* 1991), most of the world's young orogenic belts were studied (e.g. Parrish 1983; Seward & Tulloch 1991; Corrigan & Crowley 1992; Hendrix *et al.* 1994; Blythe *et al.* 1996; O'Sullivan & Currie 1996; Sorkhabi *et al.* 1996; Dunkl & Demény 1997; Kamp 1997; Sanders *et al.* 1999; Fayon *et al.* 2001; Spiegel *et al.* 2001; Glasmacher *et al.* 2002*b*; Reiners *et al.* 2002; Thomson 2002; Willet *et al.* 2003; Van der Beek *et al.* 2006; Gibson *et al.* 2007; Vincent *et al.* 2007; Glotzbach *et al.* 2008, 2009; del Río *et al.* 2009; Ruiz *et al.* 2009).

Continental rifts and passive continental margins. Continental rifts are elongated tectonic depressions that result from extension and crustal thinning caused either by a regional extensional stress field or in response to asthenospheric upwelling (cf. Olsen 1995; Ziegler & Cloething 2004). Continental separation and the onset of sea-floor spreading mark the transition from a continental rift into a passive margin. Consequently, the general morphotectonic evolution and appearance of rifts and passive margins is similar. In detail, the style of lithospheric extension, geometry of rifting, rate and amount of extension are influenced by local crust rheology coupled to erosion, which in turn is influenced by pre-existing morphology, rift-related magmatism, local drainage and climate (e.g. Gilchrist & Summerfield 1990; Kooi & Beaumont 1994; Olsen 1995; O'Sullivan & Brown 1998; Ziegler & Cloething 2004). Deconvolving these different controls has been the focus of AFT and AHe studies. Apatite thermochronological studies have been reported from most of the world's continental-size rift structures and highly extended terranes (for an overview cf. Stockli 2005) including the West Antarctic Rift System (e.g. Gleadow *et al.* 1984; Fitzgerald & Gleadow 1988; Fitzgerald 1992, 1994; Foster & Gleadow 1992*a*; Balestrieri *et al.* 1994; Fitzgerald & Stump 1997; Lisker 2002; Fitzgerald *et al.* 2006), the East African Rift System (e.g. Kohn & Eyal 1981; Omar *et al.* 1989; Foster & Gleadow 1993, 1996; Van Der Beek *et al.* 1998; Feinstein *et al.* 1996; Kohn *et al.* 1997; Balestrieri *et al.* 2005; Spiegel *et al.* 2007), and the Basin and Range Province (e.g. Armstrong *et al.* 2003; House *et al.* 2003; Colgan *et al.* 2006), as well as many other continental extension zones (e.g. Rohrmann *et al.* 1994; Van der Beek *et al.* 1996; Lisker & Fachmann 2001; Emmel *et al.* 2009). Two studies on the influence of heat-flow variation across major rifts were published by Gallagher *et al.* (1994) and Lisker *et al.* (2003).

Gallagher & Brown (1997) demonstrated how AFT data are used to study the evolution of rift-margin topography (linked to isostatic-flexural responses to erosional unloading) and to test the different geomorphic models of passive margin evolution. Most passive rift margins of the major continents have been extensively studied by apatite thermochronology, with a wealth of papers detailing different aspects of their evolution. These studies include Moore *et al.* (1986), Brown *et al.* (1990, 2002), Dumitru (1991), Kalaswad *et al.* (1993), Gallagher *et al.* (1994), Gallagher & Brown (1999), Carter *et al.* (2000), Johnson & Gallagher (2000), O'Sullivan *et al.* (2000), Persano *et al.* (2002), Gunnell *et al.* (2003), Seward *et al.* (2004), Japsen *et al.* (2006), Raab *et al.* (2005), Emmel *et al.* (2006, 2007) and Kounov *et al.* 2009). Transform margins have seen comparatively little study, with published work confined to the Ghana margin (Clift *et al.* 1997; Bouillin *et al.* 1997; Bigot-Cormier *et al.* 2005).

Cratons. Cratonic interiors were traditionally considered as tectonically and thermally stable features. They are characterized by extensive surfaces of minimal relief that infer low rates of erosion. However, thermochronological studies conducted during the last two decades on different cratons revealed that throughout the Phanerozoic many of these ancient terrains have experienced discrete episodes of kilometre-scale crustal erosion (e.g. Crowley *et al.* 1986; Wagner 1990; Noble *et al.* 1997; Harman *et al.* 1998; Cederbom *et al.* 2000; Bojar *et al.* 2002; Glasmacher *et al.* 2002*a*; Kohn *et al.* 2002; Belton *et al.* 2004; Lorencak *et al.* 2004; Soderlund *et al.* 2005; Flowers *et al.* 2006). On the other hand, cratonic areas often show apparent inconsistencies between AFT and AHe ages. These *crossover ages* are discussed controversially as either resulting from non-thermal radiation-enhanced annealing of fission tracks (e.g. Hendriks & Redfield 2005) or reflecting the dependency of He retention properties from the chemical composition of apatites (e.g. Green *et al.* 2006; Kohn *et al.* 2009).

Basin analysis

Sedimentary basins are major archives of geological history related either directly to the evolution of the basin itself (e.g. subsidence history, tectonics, climate, sea-level change) and/ or indirectly to the geological history of the sediment-source regions (e.g. provenance information, palaeo-basement, geomorphology, tectonics). Traditionally, the burial and deposition history of basin sediments is determined by investigation of the sedimentology and deposition ages derived from biostratigraphy, combined with structural and geophysical datasets. However, establishing the timing of maximum depths of burial, and the timing and extent of uplifts, is often hard to recover owing to stratigraphic gaps, poor biostratigraphic control and sediment types. In this regard, AFT chronology has proven a valuable tool.

Thermochronology becomes most useful when a basin has undergone a period of uplift or inversion, during which a section of strata has been removed by erosion (e.g. Kamp & Green 1990; Bray *et al.* 1992; Green *et al.* 1995). Specific information gained from thermochronological data include estimates of maximum palaeotemperatures, calculations of palaeogeothermal gradients and palaeo-heat flow, style (fast/slow) and time of cooling from maximum palaeotemperatures, characterization of mechanisms of heating and cooling, quantification of missing sections, stratigraphic dating, sediment provenance, and evaluation of hydrocarbon maturation (Gleadow *et al.* 1983; Green *et al.* 1989*a*; Kamp & Green 1990; Duane & Brown 1991; Mitchell 1997; Logan & Duddy 1998).

Particularly useful in basin analysis is the integration of thermochronological data with maximum palaeotemperature indicators (Fig. 5), like vitrinite reflectance (VR) and illite crystallinity

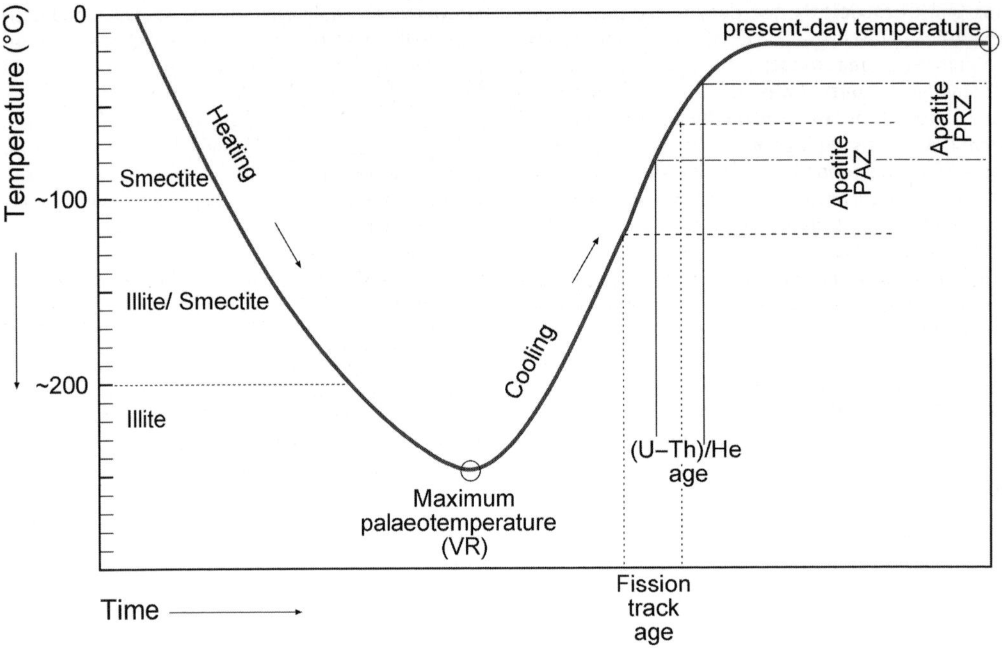

Fig. 5. Schematic to show how the thermal history of a sedimentary basin can be recovered by integration of data from multiple palaeothermometers and chronometers. Illite data constrain the burial or heating phase of a basin's thermal history, VR records maximum temperature, and combined apatite thermochronology constrains the timing of cooling and subsequent denudation (modified after Pevear 1999).

(e.g. Bray *et al.* 1992; Pagel *et al.* 1997; Duddy *et al.* 1998; Mathiesen *et al.* 2000; Ventura *et al.* 2001; Arne *et al.* 2002; Osadetz *et al.* 2002; Tingate & Duddy 2002). VR is the measure of the coalification rank of organic matter, and it is mainly dependent on temperature and time (Burnham & Sweeney 1989). VR data provide a direct estimation of maximum palaeotemperatures across the same temperature range as annealing in fission tracks in apatite, which enables the thermal history modelling of joint VR and AFT datasets to provide a more robust constraint on temperature–time histories. Basin modelling using combined thermochronological and VR constraints has become a routine, and is a valuable tool in the hydrocarbon exploration industry (Gleadow *et al.* 1983; Green *et al.* 2002; Emmerich *et al.* 2005; Underdown *et al.* 2007). Also of economic relevance is the application of apatite thermochronology to the exploration of hydrothermal ore deposits (cf. McInnes *et al.* 2005).

Tectonic processes

Thermochronological methods can be used to detect tectonic activities in two ways. In areas of substantial block uplift, thermochronological ages may be disrupted across tectonic structures. Such offsets in the palaeo-isotherm/-depth stratigraphy can be used to determine relative uplift between different blocks and the amount of throw on bounding faults (Fitzgerald & Gleadow 1988, 1990; Dumitru 1991; Foster & Gleadow 1992*a*, *b*, 1996; Fitzgerald *et al.* 1993; O'Sullivan *et al.* 1995, 2000; Johnson 1997; Rahn *et al.* 1997; Wagner *et al.* 1997; Thomson 1998; Kohn *et al.* 1999; Redfield *et al.* 2007; Ventura *et al.* 2009; Xu *et al.* 2009). Moreover, the timing of tectonic activity may be determined by the direct dating of faults, lineaments or pseudotachylites (Harman *et al.* 1998; O'Sullivan *et al.* 1998; Raab *et al.* 2002, 2009; Zwingmann & Mancktelow 2004; Tagami 2005; Timar-Geng *et al.* 2009; Yamada *et al.* 2009).

Summary

As can be seen from the papers contained in this Special Publication, apatite thermochronology has grown to become a reliable and routinely used method to helping in solving a diverse range of geological problems. Although mature, the method continues to undergo developments. Over the last decade there have been considerable improvements in AFT methodology and data interpretation, but there is still scope for further advances. Future studies to refine models of track annealing in apatite may shift away from laboratory-based experimentation to well-constrained geological experiments (e.g. Spiegel *et al.* 2007) and molecular dynamic simulations (Rabone *et al.* 2008). Further work will also need to be carried out on understanding the controls on annealing in zircons and titanites, as well as the establishment of new fission-track mineral dating systems (e.g. monazite, merrillite: cf. Wagner & Van den Haute 1992; Gleadow *et al.* 2002).

We are very grateful to A. Carter and F. Stuart for constructive reviews of the manuscript.

References

AHNERT, F. 2003. *Einführung in die Geomorphologie*. Ullmer, Stuttgart.

ARMSTRONG, P. A., EHLERS, T. A., CHAPMAN, D. S., FARLEY, K. A. & KAMP, P. J. J. 2003. Exhumation of the central Wasatch Mountains, Utah: 1. Patterns and timing of exhumation deduced from low-temperature thermochronology data. *Journal of Geophysical Research*, **108** (B3), 2172; doi:10.1029/2001JB001708.

ARNE, D. C., GRIST, A. M., ZENTILLI, M., COLLINS, M., EMBRY, A. & GENTZIS, T. 2002. Cooling of the Sverdrup Basin during Tertiary basin inversion: implications for hydrocarbon exploration. *Basin Research*, **14**, 183–206.

BALESTRIERI, M. L., BIGAZZI, G., GHEZZO, C. & LOMBARDO, B. 1994. Fission track dating of apatites from the Granite Harbour Intrusive Suite and uplift–denudation history of the Transantarctic Mountains in the area between the Mariner and David Glaciers (northern Victoria Land, Antarctica). *Terra Antartica*, **1**, 82–87.

BALESTRIERI, M. L., STUART, F. M., PERSANO, C., ABBATE, E. & BIGAZZI, G. 2005. Geomorphic development of the escarpment of the Eritrean margin, southern Red Sea from combined apatite fission-track and (U–Th)/He thermochronometry. *Earth and Planetary Science Letters*, **231**, 97–110.

BARBARAND, J., CARTER, A., WOOD, I. & HURFORD, T. 2003. Compositional and structural control of fission-track annealing in apatite. *Chemical Geology*, **198**, 107–137.

BAUMHAUER, H. A. 1894. *Die Resultate der Aetzmethode in der krystallographischen Forschung*. W. Engelmann, Leipzig.

BEAUMONT, C., KOOI, H. & WILLETT, S. 2000. Coupled tectonic-surface process models with applications to rifted margins and collisional orogens. *In*: MA, S. (ed.) *Geomorphology and Global Tectonics*. Wiley, Chichester, 29–55.

BELTON, D. X., BROWN, R. W., KOHN, B. P., FINK, D. & FARLEY, K. A. 2004. Quantitative resolution of the debate over antiquity of the central Australian landscape: implications for the tectonic and geomorphic stability of cratonic interiors. *Earth and Planetary Science Letters*, **219**, 21–34.

BENDER, M. 1973. Helium–uranium dating of corals. *Geochimica Cosmochimica Acta*, **37**, 1229–1247.

BERGER, G. W. & YORK, D. 1981. Geothermometry from dating experiments. *Geochimica et Cosmochimica Acta*, **45**, 795–811.

BERNET, M. & GARVER, J. I. 2005. Fission-track analysis of detrital zircon. *Reviews in Mineralogy and Geochemistry*, **58**, 205–237.

BERNET, M. & SPIEGEL, C. 2004. Introduction: detrital thermochronology. *In*: BERNET, M. & SPIEGEL, C. (eds) *Detrital Thermochronology – Provenance Analysis, Exhumation, and Landscape Evolution of Mountain Belts*. Geological Society of America, Special Paper, **378**, 1–6.

BERNET, M., VAN DER BEEK, P., PIK, R., HUYGHE, P., MUGNIER, J.-L., LABRIN, E. & SZULC, A. 2006. Miocene to Recent exhumation of the central Himalaya determined from combined detrital zircon fission-track and U/Pb analysis of Siwalik sediments, western Nepal. *Basin Research*, **18**, 393–412.

BIGAZZI, G. & DE MICHELE, V. 1996. New fission-track age determinations on impact glasses. *Meteoritics and Planetary Science*, **31**, 234–236.

BIGOT-CORMIER, F., BASILE, C., POUPEAU, G., BOUILLIN, J.-P. & LABRIN, E. 2005. Denudation of the Côte d'Ivoire–Ghana transform continental margin from apatite fission tracks. *Terra Nova*, **17**, 189–195; doi: 110.1111/j.1365-3121.2005.00605.x.

BLACKBURN, T. J., STOCKLI, D. F. & WALKER, D. 2007. Magnetite (U–Th)/He dating and its application to the geochronology of intermediate to mafic volcanic rocks. *Earth and Planetary Science Letters*, **259**, 360–371.

BLYTHE, A. E. & KLEINSPEHN, K. L. 1998. Tectonically versus climatically driven Cenozoic exhumation of the Eurasian plate margin, Svalbard: fission track analyses. *Tectonics*, **17**, 621–639.

BLYTHE, A. E., BIRD, J. M. & OMAR, G. I. 1996. Deformational history of the central Brooks Range, Alaska: results from fission-track and $^{40}Ar/^{39}Ar$ analyses. *Tectonics*, **15**, 440–455.

BOJAR, A.-V., FRITZ, H., KARGL, S. & UNZOG, W. 2002. Phanerozoic tectonothermal history of the Arabian Nubian shield in the Eastern Desert of Egypt: evidence from fission track and paleostress data. *Journal of African Earth Sciences*, **34**, 191–202.

BONI, M., TERRACCIANO, R., EVANS, N. J., LAUKAMP, C., SCHNEIDER, J. & BECHSTÄDT, T. 2007. Genesis of vanadium ores in the Otavi Mountainland, Namibia. *Economic Geology*, **102**, 441–469.

BOUILLIN, J. P., POUPEAU, G., LABRIN, E., BASILE, C., SABIL, N., MASCLE, J., MASCLE, G., GILLOT, F. & RIOU, L. 1997. Fission track study: heating and denudation of marginal ridge of the Ivory Coast–Ghana transform margin. *Geo-Marine Letters*, **17**, 55–61.

BRANDON, M. T. 1996. Probability density plot for fission-track grain-age samples. *Radiation Measurements*, **26**, 663–676.

BRAUN, J. 2002. Estimating exhumation rate and relief evolution by spectral analysis of age–elevation datasets: *Terra Nova*, **14**, 210–214.

BRAUN, J. 2003. Pecube: A new finite element code to solve the heat transport equation in three dimensions in the Earth's crust including the effects of a time-varying, finite amplitude surface topography. *Computer Geosciences*, **29**, 787–794.

BRAUN, J. 2005. Quantitative constraints on the rate of landform evolution derived from low-temperature thermochronology. *Reviews in Mineralogy and Geochemistry* **58**, 351–374.

BRAUN, J., VAN DER BEEK, P. & BATT, G. 2006. *Quantitative Thermochronology: Numerical Methods for the Interpretation of Thermochronological Data*. Cambridge University Press, Cambridge.

BRAY, R. J., GREEN, P. F. & DUDDY, I. R. 1992. Thermal history reconstruction using apatite fission track analysis and vitrinite reflectance; a case study from the UK East Midlands and southern North Sea. *In*: HARDMAN, R. F. P. (ed.) *Exploration Britain; Geological Insights for the Next Decade*. Geological Society, London, Special Publications, **67**, 2–25.

BROWN, R., GALLAGHER, K. & DUANE, M. 1994. A quantitative assessment of the effects of magmatism on the thermal history of the Karoo sedimentary sequence. *Journal of African Earth Sciences*, **18**, 227–243.

BROWN, R., SUMMERFIELD, M. A. & GLEADOW, A. J. W. 2002. Denudational history along a transect across the Drakensberg Escarpment of southern Africa derived from apatite fission track thermochronology. *Journal of Geophysical Research*, **107**, 2350; doi:2310.1029/2001JB000745.

BROWN, R. W., RUST, D. J., SUMMERFIELD, M. A., GLEADOW, A. J. W. & DEWIT, M. C. J. 1990. An Early Cretaceous phase of accelerated erosion on the south-western margin of Africa: evidence from apatite fission track analysis & the offshore sedimentary record. *Nuclear Tracks Radiation Measurements*, **17**, 339–350.

BROWN, R. W., SUMMERFIELD, M. A. & GLEADOW, A. J. W. 1994. Apatite fission track analysis: its potential for the estimation of denudation rates and the assessment of models of long term landscape development. *In*: KIRKBY, M. J. (ed.) *Process Models and Theoretical Geomorphology*. Wiley, Chichester, 23–53.

BURBANK, D. W. 2002. Rates of erosion and their implications for exhumation. *Mineralogical Magazine*, **66**, 25–52.

BURBANK, D. W. & ANDERSON, R. S. 2001. *Tectonic Geomorphology*. Blackwell, Oxford.

BURNHAM, A. K. & SWEENEY, J. J. 1989. A chemical kinetic model of vitrinite maturation and reflectance. *Geochimica et Cosmochimica Acta*, **53**, 2649–2657.

BURTNER, R. L., NIGRINI, A. & DONELICK, R. A. 1994. Thermochronology of lower Cretaceous source rocks in the Idaho–Wyoming thrust belt. *AAPG Bulletin*, **78**, 1613–1636.

CARLSON, W. D., DONELICK, R. A. & KETCHAM, R. A. 1999. Variability of apatite fission-track annealing kinetics; I, Experimental results. *American Mineralogist*, **84**, 1213–1223.

CARRAPA, B., DECELLES, P. G., REINERS, P. W., GEHRELS, G. E. & SUDO, M. 2009. Apatite triple dating and white mica $^{40}Ar/^{39}Ar$ thermochronology of syntectonic detritus in the Central Andes: a multiphase tectonothermal history. *Geology*, **37**, 407–410.

CARTER, A. & FOSTER, G. L. 2009. Improving constraints on apatite provenance: Nd measurement on fission-track-dated grains. *In*: LISKER, F., VENTURA, B. & GLASMACHER, U. A. (eds) *Thermochronological Methods: From Palaeotemperature Constraints to Landscape Evolution Models*. Geological Society, London, Special Publications, **324**, 57–72.

CARTER, A. & MOSS, S. J. 1999. Combined detrital-zircon fission-track and U–Pb dating: a new approach to understanding hinterland evolution. *Geology*, **27**, 235–238.

CARTER, A., ROQUES, D. & BRISTOW, C. S. 2000. Denudation history of onshore central Vietnam; constraints on the Cenozoic evolution of the western margin of the South China Sea. *Tectonophysics*, **322**, 265–277.

CEDERBOM, C., LARSON, S. A., TULLBORG, E.-L. & STIBERG, J.-P. 2000. Fission track thermochronology applied to Phanerozoic thermotectonic events in central and southern Sweden. *Tectonophysics*, **316**, 153–167.

CLIFT, P. D., LORENZO, J., CARTER, A., HURFORD, A. J. & PARTY, O. L. S. 1997. Transform tectonics and thermal rejuvenation on the Cote d'Ivoire-Ghana margin, west Africa. *Journal of the Geological Society, London*, **154**, 483–489.

COCKBURN, H. A. P., BROWN, R. W., SUMMERFIELD, M. A. & SEIDL, M. A. 2000. Quantifying passive margin denudation and landscape development using a combined fission-track thermochronology and cosmogenic isotope analysis approach. *Earth and Planetary Science Letters*, **179**, 429–435.

COLGAN, J. P., DUMITRU, T. A., MCWILLIAMS, M. & MILLER, E. L. 2006. Timing of Cenozoic volcanism and Basin and Range extension in northwestern Nevada: new constraints from the northern Pine Forest Range. *Geological Society of America Bulletin*, **118**, 126–139.

COPELAND, P., WATSON, E. B., URIZAR, S. C., PATTERSON, D. & LAPEN, T. J. 2007. Alpha thermochronology of carbonates. *Geochimica et Cosmochimica Acta*, **71**, 4488–4511.

CORRIGAN, J. D. & CROWLEY, K. D. 1992. Unroofing of the Himalayas: a view from apatite fission-track analysis of Bengal Fan sediments. *Geophysical Research Letters*, **19**, 2345–2348.

COYLE, D. A. & WAGNER, G. A. 1998. Positioning the titanite fission-track partial annealing zone. *Chemical Geology*, **149**, 117–125.

CROWLEY, K., NAESER, C. & BABEL, C. 1986. Tectonic significance of Precambrian apatite fission-track ages from the midcontinent United States. *Earth and Planetary Science Letters*, **79**, 329–336.

CROWLEY, K. D., CAMERON, M. & SCHAEFER, R. L. 1990. Annealing of etchable fission-track damage in F-, OH-, Cl- and Sr-apatite: 1. Systematics and preliminary interpretations. *Nuclear Tracks Radiation Measurements*, **17**, 409–410.

CROWLEY, K. D., CAMERON, M. & SCHAEFER, R. L. 1991. Experimental studies of annealing of etched fission tracks in fluorapatite. *Geochimica et Cosmochimica Acta*, **55**, 1449–1465.

DASZINNIES, M. C., JACOBS, J., WARTHO, J.-A. & GRANTHAM, G. H. 2009. Post Pan-African thermo-tectonic evolution of the north Mozambican basement and its implication for the Gondwana rifting. Inferences from $^{40}Ar/^{39}Ar$ hornblende, biotite and titanite fission-track dating. *In*: LISKER, F., VENTURA, B. & GLASMACHER, U. A. (eds) *Thermochronological Methods: From Palaeotemperature Constraints to Landscape Evolution Models*. Geological Society, London, Special Publications, **324**, 261–286.

DE GRAVE, J., BUSLOV, M. M., VAN DEN HAUTE, P., METCALF, J., DEHANDSCHUTTER, B. & MCWILLIAMS, M. O. 2009. Multi-method chronometry of the Teletskoye graben and its basement, Siberian Altai Mountains: new insights on its thermo-tectonic evolution. *In*: LISKER, F., VENTURA, B. & GLASMACHER, U. A. (eds) *Thermochronological Methods: From Palaeotemperature Constraints to Landscape Evolution Models*. Geological Society, London, Special Publications, **324**, 237–259.

DEL RÍO, P., BARBERO, L. & STUART, F. M. 2009. Exhumation of the Sierra de Cameros (Iberian Range, Spain): constraints from low-temperature thermochronology. *In*: LISKER, F., VENTURA, B. & GLASMACHER, U. A. (eds) *Thermochronological Methods: From Palaeotemperature Constraints to Landscape Evolution Models*. Geological Society, London, Special Publications, **324**, 153–165.

DOBSON, K. J., PERSANO, C. & STUART, F. M. 2009. Quantitative constraints on mid- to shallow-crustal processes using the zircon (U–Th)/He thermochronometer. *In*: LISKER, F., VENTURA, B. & GLASMACHER, U. A. (eds) *Thermochronological Methods: From Palaeotemperature Constraints to Landscape Evolution Models*. Geological Society, London, Special Publications, **324**, 47–56.

DONELICK, R. A. 1991. Crystallographic orientation dependence of mean etchable fission-track length in apatite: an empirical model and experimental observations. *American Mineralogist*, **76**, 83–91.

DONELICK, R. A., RODEN, M. K., MOOERS, J. D., CARPENTER, B. S. & MILLER, D. S. 1990. Etchable length reduction of induced fission tracks in apatite at room temperature, $\sim 23\,^{\circ}C$, crystallographic orientation effects and 'initial' mean lengths. *Nuclear Tracks Radiation Measurement*, **17**, 261–265.

DONELICK, R. A., O'SULLIVAN, P. B. & KETCHAM, R. A. 2005. Apatite fission-track analysis. *Reviews in Mineralogy and Geochemistry*, **58**, 49–94.

DUANE, M. & BROWN, R. 1991. Tectonic brines and sedimentary basins: further application of fission-track analysis in understanding Karoo Basin evolution (South Africa). *Basin Research*, **3**, 187–195.

DUDDY, I. R., GREEN, P. F., HEGARTY, K. A., BRAY, R. J. & O'BRIAN, G. W. 1998. Dating and duration of hot fluid flow events determined using AFTA and vitrinite reflectance-based thermal history reconstruction. *In*: PARNELL, J. (ed.) *Dating and Duration of Fluid Flow and Fluid–Rock Interaction*. Geological Society, London, Special Publications, **144**, 41–51.

DUDDY, I. R., GREEN, P. F. & LASLETT, G. M. 1988. Thermal annealing of fission tracks in apatite, 3. Variable temperature behavior. *Chemical Geology (Isotope Geoscience Section)*, **73**, 25–38.

DUMITRU, T. A. 1991. Major Quaternary uplift along the northernmost San Andreas Fault, King Range, northwestern California. *Geology*, **19**, 526–529.

DUMITRU, T. A., HILL, K. C. *ET AL*. 1991. Fission track thermochronology: application to continental rifting of south-eastern Australia. *The APEA Journal*, **31**, 131–142.

DUNKL, I. & DEMÉNY, A. 1997. Exhumation of the Rechnitz Window at the border of the Eastern Alps and Pannonian Basin during Neogene extension. *Tectonophysics*, **272**, 197–211.

DUNKL, I., DI GIULIO, A. & KUHLEMANN, J. 2001. Combination of single-grain fission-track chronology and morphological analysis of detrital zircon crystals in provenance studies – sources of the Macigno Formation (Apennines, Italy). *Journal of Sedimentary Research*, **71**, 516–525.

DUNKL, I., FRISCH, W., KUHLEMANN, J. & BRÜGEL, A. 2009. Pebble population dating as an additional tool for provenance studies – examples from the Eastern Alps. *In*: LISKER, F., VENTURA, B. & GLASMACHER, U. A. (eds) *Thermochronological Methods: From Palaeotemperature Constraints to Landscape Evolution Models*. Geological Society, London, Special Publications, **324**, 125–140.

EHLERS, T. A. & FARLEY, K. A. 2003. Apatite (U–Th)/He thermochronometry: methods and applications to problems in tectonic and surface processes. *Earth and Planetary Science Letters*, **206**, 1–14.

EHLERS, T. A., CHAUDHRI, T. *ET AL*. 2005. Computational tools for low-temperature thermochronometer interpretation. *Reviews in Mineralogy and Geochemistry*, **58**, 589–622.

EMMEL, B., GEIGER, M. & JACOBS, J. 2006. Detrital apatite fission-track ages in Middle Jurassic strata at the rifted margin of W Madagascar – indicator for a protracted resedimentation history. *Sedimentary Geology*, **186**, 27–38.

EMMEL, B., JACOBS, J., CROWHURST, P. & DASZINNIES, M. C. 2007. Combined apatite fission-track and single grain apatite (U–Th)/He ages from basement rocks of central Dronning Maud Land (East Antarctica) – Possible identification of thermally overprinted crustal segments. *Earth and Planetary Science Letters*, **264**, 72–88.

EMMEL, B., JACOBS, J. & DASZINNIES, M. C. 2009. Combined titanite and apatite fission-track data from Gjelsvikfjella, East Antarctica – another piece of a concealed intracontinental Permo-Triassic Gondwana rift basin? *In*: LISKER, F., VENTURA, B. & GLASMACHER, U. A. (eds) *Thermochronological Methods: From Palaeotemperature Constraints to Landscape Evolution Models*. Geological Society, London, Special Publications, **324**, 317–330.

EMMERICH, A., GLASMACHER, U. A., BAUER, F., BECHSTÄDT, T. & ZÜLKE, R. 2005. Meso-/Cenozoic basin and carbonate platform development in the SW-Dolomites unravelled by basin modelling and apatite FT analysis: Rosengarten and Latemar (Northern Italy). *Sedimentary Geology*, **175**, 415–438.

ENGLAND, P. & MOLNAR, P. 1990. Surface uplift, uplift of rocks, and exhumation of rocks. *Geology*, **18**, 1173–1177.

FARLEY, K. A. 2000. Helium diffusion from apatite: general behaviour as illustrated by Durango fluorapatite. *Journal of Geophysical Research*, **105**, 2903–2914.

FARLEY, K. A. 2002. (U–Th)/He dating: techniques, calibrations, and applications. *Reviews in Mineralogy and Geochemistry*, **47**, 819–844.

FARLEY, K. A., WOLF, R. A. & SILVER, L. T. 1996. The effects of long-alpha-stopping distances on (U–Th)/He ages. *Geochimica et Cosmochimica Acta*, **60**, 4223–4229.

FAYON, A. K., WHITNEY, D. L., TEYSSIER, C., GARVER, J. I. & DILEK, Y. 2001. Effects of plate convergence obliquity on timing and mechanisms of exhumation of a mid-crustal terrain, the Central Anatolian Crystalline Complex. *Earth and Planetary Science Letters*, **192**, 191–205.

FEINSTEIN, S., KOHN, B. P., STECKLER, M. S. & EYAL, M. 1996. Thermal history of the eastern margin of the Gulf of Suez, I. reconstruction from borehole temperature and organic maturity measurements. *Tectonophysics*, **266**, 203–220.

FITZGERALD, P. G. 1992. The Transantarctic Mountains of southern Victoria Land: the application of apatite fission track analysis to a rift shoulder uplift. *Tectonics*, **11**, 634–662.

FITZGERALD, P. G. 1994. Thermochronologic constraints on post-Paleozoic tectonic evolution of the central Transantarctic Mountains, Antarctica. *Tectonics*, **13**, 818–836.

FITZGERALD, P. G. & GLEADOW, A. J. W. 1988. Fission-track geochronology, tectonics and structure of the Transantarctic Mountains in Northern Victoria Land, Antarctica. *Chemical Geology (Isotope Geoscience Section)*, **73**, 169–198.

FITZGERALD, P. G. & GLEADOW, A. J. W. 1990. New approaches in fission track geochronology as a tectonic tool: examples from the Transantarctic Mountains. *Nuclear Tracks*, **17**, 351–357.

FITZGERALD, P. G. & STUMP, E. 1997. Cretaceous and Cenozoic episodic denudation of the Transantarctic Mountains, Antarctica: new constraints from apatite fission track thermochronology in the Scott Glacier region. *Journal of Geophysical Research*, **102**, 7747–7766.

FITZGERALD, P. G., BALDWIN, S. L., WEBB, L. E. & O'SULLIVAN, P. B. 2006. Interpretation of (U–Th)/He single grain ages from slowly cooled crustal terranes: a case study from the Transantarctic Mountains of southern Victoria Land. *Chemical Geology*, **225**, 91–120.

FITZGERALD, P. G., SORKHABI, R. B., REDFIELD, T. F. & STUMP, E. 1995. Uplift and denudation of the central Alaska Range: a case study in the use of apatite fission track thermochronology to determine absolute uplift parameters. *Journal of Geophysical Research*, **100**, 20,175–20,191.

FITZGERALD, P. G., STUMP, E. & REDFIELD, T. F. 1993. Late Cenozoic uplift of Denali and its relation to relative plate motion and fault morphology. *Science*, **259**, 497–499.

FLEISCHER, R. L. & PRICE, P. B. 1964. Glass dating by fission fragment tracks. *Journal of Geophysical Research*, **69**, 331–339.

FLEISCHER, R. L., PRICE, P. B. & WALKER, R. M. 1975. *Nuclear Tracks in Solids. Principles and Applications*. University of California Press, Berkeley, CA.

FLOWERS, R. M., BOWRING, S. A. & REINERS, P. W. 2006. Low long-term erosion rates and extreme continental stability documented by ancient (U–Th)/He dates. *Geology*, **34**, 925–928.

FLOWERS, R. M., KETCHAM, R. A., SHUSTER, D. L. & FARLEY, K. A. 2009. Apatite (U–Th)/He thermochronometry using a radiation damage accumulation and annealing model. *Geochimica et Cosmochimica Acta*, **73**(8), 2347–2365; doi:10.1016/j.gca.2009.01.015.

FOSTER, D. A. & GLEADOW, A. J. W. 1992*a*. Reactivated tectonic boundaries and implications for the reconstruction of southeastern Australia and northern Victoria Land, Antarctica. *Geology*, **20**, 267–270.

FOSTER, D. A. & GLEADOW, A. J. W. 1992*b*. The morphotectonic evolution of rift-margin mountains in central Kenya: constraints from apatite fission-track thermochronology. *Earth and Planetary Science Letters*, **113**, 157–171.

FOSTER, D. A. & GLEADOW, A. J. W. 1993. Episodic denudation in East Africa: a legacy of intracontinental tectonism. *Geophysical Research Letters*, **20**, 2395–2398.

FOSTER, D. A. & GLEADOW, A. J. W. 1996. Structural framework and denudation history of the flanks of the Kenya and Anza Rifts, East Africa. *Tectonics*, **15**, 258–271.

GALBRAITH, R. F. 1981. On statistical models for fission track counts. *Mathematical Geology*, **13**, 471–478.

GALBRAITH, R. F. 1988. Graphical display of estimates having differing standard errors. *Technometrics*, **30**, 271–281.

GALBRAITH, R. F. 1990. The radial plot: graphical assessment of spread in ages. *Nuclear Tracks and Radiation Measurement*, **17**, 207–214.

GALBRAITH, R. F. & GREEN, P. F. 1990. Estimating the component ages in a finite mixture. *Nuclear Tracks and Radiation Measurements*, **17**, 197–206.

GALBRAITH, R. F. & LASLETT, G. M. 1993. Statistical models for mixed fission track ages. *Nuclear Tracks and Radiation Measurements*, **21**, 459–470.

GALBRAITH, R. F., LASLETT, G. M., GREEN, P. F. & DUDDY, I. R. 1990. Apatite fission track analysis: geological thermal history analysis based on a three-dimensional random process of linear radiation damage. *Philosophical Transactions of the Royal Society, A* **332**, 419–438.

GALLAGHER, K. 1995. Evolving temperature histories from apatite fission-track data. *Earth and Planetary Science Letters*, **136**, 421–435.

GALLAGHER, K. & BROWN, R. 1997. The onshore record of passive margin evolution. *Journal of the Geological Society, London*, **154**, 451–457.

GALLAGHER, K. & BROWN, R. 1999. The Mesozoic denudation history of the Atlantic margins of southern African and Southeast Brazil and the relationship to offshore sedimentation. *In*: CAMERON, N. R., BATE, R. H. & CLURE, V. S. (eds) *The Oil and Gas Habitats of the South Atlantic*. Geological Society, London, Special Publications, **153**, 41–53.

GALLAGHER, K., BROWN, R. & JOHNSON, C. 1998. Fission track analysis and its applications to geological problems. *Annual Review of Earth and Planetary Science*, **26**, 519–572.

GALLAGHER, K., HAWKESWORTH, C. J. & MANTOVANI, M. S. M. 1994. The denudation history of the onshore continental margin of SE Brazil inferred from apatite fission track data. *Journal of Geophysical Research*, **99**, 18,117–18,145.

GALLAGHER, K., STEPHENSON, J., BROWN, R., HOLMES, C. & FITZGERALD, P. 2005. Low temperature thermochronology and modeling strategies for multiple samples 1: Vertical profiles. *Earth and Planetary Science Letters*, **237**, 193–208.

GARVER, J. I. & BRANDON, M. T. 1994. Fission-track ages of detrital zircons from Cretaceous strata, southern British Columbia: implications for the Baja BC hypothesis. *Tectonics*, **13**, 401–420.

GARVER, J. I., BRANDON, M. T., RODEN-TICE, M. & KAMP, P. J. J. 1999. Erosional denudation determined by fission-track ages of detrital apatite and zircon. *In*: RING, U., BRANDON, M. T., WILLETT, S. & LISTER, G. (eds) *Exhumation Processes: Normal Faulting, Ductile Flow, and Erosion*. Geological Society, London, Special Publications, **154**, 283–304.

GENTNER, W., KLEINMANN, B. & WAGNER, G. A. 1967. New K–Ar- and fission track ages of impact glasses and tektites. *Earth and Planetary Science Letters*, **2**, 83–86.

GIBSON, M., SINCLAIR, H. D., LYNN, G. J. & STUART, F. M. 2007. Late- to post-orogenic exhumation of the Central Pyrenees revealed through combined thermochronological data and modelling. *Basin Research*, **19**, 323–334.

GILCHRIST, A. R. & SUMMERFIELD, M. A. 1990. Differentiated denudation and flexural isostasy in formation of rifted-margin upwarps. *Nature*, **346**, 739–742.

GLASMACHER, U. A., MANN, U. & WAGNER, G. N. A. 2002*a*. Thermotectonic evolution of the Barrandian, Czech Republic, as revealed by apatite fission-track analysis. *Tectonophysics*, **359**, 381–402.

GLASMACHER, U. A., WAGNER, G. A. & PUCHKOV, V. N. 2002*b*. Thermo-tectonic evolution of the western Fold-and-Thrust belt, southern Urals, Russia, as revealed by apatite fission-track data. *Tectonophysics*, **354**, 25–48.

GLEADOW, A. J. W. & DUDDY, I. R. 1981. A natural long-term track annealing experiment for apatite. *Nuclear Tracks*, **5**, 169–174.

GLEADOW, A. J. W. & FITZGERALD, P. G. 1987. Uplift history and structure of the Transantarctic Mountains: new evidence from fission track dating of basement apatites in the Dry Valleys area, southern Victoria Land. *Earth and Planetary Science Letters*, **82**, 1–14.

GLEADOW, A. J. W. & LOVERING, J. F. 1974. Effect of weathering on fission track dating. *Earth and Planetary Science Letters*, **22**, 163–168.

GLEADOW, A. J. W., BELTON, D. X., KOHN, B. P. & BROWN, R. W. 2002. Fission track dating of phosphate minerals and the thermochronology of apatite. *In*: KOHN, M. J., RAKOVAN, J. & HUGHES, J. M. (eds) *Phosphates; Geochemical, Geobiological, and Materials Importance*. Mineralogical Society of America and Geochemical Society, Washington, DC, 579–630.

GLEADOW, A. J. W., DUDDY, I. R., GREEN, P. F. & HEGARTY, K. A. 1986*a*. Fission track lengths in the apatite annealing zone and the interpretation of mixed ages. *Earth and Planetary Science Letters*, **78**, 245–254.

GLEADOW, A. J. W., DUDDY, I. R., GREEN, P. F. & LOVERING, J. 1986*b*. Confined fission track lengths in apatite: a diagnostic tool for thermal history analysis. *Contribution Mineralogy Petrology*, **94**, 405–415.

GLEADOW, A. J. W., DUDDY, I. R. & LOVERING, J. F. 1983. Fission track analysis: a new tool for the evaluation of thermal histories and hydrocarbon potential. *The APEA Journal*, **23**, 93–102.

GLEADOW, A. J. W., GLEADOW, S., BELTON, D. X., KOHN, B. P., KROCHMAL, M. & BROWN, R. 2009. Coincidence mapping – a key strategy for the automatic counting of fission tracks in natural minerals. *In*: LISKER, F., VENTURA, B. & GLASMACHER, U. A. (eds) *Thermochronological Methods: From Palaeotemperature Constraints to Landscape Evolution Models*. Geological Society, London, Special Publications, **324**, 25–36.

GLEADOW, A. J. W., MCKELVEY, B. C. & FERGUSON, K. U. 1984. Uplift history of the Transantarctic Mountains in the Dry Valleys area, southern Victoria Land, Antarctica, from apatite fission track ages. *New Zealand Journal of Geology and Geophysics*, **27**, 457–464.

GLOTZBACH, C., REINECKER, J., DANISIK, M., RAHN, M., FRISCH, W. & SPIEGEL, C. 2008. Neogene exhumation history of the Mont Blanc massif, western Alps. *Tectonics*, **27**, TC4011; doi:10.1029/2008TC002257.

GLOTZBACH, C., SPIEGEL, C., REINECKER, J., RAHN, M. & FRISCH, W. 2009. What perturbs isotherms? An assessment using fission-track thermochronology and thermal modelling along the Gotthard transect, Central Alps. *In*: LISKER, F., VENTURA, B. & GLASMACHER, U. A. (eds) *Thermochronological Methods: From Palaeotemperature Constraints to Landscape Evolution Models*. Geological Society, London, Special Publications, **324**, 111–124.

GREEN, P. F. 1981. 'Track-in track' length measurements in annealed apatites. *Nuclear Tracks*, **5**, 121–128.

GREEN, P. F. 1985. Comparison of zeta calibration baselines for fission-track dating of apatite, zircon and sphene. *Chemical Geology*, **58**, 1–22.

GREEN, P. F. 1986. On the thermo-tectonic evolution of Northern England: evidence from fission track analysis. *Geological Magazine*, **123**, 493–506.

GREEN, P. F. 1988. The relationship between track shortening and fission track age reduction in apatite: combined influences of inherent instability, annealing anisotropy, length bias and system calibration. *Earth and Planetary Science Letters*, **89**, 335–352.

GREEN, P. F. & DURRANI, S. A. 1977. Annealing studies of tracks in crystals. *Nuclear Track Detection*, **1**, 33–39.

GREEN, P. F., CROWHURST, P. V., DUDDY, I. R., JAPSEN, P. & HOLFORD, S. P. 2006. Conflicting (U–Th)/He and fission track ages in apatite: enhanced He retention, not anomalous annealing behaviour. *Earth and Planetary Science Letters*, **250**, 407–427.

GREEN, P. F., DUDDY, I. R. & BRAY, R. J. 1995. Applications of thermal history reconstruction in inverted basins. *In*: BUCHANAN, J. G. & BUCHANAN, P. G. (eds) *Basin Inversion*. Geological Society, London, Special Publications, **88**, 149–165.

GREEN, P. F., DUDDY, I. R., GLEADOW, A. J. W. & LOVERING, J. F. 1989*a*. Apatite fission track analysis as a paleotemperature indicator for hydrocarbon exploration. *In*: NAESER, N. D. & MCCULLOH, T. H. (eds) *Thermal History of Sedimentary Basins: Methods and Case Histories*. Springer, New York, 181–195.

GREEN, P. F., DUDDY, I. R., LASLETT, G. M., HEGARTHY, K. A., GLEADOW, A. J. W. & LOVERING, J. F. 1989*b*. Thermal annealing of fission tracks in apatite: 4 – Quantitative modelling techniques and extension to geological timescales. *Chemical Geology (Isotope Geoscience Section)*, **79**, 155–182.

GREEN, P. F., DUDDY, I. R., GLEADOW, A. J. W., TINGATE, P. R. & LASLETT, G. M. 1985. Fission-track annealing in apatite; track length measurements and the form of the Arrhenius plot. *Nuclear Tracks*, **10**, 323–328.

GREEN, P. F., DUDDY, I. R., GLEADOW, A. J. W., TINGATE, P. R. & LASLETT, G. M. 1986. Thermal annealing of fission tracks in apatite, 1. A qualitative description. *Chemical Geology (Isotope Geoscience Section)*, **59**, 237–253.

GREEN, P. F., DUDDY, I. R. & HEGARTY, K. A. 2002. Quantifying exhumation from apatite fission-track analysis and vitrinite reflectance data: precision, accuracy and latest results from the Atlantic margin of NW Europe. *In*: DORÉ, A. G., CARTWRIGHT, J. A., STOKER, M. S., TURNER, J. P. & WHITE, N. (eds) *Exhumation of the North Atlantic Margin: Timing, Mechanisms and Implications for Petroleum Exploration*. Geological Society, London, Special Publications, **196**, 331–354.

GRUNDMANN, G. & MORTEANI, G. 1985. The young uplift and thermal history of the Central Eastern Alps (Austria/Italy), evidence from apatite fission track ages. *Jahrbuch der Geologischen Bundesanstalt*, **128**, 197–216.

GUNNELL, Y., GALLAGHER, K., CARTER, A., WIDDOWSON, M. & HURFORD, A. J. 2003. Denudation history of the continental margin of western peninsular India since the early Mesozoic – reconciling apatite fission-track data with geomorphology. *Earth and Planetary Science Letters*, **215**, 187–201.

HARMAN, R., GALLAGHER, K., BROWN, R., RAZA, A. & BIZZI, L. 1998. Accelerated denudation and tectonic/geomorphic reactivation of the cratons of northeastern Brazil during the Late Cretaceous. *Journal of Geophysical Research B*, **103**, 27,091–27,105.

HASEBE, N., BARBARAND, J., JARVIS, K., CARTER, A. & HURFORD, A. J. 2004. Apatite fission-track chronometry using laser ablation ICP-MS. *Chemical Geology*, **207**, 135–145.

HASEBE, N., CARTER, A., HURFORD, A. J. & ARAI, S. 2009. The effect of chemical etching on LA–ICP-MS analysis in determining uranium concentration for fission-track chronometry. *In*: LISKER, F., VENTURA, B. & GLASMACHER, U. A. (eds) *Thermochronological Methods: From Palaeotemperature Constraints to Landscape Evolution Models*. Geological Society, London, Special Publications, **324**, 37–46.

HENDRIKS, B. W. H. & REDFIELD, T. F. 2005. Apatite fission track and (U–Th)/He data from Fennoscandia: An example of underestimation of fission track annealing in apatite. *Earth and Planetary Science Letters*, **236**, 443–458.

HENDRIX, M. S., DUMITRU, T. A. & GRAHAM, S. A. 1994. Late Oligocene–early Miocene unroofing in the Chinese Tian Shan: an early effect of the India–Asia collision. *Geology*, **22**, 487–490.

HOURIGAN, J. K., REINERS, P. W. & BRANDON, M. T. 2005. U–Th zonation-dependent alpha-ejection in (U–Th)/He chronometry. *Geochimica et Cosmochimica Acta*, **69**, 3349–3365.

HOUSE, M. A., KELLEY, S. A. & ROY, M. 2003. Refining the footwall cooling history of a rift flank uplift, Rio Grande rift, New Mexico. *Tectonics*, **22**, 1060; doi:1010.1029/2002TC001418.

HURFORD, A. J. 1986. Cooling and uplift patterns in the Lepontine Alps South Central Switzerland and an age of vertical movement on the Insubric fault line. *Contributions to Mineralogy and Petrology*, **92**, 413–427.

HURFORD, A. J. 1990. Standardization of fission track dating calibration: Recommendation by the Fission Track Working Group of the I.U.G.S. Subcommission on Geochronology. *Chemical Geology*, **80**, 171–178.

HURFORD, A. J. & CARTER, A. 1991. The role of fission track dating in discrimination of provenance. *In*: MORTON, A. C., TODD, S. P. & HAUGHTON, P. D. W. (eds) *Developments in Sedimentary Provenance Studies*. Geological Society, London, Special Publications, **57**, 67–78.

HURFORD, A. J. & GREEN, P. F. 1983. The zeta age calibration of fission-track dating. *Chemical Geology*, **1**, 285–317.

HURFORD, A. J., FITCH, F. J. & CLARKE, A. 1984. Resolution of the age structure of the detrital zircon populations of two Lower Cretaceous sandstones from the Weald of England by fission track dating. *Geological Magazine*, **121**, 269–277.

HURFORD, A. J., HUNZIKER, J. C. & STOCKHERT, B. 1991. Constraints on the late thermotectonic evolution of the western Alps: evidence for episodic rapid uplift. *Tectonics*, **10**, 758–769.

ISSLER, D. R. 1996. Optimizing time step size for apatite fission track annealing models. *Computers and Geosciences*, **22**, 67–74.

JAPSEN, P., BONOW, J. M., GREEN, P. F., CHALMERS, J. A. & LIDMAR-BERGSTROM, K. 2006. Elevated, passive continental margins: Long-term highs or Neogene uplifts? New evidence from West Greenland. *Earth and Planetary Science Letters*, **248**, 330–339.

JOHNSON, C. 1997. Resolving denudational histories in orogenic belts with apatite fission-track thermochronology and structural data: an example from southern Spain. *Geology*, **25**, 623–626.

JOHNSON, C. & GALLAGHER, K. 2000. A preliminary Mesozoic and Cenozoic denudation history of the North East Greenland onshore margin. *Global and Planetary Change*, **24**, 261–274.

KALASWAD, S., RODEN, M. K., MILLER, D. S. & MORISAWA, M. 1993. Evolution of the continental margin of western India: new evidence from apatite fission-track dating. *Journal of Geology*, **101**, 667–673.

KAMP, P. J. J. 1997. Paleogeothermal gradient and deformation style, Pacific front of the Southern Alps Orogen: constraints from fission track thermochronology. *Tectonophysics*, **271**, 37–58.

KAMP, P. J. J. & GREEN, P. F. 1990. Thermal and tectonic history of selected Taranaki Basin (New Zealand) wells assessed by apatite fission track analysis. *AAPG Bulletin*, **74**, 1401–1419.

KEEVIL, N. B. 1943. The distribution of helium in rocks, V. *American Journal of Science*, **241**, 277–306.

KETCHAM, R. A. 2005. Forward and inverse modeling of low-temperature thermochronometry data. *Reviews in Mineralogy and Geochemistry*, **58**, 275–314.

KETCHAM, R. A., CARTER, A., DONELICK, R. A., BARBARAND, J. & HURFORD, A. J. 2007. Improved modelling of fission-track annealing in apatite. *American Mineralogist*, **92**, 789–798.

KETCHAM, R. A., DONELICK, R. A. & DONELICK, M. B. 2000. AFTSolve: a program for multi-kinetic modeling of apatite fission-track data. *Geological Materials Research*. **2**, 1–32.

KOHN, B. P. & EYAL, M. 1981. History of uplift of the crystalline basement of Sinai and its relation to the opening of the Red Sea as revealed by fission track dating of apatites. *Earth and Planetary Science Letters*, **52**, 129–141.

KOHN, B. P., FOSTER, D. A., STECKLER, M. S. & EYAL, M. 1997. Thermal history of the eastern Gulf of Suez, II. Reconstruction from apatite fission track and $^{40}Ar/^{39}Ar$ K-feldspar measurements. *Tectonophysics*, **283**, 219–239.

KOHN, B. P., GLEADOW, A. J. W., BROWN, R. W., GALLAGHER, K., O'SULLIVAN, P. B. & FOSTER, D. A. 2002. Shaping the Australian crust over the last 300 million years: insights from fission track thermotectonic imaging and denudation studies of key terranes. *Australian Journal of Earth Sciences*, **49**, 697–717.

KOHN, B. P., GLEADOW, A. J. W. & COX, S. J. D. 1999. Denudation history of the Snowy Mountains; constraints from apatite fission track thermochronology. *In*: KOHN, B. P. & BISHOP, P. (eds) *Long-term Landscape Evolution of the Southeastern Australian Margin; Apatite Fission Track Thermochronology and Geomorphology*. Blackwell, Melbourne, 181–198.

KOHN, B. P., LORENCAK, M., GLEADOW, A. J. W., KOHLMANN, F., RAZA, A., OSADETZ, K. G. & SORJONEN-WARD, P. 2009. A reappraisal of low-temperature thermochronology of the eastern Fennoscandia Shield and radiation-enhanced apatite fission-track annealing. *In*: LISKER, F., VENTURA, B. & GLASMACHER, U. A. (eds) *Thermochronological Methods: From Palaeotemperature Constraints to Landscape Evolution Models*. Geological Society, London, Special Publications, **324**, 193–216.

KOOI, H. & BEAUMONT, C. 1994. Escarpment evolution on high-elevation rifted margins: insights derived from a surface processes model that combines diffusion, advection, and reaction. *Journal of Geophysical Research*, **99**, 12,191–12,209.

KOONS, P. O. 1989. The topographic evolution of collisional mountain belts: a numerical look at the southern Alps, New Zealand. *American Journal of Science*, **289**, 1041–1069.

KOUNOV, A., VIOLA, G., DE WIT, M. & ANDREOLI, M. 2009. Denudation along the Atlantic passive margin: new insights from apatite fission-track analysis on the western coast of South Africa. *In*: LISKER, F., VENTURA, B. & GLASMACHER, U. A. (eds) *Thermochronological Methods: From Palaeotemperature Constraints to Landscape Evolution Models*. Geological Society, London, Special Publications, **324**, 287–306.

KUHLEMANN, J., KRUMREI, I., DANIŠÍK, M. & VAN DER BORG, K. 2009. Weathering of granite and granitic regolith in Corsica: short-term ^{10}Be versus long-term thermochronological constraints. *In*: LISKER, F., VENTURA, B. & GLASMACHER, U. A. (eds) *Thermochronological Methods: From Palaeotemperature Constraints to Landscape Evolution Models*. Geological Society, London, Special Publications, **324**, 217–235.

LASLETT, G. M., GALBRAITH, R. F. & GREEN, P. F. 1994. The analysis of projected track lengths. *Radiation Measurements*, **23**, 103–123.

LASLETT, G. M., GLEADOW, A. J. W. & DUDDY, I. R. 1984. The relationship between fission track length and track density distributions. *Nuclear Tracks*, **9**, 29–38.

LASLETT, G. M., GREEN, P. F., DUDDY, I. R. & GLEADOW, A. J. W. 1987. Thermal annealing of fission tracks in apatite, 2. A quantitative analysis. *Chemical Geology (Isotope Geoscience Section)*, **65**, 1–13.

LASLETT, G. M., KENDALL, W. S., GLEADOW, A. J. W. & DUDDY, I. R. 1982. Bias in measurement of fission-track length distributions. *Nuclear Tracks*, **6**, 79–85.

LIPPOLT, H. J., BOSCHMANN KÄTHLER, W. & ARNDT, H. 1982. Helium und Uran in Schwarzwälder Bleiglanzen, ein Datierungsversuch. *Oberrheinische Geologische Abhandlungen*, **31**, 31–46.

LIPPOLT, H. J., WERNICKE, R. S. & BÄHR, R. 1995. Paragenetic specularite and adularia (Elba, Italy): Concordant (U + Th)–He and K–Ar ages. *Earth and Planetary Science Letters*, **132**, 43–51.

LISKER, F. 2002. Review of fission track studies in northern Victoria Land – Passive margin evolution versus uplift of the Transantarctic Mountains. *Tectonophysics*, **349**, 57–73.

LISKER, F. & FACHMANN, S. 2001. Phanerozoic history of the Mahanadi region, India. *Journal of Geophysical Research*, B **106**, 22,027–22,050.

LISKER, F., BROWN, R. & FABEL, D. 2003. Denudational and thermal history along a transect across the Lambert Graben, northern Prince Charles Mountains, Antarctica derived from apatite fission track thermochronology. *Tectonics*, **22**, 1055; doi:1010.1029/2002TC001477.

LOGAN, P. & DUDDY, I. R. 1998. An investigation of the thermal history of the Ahnet–Reggane basins, central Algeria, the consequences for hydrocarbon generation accumulation. *In*: MACGREGOR, D. S., MOODY, R. T. J. & CLARK-LOWES, D. D. (eds) *Petroleum Geology of North Africa*. Geological Society, London, Special Publications, **132**, 131–155.

LONERGAN, L. & JOHNSON, C. 1998. Reconstructing orogenic exhumation histories using synorogenic detritical zircons and apatites from the Betic Cordillera, SE Spain. *Basin Research*, **10**, 353–364.

LORENCAK, M., KOHN, B. P., OSADETZ, K. G. & GLEADOW, A. J. W. 2004. Combined apatite fission track and (U–Th)/He thermochronometry in a slowly cooled terrane: results from a 3440-m-deep drill hole in the southern Canadian Shield. *Earth and Planetary Science Letters*, **227**, 87–104.

MALUSÀ, M. G., ZATTIN, M., ANDÒ, S., GARZANTI, E. & VEZZOLI, G. 2009. Focused erosion in the Alps constrained by fission-track ages on detrital apatites. *In*: LISKER, F., VENTURA, B. & GLASMACHER, U. A. (eds) *Thermochronological Methods: From Palaeotemperature Constraints to Landscape Evolution Models*. Geological Society, London, Special Publications, **324**, 141–152.

MANCKTELOW, N. S. & GRASEMANN, B. 1997. Time-dependent effects of heat advection and topography on cooling histories during erosion. *Tectonophysics*, **270**, 167–195.

MATHIESEN, A., BIDSTRUP, T. & CHRISTIANSEN, F. G. 2000. Denudation and uplift history of the Jameson Land basin, East Greenland – constrained from maturity and apatite fission track data. *Global and Planetary Change*, **24**, 275–301.

MCINNES, B. I. A., EVANS, N. J., FU, F. Q. & GARWIN, S. 2005. Application of thermochronology to hydrothermal ore deposits. *Reviews in Mineralogy and Geochemistry*, **58**, 467–498.

MEESTERS, A. G. C. A. & DUNAI, T. J. 2002*a*. Solving the production-diffusion equation for finite diffusion domains of various shapes: Part I. Implications for low-temperature (U–Th)/He thermochronology. *Chemical Geology*, **186**, 333–344.

MEESTERS, A. G. C. A. & DUNAI, T. J. 2002*b*. Solving the production-diffusion equation for finite diffusion domains of various shapes: Part II. Application to cases with [alpha]-ejection and nonhomogeneous distribution of the source. *Chemical Geology*, **186**, 347–363.

MILLER, D. S. & WAGNER, G. A. 1979. Age and intensity of thermal events by fission track analysis; the Ries impact crater. *Earth and Planetary Science Letters*, **43**, 351–358.

MITCHELL, M. M. 1997. Elevated mid-Cretaceous paleotemperatures in the western Otway Basin: consequences for hydrocarbon generation models. *The APPEA Journal*, **37**, 505–523.

MOORE, M. E., GLEADOW, A. J. W. & LOVERING, J. F. 1986. Thermal evolution of rifted continental margins: new evidence from fission tracks in basement apatites from southeastern Australia. *Earth and Planetary Science Letters*, **78**, 255–270.

MORWOOD, M. J., O'SULLIVAN, P. B., AZIZ, F. & RAZA, A. 1998. Fission-track ages of stone tools and fossils on the east Indonesian island of Flores. *Nature*, **392**, 173–176.

MURRELL, G. R., SOBEL, E. R., CARRAPA, B. & ANDRIESSEN, P. 2009. Calibration and comparison of etching techniques for apatite fission-track thermochronology. *In*: LISKER, F., VENTURA, B. & GLASMACHER, U. A. (eds) *Thermochronological Methods: From Palaeotemperature Constraints to Landscape Evolution Models*. Geological Society, London, Special Publications, **324**, 73–85.

NAESER, C. W. 1967. The use of apatite and sphene for fission track age determinations. *Geological Society of America Bulletin*, **78**, 1523–1526.

NAESER, C. W. 1979. Fission-track dating and geological annealing of fission tracks. *In*: JÄGER, E. & HUNZIKER, J. C. (eds) *Lectures in Isotope Geology*. Springer, New York, 154–169.

NOBLE, W. D., FOSTER, D. A. & GLEADOW, A. J. W. 1997. The post-Pan-African thermal and extensional history of crystalline basement rocks in eastern Tanzania. *Tectonophysics*, **275**, 331–350.

OLSEN, K. H. 1995. *Continental Rifts: Evolution, Structure, Tectonics*. Elsevier, New York.

OMAR, G. I., STECKLER, M. S., BUCK, W. R. & KOHN, B. P. 1989. Fission track analysis of basement apatites at the western margin of the Gulf of Suez rift, Egypt: evidence for synchroneity of uplift and subsidence. *Earth and Planetary Science Letters*, **94**, 316–328.

OSADETZ, K. G., KOHN, B. P., FEINSTEIN, S. & O'SULLIVAN, P. B. 2002. Thermal history of Canadian Williston basin from apatite fission-track thermochronology – implications for petroleum systems and geodynamic history. *Tectonophysics*, **349**, 221–249.

O'SULLIVAN, P. B. & CURRIE, L. D. 1996. Thermotectonic history of Mt Logan, Yukon territory, Canada: implications of multiple episodes of middle to late Cenozoic denudation. *Earth and Planetary Science Letters*, **144**, 251–261.

O'SULLIVAN, P. B. & PARRISH, R. R. 1995. The importance of apatite composition and single-grain ages when interpreting fission track data from plutonic rocks: a case study from the Coast Ranges, British Columbia. *Earth and Planetary Science Letters*, **132**, 213–224.

O'SULLIVAN, P. B., KOHN, B. P., FOSTER, D. A. & GLEADOW, A. J. W. 1995. Fission track data from the Bathurst Batholith; evidence for rapid Mid-Cretaceous uplift and erosion within the eastern highlands of Australia. *Australian Journal of Earth Sciences*, **42**, 597–607.

O'SULLIVAN, P. B., KOHN, B. P. & MITCHELL, M. M. 1998. Phanerozoic reactivation along a fundamental Proterozoic crustal fault, the Darling River Lineament, Australia; constraints from apatite fission track thermochronology. *Earth and Planetary Science Letters*, **164**, 451–465.

O'SULLIVAN, P. B., MITCHELL, M. M., O'SULLIVAN, A. J., KOHN, B. P. & GLEADOW, A. J. W. 2000. Thermotectonic history of the Bassian Rise, Australia; implications for the breakup of eastern Gondwana along Australia's southeastern margins. *Earth and Planetary Science Letters*, **182**, 31–47.

PAGEL, M., BRAUN, J. J., DISNAR, J. R., MARTINEZ, L., RENAC, C. & VASSEUR, G. 1997. Thermal history constraints from studies of organic matter, clay minerals, fluid inclusions, and apatite fission tracks at the Ardeche paleo-margin (BA1 drill hole, GPF Program), France. *Journal of Sedimentary Research, Section A: Sedimentary Petrology and Processes*, **67**, 235–245.

PARRISH, R. R. 1983. Cenozoic thermal evolution and tectonics of the Coast Mountains of British Columbia. 1. Fission-track dating, apparent uplift rates, and patterns of uplift. *Tectonics*, **2**, 601–631.

PAUL, T. A. & FITZGERALD, P. G. 1992. Transmission electron-microscopic investigation of fission tracks in fluorapatite. *American Mineralogist*, **77**, 336–344.

PERSANO, C., STUART, F. M., BISHOP, P. & BARFOD, D. N. 2002. Apatite (U–Th)/He age constraints on the development of the Great Escarpment on the southeastern Australian passive margin. *Earth and Planetary Science Letters*, **200**, 79–90.

PEVEAR, D. R. 1999. Illite and hydrocarbon exploration. *Proceedings of the National Academy of Science, USA*, **96**, 3440–3446.

PRICE, P. B. & WALKER, R. M. 1963. Fossil tracks of charged particles in mica and the age of minerals. *Journal of Geophysical Research*, **68**, 4847–4862.

RAAB, M. J., BROWN, R. W., GALLAGHER, K., CARTER, A. & WEBER, K. 2002. Late Cretaceous reactivation of major crustal shear zones in northern Namibia: constraints from apatite fission track analysis. *Tectonophysics*, **349**, 75–92.

RAAB, M. J., BROWN, R. W., GALLAGHER, K., WEBER, K. & GLEADOW, A. J. W. 2005. Denudational and thermal history of the Early Cretaceous Brandberg and Okenyenya igneous complexes on Namibia's Atlantic passive margin. *Tectonics*, **24**, TC3006; doi:10.1029/2004TC001688.

RABONE, J. A. L., CARTER, A., HURFORD, A. J. & DE LEEUW, N. H. 2008. Modelling the formation of fission tracks in apatite minerals using molecular dynamics simulations. *Physics and Chemistry of Minerals*, **35**(10), 583–596; 10.1007/s00269-008-0250-6.

RAHL, J. M., REINERS, P. W., CAMPBELL, I. H., NICOLESCU, S. & ALLEN, C. M. 2003. Combined single-grain (U–Th)/He and U/Pb dating of detrital zircons from the Navajo Sandstone, Utah. *Geology*, **31**, 761–764.

RAHN, M. K., HURFORD, A. J. & FREY, M. 1997. Rotation and exhumation of a thrust plane: apatite fission-track data from the Glarus thrust, Switzerland. *Geology*, **25**, 599–602.

REDFIELD, T. F., SCHOLL, D. W., FITZGERALD, P. G. & BECK, M. E., JR. 2007. Escape tectonics and the extrusion of Alaska: Past, present, and future. *Geology*, **35**, 1039–1042.

REINERS, P. W. & BRANDON, M. T. 2006. Using thermochronology to understand orogenic erosion. *Annual Review of Earth and Planetary Sciences*, **34**, 419–466.

REINERS, P. W. & EHLERS, T. A. (eds) 2005. *Low-temperature Thermochronology: Techniques, Interpretations, and Applications*. Reviews in Mineralogy and Geochemistry, **58**. Mineralogical Society of America, Geochemical Society.

REINERS, P. W., EHLERS, T. A., GARVER, J. I., MITCHELL, S. G., MONTGOMERY, D. R., VANCE, J. A. & NICOLESCU, S. 2002. Late Miocene exhumation and uplift of the Washington Cascades. *Geology*, **30**, 767–770.

REINERS, P. W., EHLERS, T. A. & ZEITLER, P. K. 2005. Past, present, and future of thermochronology. *Reviews in Mineralogy and Geochemistry*, **58**, 1–18.

REINERS, P. W., SPELL, T. L., NICOLESCU, S. & ZANETTI, K. A. 2004. Zircon (U–Th)/He thermochronometry: He diffusion and comparisons with $^{40}Ar/^{39}Ar$ dating. *Geochimica et Cosmochimica Acta*, **68**, 1857–1887.

RING, U., BRANDON, M. T., WILLETT, S. D. & LISTER, G. S. 1999. Exhumation processes. *In*: RING, U., BRANDON, M. T., WILLETT, S. D. & LISTER, G. S. (eds) *Exhumation Processes: Normal Faulting, Ductile Flow, and Erosion*. Geological Society, London, Special Publications, **154**, 1–27.

ROHRMANN, M., VAN DER BEEK, P. & ANDRIESSEN, P. 1994. Syn-rift thermal structure and post-rift evolution of the Oslo Rift (southeast Norway): New constraints from fission track thermochronology. *Earth and Planetary Science Letters*, **127**, 39–54.

RUIZ, G. M. H., CARLOTTO, V., VAN HEININGEN, P. V. & ANDRIESSEN, P. A. M. 2009. Steady-state exhumation pattern in the Central Andes – SE Peru. *In*: LISKER, F., VENTURA, B. & GLASMACHER, U. A. (eds) *Thermochronological Methods: From Palaeotemperature Constraints to Landscape Evolution Models*. Geological Society, London, Special Publications, **324**, 307–316.

RUTHERFORD, E. 1905. Present problems in radioactivity. *Popular Science Monthly*, May, 1–34.

SANDERS, C. A. E., ANDRIESSEN, P. A. M. & CLOETINGH, S. A. P. L. 1999. Life cycle of the East Carpathian orogen: erosion history of a doubly vergent critical wedge assessed by fission track thermochronology. *Journal of Geophysical Research*, **104**, 29,095–29,112.

SCHAER, J. P., REIMER, G. M. & WAGNER, G. A. 1975. Actual and ancient uplift rate in the Gotthard region, Swiss Alps: a comparison between precise levelling and apatite fission track age. *Tectonophysics*, **29**, 293–300.

SEWARD, D. & TULLOCH, A. J. 1991. Fission-track analysis of Tertiary uplift history of granitic basement in the Victoria Range, West Coast, New Zealand. *New Zealand Journal of Geology and Geophysics*, **34**, 115–120.

SEWARD, D., GRUJIC, D. & SCHREURS, G. 2004. An insight into the breakup of Gondwana: identifying events through low-temperature thermochronology from the basement rocks of Madagascar. *Tectonics*, **23**, TC3007; doi:10.1029/2003TC001556.

SHUSTER, D. L. & FARLEY, K. A. 2009. The influence of artificial radiation damage and thermal annealing on helium diffusion kinetics in apatite. *Geochimica et Cosmochimica Acta*, **73**(1), 183–196; doi:10.1016/j.gca.2008.10.013.

SHUSTER, D. L., FLOWERS, R. M. & FARLEY, K. A. 2006. The influence of natural radiation damage on helium diffusion kinetics in apatite. *Earth and Planetary Science Letters*, **249**, 148–161.

SHUSTER, D. L., VASCONCELOS, P. M., HEIM, J. A. & FARLEY, K. A. 2005. Weathering geochronology by (U–Th)/He dating of goethite. *Geochimica et Cosmochimica Acta*, **69**, 659–673.

SILK, E. C. H. & BARNES, R. S. 1959. Examination of fission fragment tracks with an electron microscope. *Philosophical Magazine*, **4**, 970–972.

SODERLUND, P., JUEZ-LARRE, J., PAGE, L. M. & DUNAI, T. J. 2005. Extending the time range of apatite (U–Th)/He thermochronometry in slowly cooled terranes: palaeozoic to Cenozoic exhumation history of southeast Sweden. *Earth and Planetary Science Letters*, **239**, 266–275.

SORKHABI, R. B., STUMP, E., FOLAND, K. A. & JAIN, A. K. 1996. Fission-track and $^{40}Ar/^{39}Ar$ evidence for episodic denudation of the Gangotri granites in the Garhwal Higher Himalaya, India. *Tectonophysics*, **260**, 187–199.

SPIEGEL, C., KOHN, B., BELTON, D. X., BERNER, Z. & GLEADOW, A. J. W. 2009. Apatite (U–Th–Sm)/He thermochronology of rapidly cooled samples: the effect of He implantation. *Earth and Planetary Science Letters*, doi: 10.1016/j.epsl.2009.05.045.

SPIEGEL, C., KOHN, B., RAZA, A., RAINER, T. & GLEADOW, A. 2007. The effect of long-term low-temperature exposure on apatite fission track stability: a natural annealing experiment in the deep ocean. *Geochimica et Cosmochimica Acta*, **71**, 4512–4537.

SPIEGEL, C., KUHLEMANN, J., DUNKL, I. N. & FRISCH, W. 2001. Paleogeography and catchment evolution in a mobile orogenic belt: the Central Alps in Oligo-Miocene times. *Tectonophysics*, **341**, 33–47.

STEPHENSON, J., GALLAGHER, K. & HOLMES, C. C. 2006. Low temperature thermochronology and strategies for multiple samples: 2: Partition modelling for 2D/3D distributions with discontinuities. *Earth and Planetary Science Letters*, **241**, 557–570.

STOCKLI, D. F. 2005. Application of low-temperature thermochronometry to extensional tectonic settings. *Reviews in Mineralogy and Geochemistry*, **58**, 411–448.

STÜWE, K., WHITE, L. & BROWN, R. 1994. The influence of eroding topography on steady-state isotherms. Application to fission track analysis. *Earth and Planetary Science Letters*, **124**, 63–74.

SUMMERFIELD, M. A. 1991. *Global Geomorphology*. Longman, London.

SUMMERFIELD, M. A. & BROWN, R. W. 1998. Geomorphic factors in the interpretation of fission-track data. *In*: VAN DEN HAUTE, P & DE CORTE, F. (eds) *Advances in Fission-track Geochronology*. Kluwer Academic, Dordrecht, 269–284.

TAGAMI, T. 2005. Zircon fission-track thermochronology and applications to fault studies. *Reviews in Mineralogy and Geochemistry*, **58**, 95–122.

THOMSON, S. N. 1998. Assessing the nature of tectonic contacts using fission-track thermochronology: an example from the Calabrian Arc, southern Italy. *Terra Nova*, **10**, 32–36.

THOMSON, S. N. 2002. Late Cenozoic geomorphic and tectonic evolution of the Patagonian Andes between latitudes 42°S and 76°S: an appraisal based on fission-track results from the transpressional intra-arc Liquine–Ofqui fault zone. *Geological Society of America Bulletin*, **114**, 1159–1173.

TIMAR-GENG, Z., HENK, A. & WETZEL, A. 2009. Convective heat transfer in a steeply dipping fault zone and its impact on the interpretation of fission-track data – a modelling study. *In*: LISKER, F., VENTURA, B. & GLASMACHER, U. A. (eds) *Thermochronological Methods: From Palaeotemperature Constraints to Landscape Evolution Models*. Geological Society, London, Special Publications, **324**, 87–98.

TINGATE, P. R. & DUDDY, I. R. 2002. The thermal history of the eastern Officer Basin (South Australia); evidence from apatite fission track analysis and organic maturity data. *Tectonophysics*, **349**, 251–275.

UNDERDOWN, R., REDFERN, J. & LISKER, F. 2007. Constraining the burial history of the Ghadames Basin, North Africa: an integrated analysis using sonic velocities, vitrinite reflectance data and apatite fission track ages. *Basin Research*, **19**(4), 557–578; doi: 10.1111/j.1365-2117.2007.00335.x.

VAN DER BEEK, P. A., DELVAUX, D., ANDRIESSEN, P. A. M. & LEVI, K. G. 1996. Early Cretaceous denudation related to convergent tectonics in the Baikal region, SE Siberia. *Journal of the Geological Society, London*, **153**, 515–523.

VAN DER BEEK, P., MBEDE, E., ANDRIESSEN, P. & DELVAUX, D. 1998. Denudation history of the Malawi and Rukwa Rift flanks (East African Rift System) from apatite fission track thermochronology. *Journal of African Earth Sciences*, **26**, 363–385.

VAN DER BEEK, P., ROBERT, X., MUGNIER, J.-L., BERNET, M., HUYGHE, P. & LABRIN, E. 2006. Late Miocene–Recent exhumation of the central Himalaya and recycling in the foreland basin assessed by apatite fission-track thermochronology of Siwalik sediments, Nepal. *Basin Research* **18**, 413–434.

VENTURA, B., LISKER, F. & KOPP, J. 2009. Thermal and denudation history of the Lusatian Block (NE Bohemian Massif, Germany) as indicated by apatite fission-track data. *In*: LISKER, F., VENTURA, B. & GLASMACHER, U. A. (eds) *Thermochronological Methods: From Palaeotemperature Constraints to Landscape Evolution Models*. Geological Society, London, Special Publications, **324**, 181–192.

VENTURA, B., PINI, G. A. & ZUFFA, G. G. 2001. Thermal history and exhumation of the Northern Apennines (Italy): evidence from combined apatite fission track and vitrinite reflectance data from foreland basin sediments. *Basin Research*, **13**, 435–448.

VERMEESCH, P. 2004. How many grains are needed for a provenance study? *Earth and Planetary Science Letters*, **224**, 441–451.

VINCENT, S. J., MORTON, A. C., CARTER, A., GIBBS, S. & BARABADZE, T. G. 2007. Oligocene uplift of the Western Greater Caucasus: an effect of initial Arabia–Eurasia collision. *Terra Nova*, **19**, 160–166.

WAGNER, G. A. 1966. Altersbestimmung an Tektiten und anderen natürlichen Gläsern mittels Spuren der spontanen Kernspaltung des Uran-238 ("fission track"-Methode). *Zeitschrift für Naturforschung*, **21a**, 733–745.

WAGNER, G. A. 1968. Fission track dating of apatites. *Earth and Planetary Science Letters*, **4**, 411–415.

WAGNER, G. A. 1972. The geological interpretation of fission track ages. *Transactions of the American Nuclear Society*, **15**, 117.

WAGNER, G. A. 1979. Correction and interpretation of fission track ages. *In*: JÄGER, E. & HUNZIKER, J. C. (eds) *Lectures in Isotope Geology*. Springer, New York, 170–177.

WAGNER, G. A. 1990. Apatite fission-track dating of the crystalline basement of Middle Europe: concepts and results. *Nuclear Tracks Radiation Measurements*, **17**, 277–282.

WAGNER, G. A. & REIMER, G. M. 1972. Fission track tectonics: the tectonic interpretation of fission track apatite ages. *Earth and Planetary Science Letters*, **14**, 263–268.

WAGNER, G. A. & VAN DEN HAUTE, P. 1992. *Fission-track Dating*. Ferdinand Enke, Stuttgart.

WAGNER, G. A., COYLE, D. A. *ET AL.* 1997. Post-Variscan thermal and tectonic evolution of the KTB site and its surroundings. *Journal of Geophysical Research*, **102**, B8, 18,221–18,232.

WAGNER, G. A., MILLER, D. S. & JÄGER, E. 1979. Fission track ages on apatite of Bergell rocks from central Alps and Bergell boulders in Oligocene sediments. *Earth and Planetary Science Letters*, **45**, 355–360.

WANG, W. & ZHOU, Z. 2009. Reconstruction of paleotopography from low-temperature thermochronological data. *In*: LISKER, F., VENTURA, B. & GLASMACHER, U. A. (eds) *Thermochronological Methods: From Palaeotemperature Constraints to Landscape Evolution Models*. Geological Society, London, Special Publications, **324**, 99–110.

WILLETT, S. D. 1992. Modelling thermal annealing of fission tracks in apatite. *In*: ZENTILLI, M. & REYNOLDS, P. H. (eds) *Short Course Handbook on Low Temperature Thermochronology*. Mineralogical Society of Canada, Toronto, 43–72.

WILLETT, S. D. 1997. Inverse modeling of annealing of fission tracks in apatite. 1: A controlled random search method. *American Journal of Science*, **297**, 939–969.

WILLETT, S. D. 1999. Orogeny and orography: the effects of erosion on the structure of mountain belts. *Journal of Geophysical Research*, **104**, 28,957–28,981.

WILLETT, S. D., FISHER, D., FULLER, C., EN-CHAO, Y. & CHIA-YU, L. 2003. Erosion rates and orogenic-wedge kinematics in Taiwan inferred from fission-track thermochronometry. *Geology*, **31**, 945–948.

WOLF, R. A., FARLEY, K. A. & KASS, D. M. 1998. Modeling of the temperature sensitivity of the apatite (U–Th)/He thermochronometer. *Chemical Geology*, **148**, 105–114.

WOLF, R. A., FARLEY, K. A. & SILVER, L. T. 1996. Helium diffusion and low-temperature thermochronometry of apatite. *Geochimica et Cosmochimica Acta*, **60**, 4231–4240.

XU, C., MANSY, J. L., VAN DEN HAUTE, P., GUILLOT, F., ZHOU, Z., CHEN, J. & DE GRAVE, J. 2009. Late- and post-Variscan evolution of the Ardennes in France and Belgium: constraints from apatite fission-track data. *In*: LISKER, F., VENTURA, B. & GLASMACHER, U. A. (eds) *Thermochronological Methods: From Palaeotemperature Constraints to Landscape Evolution Models*. Geological Society, London, Special Publications, **324**, 167–179.

YAMADA, R., ONGIRAD, H., MATSUDA, T., OMURA, K., TAKEUCHI, A. & IWANO, H. 2009. Fission-track analysis of the Atotsugawa Fault (Hida Metamorphic Belt, central Japan): fault-related thermal anomaly and activation history. *In*: LISKER, F., VENTURA, B. & GLASMACHER, U. A. (eds) *Thermochronological Methods: From Palaeotemperature Constraints to Landscape Evolution Models*. Geological Society, London, Special Publications, **324**, 331–337.

YAMADA, R., TAGAMI, T., NISHIMURA, S. & ITO, H. 1995. Annealing kinetics of fission tracks in zircon: an experimental study. *Chemical Geology (Isotope Geoscience Section)*, **122**, 249–258.

ZAUN, P. E. & WAGNER, G. A. 1985. Fission track stability in zircons under geological conditions. *Nuclear Tracks and Radiation Measurements*, **10**, 303–307.

ZEITLER, P. K., HERCZIG, A. L., MCDOUGALL, I. & HONDA, M. 1987. U–Th–He dating of apatite: a potential thermochronometer. *Geochimica et Cosmochimica Acta*, **51**, 2865–2868.

ZIEGLER, P. A. & CLOETHING, S. 2004. Dynamic processes controlling evolution of rifted basins. *Earth-Science Reviews*, **64**, 1–50.

ZWINGMANN, H. & MANCKTELOW, N. 2004. Timing of Alpine fault gouges. *Earth and Planetary Science Letters*, **223**, 415–425.

Coincidence mapping – a key strategy for the automatic counting of fission tracks in natural minerals

ANDREW J. W. GLEADOW[1]*, STEWART J. GLEADOW[1], DAVID X. BELTON[2], BARRY P. KOHN[1], MICHAEL S. KROCHMAL[3] & RODERICK W. BROWN[4]

[1]*School of Earth Sciences, University of Melbourne, Melbourne, Victoria 3010, Australia*

[2]*School of Physics, University of Melbourne, and CSIRO Exploration and Mining, Melbourne, Australia*

[3]*Autoscan Systems Pty Ltd, PO Box 112, Ormond, Victoria 3204, Australia*

[4]*Centre For Geosciences, University of Glasgow, Glasgow G12 8QQ, UK*

**Corresponding author (e-mail: gleadow@unimelb.edu.au)*

Abstract: We report on new image-analysis techniques that, for the first time, provide a practical solution to the problem of fully automated counting of fission tracks in natural minerals, a long-desired goal in fission-track dating. Specific challenges to be overcome have been the discrimination of fission tracks from non-track defects, polishing scratches, etc.; resolving multiple track overlaps; and reliable identification of small tracks amongst a similarly sized background of surface defects, fluid inclusions, etc. Most previous attempts at automated image analysis have failed in one or more of these tasks. The central component of our system is called 'coincidence mapping' and utilizes two images of the same tracks obtained in transmitted and reflected light. The complementary nature of the information in these two images allows a powerful discrimination of true fission tracks from most non-track features. The much smaller average track size in the reflected light image allows the resolution of most track overlaps apparent in transmitted light. The discrimination is achieved by segmenting the two images using a custom-developed thresholding routine and extracting the coincidence of features in the two binary images. The analysis is computationally efficient and takes only a few seconds to complete the processing of images that may contain up to many hundreds of tracks. Preliminary indications are that error rates are about the same as, or better than, those achieved by a human operator using normal counting conditions in transmitted light. The performance is even better at high track densities ($>10^7 cm^{-2}$) giving the potential for measuring track densities up to an order of magnitude greater than a human operator can count. Automated counting should significantly increase the speed and consistency of analysis and improve data quality in fission-track dating through better counting statistics, increased objectivity and measurement of additional track description parameters that are not currently determined.

It has long been recognized that manual counting of fission tracks in natural minerals for fission-track dating is highly labour-intensive, and requires a high level of operator skill and training in order to produce reliable results, and to minimize the possibility of observer bias. Automated counting and analysis of fission tracks, that could potentially eliminate or reduce this human element, has therefore long been desired (e.g. Gold *et al.* 1984), but has attracted surprisingly little research effort. There has been some exploration of possible approaches, and some promising developments (e.g. Wadatsumi & Masumoto 1990; Belloni *et al.* 2000; Petford *et al.* 2003), but the goal of automated fission-track counting has remained elusive, and none of these studies has produced any significant impact on the day-to-day practice of fission-track analysis. On the other hand, some success in automatic track counting has been achieved in the much simpler case of counting ion tracks in artificial track detectors (e.g. Price & Krischer 1985; Yasuda *et al.* 2005).

Image analysis for counting alpha-particle tracks in plastic track detectors, such as CR-39, has been investigated for many years (e.g. Fews 1992; Espinosa *et al.* 1996; Boukhair *et al.* 2000; Dolleiser & Hashemi-Nezhad 2002). Mostly these methods have been used for counting relatively low track densities of large and non-overlapping track etch pits, achieving some limited success for routine counting in these highly reproducible and consistent materials. However, Yasuda *et al.* (2005) recently pointed out that even the most sophisticated of these methods are usually only successful in

From: LISKER, F., VENTURA, B. & GLASMACHER, U. A. (eds) *Thermochronological Methods: From Palaeotemperature Constraints to Landscape Evolution Models*. Geological Society, London, Special Publications, **324**, 25–36.
DOI: 10.1144/SP324.2 0305-8719/09/$15.00

counting normally incident ion tracks, and are poor at discriminating tracks from non-track features and at resolving overlapping tracks. Such methods, therefore, still require a significant amount of review and supervision by a skilled human operator, and fall far short of an ideal system that would substantially relieve the operator from the tedium of track counting. For these reasons, we consider that similar methods have little prospect of success for counting tracks in natural minerals, which are optically heterogeneous and much more prone to defects of various kinds.

The optical variability of individual grains of natural minerals and the common occurrence of non-track features pose particular challenges and difficulties for the automated image analysis of spontaneous fission tracks. Digital image-analysis systems are inherently poor at complex pattern-recognition tasks compared to the human operator, and this has inhibited the development of automated track-counting methods. On the other hand, the hope has persisted that digital methods should be much better than humans in other aspects such as speed and the consistency with which they can apply counting criteria. Other, purely technical, factors have also limited the development of automated counting methods in the past. These have included inadequate microscope control systems, limited digital imaging quality and the lack of sufficient computing power, as well as limited media space for image storage. Advances in computing, digital imaging and storage technologies have now essentially eliminated these technical considerations, and open new opportunities for automated image analysis using readily available desktop computers.

Rapid advances in optical microscopy in recent years, particularly in confocal laser scanning microscopy, have also had little impact so far on the imaging of etched fission tracks in minerals. Confocal imaging is a technique for obtaining high-resolution optical images, but only under certain limited illumination conditions, especially in fluorescence microscopy. Notwithstanding claims to the contrary (e.g. Petford & Miller 1990), confocal microscopy in its current form has little benefit for the imaging of fission tracks because it remains implemented only as a reflected light technology at the present time. Under these illumination conditions some individual fission tracks are revealed superbly, but many others are rendered almost invisible. A transmitted-light confocal imaging system would overcome this problem, but such systems are much more difficult to implement technically and none are currently available. However, the emergence of confocal imaging, and other advances in optical microscopy, have spurred new developments in microscope control systems that now enable the acquisition of digital microscope images of unprecedented quality using conventional imaging modes. Our work has, therefore, concentrated on using transmitted- and reflected-light images.

Specific problems to be overcome in the automation of fission-track counting include the discrimination of fission tracks from non-track defects (polishing scratches, dislocations, etc.), resolving multiple-track overlaps, and reliable identification of small tracks amongst a similarly sized background of surface imperfections, dust particles, crystalline and fluid inclusions, etc. Most previous attempts at automated image analysis have failed in one or more of these tasks, and the goal of automated counting has remained beyond practical reach. A new approach is clearly needed. Here we report the development of a new digital image-acquisition and -processing system based on an innovative technique that we have termed 'coincidence mapping', which has successfully demonstrated the automatic counting of fission tracks in minerals for the first time. At present, we have left aside the problem of automated track-length determination, as this involves an even more challenging discrimination problem and remains readily amenable to the semi-automated methods now generally in use.

Approach

Digital image acquisition and analysis

Our system is based on the acquisition of a set of high-quality digital images of each area selected for track counting, followed by the automated analysis of these images and then by an opportunity for final review by the operator. The image-acquisition and image-processing components are separated from each other and carried out independently, which has a number of advantages, as explained later. Once the system has been set up, the image-acquisition phase is fully automated and requires no, or only minimal, human intervention, with the resulting image sets being archived to disk storage. This means that the human operator will no longer require the same degree of advanced microscopy skills that has been the case for the older manual counting techniques. The archived images constitute a high-quality, three-dimensional (3D) 'digital replica' of the tracks in each grain counted, and are available for additional study, or later analysis whenever required. The creation of such a permanent digital record will be particularly advantageous when using laser-ablation Inductively Coupled Plasma Mass Spectroscopy (ICP-MS) for uranium analysis (Hasebe *et al.* 2004), because ablation destroys the mineral surface containing the tracks.

Once acquisition is complete, processing of the images and final review of the results can be carried out offline on any suitable computer. This separation of acquisition and analysis modules (also advocated by Yasuda *et al.* 2005) allows greater flexibility, as the automated track counting no longer needs to be carried out at the microscope. One benefit of this approach will be that identical copies of standard image sets can be created and shared at will, facilitating intercalibration and standardization between laboratories. This may also be of significant benefit in training new operators. Also, the ability to analyse image resources remotely from the acquisition site means that one imaging facility could, in principle at least, service the needs of a number of counting sites, thereby increasing the accessibility of fission-track analysis by independent researchers. In anticipation of this potential, the two software modules have been written for Sun Microsystems Java™ technology to maximize their implementation across all major computer platforms. The image-processing module is computationally efficient, and automated track counting can now be carried out in a matter of a few seconds per frame, even on a laptop computer.

Coincidence mapping

Most image-analysis procedures involve three separate components: image correction for noise and background; image segmentation to select the desired features of interest; and image processing to extract and measure these features (e.g. Belloni *et al.* 2000; Russ 2002). However, this approach, when applied in a simple way to images of etched fission tracks in either transmitted or reflected light, has proved poor at discriminating tracks from non-track features and is therefore prone to significant error.

Our approach has been to learn and apply as much as possible from the techniques used by experienced human operators. In manual counting, the operator generally exploits the 3D form of fission tracks in transmitted light to assist in their identification and in resolving overlaps. The fine focus is moved up and down continually during counting at high magnification to reveal the full extent, shape, orientation and separation of the tracks. Reflected light is also used on occasion to assist in this process, and particularly in resolving multiple overlaps. Our acquisition system therefore captures a sequence of transmitted-light images closely spaced in the *z*-axis (a 'Z-stack') and a reflected-light image focused at the surface.

Our initial attempts centred on using the 3D information in the Z-stack to resolve the problem of multiple track overlaps in automated counting. Various approaches for projecting and analysing the Z-stack were tried, but, whilst some of these produced potentially useful information on track dimensions, they were processor-intensive and most were ultimately ineffective at resolving overlaps. Our attention then turned to also utilizing the information in the reflected-light image, and from this we have developed an image-analysis system that solves most of the typical track recognition and discrimination problems encountered in routine fission-track analysis in natural minerals.

The key element in this automated counting system arises from the observation that both transmitted- and reflected-light images typically include track and non-track features, but that almost completely different sets of spurious, non-track, features are found in the two images. Thus, it is only the fission tracks that are common to both images and should be resolvable using information from these two observation modes together. Mapping the areas of coincidence or overlap between a binary form of the two images, therefore, gives an intersection that contains almost exclusively the fission-track information. The objects in the resulting image can then be measured using conventional particle counting techniques. This 'coincidence mapping' technique (provisional US Patent 60/832,957) is remarkably simple in principle and is illustrated schematically in Figure 1. The actual implementation in a practical system is more complex, and will be described below.

Sample preparation

Careful sample preparation is a crucial first step for the digital image-acquisition and -analysis procedure in order to eliminate surface relief and to minimize the interference from polishing defects. Surface relief is particularly problematic, as our image-analysis procedure is only effective where the plane of focus is uniform across the captured field of view. In our procedure only diamond polishing compounds are used as these ensure both a high-quality polish with minimal polishing defects and also an optically flat surface with no relief or rounding on the grains. We avoided the use of alumina or colloidal silica compounds for the final polishing step as these have been found to produce unacceptable surface relief. A sequence of increasingly fine diamond polishing steps, although more time-consuming than those often used for preparing fission-track mounts, is, nonetheless, quite straightforward and ensures the most effective results from automated counting.

Apatite mounts are prepared in the usual way by mounting a sample of the mineral separate in epoxy (Petropoxy or Araldite) on a standard 27 × 46 mm petrographic glass slide. Pregrinding should be carried out using silicon carbide paper on a wet rotating lap to remove excess epoxy and to achieve a flat surface parallel to the underlying

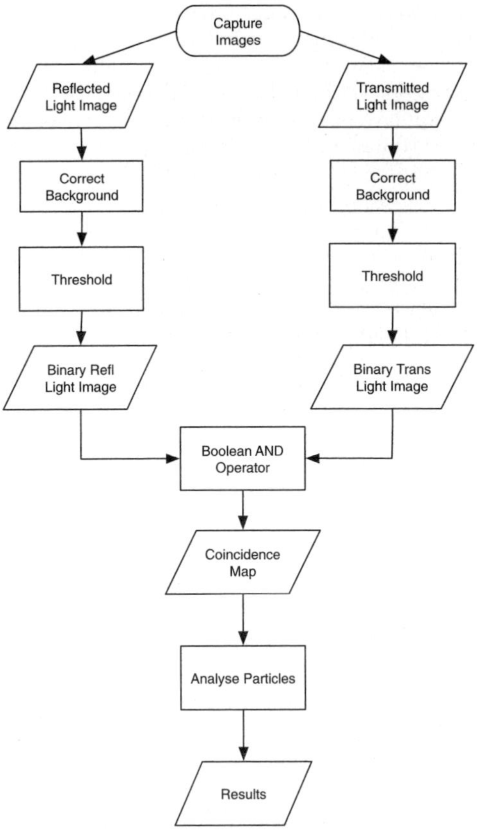

Fig. 1. Schematic diagram illustrating, in simplified form, the principle of the coincidence mapping method.

slide, and continuing until the coarsest grains are just exposed. A Struers Rotopol automated polishing machine with Rotoforce 4 polishing head is used, ensuring that polishing conditions can be precisely and consistently reproduced, as well as allowing batch processing for maximum productivity. The polishing steps used 6, 3, 1 and, finally, 0.25 μm diamond compounds on high-quality paper laps, with a rotation speed of 150 rpm and head pressure of 15 N. This usually requires only a few minutes at each step and produces the high-quality mounts required.

Image-acquisition system

Hardware implementation and image capture

Our digital image-acquisition system currently consists of a 3.2 GHz Pentium 4 desktop computer with 2 GB of random-access memory (RAM), connected via a dedicated controller unit to the microscope fitted with a digital camera. The system is based on a Zeiss Axiotron microscope with motorized stage and focus mechanism, motorized objective nosepiece and light-source shutters. All of the major microscope configuration functions are, thus, under computer control through the external control unit that handles all of the low-level function calls to the microscope. This level of control over the microscope is essential if the image acquisition is to be completed without the need for continued human supervision.

Unfortunately, the motorized focus mechanism on this microscope did not prove sufficiently accurate or reproducible at the level of about 0.1 μm required for acquisition of precisely spaced Z-stack images (although the most recent digital microscopes should be capable of this level of precision). It was therefore necessary to add a Physik Instrumente PIFOC® piezoelectric microscope-objective nanofocusing device to the 100 × objective. The PIFOC device is capable of positioning the objective lens with nanometer resolution over a total scanned range of 100 μm, with near-instantaneous response and no mechanical backlash.

The images are captured using a 3.3 megapixel Olympus ColorView 1 CCD colour digital microscope camera, with an IEEE 1394 Firewire connection directly to the computer. Captured images are stored in 24-bit TIFF or PNG format and stored to disk, initially on the control computer, and later transferred to other computers on the local network for analysis, and ultimately to DVD or network storage for permanent archiving.

Image-capture sequence

The Track Capture image-acquisition module is first used for the selection of the grain areas to be counted. These are located and labelled by the operator driving the motorized stage and focus motor using the joystick unit on the microscope controller. From this point on, the image capture sequence can be controlled autonomously by the software and requires no further supervision by the operator. The capture software controls the stage movement to visit each prelabelled point in sequence and uses an autofocus routine operating on the PIFOC device to precisely focus on the surface using the 100× objective. The software also controls camera exposure automatically throughout the capture sequence and the images are stored in 24-bit RGB format.

The capture sequence is illustrated in Figure 2. A reflected-light image focused at the surface is captured first, followed by a paired transmitted-light image and then a Z-stack sequence in transmitted light, stepping down through the surface zone

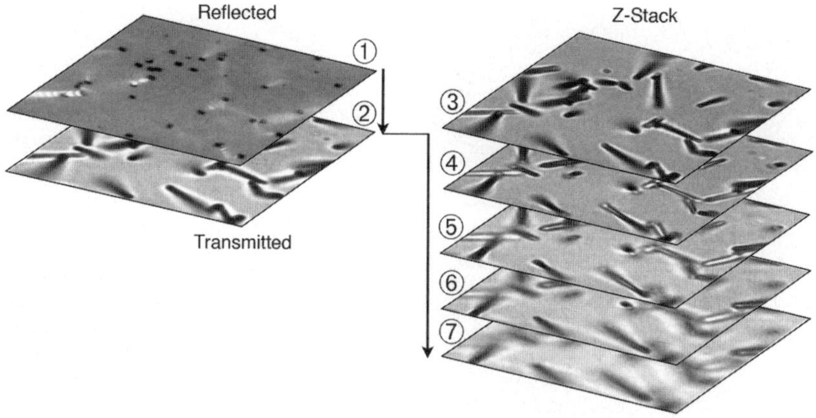

Fig. 2. Schematic diagram showing the image-capture sequence: first, a pair of surface images are captured using transmitted and reflected light, followed by a Z-stack sequence focusing progressively down into the surface of the mineral in transmitted light. In practice many more planes are captured in the Z-stack than illustrated here, to cover the full depth interval penetrated by the tracks.

containing the tracks. The spacing of the image planes in the stack is user-selectable and is typically about 0.2–0.5 μm, giving as many as 30–50 images for each field of view analysed. Typically, this takes around 2–3 min for each grain area and produces a total of about 100–150 MB of images that are then saved to the hard disk. Image-compression techniques may be used to reduce the size of the image batch, but, nonetheless, the image storage requirements of this system are quite large when 20 or more separate grain areas are counted (>2 GB per sample). At present only 'lossless' compression, using the PNG format, is used in order to maintain the highest image quality in the stored record. Only the two surface images are currently used in the image analysis, but the full Z-stack is available in scrollable form at the final review stage for comparison with results of the automatic counting. This provides the operator with a close simulation of the view seen by an observer at the microscope.

Reflected-light imaging problems

Two particular problems were encountered when capturing reflected-light images of apatite grains owing to the very low reflectivity of polished surfaces in this mineral. The first is a tendency, in many grains, for strong internal reflections to return light from below the surface, leading to an image with a partly transmitted-light character and a highly uneven background intensity. This is especially problematic for apatites that show well-faceted crystalline forms, a common occurrence for this mineral. An example is shown in Figure 3a. This phenomenon can cause significant difficulties with subsequent image processing because it includes transmitted-light information in the reflected-light image. One effective solution to this particular problem was found to be double-sided polishing of the apatite grain mounts so that the top and bottom surfaces of the grains are parallel to each other. This is relatively simple to implement, but does introduce additional processing steps, thereby making the sample processing more laborious.

A simpler procedure involved applying a very thin aluminium coating to the surface of the mount using a vacuum coating unit. This has the effect of dramatically increasing the reflectivity of the mineral surface, so that internal reflections become completely insignificant, as shown in Figure 3b. Aluminium was chosen for this purpose in preference to other commonly used coating materials, such as Au or C, as it produces the highest reflectivity, and is also easy to remove. The coating needs to be applied at just the right thickness to maximize the reflectivity, yet having only a minor effect on the intensity of the transmitted-light image, similar to using a neutral density filter. We have not yet experimented with other minerals, but the relatively much higher reflectivity of polished surfaces on zircon and titanite suggest that this procedure may not be necessary in those cases.

Common, 13 μm-thick, kitchen Al-foil is used as the source of aluminium for the coating. A 1×1 cm-square of this material is placed into a tungsten basket in a vacuum coating unit used for preparation of electron microscopy specimens. Evaporation of all of this foil at high temperature under vacuum at a distance of 9 cm from the polished surface applies a coating of 3.45 $\mu g\ cm^{-2}$,

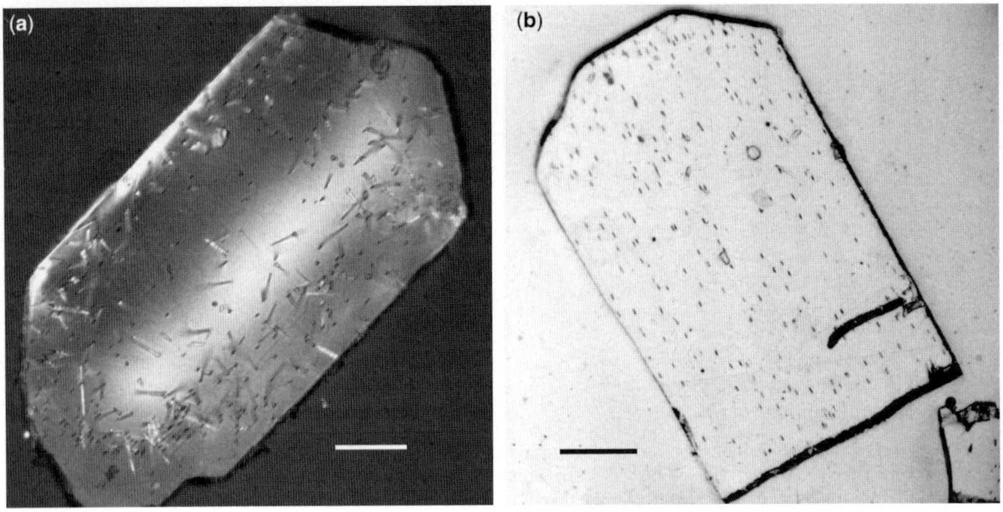

Fig. 3. In apatite the reflected-light image commonly shows internal reflections from facets and other planes below the surface, giving a bright appearance and a partially transmitted-light character, as shown in (**a**). This is a consequence of the very low surface reflectivity of apatite and the irregular shape of the crystal below the surface. A similar apatite grain in the same mount is shown in (**b**) after being coated with a thin layer of aluminium to enhance the reflectivity, and thereby suppressing internal reflections. Scale bars are 20 μm.

giving a thickness of approximately 13 nm. Twice this amount significantly impaired the transmitted-light image, and half as much did not sufficiently reduce the internal reflections. Provided that access to a vacuum coating unit is available, the Al-coating procedure is quick, simple and inexpensive, making it preferable to double-sided polishing. Also, the coating can be instantly removed by immersing the mount briefly in a dilute NaOH solution if required, with no observable damage to the polished surface, at least for apatite and mica.

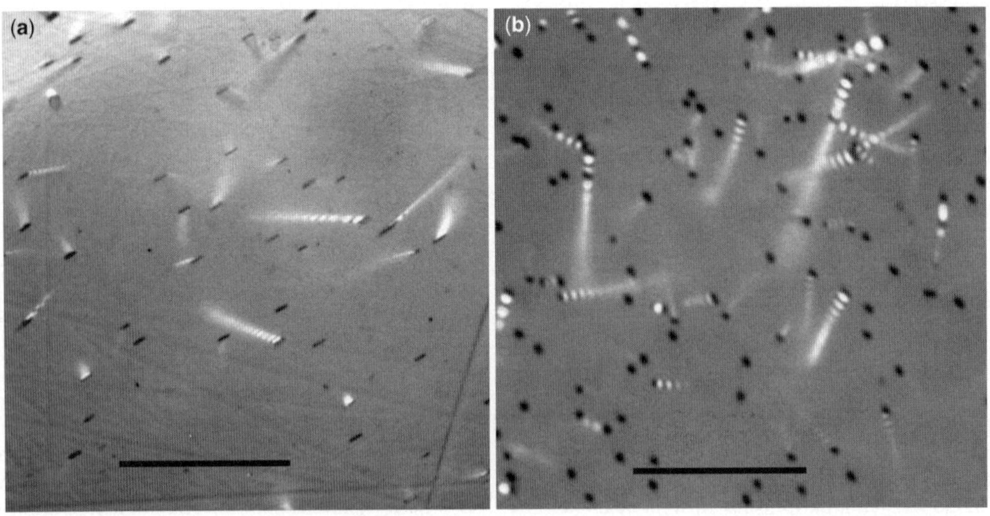

Fig. 4. A regular pattern of interference bands is observed in some shallow-dipping tracks in (**a**) apatite and (**b**) muscovite when observed in reflected light. These can cause problems for the automatic counting procedure, as described in the text, in a very small fraction (usually <3%) of the total tracks. Aluminium coating to increase the surface reflectivity (as shown in Fig. 3b), also largely eliminates this problem in both apatite and muscovite. Scale bars are 20 μm.

Another common problem, observed in reflected-light images of tracks in both apatite and muscovite, is shown in Figure 4a and b, respectively. Tracks at relatively shallow orientations to the surface are characterized by a series of interference fringes, that give them a banded 'zebra-striped' appearance. It is inferred that these bands are produced by constructive and destructive interference between rays reflected from the shallow track below the surface, and those reflected directly by the surface at that position. The spacing of the bands varies with the track dip and, for some of these tracks, the bright spot of the first interference maximum can overlap and mask the dark pit of the track entrance. Such tracks may be missed by the coincidence mapping routine and the apparent area of the track etch pit is also reduced. The problem caused by this effect is not large, but can result in a deficit in the estimated track density of up to about 3%. In principle, the number of these banded tracks that are overlooked should form a constant fraction of the total and a correction could be applied after further calibration. However, it was found that the interference bands essentially disappear after applying a thin aluminium coating, as described earlier, because the surface-reflected ray is greatly strengthened relative to the ray reflected internally from the shallow track. The use of a surface coating, therefore, should also largely eliminate this problem.

Automated fission-track counting procedure

Automatic track counting is carried out in the final Track Review software module using the pair of surface images captured in reflected and transmitted light. Some preprocessing of these colour images is required first to reduce them to 8-bit greyscale images, and to remove any background gradient that may be present. Currently, this preprocessing is carried out in a separate intermediate software module, but the intention is to include all preprocessing at the Track Capture stage before archiving to disk. The corrected images will then be loaded into the Track Review module for automatic counting and final review of results.

Background correction

The first step in processing the images is to ensure that they have a uniform background illumination, as discrimination of the tracks is made relative to the local background intensity. The ideal is that all of the images have a uniform grey background in both transmitted and reflected light, but this is frequently not realized in practice. Sometimes there may be a radial intensity gradient across the imaged field, or some persistent feature arising from dust on some out of focus optical plane within the microscope and camera. These artefacts affect all images consistently, and are best handled by taking a defocused 'background' image with no grain in the field of view and then subtracting this background from all of the captured images. The software allows for such a background image to be captured, or retrieved from memory, and subtracted before the images are stored. However, this step is frequently not needed, especially if the microscope and camera are well adjusted and maintained.

A more significant problem occurs in transmitted-light imaging of apatite grains, where the background intensity is influenced by refraction on grain irregularities below the polished surface. Most transmitted-light images show some fluctuations in background intensity across the field of view owing to this effect, as shown for an apatite grain in Figure 5a. One way to largely remove such transmitted-light effects is to double-polish the mount, as explained eariler for controlling the similar problem of internal reflections in reflected light. However, this procedure is needlessly laborious as the background variations can be removed by image processing. This is possible because the background variations are out of focus and, therefore, have a relatively low frequency, in contrast to the high-frequency information in the focused tracks. Several standard image-processing algorithms are able to remove such low-frequency background variations while retaining the high-frequency information. We use a Fast Fourier Transform bandpass filter to achieve a flattened background necessary for image thresholding and segmentation. Figure 5b shows the result of this operation on the raw image in Figure 5a.

Thresholding and segmentation

The next processing step is to segment the greyscale images by applying a threshold value to produce a pair of binary images separating the features of interest from the background. We have developed a custom image-processing routine to establish the threshold greyscale value for this segmentation after trying a number of commonly available routines, none of which worked exactly as we required. In both transmitted and reflected light, the tracks are dark (almost black) features against a uniform mid-grey background. The majority of pixels in the image actually belong to the background, so the task is, therefore, to set the threshold value at an appropriate level between the dark 'features' and the background peak in a histogram of intensity values. This must be carried out automatically and consistently, even though the average background

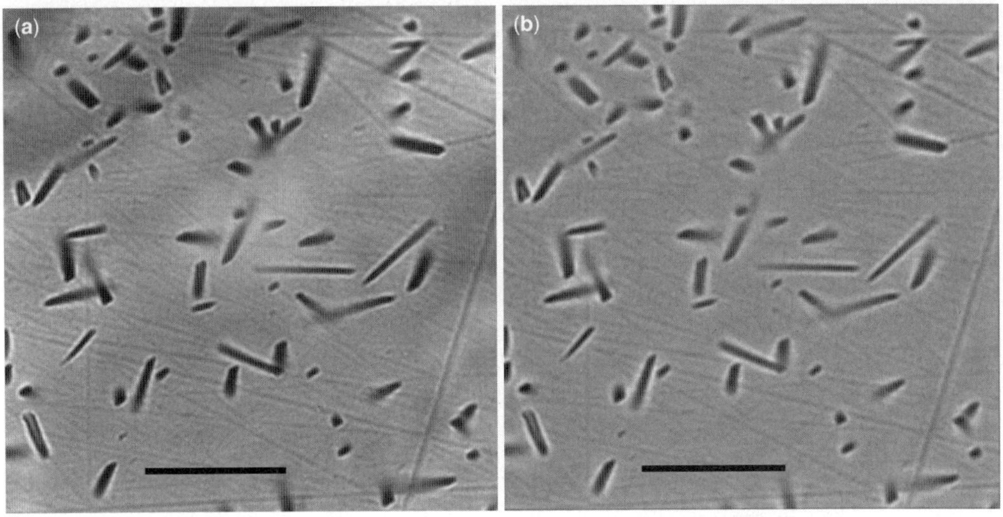

Fig. 5. Greyscale transmitted-light images of fission track in apatite: (**a**) showing the original variations in background intensity; and (**b**) the same image after application of a fast Fourier transform bandpass filter. The background has become essentially uniform in the second image. These images show the same field as the reflected-light image in Figure 4a. Scale bars are 20 μm.

intensity will vary significantly from image to image and from batch to batch, depending on such factors as the illumination level, camera exposure, transparency of the mineral, reflectivity, etc. Our routine determines the modal value of the background intensity peak and sets the threshold at a proportionate level below this peak. The resulting reflected- and transmitted-light binary images contain black 'features', including both tracks and some spurious objects, on a white background.

Coincidence mapping

The 'coincidence mapping' procedure involves combining the binary reflected-light and transmitted-light images using a Boolean AND operation, as illustrated in Figure 6. The transmitted- and reflected-light images (Fig. 6a, b, respectively) are segmented to give the binary images (Fig. 6c, d). The binary reflected-light image (Fig. 6d) clearly includes some spurious features that are not tracks. Inspection of the transmitted-light image (Fig. 6a) shows that these are actually surface dust particles, but they do not appear in the transmitted binary (Fig. 6c). The transmitted-light binary image (Fig. 6c), however, shows poor resolution of individual track features, several of which are compound objects representing two or more overlapping tracks. The intersection of the two binary images in the resulting coincidence map (Fig. 6e), however, selects for the genuine track features, and against the spurious objects (Fig. 6f), while also clearly resolving the track overlaps. The complex cluster of three overlapping tracks in the top centre of the transmitted-light image (Fig. 6a) is informative, as these are well separated and easily resolved in the coincidence map because it captures the much smaller size of the track entrance pits in reflected light.

The analysis of these images is extremely efficient and takes only a few seconds to complete, even though the images may contain hundreds of tracks. Figure 6 illustrates an example of fission tracks etched in a muscovite external detector, but the method is almost equally effective for discriminating spontaneous fission tracks in apatite, which has the additional complexity of having polishing scratches, inclusions and other defects.

Resolving track overlaps at high track densities

Coincidence mapping is particularly effective at resolving multiple track overlaps. This remains true even at high track densities, well beyond that which a human operator would normally attempt using transmitted light. At even higher track densities, where the reflected light pits themselves begin to overlap, we have used an analysis of the areas of the individual pit features to estimate the number of tracks represented by larger compound track features. In this way we have successfully been able to count track densities up to 5×10^7 cm^{-2}, well beyond the range that is reliably

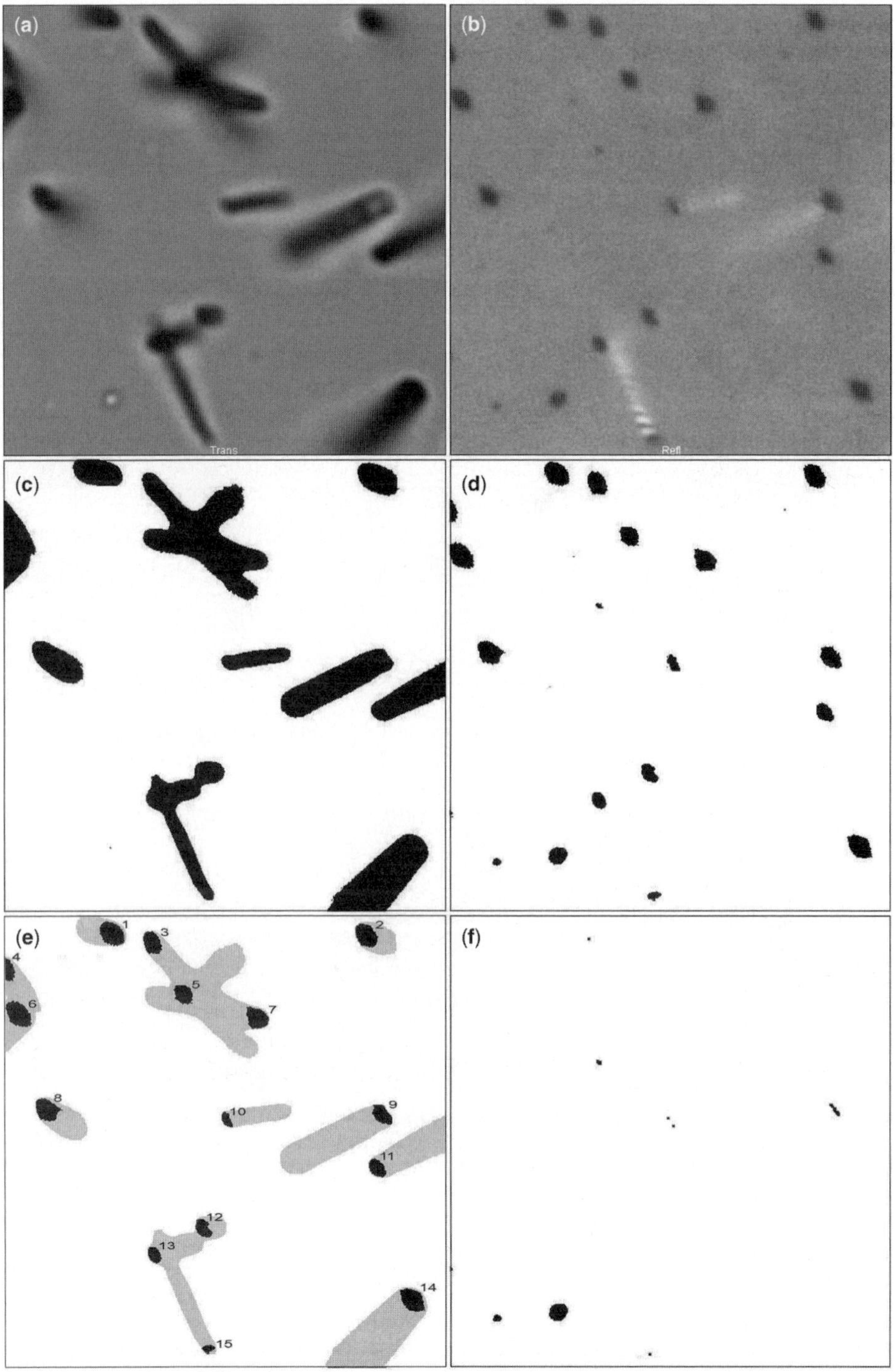

Fig. 6. Image-analysis sequence of fission tracks in muscovite mica showing: (**a**) transmitted-light image; (**b**) reflected-light image; (**c**) segmented binary image of tracks in transmitted light; (**d**) segmented binary image of tracks in reflected light; (**e**) the coincidence map showing the 15 counted tracks in black, being the intersection between the two binary images, superimposed on a grey image of the transmitted light tracks; and (**f**) spurious features rejected by the analysis. The counted fission-track features are shown as black spots with adjacent identification numbers in (e).

countable by a human operator working in transmitted light. We consider that the pit-area method should work effectively for even higher track densities, probably up to 10^8 tracks cm^{-2}, which will be discussed elsewhere in a later paper.

Final review

Coincidence mapping is extremely effective at discriminating fission tracks from surface dirt and dust, polishing scratches, and both solid and fluid inclusions. The error rate in track counts from our experimentation so far is similar to, or better than, that achieved by a human operator making repeated counts of the same view in transmitted light. The procedure is incomparably faster than a human operator, especially given that very little operator time is now needed for the image-capture process, as it can count hundreds of tracks in a few seconds. The method is also significantly better than a human operator in resolving track overlaps and counting very high track densities. However, it is not infallible, and may not discriminate fission tracks from dislocation etch pits or some deeply etched polishing defects. We have, therefore, incorporated a Track Review module that enables the operator to quickly review the automatic counting results overlaid on the transmitted light Z-stack,

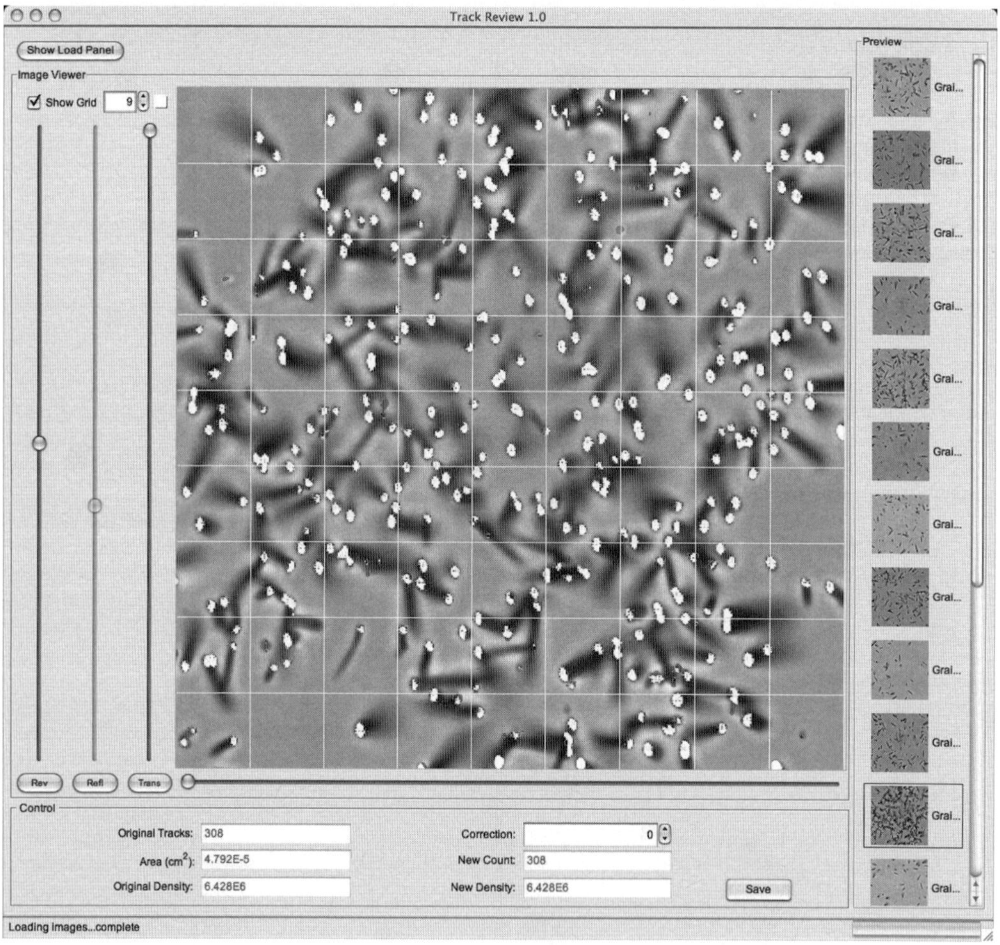

Fig. 7. The interface for the Track Review module, which constitutes a virtual microscope based on the transmitted light Z-stack that can be dynamically scrolled to examine the tracks in a way that is comparable to actual observation. On this image can be overlaid an adjustable grid, the reflected-light image and the automatic count data, all with varying degrees of transparency. Counted objects are shown here in white but can be highlighted in different colours on screen, as can the rejected objects. The count data and track density are shown at the bottom, and these can be adjusted manually if required to correct for any counting errors. Sliders on the left-hand side control the transparency of the different images, and image sets are selected from the column on the right-hand side.

which can be scrolled up and down dynamically, as well as on the original reflected-light image. The transmitted, reflected and automatic counting layers can be be turned off and on with varying degrees of transparency, and can be combined with an adjustable grid overlay. The effect is a powerful 'virtual microscope' with excellent image quality, greater flexibility and more convenience than is possible with the real microscope. Using this interface (Fig. 7) the automatic counting results can be quickly and efficiently checked, and adjusted for any errors detected. This review stage can be carried out on any computer platform (PC, Mac or Unix) and quite independently of any actual microscope.

Experimental results

To demonstrate the capabilities of this new system we show a comparison of automatic and manual counts obtained for induced fission tracks in a muscovite external detector, and for spontaneous tracks in a typical apatite sample. The apatite was separated from the Harcourt Granodiorite in central Victoria, and has a fission-track age of approximately 280 Ma. In both cases the manual count results were obtained by an experienced track counter (A. Gleadow) using the Track Review virtual microscope, the Z-stack transmitted-light images and an overlaid grid without reference to the automatically counted data or reflected-light image. This ensured that precisely the same areas and features were counted for both the manual and the automatic methods. Results are reported as raw counts, and shown in Figure 8 for the muscovite and Figure 9 for the apatite sample. In both examples the correlation between automatic and manual counts is extremely good. A single anomalous point in the apatite example (Fig. 9) was due to an unusually deep polishing scratch that was strongly revealed in both transmitted- and reflected-light images, and, therefore, erroneously included in the automatic count. Such features are readily seen and easily removed in the Track Review module, and could also be easily avoided using size and shape discriminators that will be added in later versions.

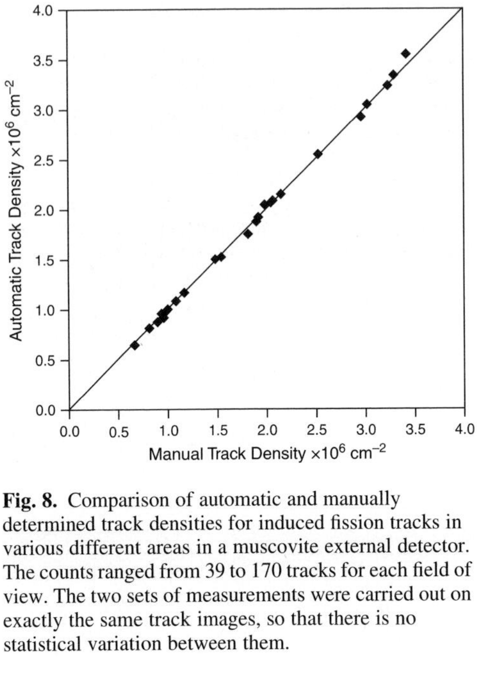

Fig. 8. Comparison of automatic and manually determined track densities for induced fission tracks in various different areas in a muscovite external detector. The counts ranged from 39 to 170 tracks for each field of view. The two sets of measurements were carried out on exactly the same track images, so that there is no statistical variation between them.

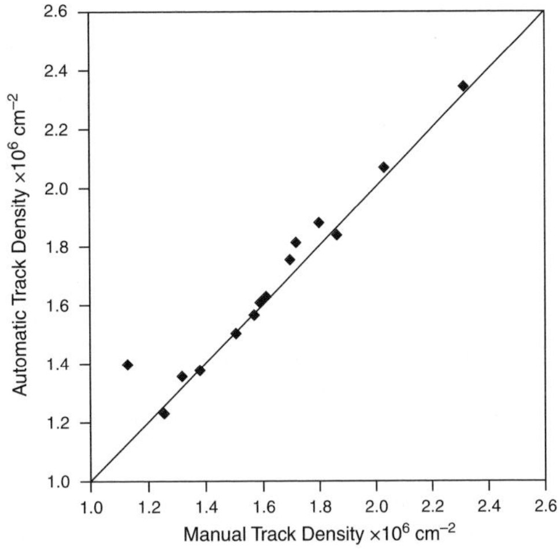

Fig. 9. Comparison of automatic and manually determined track densities for spontaneous fission tracks in different apatite grains from the Late Devonian Harcourt Granodiorite from central Victoria. The counts ranged from 52 to 99 tracks for each field of view. The one outlier on the lower left side erroneously included features from an unusually deep polishing scratch in the automatic count. The two sets of measurements were carried out on exactly the same track images, so that there is no statistical variation between them.

Conclusions

The coincidence mapping system described here provides the first truly practical solution to the problem of automatic counting of fission tracks in natural minerals. The technique is extremely fast, and promises a much more rapid sample throughput than can be achieved currently by manual counting methods, with significant implications for analytical costs and productivity. Automatic counting should also allow significantly improved data quality by obtaining larger counts and better counting statistics, as well as a more consistent application of counting criteria and increased objectivity in fission-track analysis. Other benefits will include the reduction of operator fatigue and increased accessibility of fission-track analysis to non-specialists, with much less training than is now required.

Combined with direct-uranium determination by laser-ablation ICP-MS (Hasebe *et al.* 2004), coincidence mapping should provide the basis for a fundamentally new approach to fission-track dating. The technique is still at a very early stage of its development, and clearly much more testing and elaboration of the method is needed in many areas, such as ensuring that the Al-coating does not introduce any contaminant uranium to the surface. But the results of automated fission-track counting so far are extremely promising. In future, digital image analysis should also allow measurement of additional track size, shape and orientation parameters that are not currently collected, but are potentially useful for data interpretation and modelling. Measurement of track lengths remains a more challenging problem that is not yet resolved, although several promising leads are now being investigated.

We gratefully acknowledge the financial support provided by the Australian Research Council to this project through a Linkage Project Grant with Autoscan Systems Pty Ltd as an Industry Partner. A. Raza assisted with careful sample preparation and valuable discussions throughout development of the automated counting process. We also thank R. Jonckheere, M. Wipf, B. Ventura and U. Glasmacher, whose constructive and detailed reviews significantly improved the manuscript.

References

BELLONI, F. F., KESKES, N. & HURFORD, A. J. 2000. Strategy for fission-track recognition via digital image processing, and computer-assisted track measurement. (Abstract.). *In*: *9th International Conference on Fission-track Dating and Thermochronology*. Geological Society of Australia Abstracts, **58**, 15–17.

BOUKHAIR, A., HAESSLER, A., ADLOFF, J. C. & NOURREDDINE, A. 2000. New code for digital imaging system for track measurements. *Nuclear Instruments and Methods B*, **160**, 550–555.

DOLLEISER, M. & HASHEMI-NEZHAD, S. R. 2002. A fully-automated optical microscope for analysis of particle tracks in solids. *Nuclear Instruments and Methods B*, **198**, 98–107.

ESPINOSA, G., GAMMAGE, R. B., MEYER, K. E. & DUDNEY, C. S. 1996. Nuclear track analysis by digital imaging. *Radiation Protection Dosimetry*, **66**, 363–366.

FEWS, A. P. 1992. Fully automated image analysis of etched tracks in CR-39. *Nuclear Instruments and Methods B*, **71**, 465–478.

GOLD, R., ROBERTS, J. H., PRESTON, C. C., MCNEECE, J. P. & RUDDY, F. H. 1984. The status of automated nuclear scanning systems. *Nuclear Tracks and Radiation Measurements*, **8**, 187–197.

HASEBE, N., BARBARAND, J., JARVIS, K., CARTER, A. & HURFORD, A. J. 2004. Apatite fission-track chronometry using laser ablation ICP-MS. *Chemical Geology*, **207**, 135–145

PETFORD, N. & MILLER, J. A. 1990. SLM confocal microscopy: an improved way of viewing fission-tracks. *Journal of the Geological Society, London*, **147**, 217–218.

PETFORD, N., MILLER, J. A. & BRIGGS, J. 2003. The automated counting of fission-tracks in an external detector by image analysis. *Computers and Geosciences*, **19**, 585–591.

PRICE, P. B. & KRISCHER, W. 1985. Semi-automated, three-dimensional measurement of etched tracks in solid-state nuclear track detectors. *Nuclear Instruments and Methods A*, **234**, 158–167.

RUSS, J. C. 2002. *The Image Processing Handbook*, 4th edn. CRC Press, Baton Rouge, LA.

WADATSUMI, K. & MASUMOTO, S. 1990. Three-dimensional measurement of fission-tracks: Principles and an example in zircon from the Fish Canyon Tuff. *Nuclear Tracks and Radiation Measurements*, **17**, 399–406.

YASUDA, N., NAMIKI, K., HONMA, Y., UMESHIMA, Y., MARUMO, Y., ISHII, H. & BENTON, E. V. 2005. Development of a high speed imaging microscope and new software for nuclear track detector analysis. *Radiation Measurements*, **40**, 311–315.

The effect of chemical etching on LA–ICP-MS analysis in determining uranium concentration for fission-track chronometry

NORIKO HASEBE[1]*, ANDREW CARTER[2], ANTHONY J. HURFORD[2] & SHOJI ARAI[1]

[1]*Kanazawa University, Kanazawa 920-1192, Japan*

[2]*University College London, Gower Street, London WC1E 6BT, UK*

**Corresponding author (e-mail: hasebe@kenroku.kanazawa-u.ac.jp)*

Abstract: LA–ICP-MS (laser ablation–inductively coupled plasma-mass spectrometry) has the potential to measure uranium concentration for fission-track (FT) chronometry as an alternative to thermal neutron-induced fission of ^{235}U. This study examines the effect that chemical etching, required to reveal spontaneous fission tracks of ^{238}U, has upon LA–ICP-MS analyses. Uranium concentrations were measured before and after etching for six large gem-quality apatite crystals and six zircon samples – three large crystals and three FT age standards. Comparison of the results shows no significant difference in ^{238}U concentrations measured on the etched and unetched mineral surfaces. The ^{238}U concentrations determined by the LA–ICP-MS provide reasonable FT ages for the zircon age standards, which, with the previously reported LA–ICP-MS apatite FT results, promotes the use of the LA–ICP-MS for FT chronometry.

Laser ablation – inductively coupled plasma-mass spectrometry (LA–ICP-MS) is widely accepted as a rapid method to determine the chemical composition of geological material at the micro-scale (e.g. Perkins *et al.* 1993; Bea *et al.* 1996). To convert the measured mass intensity into chemical (or isotopic) composition, appropriate standards (external standards) must be analysed together with unknown samples (e.g. Jarvis & Williams 1993; Williams & Jarvis 1993). Such an external standard should be similar (often referred to as matrix-matched) to the unknown sample in terms of material behaviour during the LA–ICP-MS measurement (e.g. aerosol formation, ionization, transport from a laser ablation cell to mass spectrometer and interference with coexisting elements) to minimize possible systematic errors (Morrison *et al.* 1995). For mineral analyses, artificial glasses such as NIST (National Institute of Standard and Technology) products have been widely used with both mineral and standard glass polished to expose flat surfaces ready for ablation by the laser (NIST620 and NIST612: Iizuka & Hirata 2004; Jeffries *et al.* 1995; Sylvester & Ghaderi 1997). Despite differences in structure (samples are usually ordered crystals, whilst glass is disordered and amorphous), the use of glass as an external standard has provided reliable results in analysis of crystalline material (Eggins *et al.* 2003).

For fission-track (FT) chronometry, the external detector technique is conventionally applied, in which the uranium concentration is determined using a neutron activation technique to induce fission in a proportion of ^{235}U atoms, recording the fission events in an external detector to give essentially a map of uranium distribution (e.g. Gleadow 1981; Hurford & Green 1982). Uranium measurement by the LA–ICP-MS method has the potential to avoid the costly and time-consuming need to irradiate samples in a nuclear reactor and the treatment of radioactive samples, and to date a sample without age standards (Hasebe *et al.* 2004; Kosler & Svojtka 2003). Possible multi-elemental analysis together with uranium concentration is also an advantage for interpreting obtained ages. However, unlike samples for other geological analyses, samples for FT chronometry are etched chemically to reveal the spontaneous fission tracks of ^{238}U before laser ablation and ICP-MS analysis to perform single-grain dating. Etching changes the surface condition of the mineral, which could induce changes in the optical characteristics against laser irradiation, in the particle distributions in the aerosol and in transport efficiency to the mass spectrometer, potentially causing a systematic over- (or under-) estimate of uranium concentration (Horn & Gunther 2002). Furthermore, chemical treatment such as etching could cause selective dissolution of a particular isotope, as has been observed for zircon (Davis & Krogh 2000). Before LA–ICP-MS is accepted for routine analysis in FT chronometry, the ability of the method to measure the uranium content of single etched mineral grains used for counting needs to be validated, in particular the effect of uranium heterogeneity must be evaluated. In this study we explore the effect of chemical etching on LA–ICP-MS measurements of ^{238}U in apatite and zircon.

From: LISKER, F., VENTURA, B. & GLASMACHER, U. A. (eds) *Thermochronological Methods: From Palaeotemperature Constraints to Landscape Evolution Models*. Geological Society, London, Special Publications, **324**, 37–46.
DOI: 10.1144/SP324.3 0305-8719/09/$15.00

Experimental strategies and analytical method

Six large gem-quality apatite crystals and six zircon samples – three large crystals and three FT age standards – (see Table 1 and Barbarand *et al.* 2003) were used in this study. In method (a) (used for all six apatite and large zircon sample TAS-1) adjacent slices were cut along the crystallographic *c*-axis using a saw (ISOMET, blade thickness of 0.3 mm) and polished using 0.3 μm diamond paste. One slice of each pair was then etched to reveal spontaneous fission tracks. Corresponding points on each slice of the sample pairs were then analysed using LA–ICP-MS. This experimental approach has the advantage that both etched and unetched slices were measured under identical analytical conditions, such as mass spectrometer sensitivity and material flow from sample chamber to mass spectrometer. However, a disadvantage is that, even though corresponding points were measured on adjacent slices, those points were separated by material removed during cutting, polishing and etching – a possible thickness of 1 mm; differences in measured uranium contents could thus result from real inhomogeneity within the sample.

In the second experimental approach, method (b), a single sample of each of the six apatite and six zircon samples was prepared and polished. Uranium content was measured at several points on each sample by LA–ICP-MS and then each sample was etched to reveal spontaneous fission tracks. After etching each sample was reanalysed using LA–ICP-MS, placing the laser spot between the previously ablated points, thus providing a direct comparison of uranium composition before and after etching from adjacent sites at a spacing of approximately 10 μm.

All analyses were made at Kanazawa University. Apatites were etched for 2 min in 0.5% HNO_3 at 33 °C (Tagami *et al.* 1988) – note that although the etching temperature, along with the etchant concentration, is chosen in function of Japanese climate, the resultant FT etch-pit size is similar to that obtained by etching in European laboratories. Zircons were etched for approximately 8–30 h in a KOH:NaOH = 1:1 (molar ratio) eutectic at 230 °C (Gleadow *et al.* 1976). LA–ICP-MS analysis used an Agilent 7500 instrument equipped with a Microlas Excimer laser ablation system (Ishida *et al.* 2004; Morishita *et al.* 2005). Details of the analytical operating conditions used for both apatite and zircon uranium measurement are summarized in Table 2. A laser-pit diameter of 20 μm and depths of *c.* 8 μm for apatite and *c.* 6 μm for zircon were planned, with the dimensions of the actual ablation pits being measured in two representative samples (apatite DUR and zircon BM4) using a Keyence laser microscope VK8500 at the Central Research Institute of Electric Power Industry, Chiba, Japan (Table 3). No systematic difference in laser-pit size between etched and unetched samples was found for either mineral, although drilling efficiency is less for zircon than for apatite and the depth of laser pits in zircon was shallower than expected.

In this study designed to examine the effect of chemical etching, ^{238}U concentration was calibrated by LA–ICP-MS measurement of NIST610 standard glass (U, $457.1 \pm 13.6\ \mu g\ g^{-1}$; Th, $457.1 \pm 27.8\ \mu g\ g^{-1}$: Pearce *et al.* 1997) as an external standard between apatite and zircon sample analyses. Uranium content is determined by use of an internal standard: an isotope of a different element whose content within the sample is known (or reasonably inferred) is measured and the mass spectrometer ion counts of uranium and calibrating element compared. Such an approach assumes knowledge of an exact value for the abundance of the internal standard, and that there is no mass discrimination in transport and measurement; however, the method does circumvent the different ablation characteristics of glass and mineral, inherent in use of external standards. In this study ^{238}U results obtained using NIST610 glass (CaO, 11.450 ± 0.231 wt%; SiO_2, 69.975 ± 0.391 wt%: Pearce *et al.* 1997) were normalized using internal standards of ^{43}Ca for apatite and ^{29}Si for zircon (Jarvis & Williams 1993). A value of 55.0 wt% is used as apatite CaO content, which is an average of EPMA (Electron Prove Micro Analyses) measurements of 13 apatites of different origin and chemistry (Barbarand *et al.* 2003). For zircon, 33.3 wt% is the average SiO_2 content determined by EPMA measurement of 100 grains from samples FCT, BM4, LMR, NST and TRG (note that zircons NST and TRG have been used for zircon annealing experiments – see Yamada *et al.* 1998). For this study, accurate CaO or SiO_2 contents are not essential because the purpose is to compare the chemical composition of two aliquots or the between analyses of the same sample. When uranium content is necessary for age determinations, it might be preferable to obtain absolute CaO or SiO_2 values for each analysed sample.

Figure 1 illustrates the stability of LA–ICP-MS measurement over a 2-year period, showing the isotope ratios determined for NIST610 during analysis of apatite ($^{238}U/^{43}Ca$) and zircon ($^{238}U/^{29}Si$) samples; NIST610 being measured after every five sample analysis spots. Differences in ratios over the analysis period reflect changes in instrumental sensitivity. NIST612 glass was also analysed over a period of 2 years, cross-calibrated against NIST610 glass, to assess the variation in uranium concentration that might reflect either

Table 1. *Comparison of ^{238}U contents measured by LA–ICP-MS on etched and unetched sections of apatite and zircon samples*

Sample	Sample description[†]	ρs[‡] ($10^6/cm^{-2}$)	Mean ^{238}U contents determined for specified sections (ppm)			
			Experimental method (a)		Experimental method (b)	
			Unetched surface $\mu g/g \pm \sigma$ (n)*	Etched surface $\mu g/g \pm \sigma$	Unetched surface $\mu g/g \pm \sigma$ (n)*	Etched surface $\mu g/g \pm \sigma$ (n)*
Apatite						
FUL	Lemon-coloured crystal from Fulford, Eagle county, Colorado, USA	0.1	7.03 ± 0.58 (12)	5.32 ± 0.35	7.10 ± 1.55 (9)	7.13 ± 1.25 (12)
DUR1, DUR2	Gem-quality, lemon-yellow apatite from Cerra de Mercado, Durango, Mexico	0.2	11.67 ± 0.61 (15) 7.10 ± 0.91 (12)	12.19 ± 0.33 5.96 ± 0.76	11.15 ± 0.20 (9)	11.56 ± 0.17 (11)
PAN	10 mm-long yellow crystal from Panasqueira Mine, Portugal	0.6	0.68 ± 0.03 (12)	0.31 ± 0.09	9.6 ± 6.1 (9)	11.01 ± 6.22 (11)
LOV	30 mm light-green crystal from near. Lovelock, Churchill County, Nevada, USA	0.7	7.52 ± 0.95 (12)	6.30 ± 0.83	11.81 ± 0.14 (9)	11.74 ± 0.16 (12)
UNK	Green-brown crystal from an unknown locality, possibly from Canada	1.0	8.84 ± 0.18 (12)	8.84 ± 0.18	10.24 ± 0.27 (9)	10.62 ± 0.15 (7)
DRV	Green fluorapatite from Ontario, Canada	High	132.6 ± 3.1 (12)	98.2 ± 20.3	132 ± 3 (9)	133 ± 3 (8)
Zircon						
BM4	Buluk Member FT age standard from Kenya	1.2			87 ± 33 (12)	84 ± 34 (11)
TAS1, TAS2	Brown *c.* 3 mm crystals from Tasmania, with a round shape, possibly by sedimentary processes	3.0	26.7 ± 5.1 (8)	36.4 ± 12.6 (8)	26.7 ± 5.1 (8) 26.0 ± 2.5 (15)	37.5 ± 8.9 (7) 25.8 ± 2.4 (21)
FCT	Fish Canyon Tuff age standard from Colorado, USA	4.8			294 ± 130 (19)	383 ± 188 (11)
LMR1, LMR2	FT secondary age standard from Australia	>20			255 ± 91 (12) 267 ± 99 (14)	222 ± 51 (7) 210 ± 42 (8)
3F	Transparant green crystal. Origin unknown	High			952 ± 18 (12)	931 ± 24 (9)
Malcan	Milky-white *c.* 3 mm crystals. Origin unknown	No tracks observed			3506 ± 771 (12)	3514 ± 738 (9)

*(n) denotes number of points analysed.
[†]See Hurford (1990) and Barbarand *et al.* (2003).
[‡]ρs, approximate mean spontaneous fission-track density.

Table 2. *Operating conditions for the LA–ICP-MS analyses*

ICP-MS	
Model	7500s (Agilent)
Forward power	1200 W
Reflected power	1 W
Carrier gas flow	1.20 l min^{-1} (Ar)
	0.3 l min^{-1} (He)
Auxiliary gas flow	1.0 l min^{-1}
Plasma gas flow	15 l min^{-1} (Ar)
Cones	Pt sample cone
	Pt skimmer cone
Laser	
Model	GeoLas Q+ (MicroLas)
Wavelength	193 nm (Excimer ArF)
Repetition rate	4 Hz (Ap) 5 Hz (Zr)
Energy density at target	8 J cm^{-2}
spot diameter	20 μm
Number of shots	40 (Ap) 30 (Zr)

Table 3. *Laser-pit size of representative samples measured by the laser microscope*

		Diameter (μm) ($\pm\sigma$)	Depth (μm) ($\pm\sigma$)
Zr (BM4)	Unetched	21.12 ± 0.34	3.72 ± 0.03
	Etched	21.51 ± 0.13	3.55 ± 0.07
Ap (DUR)	Unetched	21.71 ± 0.13	8.63 ± 0.06
	Etched	21.64 ± 0.19	8.51 ± 0.06

Each value is the average of 10 pits.
Note ± represents one standard error of the mean.

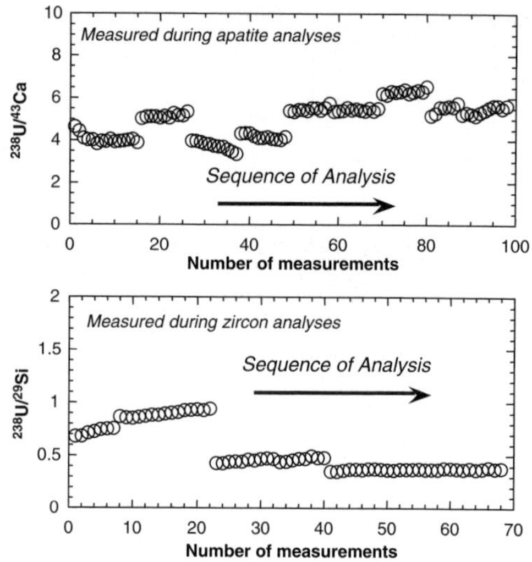

Fig. 1. Variation over the analysis period of raw signal ratios $^{238}U/^{43}Ca$ and $^{238}U/^{29}Si$ measured for NIST610. The 'jump' of values corresponds to the opening of the sample chamber.

analytical (LA–ICP-MS performance) and/or microheterogeneity of the glass. Figure 2a shows an overall average ^{238}U concentration of 37.09 ± 0.93 (1σ) μg g^{-1} for 103 measurements, consistent with a reference value of 37.15 ± 1.23 μg g^{-1} – the 'preferred average' of Pearce *et al.* (1997). These results demonstrate a reproducibility of analysis of ±5% (2σ), which is sufficient for routine FT analyses, being comparable with uncertainties of approximately 5% from the number of counted spontaneous fission tracks (Green 1981). Experimental variations between zircon and apatite analyses may result from the use of different internal standards with different mass discriminations, and also from the use of different laser repetition rates. To investigate these variation further, additional analyses of NIST612 (again calibrated against NIST610) included determination of both ^{43}Ca and ^{29}Si: Figure 2b shows for each analysis ^{238}U normalized by using both ^{43}Ca (solid symbols) and ^{29}Si (open symbols) as internal standards. Two laser repetition rates of 4 or 5 Hz were also used, shown respectively as circles and triangles in Figure 2b. Overall, the variation in experimental settings (repetition rate and the choice of internal standard) does not cause significant bias in measured uranium content.

Results – apatite

Figure 3 shows the ^{238}U concentrations measured in pairs of apatite sections using experimental method (a). Several apatite data points, especially those samples having large standard deviations in LA–ICP-MS measurement of unetched sections (Table 1), plot beneath the one-to-one correlation line, possibly the result of etching or of sample heterogeneity. Results for apatite DRV show a very wide dispersion in uranium content measured for the etched section.

Figure 4 shows the results obtained using experimental method (b), which is designed to minimize the effect of sample heterogeneity. Data for most apatites fall on the one-to-one correlation line suggesting that chemical etching has had a negligible effect on the LA–ICP-MS measurement of uranium concentration for FT dating of apatite. The exception, apatite PAN, has an extremely high standard deviation (much greater than in Fig. 3) with measured uranium content for unetched samples varying between *c.* 0 and *c.* 0–20 μg g^{-1} within *c.* 100 × 100 μm^2; such heterogeneity is also seen prominently in etched fission-track distributions.

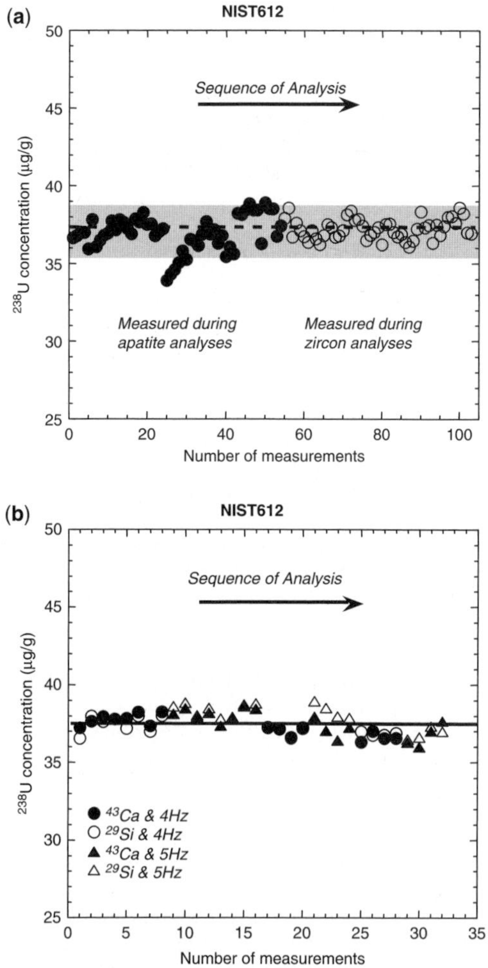

Fig. 2. (**a**) Sequential measurement of ^{238}U concentration for NIST612 standard glass. Solid circles represent measurement during apatite analyses, and open circles during zircon analyses. The overall average for 103 measurements is 37.09 $\pm$ 0.93 (1σ) (shaded area representing $\pm 2\sigma$), consistent with a reference value of 37.15 $\pm$ 1.23 (1σ) (dashed line, 'preferred average' of Pearce *et al.* 1997). (**b**) Sequential measurement of ^{238}U concentrations for NIST612 standard glass. ^{29}Si (open symbols) and ^{43}Ca (solid symbols) measured as internal standards for each analysis. Circles represent measurements using a laser repetition rate of 4 Hz and triangles those using 5 Hz. The overall average ^{238}U is 37.52 $\pm$ 0.74 (1σ).

Such large heterogeneity, at a scale smaller than the ablation pit size, means that such a sample is very difficult to analyse using LA–ICP-MS – or, indeed, by conventional FT analysis using neutron-induced fission of ^{235}U in case heterogeneity occurs vertical to an observed surface.

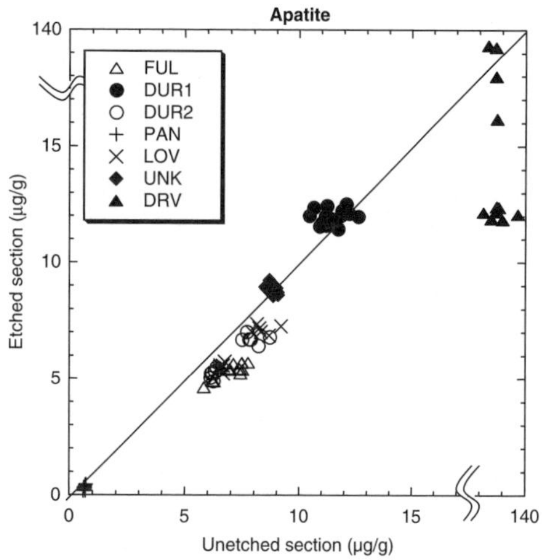

Fig. 3. Uranium content of apatite measured by experimental method (a). Several apatite data points, especially for those that have large standard deviations in the measurement of unetched samples (Table 1), are plotted under the one-to-one correlation line.

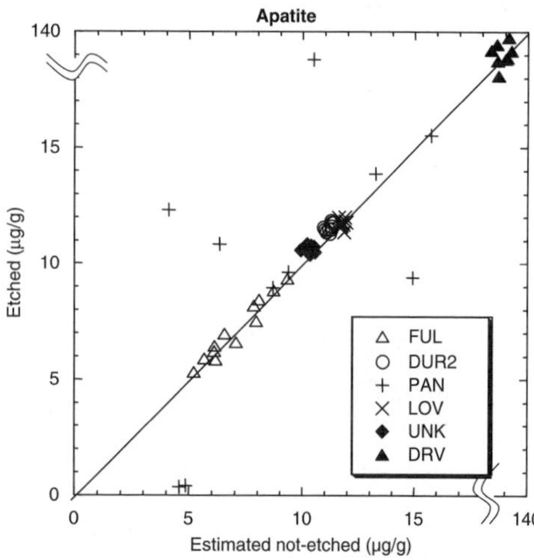

Fig. 4. ^{238}U content of apatite measured by experimental method (b). Horizontal values are estimated from the measurement of unetched samples. All data, except for PAN, fall on the one-to-one correlation line.

The coherence of uranium content in these results also indicates that analytical results are independent of any variation in instrumental sensitivity in the LA–ICP-MS over a significant period.

Results – zircon

Figure 5 shows the ^{238}U concentrations of zircons as measured using experimental method (b), better suited to minimize the effects of sample

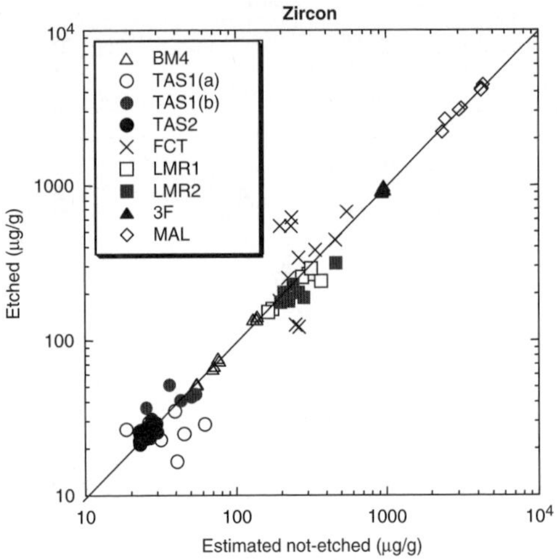

Fig. 5. ^{238}U content of zircon measured by experimental method (b), except for TAS1(a). Horizontal values are estimated from the measurement of unetched samples.

heterogeneity often encountered within zircon, as indicated by relatively large standard deviation of uranium values (see Table 1). Most zircons plot on a 1:1 correlation line, indicating that chemical etching has an insignificant effect on uranium determination by LA–ICP-MS. Data for zircon FCT scatter widely, reflecting the uranium zoning often observed in this sample. Such inhomogeneity of uranium distribution offers limits on analytical precision and requires a more exact placing of ablation pits to quarry a volume of material identical to that from which the spontaneous tracks originated.

Sample TAS-1 was also measured using method (a), and the results (open circles in Fig. 5) show greater dispersion (with uranium concentrations ranging from *c.* 20 to 60 $\mu g\ g^{-1}$) than found for values derived using method (b), where values for TAS1(b) and TAS2 (solid circles) lie close to the 1:1 correlation line.

FT age calculation

FT analyses of unknown samples using either zeta calibration, which delivers ages in correlation with age-known standards (Hurford & Green 1982), or 'absolute' calibration without age standards (see the discussion in Van den Haute *et al.* 1998; Enkelmann *et al.* 2005) are customarily validated by the analysis of samples with known ages. Hasebe *et al.* (2004) reported FT ages for apatite from three FT age standards (Fish Canyon Tuff, Durango and Mount Dromedary) determined using LA–ICP-MS-measured uranium contents. Table 4 shows individual crystal results measured in the current study for zircon age standards from the Fish Canyon Tuff and Buluk Member Tuff (Hurford 1990). Ages have been calculated using an absolute approach (Hasebe *et al.* 2004) adopting a track length of 10.5 μm (Hasebe *et al.* 1994), zircon density of $4.70\ g\ cm^{-3}$ (Morimoto *et al.* 1975) and decay constant of $8.46 \cdot 10^{-17}\ year^{-1}$ (Spadavecchia & Hahn 1967; Bigazzi 1981; Holden & Hoffman 2000; Guedes *et al.* 2003; Yoshioka *et al.* 2005). The track etching and counting efficiency is assumed to be 1. The resulting age of each zircon age standard is in reasonable agreement with the independent reference ages: for the Fish Canyon Tuff 26.0 ± 2.0 Ma (2σ) compares with $^{40}Ar/^{39}Ar$ ages of around 28 Ma (Backmann *et al.* 2007); for the Buluk Member 16.7 ± 1.4 Ma (2σ) compares with the 16.2 ± 0.2 Ma K–Ar high-temperature alkali feldspar ages (Hurford 1990).

These results were obtained using NIST610 glass as an external standard for calibration and assuming a generalized silicon content of 33.3%. Use of a standard zircon (Black *et al.* 2004; Goolaerts *et al.* 2004) as an LA–ICP-MS external standard to minimize the matrix effect (Stirling *et al.* 2000; Tiepolo 2003; Woodhead *et al.* 2004; Gagnevin *et al.* 2005) and specific knowledge of the SiO_2 content of the dated grains would permit a more accurate estimation of uranium contents and resulting FT ages.

The calculation of FT age using the absolute approach adopted above assumes track length and fission fragment registration factors, and requires a value for the ^{238}U spontaneous fission decay constant, λ_f. To avoid such a selection of factors it may be more appropriate to adopt a comparative, zeta-type approach to FT analysis using LA–ICP-MS, determining zeta-type values for age standards analysed in each LA–ICP-MS session (see discussion in Hasebe *et al.* 2004).

$^{238}U/^{235}U$ and $^{238}U/^{232}Th$ ratios

Although the main focus of this paper is to determine the reliability of ^{238}U measurement on etched surfaces, ^{235}U and ^{232}Th were also measured and provide some observations for the LA–ICP-MS measurement of U and Th in dated minerals. Table 5 lists average concentrations of the three isotopes and their ratios for NIST glasses, and apatite and zircon samples; mean values are calculated from all the spot analyses, combining data from multiple grains and from etched and unetched surfaces. Broadly similar numbers of ablated spots were measured for each sample and standard. Standard deviations of those samples expected to have homogeneous elemental composition (e.g. NIST glass) should be the lowest and effectively represent

Table 4. *Age calculation of zircon age standards*

Grain number	N_s	ρs (10^6 cm^{-2})	Number of ablation pits	$^{238}C \pm \sigma$ (μg g^{-1})	$T \pm \sigma$ (Ma)
FCT: Fish Canyon Tuff					
1	70	7.79	1	643 ± 64	22.9 ± 3.6
2	66	8.27	2	511 ± 17	30.5 ± 3.9
3	147	6.14	2	359 ± 90	32.2 ± 8.5
4	153	6.39	3	492 ± 47	24.5 ± 3.1
5	47	2.94	2	190 ± 2	29.2 ± 4.3
6	57	6.35	1	381 ± 38	31.5 ± 5.2
7	42	2.34	2	256 ± 3	17.3 ± 2.7
8	48	2.29	3	143 ± 4	30.2 ± 4.5
9	71	7.90	1	834 ± 83	17.9 ± 2.8
10	98	6.14	3	530 ± 5	21.9 ± 2.2
11	84	4.01	3	338 ± 11	22.3 ± 2.5
12	80	8.02	2	462 ± 18	32.8 ± 3.9
13	48	4.01	1	234 ± 23	32.2 ± 5.7
Weighted mean					26.0 ± 1.0
BM: Buluk Member Tuff					
1	44	1.26	6	129 ± 3	18.4 ± 2.8
2	36	0.90	4	127 ± 8	13.4 ± 2.4
3	32	1.28	4	114 ± 3	21.2 ± 3.8
4	36	1.03	6	124 ± 6	15.7 ± 2.7
5	42	1.40	4	137 ± 5	19.4 ± 3.1
6	32	0.64	4	78 ± 3	15.5 ± 2.8
7	56	1.15	4	146 ± 6	14.9 ± 2.1
8	57	1.14	4	136 ± 2	15.9 ± 2.1
9	60	1.00	6	119 ± 7	15.9 ± 2.2
10	45	1.29	4	142 ± 8	17.2 ± 2.7
11	60	1.43	3	163 ± 18	16.6 ± 2.8
12	55	1.38	4	133 ± 0	19.6 ± 2.6
13	26	0.72	3	93 ± 2	14.7 ± 2.9
14	40	1.25	2	131 ± 3	18.0 ± 2.9
15	28	0.80	2	148 ± 7	10.2 ± 2.0
16	41	0.68	3	68 ± 5	19.2 ± 3.4
17	23	0.66	2	79 ± 3	15.8 ± 3.3
Weighted mean					16.7 ± 0.7

N_s, number of spontaneous tracks counted; ρs spontaneous track density; ^{238}C average uranium content of a grain obtained by LA–ICP-MS analyses of several spots within area where spontaneous tracks were counted. Where only one spot was measureable, 10% of an average was given as an error.

best analytical precision. For other samples the standard deviations reflect compositional variation between grains (or between analysed spots). From the NIST glass results, concentrations of ^{235}U as low as 0.1 μg g^{-1} are readily measurable by the LA–ICP-MS within an uncertainty of approximately 10%.

The mean $^{238}U/^{235}U$ ratio of 421.6 ± 39.0 calculated for NIST612 from measured ^{238}U and ^{235}U concentrations agrees well with the value of approximately 417 given without uncertainty by Carpenter & Reimer (1974) – note that NIST612 was prepared with depleted uranium. For mineral samples, the mean $^{238}U/^{235}U$ ratios (139.5 for apatite and 137.9 for zircon) compare well with the natural abundance ratio of 137.9 (Steiger & Jager 1977). These ratios calculated from the concentration of each isotope agree well with ratios obtained by the MS signal intensities. No discrimination or fractionation occurred between ^{238}U and ^{235}U during measurements, implying that the known concentration of one isotope is sufficient to estimate the concentrations of all the uranium isotopes.

The concentration of ^{232}Th measured for the NIST612 glass (37.8 μg g^{-1} for NIST612 measured with apatite and zircon) shows good agreement with the reference value of 37.23 μg g^{-1} (Pearce *et al.* 1997). Similarly, the $^{238}U/^{232}Th$ ratios are in accord with reference values. However, for apatite and zircon $^{238}U/^{232}Th$ ratios obtained by calculated concentrations and by LA–ICP-MS signal intensities are discordant compared with reference values. This is particularly true for zircons.

Table 5. *LA–ICP-MS measured concentrations of the isotopes ^{238}U, ^{235}U and ^{232}Th and their ratios*

Sample code	*N*	^{238}U (μg/g)	SD (^{238}U) (%)	^{235}U (μg/g)	SD (^{235}U) (%)	^{232}Th (μg/g)	SD (^{232}Th) (%)	$^{238}U/^{235}U$	SD ppm ($^{238}U/^{235}U$) (%)	$^{238}U/^{235}U$ int/int	SD int ($^{238}U/^{235}U$) (%)	$^{238}U/^{232}Th$	SD ppm ($^{238}U/^{232}Th$) (%)	$^{238}U/^{232}Th$ int/int	SD int ($^{238}U/^{232}Th$) (%)
NIST612(Ap)	47	37.3	2.5	0.089	9.8	37.8	2.6	420.9	10.0	415.6	9.9	0.987	3.2	1.027	7.9
NIST612(Zr)	49	37.2	1.7	0.089	8.8	37.8	1.6	422.3	8.5	421.8	9.2	0.986	2.3	1.358	3.7
Reference values (Carpenter & Reimer 1974; (Pearce et al. *1997)*		*37.1*	*3.4*	*0.088*	*3.4*	*37.2*	*1.9*	*417.1*							
Apatite															
FUL	45	6.6	19.3	0.048	20.4	49.0	7.7	138.7	11.9	139.1	11.7	0.134	13.0	0.139	19.2
DUR	75	10.0	25.1	0.073	30.6	185.6	45.9	140.5	16.0	138.3	15.5	0.066	39.9	0.070	43.1
LOV	45	9.2	27.8	0.066	28.9	121.6	52.2	141.6	12.1	140.5	12.0	0.088	27.8	0.093	24.3
UNK	40	9.5	8.5	0.068	11.7	54.9	8.1	139.5	7.9	139.3	7.9	0.172	2.0	0.187	4.2
DRV	39	122.6	15.9	0.895	16.9	841.2	23.0	137.3	21.2	136.5	21.1	0.150	0.0	0.156	0.0
Mean apatite								139.5							
Zircon															
BM4	23	85.4	39.9	0.615	41.6	50.8	43.5	139.8	2.5	140.0	2.1	1.705	5.6	2.456	5.5
TAS	45	28.9	27.3	0.209	26.8	8.2	31.1	139.1	3.4	138.8	3.8	3.571	7.7	4.817	9.0
FCT	30	326.8	48.9	2.386	48.5	146.6	74.3	136.9	3.2	137.6	1.7	2.969	60.7	3.951	61.2
LMR	43	238.1	36.8	1.720	36.5	83.8	77.0	138.3	2.3	137.4	1.8	4.125	60.0	5.741	61.0
3F	21	942.9	2.5	6.916	3.1	117.0	3.1	136.4	1.8	134.2	0.7	8.062	2.6	11.249	3.0
Malcan	21	3527.0	22.2	25.715	21.9	462.4	28.8	137.1	1.5	135.8	0.7	7.883	15.7	11.236	15.1
Mean zircon								137.9							
Natural ratios from atom% (Steiger & Jager, 1977)								*137.9*							

int, signal intensity of mass spectrometry.

These results indicate: (1) the different behaviour of U and Th through zircon LA–ICP-MS measurement; (2) the effect of the internal standard (Ca or Si); and/or (3) the role of instrumental setting (such as 4 or 5 Hz laser repetition rate). Additional experiments would be necessary to determine which elements are key to understanding these observations.

Summary

- Chemical etching of fission tracks in apatite and zircon for FT chronometry causes no significant effect on measurement of uranium concentration by LA–ICP-MS, giving support to its employment as an alternative to thermal neutron-induced fission of ^{235}U at nuclear reactors. This study used an Excimer laser, but, because laser ablation efficiency depends largely on laser wavelength (Gunther *et al.* 1999), different laser systems might show different results.
- Ages calculated for zircon standards show good agreement with their reference ages. NIST610 glass appears an effective external standard for zircon as well as for apatite (Hasebe *et al.* 2004), and use of an average SiO_2 content of 33.3% is reasonable for age calculation of zircon at a first level of precision.

This study is supported by the Mitani Research and Development Foundation. The LA–ICP-MS measurements were helped by Dr A. Tamura. Laser microscope observation was assisted by Dr H. Ito. Prof. G. A. Wager and Dr R. Jonckheere gave helpful comments and suggestions to improve the manuscript. We are grateful to Dr B. Ventura, the editor, for her additional comments and guidance.

References

BACHMANN, O., OBERLI, F., DUNGAN, M. A., MEIER, M., MUNDIL, R. & FISCHER, H. 2007. Ar/Ar and U–Pb dating of the Fish Canyon magmatic system, San Juan Volcanic field, Colorado: evidence for an extended crystallization history. *Chemical Geology*, **236**, 134–166.

BARBARAND, J., CARTER, A., WOOD, I. & HURFORD, A. 2003. Compositional and structural control of fission-track annealing in apatite. *Chemical Geology*, **198**, 107–137.

BEA, F., MONTERO, P., STROH, A. & BAASNER, J. 1996. Microanalysis of minerals by an Excimer UV-LA-ICP-MS system. *Chemical Geology*, **133**, 145–156.

BIGAZZI, G. 1981. The problem of the decay constant λ_f of ^{238}U. *Nuclear Tracks*, **5**, 35–44.

BLACK, L. P., KAMO, S. L. ET AL. 2004. Improved $^{206}Pb/^{238}U$ microprobe geochronology by the monitoring of a trace-element-related matrix effect; SHRIMP, ID-TIMS, ELA-ICP-MS and oxygen isotope documentation for a series of zircon standards. *Chemical Geology*, **205**, 115–140.

CARPENTER, B. S. & REIMER, G. M. 1974. *Calibrated Glass Standards for Fission Track Use.* National Bureau of Standards, Special Publications, 260–49.

DAVIS, D. W. & KROGH, T. E. 2000. Preferential dissolution of U^{234} and radiogenic Pb from α-recoil-damaged lattice sites in zircon: implications for thermal histories and Pb isotopic fractionation in the near surface environment. *Chemical Geology*, **172**, 41–58.

EGGINS, S., GRÜN, R., PIKE, A. W. G., SHELLEY, M. & TAYLOR, L. 2003. ^{238}U, ^{232}Th profiling and U-series isotope analysis of fossil teeth by laser ablation–ICPMS. *Quaternary Science Reviews*, **22**, 1373–1382.

ENKELMANN, E., JONCKHEERE, R. & WAUSCHKUHN, B. 2005. Independent fission-track ages (ϕ-ages) of proposed and accepted apatite age standards and a comparison of ϕ-, Z-, ζ- and ζ_0-ages: implications for method calibration. *Chemical Geology*, **222**, 232–248.

GAGNEVIN, D., DALY, J. S., WAIGHT, T. E., MORGAN, D. & POLI, G. 2005. Pb isotopic zoning of K-feldspar megacrysts determined by Laser Ablation Multi-Collector ICP-MS: insights into granite petrogenesis. *Geochimica et Cosmochimica Acta*, **69**, 1899–1915.

GLEADOW, A. J. W. 1981. Fission-track dating methods: what are the real alternatives? *Nuclear Tracks*, **5**, 3–14.

GLEADOW, A. J. W., HURFORD, A. J. & QUAIFE, R. D. 1976. Fission track dating of zircon: improved etching conditions. *Earth and Planetary Science Letters*, **33**, 273–276.

GOOLAERTS, A., MATTIELLI, N., JONG, J. D., WEIS, D. & SCOATES, J. S. 2004. Hf and Lu isotopic reference values for the zircon standard 91500 by MC-ICP-MS. *Chemical Geology*, **206**, 1–9.

GREEN, P. F. 1981. A new look at statistics in fission-track dating. *Nuclear Tracks*, **5**, 77–86.

GUEDES, S., HADLER, J. C., IUNES, P. J., ZUNIGA, A., TELLO, C. A. & PAULO, S. R. 2003. The use of the U(n,f) reaction dosimetry in the determination of the λf value through fission-track techniques. *Nuclear Instruments and Methods in Physics Research Section A*, **436**, 215–221.

GUNTHER, D., JACKSON, S. E. & LONGERICH, H. P. 1999. Laser ablation and arc/spark solid sample introduction into inductively coupled plasma mass spectrometers. *Spectrochimica Acta Part B: Atomic Spectroscopy*, **54**, 381–409.

HASEBE, N., BARBARAND, J., JARVIS, K., CARTER, A. & HURFORD, A. J. 2004. Apatite fission-track chronometry using laser ablation ICP-MS. *Chemical Geology*, **207**, 135–145.

HASEBE, N., TAGAMI, T. & NISHIMURA, S. 1994. Towards zircon fission track thermochronology: reference framework for confined track length measurements. *Chemical Geology (Isotope Geoscience Section)*, **112**, 169–178.

HOLDEN, N. E. & HOFFMAN, D. C. 2000. Spontaneous fission half-lives for ground-state nuclides (Technical Report). *Pure and Applied Chemistry*, **72**, 1525–1562.

HORN, I. & GUNTHER, D. 2002. Transport efficiency and particle size distribution of laser induced aerosols.

In: *Abstracts of the 12th Annual V. M. Goldshdmit Conference, Davos, Switzerland, Geochimica et Cosmochimica Acta*, Special Supplement, **66**, A341.

HURFORD, A. J. 1990. Standardization of fission track dating calibration: Recommendation by the Fission Track Working Group of the I.U.G.S. Subcommission on Geochronology. *Chemical Geology (Isotope Geoscience Section)*, **80**, 171–178.

HURFORD, A. J. & GREEN, P. F. 1982. A user's guide to fission track dating calibration. *Earth & Planetary Science Letters*, **59**, 343–354.

IIZUKA, T. & HIRATA, T. 2004. Simultaneous determinations of U–Pb age and REE abundances for zircons using ArF excimer ablation-ICPMS. *Geochemical Journal*, **35**, 229–241.

ISHIDA, Y., MORISHITA, T., ARAI, S. & SHIRASAKA, M. 2004. Simultaneous in-situ multi-element analysis of minerals on thing section using LA–ICP-MS. *Science Reports of Kanazawa University*, **48**, 31–42.

JARVIS, K. E. & WILLIAMS, J. G. 1993. Laser ablation inductively coupled plasma mass spectrometry (LA-ICP-MS): a rapid technique for the direct, quantitative determination of major, trace and rare-earth elements in geological samples. *Chemical Geology*, **106**, 251–262.

JEFFRIES, T. E., PERKINS, W. T. & PEARCE, N. J. G. 1995. Measurements of trace elements in basalts and their phenocrysts by laser probe microanalysis inductively coupled plasma mass spectrometry (LPMA-ICP-MS). *Chemical Geology*, **121**, 131–144.

KOSLER, M. & SVOJTKA, M. 2003. Present trends and the future of zircon in geochronology: laser ablation ICPMS. *Reviews in Mineralogy and Geochemistry*, **54**, 243–275.

MORIMOTO, N., SUNAGAWA, K. & MIYASHIRO, A. 1975. *Mineralogy*. Iwanami Publishing, Tokyo.

MORISHITA, T., ISHIDA, Y., ARAI, S. & SHIRASAKA, M. 2005. Determination of multiple trace element compositions in thin ($<30\ \mu m$) layers of NIST SRM 614 and 616 using laser ablation–inductively coupled plasma-mass spectrometry (LA–ICP-MS). *Geostandards and Geoanalytical Research*, **29**, 107–122.

MORRISON, C. A., LAMBERT, D. D., MORRISON, R. J. S., AHLERS, W. W. & NICHOLLS, I. A. 1995. Laser ablation-inductively couple plasma-mass spectrometry: an investigation of elemental responses and matrix effects in the analysis of geostandard materials. *Chemical Geology*, **119**, 13–29.

PEARCE, N. J. G., PERKINS, W. T., WESTGATE, J. A., GORTON, M. P., JACKSON, S. E., NEAL, C. R. & CHENERY, S. P. 1997. A compilation of new and published major and trace element data for NIST SRM 610 and NIST SRM 612 glass reference materials. *Geostandards Newsletter*, **21**, 115–144.

PERKINS, W. T., PEARCE, N. J. G. & JEFFRIES, T. E. 1993. Laser ablation inductively coupled plasma mass spectrometry: a new technique for the determination of trace and ultra-trace elements in silicates. *Geochimica et Cosmochimica Acta*, **57**, 475–482.

STEIGER, R. H. & JAGER, E. 1977. Subcommission on Geochronology: Convention on the use of decay constraints in geo- and cosmochronology. *Earth and Planetary Science Letters*, **36**, 359–362.

SPADAVECCHIA, A. & HAHN, B. 1967. Die Rotationskammer und einige anwendungen. *Helvetica Physica Acta*, **40**, 1063–1079.

STIRLING, C. H., LEE, D. C., CHRISTENSEN, J. N. & HALLIDAY, A. N. 2000. High-precision in situ ^{238}U–^{234}U–^{230}Th isotopic analysis using laser ablation multiple-collector ICPMS. *Geochimica et Cosmochimica Acta*, **64**, 3737–3750.

SYLVESTER, P. J. & GHADERI, M. 1997. Trace element analysis of scheelite by excimer laser ablation–inductively coupled plasma-mass spectrometry (ELA–ICP-MS) using a synthetic silicate glass standard. *Chemical Geology*, **141**, 49–65.

TAGAMI, T., LAL, N., SORKHABI, R. B., ITO, H. & NISHIMURA, S. 1988. Fission track dating using external detector method: a laboratory procedure. *Memoirs of the Faculty of Science, Kyoto University. Series of Geology & Mineralogy*, **53**, 14–30.

TIEPOLO, M. 2003. *In situ* Pb geochronology of zircon with laser ablation–inductively coupled plasma-sector field mass spectrometry. *Chemical Geology*, **199**, 159–177.

VAN DEN HAUTE, P., DE CORTE, F., JONCKHEERE, R. & BELLMANS, F. 1998. The parameters that govern the accuracy of fission-track age determinations: a reappraisal. *In*: VAN DEN HAUTE, P. & DE CORTE, F. (eds) *Advances in Fission-Track Geochronology*. Kluwer, Dordrechts, 33–46.

WILLIAMS, I. S. & JARVIS, K. 1993. Preliminary assessment of laser ablation inductively coupled plasma mass spectrometry for quantitative multi-element determination in silicates. *Journal of Analytical Atomic Spectrometry*, **8**, 25–34.

WOODHEAD, J., HERGT, J., SHELLEY, M., EGGINS, S. & KEMP, R. 2004. Zircon Hf-isotope analysis with an excimer laser, depth profiling, ablation of complex geometries, and concomitant age estimation. *Chemical Geology*, **209**, 121–135.

YAMADA, R., YOSHIOKA, T., WATANABE, K., TAGAMI, T., NAKAMURA, H., HASHIMOTO, T. & NISHIMURA, S. 1998. Comparison of experimental techniques to increase the number of measurable confined fission tracks in zircon. *Chemical Geology*, **149**, 99–107.

YOSHIOKA, T., TSURUTA, T., IWANO, H. & DANHARA, T. 2005. Spontaneous fission decay constant of ^{238}U determined by SSNTD method using CR-39 and DAP plates. *Nuclear Instruments and Methods in Physics Research A*, **555**, 386–395.

Quantitative constraints on mid- to shallow-crustal processes using the zircon (U–Th)/He thermochronometer

KATHERINE J. DOBSON[1,2]*, CRISTINA PERSANO[1] & FINLAY M. STUART[2]

[1]*Geographical & Earth Sciences, University of Glasgow, Gregory Building, Glasgow G12 8QQ, UK*

[2]*Scottish Universities Environmental Research Centre (SUERC), Rankine Avenue, Scottish Enterprise Technology Park East Kilbride G75 0QF, UK*

**Corresponding author (e-mail: kate.dobson@ges.gla.ac.uk)*

Abstract: Despite the potential of zircon He thermochronometry for constraining rock thermal histories, it remains less commonly exploited than the apatite He chronometer. In part, this is due to the more challenging analytical techniques required to extract He, U and Th. Here we present a new method for the routine determination of zircon (U–Th)/He ages, and demonstrate how it can be used to constrain thermal histories and to quantify cooling in different tectonic settings. We present zircon (U–Th)/He ages that place a firm upper limit on the extent of denudation-induced cooling (*c.* 3 km) on the SE Australian passive margin; a region where synrift apatite fission-track and apatite (U–Th)/He ages have previously prevented quantitative constraint. We have also used the zircon (U–Th)/He thermochronometer to quantify the cooling of early Tertiary mafic plutons from Skye, Scotland, where the rate and timing of cooling cannot be determined using other thermochronometers.

The majority of studies that apply low-temperature thermochronology to continental-scale processes aim to resolve the rate, timing and distribution of uplift and denudation; typically in the context of orogenesis (e.g. Bigot-Cormier *et al.* 2000; Blythe *et al.* 2000), the evolution of passive rifted margins (e.g. Gallagher & Brown 1997; Lisker 2002) or the burial and exhumation of sedimentary basins (e.g. Duddy & Gleadow 1985). In these settings the combination of apatite fission-track (AFT) and apatite (U–Th)/He (AHe) thermochronometry has provided temporal constraints on many shallow-crustal processes (Persano *et al.* 2005; Spotila 2005; Stockli 2005). However, to date, the quantitative constraint on the timing of cooling through higher temperatures has often been prevented by the poor understanding of crystal specific annealing behaviour of fission tracks in zircon (ZFT) (Hasebe *et al.* 1994; Yamada *et al.* 1995; Bernet & Garver 2005; Garver *et al.* 2005). The inability to accurately determine the timing of cooling between approximately 350 °C (the closure temperature of the K/Ar system in mica) and *c.* 110 °C (the AFT closure temperature) prevents the resolution of the timing and rate of shallow- and mid-crustal processes, and limits our understanding of the interplay between plate-scale tectonics and landscape evolution.

The zircon (U–Th)/He (ZHe) thermochronometer has a closure temperature of 170–190 °C for typical cooling rates and crystal sizes (Reiners *et al.* 2002, 2004), and has the potential to provide time–temperature constraints unavailable from existing thermochronometers. The common occurrence of zircon as an accessory phase in many igneous and metamorphic rocks, and its robustness in the geological record means that it can be applied to a wide range of studies. ZHe has previously been used to constrain the cooling of mid-crustal plutonic rocks (Reiners & Spell 2002; Reiners *et al.* 2002, 2004; Kirstein *et al.* 2006), for dating volcanic tuff sequences (Reiners *et al.* 2002, 2004; Tagami *et al.* 2003) and for determining sediment provenance using detrital zircon populations (Hourigan *et al.* 2003; Reiners *et al.* 2005). However, the application of the technique has been limited because of the analytical considerations. In particular, the time, effort and cost of U and Th extraction from zircon is considerably greater than from apatite (Tagami *et al.* 2003; Reiners 2005; Reiners & Nicolescu 2006), and the prevailing analytical protocols are not without their complications (Reiners 2005; Reiners & Nicolescu 2006). Here we present a technique that allows the routine determination of (U–Th)/He ages from single zircon crystals, and illustrate the strength of the new technique with two short studies.

From: LISKER, F., VENTURA, B. & GLASMACHER, U. A. (eds) *Thermochronological Methods: From Palaeotemperature Constraints to Landscape Evolution Models*. Geological Society, London, Special Publications, **324**, 47–56.
DOI: 10.1144/SP324.4 0305-8719/09/$15.00

Analytical technique

Zircons were hand-picked to be free from visible fluid/mineral inclusions, cracks and fractures using a stereoscopic microscope at ×500 magnification. The crystal length, termination lengths and orthogonal crystal widths were determined using a graticule. Individual crystals were packed into 2.0 × 0.5 mm Pt-foil tubes that were crimped closed at each end. Helium extraction is now routinely performed using a diode laser (Foeken *et al.* 2006), but early experiments used a double-walled resistance furnace previously used for noble gas extraction at SUERC. For furnace extraction, Pt-foil tubes were wrapped in small Mo-foil envelopes to ease their removal after helium extraction. For laser heating the Pt-foil packets were loaded directly into 3 mm-deep pits in a Cu laser pan. Complete helium extraction was achieved by heating at approximately 1200 °C for 25 min. Laser helium blank levels (1.5×10^{-12} cm^{-3} STP ^{4}He) are significantly lower than furnace blanks (1.4×10^{-11} cm^{-3} STP ^{4}He). Helium re-extractions were performed routinely and were usually indistinguishable from the preceding hot blank, and generally less than 0.5% of the ^{4}He. Full details of gas extraction, clean up and He measurement are given in Foeken *et al.* (2006).

The Pt-foil packets were removed from the pan after He extraction. To prevent possible crystal loss and the incomplete extraction of volatilized U and Th, which may have condensed onto the internal surface of the Pt-foil, the unopened degassed foil packets were placed in 0.35 ml Parrish-type Teflon microcapsules with 30 μl of 11.2 M HCl and approximately 2.28 ng ^{230}Th and *c.* 0.93 ng ^{235}U spikes dissolved in 300 μl of 5% HNO_3. The sample solution was reduced by 50% on a hotplate at 80 °C, and the microcapsules sealed and refluxed overnight to completely dissolve the Pt-foil. The solutions were then evaporated to dryness, and rehydrated with 15 μl of 16 M HNO_3 and 180 μl of 27.6 M HF. Eight microcapsules were loaded into a 125 ml capacity Parr^{TM} bomb with 180 μl of 16 M HNO_3 and 9 ml of 27.6 M HF. The assembled bomb was heated for 48 h at 230 ± 1 °C in a thermostatically controlled oven. To ensure that the refractory fluoride salts were fully dissolved, the microcapsules were removed from the oven, evaporated to dryness and rehydrated with 195 μl of 3 M HCl. The microcapsules were then returned to the bomb with 9 ml of 3 M HCl, and heated for a further 14 h at 230 °C. This procedure is sufficient to dissolve the majority of zircons (Parrish 1987). The initial dissolution cycle can be extended if necessary.

The dissolution of the Pt-foil packets introduced approximately 200 μg of Pt into the sample solution. The dissolved Pt forms $PtAr^{+}$ during ionization in the plasma that generated peaks which interfere at several of the masses that are routinely measured for U and Th concentration determinations. In particular, $^{198}Pt^{40}Ar^{+}$ (mass 238) and $^{195}Pt^{40}Ar^{+}$ (mass 235) restrict using the Pt-bearing solutions for U concentration measurement by conventional isotopic dilution techniques using quadrupole inductively coupled plasma mass spectrometers (ICP-MS). For routine application the Pt is removed prior to analysis by anion-exchange column chemistry. The purification of the sample and the reduction in the concentration of contaminant prevents $PtAr^{+}$ interference and removes the approximately 35% reduction in ICP-MS sensitivity that results from the high volume of dissolved species (Reiners & Nicolescu 2006). The sample solutions were transferred from the microcapsules to 2 ml Teflon beakers. The solutions were evaporated to dryness then equilibrated with 1 ml of 1.5 M HNO_3. Purification of the U and Th was then achieved using Teflon columns loaded with 1 ml of Eichrom TRU Resin (Blue). The columns were rinsed with 9 ml of 0.2 M HCl and 9 ml of 0.1 M HCl–0.3 M HF, then preconditioned using 9 ml of 1.5 M HNO_3 before the equilibrated sample solution was introduced. The Pt and other contaminants were removed with rinses of 12 ml of 1.5 M HNO_3 and 2.5 ml of 3 M HCl. Elution of U and Th was achieved using a rinse of 12 ml of 0.1 M HCl–0.3 M HF. The U- and Th-bearing elute was evaporated to dryness, and equilibrated with 2 ml of 5% HNO_3 and trace HF mixture prior to ICP-MS analysis. The microcapsules were cleaned by refluxing with 0.1 M HCl–0.3 M HF on a hotplate at 130 °C. Next 15 μl of 16 M HNO_3 and 180 μl of 27.6 M HF were added to each microcapsule, and the microcapsules loaded into the bomb and heated at 235 °C for a further 48 h. This procedure was sufficient, as blanks performed routinely yield levels of U and Th indistinguishable from background ICP-MS measurements.

U and Th analyses were performed on a VG PQ2plus ICP-MS. Isotopic fractionation was monitored using a certified U500 standard solution. The U and Th measurements were replicated five times. The Pt-foil tubes used during this study contained measurable U and Th, which generated a procedural blank of 0.1067 ± 0.0149 ng U ($n = 11$) and 0.0997 ± 0.0110 ng Th ($n = 11$). Although analysis of single crystals is possible, until the U and Th content of a sample is established, aliquots of two or three crystals are preferred, in order to ensure that the uncertainty in the U and Th blank has a minimal effect on the measured U and Th concentrations and the calculated ZHe age.

Zircon (U–Th)/He age standards

Zircon from the Fish Canyon Tuff (FCT) has been adopted as the ZHe age standard mineral (Reiners *et al.* 2002; Tagami *et al.* 2003; Reiners 2005). Unlike the Durango apatite standard, (U–Th)/He ages of FCT zircon were determined on complete crystals and therefore all ages required correction for the recoil loss of ^{4}He at the crystal boundaries (Farley 2002). In this study, the recoil correction was based on measured grain dimensions and used the equations developed by Hourigan *et al.* (2005). The average (U–Th)/He age of the 16 FCT zircon aliquots analysed as part of this study (27.6 ± 3.3 Ma; Table 1) is indistinguishable from the He ages of 28.3 ± 2.6 Ma ($n = 83$) reported by Reiners (2005), and the average age of 127 analyses from all laboratories (28.3 ± 3.1 Ma, Tagami *et al.* 2003; Reiners 2005; Pik pers. comm.) (Fig. 1).

The age reproducibility of this standard (±11%) was significantly poorer than the analytical precision (typically ±2%: Dobson 2006; Reiners & Nicolescu 2006), and the He age reproducibility of the Durango apatite standard (typically ±6–8%: Farley 2002; Boyce & Hodges 2005). This is thought to be because of heterogeneous and variable U and Th distributions within the zircons, which can exhibit up to a factor of 20 change in the concentration gradients in the outer 20 μm of some crystals (Dobson 2006). The uncertainty in the age of unknown samples is generally calculated from the 2σ reproducibility of relevant mineral standards (e.g. Reiners 2005). Using the FCT zircon as an age standard limits the precision with which dating can be performed, and the variable zonation of U and Th in FCT zircon also prevents a true estimation of analytical precision (Dobson 2006).

The Muck Tuff (MT) is a sequence of zircon-bearing crystal lithic tuffs that are preserved on the island of Muck, western Scotland. These tuffs mark the earliest Palaeogene volcanism on the European Atlantic margin (Saunders *et al.* 1997), and field relationships indicate it has been within 500 m of the surface since approximately 58.7 Ma (Chambers *et al.* 2005). Five single zircon aliquots from Camas Mor, Muck, have an average He age of 58.8 ± 3.5 Ma (Table 1). This is within error of the zircon U–Pb age (61.15 ± 0.25 Ma: Chambers *et al.* 2005) and the sanidine ^{40}Ar–^{39}Ar age (62.8 ± 0.6 Ma, Pearson *et al.* 1996, to 60.45 ± 0.03 Ma, Chambers *et al.* 2005) obtained from this sample, implying that the Muck Tuff cooled to <170 °C rapidly after eruption and experienced no reheating. The age reproducibility of the Muck Tuff is significantly better (±*c.* 6%) than that determined for the Fish Canyon Tuff zircons (±*c.* 11%), and is approximately the same as reported for the Durango apatite standard. The improved reproducibility of the MT may be due to less inter-crystal variation of U and Th zonation, therefore improving the accuracy of the alpha-recoil correction. We suggest that the age reproducibility of the MT zircons reflects better the true precision with which ZHe ages can be determined, and that it may be a more appropriate age standard than FCT zircons. Further work constraining the effect of variable U–Th zonation in crystal populations on the alpha-recoil correction is ongoing (Dobson 2006; Dobson *et al.* 2008).

Applications of zircon (U–Th)/He thermochronology

Constraining exhumation at passive margins

AFT thermochronology is sensitive to cooling from approximately 110 to 70 °C (Laslett *et al.* 1987; Ketcham *et al.* 1999). This corresponds to the removal of a crustal section of less than 4 km when geothermal gradients are 25–30 °C km^{-1}. The apatite (U–Th)/He thermochronometer is sensitive to cooling from approximately 75 to 35 °C (House *et al.* 1999; Farley 2000), and can be used to determine the timing of removal of 1–2 km of crust. Combining the AHe and AFT thermochronometers in samples from the SE Australian and the Eritrean high-elevation passive margins has demonstrated that continental break-up-driven denudation varied systematically across both margins. Denudation decreases from a maximum at the present coastline, where synrift AFT and AHe ages are similar, to the continental interior, where ages may be 100s of Ma older (Persano *et al.* 2002, 2005; Balestrieri *et al.* 2005). The synrift AFT ages at the coast prevent a precise determination of the amount of denudation, and a higher temperature thermochronometer is necessary to in order to: (i) constrain the maximum amount of break-up-driven denudation at the rifted margin; and (ii) determine the volume of sediments delivered to the offshore basins. Constraining the maximum amount of break-up-driven denudation experienced by a continental margin has important implications for the understanding of the flexural properties of the crust that allow erosion-driven isostatic rebound to occur. Numerical models of landscape evolution at the eastern Australian passive margin indicate that for typical values of crustal elastic thickness ($T_e = 10$–15 km), denudation at the coast could not have exceeded 2 km (Braun & Van der Beek 2004). This implies that the geothermal gradient at the time of rifting was at least 60 °C km^{-1}. If the geothermal gradient has remained constant at 25 °C km^{-1} since before break up, the minimum amount of denudation derived from the synrift

Table 1. *He, U and Th data from the Fish Canyon Tuff zircon*

Sample	Number of crystals	^{4}He (cm^3)	^{238}U (ng)	^{232}Th (ng)	Th/U	Uncert.	Raw age (Ma)	F_T*	Corrected age (Ma)	2σ uncert. (Ma)
FCT #1[†]	1	2.12E-08	6.68	3.37	0.50	3.9%	23.2	0.85	27.4	3.1
FCT #2[†]	2	4.59E-09	1.58	1.04	0.66	2.7%	19.8	0.75	26.5	3.0
FCT #3[†]	2	4.96E-09	1.62	2.44	1.50	2.8%	18.5	0.70	26.4	3.0
FCT #4[†]	1	4.69E-09	1.51	0.83	0.55	2.4%	22.6	0.80	28.3	3.2
FCT #5[†]	2	1.1E-09	0.47	0.22	0.46	6.2%	17.1	0.70	24.4	2.7
FCT #6[†]	2	1.84E-09	0.66	0.30	0.46	8.6%	20.6	0.70	29.3	3.3
FCT #7[†]	2	1.33E-09	0.53	0.23	0.43	6.9%	18.6	0.68	27.2	3.1
FCT #8[†]	1	2.11E-09	0.66	0.39	0.59	4.2%	23.0	0.84	27.3	3.1
FCT #9[†]	1	1.89E-09	0.69	0.32	0.47	4.5%	20.3	0.69	29.5	3.3
FCT #10[†]	1	1.31E-09	0.41	0.25	0.61	6.6%	22.9	0.82	27.9	3.1
FCT #11[†]	1	7.73E-10	0.40	0.23	0.58	6.6%	14.1	0.58	24.4	2.7
FCT #12[†]	1	9.7E-10	0.60	0.52	0.86	4.0%	11.0	0.66	16.6[§]	1.9[§]
FCT #13[†]	3	5.12E-09	1.68	1.23	0.73	1.9%	21.3	0.76	28.1	3.2
FCT #14[‡]	3	4.19E-09	1.44	0.89	0.61	2.3%	20.8	0.72	28.9	3.3
FCT #15[‡]	3	1.15E-08	4.08	2.38	0.59	2.2%	20.2	0.71	28.6	3.2
FCT #16[‡]	3	9.50E-09	3.32	1.82	0.56	2.2%	20.6	0.70	29.3	3.3
FCT #17[‡]	3	4.43E-09	1.62	1.12	0.71	2.4%	19.3	0.65	29.7	3.3
FCT #18[‡]	3	4.84E-09	1.76	1.11	0.65	2.6%	19.7	0.69	28.6	3.2
FCT #19[‡]	3	1.16E-08	3.83	3.08	0.81	2.4%	20.8	0.77	27.2	3.1
Average							*19.5*	*0.73*	*27.7*	*3.1 (11.3%)*[¶]

*F_T – correction for alpha-recoil loss, after Farley (2002) using the parameters for tetragonal pyramid-terminated crystals determined by Hourigan *et al.* (2005) assuming U–Th homogeneity; 2σ uncertainties of each replicate represent the analytical uncertainty (uncert.) of the individual U–Th–He measurements. # – replicate number; [†]furnace He extraction; [‡]diode laser He extraction; [§]data not included in the calculation of the average ZHe age, as this single crystal aliquot is thought to have atypical and extreme zonation (Dobson 2006); [¶]1σ standard deviation of the mean FCT ZHe age.

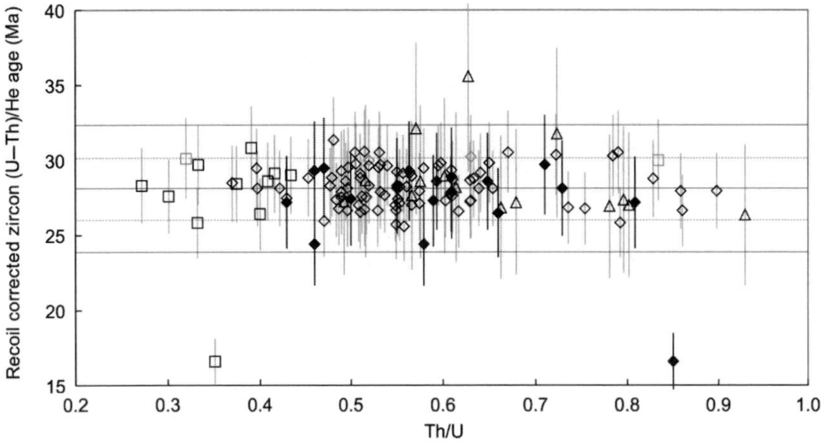

Fig. 1. The zircon (U–Th)/He ages of the Fish Canyon Tuff. Data from all laboratories routinely making ZHe age determinations are shown: open diamonds, Reiners (2005); open squares, Tagami *et al.* (2003); open triangles, Pik (pers. comm.); filled diamonds, this study. All error bars are 2σ. The two ZHe ages younger than 20 Ma are not included in the average age calculation. These samples may be affected by an extreme form of U and Th zonation (Tagami *et al.* 2003; Dobson 2006). The average age of the FCT zircons is 28.3 ± 3.1 Ma (2σ, $n = 127$).

AFT ages (3–4 km) can only be accommodated by an elastic thickness of 8 km. Although this is an acceptable value, it is at the low end of the range determined for continental crust (Braun & Van der Beek 2004). If the amount of synrift denudation was significantly greater, that is if the samples now at the surface were at more than 110 °C at the onset of continental break-up, a constant geothermal gradient of 25 °C km^{-1} would require unrealistically low values of T_e. If faulting can be discounted, cooling from significantly more than 110 °C implies a significantly higher synrift geothermal gradient. An accurate estimation of the maximum amount of cooling during rifting has the potential to constrain this palaeogeothermal gradient.

Along the SE coast of Australia AFT and AHe ages are indistinguishable from each other, and

Table 2. *He, U and Th data from zircons from SE Australia and the Hebridean Igneous Province, Scotland*

Sample	Number of crystals	^{4}He (10^{-8} cm^3)	^{238}U (ng)	^{232}Th (ng)	Th/U	Uncorrected age (Ma)	F_T*	Corrected age (Ma)
Australia								
99OZ-12 #1	2	3.79	1.357	0.738	0.54	217.4	0.76	285.5 ± 16.9
99OZ-12 #2	2	2.24	0.810	0.625	0.77	217.2	0.73	297.7 ± 17.6
Muck Tuff								
MT#1	1	3.20	3.95	4.3	1.09	52.8	0.90	58.7 ± 3.5
MT#2	1	0.40	0.47	0.55	1.16	54.1	0.95	56.9 ± 3.4
MT#3	1	0.58	0.6	0.95	1.57	57.3	0.93	61.6 ± 3.7
MT#4	1	2.17	2.49	4.08	1.64	51.5	0.89	57.9 ± 3.5
MT#5	1	1.99	2.42	3.01	1.24	52.0	0.88	59.1 ± 3.5
Average						53.5	0.91	58.8 ± 3.5¶
Mull								
ML3 #1	1	0.23	0.42	0.22	0.54	39.2	0.67	58.9 ± 1.3
ML3 #2	1	0.16	0.36	0.18	0.5	33.5	0.58	57.8 ± 1.3
ML5 #1	3	0.33	0.60	0.71	1.19	35.2	0.61	57.7 ± 1.3
Average						36.0	0.64	58.1 ± 1.3¶

*F_T correction for alpha-recoil loss, after Farley (2002) using the parameters for tetragonal pyramid-terminated crystals determined by Hourigan *et al.* (2005). All ages quoted with 2σ uncertainties representing the analytical uncertainty of the individual U–Th–He measurements.

¶1σ standard deviation of the mean FCT ZHe age.

from the age of rifting (Persano *et al.* 2002, 2005) (sea-floor spreading began at *c.* 85 Ma: Weissel & Hayes 1977). In a preliminary study we determined zircon He ages from the Bega granite (crystallization age of *c.* 400 Ma: Chappell & Stevens 1988; Williams 2001) within 5 km of the coast (99-OZ-12; Table 2) in order to constrain the maximum amount of denudation at the SE Australian passive margin. This sample previously yielded an AFT age of 135 ± 5 Ma and an AHe age of 90 ± 9 Ma (Persano *et al.* 2005), and on the basis of remagnetization of pyrrhotite Dunlop *et al.* (2000) have argued that pre-break-up temperatures were 165 ± 30 °C. Two aliquots of two euhedral zircon crystals yielded reproducible recoil-corrected ZHe ages of 286 and 298 Ma, respectively (Table 1). These are almost 200 Ma older than break-up. By combining the zircon He ages with the existing AFT thermochronology it is possible to constrain the pre-break-up thermal history of the margin and determine the amount of cooling during the isostatic uplift after break-up (Persano *et al.* 2005). In Figure 2 we show the He ages (uncorrected for alpha-recoil) predicted for a range of thermal histories for zircons with the same surface area-to-volume ratio as the analysed crystals. The He ages vary from 70 to 280 Ma depending on the cooling rate (Fig. 2B). Assuming the relatively simple thermal histories modelled here, the measured ZHe ages are consistent with a relatively rapid post-crystallization cooling (*c.* 7 °C Ma^{-1}), followed by a period of slower cooling (*c.* 0.1 °C Ma^{-1}) until the time of break-up, when the cooling rate dramatically increased (Persano *et al.* 2005). High rates of cooling in the early Devonian are consistent with typical plutonic cooling profiles and indicate that the granite cooled to approximately 200 °C in less than 50 Ma. This suggests that this pluton was intruded at relatively shallow depth (between about 5 and 8 km for palaeogeothermal gradients of 20–35 °C km^{-1}) or was exhumed to this depth immediately after intrusion. A long period of slow cooling is consistent with the old apatite fission track (250–350 Ma: Dumitru *et al.* 1991; Gleadow 2000; Persano *et al.* 2005) and He ages (200–250 Ma: Persano *et al.* 2005) and mixed fission-track length distribution found where the SE Australian highlands were not subjected to break-up-related denudation (Persano *et al.* 2005).

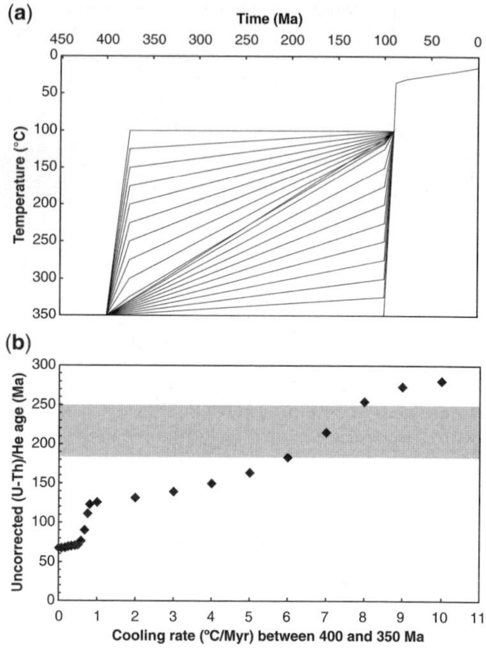

Fig. 2. (**a**) The possible thermal histories for the coastal samples from SE Australia. The time–temperature constraint at approximately 400 Ma is derived from the emplacement age of the Bega batholith (Chappell & Stevens 1988), 90 Ma is the latest time at which the sample could have passed through 100 °C (Persano *et al.* 2005). These thermal histories were used to predict the zircon (U–Th)/He ages shown in (b). (**b**) The uncorrected zircon He ages predicted from the thermal histories in (a) plotted as a function of the initial cooling rate from 400 Ma. Zircon He ages were predicted using DECOMP (Meesters & Dunai 2002) using the He diffusion parameters in zircon, $D_0 = 0.46$ cm s^{-1} and $E_a = 40.4$ kcal mol^{-1} (Reiners *et al.* 2004) and a stopping distance of 17 μm (Hourigan *et al.* 2005). 'Uncorrected' ZHe ages are used because these are the form of the data output by DECOMP, and previous studies have shown that applying an alpha-recoil correction is unjustified unless cooling is rapid (Meesters & Dunai 2002). The grey region represents the 'uncorrected' measured He age 217 Ma with a ±15% (±33 Ma) uncertainty, as calculated from the reproducibility of the uncorrected He ages of the FCT (Table 1). A cooling rate of between 6 and 8 °C Ma^{-1} best fits the measured uncorrected ages, this means that the sample was between 170 and 200 °C at 370 Ma.

In order to constrain the amount of break-up denudation we need to determine the temperature of sample 99-OZ-12 at approximately 100 Ma. To achieve this we calculated the He age of zircons for a suite of thermal histories (Fig. 3) that span the range predicted by preliminary models. These thermal histories include a period of rapid cooling from more than 350 °C, and the samples cool to 170–200 °C at 370 Ma, followed by variably slow cooling monotonic until 100 Ma. The predicted ZHe ages are indistinguishable from the measured age if the sample cooled to 100 °C (for slow cooling starting at 200 °C) and 125 °C (for slow cooling starting at 170 °C) at about 100 Ma (Fig. 3). These results indicate that the rocks now at the coast were close to, or slightly beyond, the

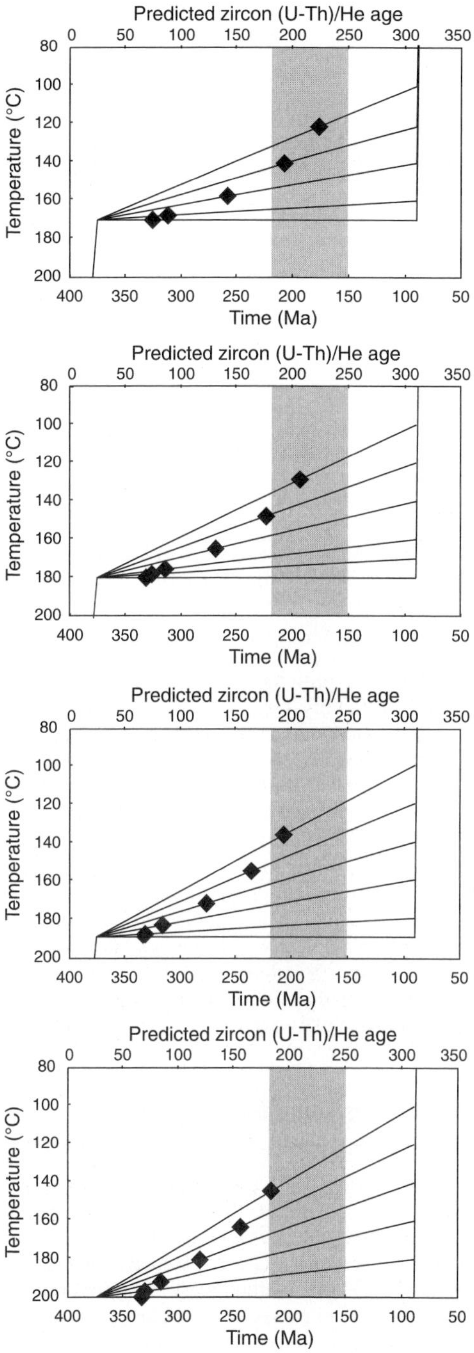

Fig. 3. Modelled time–temperature paths of the coastal samples at the onset of rifting, where cooling starts at 370 Ma at 170 °C (top panel) to 200 °C (bottom panel), using the constraints provided by Figure 2b and the latest time at which the sample could have passed through 100 °C (Persano *et al.* 2005). The DECOMP-derived ZHe ages for each thermal history are represented by black diamonds on the corresponding time–temperature path. The acceptable time–temperature paths are those where the predicted ages fall within the shaded region, which represents the uncorrected measured ZHe age of the coastal sample 99-OZ-12 (as on Fig. 2b). The plots show that this sample cannot have been at temperatures in excess of 125 °C at the onset of rifting (100–90 Ma).

base of the apatite partial annealing zone at the onset of break-up. Consequently, the amount of denudation derived from the AFT–AHe data (2–4 km, depending on the geothermal gradient) is close to the maximum amount experienced by the margin. The ZHe ages are inconsistent with the suggestion that remagnetization indicates break-up temperatures of 165 ± 30 °C (Dunlop *et al.* 2000). Higher temperatures were either very localized or were short-lived and therefore did not affect the He diffusion in zircon.

Constraining the cooling of plutonic systems

While the cooling history of acid igneous rocks can generally be constrained using a number of mineral thermochronometers, the cooling history of mafic and ultra-mafic rocks tends to be more difficult to extract because of a paucity of mineral phases suitable for radiometric dating. This generally limits the thermal constraint to crystallization ages determined from U/Pb analysis of zircon. Poor constraints on the thermal history prevent determination of the depth of emplacement, the timescale of hydrothermal activity and associated mineralization, and establishing accurately the duration of magmatic activity. Even for plutonic units where the application of $^{40}Ar/^{39}Ar$ and FT thermochronology has allowed the partial constraint on the cooling history, the ZHe thermochronometer can provide additional and more specific time–temperature constraint (Reiners & Spell 2002; Reiners *et al.* 2004).

The Palaeogene extrusive and intrusive volcanic sequences of the European North Atlantic margin were generated in response to the impact of the proto-Iceland plume (Saunders *et al.* 1997). The Hebridean Igneous Province (HIP) (Fig. 4) is one of the earliest igneous provinces on the European rift flank, and lies along the west coast of Scotland. The HIP is characterized by three fissure-fed basaltic lava fields and four plutonic complexes. These plutonic complexes represent the root zones of Palaeogene volcanoes (Bell & Williamson 2002) and are thought to have been emplaced at depths of 2–3 km (Holness 1999; Bell & Williamson 2002). Post-magmatic denudation has removed much of the basaltic sequence, and the timing and volume of material removed during this denudation episode is poorly constrained. Recent radiometric

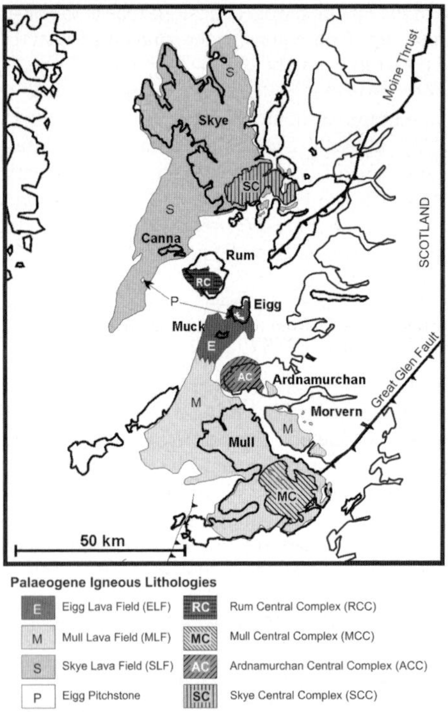

Fig. 4. A map of the Hebridean Igneous Province, NW Scotland (modified after Emeleus & Bell 2005).

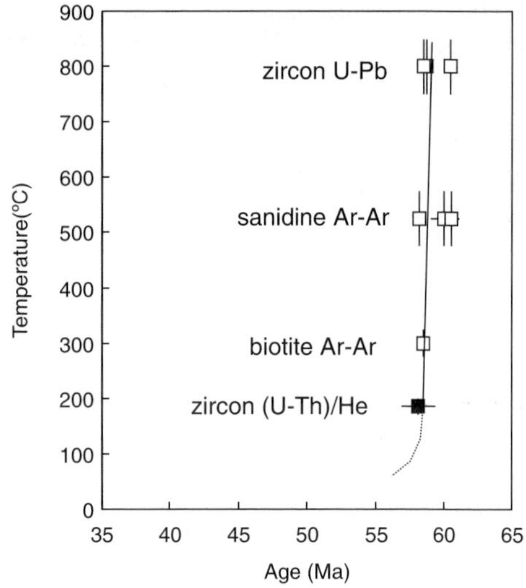

Fig. 5. The U/Pb (in Emeleus & Bell 2005), Ar/Ar (Chambers & Pringle 2001) and zircon (U–Th)/He ages determined on plutons from the Mull plutonic complex.

dating has provided high precision U–Pb (zircon) and Ar/Ar (biotite and sanidine) ages for several of the intrusive units across the HIP, but constraint of the plutonic cooling through lower temperatures is limited to ZFT from the plutonic units of the Skye plutonic complex (Lewis *et al.* 1992). These data suggest that there was prolonged heat flow through the shallow-level plutonic complex (Lewis *et al.* 1992). The source of this heat remains unidentified, and continued heat flow is inconsistent with field evidence for rapid synmagmatic denudation (Brown 2003) and short-lived magmatic activity (Bell & Williamson 2002). No low-temperature constraints have been placed on the plutonic cooling of the other plutonic complexes.

In order to constrain the cooling of the plutonic complex on the Isle of Mull (Fig. 4), zircon He (ZHe) thermochronometry was performed on euhedral zircons from two samples. ML5 is a gabbro which, from field relationships, was emplaced during the main phase of intrusion (58.3 Ma, U/Pb: Hamilton in Emeleus & Bell 2005); whereas ML3 is from the youngest intrusion in the complex, the Loch Ba Felsite (58.5 Ma, U/Pb: Hamilton in Emeleus & Bell 2005). The ZHe ages from both samples are indistinguishable, with an average age of 58.1 $\pm$ 6.6 Ma (Table 2), and show that the Mull plutonic complex cooled to below about 170 °C very rapidly (at in excess of 200 °C Ma^{-1}), immediately after intrusion (Fig. 5). This is in sharp contrast to the ZFT data from the Skye plutonic complex, which suggests that temperatures remained at approximately 250 °C until about 47 Ma (Lewis *et al.* 1992). The cooling of the Mull plutonic complex is consistent with the short period of intrusion suggested from the radiometric crystallization ages, and field evidence for rapid unroofing immediately after the cessation of magmatic activity (Emeleus & Bell 2005). A more complete investigation of the cooling of the HIP will be presented elsewhere.

Concluding thoughts

We have developed a methodology for the determination of zircon He ages that enables routine age determination without the possibility of crystal loss, parent-element loss, mass interference or loss of sensitivity during U and Th measurement. The application of this methodology is limited only by the time required for the anion-exchange chemistry, and by U and Th blanks in the Pt-foils. The technical developments presented here have allowed the identification of several issues that require further investigation. Most important is the need to accurately correct for He-recoil loss when crystals exhibit U and Th zonation, and to fully understand the effects of zonation on helium diffusion. However, we have shown that the application of the ZHe thermochronometer can provide constraints on the rates, timings and amounts of rock cooling in the mid- to shallow crust that are generally unavailable with other techniques.

This study forms part of K. J. Dobson's Ph.D. supported by NERC (NER/S/A/2002/10370) and STATOIL. We thank J. Foeken, V. Olive and D. Vilbert for invaluable assistance in the laboratory, and L. Chambers and M. Hamilton for supplying samples. SUERC is supported by the Scottish Universities.

References

BALESTRIERI, M. L., STUART, F. M., PERSANO, C., ABBATE, E. & BIGAZZI, G. 2005. Geomorphic development of the escarpment of the Eritrean margin, southern Red Sea from combined apatite fission-track and (U–Th)/He thermochronometry. *Earth and Planetary Science Letters*, **231**, 97.

BELL, B. R. & WILLIAMSON, I. T. 2002. Tertiary igneous activity. *In*: TREWIN, N. H. (ed.) *The Geology of Scotland*. Geological Society, London, 371–430.

BERNET, M. & GARVER, J. I. 2005. Fission-track analysis of detrital zircon. *In*: REINERS, P. W. & EHLERS, T. A. (eds) *Thermochronology*. Reviews in Mineralogy and Geochemistry, Mineralogical Society of America, **58**, 205–238.

BIGOT-CORMIER, F., POUPEAU, G. & SOSSON, M. 2000. Differential denudations of the Argentera Alpine external crystalline massif (SE France) revealed by fission track thermochronology (zircons, apatites). *Comptes Rendus de l'Academie des Sciences*, **330**, 363–370.

BLYTHE, A. E., BURBANK, D. W., FARLEY, K. A. & FIELDING, E. J. 2000. Structural and topographic evolution of the central Transverse Ranges, California, from apatite fission-track, (U–Th)/He and digital elevation model analyses. *Basin Research*, **12**, 97–114.

BOYCE, J. W. & HODGES, K. V. 2005. U and Th zoning in Cerro de Mercado (Durango, Mexico) fluorapatite: insights regarding the impact of recoil redistribution of radiogenic ^{4}He on (U–Th)/He thermochronology. *Chemical Geology*, **219**, 261–274.

BRAUN, J. & VAN DER BEEK, P. A. 2004. Evolution of passive margin escarpments: what can we learn from low-temperature thermochronology? *Journal of Geophysical Research*, **109**, F4009, doi: 10.1029/2004JF000147.

BROWN, D. J. 2003 *The Nature and Origin of Breccias Associated with Central Complexes and Lava Fields of the British Tertiary Igneous Province*. Ph.D. thesis, University of Glasgow.

CHAMBERS, L. M. & PRINGLE, M. S. 2001. Age and duration of activity at the Isle of Mull Tertiary igneous centre, Scotland, and confirmation of the existence of subchrons during Anomaly 26r. *Earth and Planetary Science Letters*, **193**, 333–345.

CHAMBERS, L. M., PRINGLE, M. S. & PARRISH, R. R. 2005. Rapid formation of the Small Isles Tertiary centre constrained by precise ^{40}Ar/^{39}Ar and U–Pb ages. *Lithos*, **79**, 367–384.

CHAPPELL, B. W. & STEVENS, W. E. 1988. Origin of intracrustal (I-type) granite magmas. *Transactions of the Royal Society, Edinburgh*, **79**, 71–86.

DOBSON, K. J. 2006. *The Zircon (U–Th)/He Thermochronometer: Development and Application of Thermochronometers in Igneous Provinces*. PhD thesis, University of Glasgow.

DOBSON, K. J., STUART, F. M., DEMPSTER, T. J. & EIMF. 2008. U and Th zonation in Fish Canyon Tuff zircons: Implications for a zircon (U–Th)/He standard. *Geochimica et Cosmochimica Acta*, **72**, 4745–4755.

DUDDY, I. R. & GLEADOW, A. J. W. 1985. The application of fission track thermochronology to sedimentary basins. 2 Case studies. *Nuclear Tracks & Radiation Measurements*, **10**, 408–408.

DUMITRU, T. A., HILL, K. C. *ET AL*. 1991. Fission track thermochronology: application to continental rifting of south-eastern Australia. *Australian Petroleum Exporation Association Journal*, **31**, 131–142.

DUNLOP, D. J., OZDEMIR, O., CLARK, D. A. & SCHMIDT, P. W. 2000. Time–temperature relations for the remagnetization of pyrrhotite (Fe_7S_8) and their use in estimating paleotemperatures. *Earth and Planetary Science Letters*, **176**, 107–116.

EMELEUS, C. H. & BELL, B. R. 2005. *The Palaeogene Volcanic Districts of Scotland*, (4th edn). British Geological Survey, Nottingham.

FARLEY, K. A. 2000. Helium diffusion from apatite: General behavior as illustrated by Durango fluorapatite. *Journal of Geophysical Research*, **105**, 2903–2914.

FARLEY, K. A. 2002. (U–Th)/He dating: techniques, calibrations, and applications. *In:* PORCELLI, D., BALLANTINE, C. K. & WIELER, R. (eds) *Noble Gases in Geochemistry and Cosmochemistry*. Reviews in Mineralogy and Geochemistry, Mineralogical Society of America, **47**, 819–843.

FOEKEN, J. P. T., STUART, F. M., DOBSON, K. J., PERSANO, C. & VILBERT, D. 2006. A diode laser system for heating minerals for (U–Th)/He chronometry. *Geochemistry Geophysics Geosystems*, **7**, Q04015, doi: 10.1029/2005GC001190.

GALLAGHER, K. & BROWN, R. 1997. The onshore record of passive margin evolution. *Journal of the Geological Society, London*, **154**, 451–457.

GARVER, J. I., REINERS, P. W., WALKER, L. J., RAMAGE, J. M. & PERRY, S. E. 2005. Implications for timing of Andean uplift from thermal resetting of radiation-damaged zircon in the Cordillera Huayhuash, northern Peru. *Journal of Geology*, **113**, 117–138.

GLEADOW, A. J. W. 2000. *Australian FT Summary Database*. World wide web address: http://web.earthsci.unimelb.edu.au/Gleadow/AustraliaAFTDatabase.xls.

HASEBE, N., TAGAMI, T. & NISHIMURA, S. 1994. Towards zircon fission track thermochronology; reference framework for confined track length measurements. *Chemical Geology*, **112**, 169–178.

HOLNESS, M. B. 1999. Contact metamorphism and anatexis of Torridonian arkose by minor intrusions of the Rum Igneous Complex, Inner Hebrides, Scotland. *Geological Magazine*, **136**, 527–542.

HOURIGAN, J. K., REINERS, P. W. & BRANDON, M. T. 2005. U–Th zonation-dependent alpha-ejection in

(U–Th)/He chronometry. *Geochimica et Cosmochimica Acta*, **69**, 3349–3365.

HOURIGAN, J. K., REINERS, P. W., MATTINSON, J. M., CAMPBELL, I. S., NICOLESCU, S. & ALLEN, C. 2003. Zircon He–Pb* dating: approaches and applications. *Geochimica et Cosmochimica Acta*, **67**, A159.

HOUSE, M. A., FARLEY, K. A. & KOHN, B. P. 1999. An empirical test of helium diffusion in apatite: borehole data from the Otway basin, Australia. *Earth and Planetary Science Letters*, **170**, 463–474.

KETCHAM, R. A., DONELICK, R. A. & CARLSON, W. D. 1999. Variability of apatite fission-track annealing kinetics: III. Extrapolation to geological time scales. *American Mineralogist*, **84**, 1235–1255.

KIRSTEIN, L. A., SINCLAIR, H., STUART, F. M. & DOBSON, K. 2006. Rapid Miocene exhumation of the Ladahk batholith, western Himalaya. *Geology*, **34**, 1049–1052.

LASLETT, G. M., GREEN, P. F., DUDDY, I. R. & GLEADOW, A. J. W. 1987. Thermal annealing of fission tracks in apatite II: a quantitative analysis. *Chemical Geology*, **65**, 1–13.

LEWIS, C. L. E., CARTER, A. & HURFORD, A. J. 1992. Low temperature effects of the Skye Tertiary intrusions on Mesozoic sediments in the Sea of Hebrides basin. *In*: PARNELL, J. (ed.) *Basins on the Atlantic Seaboard: Petroleum Geology, Sedimentology and Basin Evolution.* Geological Society, London, Special Publications, **62**, 175–188.

LISKER, F. 2002. Review of fission track studies in northern Victoria Land, Antarctica – passive margin evolution versus uplift of the Transantarctic Mountains. *Tectonophysics*, **349**, 57–73.

MEESTERS, A. G. C. A. & DUNAI, T. J. 2002. Solving the production – diffusion equation for finite diffusion domains of various shapes – Part I. Implications for low-temperature (U–Th)/He thermochronology. *Chemical Geology*, **186**, 333–344.

PARRISH, R. R. 1987. An improved micro-capsule for zircon dissolution in U–Pb geochronology. *Chemical Geology*, **66**, 99–102.

PEARSON, D. G., EMELEUS, C. H. & KELLEY, S. P. 1996. Precise $^{40}Ar/^{39}Ar$ age for the initiation of Palaeogene volcanism in the Inner Hebrides and its regional significance. *Journal of the Geological Society, London*, **153**, 815–818.

PERSANO, C., BISHOP, P. & BARFOD, D. N. 2002. Apatite (U–Th)/He age constraints on the development of the Great Escarpment on the southeastern Australian passive margin. *Earth and Planetary Science Letters*, **200**, 79–90.

PERSANO, C., STUART, F. M., BISHOP, P. & DEMPSTER, T. 2005. Deciphering continental breakup in eastern Australia by combining apatite (U–Th)/He and fission track thermochronometers. *Journal of Geophysical Research*, **110**, 12, 405.

REINERS, P. W. 2005. Zircon (U–Th)/He thermochronometry. *In*: REINERS, P. W. & EHLERS, T. A. (eds) *Thermochronology.* Reviews in Mineralogy and Geochemistry, Mineralogical Society of America, **58**, 151–179.

REINERS, P. W. & NICOLESCU, S. 2006. *Measurement of parent nuclides for (U–Th)/He chronometry by solution sector ICP-MS*, ARHDL Report 1, http://www.geo.arizona.edu/~reiners/arhdl/arhdl.htm.

REINERS, P. W. & SPELL, T. L. 2002. Intercalibration of zircon (U–Th)/He and K-feldspar $^{40}Ar/^{39}Ar$ thermochronometry. *Geochimica et Cosmochimica Acta*, **66**, 631.

REINERS, P. W., CAMPBELL, I. H. *ET AL.* 2005. (U–Th)/(He–Pb) double dating of detrital zircons. *American Journal of Science*, **305**, 259–311.

REINERS, P. W., FARLEY, K. A. & HICKES, H. J. 2002. He diffusion and (U–Th)/He thermochronometry of zircon: initial results from Fish Canyon Tuff and Gold Butte. *Tectonophysics*, **349**, 297–308.

REINERS, P. W., SPELL, T. L., NICOLESCU, S. & ZANETTI, K. A. 2004. Zircon (U–Th)/He thermochronometry He diffusion and comparisons with $^{40}Ar/^{39}Ar$ dating. *Geochimica et Cosmochimica Acta*, **68**, 1857–1887.

SAUNDERS, A. D., FITTON, J. G., KERR, A. C., NORRY, M. J. & KENT, R. W. 1997. The North Atlantic Igneous Province. *In*: MAHONEY, J. J. & COFFIN, M. F. (eds) *Large Igneous Provinces. Continental, Oceanic, and Planetary Flood Volcanism.* American Geophysical Union, Geophysical Monograph, **100**, 45–93.

SPOTILA, J. A. 2005. Applications of low-temperature thermochronometry to quantification of recent exhumation in mountain belts. *In*: REINERS, P. W. & EHLERS, T. A. (eds) *Thermochronology.* Reviews in Mineralogy and Geochemistry, Mineralogical Society of America, **58**, 449–466.

STOCKLI, D. 2005. Application of low-temperature thermochronometry to extensional tectonic settings. *In*: REINERS, P. W. & EHLERS, T. A. (eds) *Thermochronology.* Reviews in Mineralogy and Geochemistry, Mineralogical Society of America, **58**, 411–448.

TAGAMI, T., FARLEY, K. A. & STOCKLI, D. F. 2003. (U–Th)/He geochronology of single zircon grains of known Tertiary eruption age. *Earth and Planetary Science Letters*, **207**, 57–67.

WILLIAMS, I. S. 2001. Response of detrital zircon and monazite, and their U–Pb isotopic systems, to regional metamorphism and host-rock partial melting, Cooma Complex, southeastern Australia. *Australian Journal of Earth Sciences*, **48**, 557–580.

WEISSEL, K. J. & HAYES, D. E. 1977. Evolution of the Tasman Sea reappraised. *Earth and Planetary Science Letters*, **36**, 77–84.

YAMADA, R., TAGAMI, T., NISHIMURA, S. & ITO, H. 1995. Annealing kinetics of fission tracks in zircon; an experimental study. *Chemical Geology*, **122**, 249–258.

Improving constraints on apatite provenance: Nd measurement on fission-track-dated grains

A. CARTER[1]* & G. L. FOSTER[2]

[1]*School of Earth Sciences, Birkbeck, University of London, Malet Street, London WC1E 7HX, UK*

[2]*Bristol Isotope Group, Department of Earth Sciences, Bristol University, Queens Road, Bristol BS8 1RJ, UK*

**Corresponding author (e-mail: a.carter@ucl.ac.uk)*

Abstract: Detrital thermochronology is commonly used to locate source and improve the understanding of a region's long-term development. Provenance detrital age data typically comprise a number of distinct age modes representing contributions from a number of different sources. For methods such as apatite fission track (FT) this can present a challenge in that direct assignment of age modes to particular lithological or tectono-stratigraphic units is rarely possible, particularly when ancient orogenic sediments are examined. A new approach described here is based on measuring Nd isotopic data on single apatites by laser ablation ICPMS where the $^{143}Nd/^{144}Nd$ ratios are diagnostic of grain source. By combining Nd isotopic measurement with apatite FT analysis it is possible to link detrital apatite FT ages to specific rock unit sources. The methodology and wider benefits of this new approach are discussed using examples from the Himalaya and Andaman Islands.

There has been significant growth in the number of studies using sediments either to learn more about a source region's long-term geodynamic development or to characterize a source region's current behaviour in terms of rock uplift and erosion (e.g. DeCelles *et al.* 2004; Najman *et al.* 2005; van der Beek *et al.* 2006). Whilst the common provenance tools, such as petrographical and geochemical data, are useful for identifying where sediment has come from they rarely offer insight into how the source region has developed (mostly topographically) in response to the interaction between tectonics and climate erosion. Extraction of this type of information is best obtained using low-temperature thermochronometry (FT and $^{40}Ar-^{39}Ar$) applied to single detrital grains, typically zircon, apatite and mica. Conceptually, detrital thermochronometry has the potential to provide source behavioural information in terms of temporal and spatial variations in rock uplift and exhumation, although in practice this is often difficult to achieve because a detrital age alone is not sufficiently unique.

In this study we consider a new methodology aimed at improving constraints on grain source through a combination of FT thermochronometry and Nd geochemical analysis (Foster & Carter 2007). Such an approach has the potential to yield more site-specific information about from where in the hinterland the apatite grain has come. This information is needed because in active geodynamic settings where exhumation rates are high it is commonplace to obtain similar FT ages from different rock units. A good example of this occurs in the Himalaya where apatite FT bedrock ages from the orogen front, where erosion is localized, rarely differ by more than 2–3 Ma (Fig. 1). Here, the focus of recent research has been to map range-scale behaviour to test a conceptual linkage between erosion and deformation based on numerical models that argue for deformation being focused where erosion is persistently high (e.g. Willett 1999; Beaumont *et al.* 2001). To study for such effects requires examination of both the long-term exhumation record based on bedrock and foreland basin sediments, and detailed studies of small catchments to see how the modern detrital outputs of such catchments capture variations in local erosion rate that might reflect differences in structure, lithology, relief and bedrock mineral concentrations (Brewer *et al.* 2003; Amidon *et al.* 2005). The aim of this paper is to consider the merits of measuring Nd ratios on FT-dated apatite grains as a means of providing more site-specific information about from where an apatite grain has come and to consider this approach alongside more conventional Nd isotopic analysis based on whole-rock (WR) samples. Using examples we aim to demonstrate the utility of apatite Nd isotopic data taking into account analytical uncertainties and any practical limitations.

From: LISKER, F., VENTURA, B. & GLASMACHER, U. A. (eds) *Thermochronological Methods: From Palaeotemperature Constraints to Landscape Evolution Models*. Geological Society, London, Special Publications, **324**, 57–72.
DOI: 10.1144/SP324.5 0305-8719/09/$15.00

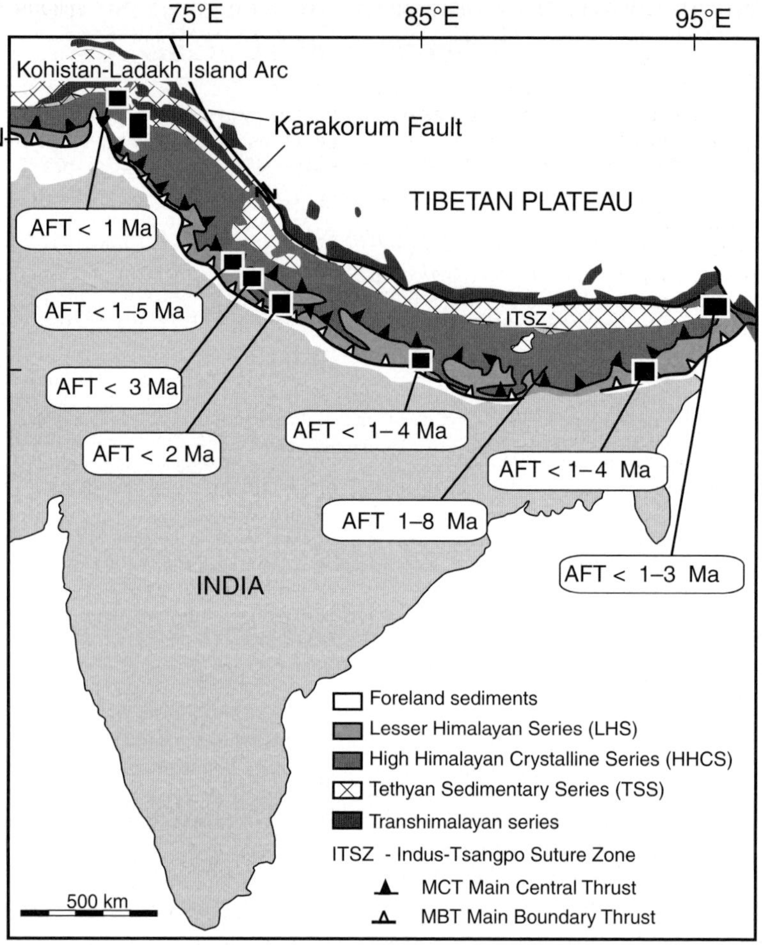

Fig. 1. Geological map of the Himalayas illustrating principal structures and main tectono-stratigraphic units. Also shown are examples of the age range for bedrock AFT ages across the Himalayan arc (data sources: Treloar *et al.* 2000; Burbank *et al.* 2003; Schlup *et al.* 2003; Vannay *et al.* 2004; Grujic *et al.* 2006; Blythe *et al.* 2007; Carter unpublished data).

Sm–Nd geochemistry

The bulk Nd isotopic composition of sedimentary rocks is widely used to fingerprint sediment source (e.g. France-Lanord *et al.* 1993; Galy *et al.* 1996). Based on the alpha decay of ^{147}Sm to ^{143}Nd measured, ^{143}Nd/^{144}Nd ratios generally represent weighted averages of when the sources of the sediment were extracted from the mantle. Sm and Nd generally behave as immobile elements, so that model ages and initial ratios are considered relatively insensitive to the effects of weathering and metamorphism. In practice this underlying assumption may be wrong as chemical weathering, diagenesis or low-grade metamorphism can cause mobilization (e.g. Ehrenberg & Nadeau 2002), but the impact of any mobilization on model ages is only significant if mobilization were to happen long after crust formation (e.g. Ohlander *et al.* 2000), as is the case for the Himalayas. In sedimentary provenance studies the emphasis is not to obtain a model age, but to simply compare sediment sources in terms of their Nd isotopic composition. Provided the sediment sources have a distinct geological history and known Nd isotopic composition, which is often the case, the sediment provenance can be extracted.

For convenience bulk-rock and ^{143}Nd/^{144}Nd isotope data are generally normalized to the modern Bulk Earth value and expressed in epsilon (ε) units, which are parts per 10 000 deviation from the chondritic uniform reservoir

($^{143}Nd/^{144}Nd = 0.512638$: Jacobsen & Wasserburg 1980), where by definition the ε_{Nd} of CHUR=0 (a single epsilon unit is equivalent to a difference in $^{143}Nd/^{144}Nd$ ratio at the fourth digit). For detrital sediments that comprise a mixture of minerals derived from a number of sources the ε_{Nd} reflects not only the time since deposition (i.e. the amount of ^{147}Sm that has decayed) but it also depends on the amount of time since the Nd within the constituents of the sediment separated from the mantle. For example, sediments predominantly derived from young volcanic material will tend to have a bulk-rock ε_{Nd} similar to that of modern volcanic rocks where the Nd has only recently been extracted from the mantle (e.g. $\varepsilon_{Nd} = +10$ to -5). Sediments predominantly derived from ancient crustal sources, where the Nd separated from CHUR a few billion years ago, will have ε_{Nd} in the range -20 to -40, reflecting the amount of time that they have been depleted in Sm relative to Nd compared to CHUR (see Fig. 2). Sediment derived from a mixture of these two end members will have an ε_{Nd} somewhere in between, depending on the exact proportions of each contribution and its associated Nd concentration.

However, the exact interpretation of bulk sediment Nd data is more complicated than this simple treatment suggests. For instance, differences in the grain size of the sediment can give different bulk-rock values, normally attributed to variations in the proportions of rare earth-bearing minerals such as apatite, zircon and monazite. Confinement to clay-sized fractions reduces such effects. Furthermore, it is unclear what bulk-rock values (weighted average) mean in terms of source, especially if a sample comes from a large basin such as the Indus or Bengal Fan fed by rivers that drain large areas (>10^5 km^2) where the erosional flux is variable at a variety of spatial and temporal scales and considerable (>1 Ma) sediment storage can occur. In this regard single detrital grain Nd data has the potential (for some units) to offer a more detailed insight into sediment source(s) by giving a more direct insight into mixing proportions. The Sm/Nd ratios of apatites, which typically range from 0.2 to 0.5 (Belousova *et al.* 2002), are similar to those of average continental crust (*c.* 0.2: Taylor & McLennan 1985). As apatite is both a reactant and product in a variety of accessory-mineral-forming reactions (e.g. Bingen *et al.* 1996; Pyle & Spear 2000; Pyle *et al.* 2001; Wing *et al.* 2003) it implies that at any one time the $^{143}Nd/^{144}Nd$ of metamorphic apatite will maintain isotopic equilibrium with its whole rock, that is apatite will lie on an isochron defined by the whole rock and the other Nd-bearing phases such as garnet. Similarly, magmatic apatite is also likely to have crystallized along with the major phases of rock, thereby also maintaining isotopic equilibrium. Apatite has been shown to be relatively stable with regard to rare earth element (REE) diffusion, and a closure temperature for Nd has been estimated as approximately 750 °C for cooling rates of 10 °C Ma and a grain radius of 250 μm (Cherniak 2000).

Igneous and metamorphic apatites are unlikely to develop significantly higher $^{143}Nd/^{144}Nd$ ratios than the rock in which they are hosted provided that a relatively short period has elapsed since they were last equilibrated with the whole rock. This is demonstrated in Figure 3, which plots whole rock, garnet and apatite for 10, 50, 100 and 500 Ma durations since equilibration, where clear significant differences in Nd isotopic compositions are only achieved in apatite after 50 Ma or so. For provenance studies an ideal situation would be to compare detrital apatite Nd values directly

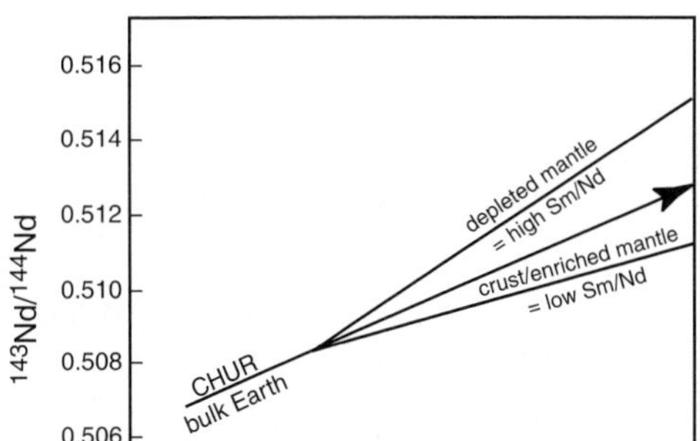

Fig. 2. Growth curves to show the Nd isotope evolution in mantle and crust. The bold line marks the evolution of the bulk earth or CHUR (chondritic uniform reservoir).

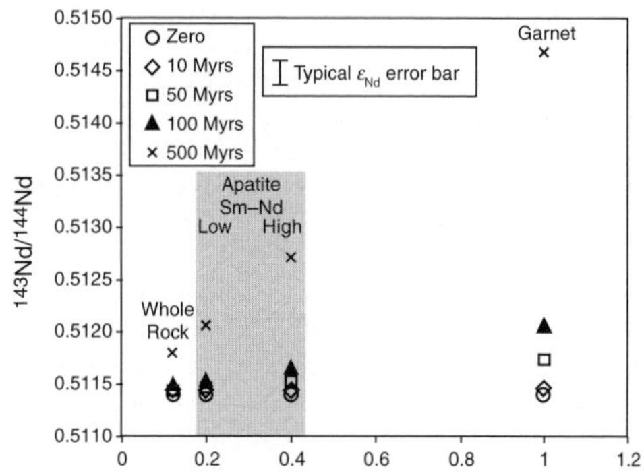

Fig. 3. Plot to show how apatite Nd evolution through time compares to whole rock and garnet. Apatite Nd values only differ significantly from their whole rock after 100 Ma.

with source apatite Nd data, but currently such information is unlikely to be available. This is not a significant problem, however, as the plot in Figure 3 shows that where there is no source-rock apatite data, useful information can still be obtained using source whole-rock values provided that the detrital apatites grew comparatively recently, that is in the Late Mesozoic–Cenozoic. This suggests that in young mountain belts, like the Himalaya, where there is a largely pervasive Cenozoic–aged metamorphism (Hodges 2000 and references therein), the non-age corrected $^{143}Nd/^{144}Nd$ ratio of apatites will closely reflect the Nd isotopic composition of their parent whole rock. For sediment provenance studies this is an important observation as it means that detrital apatites can be directly linked to their sources using either apatite or, where this is lacking, bulk-rock data.

Approach and methodology

Measurement of the Nd isotopic composition of single apatites requires a technique able to measure small quantities of Nd hosted in the single apatite crystals and, crucially, it must be able to determine $^{143}Nd/^{144}Nd$ to within about 100 ppm of the mean (1 epsilon unit). Our goal is to be able to measure Nd on FT-dated grains within their FT mount, which is best achieved *in situ* using laser ablation protocols. This methodology, outlined in Foster & Vance (2006), is based on laser ablation multicollector inductively coupled plasma mass spectrometry (LA–MC-ICPMS). Measurement was undertaken at the University of Bristol with a 193 nm homogenized ArF New Wave/Merchantek laser ablation system linked to a Thermofinigan Neptune ICPMS. All ablation was carried out in a He environment, and mixed with Ar and N after the ablation cell in a gas-mixing bulb. Laser repetition rates used were 4 Hz, and spot size (65–90 μm) and laser power (1–8 J cm^{-2}) was varied according to grain size. Where apatite grains were small, spot sizes and power were decreased; where grains had low Nd concentrations, mean internal precisions were often worse than 0.5 but generally better than 2 epsilon units. NIST 610 was used to correct for mass bias in Sm–Nd. During each analysis Durango apatite was used as a consistency standard and the overall reproducibility was better than 0.5 epsilon units. No degradation of data was observed between measurements made on polished grains or apatites etched for FT analysis. Grain size proved to be a limiting factor as the energy of the laser ablation in some cases caused grain disintegration. The analysis procedure is rapid, taking on average 3–5 min per grain, which means that a large number of analyses can be performed in a typical analytical session.

Regional geology

The viability of using Nd in apatite grains to constrain source is demonstrated here using rocks from, or associated with, the Himalaya that are well characterized in terms of their WR Nd values. The crystalline rocks of the main Himalayan chain can be divided (from north to south) into six tectono-stratigraphic units (Fig. 1): the Trans-Himalaya (TH; here also including southern Tibetan units such as the Lhasa Block); the Indus-Tsangpo Suture Zone (ITSZ); the Tethyan Sedimentary Series (TSS); the High Himalayan Crystalline Series (HHCS); the Lesser Himalayan Series (LHS); and the foreland sediments of the Sub-Himalaya. In Pakistan and the Western Himalaya there are two additional units: the Kohistan–Ladakh Island Arc; and the Karakorum Belt. In terms of metamorphic and structural evolution these units are largely distinct (see Hodges 2000 for a review); however, what is of relevance here is the isotopic contrast that exists between the different units. Figure 4 is a histogram showing the range in Nd isotopic composition exhibited by the rocks belonging to each unit. These data are from a variety of sources: Deniel *et al.* (1987); France-Lanord *et al.* (1993); Inger & Harris (1993); Parrish & Hodges (1996); Ayres (1997); Prince (1999); Whittington *et al.* (1999); Ahmad *et al.* (2000); Miller *et al.* (2001) and Robinson *et al.* (2001). Data for the Lhasa Block are from Debon *et al.* (1986); for the Trans-Himalaya from Allegre & Ben Othman (1980); for Kohistan from Petterson *et al.* (1993); Khan *et al.* (1997) and Jagoutz *et al.* (2006); and for the Karakoram from Schärer *et al.* (1990). It is clear from Figure 4 that resolvable differences exist between most of these units, which if replicated in apatite grains provide a simple means of discerning source. Exceptions are the TSS and HHCS, which cannot be distinguished by WR Nd model ages alone (Robinson *et al.* 2001).

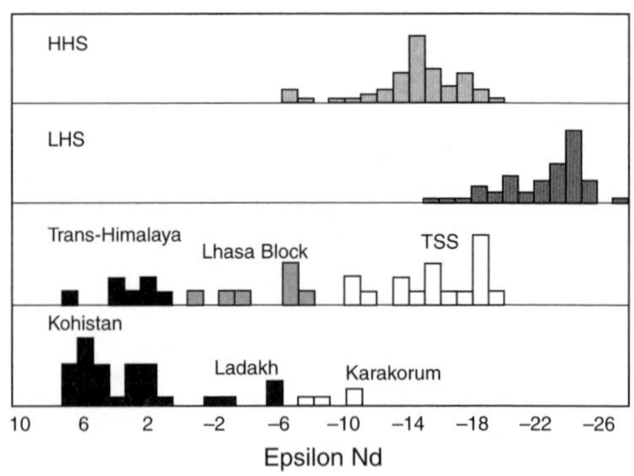

Fig. 4. Histogram showing the range of WR Nd values for the principal rock units that comprise the Himalaya.

Application

To use the Nd isotopic composition of apatite in detrital sediments as an indicator of sediment provenance requires the variation in Nd isotopic composition of apatites in potential source rocks to be known *a priori*. Foster & Carter (2007) examined the utility of apatite as a provenance indicator by examining apatite separated from modern river sands in Bhutan, where the catchments have a well-defined geology with well-characterized apatite Sm–Nd systematics. The location and details of these samples are reported in Foster & Carter (2007). Here we summarize the principal results of this study. Three sand samples were analysed for both Nd and apatite FT: sample 02g26 had a drainage that lay exclusively within the HHCS unit; sample 02g27 had a drainage that lay predominantly within the Chekha Formation, but also incorporated some of the HHCS; and 02g42 had the largest drainage, which incorporated the HHCS, TSS and LHS (see Fig. 2 in Foster & Carter 2007). The petrological composition of each sand sample reflects their varied sources. Garzanti *et al.* (2004) noted that these sand samples consist mainly of metamorphic rock fragments, mica, hornblende, sillimanite and garnet largely derived from the HHCS, indicating erosion is highest in this unit for each catchment, and hence the apatites in each should largely be derived from this unit (although small differences in catchment size might also have an influence). Sample 02g42, the Kuru Chu, also contains slate and metacarbonate grains, indicating a significant contribution of the TSS unit that crops out in its headwaters. The LHS rocks, through which the Kuru Chu passes prior to the sample site for 02g42, comprises quartzites and schists of the Shumar–Daling unit (Gansser 1983), but these leave little mark on the sand petrology (E. Garzanti pers. commun.). We found that apatite concentrations were very low in the quartzites of the LHS in the Kuru Chu catchment.

Figure 5 shows the distribution of bedrock apatite $^{143}Nd/^{144}Nd$ ratios used in the study by Foster & Carter (2007), with the bulk-rock data from Figure 4. Comparison of these data with detrital apatites from the modern river sediments shows that a predominant HHCS source for sample 02g26 is entirely consistent with the *in situ* apatite isotopic data (Fig. 6). There is clear overlap between the majority of apatites analysed in this sand and the apatites separated from HHCS bedrock. For 02g27 apatites, although the $^{147}Sm/^{144}Nd$ ratios exhibits a greater range, the $^{143}Nd/^{144}Nd$ ratios are similar to the local bedrock apatite sample BH74 (Fig. 6), indicating that the

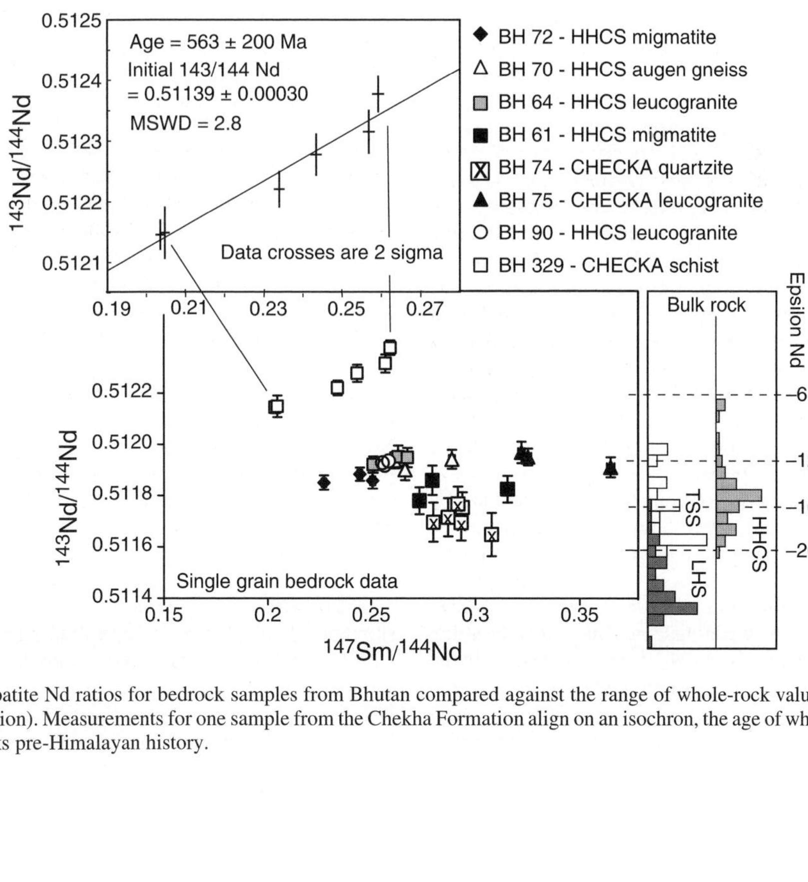

Fig. 5. Apatite Nd ratios for bedrock samples from Bhutan compared against the range of whole-rock values (grey shaded region). Measurements for one sample from the Chekha Formation align on an isochron, the age of which relates to the rocks pre-Himalayan history.

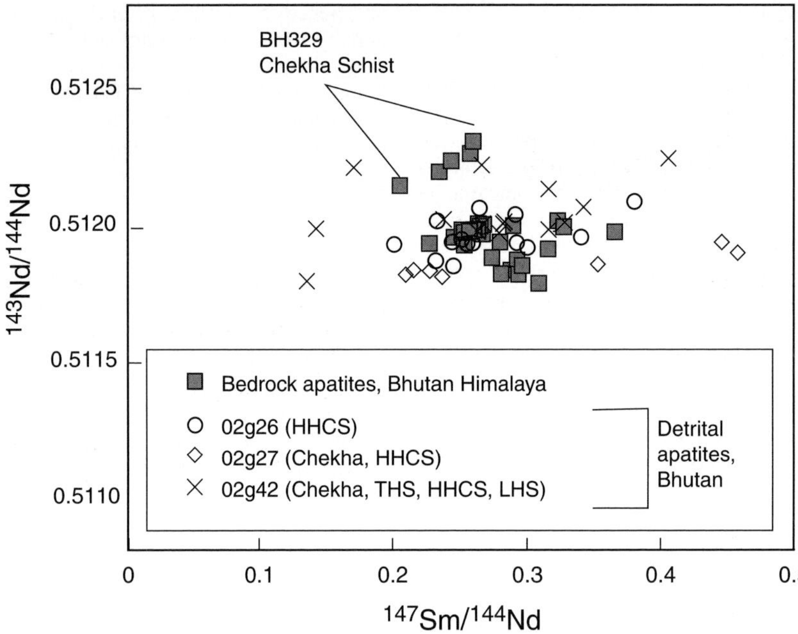

Fig. 6. Plot comparing bedrock apatites from Bhutan with detrital apatites extracted from river sediments in catchments with restricted geology in terms of Himalayan tectono-stratigraphic units.

apatites in this sample were probably derived from a Chekha source. For 02g42 the majority of the grains overlap with the HHCS bedrock, whereas some of the grains show similar values to the apatites from the Chekha Formation schist sample BH329 (Figs 5 and 6). This interpretation of a significant involvement of the Chekha in the source of 02g42 is consistent with the sand petrology (see earlier) and the geology of its catchment, assuming that the Chekha Formation has a similar isotopic composition to the TSS (Fig. 4). We did not analyse any apatites from bedrock samples belonging to the LHS; however, given the relatively low $^{143}Nd/^{144}Nd$ ratio of the LHS (Fig. 4), the absence of similarly low $^{143}Nd/^{144}Nd$ apatites in 02g42 is also consistent with the petrology of the sand, which suggests a minimal involvement of the LHS.

The approach and data presented above indicate that Nd values measured in detrital apatites can yield useful information concerning sediment provenance. However, as yet there exists only a limited database of apatite Nd for the Himalaya. In the absence of apatite data it is instructive to compare apatite Nd to the much more extensive bulk-rock Nd database for the Himalaya. This is illustrated in Figure 5, where the $^{143}Nd/^{144}Nd$ of apatites from bedrock samples from Bhutan are clearly similar to bulk-rock values, despite different Sm/Nd ratios by virtue of their young age (<50 Ma). This is also likely to be true of much of the LHS (particularly the upper parts nearest the MCT where metamorphic grade is highest: Bollinger *et al.* 2004), TSS, Kohistan and the Karakorum owing to the pervasive <100 Ma-aged metamorphism and the fact that most of the igneous rocks of these units are also <100 Ma old (e.g. Hodges 2000 and references therein). One sample from the higher structural levels of the Chekha Formation of Bhutan (thought to be the equivalent of the Hiamanta of the Eastern Himalaya and Everest Series of Nepal) had not been affected by Cenozoic metamorphism as the apatites in this sample were observed to align on an approximately 560 Ma-isochron (Fig. 5), which relates to the rock's pre-Himalayan history.

We now consider some examples that demonstrate the potential of this methodology by linking Nd isotopic measurement with individual apatite FT age. By comparing the detrital apatite Nd with the current apatite Nd and bulk-rock database it is possible to directly link each age peak identified by FT with an isotopically distinct source region. Figure 7 compares single-grain FT and Nd isotopic data for samples 02g26 and 02g42, as well as showing a histogram comparing sample grain $^{143}Nd/^{144}Nd$ ratios with the bedrock data. For sample 02g26 the Nd isotopic data confirm a HHCS source and, therefore, the observed different FT-age modes must relate to different locations (and

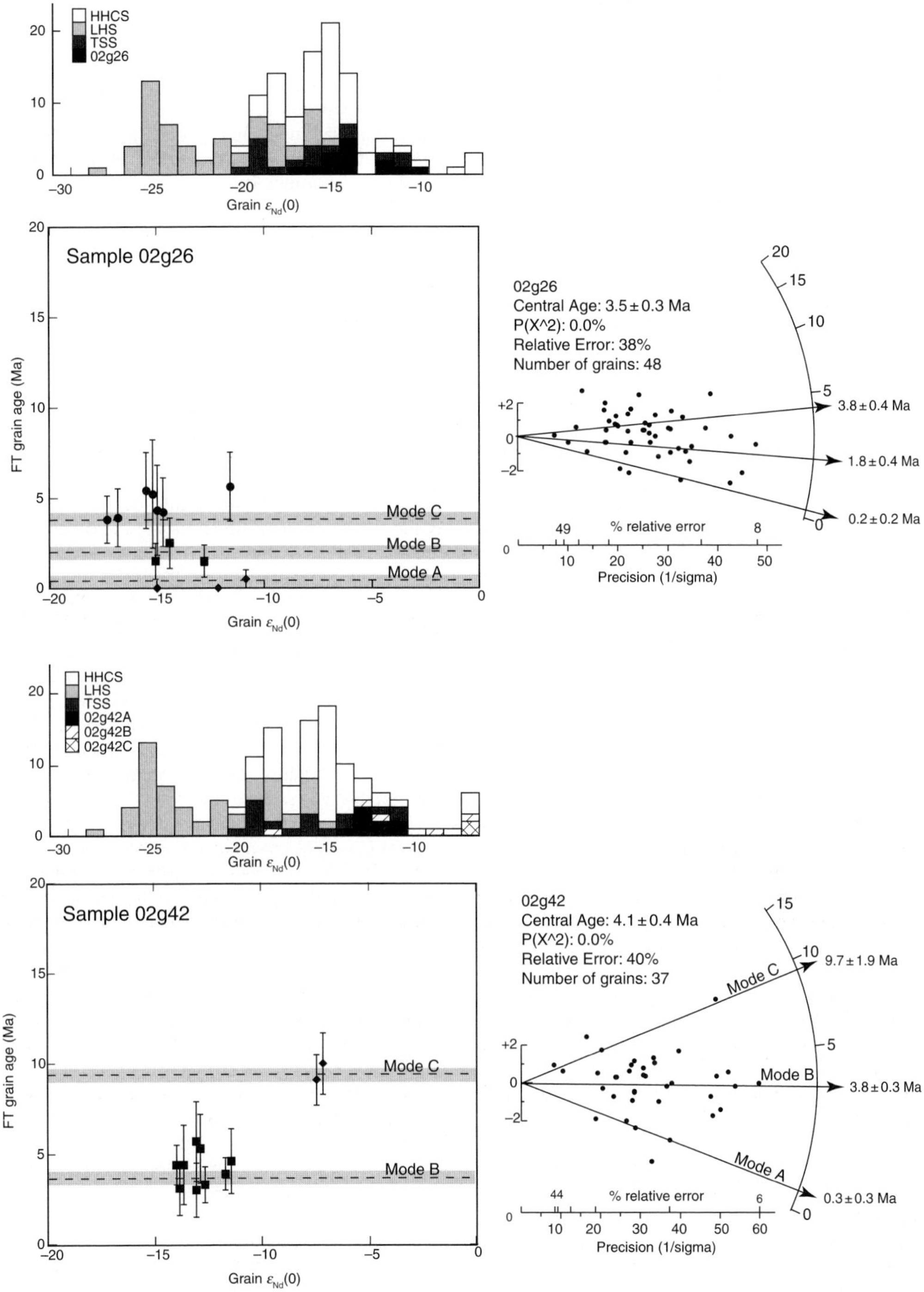

Fig. 7. Plots comparing FT and Nd data measured on the same grains. The histograms compare the sample Nd data with bedrock Nd isotopic values that define the range of values present in the principal Himalayan tectono-stratigraphic units.

exhumation history) within the HHCS source area. Reworking of stored sediments is not considered likely given the proximity of the sample to the bedrock source and lack of any significant sediment cover in the catchment. The data for sample 02g42 contain two apatites with much older FT ages at 9.7 ± 1.9 Ma. Nd isotopes measured on these grains show less negative epsilon values (but within the tail end of HHCS values) that are distinctly different from the data obtained from the dominant age mode at 3.8 ± 0.3 Ma.

Figure 8 shows the results for sample 02g27. This sample is extremely young, and there are insufficient grains and track counts to obtain a good constraint on the true FT age. Nevertheless, the FT-dated grains show $^{143}Nd/^{144}Nd$ ratios that are similar to the local bedrock apatite sample BH74 (Fig. 6), indicating that the apatites in this sample were probably derived from a Chekha source that is consistent with the sampled rivers' drainage that lies predominantly within the Chekha Formation. Whilst the data indicate a Chekha Formation source, this cannot be the same as bedrock sample BH329, which has a 560 Ma-isochron. The $^{143}Nd/^{144}Nd$ ratios from the FT-dated apatites align on an 89 Ma-isochron (Fig. 8), indicating a part of the Chekha Formation that has been reset by metamorphism.

Recent sediment record

The present-day drainage of the River Ganga suggests that sediment in the surface floodplain of the Bengal Delta region will be mostly derived from the Higher Himalayan crystalline units comprising staurolite–sillimanite-bearing schists and gneisses, diopside-bearing banded marbles and amphibolites (Garzanti *et al.* 2004). There is no evidence to suggest that sediment routing was any different in the Late Quarternary, except that there are major changes in sediment flux associated with dramatic climate change. Goodbred *et al.* (2003) found that between 11 and 7 ka there was a threefold increase in sediment flux to the Bengal Delta and offshore fan. The peak flux occurred at approximately 9 ka, coincident with a climatic peak in regional insolation (Fig. 9). Before 15 ka sediment flux was much lower and is associated with the dominant NE arid monsoon. These changes occurred as sea level was steadily rising, so that changes in sediment flux cannot be attributed to sea-level change. The source of sediment that contributed to the dramatic rise in accumulation is either due to an increase in erosion of Himalayan bedrock and/or to the reworking of stored sediment on the delta plain. If the sediment source region remained the same, an indication of reworking of delta-plain material would be a shift to older apatite FT ages with decreasing depositional age as the expectation would be that stored sediment contains earlier exhumation signals. To test for this a shallow borehole in the delta plain was drilled from which samples were collected at depth intervals that span the changes in Late Quarternary climate. Depositional age control was by optical luminescence (OSL) dating (Hudson-Edwards *et al.* 2004). Table 1 details the apatite FT data, analytical details and OSL ages, and the results are plotted in Figure 9 together with climate and sediment-accumulation plots from Goodbred *et al.* (2003).

There is no major change in apatite FT age over the period from 23 to 8 ka, although there is a small increase in FT age associated with the two youngest samples, but this post-dates peak insolation and sediment flux. This change seems contrary to the

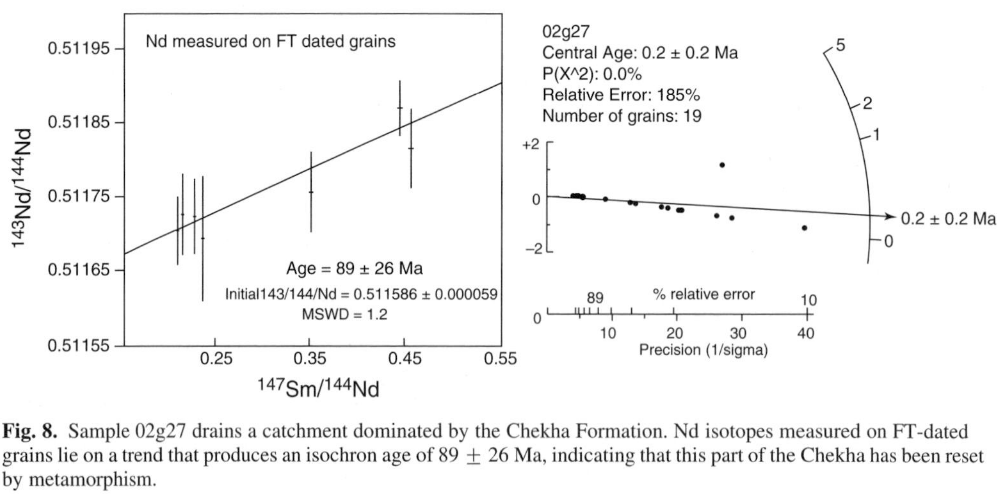

Fig. 8. Sample 02g27 drains a catchment dominated by the Chekha Formation. Nd isotopes measured on FT-dated grains lie on a trend that produces an isochron age of 89 ± 26 Ma, indicating that this part of the Chekha has been reset by metamorphism.

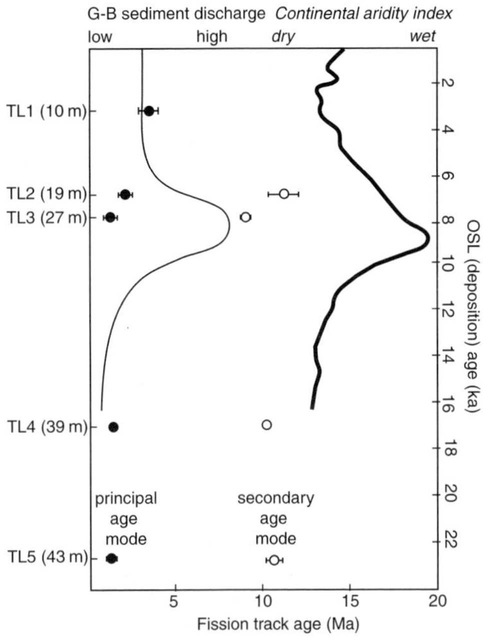

Fig. 9. Plot showing the relationship between apatite FT detrital age modes for samples from a borehole into the Late Quarternary sediments of the Bengal Delta at Joypur in West Bengal. Also on the plot are curves showing Late Quarternary relative changes in insolation, and sediment discharge by the Ganges Brahmaputra rivers complied by the study of Goodbred *et al.* (2003).

expected response to increased wetness that, in theory at least, should increase source erosion rates; but increased wetness also raises river levels, hence reworking of older sediments (containing older AFT ages) would also be expected. Sample TL3, which has a depositional age of approximately 8 ka, overlaps with the wettest period. It comes from a unit rich in quartz and mica (biotite, muscovite and chlorite) with subordinate ferro-hornblende, calcite, dolomite, K-feldspar, albite, smectite and illite, and minor monazite, zircon, illmenite, garnet and rutile, an assemblage that is consistent with High Himalayan crystalline sources. The sample has three age populations – A (1.5 ± 0.3 Ma; 61%), B (9.0 ± 1.0 Ma; 34%) and C (two grains ≥ 140 Ma) – detected in 59 single-grain ages. Reported Nd WR data from Late Quarternary sediments in the coastal parts of the Bengal Fan closest to the sample location record values for ε_{Nd} of between -11 and -16 and confirm a Himalayan provenance (Pratima *et al.* 2005), but since this signal is only an averaged image of source it is unclear if this relates to HHCS and/or TSS sources (bulk-rock Nd data cannot distinguish between these two). A LHS source would give ε_{Nd} values in the range -20 to -26, and the minor involvement of this source is not precluded by the bulk-rock Nd data. Foster & Carter (2007) measured Nd isotopic composition on representatives of the FT-dated grains for sample TL3 (some grains were too small to analyse) that enabled grains to be attributed to a source tectono-stratigraphic unit. A good correspondence between examples of bedrock apatites from the main Himalayan units in Bhutan (taken as being representative of the HHCS and TSS of the main Himalayan orogen) and Bengal Delta apatites can be seen in Figure 10.

Based on current understanding of exhumation rates and the distribution of bedrock apatite FT ages in the Himalaya, a source interpretation made on the detrital FT ages alone would infer a HHCS source for age mode A, a LHS source for mode B and the slowly eroding Trans-Himalayan region for the old ages in mode C. The apatite Nd results are consistent with this interpretation, except for mode B where the Nd ratios indicate HHCS sources. This finding is important as it means that we can now attribute different ages to the same unit, that is both 1.5 and 9 Ma ages coming from the same (HHCS) tectono-stratigraphic unit. Thus, by combining AFT data with Nd on the same grains it is possible to extract more detail about apatite sediment provenance (and/or reworking) than has been possible using apatite FT data alone.

Ancient sediments

The study of ancient sediments to reconstruct a source region's past behaviour in terms of rock uplift and exhumation–erosion has been an area of increased research, particularly for the Himalaya where a key objective is to constrain early growth history and to identify the point in time when exhumation rates accelerated in response to crust thickening and topographic growth (e.g. DeCelles *et al.* 2004; Najman *et al.* 2005). The adjacent foreland basin record has been shown to be incomplete, with most of the key Palaeogene sections missing (Najman 2006). More complete stratigraphic sections may lie beneath the Indus or Bengal Fan, but most of the early sediments are inaccessible to ocean drilling because they are either deeply buried or lie in deep waters. One possible candidate for research lies on the Andaman Islands located in the Bay of Bengal (Fig. 11). Here, a Palaeogene forearc-accretionary sequence has been uplifted and exposed. Part of this uplifted sequence includes a turbidite sequence known as the Andaman Flysch, whose origin has been debated for over 20 years. It has been variously proposed that the Andaman Flysch sediment has been derived from a palaeo-Irrawaddy River (Pal *et al.* 2003) or, alternatively, from Bengal Fan material shed from the

Table 1. *Fission-track results for samples of Late Quarternary Bengal Delta sands from Joypur, West Bengal, India. Sample depositional age is constrained by OSL ages from Hudson-Edwards* et al. *(2004)*

Sample	OSL age (ka)	Depth (m)	No. of grains	Dosimeter		Spontaneous		Induced		Age dispersion		Central age $\pm 1\sigma$	Age groups (No. of grains)			
				ρd	Nd	ρs	Ns	ρi	Ni	$P\chi^2$	RE (%)					
West Bengal, Shallow borehole onshore Bengal delta at 22°44.43′N, 88°29.45′E																
TL1	3.8 ± 1.3	10.0–10.5	42	1.264	7010	0.078	84	2.254	2431	0	65	6.2 ± 2.9	3.6 ± 0.6 (41)			231 ± 51 (1)
TL2	7.1 ± 0.4	19.0–19.5	40	1.264	7010	0.093	115	2.65	3273	0	263	8.1 ± 3.5	2.3 ± 0.4 (30)	12 ± 3 (7)	38 ± 8 (1)	317 ± 85 (1)
TL3	7.6 ± 0.4	27.5–28.0	59	1.264	7010	0.061	167	2.639	7279	0	100	6.9 ± 2.4	1.5 ± 0.3 (36)	9 ± 1 (20)	140 ± 74 (1)	236 ± 58 (1)
TL5	17.5 ± 1.2	39.0–39.5	51	1.264	7010	0.066	235	2.369	8356	0	265	10.7 ± 3.9	1.5 ± 0.2 (37)	10 ± 1 (11)	150 ± 34 (1)	300 ± 100 (2)
TL6	23.4 ± 1.5	43.0–43.5	35	1.264	7010	0.038	112	1.974	5798	0	125	3.2 ± 0.8	1.5 ± 0.3 (28)	11 ± 2 (7)		
Modern river sediments, Bhutan																
	River	Location														
02g26	Yangste	27°36.74′N 91°29.61′E	48	1.034	2762	0.042	204	2.175	10687	0	37.6	3.5 ± 0.3	0.2 ± 0.2 (7)	1.8 ± 0.4 (7)	3.4 ± 0.4 (34)	
02g27	Radi	27°20.83′N 91°37.07′E	19	1.038	2762	0.008	2	0.685	1700	7	184	0.2 ± 0.2				
02g42	Kuru Chu	27°15.90′N 91°11.68′E	37	1.048	2762	0.061	293	2.596	12481	<1	40.4	4.1 ± 0.4	0.3 ± 0.3 (5)	3.8 ± 0.3 (30)	10 ± 2 (2)	

(i) Track densities are ($\times 10^6$ tr cm^{-2}) numbers of tracks counted (N) shown in brackets; (ii) analyses by external detector method using 0.5 for the $4\pi/2\pi$ geometry correction factor; (iii) ages calculated using dosimeter glass CN-5; (apatite) $\zeta_{CN\text{-}5} = 338 \pm 4$; calibrated by multiple analyses of IUGS apatite age standards (Hurford 1990); (iv) $P\chi^2$ is probability for obtaining χ^2 value for v degrees of freedom, where v is the number of crystals -1; (v) central age is a modal age, weighted for different precisions of individual crystals (see Galbraith & Laslett 1993); (vi) mixed data deconvolved using the binomial peak-fitting algorithm of Galbraith & Green (1990); (vii) OSL ages from Hudson-Edwards *et al.* (2004).

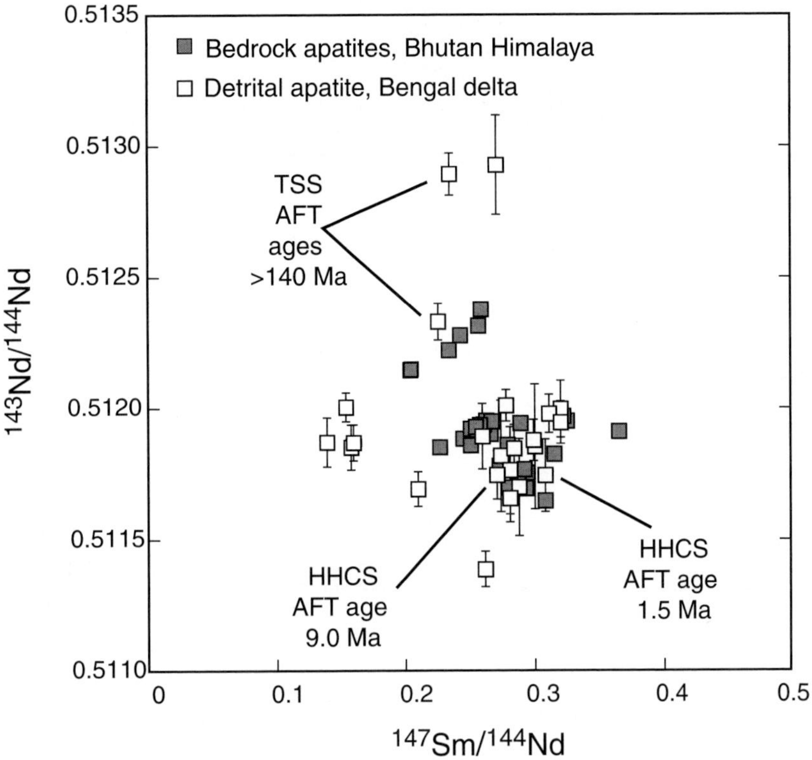

Fig. 10. Plot to compare Nd data measured on individual FT-dated grains with bedrock apatites from the main Himalayan units in Bhutan and the Bengal Delta.

nascent Himalaya and sourced directly by submarine fans or by tectonic emplacement as an allochthon caused by oblique subduction (Curray 2005). The latter model is of major interest as these sediments may represent all or part of the missing record of early Himalayan growth history.

Much of the sediment exposed on the Andaman Islands has a volcanic arc provenance or is reworked from the uplifted ophiolite (Allen *et al.* 2007). Typical of the earliest Palaeogene sediments is the Namunagargh Grit Formation, which contains abundant fresh andesitic material including pumice and undeformed glass shards diagnostic of direct volcanic input. The Andaman Flysch, a siliciclastic turbidite sequence deposited on a submarine fan, is in marked contrast to the relatively quartz-free volcanic sediments and suggests a very different provenance, likely to be continental, and this raises the prospect that the Andaman Flysch might contain grains derived from the nascent Himalaya, as has been suggested by Curray (2005). Fission-track data for this sample, reported by Allen *et al.* (2007), record partial resetting (although the amount of grain-age dispersion is not very high) and, thus, the apatite grain ages do not directly record their provenance.

To constrain their source the FT-dated Andaman Flysch grains were subjected to *in situ* Nd analysis. The majority (*c.* 60%) of grains produced ε_{Nd} values of between -5 and $+5$, typical of juvenile volcanic source. In contrast, whole-rock values for samples from the same locality (Allen *et al.* 2007) give ε_{Nd} values of between -11.1 and -8.2. The restricted range of ε_{Nd} values for most of the AFT-dated apatites indicate a dominant volcanic source in contrast to the WR data that imply some input from older less radiogenic (continental) material, that is far-field transport of mud derived possibly from the Himalaya but more likely from the east Asian continental margin (present-day eastern Burma and Thailand). If the provenance of the apatites was based on the sample WR Nd data we would get an incorrect image of their source, the implication being that the apatite came from a continental region such as the Himalaya. By contrast, the image from direct Nd analysis of the apatites shows that they are derived locally and are volcanic in origin.

Further indication that the FT-dated apatites from the Andaman Flysch do not come from the metamorphic Himalaya is evident through a comparison between these single-grain Nd data with

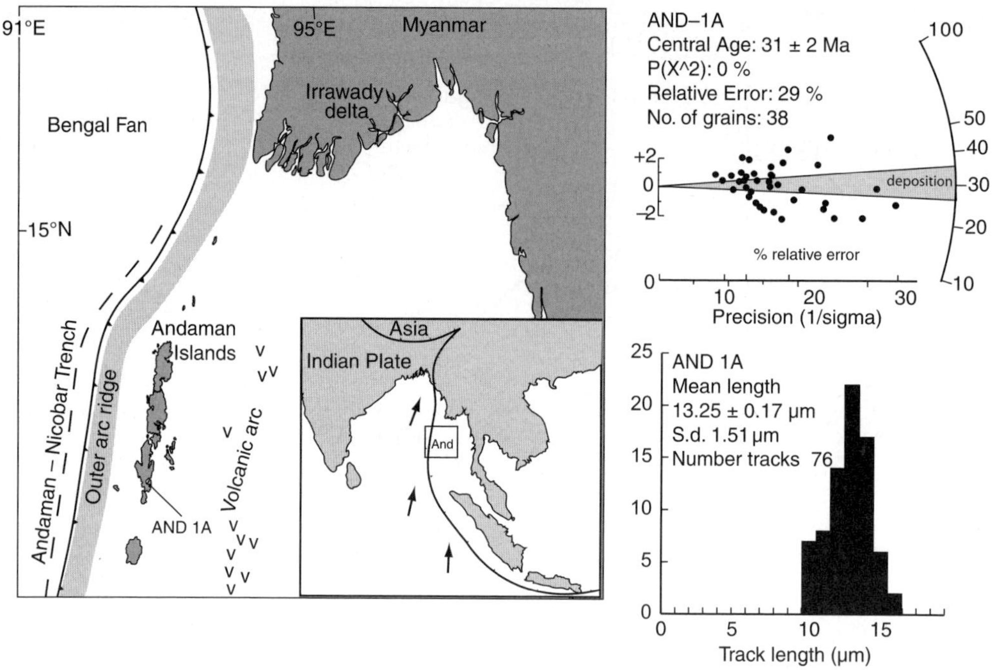

Fig. 11. Map to show the location of the Andaman Islands and sample 1A used in this study. Also shown are the radial plots of individual apatite grain ages and their track-length distribution. The more complete dataset and analytical details can be found in Allen *et al.* (2007).

apatites from the modern Bengal Delta. If the Andaman Flysch were an uplifted part of the Bengal Fan, sourced at a time when erosion and exhumation rates were accelerating with major crust thickening as predicted for the Palaeogene (Beaumont *et al.* 2001), we would expect Nd values similar to apatites in the present-day fan/delta sediments (although AFT ages would be very different). Figure 12 shows a plot of the Nd Andaman Flysch apatites against data obtained

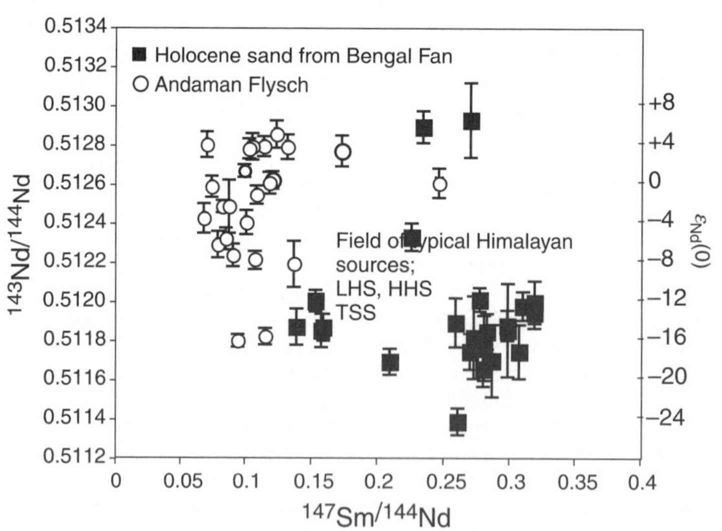

Fig. 12. Plot comparing *in situ* apatite Nd ratios for samples from Joydapur, West Bengal, in the Bengal Basin and the Andaman Flysch, South Andaman.

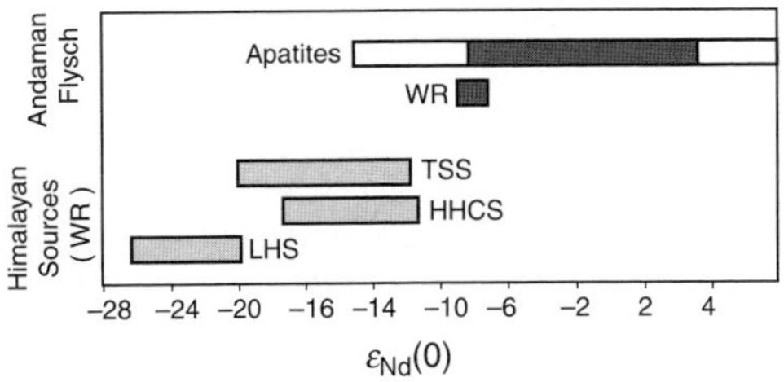

Fig. 13. Nd isotopic data from the Andaman Island Palaeogene sediments compared with WR data from the principal Himalayan tectono-stratigraphic units.

from Holocene sand collected from the Bengal delta at Joypur, West Bengal, whilst Figure 13 shows the Andaman data with WR values for the principal Himalayan units (Robinson *et al.* 2001; Clift *et al.* 2002; Richards *et al.* 2006). It is clear from both Figures 12 and 13 that the predominant source of the Andaman apatites is quite different from the Himalayan sources.

Discussion

The two examples briefly described earlier demonstrate both the utility and benefit of combining *in situ* Nd isotopic data with FT analysis on the same apatite grains. For provenance studies detrital apatite FT ages can be linked to a source by using sample heavy mineral signatures (if diagnostic minerals are present) or by comparison with likely source bedrock FT ages (if there is some prior knowledge of source exhumation history). The latter approach is only practical for recent sediments as over the long term erosion would have removed any earlier exhumation record, that is it is not meaningful to compare a 35 Ma detrital AFT age with present-day bedrock FT ages unless there has been little, if any, erosion since the Oligocene or there are known 35 Ma bedrock sources. There are not many examples of apatite FT provenance studies in the literature because partial–total resetting by post-deposition burial commonly overprints any provenance signal. Measuring each grain's constituent $^{143}Nd/^{144}Nd$ ratio can now circumvent the problem of resetting, as has been demonstrated for the Andaman Island apatites. Comparison between Andaman apatites and Bengal Delta apatites known to be derived from the Himalayas provides unambiguous evidence that the Andaman apatites are not 'Himalayan' in origin, instead they are probably derived from juvenile arc-type magmatic sources with some minor continental input from the nearby east Asian margin.

For study areas such as the Himalaya a key objective is to find ways to monitor orogenic growth from early collision and erosion of cover rocks to later exhumation of the deeper metamorphic units produced by crust thickening. It is likely that this type of record exists within the sediment archive of the Bengal or Indus Fan. ODP Leg 116 drilled parts of the Bengal Fan in the late 1980s, and Nd WR isotopic data were used to map provenance and to identify when particular tectono-stratigraphic units were first eroded (Bouquillon *et al.* 1989). Such studies used the finest sediment fraction, as this material was considered to be the most sensitive to oceanic sources (McLennan *et al.* 1989). Since this time Nd WR rock data have been widely used to characterize sediment provenance, even though ε_{Nd} only represents the weighted average of when the sediment sources were extracted from the mantle.

Few studies have examined in detail the effect of grain size on whole-rock sedimentary samples. Galy *et al.* (1996) revisited the Bengal Fan sediments from ODP Leg 116 and performed analyses on different grain sizes. They found that clay fractions ($<2\ \mu m$) had slightly higher ε_{Nd} than coarser samples. Although this study was unable to identify the precise cause, it was implied that a small contribution of weathered products from the radiogenic Deccan traps was the most likely culprit, that is a small amount of material can potentially have a big impact on WR ε_{Nd}. More recently, Singh & France-Lanord (2002) studied sediments in the modern Brahmaputra, and compared Nd concentrations in suspended and riverbank sediments. They found that the $^{143}Nd/^{144}Nd$ ratio in riverbank sediments decreased as SiO_2 increased, implying a control by heavy minerals. The suspended load shows no dependence of Nd isotopic composition on SiO_2. Furthermore, clay fractions in the Brahmaputra River comprise up to 20% more HHCS material than in the silt fraction (Galy 2004), implying that the clay fraction is biased towards the HHCS unit. Galy (2004) proposed that this bias arises because the upper regions of the HHCS are subjected to glacial abrasion, producing more clay-size particles relative to non-glaciated regions. What these examples demonstrate is that Nd isotopic data in sedimentary rocks do not necessarily provide a reliable image of the source region from which the sediments are derived because the data are vulnerable to distortion through weathering, erosion and transport processes (and, on occasion, diagenesis). Measurement of Nd isotopes on single grains of FT-dated apatite does not suffer from such problems as it provides a direct record of the grains' source unit.

The equivalence of apatite Nd isotopic data to their parent whole-rock values depends on the time since the apatite equilibrated with its host rock. This condition will not apply to some Himalayan units that have been relatively unaffected by Himalayan metamorphism. An example mentioned earlier is the Chekha Formation of Bhutan, which is separated from the HHCS by an extensional detachment. Apatite Nd isotopic data from the upper Chekha Formation were observed to align on an approximately 560 Ma-isochron (Fig. 5), which relates to the rock's pre-Himalayan history. A similar situation is likely to exist in the other units that structurally overlie the HHCS such as the TSS and the Haimanta Group of the Western Himalaya. Conventional Nd WR studies are unable to distinguish between a TSS-like source and a HHCS source, (e.g. Fig. 4), however, as the Chekha example shows by measuring Nd isotopes on apatite grains, this is now possible, so raising the prospect of being able to track exhumation of TSS units in much more detail; although, as a cautionary note, we did find one Chekha sample that had been reset and, usefully, it was possible to detect this on FT-dated grains in sample 02g27.

Whilst Nd isotopic measurements on apatite grains alone can significantly enhance provenance interpretations, their combination with FT analysis raises further the potential level of source resolution. Consider the Holocene sand sample from the Bengal Delta, which by virtue of its young depositional age enables a direct comparison between the FT data and present-day bedrock FT ages. The sample yielded three age components: Nd analysis on the dated grains identified age modes A and B as being from HHCS sources. The youngest mode, A (1.5 ± 0.3 Ma), is typical of the HHCS and could be from sources along the strike of much of the Himalayan range. However, mode B (9.0 ± 1.0 Ma) is significantly older than most known bedrock AFT ages, the exception being Bhutan where AFT ages in some of the HHCS extends back as far as about 8 Ma (Grujic *et al.* 2006). But it is unlikely Bhutan was ever a significant source of sediment, since the contribution of sediment from any area will be proportional to erosion rate. In Bhutan FT ages are older than elsewhere and erosion rates lower. Given the likelihood that the apatites do not come from Bhutan, an alternative explanation is that the Holocene sands include some reworked older sediment either from parts of the delta or further afield from the Siwalik basins (e.g. van der Beek *et al.* 2006). Although this is only one sample and so we cannot be certain if the 9 Ma age represents either a Bhutan source or reworking, this example does serve to illustrate the potential for extracting more detailed provenance information using this approach. Future single-grain Nd studies on deltas and their catchments may be able to reveal more detail as to the true extent of sediment storage and reworking.

With detrital thermochronometric datasets such as FT it is common practice to extract the main age modes using statistical procedures. These deconvolution methods generally apply Poissonian models and goodness-of-fit criteria to identify the likelihood of single grains belonging to any particular age mode. However, this approach can have its limitations, especially in areas, such as in large orogens, where different sources (spatially and temporally) might produce similar FT ages. It is entirely conceivable that within a single age mode there might be a mixture of different source areas, partly reflecting the drainage system but possibly also due to local reworking of sediment. Statistical deconvolution of the data will not be sensitive to detect for this effect, but the inclusion of Nd isotopic data measured on the FT-dated grains will enable identification of any different source inputs that have the same age.

Conclusions

We have shown here how it is possible to measure Nd on single apatite grains previously etched for FT analysis, enabling their Nd isotopic composition to be attributed to a sample's detrital FT age structure. Using examples from the Andaman Islands and the Himalaya, Nd measurements on apatite grains have been shown to be a powerful provenance tool that in future will enable more detailed interrogation of a sediment source. When combined with FT analysis the methodology has the potential to extend source interpretation to a level of detail beyond our current capabilities, including providing new insights into sediment routing systems, levels of sediment reworking and duration of storage.

We are grateful to the Natural Environment Research Council for providing funding for this contribution in the form of Joint Infrastructure Fund grant GR3/JIF/46, NERC fellowship to G. L. Foster and award NE/B503192/1 to A. Carter. Constructive reviews were provided by M. Bernet and A. Gleadow.

References

AHMAD, T., HARRIS, N. B. W., BICKLE, M., CHAPMAN, H., BUNBURY, J. & PRINCE, C. I. 2000. Isotopic constraints on the structural relationships between the Lesser Himalayan Series and the High Himalayan Cyrstalline Series, Garhwal Himalaya. *Geological Society of America Bulletin*, **112**, 467–477.

ALLEGRE, C. J. & BEN OTHMAN, D. 1980. Nd–Sr isotopic relationship in granitoid rocks and continental crust development: a chemical approach to orogenesis. *Nature*, **286**, 335–342.

ALLEN, R., CARTER, A. *ET AL.* 2007. New constraints on the sedimentation and uplift history of the Andaman–Nicobar accretionary prism, South Andaman Island. *In*: DRAUT, A. E., CLIFT, P. D. & SCHOLL, D. W. (eds) *Formation and Applications of the Sedimentary Record in Arc Collision Zones*. Geological Society of America, Special Publications, **436**, Chap. 11.

AMIDON, W. H., BURBANK, D. W. & GEHRELS, G. E. 2005. U–Pb zircon ages as a sediment mixing tracer in the Nepal Himalaya. *Earth and Planetary Science Letters*, **235**, 244–260.

AYRES, M. W. 1997. *Trace-element Behaviour During High-grade Metamorphism and Anatexis of the Himalayas*, PhD thesis, Open University.

BEAUMONT, C., JAMISON, R. A., NGUYEN, M. H. & LEE, B. 2001. Himalayan tectonics explained by extrusion of a low-viscosity crustal channel coupled to focused surface denudation. *Nature*, **414**, 738–742.

BELOUSOVA, E. A., GRIFFIN, W. L., O'REILLY, S. Y. & FISHER, N. I. 2002. Apatite as an indicator mineral for mineral exploration: trace-element compositions and their relationship to host rock type. *Journal of Geochemical Exploration*, **76**, 45–69.

BINGEN, B., DEMAIFFE, D. & HERTOGEN, J. 1996. Redistribution of Rare-Earth Elements, Thorium, and Uranium over accessory minerals in the course of amphibolite to granulite-facies metamorphism – the role of apatite and monazite in orthogneisses from Southwestern Norway. *Geochimica et Cosmochimica Acta*, **60**, 1341–1354.

BLYTHE, A. E., BURBANK, D. W., CARTER, A., SCHMIDT, K. & PUTKONEN, J. 2007. Exhumational history of the Annapurna region in the central Nepalese Himalaya from apatite and zircon fission track and apatite (U–Th)/He analyses. *Tectonics*, **26**, TC3002, doi:10.1029/2006TC001990.

BOLLINGER, L., AVOUAC, J.-P. *ET AL.* 2004. Thermal structure and exhumation history of the Lesser Himalaya in central Nepal. *Tectonics*, **23**, (TC5015), doi:10.1029/2003TC001564.

BOUQUILLON, A., FRANCE-LANORD, C., MICHARD, A. & TIERCELIN, J.-J. 1990. Sedimentology and isotopic chemistry of the Bengal fan sediments: the denudation of the Himalaya. *Proceedings of the Ocean Drilling Program, Scientific Results*, **116**, 43–58.

BREWER, I. D., BURBANK, D. W. & HODGES, K. V. 2003. Modelling detrital cooling-age populations: insights from two Himalayan catchments. *Basin Research*, **15**, 305–320.

BURBANK, D. W., BLYTHE, A. E. *ET AL.* 2003. Decoupling of erosion and precipitation in the Himalayas. *Nature*, **426**, 652–655.

CHERNIAK, D. J. 2000. Rare earth element diffusion in apatite. *Geochimica et Cosmochimica Acta*, **64**, 3871–3885.

CLIFT, P. D., LEE, J. I. *ET AL.* 2002. Nd and Pb isotope variability in the Indus River System: implications for sediment provenance and crustal hetrogeneity in the Western Himalaya. *Earth and Planetary Science Letters*, **200**, 91–106.

CURRAY, J. R. 2005. Tectonics and history of the Andaman Sea region. *Journal of Asian Earth Sciences*, **25**, 187–232.

DEBON, F., LE FORT, P., SHEPPARD, S. M. F. & SONET, J. 1986. The four plutonic belts of the Transhimalaya-Himalaya: a chemical, mineralogical, isotopic, and chronological synthesis along a Tibet-Nepal section. *Journal of Petrology*, **27**, 219–250.

DECELLES, P. G., GEHRELS, G. E., NAJMAN, Y., MARTIN, A. J., CARTER, A. & GARZANTI, E. 2004. Detrital geochronology and geochemistry of Cretaceous–Early Miocene strata of Nepal: Implications for timing and diachroneity of initial Himalayan orogenesis. *Earth and Planetary Science Letters*, **227**, 313–330.

DENIEL, C., VIDAL, P., FERNANDEZ, A., LE FORT, P. & PEUCAT, J. J. 1987. Isotopic study of the Manaslu granite (Himalaya, Nepal): inferences on the age and source of Himalayan leucogranites. *Contributions in Mineralogy and Petrology*, **96**, 78–92.

EHRENBERG, S. N. & NADEAU, P. H. 2002. Postdepositional Sm/Nd fractionation in sandstones: implications for neodymium-isotope stratigraphy. *Journal of Sedimentary Research*, **72**, 304–315.

FOSTER, G. L. & CARTER, A. 2007. Improving the understanding of the location and long-term rates of Himalayan erosion; a combined fission track and *In situ* Nd isotopic study of detrital apatite. *Earth and Planetary Science Letters*, **257**, 407–418.

FOSTER, G. L. & VANCE, D. 2006. *In situ* Nd isotopic analysis of geological materials by laser ablation MC-ICP-MS. *Journal of Analytical Atomic Spectrometry*, **21**, 288–296.

FRANCE-LANORD, C., DERRY, L. & AMICHARD, A. 1993. Evolution of the Himalaya since Miocene time: isotopic and sedimentological evidence from the Bengal Fan. *In*: TRELOAR, P. J. & SEARLE, M. P. (eds) *Himalayan Tectonics*. Geological Society, London, Special Publications, **74**, 605–622.

GALBRAITH, R. F. & GREEN, P. F. 1990. Estimating the component ages in a finite mixture. *Nuclear Tracks and Radiation Measurement*, **17**, 197–206.

GALBRAITH, R. F. & LASLETT, G. M. 1993. Statistical models for mixed fission track ages. *Nuclear Tracks and Radiation Measurement*, **21**, 459–470.

GALY, A. 2004. Fingerprinting the source of fluvial deposits: lessons learnt from the study of modern Himalayan sediments. *In*: *William Smith Meeting*. Geological Society, London, Abstract Volume.

GALY, A., FRANCE-LANORD, C. & DERRY, L. A. 1996. The Late Oligocene–Early Miocene Himalayan belt constraints deduced from isotopic compositions of Early Miocene turbidites in the Bengal Fan. *Tectonophysics*, **260**, 109–118.

GANSSER, A. 1983. *Geology of the Bhutan Himalaya*. Birkhauser Verlag, Basel.

GARZANTI, E., VEZZOLI, G., ANDO, S., FRANCE-LANORD, C., SINGH, S. K. & FOSTER, G. L. 2004. Sand petrology and focused erosion in collision orogens: the Brahmaputra case. *Earth and Planetary Science Letters*, **220**, 157–174.

GOODBRED, S. L., KUEHL, S. A., STECKLER, M. S. & SARKER, M. H. 2003. Controls on facies distribution and stratigraphic preservation in the Ganges–Brahmaputra delta sequence. *Sedimentary Geology*, **155**, 301–316.

GRUJIC, D., COUTAND, I., BOOKHAGEN, B., BONNET, S., BLYTHE, A. & DUNCAN, S. 2006. Climatic forcing of erosion, landscape, and tectonics in the Bhutan Himalayas. *Geology*, **34**, 801–804.

HODGES, K. V. 2000. Tectonics of the Himalaya and southern Tibet from two perspectives. *Geological Society of America Bulletin*, **112**, 324–350.

HUDSON-EDWARDS, K. H., BANERJEE, R. *ET AL.* 2004. A sedimentary framework for arsenic-contaminated groundwater in West Bengal. *Geochimica et Cosmochimica Acta*, **68**, A515.

HURFORD, A. J. 1990. Standardization of fission track dating calibration: Recommendation by Fission Track Working Group of IUGS Subcommission on Geochronology. *Chemical Geology*, **80**, 171–178.

INGER, S. & HARRIS, N. 1993. Geochemical constraints on leucogranite magmatism in the Langtang Valley, Nepal Himalaya. *Journal of Petrology*, **34**, 345–368.

JACOBSEN, S. B. & WASSERBURG, G. J. 1980. Sm–Nd isotopic evolution of chondrites. *Earth and Planetary Science Letters*, **50**, 139–155.

JAGOUTZ, O., MUNTENER, O., BURG, J.-P., ULMER, P. & JAGOUTZ, E. 2006. Lower continental crust formation through focused flow in km-scale melt conduits: The zoned ultramafic bodies of the Chilas Complex in the Kohistan island arc (NW Pakistan). *Earth and Planetary Science Letters*, **242**, 320–342.

KHAN, M. A., STERN, R. J., GRIBBLE, R. F. & WINDLEY, B. F. 1997. Geochemical and isotopic constraints on subduction polarity, magma sources, and paleogeography of the Kohistan intra-oceanic arc, northern Pakistan Himalaya. *Journal of the Geological Society, London*, **154**, 935–946.

MCLENNAN, S. M., MCCULLOCH, M. T., TAYLOR, S. R. & MAYNARD, J. B. 1989. Effect of sedimentary sorting on neodymium isotopes in deep-sea turbidites. *Nature*, **337**, 547–549.

MILLER, C., THONI, M., FRANK, W., GRASEMANN, B., KLOTZLI, U., GUNTLI, V. & DRAGANITS, E. 2001. The early Palaeozoic magmatic event in the Northwest Himalaya, India: source, tectonic setting and age of emplacement. *Geology Magazine*, **138**, 237–251.

NAJMAN, Y. 2006. The detrital record of orogenesis: A review of approaches and techniques used in the Himalayan sedimentary basins. *Earth-Science Reviews*, **74**, 1–72.

NAJMAN, Y., CARTER, A., OLIVER, G. & GARZANTI, E. 2005. Provenance of early foreland basin sediments, Nepal: constraints to the timing and diachroneity of early Himalayan orogenesis. *Geology*, **33**, 309–312.

OHLANDER, B., INGRI, J., LAND, M. & SCHOBERG, H. 2000. Change of Sm–Nd isotope composition during weathering of till. *Geochimica et Cosmochimica Acta*, **64**, 813–820.

PAL, T., CHAKRABORTY, P. P., GUPTA, T. D. & SINGH, C. D. 2003. Geodynamic evolution of the outer-arc–forearc belt in the Andaman Islands, the central part of the Burma–Java subduction complex. *Geology Magazine*, **140**, 289–307.

PARRISH, R. R. & HODGES, K. V. 1996. Isotopic constraints on the age and provenance of the Lesser and Greater Himalayan sequences, Nepalese Himalaya. *Geological Society of America Bulletin*, **108**, 904–911.

PETTERSON, M. G., CRAWFORD, M. E. & WINDLEY, B. F. 1993. Petrogenetic implications of neodymium isotope data from the Kohistan batholith, North Pakistan. *Journal of the Geological Society, London*, **150**, 125–129.

PRATIMA, M., KESSARKAR, V. *ET AL*. 2005. Changing sedimentary environment during the Late Quaternary: Sedimentological and isotopic evidence from the distal Bengal Fan. *Deep-Sea Research*, **52**, 1591–1615.

PRINCE, C. I. 1999. *The timing of Prograde Metamorphism in the Garhwal Himalaya, India*. PhD thesis, Open University.

PYLE, J. M. & SPEAR, F. S. 2000. An empirical garnet (YAG)-xenotime thermometer, Contributions. *Mineralogy and Petrology*, **138**, 51–58.

PYLE, J. M., SPEAR, F. S., RUDNICK, R. L. & MCDONOUGH, W. F. 2001. Monazite-xenotime-garnet equilibrium in metapelites and a new monazite-garnet thermometer. *Journal of Petrology*, **42**, 2083–2107.

RICHARDS, A., PARRISH, P. P., HARRIS, N. B. W., ARGLES, T. W. & ZHANG, L. 2006. Correlation of lithotectonic units across the eastern Himalaya, Bhutan. *Geology*, **34**, 341–344.

ROBINSON, D. M., DECELLES, P. G., PATCHETT, P. L. & GARZIONE, C. N. 2001. The kinematic evolution of the Nepalese Himalaya interpreted from Nd isotopes. *Earth and Planetary Science Letters*, **192**, 507–521.

SCHÄRER, U., COPELAND, P., HARRISON, T. M. & SEARLE, M. P. 1990. Age, cooling history, and origin of post-collisional leucogranites in the Karakoram Batholith; a multi-system isotope study. *Journal of Geology*, **98**, 233–251.

SCHLUP, M., CARTER, A., COSCA, M. & STECK, A. 2003. Exhumation history of eastern Ladakh revealed by $^{40}Ar/^{39}Ar$ and fission-track ages: the Indus River–Tso Morari transect, NW Himalaya. *Journal of the Geological Society, London*, **160**, 385–399.

SINGH, S. K. & FRANCE-LANORD, C. 2002. Tracing the distribution of erosion in the Brahmaputra watershed from isotopic compositions of stream sediments. *Earth and Planetary Science Letters*, **202**, 645–662.

TAYLOR, S. R. & MCLENNAN, S. M. 1985. *The Continental Crust: Its Composition and Evolution*. Blackwell, London.

TRELOAR, P. J., REX, D. C., GUISE, P. G., WHEELER, J., HURFORD, A. J. & CARTER, A. 2000. Geochronological constraints on the evolution of the Nanga Parbat syntaxis, Pakistan Himalaya. *In*: KHAN, M. A., TRELOAR, P. J., SEARLE, M. P. & JAN, M. Q. (eds) *Tectonics of the Nanga Parbat Syntaxis and the Western Himalaya*. Geological Society, London, Special Publications, **170**, 137–162.

VAN DER BEEK, P. A., ROBERT, X., MUGNIER, J. L., BERNET, M., HUYGHE, P. & LABRIN, E. 2006. Late Miocene–Recent exhumation of the central Himalaya and recycling in the foreland basin assessed by apatite fission-track thermochronology of Siwalik sediments, Nepal. *Basin Research*, **18**, 413–434.

VANNAY, J. C., GRASEMANN, B., RAHN, M., CARTER, A., BAUDRAZ, V. & COSCA, M. 2004. Miocene to Holocene exhumation of metamorphic crustal wedges in the NW Himalaya: Evidence for tectonic extrusion coupled to fluvial erosion. *Tectonics*, **23**, 1014–1029.

WHITTINGTON, A. G. 1996. Exhumation overrated at Nanga Parbat, Northern Pakistan. *Tectonophysics*, **206**, 215–226.

WILLETT, S. D. 1999. Orography and orogeny: the effects of erosion on the structure of mountain belts. *Journal of Geophysical Research*, **104**, 28,957–29,981.

WING, B. A., FERRY, J. M. & HARRISON, T. M. 2003. Prograde destruction and formation of monazite and allanite during contact and regional metamorphism of pelites: petrology and geochronology. *Contributions in Mineralogy and Petrology*, **145**, 228–250.

Calibration and comparison of etching techniques for apatite fission-track thermochronology

G. R. MURRELL[1,2], E. R. SOBEL[3]*, B. CARRAPA[3,4] & P. ANDRIESSEN[1]

[1]*Faculteit der Aard- en Levenswetenschappen, Vrije Universiteit, De Boelelaan 1085, 1081 HV Amsterdam, The Netherlands*

[2]*Present address: Enhanced Oil Recovery Institute, University of Wyoming, Laramie, WY 82071-3006, USA*

[3]*Institut für Geowissenschaften, Universität Potsdam, Karl-Liebknecht-Strasse 24, 14476, Golm, Germany*

[4]*Present address: Department of Geology and Geophysics, University of Wyoming, Laramie, WY 82071-3006, USA*

**Corresponding author (e-mail: sobel@rz.uni-potsdam.de)*

Abstract: Understanding time–temperature histories using apatite fission-track thermochronology involves sample preparation, analysis and then thermal modelling using an appropriate annealing algorithm. A subtle point in this sequence is ascertaining that the sample preparation utilized is compatible with the methodology used in obtaining the data for constructing the annealing data set. This issue is important if one wishes to utilize the relatively new multikinetic annealing algorithm of Ketcham *et al.* that is implemented in their AFTSolve and HeFTy models which is based on a different etching recipe than those previously used. A preliminary calibration step involves comparing published etch pit diameters for a suite of samples with those analysed by an operator. Results show that the operator can reliably reproduce the calibration data set. We then report a laboratory experiment using samples from Finland and Spain that compares the results obtained using two different etching methodologies (7% nitric acid with qualitative etching conditions and 5.5 M nitric acid at constant conditions). The two raw data sets yield variable results. Comparing the two etching methodologies reveals the influence of this procedure on the kinetic parameter D_{par}.

A recurring issue in thermochronology involves incompatibilities between thermal modelling results and geological observations. Murrell (2003) revealed some discrepancies between thermal histories produced through numerical modelling of apatite fission-track thermochronology (AFTT) data and other available thermal indicators and physical data sets in a study based in Finland. One explanation is that the AFTT thermal models overestimated the amount of heating that has occurred due either to the application of an inappropriate annealing model, the inability of the annealing model to resolve low-temperature annealing over long timespans or a combination of both. In the first case, the inappropriateness of the applied model may arise as a result of the influence that apatite chemical composition has on fission-track annealing. A particular example is the observation that fluorine-rich apatites are less resistant (anneal more easily and thus record cooler temperatures) than chlorine-rich apatites (Green *et al.* 1986). The Laslett *et al.* (1987) monocompositional annealing model used in Murrell (2003) is based on experimental data derived from the analysis of the Durango apatite (Green *et al.* 1986), a relatively chlorine-rich sample, whereas microprobe data show that the samples analysed by Murrell (2003) are predominantly fluorine-rich apatites; thus, the thermal histories produced predict hotter temperatures than may have occurred. A further problem arises owing to potential variations in apatite chemistry both between and within samples. Rigorous treatment of such samples requires the use of an annealing model appropriate for individual samples as well as samples with intrasample chemical variations (Murrell 2003).

Newer annealing models attempt to take into account these variations (e.g. Laslett & Galbraith 1996). One of the latest advances has been documented in a series of papers published in 1999 (Carlson *et al.* 1999; Donelick *et al.* 1999; Ketcham *et al.* 1999). This work seeks to utilize the dependence of fission-track annealing rate on apatite composition in order to enhance AFTT

From: LISKER, F., VENTURA, B. & GLASMACHER, U. A. (eds) *Thermochronological Methods: From Palaeotemperature Constraints to Landscape Evolution Models*. Geological Society, London, Special Publications, **324**, 73–85.
DOI: 10.1144/SP324.6 0305-8719/09/$15.00

thermal modelling (Ketcham *et al.* 2000, 2005). The workers observed that quantifying this dependence allows the range of temperatures over which apatite fission tracks serve as recorders of thermal history to be extended and the precision of thermal histories to be improved, with more resistant apatites being used to probe the higher temperature portions of the thermal history while less resistant apatites address the lower temperature portions. They introduce a multikinetic annealing model based on the experimental analysis of track annealing in apatites of variable composition. The annealing rate is expressed as a function of measurable chemical or crystallographic parameters, allowing an estimate of the annealing characteristics to be made and included in the modelling procedure.

The benefit of using such a multicompositional model is the production of more accurate thermal models. No longer is it necessary to estimate (guess) the effect of modelling F-apatites with a Durango apatite based on an annealing model, nor is it necessary to select an annealing model that best fits the majority of the samples to the potential detriment of others. It is not necessary to exclude samples with intergrain chemical variations from the modelling exercise.

Prior to using a specific modelling procedure, one must verify that the sample input data were collected in a manner that is compatible with the data set upon which the model was constructed. When methodologies vary, ideally one should conduct a calibration exercise to examine the differences between the two different methodologies. In the case of AFTT, a fundamental methodological issue concerns sample etching. Apatites are etched with nitric acid prior to irradiation to reveal spontaneous fission tracks; post-irradiation, the mica external detector is etched in hydrofluoric acid to reveal the reactor-induced fission tracks (e.g. Wagner & Van der Haute 1992). At present, no unique etching protocol exists. Although many laboratories use constant etch time and temperature, the fission-track community routinely uses several different nitric acid etches (e.g. 5, 5.5 and 1.6 M [7%]), with variations in acid strength, etch time and etchant temperature. The available published annealing models are also based on several different etching recipes.

Herein we report on an exercise to compare the effect of two different etching recipes with the goal of quantifying differences in raw AFTT data, which is one of the principal inputs into modelling programs. The project began as an effort to assess whether an existing data set could be used with a different annealing model and progressed to a more thorough examination of the differences that result from using these different recipes. We compare laboratory experiments using two different etching methods on a well-characterized suite of samples (Carlson *et al.* 1999; Donelick *et al.* 1999) in order to develop a calibration system. An additional data set is examined using the same two etching procedures on samples from Finland (Murrell 2003) and Spain (de Bruijne 2001). Although one of the etching procedures examined herein has now fallen out of favour, it is nevertheless worthwhile to present a direct comparison of the two methodologies, as the principles examined can also be applied to other etching recipes. Indeed, several new etching protocols have been used in papers published in the last several years; the approach presented here could be used to examine and calibrate various recipes.

Laboratory procedures and system calibration

Ketcham *et al.* (1999) stressed the important influence that the applied etching method has on the derivation of the data used as the basis of the annealing model. They also point out that the etching procedure used in the study is not commonly applied in the FT community and state that if alternative etching techniques are used then calibration is required. However, to date, quantitative reports on the effect of using different etching procedures on the final apatite fission-track ages and modelling results are lacking. We report results from a laboratory exercise that quantifies the effects of two different, widely used, fission-track methodologies.

The first one, referred to here as Method I, applies a non-quantitative etching procedure where the fission-track mounts were etched in 7% (1.6 M) HNO_3 at room temperature (18–22 °C) for between 25 and 40 s until the tracks were clearly visible and of 'equal' scale. Fission-track analysis did not include *c*-axis length correction, as *c*-axis angles were not measured. Separate mounts were prepared for single-grain age and track-length determinations; all mounts were etched individually. Modelling is based on the model of Laslett *et al.* (1987). The goal of this etching procedure is to produce tracks of a uniform appearance so that the analyst will analyse all samples similarly. This procedure was often applied in the past, prior to the introduction of multikinetic annealing algorithms and the increased availability of temperature-controlled water baths.

The second methodology, referred to here as Method II, uses a quantitative etching procedure involving etching in 5.5 M HNO_3 for 20 s at a controlled 21 °C. This etching procedure matches that used by Carlson *et al.* (1999) to produce the database for the annealing model of Ketcham *et al.*

(1999). Suites of samples were all etched together for a constant, consistent time. Calibration with the results of Carlson *et al.* (1999) was possible using an etching experiment conducted on a split of the original samples analysed by Carlson *et al.* (1999) kindly provided by R. A. Donelick. Samples were etched according to this procedure and the etch-pit sizes compared with those reported in Carlson *et al.* (1999) (their Appendix, see Table 1). *c*-axis angles were measured. All measurements in this study, with the exception of the data published by de Bruijne (2001), were obtained by a single operator (G. R. Murrell) using a single microscope. Tracks were counted under 1000× magnification with a 100× dry objective and a numerical aperture of 0.90.

The method used to etch the mica external detectors also varied. For both Method I and Method II, micas were etched using 48% HF for 12 min at room temperature (*c.* 21 °C). In the former case, micas were etched individually and, therefore, the etch time could vary slightly between samples; in the latter case, all samples were etched together for the same time.

D_{par} is the arithmetic mean maximum diameter of fission-track etch pits (Donelick 1993; Burtner *et al.* 1994) (Fig. 1). This parameter is measured in the same manner as the track lengths, and is easy, quick and inexpensive to obtain. D_{par} is used as the calibration parameter as it is a direct measure of the apatite solubility and, in turn, the etch characteristic (Honess 1927). To avoid confusion, the D_{par} values reported by Carlson *et al.* (1999) are here referred to as D_C, while the values produced with the variable etch (Method I) during the calibration exercise are referred to as D_I. Values reported herein produced using the Donelick *et al.* (1999) etch (Method II) are referred to as D_{II}.

The values of D_I, D_{II} and D_C reported for each apatite species (Table 1 and Fig. 2) are arithmetic means determined from at least 25 measurements. The reported errors on the means are standard arithmetic errors. To aid comparison, the values of D_I and D_{II} were normalized against the corresponding values of D_C (Fig. 2). The results show that the D_I values are all too large, ranging between +4.3 and +39.3% of the corresponding D_C values; the average being +25.5%. In contrast, D_{II} values are between −2.4 and +5.4% of the corresponding D_C values, with the average being +1.1%. One specimen (B3) produced D_{II} values 24.8% less than the corresponding D_C; however, these measurements have been excluded because they were taken from non-*c*-axis-parallel surfaces.

The importance of the 1.1% difference in the values of D_C and D_{II} can be assessed with respect to its effect on initial track length (l_0). The l_0 parameter has a large influence on the derivation of annealing models, as discussed in great detail by Carlson *et al.* (1999). Indeed, an obvious influence is present in the Laslett *et al.* (1987) model, which predicts ubiquitously the presence of a late cooling event in modelled thermal histories. This effect is, in part, due to the model having an inappropriately large initial track-length parameter (Gallagher *et al.* 1998). Carlson *et al.* (1999) show that initial

Table 1. *Comparison between the* D_{par} *values obtained with Method I* (D_I)*, Method II* (D_{II}) *and the* D_{par} *values reported by Carlson* et al. *(1999)* (D_C)

Sample name	D_I (μm)	D_{II} (μm)	D_C (μm)	D_I/D_C	D_{II}/D_C
AY	2.72 ± 0.24	2.06 ± 0.03	1.95 ± 0.03	1.3927 ± 0.0259	1.0545 ± 0.0214
B2	5.50 ± 0.56	4.47 ± 0.08	4.58 ± 0.06	1.2008 ± 0.0248	0.9767 ± 0.0211
B3	5.21 ± 0.19	3.75 ± 0.06	4.99 ± 0.06	1.0432 ± 0.0159	0.7522 ± 0.0184
DR	2.40 ± 0.25	1.84 ± 0.03	1.83 ± 0.02	1.3089 ± 0.0226	1.0071 ± 0.0213
HS	3.59 ± 0.32	2.99 ± 0.06	3.00 ± 0.06	1.1979 ± 0.0306	0.9962 ± 0.0292
KP	2.71 ± 0.23	2.13 ± 0.05	2.06 ± 0.04	1.3178 ± 0.0296	1.0321 ± 0.0305
OL	3.21 ± 0.31	2.42 ± 0.03	2.35 ± 0.02	1.3646 ± 0.0190	1.0294 ± 0.0147
PC	2.60 ± 0.25	2.16 ± 0.03	2.15 ± 0.03	1.2111 ± 0.0255	1.0053 ± 0.0206
PQ	2.06 ± 0.20	1.60 ± 0.04	1.59 ± 0.02	1.2950 ± 0.0228	1.0041 ± 0.0267
RN	1.97 ± 0.17	1.63 ± 0.03	1.65 ± 0.03	1.1963 ± 0.0261	0.9897 ± 0.0241
SC	2.06 ± 0.18	1.70 ± 0.03	1.71 ± 0.03	1.2043 ± 0.0253	0.9930 ± 0.0248
TI	3.17 ± 0.27	2.44 ± 0.03	2.45 ± 0.04	1.2955 ± 0.0243	0.9952 ± 0.0204
UN	2.20 ± 0.18	1.93 ± 0.05	1.89 ± 0.02	1.1641 ± 0.0197	1.0228 ± 0.0266
WK	2.17 ± 0.24	1.94 ± 0.03	1.87 ± 0.03	1.1604 ± 0.0251	1.0387 ± 0.0224
Mean				1.239	0.993
Mean (excluding B3)				1.255	1.011

$D_I = D_{par}$ in apatite etched using the variable etch.
$D_{II} = D_{par}$ in apatite etched using the Carlson *et al.* (1999) etch.
$D_C = D_{par}$ reported by Carlson *et al.* (1999).

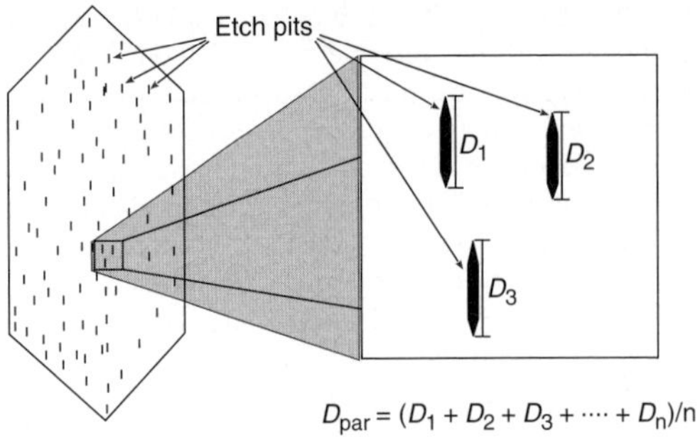

Fig. 1. Schematic describing measurement and calculation of D_{par} kinetic parameter.

track length appears systematic with respect to certain measurable parameters including D_{par} and present an equation describing this relationship: $l_0 = 15.63 + 0.283\ D_{par}$. This equation can only be strictly applied for samples etched in the same way as demonstrated by Carlson *et al.* (1999), as we have attempted with our Method II etch. Using this equation, an apatite specimen with $D_{par} = 2.00\ \mu m$ would produce an initial track-length parameter $l_0 = 16.196\ \mu m$. The 1.1% difference produced in the calibration exercise would imply that the particular apatite would have a $D_{par} = 2.022\ \mu m$ if it was etched with Method II. This would in turn produce an $l_0 = 16.202\ \mu m$, just 0.008 μm greater than the 'real' value. Carlson *et al.* (1999) stated that a 0.5 μm variation in l_0 leads to a difference (overestimation) of 10–15 °C in predicted temperature along the later, cooler portions of a time–temperature path. This would imply that the 1.1% difference in D_{par} would equal just *c.* 0.2 °C difference (overestimation) in this temperature. This is a negligible amount and falls well within the errors of the modelling process and, consequently, can be ignored. Thus, the calibration procedure has shown that the etching procedure of Method II satisfactorily matches that used in the derivation of the data reported by Carlson *et al.* (1999); therefore, the annealing model presented by Ketcham *et al.* (1999) can be applied.

The calibration exercise conducted here has probably been overvigorous, as there is no great variability in normalized D_{par} between the different apatite species. Therefore, for other laboratories that use the Ketcham *et al.* (1999) annealing model, calibration with a smaller set of freely available apatite specimens (e.g. the Durango and Fish Canyon Tuff standards) is probably sufficient.

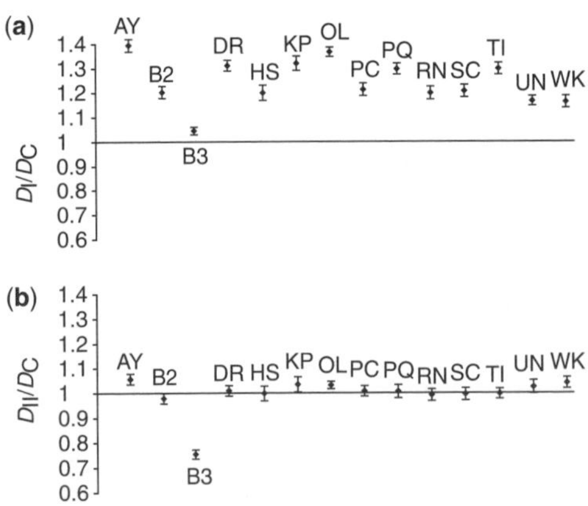

Fig. 2. Plots of normalized D_{par} for each of the apatite specimens, showing relative difference in amount of etching that has occurred due to etching with (**a**) Method I and (**b**) Method II v. the etching used in the derivation of the annealing data of Carlson *et al.* (1999).

Case study comparison

A major goal of AFTT is to obtain data for thermal modelling using an annealing model. In order to assess the influence of etching procedure on the raw AFTT data that would be input into such a model, we present results comparing 12 real samples from Finland and Spain. The six samples from Finland (98005, 98012, 98017, 98021, 98033 and 98048) were selected because they span the range of age results reported by Murrell (2003): from 400 to 1000 Ma. The samples from Spain (clo3, scs3, scs22, scs34, scs37 and scs45) were reported by de Bruijne (2001), and span a range of ages considerably younger than the Finnish examples: from 15 to 215 Ma. The two data sets combined provide a selection spanning a great range and variety of AFTT data. All samples were originally prepared following Method I; these analytical results are available in de Bruijne (2001) and Murrell (2003). Splits from the original mineral separates were processed and analysed with Method II. The raw AFTT results (Central age, Mean track length, D_{par}, etc.) of Method II are reported in Table 2 and shown in Figure 3. To aid comparison, the age and length data produced by Method II, following the Carlson *et al.* (1999) procedure, have been normalized against the results produced by Method I (Table 3).

There is a general variation between the two data sets (Table 3 and Fig. 4). The majority of samples produce considerably older ages with the new etch procedure (e.g. 98017 and 98048) while several are considerably younger (e.g. scs34). Samples 98017, scs34 and 98012 contain bimodal age distributions (Fig. 3), and, therefore, it is possible that the difference in ages between the two data sets in these

Table 2. *AFTT data from the re-analysis of selected samples from Murrell (2003) and some samples reported by de Bruijne (2001)*

Sample number*	Elv. (m.a.s.l.)	Stratigraphic description *Lithology*	Lat. (°N) Long. (°E)	$\rho^{d\dagger}$ ($\times 10^5$)	$\rho^{s\dagger}$ ($\times 10^5$)	$\rho^{i\dagger}$ ($\times 10^5$)	χ^2 (%)	Age (Ma $\pm$ 1σ)	Length statistics†
clo3	1700	SCS Hercynian Basement	354–357	8.61	41.349	32.717	82.7	187.3 ± 7.8	11.7 ± 0.2
18		*Gneiss*	40–41 30.0	*13345*	*1782*	*1410*			(1.6) *80*
scs3	880	SCS Hercynian Basement	354–357	8.61	5.221	44.587	57.5	17.6 ± 0.9	13.2 ± 0.2
24		*Granodiorite*	40–41 30.0	*13345*	*497*	*4244*			(1.7) *100*
scs22	1460	SCS Hercynian Basement	354–357	8.61	20.092	31.213	5.8	96.0 ± 4.0	11.9 ± 0.2
25		*Granite*	40–41 30.0	*13345*	*2009*	*3121*			(1.9) *100*
scs34	550	SCS Hercynian Basement	354–357	8.61	8.38	25.616	0.1	51.0 ± 3.1	12.6 ± 0.2
25		*Granite*	40–41 30.0	*13345*	899	2748			(2.0) *100*
scs37	680	SCS Hercynian Basement	354–357	8.61	5.005	34.792	5.5	21.6 ± 1.1	13.5 ± 0.2
25		*Granite*	40–41 30.0	*13345*	540	3754			(1.7) *71*
scs45	400	SCS Hercynian Basement	354–357	8.61	5.04	35.122	44.8	21.6 ± 1.2	13.3 ± 0.1
25		*Granite*	40–41 30.0	*13345*	425	2962			(1.4) *100*
98005	95	SF Collision Rel. Intr.	61 36.381	8.61	26.221	3.758	82.4	972.2 ± 55.4	13.3 ± 0.2
27		*Gabbro*	29 41.875	*13345*	*2882*	*413*			(1.6) *100*
98012	155	Karelian L. Oro. Intr.	65 20.336	8.61	11.75	2.838	34.2	594.3 ± 34.5	11.8 ± 0.2
26		*Granite*	27 24.806	*13345*	*1776*	*429*			(1.7) *100*
98017	10	SF Suprecrustal	63 38.964	8.61	38.513	13.869	0.5	404.2 ± 23.3	10.7 ± 0.2
27		*Mica Schist*	22 53.543	*13345*	*2627*	*946*			(1.8) *120*
98021	115	SF Collision Rel. Intr.	62 14.382	8.61	*36.211*	*9.59*	96.3	544.2 ± 24.9	11.9 ± 0.1
25		*Gabbro*	23 53.571	*13345*	*2964*	*785*			(1.5) *105*
98033	100	SF Collision Rel. Intr.	60 40.031	8.61	*27.144*	*4.575*	98.3	835.7 ± 55.3	13.8 ± 0.2
15		*Gabbro*	24 39.812	*13345*	*1780*	*300*			(1.5) *100*
98048	25	SF L. Oro. Intr.	60 07.227	8.61	*26.659*	*9.25*	95.2	419.4 ± 15.3	11.7 ± 0.2
26		*Granite*	24 32.160	*13345*	*4542*	*1576*			(1.7) *100*

*Italics show number of grains analysed. †Mean track length stated in μm ± standard error, values in brackets state 1σ standard deviation in μm. Italics show the number of tracks counted/measured. Standard (ρ^d) and induced (ρ^i) track densities (tracks cm^{-2}) were measured on mica external detectors, and spontaneous track densities (ρ^s) were measured on internal mineral surfaces. Sample preparation was carried out following Method II. Elv., elevation; m.a.s.l., metres above sea level. SF, Svecofennian; AGC, Achaean Gneiss Complex; SCS, Spanish Central System; L. Oro. Intr., Late Orogenic Intrusion; Rel. Intr., Related Intrusion. Ages were calculated using $\zeta = 349 \pm 7$ for dosimeter glass CN5 for G. R. Murrell.

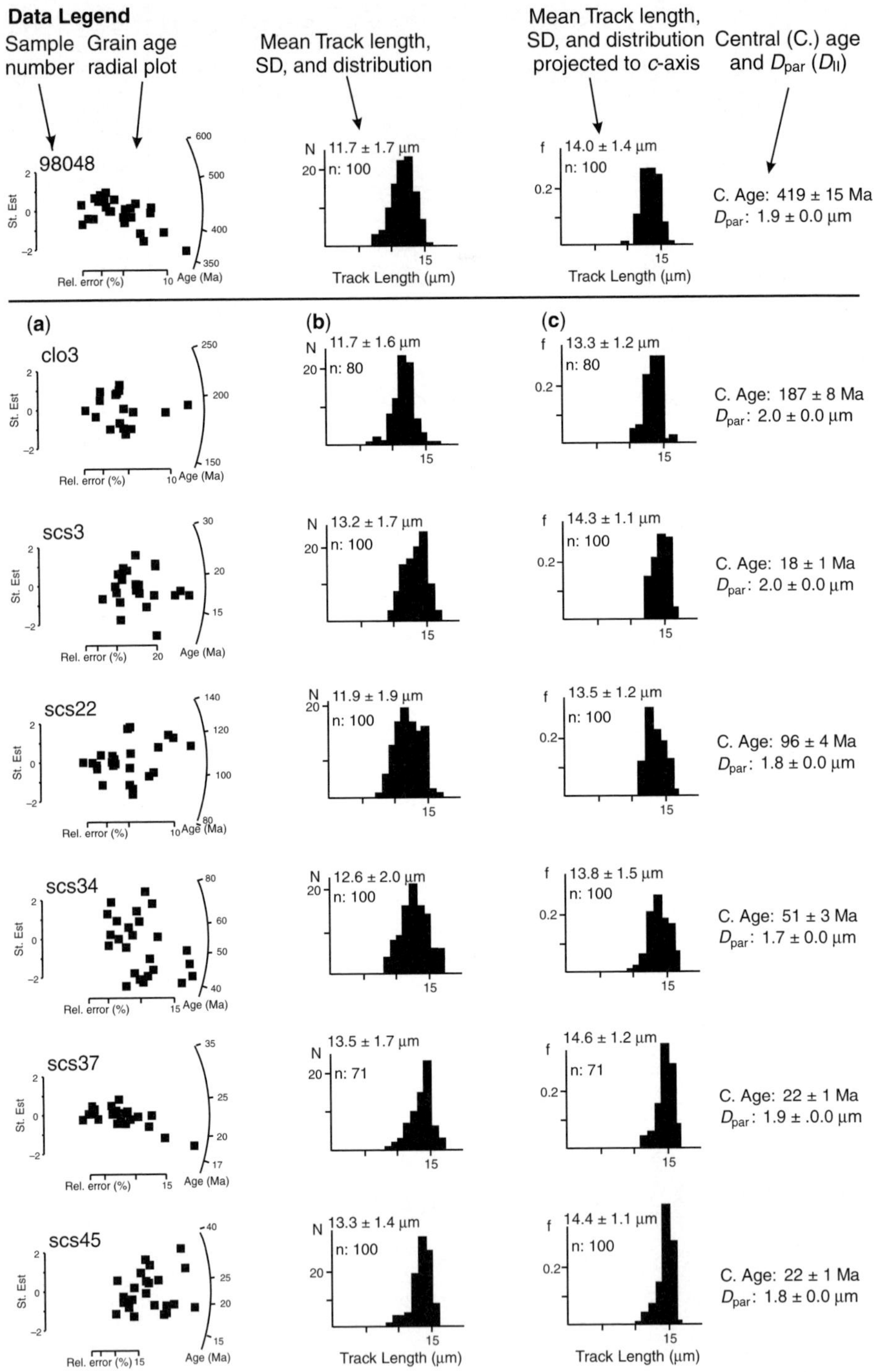

Fig. 3. Multikinetic AFTT raw data for each sample prepared using Method II. (**a**) Grain age radial plots, (**b**) length distributions and (**c**) corrected length distributions and D_{par}. Standard errors for D_{par} have been rounded and are typically *c.* 3%; values less than 0.05 are rounded to 0.0.

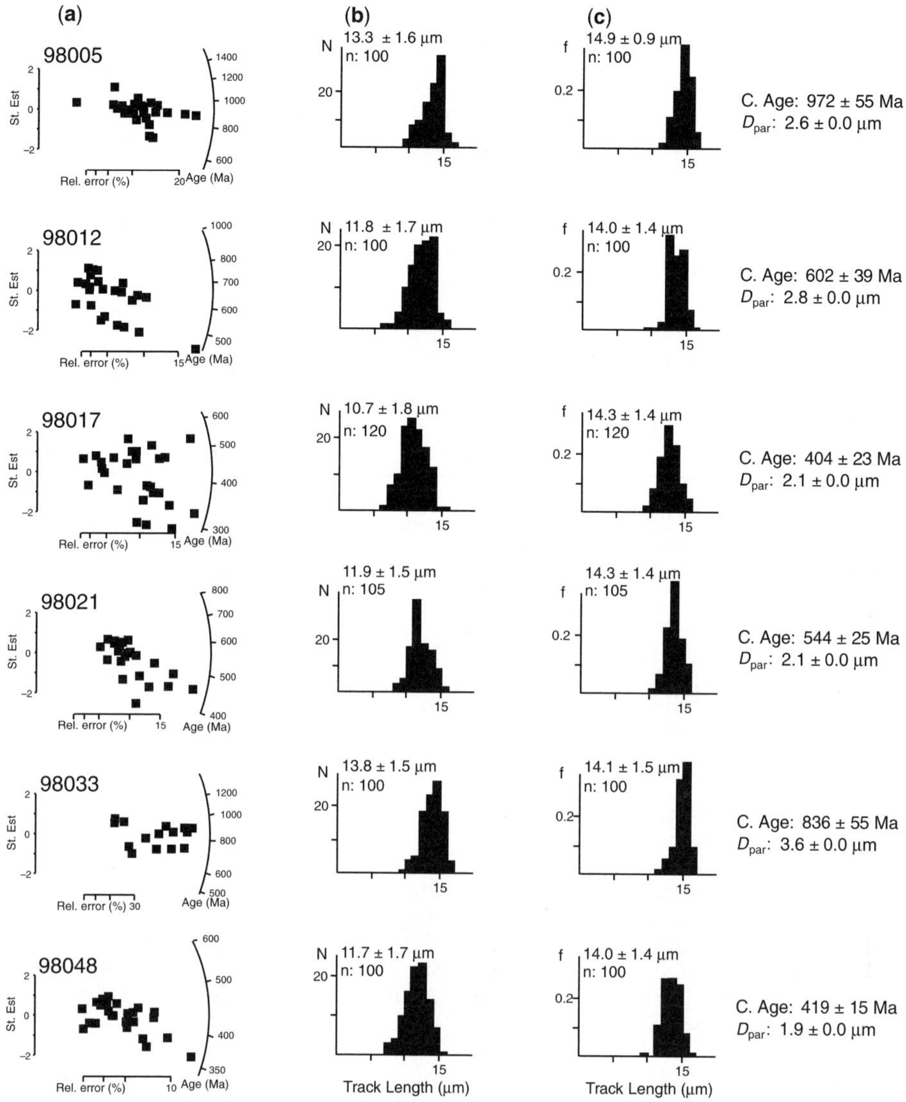

Fig. 3. (*Continued*).

cases may be due to counting more grains of one age mode than the other in the two mounts. However, this is not the case for the other samples.

Method I produced longer track lengths than Method II in seven of 12 samples and shorter in two of 12 sample. Only three of the 12 samples produced replicate lengths that agreed within 1 standard error. The reason for the discrepancy of mean track length between the two data sets is, in part, externally derived: under Method I the length data are taken from separate sample mounts that are etched until the confined tracks are all of similar 'cigar' shape. This invariably meant that the length mounts were etched for a longer period of time than the age mounts. This means that it is even more difficult to compare the raw AFTT data produced by each method. Unfortunately, as *c*-axis data were not collected for Method I, it is not possible to determine if the track-length differences between the two methods reflect track-angle observational biases.

The variation in AFTT age is potentially due to variations in the amount of etching that each procedure has induced in each sample. As stated earlier, Method I involved varying the etch time for each sample so that they appeared to have experienced

Table 3. *Comparison of AFTT data from samples prepared with both Method I and Method II*

Sample	Central age			Mean track length (μm) ± SE		
	Method II (Ma) ± (1σ)	Method I (Ma) ± (1σ)	Normalized (II/I)	Method II (μm) ± SE	Method I (μm) ± SE	Normalized (II/I)
clo3	187.3 ± 7.8	214 ± 28	0.88 ± 0.12	11.7 ± 0.2	11.9 ± 0.2	0.98 ± 0.02
scs3	17.6 ± 0.9	15 ± 2	1.17 ± 0.16	13.2 ± 0.2	13.9 ± 0.1	0.95 ± 0.02
scs22	96.0 ± 4.0	117 ± 12	0.82 ± 0.09	11.9 ± 0.2	13.2 ± 0.1	0.90 ± 0.02
scs34	51.0 ± 3.1	137 ± 14	0.37 ± 0.07	12.6 ± 0.2	13.2 ± 0.2	0.95 ± 0.02
scs37	21.6 ± 1.1	20 ± 2	1.08 ± 0.12	13.5 ± 0.2	14.3 ± 0.1	0.94 ± 0.02
scs45	21.6 ± 1.2	22 ± 2	0.98 ± 0.11	13.3 ± 0.1	13.0 ± 0.2	1.02 ± 0.02
98005	972.2 ± 55.4	869.5 ± 38.3	1.12 ± 0.08	13.3 ± 0.2	13.3 ± 0.2	1.00 ± 0.02
98012	602.1 ± 38.5	559.0 ± 24.6	1.08 ± 0.08	11.8 ± 0.2	12.2 ± 0.1	0.97 ± 0.02
98017	404.2 ± 23.3	300.8 ± 15.8	1.34 ± 0.09	10.7 ± 0.2	11.1 ± 0.1	0.96 ± 0.02
98021	544.2 ± 24.9	510.8 ± 20.8	1.07 ± 0.06	11.9 ± 0.1	12.2 ± 0.1	0.98 ± 0.01
98033	835.7 ± 55.3	665.1 ± 32.3	1.26 ± 0.09	13.8 ± 0.2	12.5 ± 0.1	1.10 ± 0.02
98048	419.4 ± 15.3	374.8 ± 13.5	1.12 ± 0.05	11.7 ± 0.2	11.6 ± 0.2	1.01 ± 0.02

Central ages are reported in order to be able to include scs34 and 98017 in the subsequent comparison.
Method II, samples prepared with new etch procedure (Carlson *et al.* 1999).
Method I, samples prepared with Method I (variable etching).
SE, standard error.

a similar amount of etching. The etch time was recorded for consideration later. Method II has a different approach in that all samples are etched quantitatively with set time and temperature. Thus, samples can display vastly different degrees of etching due to their compositionally induced susceptibility to the applied etchant; an apatite specimen prepared with both methods can potentially have had largely different amounts of etching. Measuring D_{par} for both data sets can help to assess this possibility as it is a measure of the crystals susceptibility to the etchant and, in turn, the degree of etching that has occurred. The D_{par} measurements presented in Table 4 and Figure 5a show that Method I was largely unsuccessful at producing generally similar degrees of etching for each sample as D_I varies significantly between 1.01 and 2.45 μm. This was not totally unexpected as the analyst judged the degree of etching visually. The quantitative etching method also shows considerable variation, with D_{II} varying from 1.70 to 3.58 μm. However, the variation present in the qualitative method is due to differences in the etch procedure, since the (imperfectly achieved) goal of this was create uniform etch pits, whereas the variations present in the quantitative method are due to compositional differences in the apatite crystals. For comparison, the available wt% Cl is also reported in Table 4. Normalizing the D_{par} of Method II (D_{II}) against the D_{par} of Method I (D_I) produces a comparison of the degree of etching that each sample has experienced from each etch method (Fig. 5b). This shows that some samples have experienced considerably more etching with Method II compared with Method I (e.g. clo3, 98005 and 98033), while the reverse has also occurred (e.g. scs22).

Any potential correlation between the degree of etching that has occurred (i.e. normalized D_{par} [D_{II}/D_I]) and the variation in Central ages (i.e. normalized C. age [C.age$_{II}$/C.age$_I$]) can be tested simply by plotting the two variables against each other (Fig. 6). If the amount of etching that a sample had experienced was independent of the AFTT age obtained, then the data presented here would all plot along the C.age$_{II}$/C.age$_I$ = 1 line. That is to say that despite any difference in the degree of etching, the ages obtained would be the same. This is clearly not the case: when the data are plotted it can be seen that there is a positive correlation between the degree of etching and the age obtained. This means that normalized Central age is >1 (<*1*) when normalized D_{par} is >1 (<*1*).

There are two significant exceptions to this correlation (clo3 and scs34). These samples show a higher degree of etching after preparation with Method II but younger ages. The reason for this discrepancy becomes apparent when the comparative etching of the external detector is taken into account. The external detectors (mica slices) of the samples prepared with Method I were etched individually. This allowed human error (in the form of variations in etching time) to enter the procedure, resulting in variable etching conditions for the external detector of each sample. This can be clearly seen in Figure 5c, where D_{par} values for the etch pits in the mica are shown. The D_{par} values for the qualitative etching show great

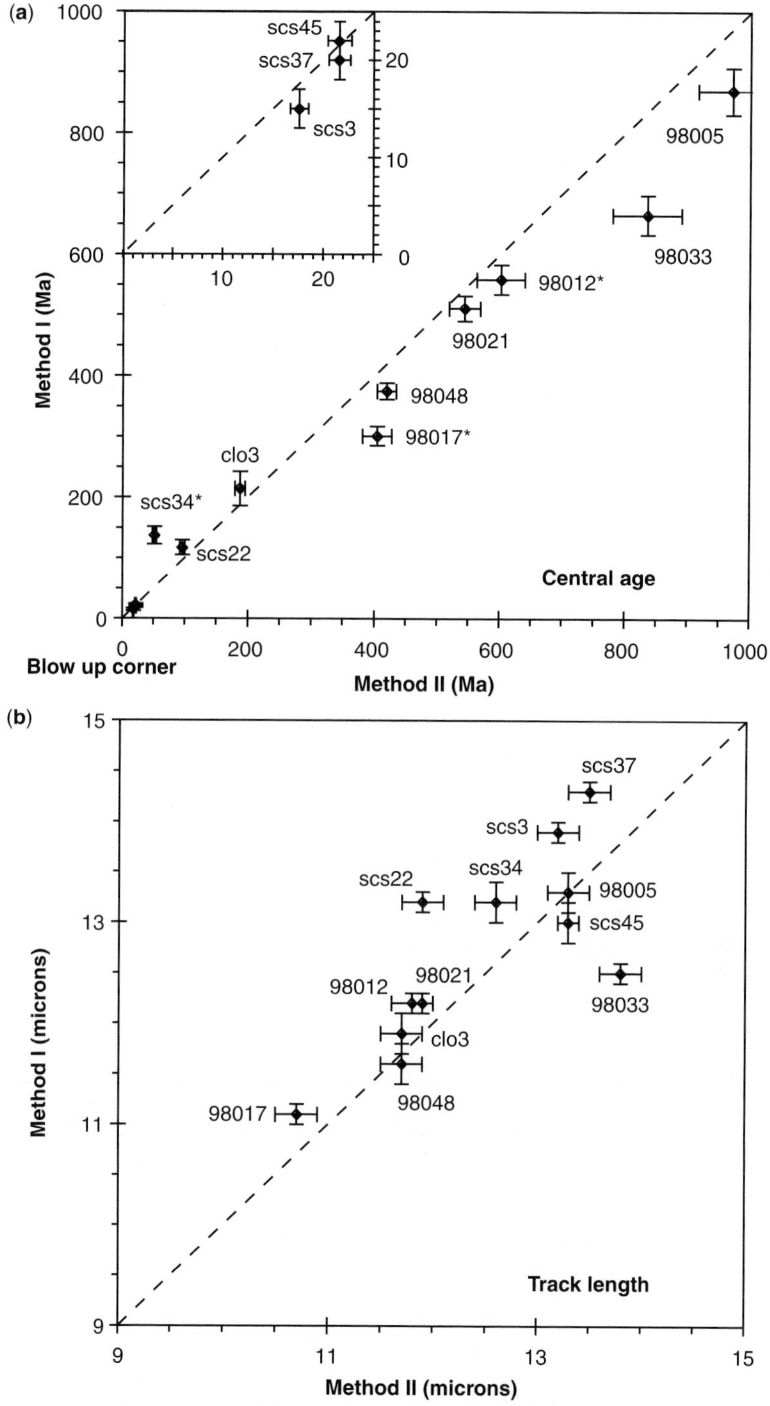

Fig. 4. Comparison between the raw AFTT data obtained from samples prepared with Method I and Method II as outlined by Carlson *et al.* (1999). C. age, Central age. Samples marked with * have bimodal single-grain age distributions, as shown in Figure 3.

Table 4. D_{par} *values for both apatite and external detector for the samples prepared with the two etch procedures*

Sample	D_{par} apatite D_{II} (μm) ± SE	D_I (μm) ± SE	D_{II}/D_I	D_{IIm} (μm) ± SE	D_{par} mica D_{Im} (μm) ± SE	D_{IIm}/D_{Im}	Apatite Cl (wt%)
clo3	2.0 ± 0.0	1.0 ± 0.0	2.0 ± 0.0	1.8 ± 0.1	1.1 ± 0.1	1.7 ± 0.1	
scs3	2.0 ± 0.0	1.6 ± 0.0	1.2 ± 0.0	1.8 ± 0.1	??		
scs22	1.8 ± 0.0	2.5 ± 0.0	0.8 ± 0.0	1.8 ± 0.0	1.5 ± 0.0	1.2 ± 0.1	
scs34	1.7 ± 0.0	1.4 ± 0.0	1.3 ± 0.0	1.8 ± 0.0	0.9 ± 0.0	2.0 ± 0.1	
scs37	1.9 ± 0.0	1.7 ± 0.0	1.1 ± 0.0	1.8 ± 0.1	1.6 ± 0.0	1.1 ± 0.0	
scs45	1.8 ± 0.0	1.6 ± 0.0	1.1 ± 0.0	1.8 ± 0.0	2.1 ± 0.1	0.9 ± 0.0	
98005	2.6 ± 0.0	1.8 ± 0.0	1.5 ± 0.0	1.8 ± 0.1	1.6 ± 0.0	1.1 ± 0.0	0.46
98012	2.8 ± 0.0	1.9 ± 0.0	1.5 ± 0.0	1.9 ± 0.1	2.1 ± 0.1	0.9 ± 0.0	0.08
98017	2.1 ± 0.0	1.6 ± 0.0	1.4 ± 0.0	1.9 ± 0.0	1.8 ± 01	1.0 ± 0.0	
98021	2.1 ± 0.0	2.1 ± 0.0	1.0 ± 0.0	1.8 ± 0.0	1.7 ± 0.0	1.0 ± 0.0	0.07
98033	3.6 ± 0.0	2.4 ± 0.0	1.5 ± 0.0	1.9 ± 0.0	1.8 ± 0.0	1.0 ± 0.0	1.92
98048	1.9 ± 0.0	1.6 ± 0.0	1.2 ± 0.0	1.9 ± 0.0	1.9 ± 0.0	1.0 ± 0.0	0.00

Values are arithmetic means and SE is the standard error on the mean. The mount for scs3 was missing and D_{par} for the external detector could not be measured.
D_{II}, apatite D_{par} for samples prepared with Method II.
D_I, apatite D_{par} for samples prepared with Method I (variable etch procedure).
D_{IIm}, external detector D_{par} for samples etched with Method II.
D_{Im}, external detector D_{par} for samples etched with Method I.
Cl wt% measured with electron microprobe, as described in Murrell (2003).
Errors have been rounded and are approximately 3%; values less than 0.05 are rounded to 0.0.

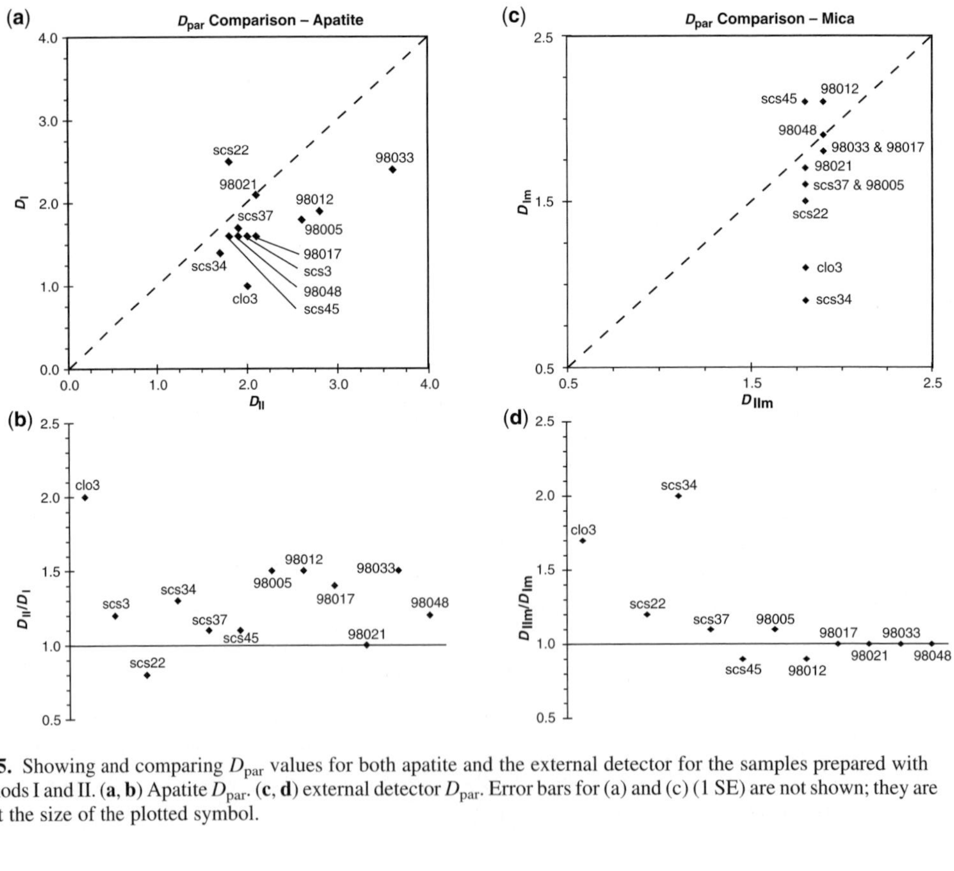

Fig. 5. Showing and comparing D_{par} values for both apatite and the external detector for the samples prepared with Methods I and II. (**a**, **b**) Apatite D_{par}. (**c**, **d**) external detector D_{par}. Error bars for (a) and (c) (1 SE) are not shown; they are about the size of the plotted symbol.

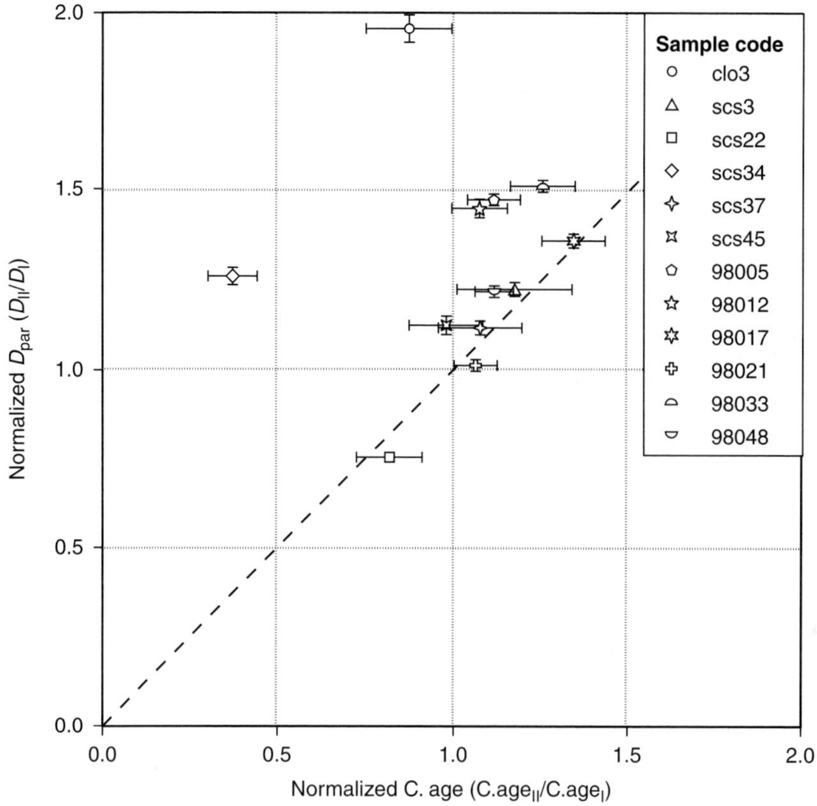

Fig. 6. Plot of the relationship between the degree of etching and the relative difference in Central age for each etch procedure.

variation, between approximately 0.9 and 2.1 μm. In contrast, the external detectors of the samples prepared with Method II were etched collectively, thus ensuring that all samples were etched for the same time. As a result, D_{par} for these samples' external detector etch pits show a considerably more consistent degree of etching, with D_{par} clustering around 1.8 μm. Normalizing the mica etch pit D_{par} between the two data sets (mica D_{IIm}/mica D_{Im}) (Fig. 5d) shows that the external detectors of samples clo3 and scs34 had been etched for greatly differing amounts. Furthermore, the nature of this difference is directly correlated (although in an inverse sense, with the correlation observed between degree of etching and measured age) with the comparison of the ages; samples with underetched external detectors producing older AFTT ages.

It is clear that the amount of etching that a sample experiences influences the AFTT age obtained. This effect not only applies to etching of the apatite but also of the mica external detector. Early studies of how etched track geometry quantitatively affects observable track parameters and, in turn, track density present some possible explanations for why this occurs (see Fleischer *et al.* 1975; Wagner & Van den Haute 1992). These results are in agreement with previous studies suggesting that the effect of prolonged etching is to increase etch efficiency, which is a measure of the number of tracks revealed (actually measured) with respect to the number of tracks that intersect the surface (Fleischer *et al.* 1975, p. 63). In particular these authors observe that: 'longer etches reveal more pits for three main reasons: 1) very small features that were not originally perceptible are enlarged; 2) new tracks that began and ended beneath the original surface are encountered as material is removed; and 3) etching along a track having a low dip angle and damage that increases downward reveals tracks part way along their lengths'.

The first mechanism is essentially anthropogenic. An individual's ability to perceive and count a track that is very small is entirely personal, and is dependent on experience and one's individual criteria and skill. It follows that as a track is enlarged that an operator will be more likely to count it and include it in their measurement (Wagner & Van den Haute 1992). This can be referred to as an observation factor and is incorporated into the development of individual ζ values, which are applied in the derivation of AFTT ages.

The second mechanism refers to the bulk etch rate (V_s) of the apatite, which is very low with respect to preferential etching along the track (V_t). This process has been shown to contribute to a slow increase in track density, which cannot be neglected. This observation has led to the definition of an optimum degree of etching (Gleadow & Lovering 1977; Wagner & Van den Haute 1992) and reflects the earlier conclusion of Fleischer *et al.* (1975, p. 64) that 'use of fixed etching conditions and time is advisable so that the thickness ... of removed material is standardised. Otherwise the total number of pits will increase monotonically'.

The third mechanism refers to the case where V_t varies along the track, but, as normally $V_t \gg V_s$ in minerals, the effect is largely overwhelmed by the influence of the first two mechanisms.

Thus, the degree of etching has an influence on track density and, consequently, sample age. The influence of mechanism 2 had been accounted for by etching to 'optimum' levels. Method I was based on this concept. Method II sacrifices this principle in an attempt to extract compositional information for a more accurate interpretation of annealing kinetics.

The first mechanism was essentially removed as a result of the adaptation of the ζ age determination method. However, the derivation of an individuals personal ζ value is based on the repetitive analysis of age standards, which have been prepared in the same manner as the samples to be analysed (Hurford & Green 1983; Wagner & Van den Haute 1992). This raises a problem because the quantitative etch method produces samples with varying degrees of etching while each age standard only reflects one degree of etching. It therefore follows that for the quantitative etch method to produce *raw* AFTT data, which can be compared with other externally derived AFTT data sets, a series of ζ values for each standard are required, each corresponding to a different degree of etching.

It can, therefore, be concluded that samples prepared with the two methods can produce significantly different primary AFTT data and that the cause of this difference is the corresponding variations in the degree of etching that occurred. Together, this implies that direct comparison between data sets must be carried out very carefully and with great care taken to remove any potential for procedural variations to enter the analysis (i.e. apatite and external detector etching and ζ considerations).

Conclusions

The results presented here demonstrate that different fission-track methodologies involving different etching, length measurement and, ultimately, modelling procedures can lead to important differences in the final apatite fission-track age and model results. Utilization of the recent multikinetic annealing model of Ketcham *et al.* (1999) provides the opportunity to eliminate many of these issues; however, in order to apply it correctly (or at least with confidence) a strong understanding of it, its development and its core data set, as well as calibration against the procedures applied in the derivation of that data set, is essential. It is apparent that Method II coupled with the annealing model of Ketcham *et al.* (1999) is an advance on the variable etching method (Method I). Furthermore, the data set presented herein has been obtained with the primary objective of characterizing annealing behaviour with respect to measurable crystallographic or chemical parameters while removing as much potential for systematic error as possible. In this study, Method II produces D_{par} values within 1.1% of those stated by Carlson *et al.* (1999). The effect of this on initial track length is discussed and shown to be negligible.

A case study comparing samples from Finland and Spain shows that, in the derivation of the primary AFTT data, the relative amounts of etching that have occurred (both of the apatite and the external detector) play a large part in accounting for any differences present. This observation implies that raw AFTT data comparison can only be made if a series of ζ values corresponding to variable degrees of etching of the age standards is available.

This exercise provides a quantitative example of the influence of sample preparation methodology on fission-track data and thermal modelling results. While it is, perhaps, not surprising that such an influence exists, it has so far not been well documented and quantitatively constrained. Indeed, methodological practices in thermochronology evolve over time in response to published studies. The goal of this study is to present a laboratory exercise that can be used as an example for future fission-track studies to be conducted with confidence and accuracy; the advantage being improved modelling and, therefore, better agreement between thermal models and corresponding geological observations.

The authors would like to thank K. de Bruijne for supplying splits of her samples and R. Donelick for providing splits of his original calibration apatites. R. Ketcham, J. Barbarand and F. Lisker provided constructive reviews. NWO and the Vrije Universtieit Amsterdam provided financial support.

References

BURTNER, R. L., NIGRINI, A. & DONELICK, R. A. 1994. Thermochronology of Lower Cretaceous source rocks in the Idaho–Wyoming thrust belt. *AAPG Bulletin*, **78**(10), 1613–1636.

CARLSON, W. D., DONELICK, R. A. & KETCHAM, R. A. 1999. Variability of apatite fission-track annealing kinetics: I. Experimental results. *American Mineralogist*, **84**(9), 1213–1223.

DE BRUIJNE, C. H. 2001. *Denudation, Intraplate Tectonics and Far Field Effects*. PhD thesis, Vrije Universiteit.

DONELICK, R. A. 1993. A method of fission track analysis utilizing bulk chemical etching of apatite. Patent No. 5 267 274.

DONELICK, R. A., KETCHAM, R. A. & CARLSON, W. D. 1999. Variability of apatite fission-track annealing kinetics: II. Crystallographic orientation effects. *American Mineralogist*, **84**(9), 1224–1234.

FLEISCHER, R. L., PRICE, P. B. & WALKER, R. M. 1975. *Nuclear Tracks in Solids: Principles and Applications*. University of California Press, Berkeley, CA.

GALLAGHER, K., BROWN, R. & JOHNSON, C. 1998. Fission track analysis and its applications to geological problems. *Annual Reviews in Earth and Planetary Sciences*, **26**, 519–572.

GLEADOW, A. J. W. & LOVERING, J. F. 1977. Geometry factor for external detectors in fission track dating. *Nuclear Track Detection*, **1**, 99–106.

GREEN, P. F., DUDDY, I. R., GLEADOW, A. J. W., TINGATE, R. R. & LASLETT, G. M. 1986. Thermal annealing of fission tracks in apatite. 1. A qualitative description. *Chemical Geology, Isotope Geoscience Section*, **59**, 237–253.

HONESS, A. P. 1927. *The Nature, Origin And Interpretation of the Etch Figures on Crystals*. Wiley, Chichester.

HURFORD, A. J. & GREEN, P. F. 1983. The zeta age calibration of fission-track dating. *Chemical Geology*, **41**(4), 285–317.

KETCHAM, R. A. 2005. Forward and inverse modeling of low-temperature thermochronometry data. *Mineralogical Society of America, Reviews in Mineralogy and Geochemistry*, **58**, 275–314.

KETCHAM, R. A., DONELICK, R. A. & CARLSON, W. D. 1999. Variability of apatite fission-track annealing kinetics: III. Extrapolation to geological time scales. *American Mineralogist*, **84**(9), 1235–1255.

KETCHAM, R. A., DONELICK, R. A. & DONELICK, M. B. 2000. AFTSolve: a program for multi-kinetic modeling of apatite fission-track data. *Geological Materials Research*, **2**(1), 1–32.

LASLETT, G. M. & GALBRAITH, R. F. 1996. Statistical modelling of thermal annealing of fission tracks in apatite. *Geochimica et Cosmochimica Acta*, **60**(24), 5117–5131.

LASLETT, G. M., GREEN, P. F., DUDDY, I. R. & GLEADOW, A. J. W. 1987. Thermal annealing of fission tracks in apatite, 2: a quantitative analysis. *Chemical Geology, Isotope Geoscience Section*, **65**, 1–13.

MURRELL, G. R. 2003. *The long-term thermal evolution of Central Fennoscandia*. PhD thesis, Vrije Universiteit.

WAGNER, G. & VAN DER HAUTE, P. 1992. Fission-track Dating. Solid Earth Sciences Library, **6**. Kluwer, Dordrecht.

Convective heat transfer in a steeply dipping fault zone and its impact on the interpretation of fission-track data – a modelling study

ZOLTAN TIMAR-GENG[1], ANDREAS HENK[1]* & ANDREAS WETZEL[2]

[1]*Geologisches Institut, Albert-Ludwigs-Universität Freiburg, Albertstrasse 23b, D-79104 Freiburg, Germany*

[2]*Geologisch-Paläontologisches Institut, Universität Basel, Bernoullistrasse 32, CH-4056 Basel, Switzerland*

**Corresponding author (e-mail: henk@uni-freiburg.de)*

Abstract: The effects of convective heat transfer by hydrothermal fluid flow on fission-track (FT) thermochronology are studied using numerical modelling techniques. Parameter studies are carried out on two-dimensional crustal segments with a steeply dipping fault zone exposed to constant denudation to evaluate the relative importance of different variables, including denudation rate as well as hydraulic and material properties. Time–temperature histories of particle points are calculated in the vicinity and also a few kilometres away of the fault zone. These time–temperature paths are then used in a forward-modelling approach to determine the expected FT cooling ages and track-length distributions.

Modelling results indicate that hydrothermal fluid flow can significantly disturb the background conductive thermal state of the upper crust, and the interpretation of FT data using a steady-state geothermal gradient can result in erroneous denudation rates that overestimate the true erosion rates by more than 80%. A pattern of highly varied FT cooling ages from samples at the same elevation does not necessarily ask for differential tectonic movements, instead it can be generated by deep circulation of groundwater within a few million years (Ma). Denudation rates inferred from FT cooling age–elevation plots are likewise inaccurate in a hydrothermally active area because the important assumption about closure temperature isotherms being horizontal or at a constant depth below the surface is not met.

Fission-track (FT) thermochronology is a tool routinely used for studies of surface denudation because of its sensitivity to the low temperatures found in the uppermost part of the crust. FT ages and associated track-length distributions are regularly interpreted in terms of erosion rates assuming a steady-state temperature field and conductive heat transfer only. However, such a simplified interpretation of thermochronological studies may lead to invalid conclusions if the temperatures at a certain depth had actually varied with time. For example, the convective transfer of heat by hydrothermal fluid flow can cause transient thermal events within the upper crust. In particular, fluid circulation along fault zones can result in substantial temperature anomalies in the adjacent rocks (Zuther & Brockamp 1988; Fleming *et al.* 1998; Lampe & Person 2000; Bächler *et al.* 2003). Hot springs frequently associated with faults are the obvious surface manifestations of these deeply reaching hydrothermal circulation patterns. Topographically driven groundwater flow can also cause a significant disturbance of the upper-crustal temperature field in mountainous terrains (Forster & Smith 1989). In particular, the existence of high-permeability fault zones acting as preferential pathways for fluid flow enable the development of geothermal systems and associated thermal anomalies in the vicinity of the fault zones (Fleming *et al.* 1998; Ehlers & Chapman 1999).

The main objective of this study is to assess quantitatively to what extent hydrothermal fluid flow and convective heat transport can alter the upper-crustal temperature field and how these processes modify the cooling ages and track-length distributions observed in apatite FT data. The numerical simulations utilize finite-element techniques using the subsurface flow and transport simulation software FEFLOW® (Diersch 2002). The models are two-dimensional and describe a cross-section through the upper crust in the vicinity of a major, steeply dipping fault zone. Parameter studies covering a wide range of denudation rates, hydraulic potential differences as well as hydraulic and thermal material properties are carried out in order to gain a thorough understanding of the main

From: LISKER, F., VENTURA, B. & GLASMACHER, U. A. (eds) *Thermochronological Methods: From Palaeotemperature Constraints to Landscape Evolution Models*. Geological Society, London, Special Publications, **324**, 87–98.
DOI: 10.1144/SP324.7 0305-8719/09/$15.00

factors controlling convective heat transport. For each parameter study a corresponding set of FT parameters are produced, thus providing a catalogue of FT ages and track-length distributions that will help to interpret real data sets.

Modelling approach

Several numerical studies of coupled fluid flow and heat transfer in regional-scale flow systems examine the impact of fluid flow on thermal regimes and characterize some of the parameters that affect the system's behaviour (Forster & Smith 1989; López & Smith 1995, 1996; Wisian & Blackwell 2004).

The numerical simulation used in this study describes fluid-flow and heat-transport processes as coupled phenomena (e.g. Smith & Chapman 1983; Clauser & Villinger 1990; Diersch 2002). Two-dimensional fluid flow through a porous medium is described by:

$$S_s \frac{\partial h_0}{\partial t} = \frac{\partial}{\partial x}\left(K_x \frac{\partial h_0}{\partial x}\right) + \frac{\partial}{\partial z}\left(K_z \frac{\partial h_0}{\partial z} + \rho_r(T)\right) \quad (1)$$

where h_0 is hydraulic potential, t is time, x is horizontal distance, z is the elevation above a reference depth, S_s is the specific storage coefficient and $\rho_r(T)$ is the temperature-dependent density contrast of the fluid with respect to a reference density. K_x and K_z are the hydraulic conductivities in the horizontal and vertical directions defined as:

$$K_x = \frac{\rho(T)g}{\mu(T)} k_x \quad (2)$$

and

$$K_z = \frac{\rho(T)g}{\mu(T)} k_z \quad (3)$$

where k_x and k_z are the permeabilities in the horizontal and vertical directions, g is the gravitational acceleration, and $\rho(T)$ and $\mu(T)$ are the temperature-dependent fluid density and viscosity.

Time-dependent heat transport in two dimensions with its components conduction, convection and production is described by:

$$c\rho(T)\frac{\partial T}{\partial t} = \lambda_x \frac{\partial^2 T}{\partial x^2} + \lambda_z \frac{\partial^2 T}{\partial z^2} - c\rho(T)\, v_x \frac{\partial T}{\partial x} - c\rho(T)\, v_z \frac{\partial T}{\partial z} + H \quad (4)$$

where T is temperature, c is specific heat capacity, λ_x and λ_z are the thermal conductivities in the horizontal and vertical directions, and H is radiogenic heat production. The Darcy velocities in the horizontal and vertical directions, v_x and v_z are defined as:

$$v_x = -K_x \frac{\partial h_0}{\partial x} \quad (5)$$

and

$$v_z = -K_z \left(\frac{\partial h_0}{\partial z} + \rho_r(T)\right). \quad (6)$$

These equations describing fluid flow and heat transfer are solved using finite-element techniques and the software package FEFLOW®. Radiogenic heat production and the coupling of temperature and viscosity are incorporated, as well as the coefficient of thermal expansion and compressibility, which allow the fluid density to be reproduced over a wide temperature and pressure range (Magri 2004).

The geometric model for coupled fluid flow and heat transfer used in this study is shown in Figure 1. The modelled cross-section covers an area of 20 km width and 10 km depth, representing a section through the upper part of the crust. The finite-element mesh is composed of 62 699 nodes and 124 604 three-noded triangular elements. Boundary conditions assigned to the vertical sides of the thermal model do not allow lateral heat transport. A constant surface temperature and a constant heat flow are assigned to the top and base of the model, respectively. For fluid-flow modelling no horizontal flow is allowed through the sides of the model. Similarly, a no-flow boundary is assigned to the base of the model. Constant head-boundary conditions are applied to the nodes at the top of the fluid-flow model. In order to initiate fluid flow and hydrothermal circulation along the fault zone a hydraulic head difference, $h_m - h_f$, is assumed between the model boundaries and the fault zone. The lower part of the model is characterized by a low hydraulic conductivity, C_2, and a low porosity, P_2, in order to represent a domain of strongly reduced fluid flow and mainly conductive heat transfer. Groundwater circulation predominantly takes place in the upper part of the model with one or two orders of magnitude higher values of hydraulic conductivity, C_1, and porosity, P_1. The hydraulic properties assigned to these two layers represent values typically observed in crystalline basement rocks (Stober 1995). The fault zone is represented by a narrow (100 m) zone where tectonic activity has resulted in an enhanced hydraulic conductivity, C_f, and porosity, P_f. The 'fault zone' is tectonically not active; it serves solely as a high permeable unit enabling enhanced fluid flow and heat convection.

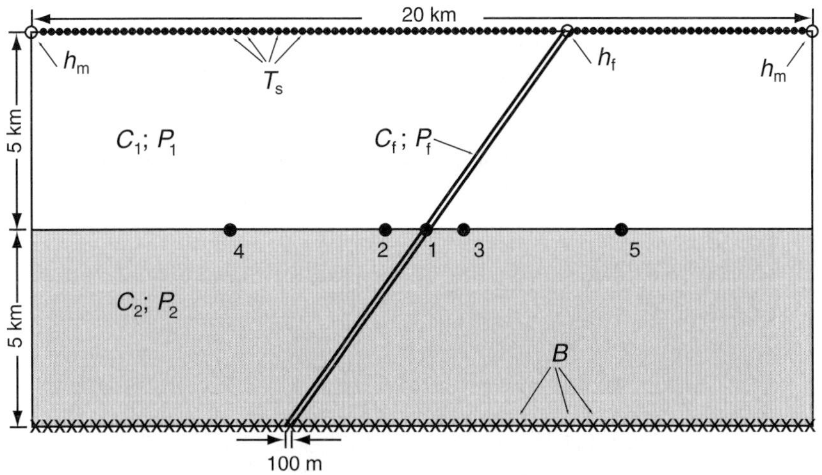

Fig. 1. Model geometry and base configuration for two-dimensional numerical models of convective heat transfer in an upper-crustal segment with a steeply dipping fault zone. T_s represents the surface temperature as a boundary condition; h_m and h_f, the hydraulic head at the upper margins of the model and at the top of the fault zone; B, the basal heat flux; C and P, different values for hydraulic conductivity and porosity; 1–5, observation points for tracking the time–temperature histories of the respective particle points.

The numerical simulation starts with a steady-state flow and heat-transport model based on the geometry and boundary conditions described above. This initial model is then subject to erosion, and the transient effects on fluid flow and heat transfer are studied. Denudation is simulated by a downward displacement of the surface temperature boundary condition, T_s, moving from the top into the model (Fig. 2). Simultaneously, the higher-permeable upper part of the model (with the material parameters C_1, P_1 and C_f, P_f) is also shifted downwards, thus providing a constant thickness of the generated geothermal system throughout the model run. Similarly, the constant head-boundary conditions are moved downwards and the hydraulic conductivities of the 'eroded' section are changed to several orders of magnitude lower values, so that hydrothermal fluid flow is confined to the model section below.

The evolving temperature field is monitored with the help of predefined observation points (1–5 in Figs 1 and 2). All points are located at the same elevation and experience the same exhumation (decompression) history. The time–temperature histories of these particle points are tracked as denudation moves them closer to the surface until they finally reach Earth's surface. The resulting time–temperature paths are then used in a forward-modelling approach to determine the expected FT age and length distributions (see below). Thus, these predefined observation points are comparable to the samples collected at Earth's surface and used for FT analysis.

Fission-track modelling

Roughly two decades ago the quantitative modelling of a rock's thermal history based upon the thermal annealing behaviour of fission tracks in apatite was introduced (Green *et al.* 1986, 1989; Laslett *et al.* 1987; Duddy *et al.* 1988). Since then, the technique has been applied in a large number of thermochronological studies (e.g. Tingate & Duddy 2002; Spiegel *et al.* 2004; Senglaub *et al.* 2005; Timar-Geng *et al.* 2006). These studies utilize some theoretical annealing model that predicts how the FT system develops as a function of time and temperature in the inverse sense: a set of possible thermal histories is determined and statistically evaluated to be consistent with measured FT data. Thus, inverse model results do not provide a simple time–temperature path, but a range of reliable thermal histories. In contrast, the forward-modelling approach, that is the prediction how the FT system will evolve at a given time–temperature history, provides a unique theoretically expected FT dataset (see also Stüwe *et al.* 1994).

The distribution of FT lengths observed in a sample preserve an integrated thermal history, since fission tracks form continuously over time and subsequently anneal during their residence below the total annealing temperature (Ketcham 2005). Several annealing models for apatite have been developed to characterize FT annealing as a function of time and temperature (e.g. Laslett *et al.* 1987; Carlson 1990; Crowley *et al.* 1991; Laslett & Galbraith 1996; Ketcham *et al.* 1999).

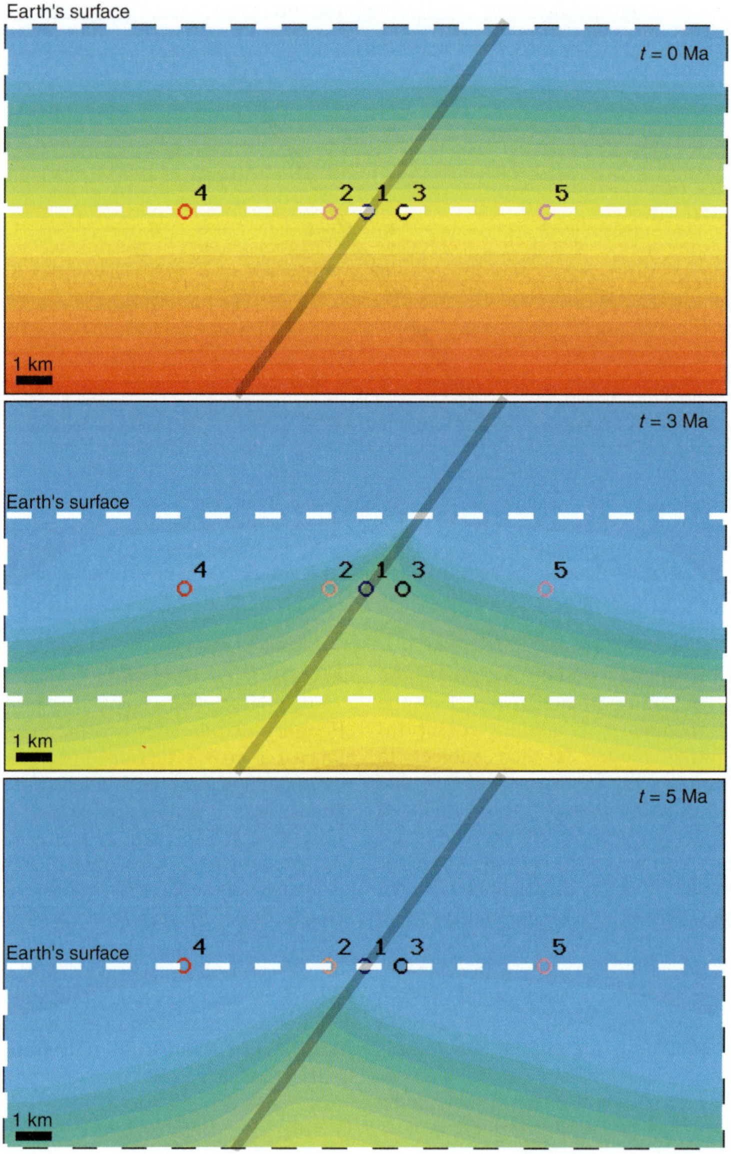

Fig. 2. The starting point ($t = 0$ Ma), an intermediate step ($t = 3$ Ma) and the final stage ($t = 5$ Ma) of a transient model run illustrating the evolution of the temperature field in a geothermal system with ascending fluids along a fault zone, involving surface denudation. The bottom temperature at the beginning of the model run is 400 °C; surface temperature is 10 °C.

They can produce a FT age and length distribution, and require only a time–temperature path as input. A number of computer programs have been developed that implement such numerical FT annealing models, one of which being AFTSolve® (Ketcham *et al.* 2000) that utilizes the annealing model of Laslett *et al.* (1987). The time–temperature paths derived from the coupled fluid-flow and heat-transport modelling described earlier are used in a forward-modelling sense as input for the AFTSolve® software (Ketcham *et al.* 2000).

Results

Reference model

The reference model involves the cooling of rocks, which are exhumed at a constant erosion rate of

Table 1. *Model parameters and boundary conditions for the reference model*

Model parameter/boundary condition	Value
Denudation rate (mm year^{-1})	1
Flow material parameters:	
Conductivity, C_1 (m s^{-1})	5×10^{-9}
Conductivity in the fault zone, C_f (m s^{-1})	1×10^{-6}
Heat material parameters:	
Porosity, P_1	0.01
Porosity in the fault zone, P_f	0.1
Capacity fluid (10^6 J/m^3/K)	4.2
Capacity solid (10^6 J/m^3/K)	2.52
Conductivity fluid (J/m/s/K)	0.65
Conductivity solid (J/m/s/K)	2.8
Radiogenic heat production (μW m^{-3})	4
Boundary conditions:	
Hydraulic head difference, $h_m - h_f$ (m)	1000
Temperature at the surface, T_s (°C)	10
Basal heat flux, B (mW m^{-2})	85

1 mm year^{-1}. The model covers 5 Ma, that is the timespan it takes for the rocks to pass through the partial annealing zone (PAZ; the temperature range in which fission tracks gradually shorten but do not disappear completely: Wagner 1979) and reach the surface. Table 1 lists the model parameters and boundary conditions used for the reference model.

The comparatively high erosion rates result in transient temperature fields. All time–temperature paths monitored by the observation points (Fig. 3) have a similar shape. However, the particle points in the vicinity of the fault zone cool significantly later. By the time the fault-zone sample enters the PAZ, the samples 5 km further away (observation points 4 and 5) are already approximately 30 °C cooler. In spite of the different timing when the samples enter the PAZ, the style of the cooling history, that is relatively rapid cooling through the PAZ over a period of 1–2 Ma, is similar for all samples. This gives rise to very similar expected FT length distributions (Fig. 3) about a mean value of slightly below 14 μm. This also holds for all subsequent model runs.

Despite the same depth of interest and the same denudation history of the observed particle points, the modelled FT ages exhibit significant differences between the samples in the vicinity of the fault zone and those further away (Fig. 4). The sample directly from the fault zone has a modelled FT age of 2.1 Ma, whereas the most distant sample from the hanging wall is significantly older (2.9 Ma). Although the observation points are equidistant on both sides of the fault zone, the modelled FT ages are slightly younger in the footwall than in the hanging wall. This is the consequence of the fault-zone geometry (dip angle) and the resulting asymmetric temperature field around it (see also Fig. 2).

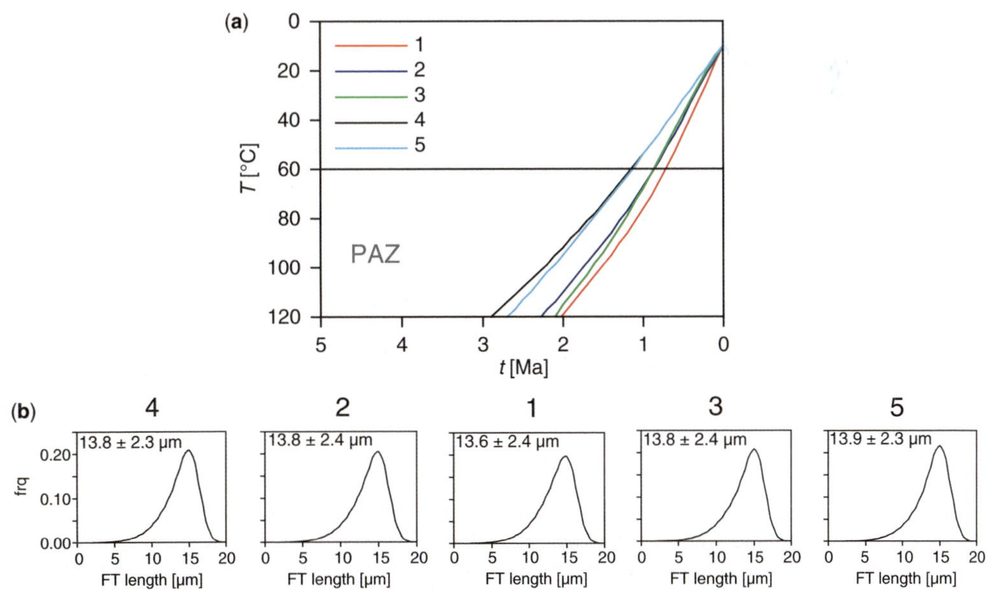

Fig. 3. (**a**) Time–temperature paths of the observation points (1–5) for the reference model; (**b**) modelled FT length distributions for the observation points (1–5) in the reference model.

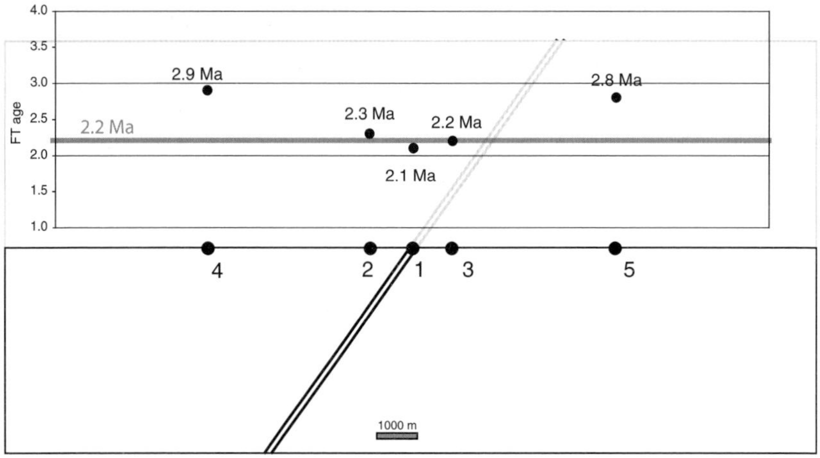

Fig. 4. Modelled FT ages for the observation points (1–5) in the reference model.

Parameter studies

In this section we analyse how parameters different from those in the reference model influence heat transfer and the evolving FT system. These parameter studies include denudation rate, fault-zone conductivity, hydraulic head difference and basal heat flux.

Denudation rate. In this parameter study the temperature effects arising from lower denudation rates than in the reference model (0.5 and 0.1 mm year^{-1}, Fig. 5) are investigated. Significantly longer exhumation periods of 10 Ma (0.5 mm year^{-1}) and 50 Ma (0.1 mm year^{-1}) are the direct consequence of decreasing denudation rates, which allow for approaching a quasi-steady state temperature field during erosion. As a result the trend of the time–temperature paths of particle points are very similar, and particularly for the lowest denudation rate nearly linear in shape (Fig. 6). Thus, the observed differences in the modelled FT ages for the reference model are merely amplified in absolute terms for lower denudation rates (longer exhumation periods), but the relative differences are somewhat lowered. The distribution of the FT ages with decreasing denudation rates predominantly reflects the geometry of the temperature field, which was perturbed but only slightly affected by transient effects with respect to depth throughout the simulations.

Fault-zone conductivity. An increase of the fault-zone conductivity up to one order of magnitude has a great impact on fluid convection within the fault zone, resulting in an enhanced convective heat transfer and strongly modified time–temperature paths for individual particle points in comparison to the reference model (Fig. 5). The increased convective heat transfer causes a larger temperature contrast between the observation points in the vicinity of the fault zone and those positioned further away. In the case of increased fault-zone conductivity particle points 4 and 5 cool earlier into the PAZ, whereas particle point 1 from within the fault zone reaches this temperature interval significantly later than in the reference model. The trends of the time–temperature paths are also considerably different from each other, reflecting a more dynamic evolution of the temperature field during the model runs (Fig. 6). The strong temperature disturbance within the modelled geothermal system is also visible in the resulting large spread of FT ages ranging from 1.9 Ma in the fault zone to 3.7 Ma 5 km further away (Fig. 5). It is interesting to note that there is no linear relationship between the fault-zone conductivity (or the conductivity contrast between the fault zone and outside of the fault zone) and the FT ages of the particle points. There is obviously a threshold conductivity value, above which the convective heat transfer does not significantly change (see the minor differences between Fig. 6a and b in spite of the additional increase in the fault-zone conductivity).

Hydraulic head difference. Doubling the imposed hydraulic difference from 1000 m to 2000 m has only a minor effect on the evolving temperature field, as is indicated in the corresponding time-temperature paths (Figs 3 and 5). It generates a slight increase in the temperature contrast between the observation points, giving rise to a minor enlargement of the overall timespan in which all observation points enter the PAZ. Reduction of the hydraulic head difference from 1000 to 500 m has the opposite effect, again with only very minor

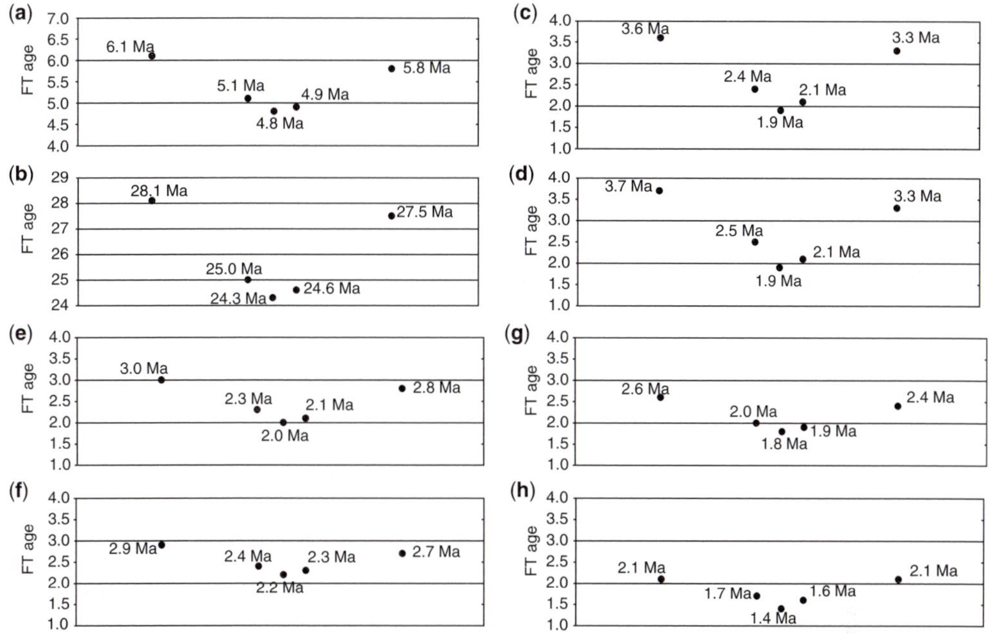

Fig. 5. Modelled FT ages against horizontal distance. For different denudation rates: (**a**) 0.5 mm year^{-1}; (**b**) 0.1 mm year^{-1}. For different fault-zone conductivities: (**c**) 5×10^{-6} m^{-1}; (**d**) 1×10^{-5} m s^{-1}. For different hydraulic head differences between the top margins of the model and the fault zone: (**e**) 2000 m; (**f**) 500 m. For different heat flux values: (**g**) 100 mW m^{-2}; (**h**) 120 mW m^{-2}.

changes in the convective heat transfer (Fig. 5). The modelled FT ages reflect these minor changes with an age distribution very similar to the reference model.

Basal heat flux. The increase in the basal heat flux has a direct impact on the time–temperature paths by significantly raising the conductive component of the heat transfer within the geothermal system. The consequence is that all observation points enter the PAZ progressively later with increasing heat flux values (Fig. 5). The temperature contrasts during the simulations remain roughly the same as in the reference model. Thus, with increasing basal heat flux values all modelled FT ages tend to become progressively younger.

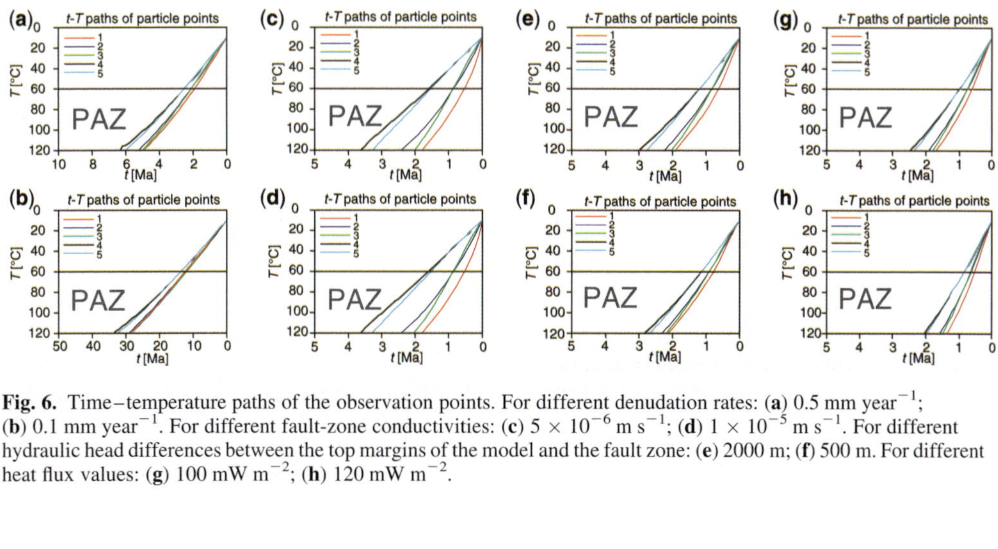

Fig. 6. Time–temperature paths of the observation points. For different denudation rates: (**a**) 0.5 mm year^{-1}; (**b**) 0.1 mm year^{-1}. For different fault-zone conductivities: (**c**) 5×10^{-6} m s^{-1}; (**d**) 1×10^{-5} m s^{-1}. For different hydraulic head differences between the top margins of the model and the fault zone: (**e**) 2000 m; (**f**) 500 m. For different heat flux values: (**g**) 100 mW m^{-2}; (**h**) 120 mW m^{-2}.

Discussion

The various parameter studies were carried out as a sensitivity analysis to assess the effect of different variables on the system behaviour. The sensitivity of the system to each of the input parameters varies to some degree. Among the material parameter variations, the increase in the fault-zone conductivity (relative to a given host-rock conductivity) had the largest impact on the modelled FT data. The system behaviour cannot be treated, however, without the inclusion of the bulk hydraulic conductivity of the host rock. Indeed, several numerical studies (Forster & Smith 1989; Wisian & Blackwell 2004) found that bulk permeability of the host rock is the key factor that influences fluid flow. In addition, fault-zone-related geothermal systems can only exist in a relatively narrow range of host-rock (bulk) permeability ($10^{-15}-10^{-16}$ m^2: Wisian & Blackwell 2004). Outside of this permeability 'window' the system is either dominated by conductive heat transfer without significant fluid flow at low permeabilities or it begins to 'wash-out' at high permeabilities, and fault-zone temperatures decrease rapidly despite increasing fluid flow (Wisian & Blackwell 2004).

The selected denudation rate of 1 mm year^{-1} for the reference model is rather high and does not represent the most common geological scenario. The reason for our choice is twofold. First, the generated FT cooling age distribution for low denudation rates (e.g. Fig. 5b) may be difficult to verify because of the commonly large relative errors of measured FT ages. Even in the case of younger FT ages, as expected from the reference model, lack of spontaneous tracks and low U content could result in very large relative errors (in excess of 10%) and cause difficulties by the verification of the modelling results in a potential case study. The second reason is that geothermal systems are relatively short-lived geological features that make a simulation period of 5 Ma more reasonable than the longer periods needed for exhumation of the particle points at lower denudation rates. An alternative and reasonable modelling approach could be to exhume an extinct palaeogeothermal system after a few million years of activity at low denudation rate.

The results of the modelling study are of particular importance for low-temperature thermochronological studies that involve rapid exhumation and fault-controlled fluid flow, e.g. rift shoulder uplift. The simplification commonly made in FT studies assuming a steady-state temperature field and only conductive heat transfer may often be adequate, especially in the absence of preferential pathways for rapid, ascending fluid motion like high-permeability fault zones. However, particularly in high-relief mountainous terrains, hydraulic head differences can trigger substantial fluid-circulation systems, and positive heat-flow disturbances may reach up to 1.5–1.8 times the basal heat-flow value (Bodri & Rybach 1998). The significant relief may enhance the vertical component of groundwater flow and cause a major convective disturbance of the temperature field (Forster & Smith 1989). Because of the widespread existence of palaeo- and recent thermal anomalies in extensional geological settings, like the Basin and Range region of the western United States or the European Cenozoic rift system (e.g. Wisian & Blackwell 2004; Timar-Geng *et al.* 2004, 2006), neglecting the convective part of the regional heat flow in FT studies would inevitably result in erroneous interpretations. For example, the basement rocks of the southern Upper Rhine Graben area (the central segment of the European Cenozoic rift system) experienced an ongoing hydrothermal activity throughout the Mesozoic and mainly in the early Tertiary related to the main rifting stage (e.g. Werner & Franzke 2001; Wetzed *et al.* 2003). Thus, statements about the denudation of the rift flanks from FT data can only be made if reliable estimates about the palaeotemperature field are available (Timar-Geng *et al.* 2006).

There is one common insight about modelling studies in general and it also applies for our models: the simulation results are only as good as the model itself. Every model has severe limitations and the results are still only estimates. The most important limitations in our models are the lack of lateral heat transport, relief and the third dimension, as well as the explicit specification of the position of the water table by a constant hydraulic head difference between the top margin of the model and the fault zone. For more realistic simulation results it is desirable to incorporate some kind of landscape evolution model, which also influences the geometry of the temperature field. A fixed water-table position means that the model must recharge as much water as needed to balance the fault discharge, which may yield unrealistic values for groundwater recharge. Despite these limitations, the modelling results clearly show the severe impact of the convective heat transfer on the FT system and the well-defined trends in the distribution of FT cooling ages around a geothermal system.

Implications for the interpretation of AFT data

Modelling results clearly illustrate that convective heat transfer by fluid flow can significantly disturb the background conductive thermal state of the crust. As the observed FT age distribution is a direct consequence of the spatially and temporally changing temperatures in the upper crust, any estimate for the amount and rate of denudation from FT studies requires a thorough knowledge of the palaeogeothermal gradients. If vertical sampling profiles are available, they can be estimated directly from the observed FT data. Otherwise, they have to be assumed. Many geological processes, such as magmatism, fault motion as well as fluid flow, contribute to the significant natural variability in crustal thermal gradients. Thus, denudation estimates based on assumed *constant* palaegeothermal gradients could lead to erroneous results.

The possible magnitude of such errors can be quantified, as shown in the presented modelling study for a specific geometry. Assuming that the calculated time–temperature histories from this study (for example, Fig. 3) were best-fit inverse-modelling results from a measured regional FT dataset, one could infer from this cooling history and a *postulated* temporally constant palaeogeothermal gradient of 30 °C km^{-1} that particle point 4 experienced a denudation rate of 1.26 mm year^{-1} and particle point 1 a denudation rate of 1.83 mm year^{-1} (denudation rate = cooling rate/geothermal gradient). However, the erosion rate for all points was actually 1 mm year^{-1}. Thus, calculation of the denudation rate based on such an assumed geothermal gradient would have lead to an error in excess of 80% and a significant overestimation of the amount of eroded section. It also has to be noted that at higher denudation rates (≥0.2 mm year^{-1}) the exhumation process in itself compresses the isotherms and increases the geothermal gradients significantly (e.g. Ehlers *et al.* 2001). Thus, convective heat transfer by fluid flow further magnifies this effect in the vicinity of the fault zone.

In addition, the assumption of a regionally constant geothermal gradient for the calculated FT age distribution would indicate differential denudation and require a tectonic interpretation with an active fault somewhere between the two samples. The magnitude of vertical displacement along this required fault would amount to up to 1 km (Fig. 7). Consequently, neglecting the significance of changing geothermal gradients and applying a pure tectonic interpretation to the FT data can lead to erroneous conclusions and significant mistakes in estimates for denudation rates.

Conversely, differences in FT cooling ages from samples collected at the same elevation of a region do not necessarily imply fault-block tectonics and the related differential tectonic movements. A basin and range type extensional geothermal system with deep circulation of groundwater is sufficient to generate this pattern of varying FT cooling ages.

Another frequently used approach to obtain information about denudation rate is to plot the FT cooling age against elevation. Under certain assumptions the slope of the best-fit line through the data equals the denudation rate (Fig. 8). The thermal models associated with the interpretation of FT data from age–elevation plots incorporate the important hypothesis that all samples pass through the closure temperature at the same elevation or at a constant depth below the surface (if closure isotherms mimic the topography) (Ehlers 2005). Samples collected in a vertical profile, for example from a borehole, would have passed through the closure temperature at the same depth. This approach can be tested with the help of our models by including additional observation points in a vertical array and calculating the corresponding expected FT cooling ages (Fig. 8). The denudation rates inferred from the age–elevation plots show a much better agreement with the true denudation rate than the values deduced from any single sample dataset (Fig. 7). However, this method would also suggest differential denudation between the two borehole locations and result in an underestimation of the true denudation rate for particle points 4–4d. It is obvious that in an active geothermal system with considerable convective heat transfer the depth to the closure

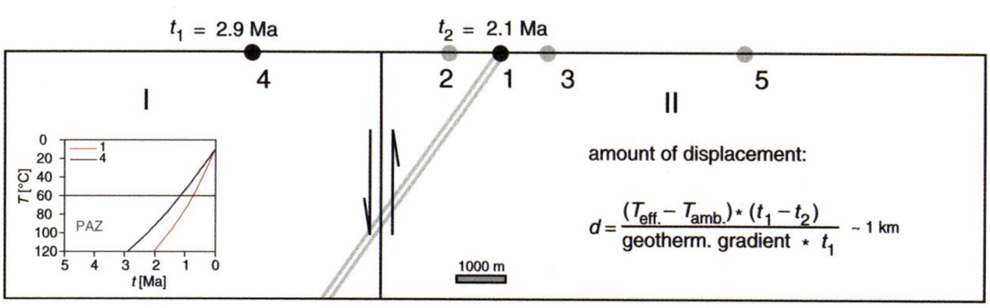

Fig. 7. Possible tectonic interpretation of two modelled FT cooling ages of the reference model (see also Fig. 4) assuming a constant geothermal gradient (30 °C km^{-1}) and a constant uplift rate for block I. $T_{eff.}$ is the effective retention temperature (120 °C), $T_{amb.}$ is the ambient surface temperature (10 °C) and d is the amount of vertical displacement on the postulated fault. After Wagner & Van den Haute (1992).

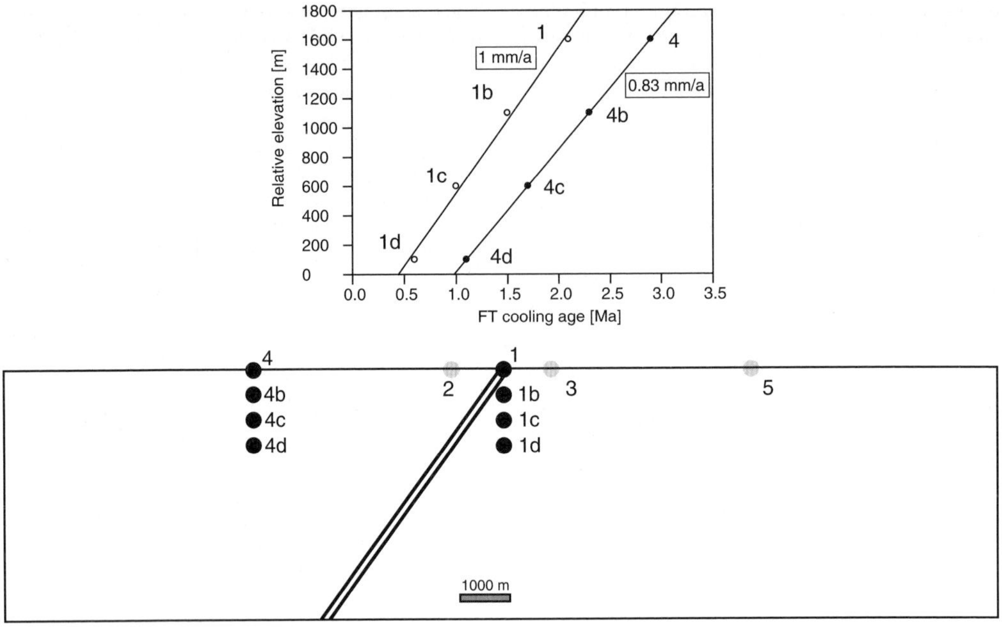

Fig. 8. FT age–elevation plots from two vertical profiles (4–4d and 1–1d) and the inferred denudation rates assuming a constant depth of the closure temperature isotherms below the surface. The model configuration is the same as in the reference model.

isotherm (i.e. the geothermal gradient) is temporally and spatially significantly variable. In such a geological setting more sophisticated numerical modelling techniques are required to assess transient and spatial perturbations to the crustal temperature field, which influence the cooling histories of rock samples.

Conclusions

In this study we quantify the impact of convective heat transfer by fluid flow on the interpretation of apatite FT data using finite-element numerical techniques on cross-sectional models of a geothermal system exposed to denudation. Modelling results show that groundwater flow can significantly disturb the thermal regime of the upper crust.

In a series of parameter studies the relative importance of different variables, including denudation rate as well as hydraulic and material properties, has been evaluated. All model configurations resulted in a significant distribution of FT cooling ages despite the uniform denudation history. For a given host-rock hydraulic conductivity, an increase in the fault-zone conductivity has the greatest effect on the distribution of the FT cooling ages.

It has been shown that denudation rates inferred from FT thermochronology are dependent on spatial and temporal variations in crustal palaeotemperature fields, and the interpretation of FT data using a steady-state geothermal gradient can result in erroneous denudation rates with an overestimation of more than 80%. Significant variations in FT cooling ages around an exhumed geothermal system in samples from the same elevation can be generated by the deep circulation of groundwater within a few million years. Differential tectonic movements are not necessarily needed. Denudation rates derived from FT cooling age–elevation plots are likewise imprecise in a region dominated by deep groundwater circulation because the important hypothesis about closure temperature isotherms being horizontal or at a constant depth below the surface is not satisfied.

This work has been supported by the Deutsche Forschungsgemeinschaft. We gratefully acknowledge the critical reading of the manuscript and the constructive comments by K. Stüwe and an anonymous reviewer.

References

BÄCHLER, D., KOHL, T. & RYBACH, L. 2003. Impact of graben-parallel faults on hydrothermal convection; Rhine Graben case study. *Physics and Chemistry of the Earth*, **28**, 431–441.

BODRI, B. & RYBACH, L. 1998. Influence of topographically driven convection on heat flow in the Swiss Alps: a model study. *Tectonophysics*, **291**, 19–27.

CARLSON, W. D. 1990. Mechanisms and kinetics of apatite fission-track annealing. *American Mineralogist*, **75**, 1120–1139.

CLAUSER, C. & VILLINGER, H. 1990. Analysis of conductive and convective heat transfer in a sedimentary basin, demonstrated for the Rheingraben. *Geophysical Journal International*, **100**, 393–414.

CROWLEY, K. D., CAMERON, M. & SCHAEFER, R. L. 1991. Experimental studies of annealing etched fission tracks in fluorapatite. *Geochimica et Cosmochimica Acta*, **55**, 1449–1465.

DIERSCH, H.-J. G. 2002. *WASY Software FEFLOW®. Reference Manual.* WASY GmbH, Berlin.

DUDDY, I. R., GREEN, P. F. & LASLETT, G. M. 1988. Thermal annealing of fission tracks in apatite 3. Variable temperature behaviour. *Chemical Geology*, **73**, 25–38.

EHLERS, T. A. 2005. Crustal thermal processes and the interpretation of thermochronometer data. *Reviews in Mineralogy and Geochemistry*, **58**, 315–350.

EHLERS, T. A. & CHAPMAN, D. S. 1999. Normal fault thermal regimes: conductive and hydrothermal heat transfer surrounding the Wasatch fault, Utah. *Tectonophysics*, **312**, 217–234.

EHLERS, T. A., ARMSTRONG, P. A. & CHAPMAN, D. S. 2001. Normal fault thermal regimes and the interpretation of low-temperature thermochronometers. *Physics of the Earth and Planetary Interiors*, **126**, 179–194.

FLEMING, C. G., COUPLES, G. D. & HAZELDINE, R. S. 1998. Thermal effects of fluid flow in steep fault zones. *In*: JONES, G., FISHER, Q. J. & KNIPE, R. J. (eds) *Faulting, Fault Sealing and Fluid Flow in Hydrocarbon Reservoirs.* Geological Society, London, Special Publications, **147**, 217–229.

FORSTER, C. & SMITH, L. 1989. The influence of groundwater flow on thermal regimes in mountaineous terrain: a model study. *Journal of Geophysical Research*, **94**, (B7), 9439–9451.

GREEN, P. F., DUDDY, I. R., GLEADOW, A. J. W., TINGATE, P. R. & LASLETT, G. M. 1986. Thermal annealing of fission tracks in apatite 1. A qualitative description. *Chemical Geology*, **59**, 237–253.

GREEN, P. F., DUDDY, I. R., LASLETT, G. M., HEGARTY, K. A., GLEADOW, A. J. W. & LOVERING, J. F. 1989. Thermal annealing of fission tracks in apatite 4. Quantitative modeling techniques and extension to geological time scales. *Chemical Geology*, **79**, 155–182.

KETCHAM, R. A. 2005. Forward and inverse modeling of low-temperature thermochronometry data. *Reviews in Mineralogy and Geochemistry*, **58**, 275–314.

KETCHAM, R. A., DONELICK, R. A. & CARLSON, W. D. 1999. Variability of apatite fission-track annealing kinetics III: extrapolation to geological time scales. *American Mineralogist*, **84**, 1235–1255.

KETCHAM, R. A., DONELICK, R. A. & DONELICK, M. B. 2000. AFTSolve: a program for multi-kinetic modelling of apatite fission track data. *Geological Materials Research*, **2**, 1–18.

LAMPE, C. & PERSON, M. 2000. Episodic hydrothermal fluid flow in the Upper Rhinegraben (Germany). *Journal of Geochemical Exploration*, **69/70**, 37–40.

LASLETT, G. M. & GALBRAITH, R. F. 1996. Statistical modelling of thermal annealing of fission tracks in apatite. *Geochimica et Cosmochimica Acta*, **60**, 5117–5131.

LASLETT, G. M., GREEN, P. F., DUDDY, I. R. & GLEADOW, A. J. W. 1987. Thermal annealing of fission tracks in apatite 2. A quantitative analysis. *Chemical Geology*, **65**, 1–13.

LÓPEZ, D. L. & SMITH, L. 1995. Fluid flow in fault zones: Analysis of the interplay of convective circulation and topographically driven groundwater flow. *Water Resources Research*, **31**(6), 1489–1503.

LÓPEZ, D. L. & SMITH, L. 1996. Fluid flow in fault zones: Influence of hydraulic anisotropy and heterogeneity on the fluid flow and heat transfer regime. *Water Resources Research*, **32**(10), 3227–3235.

MAGRI, F. 2004. Derivation of the coefficients of thermal expansion and compressibility for use in FEFLOW. *In*: DIERSCH, H.-J. G. (ed.) *WASY Software FEFLOW, White Papers III*, WASY GmbH, Berlin, 13–23.

SENGLAUB, Y., BRIX, M. R., ADRIASOLA, A. C. & LITTKE, R. 2005. New information on the thermal history of the southwestern Lower Saxony Basin, northern Germany, based on fission track analysis. *International Journal of Earth Sciences*, **94**, 876–896.

SMITH, L. & CHAPMAN, D. S. 1983. On the thermal effects of groundwater flow – Regional scale systems. *Journal of Geophysical Research*, **88**, 593–608.

SPIEGEL, C., SACHSENHOFER, R. F., PRIVALOV, V. A., ZHYKALYAK, M. V. & PANOVA, E. A. 2004. Thermotectonic evolution of the Ukrainian Donbas Foldbelt: evidence from zircon and apatite fission track data. *Tectonophysics*, **383**, 193–215.

STÜWE, K., WHITE, L. & BROWN, R. 1994. The influence of eroding topography on steady-state isotherms; application to fission track analysis. *Earth and Planetary Science Letters*, **124**, 63–74.

STOBER, I. 1995. *Die Wasserführung des kristallinen Grundgebirges.* Enke, Stuttgart.

TIMAR-GENG, Z., FÜGENSCHUH, B., SCHALTEGGER, U. & WETZEL, A. 2004. The impact of the Jurassic hydrothermal activity on zircon fission track data from the southern Upper Rhine Graben area. *Schweizerische Mineralogische und Petrographische Mitterlungen*, **84**, 257–269.

TIMAR-GENG, Z., FÜGENSCHUH, B., WETZEL, A. & DRESMANN, H. 2006. Low-temperature thermochronology of the flanks of the southern Upper Rhine Graben. *International Journal of Earth Sciences*, **95**, 685–702.

TINGATE, P. R. & DUDDY, I. R. 2002. The thermal history of the eastern Officer Basin (South Australia): evidence from apatite fission track analysis and organic maturity data. *Tectonophysics*, **349**, 251–275.

WAGNER, G. A. 1979. Correction and interpretation of fission track ages. *In*: JÄGER, E. & HUNZIKER, J. C. (eds) *Lectures in Isotope Geology.* Springer, Berlin, 170–177.

WAGNER, G. A. & VAN DEN HAUTE, P. 1992. *Fission-track Dating*. Enke, Stuttgart.

WERNER, W. & FRANZKE, H. J. 2001. Postvariszische bis neogene Bruchtektonik und Mineralisation im südlichen Zentralschwarzwald. *Zeitschrift der Deutschen Geologischen Gesellschaft*, **152**, 405–437.

WETZEL, A., ALLENBACH, R. & ALLIA, V. 2003. Reactivated basement structures affecting the sedimentary facies in a tectonically 'quiescent' epicontinental basin: an example from NW Switzerland. *Sedimentary Geology*, **157**, 153–172.

WISIAN, K. W. & BLACKWELL, D. D. 2004. Numerical modeling of Basin and Range geothermal systems. *Geothermics*, **33**, 713–741.

ZUTHER, M. & BROCKAMP, O. 1988. The fossil geothermal system of the Baden-Baden trough (Northern Black Forest, Germany). *Chemical Geology*, **71**, 337–353.

Reconstruction of palaeotopography from low-temperature thermochronological data

WANG WEI* & ZHOU ZUYI

State Key Laboratory of Marine Geology, Tongji University, Shanghai, 200092, China

**Corresponding author (e-mail: ww200211@gmail.com)*

Abstract: Thermochronology can provide information on palaeotopography and its evolution. We present a new method to reconstruct the shape of the palaeotopography from the palaeoisotherm derived from low-temperature thermochronological data. While deriving palaeoisotherm and reconstructing palaeotopography, the exhumation rate may also be constrained. The proposed method is independent of the relationship between the ancient and modern topography; however, it requires that the palaeotopography maintains its shape and the subsurface thermal field is invariant with time during the period when a set of samples passed through the closure temperature. It is shown that the inherent uncertainties in sample age and the limited sampling density will inevitably induce errors in the reconstructed topography, and these errors can be reduced by eliminating the noise in the derived isotherm. Increasing the number of samples helps to reduce the noise, and if, before sampling, we can make rough estimates about the geothermal gradient, the exhumation rate, the uncertainties in sample age, and the wave components and the relief of the palaeotopography to be reconstructed, the number of samples can be tentatively decided. It is also shown that the reconstructed topography is sensitive to exhumation rate and geothermal gradient, and a concrete sensitivity analysis is needed for a given dataset. By reinterpreting the (U–Th)/He data from the Sierra Nevada, California, we show that our method can reveal palaeorelief as well as other valuable information on the palaeotopography of the studied area.

The Earth's surface and its evolution are among the most important factors that affect the thermal structure of the uppermost crust, to which low-temperature thermochronological systems, such as (U–Th)/He or fission-track dating in apatite and zircon, are most sensitive (e.g. Turcotte & Schubert 1982; Stüwe *et al.* 1994; Mancktelow & Grasemann 1997; Braun 2002*a*, *b*). Although this complicates the interpretation of low-temperature thermochronological data, it enables some of the important attributes of the palaeotopography to be constrained. These attributes may include topographic relief or the form, location and scale of topographic relief in the past, but do not include palaeoelevation because thermochronology is only concerned with vertical movement of rocks towards the Earth's surface, which is a moving boundary relative to sea level, etc. (Reiners 2007).

In the last decade, there have been several studies to constrain palaeotopography using data from low-temperature thermochronology (e.g. House *et al.* 1998, 2001; Braun 2002*a*, *b*; Persano *et al.* 2002; Ehlers *et al.* 2003; Reiners *et al.* 2003; Braun & Roberts 2005). Some of these studies are based on numerical modelling and require a topography evolution model that links the palaeotopography with the present-day topography. For example, the model by Braun (2002*b*) hypothesized that the ancient and the current topography had the same shape but different relief (or unevenness). However, so far no model can offer a detailed description of the long-term topography evolution because many factors that may influence the evolution of topography are difficult to constrain.

Here we present a new method to reconstruct the shape of palaeotopography from the thermochronological data. This method is independent of the relationship between the ancient and the present-day topography, but requires that the palaeotopography maintains its shape and the subsurface thermal field is invariant with time during a period in the past; for example, during the period when a set of samples are passing through the closure temperature.

Reconstructing palaeotopography from palaeoisotherm

Surface topography can be expanded in an infinite Fourier series. In practice, the sum of a finite number of cosine and sine is adequate to approximate surface topography:

$$h(x) \approx \sum_{n=0}^{N} [a_n \cos(2n\pi x/\lambda) + b_n \sin(2n\pi x/\lambda)] \tag{1}$$

From: LISKER, F., VENTURA, B. & GLASMACHER, U. A. (eds) *Thermochronological Methods: From Palaeotemperature Constraints to Landscape Evolution Models*. Geological Society, London, Special Publications, **324**, 99–110.
DOI: 10.1144/SP324.8 0305-8719/09/$15.00

where $h(x)$ is the height of ground surface at the horizontal position x, λ is the length of a given transect, and a_n and b_n are amplitudes of cosine and sine functions whose wavelength are λ/n. Equation (1) can be expressed in another form:

$$h(x) \approx \sum_{n=0}^{N} c_n \cos(2n\pi x/\lambda + \phi_n) \quad (2)$$

where $c_n = \sqrt{[(a_n)^2 + (b_n)^2]}$, c_n and ϕ_n are amplitude and phase, respectively, of the n-order harmonic whose wavelength is λ/n.

Because heat conduction in solids can be approximated as a linear physical process, the total perturbation of the thermal field caused by the topography can be adequately approximated by the sum of the perturbations caused by each of the components of the Fourier series. The perturbation caused by a periodic topography with small amplitude has been shown to be proportional to the amplitude of the topography, and to decay exponentially with depth (Turcotte *et al.* 1982). The steady-state solution for topography of constant geometry can therefore be expressed as (Mancktelow & Grasemann 1997):

$$T(x, y) = T_0(y) + \sum_{n=0}^{N}\left[a_n \cos\left(\frac{2n\pi x}{\lambda}\right) + b_n \sin\left(\frac{2n\pi x}{\lambda}\right)\right] e^{ym_n}\left(\frac{\partial T_0(y)}{\partial y_{y=0}} - \phi\right) \quad (3)$$

$$m_n = \frac{1}{2}\left(-u/\kappa - \sqrt{(u/\kappa)^2 + (4n\pi/\lambda)^2}\right) \quad (4)$$

where u is exhumation rate, κ is thermal diffusivity and ϕ is the lapse rate of surface temperature with altitude ($-4.5\ ^\circ\mathrm{C\ km^{-1}}$ altitude). $T_0(y)$ describes the geotherm in the absence of any surface perturbations, and can be fixed according to the temperature at the surface and at the lower boundary of a slab (see equation 23 of Mancktelow & Grasemann 1997).

During the period when a set of samples are passing through their closure temperature (T_c) – if the palaeotopography maintains its shape and the subsurface thermal field is invariant with time – the isotherm of T_c maintains its shape and depth. Assuming the exhumation rate is constant, the difference in the depth of palaeoisotherm of T_c at the horizontal position of two samples, Sample$_1$ (Age$_1$, Elevation$_1$) and Sample$_2$ (Age$_2$, Elevation$_2$), is:

$$D_h = u \cdot (\mathrm{Age}_1 - \mathrm{Age}_2) - (\mathrm{Elevation}_1 - \mathrm{Elevation}_2). \quad (5)$$

The shape of palaeoisotherm is determined by equation (5) and the average depth of palaeoisotherm (y) can be estimated through the following equation:

$$T_0(y) = T_c. \quad (6)$$

Thus, the depth of palaeoisotherm at the horizontal position of each sample can be obtained, that is for each sample we have:

$$T(x_i, y_i) = T_c \quad (7)$$

where x_i and y_i are the horizontal position and the depth of palaeoisotherm at the position of the i-th sample, respectively. Expanding equation (7), a system of linear equations with unknown values of a_n and b_n ($0 \le n \le N$) as variables is obtained. These unknown values can be fitted by multiple linear regression. Then a_n and b_n ($0 \le n \le N$) can be used in equation (1) and the palaeotopography is reconstructed.

A vital issue is how to select a suitable value of N. Obviously, the greater the value of N, the more detailed the topography to be reconstructed, so we have called λ/N *the precision of topography reconstruction* (PTR). According to the theory of linear regression, the maximal value of N cannot exceed half of the number of samples. In reality, the value of N should be smaller than its maximum owing to data uncertainties.

Uncertainties in the calculated palaeoisotherm and selection of the PTR for a given dataset

A synthetic example

Here, a synthetic example is used to test the validity of the method and to illustrate how uncertainties in the calculated isotherm affect the reconstruction of palaeotopography. In our experiment, we use a 192 km-long topographic profile from a 1 km-resolution DEM (GTOPO30) of Dabie Shan, eastern China.

We first calculated the steady-state thermal field using the finite-difference method and accepted it as the 'real' thermal field under the topography. In this forward model, the parameters were set as: the surface temperature $T_s = 0\ ^\circ\mathrm{C}$, the thickness of the layer being exhumed $L = 100$ km, the constant exhumation rate $u = 50\ \mathrm{m\ Ma^{-1}}$, the heat diffusivity $\kappa = 10^{-6}\ \mathrm{m^2\ s^{-1}}$ and the lapse rate of surface temperature with altitude $\phi = -4.5\ ^\circ\mathrm{C\ km^{-1}}$. Because heat production is not considered in this model, the temperature at the low boundary is set as $T_L = 2500\ ^\circ\mathrm{C}$ to ensure a near-surface geothermal gradient of approximately $25\ ^\circ\mathrm{C\ km^{-1}}$. This basal temperature is clearly much too large, but it

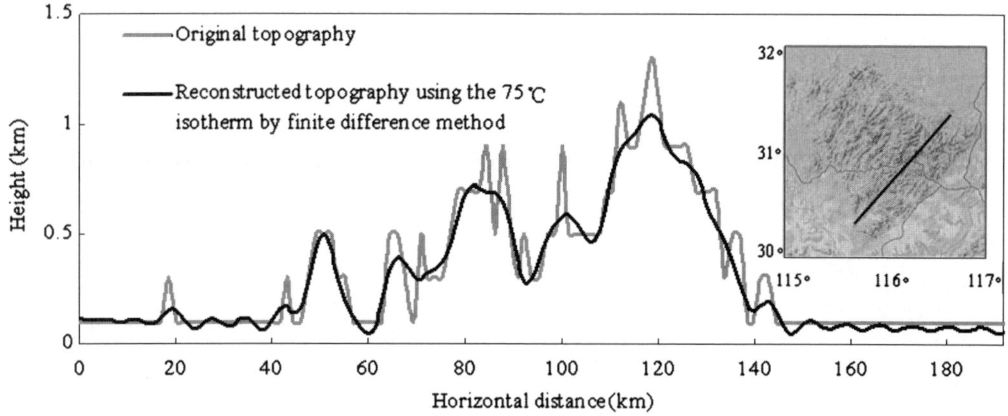

Fig. 1. A synthetic example for testing the validity of our method. The original topography is from Dabie Shan, eastern China, and is reconstructed using our method from the 75 °C isotherm obtained by finite-difference method.

is the selection of the near-surface geothermal gradient, instead of the basal temperature, that affects the result.

The isotherms obtained by the finite-difference method are then sampled at equal intervals to form a dataset (DS1) of 128 data pairs. Each data pair contains the horizontal position and the depth of isotherm at that position. Finally, the topography is reconstructed with this dataset using the method that we have described above, with the value of the PTR set as 6 km.

As shown in Figure 1, taking the 75 °C isotherm as an example, the reconstructed topography is almost identical to the original topography. Therefore, if the isotherm could be accurately derived from the low-temperature thermochronological data, the topography can be accurately reconstructed given that the parameters, such as the exhumation rate and the near-surface geothermal gradient, are known.

However, uncertainties are inherent in analysing the cooling age of rocks. They will lead to uncertainties in the calculated isotherm and errors in the reconstructed topography. According to equation (5), the error of the calculated isotherm at the horizontal position of the i-th sample is:

$$\varepsilon_i = u \cdot \varepsilon_{ai} \tag{8}$$

where ε_{ai} is the error in the i-th sample age. If the error in all sample ages obeys the normal distribution with a mean of zero and one standard error (1 SE) of σ, then ε_i obeys the normal distribution with a mean of zero and a deviation of $u\sigma$.

To illustrate how uncertainties in the calculated isotherm affect the reconstruction of palaeotopography, random error is added to the depth of isotherm of every data pair of DS1 to form a new dataset (DS2). The random errors are generated from a normal distribution with a mean of zero and 1 SE of 150 m to simulate the situation in which 1 SE of sample age is 3 Ma and the exhumation rate is 50 m Ma^{-1}.

Using DS2, the wave components whose wavelengths exceed 8, 32, 48, 96 and 192 km (i.e. $\lambda = 192$ km, and $n = 24, 6, 3, 2$ and 1, respectively, see equation 2) are regressed respectively, and the reconstructed topographies are shown in Figure 2. In order to quantify the 'goodness' of topography reconstruction, average deviations between the reconstructed and the original topography are calculated at every 1.5 km.

As shown in Figure 2, the topography reconstructed with long wave components (Fig. 2a) exhibits a similar overall trend as the original topography, but lacks detail. More details are revealed with decreasing wavelength (Fig. 2b). However, with a further decrease in wavelength, drastic oscillations of short wavelength appear in the reconstructed topography (Fig. 2c). Correspondingly, with a decrease in wavelength, the deviation reduces (Fig. 2b), but this deviation gradually widens with a further decrease in wavelength.

Obviously, owing to uncertainties in sample ages and errors in calculated isotherm, only the long wave component of the topography can be generally reconstructed. However, the problem is how to select an appropriate wavelength and to reconstruct the topography given the dataset and the uncertainties in sample ages.

The PTR for a given dataset

To understand how uncertainties in the calculated isotherm affect the reconstruction of palaeotopography, isotherms obtained by the above two datasets

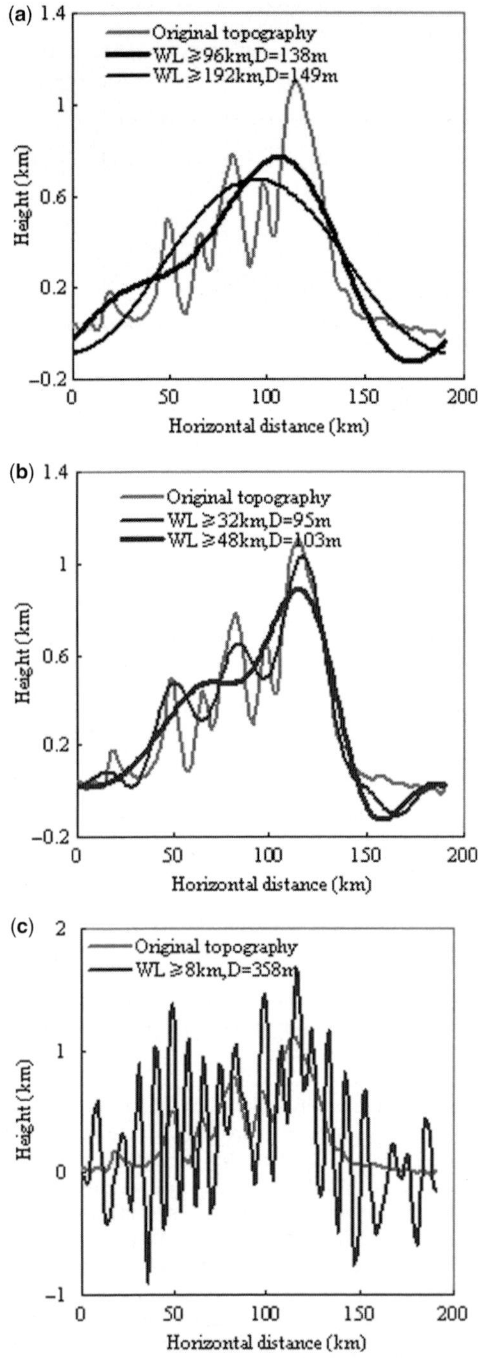

Fig. 2. The wave components whose wavelengths exceed 8, 32, 48, 96 and 192 km are reconstructed for topography in Figure 1, respectively, from the 75 °C isotherm with the sampling interval set at 1.5 km and the 1 SE of sample age at 3 Ma. The topography cannot be accurately reconstructed owing to the presence of errors in the isotherm. WL, wavelength; D, deviations.

(DS1 and DS2) are broken down into different wave components by Fast Fourier Transform (FFT) procedure. The amplitude of each wave components is shown in Figure 3. Where c_{tn} (the actual amplitude) is for DS1 and c_n (the calculated amplitude with errors) is for DS2.

As shown in Figure 3, for wave components whose wavelength are longer than 27.4 km (i.e. $n \leq 6$), the calculated amplitudes are near the actual amplitudes and so the corresponding wave components in the topography are well reconstructed, as shown in Figure 2a and b. For the wave components whose wavelength are shorter than 32 km (i.e. $n > 6$), the actual amplitudes are near zero; however, the corresponding calculated amplitudes are much greater than zero and thus lead to short-wavelength oscillations in the reconstructed topography, as shown in Figure 2c.

In fact, if the actual amplitude is zero, the uncertainties of sample ages may lead to calculated amplitudes much greater than zero. It can be proven that, under this circumstance:

$$\left(\frac{C_n}{u\sigma\sqrt{2/N_s}}\right) \approx \chi^2(2), \text{ while } C_{tn} = 0 \qquad (9)$$

where $\chi^2(2)$ is the Chi-square distribution with a mean of 2. Therefore, if a calculated c_n is small, they may be noise induced completely by uncertainties in the sample ages and limited sampling density. However, it is very difficult to tell whether a value of c_n is small enough to be regarded as noise and should be eliminated from the calculated isotherm while reconstructing palaeotopography.

Here, based on our model experiments and considering the statistic characteristics of the calculated amplitudes, we select a threshold value of:

$$\mathrm{TV} = 3.5u\sigma/N_s. \qquad (10)$$

This threshold value (TV) is used to help us discern whether a wave component of the isotherm is noise or not. In this example the threshold value is 0.046 km, and the amplitudes for wave components of 192, 96, 64, 38.5 and 32 km exceed the threshold value (Fig. 3). Therefore, these wave components of the isotherm can be discerned from the noise, and the corresponding wave components of the topography should be reconstructed.

In practice, we reconstruct all the wave components of the topography whose wavelength exceeds that of the shortest discernible wave component of the isotherm. In this example the calculated amplitude of the 48 km-wave component of the isotherm is less than the threshold value. Nonetheless, for the sake of convenience, the 48 km-wave component of the topography is reconstructed.

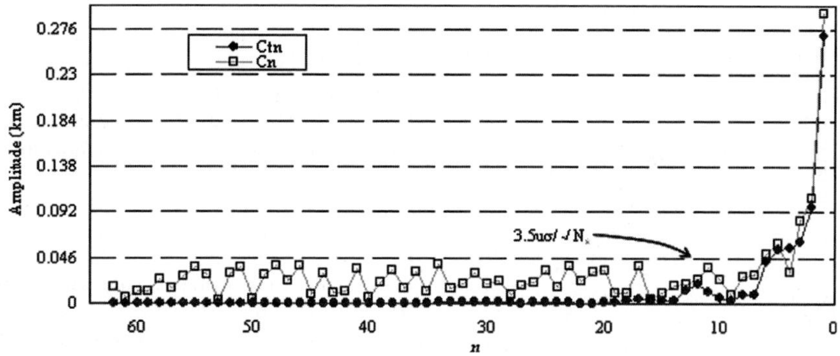

Fig. 3. The amplitudes of each wave component of the 75 °C isotherm for the topography in Figure 1. n is the n-order harmonic whose wavelength is λ/n. c_{tn} (the actual amplitude) is for DS1 and c_n (the calculated amplitude with uncertainties) is for DS2. See the text for a detailed explanation.

Therefore, in our paper, the PTR is defined as the wavelength of the shortest discernible wave component of the isotherm and all wave components whose wavelengths exceed the PTR are reconstructed. In this example, the PTR is 32 km. It should be mentioned that:

- the threshold value is a statistical concept, so the noise may exceed it; in fact, other threshold values may be selected and it is suggested its value should not be larger than $5u\sigma/\sqrt{N_s}$;
- in most of our model experiments the deviation between the original topography and the topography reconstructed using the PTR thus selected is the least.

Some factors that affect the isotherm and the selection of sampling density

Whether a wave component of the palaeotopography can be reconstructed is dependent on two facts:

- the wave component can cause remarkable perturbation on the isotherm;
- this perturbation can be discerned from the noise given the number of the samples and the uncertainties in the sample ages.

These two factors can be linked by the following inequality:

$$C_{tn} > 3.2u\sigma/\sqrt{N_s}. \quad (11)$$

That is to say, if the wave component of the topography could cause an undulation whose amplitude exceeds $3.2u\sigma/\sqrt{N_s}$ in the isotherm, the corresponding calculated amplitude would have a good chance (50%) of exceeding the threshold value of $3.2u\sigma/\sqrt{N_s}$ and to be discerned from the noise. Then what factors affect the magnitude of the perturbation on the isotherm induced by topography?

It has been shown that the perturbation is proportional to the amplitude of the topography. As shown in Figure 4, for a periodic topography, the isotherm follows the shape of the topography but with different amplitude, and the ratio α (Braun 2002*b*) between the amplitude of the topography and that of the isotherm is almost constant, only weakly sensitive to topographic amplitude (Reiners 2007).

However, the amplitude of the topography is just one of the factors that affect the magnitude of the perturbation. Other factors include the geothermal gradient, the wavelength of the topography and the temperature of the isotherm. Figure 5 shows the ratios (α) for the topographies of different wavelengths and the isotherms of different temperatures when the geothermal gradient is 10, 25

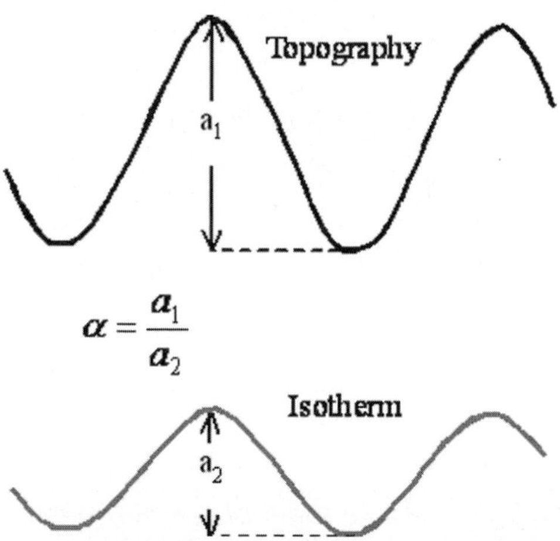

Fig. 4. For a periodic topography, the ratio (α) between the amplitude of the topography and that of the isotherm is almost irrelevant to the amplitude of the topography.

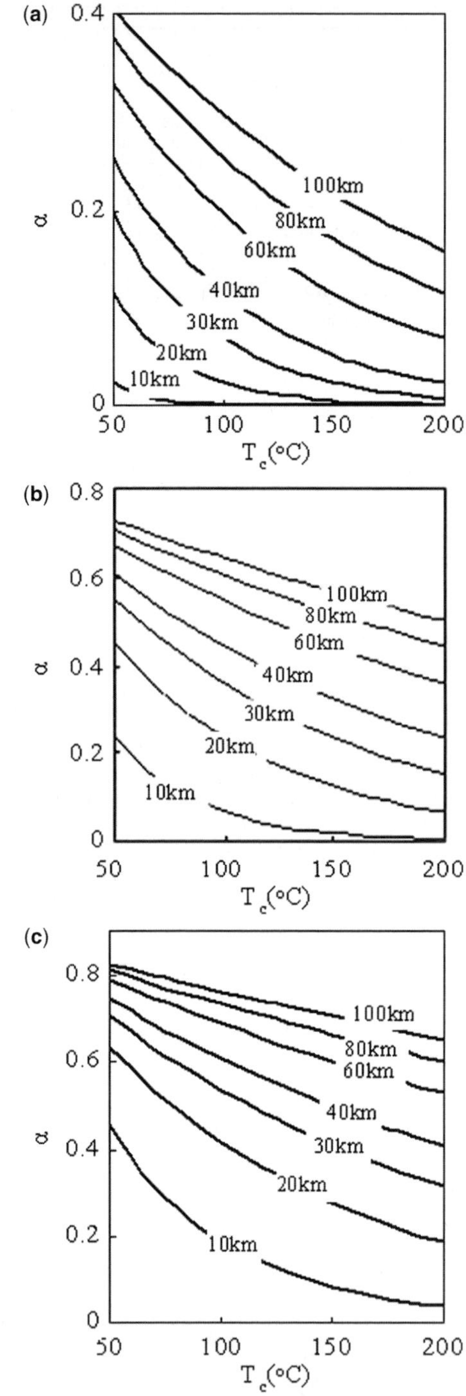

Fig. 5. The α ratios (as defined in Fig. 4) are calculated for periodic topographies of different wavelengths (in km) and the isotherms of different temperatures when the geothermal gradient (dT/dz) is 10, 25 or 40 °C km^{-1}. The heat diffusivity is 10^{-6} m^2 s^{-1} and heat production is not considered.

or 40 °C km^{-1}, where the heat diffusivity is 10^{-6} m^2 s^{-1} and heat production is not considered. It is shown in Figure 5 that, as the wavelength of the topography and the near-surface geothermal gradient decrease, the ratio α decreases. In contrast, the ratio α increases as the temperature of the isotherm decreases.

For topographies of the same amplitude, a smaller α means that the perturbation, induced by the topography, is more difficult to discern from the noise in the calculated isotherm. Nonetheless, the corresponding topography could be reconstructed, if we were able to reduce the noise by increasing the number of samples.

However, it is difficult to decide how many samples are sufficient for our purpose, because such a decision relies on our knowledge of the geothermal gradient, the exhumation rate, and, in particular, the wave components and the relief of the topography to be reconstructed. If we do have some vague knowledge about these factors, then Figure 5 can help us tentatively to decide the number of the samples. In the case shown in Figure 5a, assuming the main wave component of the palaeotopography is 40 km, its amplitude is 1 km and the temperature of the isotherm is 105 °C, then the amplitude of isotherm is approximately 0.13 km. If the 1 SE of the sample age is 3 Ma, then, according to inequality (11),

$$N_s > (3.2u\sigma/C_{tr})^2 = (3.2 \times 3 \times 0.05/0.13)^2 \approx 14.$$

Therefore, 14 samples are needed on a 40 km-long transect. It should be mentioned that inequality (11) is only applicable in the situation where the number of the samples is a multiple of 2 and the samples are sampled at equal intervals.

In practice, the topography contains a series of wave components, and numerical modelling is needed to calculate the 'conjectural' isotherm and break it into different wave components. Then the number of samples can be decided according to the amplitudes of the wave components in the isotherm. In the case of a moderate exhumation rate of 50 m Ma^{-1} and a geothermal gradient of 25 °C km^{-1} and one standard error of 3 Ma or so in apatite (U–Th)/He ages, and assuming the topography in Figure 1 is a rough estimation of the topography to be reconstructed, the wave components of the isotherm are shown in Figure 3. According to the amplitude of each component and inequality (11), four samples are needed for the reconstruction of the 192 km-wave component of the topography, 32 samples are needed for the reconstruction of the 96 km-wave component, 64 samples are needed for the reconstruction of the 38.5, 48 and 64 km-wave components, and 128 samples needed to reconstruct the 32 km-wave component.

Sensitivity to exhumation rate and geothermal gradient

For a given dataset, the sensitivity of reconstructed topography to the exhumation rate is related to the age and height distribution of the samples. According to equation (5), if the samples come from locations of the same height, the isotherm is determined completely by the age distribution of the samples. In such case, the relief of the isotherm and then the reconstructed topography are proportional to the exhumation rate. However, if all the sample ages are identical, the isotherm is determined completely by the height distribution of the samples, and the exhumation rate has no influence on the reconstructed topography.

In practice, a concrete analysis is needed. As shown in Figure 6a, different exhumation rates are used to reconstruct the topography in Figure 1 with DS2, and the reconstructed topography is not sensitive to exhumation rate.

The near-surface geothermal gradient is another parameter that has a vital influence on the reconstructed topography. According to Figure 5, this influence varies with the wavelength of the topography to be reconstructed. Supposing the exhumation rate is 50 m Ma^{-1} and the actual geothermal gradient is 40 °C km^{-1}, an observed amplitude of 0.5 km for the 30 km-wave component of the 100 °C isotherm means a relief of 1 km or so in the topography (see Fig. 4c). However, if we reconstruct the topography with a geothermal gradient of 10 °C km^{-1}, we will get a relief of 8 km or so in the topography (see Fig. 4a).

In practice, limited samples are only sufficient to reconstruct the long wavelength of the topography. Under this circumstance, the reconstructed topography is not as sensitive to geothermal gradient. As shown in Figure 6b, different geothermal gradients were used to reconstruct the topography in Figure 1 with DS2, where the 'real' geothermal gradient is 25 °C km^{-1}.

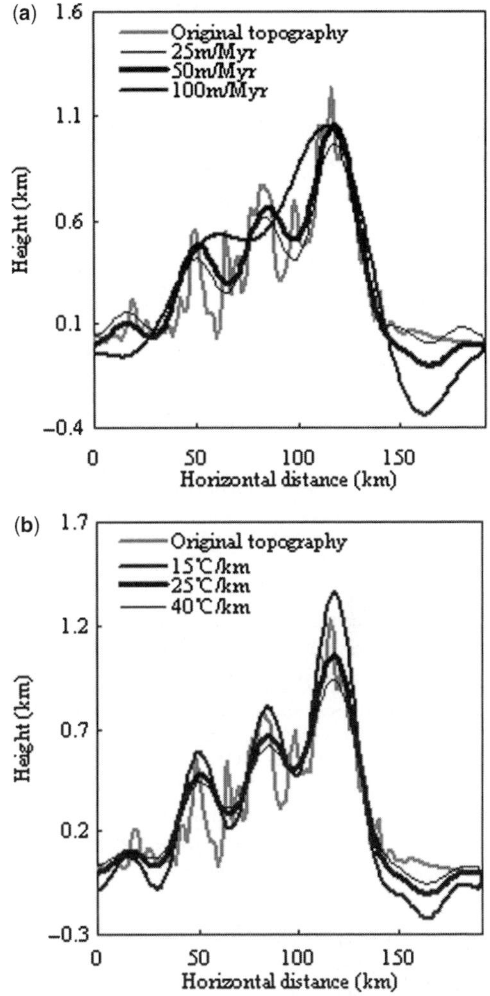

Fig. 6. (**a**) Different exhumations and (**b**) geothermal gradients are used to reconstruct the topography with DS2 to test the parameter sensitivity.

Constraining exhumation rate while reconstructing palaeotopography

Exhumation rate can be constrained by the slope of the 'age–elevation relationship'. This method requires that all the samples have undergone the closure temperature at constant depth below a mean surface. If the samples are not come from a vertical profile, the influence of topography on the isotherm should be taken into account, and a topographic correction is needed (Stüwe *et al.* 1994). However, the palaeotopography is unknown. Here, we propose a new method to constrain the exhumation rate while reconstructing palaeotopography.

According to our model experiments, if the samples do not come from locations of the same height, an exhumation rate other than the real value is *inclined* to boost up the short wave components much more than that of long wave components of the isotherm. This will lead to a 'rougher' calculated isotherm than the real one, as shown in Figure 7, where the actual exhumation rate is 50 m Ma^{-1}.

Therefore, a smoother isotherm is usually preferred. For a given dataset and different exhumation rates (e.g. u_1, u_2), the corresponding PTRs (e.g. PTR_1, PTR_2) can be determined by the threshold value. After eliminating from the calculated isotherms the wave components whose wavelengths

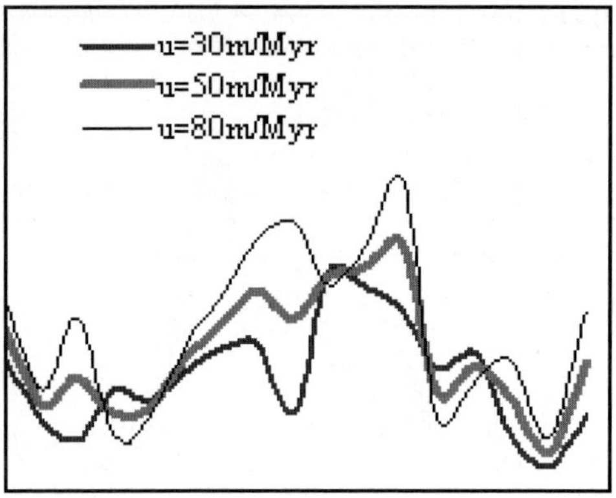

Fig. 7. If the samples do not come from locations of the same height, an exhumation rate (u) other than the real one (50 m Ma^{-1}) is inclined to boost up the amplitude of short wave components much more than that of long wave components of the isotherm, and leads to the calculated isotherm being rougher than the real isotherm.

are shorter than the PTR, using equations (5) and (6), smoothed isotherms are obtained.

- If $PTR_1 > PTR_2$, the smoothed isotherm with u_1 contains less short wave components and is therefore smoother, and so u_1 is preferred.
- If $PTR_1 = PTR_2 = PTR$, reconstructing the topography with u_1 and u_2, respectively, and calculating the corresponding root mean square (RMS) of the difference between the predicted and the observed ages:

$$\mathrm{RMS} = \sqrt{\frac{\sum_{1}^{N} (\mathrm{Age}_{pi} - \mathrm{Age}_{oi})^2}{N_s}}$$

where Age_{pi} and Age_{oi} are the predicted and observed ages for the i-th sample, respectively. Note that RMS is the 'distance' between the calculate isotherm and the smoothed isotherm (not the real isotherm); a small RMS means that the short wave components (<PTR) in the calculated isotherm are less remarkable and the calculated isotherm is smoother, so the corresponding exhumation rate is more preferred.

Here we provide a synthetic example. In this example, the sampling interval is 12 km, exhumation rate is 50 m Ma^{-1}, and 1 SE of sample age is 3 Ma. Through a forward modelling, a synthetic dataset was obtained, as shown in Figure 8a. The isotherm is then calculated using different exhumation rates between 10 and 100 m Ma^{-1}, and the PTR was determined. The PTR is 192 km for exhumation rates of between 40 and 100 m Ma^{-1}, and is shorter than 192 km for exhumation rates of between 10 and

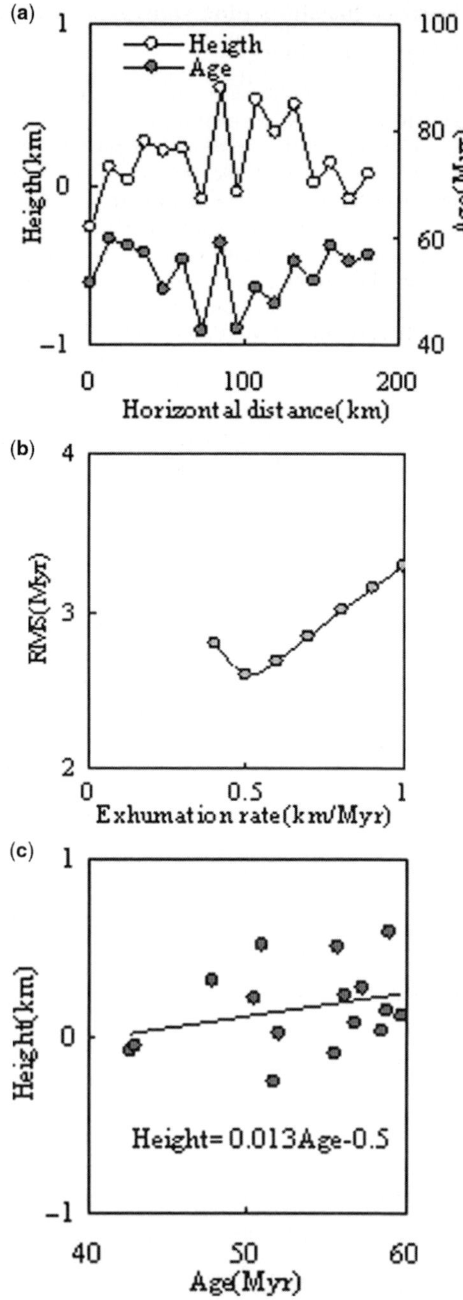

Fig. 8. A synthetic example. (**a**) The age and the height distribution of the samples; (**b**) RMS for different exhumation rates; and (**c**) the age–elevation relationship.

30 m Ma^{-1}, so an exhumation rate larger than 30 m Ma^{-1} is preferred. For different exhumation rates from 40 to 100 m Ma^{-1}, topographies are reconstructed and RMSs are calculated. As shown

in Figure 8b, the RMS is least when the exhumation rate is 50 m Ma^{-1}.

Because the sampling interval is too large and the height difference between samples is too small, it is not suitable to use the slope of the 'age–elevation relationship' to constrain the exhumation rate with this synthetic dataset, as shown in Figure 8c. In fact, under such circumstances, the slope of the 'age–elevation relationship' tends to underestimate the exhumation rate because of the influence of the topography.

The example of the Sierra Nevada, California

One of the most successful attempts to directly constrain the palaeotopography by low-temperature thermochronology was undertaken by House *et al.* (1998, 2001) in the Sierra Nevada of northern California. This region is a relatively simple west-tilted block with no evidence of significant folding or faulting. These features make it an ideal place to study the evolution of topography.

In the study by House *et al.* (1998), 36 samples were collected at the approximately same height of 2 km along a 200 km range-parallel transect. As shown in Figure 9a, the average sampling interval is about 6 km, and most of the mean apatite (U–Th)/He ages range from 50 to 80 Ma. The 1 SE of sample age is approximately 3 Ma. House *et al.* (1998) showed that if the geothermal gradient is close to 20 °C km^{-1} and radioactive heat production is considered for (U–Th)/He samples taken from the same elevation, the age distribution is mainly determined by the relief and less affected by the wavelength of the topography. According to

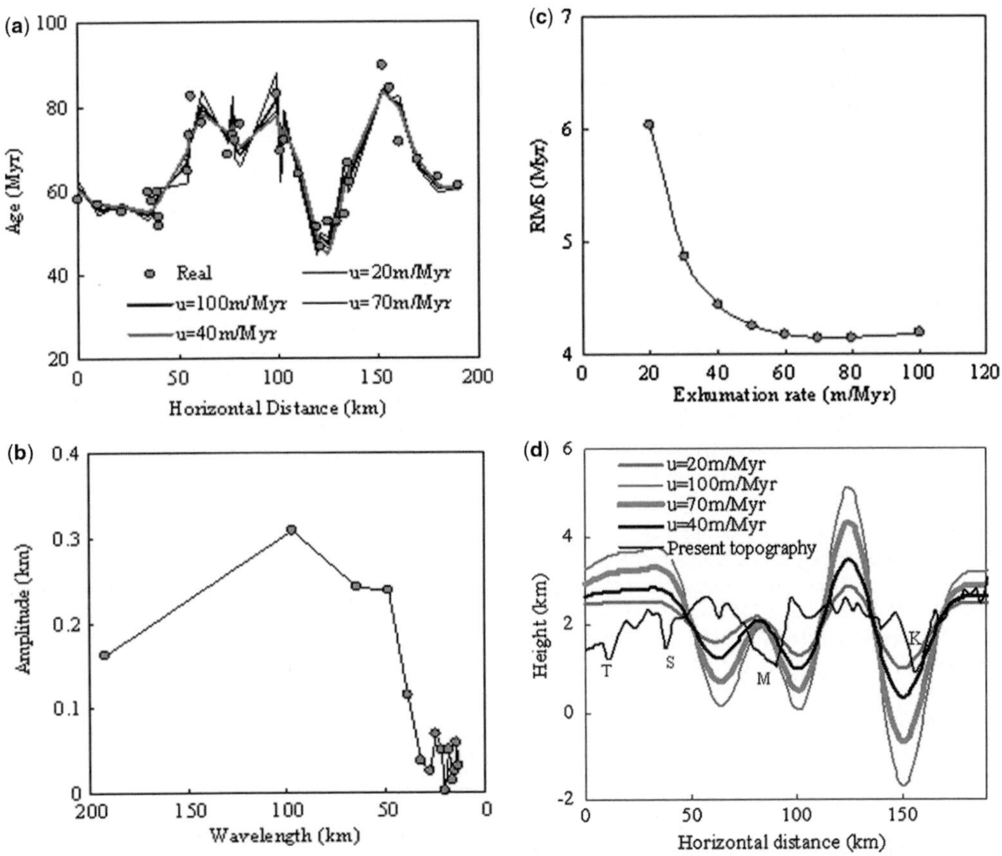

Fig. 9. Application to the Sierra Nevada. (**a**) Apatite (U–Th)/He ages from a transect through the Sierra Nevada (House *et al.* 1998) and the predicted ages for different exhumation rates. (**b**) The wave components of the 75 °C isotherm calculated from the (U–Th)/He age and a exhumation rate of 40 m Ma^{-1}. (**c**) The RMS of the difference between the predicted and the observed ages for different exhumation rates. (**d**) The present-day topography and the reconstructed topographies for different exhumation rates. T, M, S and K are four valleys in the transect.

their calculations, the observed 20–30 Ma variation in (U–Th)/He ages correspond to a long-wavelength relief of 2–4 km.

We reinterpreted the data of House *et al.* (1998) using our method. The parameters were set as: $L = 100$ km, $T_s = 0$ °C, $T_L = 2500$ °C, $\kappa = 10^{-6}$ m^2 s^{-1}, $\phi = -4.5$ °C. Therefore, the near-surface geothermal gradient is 27 °C km^{-1}, a little larger than that suggested by House *et al.* (1997). We assumed that the closure temperature of apatite (U–Th)/He thermochronological system is 75 °C (e.g. Wolf *et al.* 1996).

The exhumation rate in this area is between 40 and 100 m Ma^{-1} (House *et al.* 1998; Braun 2002*a*). We first selected an exhumation rate of 40 m Ma^{-1} and calculated the 75 °C isotherm, as shown in Figure 9b. Under this circumstance, the threshold value is 0.074 km, so the PTR is 38.5 km. The same value of the PTR is also obtained when the exhumation rate is set anywhere between 10 and 100 m Ma^{-1}.

We tried to constrain the exhumation rate while reconstructing the palaeotopography. As shown in Figure 9c, the exhumation rate should not be less than 40 m Ma^{-1}. However, the RMS is not affected significantly by the change in exhumation rate when the rate is greater than 40 m Ma^{-1}. Therefore, the exhumation rate cannot be well constrained with this dataset.

The topography at 50–80 Ma was reconstructed for different exhumation rates, as shown in Figure 9d. If the exhumation rate is 40 m Ma^{-1}, the palaeorelief would be 2–3 km, and an exhumation rate of 100 m Ma^{-1} would give a relief of 4–6 km. Nevertheless, the reconstructed topography has a larger relief compared to the present-day topography. This suggests that the current topography is probably the result of long-term erosion of the mountain belts. In addition to palaeorelief, other valuable information on the palaeotopography of the studied area can be revealed. For example, by comparing the current topography and the reconstructed topography, we note that the ancient mountain ridge between valley S and K is narrower than its current state. This may be related to a lateral migration of the drainage system in this area after 50 Ma.

Conclusions

- In a tectonically stable region, assuming the palaeotopography maintains its shape and the thermal field is invariant with time during the period when a set of samples is passing through the closure temperature, the shape of palaeotopography can be reconstructed from the isotherm derived from the low-temperature thermochronological data. By reinterpreting the (U–Th)/He data from the Sierra Nevada, California, we show that this method can reveal more details, in addition to the relief (i.e. the unevenness), of the palaeotopography.
- Because of the inherent uncertainties in sample age and the limited sampling density, errors are inevitable in the calculated isotherm. If the amplitudes of the wave components of the calculated isotherm are small, they may be noise induced completely by uncertainties in the sample ages and should be eliminated from the calculated isotherm while reconstructing palaeotopography. A threshold value, based on model tests and the statistic characteristics of the calculated amplitudes, is needed to help us discern whether a wave component of the isotherm is noise or not.
- It is difficult to decide how many samples are sufficient for our purpose because such a decision relies on our knowledge of the geothermal gradient, the exhumation rate, the uncertainties in sample ages, and, in particular, the wave components and the relief of the topography to be reconstructed. However, if we do have some knowledge about these factors, then the number of the samples can be tentatively decided.
- The reconstructed topography is sensitive to exhumation rate and geothermal gradient. The sensitivity varies with many factors and cannot be generalized. Therefore, a concrete sensitivity analysis is needed for a given dataset.
- The exhumation rate can be constrained while deriving palaeoisotherm and reconstructing palaeotopography. This method can be used in the situation where the slope of the 'age–elevation relationship' is not applicable to constrain the exhumation rate when the sampling interval is too large and the height difference between samples is too small.
- Owing to the presence of many factors that are hard to constrain, it is difficult to establish models of long-term topography evolution. Our method is independent of the relationship between the ancient and the present-day topography, and can be used to constrain models of long-term topography evolution.

This work has been supported by Chinese National Science Foundation (Programs 40572075 and 40621063). We are grateful to J. Braun and C. Persano for their constructive reviews that greatly improved the manuscript.

Appendix: A statistic analysis for the uncertainties in the calculated isotherm

The depth of the isotherm in the horizontal position of the *i*-th sample can be denoted as:

$$y_i = y_{ti} + \varepsilon_i \quad \text{(A1)}$$

where y_i is the calculated depth from equations (5) and (6), and y_{ti} and ε_i are the true value and the error of y_i, respectively. According to equation (5), the error of the isotherm in the position of the i-th sample is:

$$\varepsilon_i = u\varepsilon_{ai} \tag{A2}$$

where u is the exhumation rate and ε_{ai} is the error in the i-th sample age. Supposing the error in all sample ages obeys the normal distribution with a mean of zero and a deviation of σ^2, and then ε_i obeys the normal distribution with a mean of zero and a deviation of $u^2\sigma^2$.

If the number of the samples is a multiple of 2 and the samples are sampled at equal intervals, then the calculated isotherm can be broken down into a series of cosine and sine functions (see equation 1) by Fast Fourier Transform. The amplitudes of the cosine and sine functions, whose wavelength are λ/n, consist of two elements:

$$a_n = a_{tn} + a_{en} = a_{tn} + \frac{2}{N_s}\sum_{k=0}^{N_s-1} \varepsilon_i \cos\left(\frac{nk\pi}{N_s}\right) \tag{A3}$$

$$b_n = b_{tn} + b_{en} = b_{tn} + \frac{2}{N_s}\sum_{k=0}^{N_s-1} \varepsilon_i \sin\left(\frac{nk\pi}{N_s}\right) \tag{A4}$$

where a_{tn} and b_{tn} are Fourier transformations of y_{ti}, a_{en} and b_{en} are Fourier transformations of ε_i, and N_s is the number of samples. The observed value and the true value of the amplitude of the n-order wave component of the isotherm are:

$$C_n = \sqrt{(a_{tn} + a_{en})^2 + (b_{tn} + b_{en})^2} \tag{A5}$$

$$C_{tn} = \sqrt{a_{tn}^2 + b_{tn}^2} \tag{A6}$$

If errors in different sample ages are independent of each other, it can be proven that a_{en} and b_{en} obey the normal distribution, and:

$$E(a_{en}) = E(b_{en}) = 0 \tag{A7}$$

$$D(a_{en}) = D(b_{en}) = \frac{2u^2\sigma^2}{N_s} \tag{A8}$$

Supposing a_{tn} and b_{tn} are zero, then $C_n^2 = a_{en}^2 + b_{en}^2$. Under this circumstance, the errors in the calculated isotherm result completely from the uncertainties of the sample ages. Because a_{en} and b_{en} obey the normal distribution and are independent of each other, then:

$$\left(\frac{C_n}{u\sigma\sqrt{2/N_s}}\right)^2 \approx \chi^2(2), \quad \text{while } C_{tn} = 0 \tag{A9}$$

where $\chi^2(2)$ is Chi-square distribution with a mean of 2. The conditional probability of $P(C_n < C|C_{tn} = 0)$ can be readily obtained, as shown in Figure 10a. According to Figure 10a, $P(C_n > u\sigma/\sqrt{N_s}|C_{tn} = 0) = 1 - P(C_n < u\sigma/\sqrt{N_s}|C_{tn} = 0) = 0.8$. This is to say, if the actual amplitude is zero, then the calculated amplitude would exceed $u\sigma/\sqrt{N_s}$, with a probability of 80%. Therefore, if a calculated c_n is small (e.g. smaller than $u\sigma/\sqrt{N_s}$) it may be induced completely by uncertainties in the sample ages and should be regarded as noise and eliminated from the calculated isotherm while reconstructing the palaeotopography. However, it is not easy to tell whether a value of c_n is small enough to be regarded as noise.

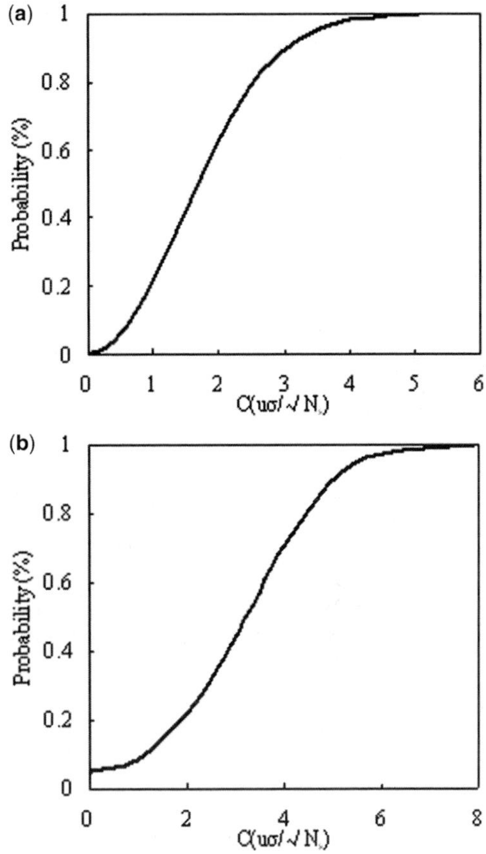

Fig. 10. The probability for (**a**) $C_n < C$ while $C_{tn} = 0$ and (**b**) $C_n > 3.5u\sigma/\sqrt{N_s}$ while $C_{tr} = C$.

Based on our model experiments, and considering $P(C_n < 3.5u\sigma/\sqrt{N_s}|C_{tn} = 0) = 95\%$, we selected a threshold value of $3.5u\sigma/\sqrt{N_s}$ to help us to discern whether a wave component of the isotherm is noise or not. It should be mentioned that the noise may exceed this threshold value and this might cause intolerable errors in the reconstructed topography. In fact, a larger threshold value can be selected to avoid such a situation and it is suggested that its value should not be greater than $5u\sigma/\sqrt{N_s}$.

Then, a related question is how large the actual amplitude of the wave components in the isotherm should be to ensure that the calculated amplitude exceeds the threshold value of $3.5u\sigma/\sqrt{N_s}$. In order to understand this question the probability of $P(C_n > 3.5u\sigma/\sqrt{N_s}|C_{tr} = C)$ was calculated using a Monte Carlo sampling of equation (5). As shown in Figure 10b, if the actual amplitude is $3.2u\sigma/\sqrt{N_s}$ then the calculated amplitude would exceed $3.5u\sigma/\sqrt{N_s}$ with a probability of 50%.

References

BRAUN, J. 2002*a*. Estimating exhumation rate and relief evolution by spectral analysis of age-elevation datasets. *Terra Nova*, **14**, 210–214.

BRAUN, J. 2002*b*. Quantifying the effect of recent relief changes on age–elevation relationships. *Earth and Planetary Science Letters*, **200**, 331–343.

BRAUN, J. & ROBERT, X. 2005. Constraints on the rate of post-orogenic erosional decay from thermochronological data: example from the Dabie Shan, China. *Earth Surface Processes and Landforms*, **30**, 1203–1225.

EHLERS, T., WILLETT, S., ARMSTRONG, P. & CHAPMAN, D. 2003. Exhumation of the central Wasatch Mountains: 2. Thermokinematic models of exhumation, erosion and low-temperature thermochronometer interpretation. *Journal of Geophysical Research*, **108**, 2173, doi:10.1029/2001JB001723.

HOUSE, M., WERNICKE, B. & FARLEY, K. 2001. Paleo-geomorphology of the Sierra Nevada, California, from (U–Th)/He ages in apatite. *American Journal of Science*, **301**, 77–102.

HOUSE, M. A., WERNICKE, B. P. & FARLEY, K. A. 1998. Dating topography of the Sierra Nevada, California, using apatite (U–Th)/He ages. *Nature*, **396**, 66–69.

HOUSE, M. A., WERNICKE, B. P., FARLEY, K. A. & DUMITRU, T. A. 1997. Cenozoic thermal evolution of the central Sierra Nevada, CA from (U–Th)/He thermochronometry. *Earth and Planetary Science Letters*, **151**, 167–179.

MANCKTELOW, N. S. & GRASEMANN, B. 1997. Time-dependent effects of heat advection and topography on cooling histories during erosion. *Tectonophysics*, **270**, 167–195.

PERSANO, C., STUART, F. M., BISHOP, P. & BARFORD, D. N. 2002. Apatite (U–Th)/He age constraints on the development of the Great Escarpment on the southeastern Australian passive margin. *Earth and Planetary Science Letters*, **200**, 79–90.

REINERS, P., ZHOU, Z., EHLERS, T., XU, C., BRANDON, M., DONELICK, R. & NICOLESCU, S. 2003. Post-orogenic evolution of the Dabie Shan, eastern China, from (U–Th)/He and fission-track thermochronology. *American Journal of Science*, **303**, 489–518.

REINERS, P. W. 2007. Thermochronologic approaches to paleotopography. *Reviews in Mineralogy & Geochemistry*, **66**, 243–267.

STÜWE, K., WHITE, L. & BROWN, R. 1994. The influence of eroding topography on steady-state geotherms. Application to fission track analysis. *Earth and Planetary Science Letters*, **124**, 63–74.

TURCOTTE, D. L. & SCHUBERT, G. 1982. *Geodynamics: Applications of Continuum Physics to Geological Problems*, 1st edn. Wiley, New York.

WOLF, R. A., FARLEY, K. A. & SILVER, L. T. 1996. Helium diffusion and low temperature thermochronometry of apatite. *Geochimica et Cosmochimica Acta*, **60**, 4231–4240.

What perturbs isotherms? An assessment using fission-track thermochronology and thermal modelling along the Gotthard transect, Central Alps

C. GLOTZBACH[1,2]*, C. SPIEGEL[1,3], J. REINECKER[1], M. RAHN[4] & W. FRISCH[1]

[1]*Institute of Geoscience, University Tübingen, Sigwartstrasse 10, 72076 Tuebingen, Germany*

[2]*Present address: Laboratoire de Géodynamique des Chaînes Alpines, Université Joseph Fourier, BP53, 38400 Grenoble, France*

[3]*Present address: FB 5 – Geoscience, University Bremen, Postfach 330 440, 28334 Bremen, Germany*

[4]*Institute of Mineralogy – Geochemistry, University Freiburg, Albertstrasse 23b, 79104 Freiburg, Germany*

**Corresponding author (e-mail: christoph.glotzbach@bvra.ujf-grenoble.fr)*

Abstract: Interpretation of low-temperature thermochronological data usually relies on assumptions on the shape of isotherms. Recently, a number of thermal modelling approaches investigate and predict the theoretical influence of topography on isotherms. The application and proof of these predictions is not well confirmed by measured data. Here we present apatite fission-track (AFT) data from samples collected along the Gotthard road tunnel and its corresponding surface line to test these predictions. AFT ages broadly cluster around 6 Ma along the tunnel. No correlation of tunnel ages with superimposed topography is seen, which means that topography-induced perturbation of isotherms under given boundary conditions (topographic wavelength 12 km; relief 1.5 km; exhumation rate 0.45 km Ma^{-1}) can be neglected for the interpretation of AFT ages. Thus, in areas characterized by similar topographies and exhumation rates, apparent exhumation rates deduced from the age–elevation relationship (AER) of AFT data need no correction for topography-induced perturbation of isotherms. Three-dimensional (3D) numerical thermal modelling was carried out incorporating thermally relevant parameters and mechanisms, such as topography, geology, thermal conductivities and heat production. Modelling reveals a strong influence on the shape of isotherms caused by spatially variable thermal parameters, especially heat production rates. Therefore, not only topography has to be considered for interpreting low-temperature thermochronological data, but also other parameters like heat production rates.

Supplementary material: 1. Electron microprobe analyses, 2. Topography and model extend, 3. Model parameters are all available online at http://www.geolsoc.org.uk/SUP18380.

The first geological applications of low-temperature thermochronology commonly assumed that critical isotherms (e.g. 110 °C for apatite fission track) remain horizontal in relation to topography (Schaer *et al.* 1975). Thermal modelling, however, predicts that in areas with pronounced relief near-surface isotherms will be influenced by topography with compressed isotherms beneath valleys and wider-spaced isotherms beneath ridges (e.g. Stüwe *et al.* 1994). The impact for the interpretation of low-temperature thermochronological data, especially apatite fission track (AFT) and (U–Th)/He data, was investigated recently (Stüwe *et al.* 1994; Mancktelow & Grasemann 1997; House *et al.* 1998, 2001; Stüwe & Hintermüller 2000; Braun 2002; Foeken *et al.* 2007). On the other hand, bending of isotherms carries potential information to reveal minimum ages of topography and the evolution of palaeorelief (House *et al.* 1998; House *et al.* 2001; Braun 2002; Foeken *et al.* 2007).

A widely used technique to derive exhumation rates utilizes the correlation of ages of a single isotopic system, e.g. AFT or apatite (U–Th)/He, with elevation, the so-called age–elevation relationship (AER) (Wagner & Reimer 1972; Schaer *et al.* 1975; Stüwe *et al.* 1994). The great advantage, in contrast to the mineral-pair method (Wagner *et al.* 1977), is that no estimation of the geothermal gradient has to be made. However, this proxy is only valid under certain assumptions (e.g. Parrish 1983): (1) during and after passing the closure isotherm all samples followed a vertical exhumation pathway and no tectonic displacement exists between samples; (2) all samples are kinetically uniform;

From: LISKER, F., VENTURA, B. & GLASMACHER, U. A. (eds) *Thermochronological Methods: From Palaeotemperature Constraints to Landscape Evolution Models*. Geological Society, London, Special Publications, **324**, 111–124.
DOI: 10.1144/SP324.9 0305-8719/09/$15.00

and (3) the closure isotherm is fixed with respect to some geographical reference horizon, e.g. sea level.

A drawback of many studies is that sample profiles are more horizontal than vertical, and thus deduced exhumation rates from such profiles are affected by topography-induced perturbation of isotherms. The magnitude of isotherm perturbation increases with topographic wavelength, relief and exhumation rate (Stüwe *et al.* 1994). Hence, in mountainous regions the assumption that AFT relevant isotherms are flat is probably not fulfilled. In this case only vertical profiles (along a borehole or steep cliff) give correct values of exhumation rates. Resulting apparent exhumation rates of AER plots from 'near'-vertical profiles have to be corrected for the effect of isotherm perturbation due to topography. This leads to the question of the geomorphic and structural–kinematic boundary conditions that cause isotherm perturbation. Based on numerical modelling and analytical solutions (Stüwe *et al.* 1994; Mancktelow & Grasemann 1997; Stüwe & Hintermüller 2000; Braun 2002), exhumation rates of $\geq$0.5 km Ma^{-1} along with a topographic wavelength of $\geq$10 km and a relief of $\geq$2 km are threshold values for perturbing the 110 °C isotherm.

The aim of this study is to estimate the perturbation of isotherms under a given framework of exhumation rate, topographic wavelength, relief amplitude, rock properties and geothermal gradient by applying AFT thermochronology. AFT age variations along the tunnel transect give evidence as to whether the 110 °C isotherm was perturbed and, if so, to what extent. For this we sampled the Gotthard road tunnel as well as the corresponding surface line directly above the tunnel. The Gotthard road tunnel is located in central Switzerland and has a length of 16.3 km (Fig. 1). Present-day uplift rates range between 0.7 mm year^{-1} in the north and 1.1 mm year^{-1} in the south of the tunnel (Kahle 1997) (Figs 1b & 2a). The Gotthard transect is characterized by more or less ENE–WSW-trending ridges and valleys with a topographic wavelength of 12 km and relief of up to 1.5 km (Fig. 1b). Published exhumation rates constrained by AFT data from the Gotthard Massif range between 0.45 and 0.5 km Ma^{-1} between 10 and 6 Ma (Schaer *et al.* 1975; Wagner *et al.* 1977; Michalski & Soom 1990). These values suggest that the Gotthard transect is close to the lower limit of the proposed parameter values given for a 110 °C isotherm perturbation. Thus, this study provides a natural benchmark to verify topography-induced perturbation of isotherms predicted by existing modelling approaches.

In addition, a new three-dimensional (3D) finite-difference thermal model for predicting the positions of isotherms beneath topography is presented. Compared to previously published analytical solutions (Stüwe *et al.* 1994; Mancktelow & Grasemann 1997; Stüwe & Hintermüller 2000) our model provides the following advantages: (1) it incorporates 3D topography; and (2) it includes temporally and spatially variable parameters (e.g. thermal conductivity, internal heat production, exhumation rate and topography). Thus, the model provides information about the influence of different thermal parameters on predictions of the shape of near-surface isotherms, comparable to the program 'Pecube' of Braun (2003).

Geological setting

The Gotthard road tunnel cross-cuts the entire Gotthard Massif (GM) and the southern part of the Aar Massif (AM). The GM forms an ENE–WSW-trending mountain range 80 km long and up to 12 km wide (Fig. 1a). The GM and AM belong to the external Massifs of the Alps. They consist of pre-Variscan polymetamorphic basement intruded by late Variscan granitoids and covered by Late Palaeozoic–Mesozoic sedimentary rocks (Fig. 1) (e.g. Labhart 1977, 1999; Schaltegger 1994). The boundary between GM and AM is marked by the heavily tectonized Urseren–Garvera zone, built up by steeply dipping Permo-Carboniferous and Mesozoic metasediments. The sedimentary cover of the GM and AM was mostly detached during the Alpine orogeny, forming parts of the Helvetic nappes.

Peak Alpine metamorphic conditions along the tunnel were reached at around 35–30 Ma, with greenschist-facies conditions in the north and amphibolite-facies conditions in the south (Frey & Ferreiro Mählmann 1999). The post-metamorphic evolution of the study area was mainly controlled by the northward movement of the Adriatic indenter and related thrusting in the external Alps (e.g. Schmid *et al.* 1996). Thrusting during the 'Grindelwald stage' (22–12 Ma) propagated towards the foreland (Schmid *et al.* 1996), resulting in thrusting of the Gotthard Massif upon the Tavetsch and Aar Massif around 20 Ma (ductile deformation) and subsequent steepening of these structures (brittle deformation) (Wyder & Mullis 1998). It is likely that brittle deformation continues up to the present, as indicated by post-glacial vertical movements that are common in the study area (Persaud & Pfiffner 2004; Ustaszewski *et al.* 2008).

Methods

Sample sets along the Gotthard transect were collected and 31 AFT ages measured (Fig. 2a and Table 1). Minerals were separated using standard magnetic and heavy liquid techniques. Apatites were mounted in epoxy, and their internal surfaces

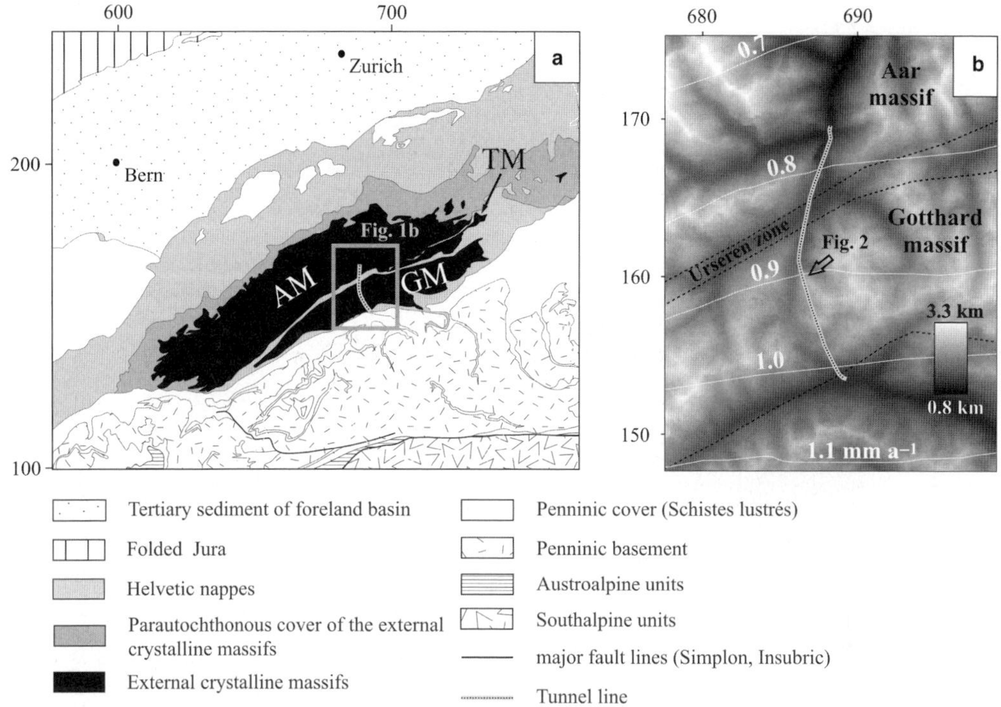

Fig. 1. (**a**) Geological sketch map of central Switzerland in Swiss coordinate system. The Gotthard road tunnel is located in the centre, crossing the southern part of the Aar Massif (AM) and the Gotthard Massif (GM). (**b**) Shuttle Radar Topography Mission (SRTM) digital elevation model (DEM) of the study area. Contours are measured recent rock uplift rates in mm year^{-1} from precise levellings (Kahle 1997).

were excavated by grinding and polishing. Afterwards the mounts were etched by 5 M HNO_3 for 20 s at 20 °C. Irradiation was carried out at the FRM-II reactor in Garching (TU München, Germany). Mica detectors were etched to reveal induced tracks using 40% HF at 20 °C for 40 min. Fission-track counting was carried out with an optical microscope (Axioscope 1, Zeiss) under 1000× magnification using dry lenses. Samples were dated using the external detector method (Naeser 1978; Gleadow 1981) using the zeta calibration approach (Hurford & Green 1982, 1983), with a zeta of 354 ± 7 years cm^{-2} for dosimeter glass CN5 and Durango and Fish Canyon Tuff apatite age standards. Age determination, visualization and statistics were calculated and performed with Trackkey 4.2 g (Dunkl 2000). All AFT ages are displayed as central ages and errors as $\pm 1\sigma$ (Galbraith & Laslett 1993).

Measured D_{par} values were used as kinetic parameters (e.g. Burtner *et al.* 1994), complemented by electron microprobe analysis on selected crystals, using a JEOL Superprobe with a beam current of 30 nA, an acceleration voltage of 15 kV and a beam diameter of 10 μm.

Inferred shape of isotherms from AFT data

The aim of this section is to investigate the perturbation of measured AFT ages under the given boundary conditions, especially with respect to topography.

Table 1 summarizes the results of AFT dating along the tunnel and the corresponding surface line. Previously published AFT data for the Gotthard area (Schaer *et al.* 1975; Wagner *et al.* 1977; Michalski & Soom 1990) are in good agreement with our data and are therefore included in our interpretations (cf. Fig. 2b).

To make sure that measured variations of AFT ages are, in fact, the result of the shape of the palaeo-isotherm, the kinetic homogeneity of our apatites was tested by etch-pit diameter (D_{par}) measurements. D_{par} values obtained for measured samples are small and relatively uniform, varying between 1.1 and 1.6 μm (Table 1). Microprobe analyses of 15 samples from different lithologies were carried out, demonstrating that all samples are close to the F-apatite end member, with Cl contents of less than 0.1 wt%. Furthermore, analysed elements such as Si, Mn, Ce and Sr do not show

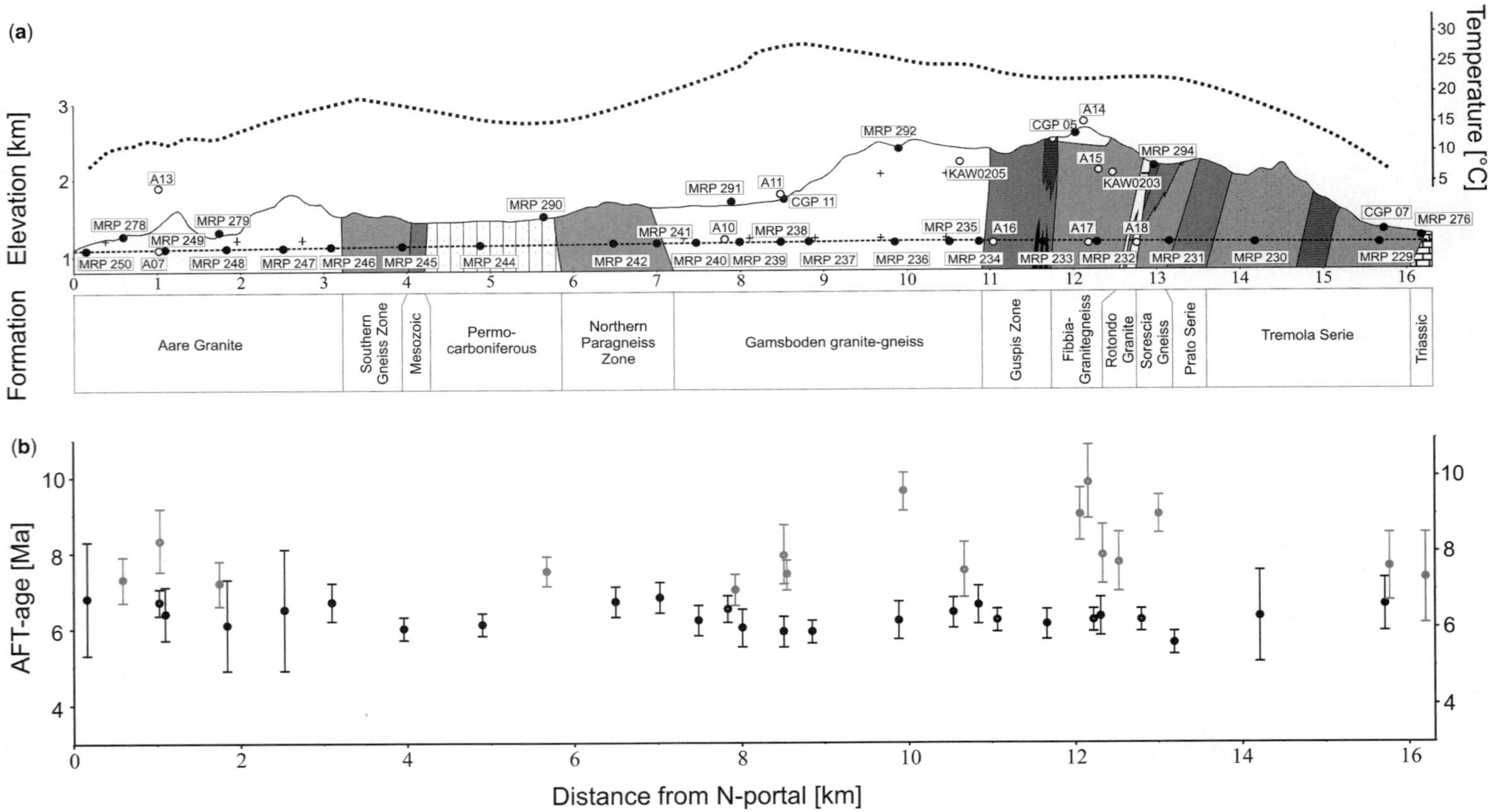

Fig. 2. (**a**) Geological profile along the Gotthard road tunnel (modified after Keller *et al.* 1987) with published (open circles) and own samples (black dots) (Schaer *et al.* 1975; Wagner *et al.* 1977). All samples are directly located on the plane of the transect, except some samples plotted between the tunnel and surface line, and above the surface line. Their distance to the plane is: 750 m (A13), 280 m (A14), 890 m (A15), 300 m (A11), 780 m (KAW0203) and 420 m (KAW0205). The dotted line shows interpolated *in situ* rock temperatures measured during excavation of the tunnel (Keller *et al.* 1987). (**b**) AFT ages along the tunnel (black) and on the surface (grey).

Table 1. *Apatite fission-track data from the Gotthard road tunnel and a surface line directly above the tunnel*

Sample	Geological unit	Swiss grid (CH1903) (m) X	Swiss grid (CH1903) (m) Y	Elevation (m)	Number of grains	ρ_s	N_s	ρ_i	N_i	ρ_d	N_d	$P(x^2)$ (%)	Dispersion	Central age $\pm 1\sigma$ (Ma)	D_{par} (μm)	U (ppm)
Tunnel																
MRP 250	Aare Granite	688290	168840	1082	20	0.800	27	21.100	671	10.45	4998	13.2	0.30	6.8 ± 1.5	1.31	16
MRP 249	Aare Granite	688020	167940	1093	30	0.501	104	8.899	1846	6.407	3310	93.8	0.00	6.4 ± 0.7	1.34	16
MRP 248	Aare Granite	687770	167240	1102	10	0.429	30	8.000	559	6.424	3310	78.5	0.01	6.1 ± 1.2	1.12	16
MRP 247	Aare Granite	687530	166590	1110	12	0.230	17	4.037	298	6.441	3310	97.6	0.00	6.5 ± 1.6	1.13	8
MRP 246	Aare Granite	687340	166060	1117	60	0.527	184	9.012	3145	6.45	3118	99.9	0.00	6.7 ± 0.5	1.24	17
MRP 245	Southern Gneiss Zone	687060	165240	1127	49	1.942	434	37.985	8491	6.655	3311	47.2	0.04	6.0 ± 0.3	1.63	77
MRP 244	Permo-Carboniferous	686850	164330	1138	50	1.604	571	30.145	10734	6.474	3310	84.1	0.04	6.1 ± 0.3	1.32	57
MRP 242	Northern Paragneiss Zone	686530	162770	1157	50	0.770	360	13.266	6202	6.491	3310	51.4	0.07	6.7 ± 0.4	1.41	23
MRP 241	Northern Paragneiss Zone	686430	162250	1163	50	0.763	328	13.140	5649	6.6	3311	92.3	0.00	6.8 ± 0.4	1.47	22
MRP 240	Gamsboden Granite Gneiss	686340	161790	1168	47	1.133	315	21.050	5852	6.516	3118	99.8	0.00	6.2 ± 0.4	1.41	37
MRP 239	Gamsboden Granite Gneiss	686240	161280	1174	33	0.754	215	14.658	4182	6.608	3311	25.6	0.04	6.0 ± 0.5	1.25	25
MRP 238	Gamsboden Granite Gneiss	686280	160790	1173	35	0.900	234	17.538	4559	6.508	3310	99.4	0.00	5.9 ± 0.4	1.21	32
MRP 237	Gamsboden Granite Gneiss	686360	160460	1172	50	1.165	431	23.019	8519	6.549	3118	97.8	0.00	5.9 ± 0.3	1.41	41
MRP 236	Gamsboden Granite Gneiss	686640	159460	1168	50	0.863	161	16.091	3003	6.524	3310	74.7	0.02	6.2 ± 0.5	1.29	28
MRP 235	Gamsboden Granite Gneiss	686820	158830	1166	50	0.754	277	13.656	5016	6.541	3310	77.8	0.03	6.4 ± 0.4	1.41	25
MRP 234	Gamsboden Granite Gneiss	686920	158500	1165	50	0.821	192	15.543	3637	7.055	3397	81.3	0.00	6.6 ± 0.5	1.41	26
MRP 233	Guspis Zone	687160	157770	1162	50	0.419	240	7.964	4562	6.558	3310	99.5	0.00	6.1 ± 0.4	1.57	15
MRP 232	Fibbia Granite Gneiss	687370	157150	1160	51	0.743	212	13.843	3951	6.574	3310	66.4	0.05	6.3 ± 0.5	1.34	25
MRP 231	Sorescia Gneiss	687670	156340	1157	50	1.755	725	36.502	15083	6.591	3310	60.8	0.09	5.6 ± 0.3	1.26	66
MRP 230	Tremola Serie	688030	155380	1153	25	0.123	27	2.304	504	6.582	3118	90.5	0.00	6.3 ± 1.2	1.51	4
MRP 229	Tremola Serie	688730	154050	1147	50	0.223	90	3.941	1589	6.615	3118	95.8	0.00	6.6 ± 0.7	1.42	7
Surface																
MRP 278	Aare Granite	688180	168400	1265	50	0.806	168	16.205	3376	8.277	4059	50.4	0.11	7.3 ± 0.6	1.32	23
MRP 279	Aare Granite	687800	167320	1315	50	0.821	171	16.885	3516	8.303	4059	7.0	0.27	7.2 ± 0.6	1.20	23
MRP 290	Permo-Carboniferous	686730	163585	1510	50	0.994	420	19.662	8305	8.354	4059	4.1	0.20	7.5 ± 0.4	1.26	27
MRP 291	Gamsboden Granite Gneiss	686150	161390	1690	50	1.348	499	21.959	8131	6.391	3310	71.8	0.02	7.0 ± 0.4	1.26	40
CGP 11	Gamsboden Granite Gneiss	686240	160740	1725	44	1.230	467	19.488	7397	6.608	3310	32.5	0.10	7.4 ± 0.4	1.28	33
MRP 292	Gamsboden Granite Gneiss	686650	159420	2380	50	1.446	423	22.012	6439	8.430	4059	71.3	0.06	9.6 ± 0.5	1.31	30
CGP 05	Rotondo Granite	687129	157340	2580	36	0.898	197	11.761	2579	6.658	3310	59.6	0.09	9.0 ± 0.7	1.30	21
MRP 294	Sorescia Gneiss	687640	156530	2140	50	0.896	389	11.303	4905	6.374	3310	99.9	0.00	9.0 ± 0.5	1.28	21
CGP 07	Tremola Serie	688851	154073	1320	23	0.459	83	7.140	1291	6.624	3310	72.7	0.16	7.6 ± 0.9	1.31	13
MRP 276	Tremola Serie	689065	153820	1220	17	0.525	43	10.755	881	8.405	4059	54.6	0.01	7.3 ± 1.2	1.59	14

ρ_s (ρ_i) are spontaneous (induced) track densities (10^5 tracks cm^{-2}); N_s (N_i) is number of counted spontaneous (induced) tracks; ρ_d is dosimeter track density (10^5 tracks cm^{-2}); N_d is number of tracks counted on dosimeter; $P(X^2)$ is the probability obtaining Chi-square value (X^2) for n degrees of freedom (where n is the number of crystals minus 1); dispersion is a real number, that gives an idea of the variability of the single grain age distribution, it is zero if all the data are identical, and increases as the data become more diverse; age ± 1 s is the central age ± 1 standard error (Galbraith & Laslett 1993); D_{par} is the etch-pit diameter of fission tracks, averaged from four measurements per analysed grain. Ages were calculated using the zeta calibration method (Hurford & Green 1983), glass dosimeter CN-5, and a zeta value of 354.92 ± 7.03 years cm^{-2}.

any significant variation, suggesting that apatites are kinetically uniform (for details see p. 1 of the supplementary material for this paper).

AFT ages of samples collected at the same elevation can be used to predict the palaeoshape of the critical isotherm (House *et al.* 1998). All samples along the Gotthard tunnel are at approximately 1.1 km elevation and yield AFT ages of 6.2 ± 0.6 Ma, overlapping within 1σ-error limits (Fig. 2b). Individual samples characterized by low U-content show larger 1σ-error, especially those from the northern and southern ends of the transect.

Surface samples yield AFT ages ranging from 7.0 to 9.6 Ma. Surface samples collected at approximately 1.3 km elevation along the Gotthard transect (CGP 07, MRP 276, MRP 278 and MRP 279) exhibit the same ages of 7.4 ± 0.2 Ma. Thus, AFT age patterns on two structural elevation levels (*c.* 1.1 and *c.* 1.3 km) indicate that between 6.2 and 7.4 Ma the 110 °C isotherm was flat, suggesting that topography has no visible effect on the AFT age pattern along the tunnel. Small variations in ages partly coincide with stratigraphic–tectonic boundaries (Fig. 2b), which can be explained by kinetic or physical differences (such as heat production and thermal conductivities).

Surface AFT ages show an increase with elevation (Figs 2b & 3). Slopes of AER were calculated by weighting the ages according to their errors. Slopes and uncertainties were inverted to estimate exhumation rates with defined uncertainties (Reiners *et al.* 2003).

Regression bands were plotted with a 95% confidence interval, which allows prediction of the error of the regression lines. The AER plot for samples of the northern flank of the Gotthard Massif (MRP 237, MRP 238, CGP 11 and MRP 292) yields an apparent exhumation rate of 0.35 ± 0.1 km Ma^{-1} for the period between 10 and 5 Ma (Fig. 3a). The horizontal and vertical distance between the samples is 1.5 and 1.2 km, respectively. Samples of the southern flank (MRP 229, MRP 276, MRP 294, CGP 07 and A14 of Schaer *et al.* 1975) yield an apparent exhumation rate of 0.5 ± 0.2 km Ma^{-1} for the period between 10 and 6 Ma (Fig. 3b). Here the horizontal and vertical distance between the samples is 4 and 1.6 km, respectively. Deduced apparent exhumation rates are in good accordance with previously published data (Schaer *et al.* 1975; Michalski & Soom 1990). The presented data suggest that critical isotherms are flat under the Gotthard transect and a correction for topography is not necessary.

Modelling the shape of isotherms

Although we demonstrated that in the specific case of the Gotthard transect no topographic correction is required, in the following section we test the theoretical influence of topography and other thermally relevant parameters on near-surface isotherms. We present numerical 3D finite-difference thermal models developed for a crustal block of 14 × 26 × 10 km around the Gotthard road tunnel (an overview is provided on p. 2 of the supplementary material). Modelling incorporates spatial and/or temporal varying parameters, namely topography, exhumation rates, thermal conductivity and heat production (for details see p. 3 of the supplementary material). The parameter values are either taken from this study (exhumation rates) or from geophysical measurements (e.g. Busslinger & Rybach 1999). In the following, all mentioned isotherm perturbation values refer to the location of the Gotthard road tunnel transect, and are defined here as the maximum vertical deflection of the corresponding isotherm.

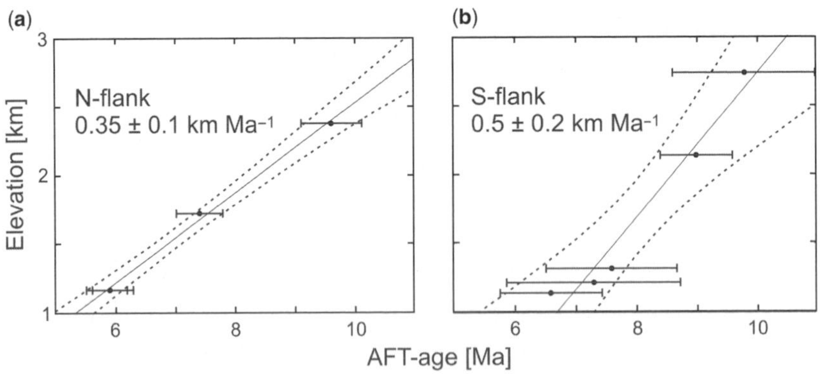

Fig. 3. AFT age–elevation relationship of surface samples (**a**) of the northern flank (MRP 292, CGP 11, MRP 237, MRP 238) and (**b**) of the southern flank (A14 from Schaer *et al.* 1975: MRP 294, MRP 276, CGP 07, MRP 229) of the Gotthard transect.

For simplification in the first step it is assumed that heat transfer in the crust is controlled by conduction only (Stüwe 2007). In this case Fourier's law of heat conduction can be used:

$$\frac{\partial T}{\partial t} = \kappa \cdot \nabla^2 T \qquad (1)$$

where κ is the thermal diffusivity with typical values of around 10^{-6} m^2 s^{-1}, T is temperature, t is time and ∇ is the nabla operator, describing here the spatial change of temperature in 3D (e.g. Stüwe 2007). Equation (1) is modified to account for the effect of exhumation and internal heat production due to radioactive decay, resulting in:

$$\frac{\partial T}{\partial t} = \kappa \cdot \nabla^2 T + u \cdot \frac{\partial T}{\partial z} + \frac{S}{\rho \cdot c_p} \qquad (2)$$

where z is depth, S the heat production rate, ρ the rock density and c_p the specific heat. Initial and boundary conditions necessary for the solution of the differential equation (2) are as follows.

- The lower boundary of the model was placed at -10 km from the surface and was fixed with a constant vertical geothermal gradient, for which a value of 20 °C km^{-1} was used based on borehole and tunnel temperature measurements in or near the Gotthard road tunnel (Keller *et al.* 1987). To verify that this relatively 'shallow' lower boundary allows the deflection of isotherms caused by thermal relevant parameters (e.g. topography and exhumation), a simple 2D sinus-shaped topography was modelled analytically using the program 'TERRA' (Ehlers *et al.* 2005) (Fig. 4). Used parameters fit those from the Gotthard transect. Position of isotherms are unaffected by the depth of the lower boundary of the model.
- Topography was used as an upper-boundary condition, discretized with a rectangular grid and converted into air temperature, depending on elevation (Busslinger & Rybach 1999):

$$T_{Air} = T_0 - \alpha \cdot h \qquad (3)$$

with T_{Air} being the mean air temperature at surface, T_0 the air temperature at sea level, α the atmospheric lapse rate and h the elevation. Kohl *et al.* (2001) showed that ground surface temperature (GST) data collected from high elevations in Alpine terrain are best described by the model of Niethammer (1910), with $T_0 = 12$ °C and $\alpha = 4.6$ °C km^{-1}. These values were used to calculate GST for each node based on a Shuttle Radar Topography Mission (SRTM) digital elevation model (DEM). We neglected the difference between air temperature and the corresponding GST depending on exposition (slope and orientation), vegetation, snow cover and rock surface properties (e.g. Šafanda 1999). Consequently, each node of the upper boundary is composed of an elevation and a corresponding GST.
- Lateral boundaries were fixed with no horizontal heat flow. For simplification a uniform distance from node to node ($\Delta x = \Delta y = \Delta z$) of 132.9 m was used, according to the spatial resolution of the input DEM. Thus, the total mesh was built up by 1.55×10^6 nodes. A simplified geological model was used for the modelling, which considered five geological units: Aar Massif, Urseren Zone, Tavetsch Massif, Gotthard Massif and Schistes lustrés (Fig. 1 and Table 2). Vertical boundaries between individual geological units were assumed, consistent with high dip angles observed at the surface and during tunnel construction (Fig. 2a). The geological units are characterized by different physical and thermal properties. The foliation developed during Alpine metamorphism results in a pronounced anisotropic thermal conductivity of all geological units, with ratios of up to 1.5 between maximum ($k_{\parallel}$) and minimum ($k_{\perp}$) values within one unit (Kappelmeyer & Haenel 1974; Kohl *et al.* 2001). Input thermal conductivities are listed in Table 2. Rock radioactivity measurements within the tunnel (Keller *et al.* 1987) were used to calculate the total heat production related to the radioactive decay of uranium, thorium and potassium (Turcotte & Schubert 1982, equations 4–6). Resulting heat production values ranged from 7.54×10^{-11} to 2.56×10^{-9} W kg^{-1} for individual geological

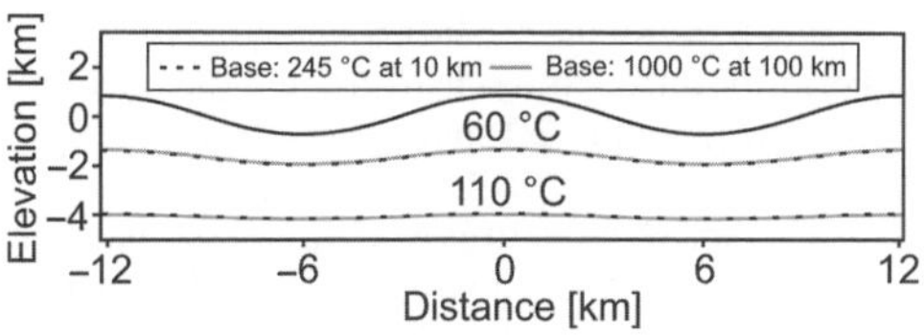

Fig. 4. Position of the 60 and 110 °C isotherms depending on the depth of the lower boundary. Model parameters are: steady-state topography with a wavelength of 12.5 km and an amplitude of 0.5 km; exhumation rate of 0.45 km Ma^{-1}; lower boundary at -10 km/-100 km with a temperature of 245 °C/1000 °C; upper boundary fixed by topography and surface temperature calculated with 12 °C at 0 m and a lapse rate of 4.6 °C km^{-1}; thermal diffusivity 1×10^{-6} m^2 s^{-1}; density 2700 kg m^{-3}; specific heat 1000 J kg^{-1} K^{-1}; heat production 3×10^{-6} W m^{-3} with a skin depth of 10 km.

Table 2. *Geological units with their anisotropic thermal conductivities* (k) *parallel and perpendicular to foliation and heat production values*

Geological unit	$k_{\parallel}$ ($W\ m^{-1}\ K^{-1}$)	$k_{\perp}$ ($W\ m^{-1}\ K^{-1}$)	Heat production ($10^{-9}\ W\ kg^{-1}$)
Aar Massif	3.66*	2.83*	1.57–1.92
Gotthard Massif	4.43*	2.95*	0.51–1.50
Tavetsch Massif	3.61*	2.79*	1.92
Urseren-Garvera zone	3.10*	2.45*	1.17
Schistes lustrés	3.10†	2.40†	0.51

*Kohl *et al.* (2001); †Kappelmeyer & Haenel (1974).

units, with $1.22 \times 10^{-9}\ W\ kg^{-1}$ as an average value. Modelling was carried out either with individual values or an averaged heat production rate extrapolated to the surroundings of the Gotthard road tunnel (Table 2). Tunnelling and geophysical investigations revealed that the Aar and Gotthard Massifs continue towards depth, possibly by more than 10 km (Pfiffner & Heitzmann 1997). Therefore, heat production rates are assumed to be constant with depth. Modelling was conducted assuming both a steady-state and time-varying topography, and a constant exhumation rate of $0.45\ km\ Ma^{-1}$ (mean value of the AER plots in Fig. 3). To account for a time-varying topography nodes were added/removed and temperature adjusted (according to equation 3) at the upper boundary during the modelling progress. For instance, a change in elevation from 0 to 2000 m in 2 Ma, is implemented by adding one node (132.9 m) every 0.133 Ma (reaching an elevation of 1994 m after 2 Ma). A simple 2D thermal model was numerically calculated and compared with an analytical solution calculated with the program 'TERRA' (Ehlers *et al.* 2005) (Fig. 5), proving the correctness of the used finite-difference thermal model approach.

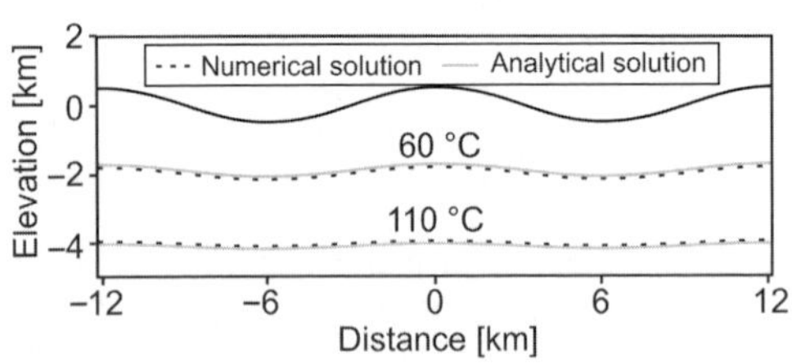

Fig. 5. Comparison between an analytical solution from thermal modelling program TERRA (Ehlers *et al.* 2005) and the numerical model presented in this study for a simple 2D model. Model parameters are the same as in Figure 4, except a lower boundary at −10 km with a temperature of 220 °C.

From several performed model runs we present five solutions, which differ by the assumed thermal conductivity, heat production and topographic evolution (Table 3). Models 1–3 were calculated until reaching a steady-state temperature field, which takes around 5 Ma depending on the parameterization. Models 4 and 5 were calculated with an initial steady-state temperature field, subsequently adapted during the modelling progress because of a changing topography.

To evaluate the influence of thermally relevant parameters (e.g. topography) on low-temperature thermochronology in the following models, closure temperatures of 110 and 60 °C were assumed for the apatite fission-track (e.g. Rahn & Grasemann 1999) and (U–Th)/He system (e.g. Ehlers & Farley 2003), respectively. AFT age differences along the tunnel are predicted, using forward modelling of AFT ages with the program HeFTy (Ketcham *et al.* 2007) based on measured kinetic parameters (D_{par}) and generated time–temperature (*tT*) path for selected localities. Therefore, during modelling progress the temperatures of selected localities were read out every 10 ka, yielding nearly continuous *tT* paths.

Figure 6 shows the predicted steady-state 110 °C isotherm for different model runs, and the contours of the corresponding 2D shape of the 60 and 110 °C isotherms along the Gotthard road tunnel.

The benefit of model 1 is that we can estimate the net influence of the topography on a chosen isotherm because all other input parameters are spatially uniform. The modelled 110 °C isotherm clearly demonstrates that small topographic features have no influence on the isotherm (Stüwe *et al.* 1994; Mancktelow & Grasemann 1997). Thus, according to model 1, the shape of the isotherm follows the topography in a strongly dampened fashion (Stüwe *et al.* 1994) and is clearly deformed beneath the Gotthard Massif, with a perturbation of 250 m. Resulting maximum AFT age difference along the tunnel is 0.5 Ma (modelled ages range from 6 to 5.5 Ma).

Table 3. *Input parameters of individual model runs*

Model No.	Topography	Thermal conductivity (W K^{-1} m^{-1})	Heat production (10^{-9} W kg^{-1})	Rock uplift rate (km Ma^{-1})	Surface uplift rate (km Ma^{-1})	Exhumation rate (km Ma^{-1})
1	Recent, steady state	2.7	1.22	0.45	0	0.45
2	Recent, steady state	Anisotropic, cf. Table 2	1.22	0.45	0	0.45
3	Recent, steady state	Anisotropic, cf. Table 2	Anisotropic, cf. Table 2	0.45	0	0.45
4	0 m to recent, increasing	Anisotropic, cf. Table 2	Anisotropic, cf. Table 2	0.45	0.10–0.27	0.18–0.35
5	Decreasing	2.7	1.22	0.45	−0.10 to −0.22	0.55–0.67

Incorporating spatially variable and anisotropic thermal conductivities (model 2) results in an overall deeper 110 °C isotherm, because average input thermal conductivities are higher than those used for model 1. The shape of the 110 °C isotherm in model 2 is similar to that from model 1, apart from a smaller isotherm perturbation of only 150 m (Fig. 6). Areas characterized by low thermal conductivities (Urseren zone and Schistes lustrés, cf. Fig. 1) accumulate heat that bulge the 110 °C isotherm. We conclude that the observed differences and anisotropy of thermal conductivities along the Gotthard road tunnel (Table 2) do only weakly influence the shape of the near-surface isotherms. For models 1 and 2, the corresponding perturbation of the 60 °C isotherm is more or less twice

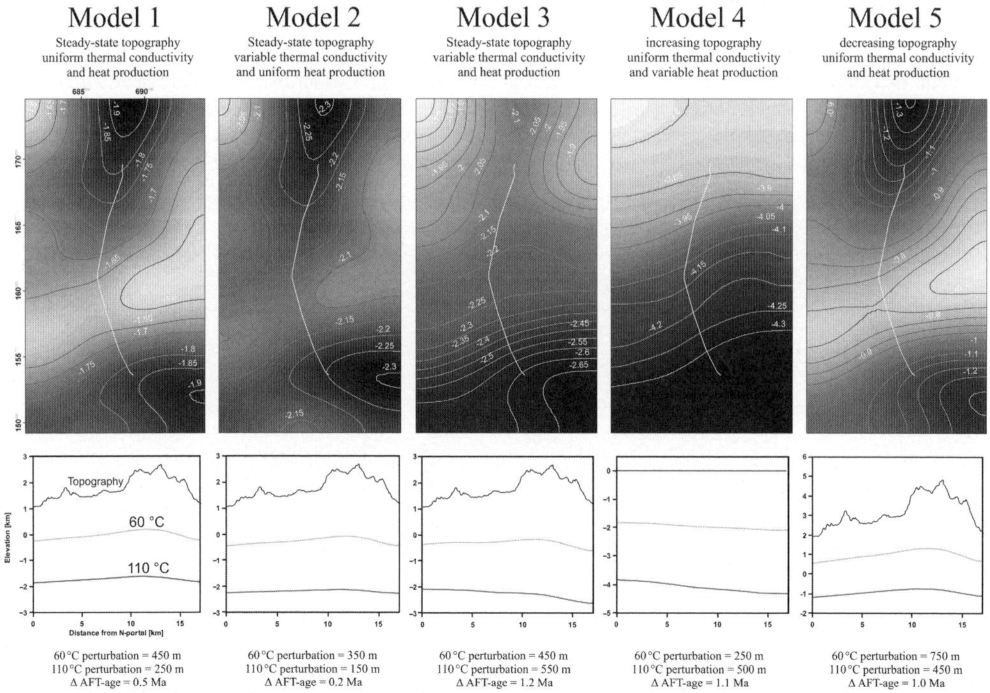

Fig. 6. Three-dimensional models of the 110 °C isotherms with contours in metres below sea level calculated with parameters listed in Tables 2 and 3. The Gotthard road tunnel is marked by the white line. The corresponding position of the 60 and 110 °C isotherms with topography is shown in the graphs below. Δ AFT describes the variation of AFT ages along the tunnel transect, predicted by the different model runs.

as high compared to the perturbation of the 110 °C isotherm. However, the resulting AFT age difference along the tunnel is only 0.2 Ma (modelled ages range from 6.7 to 6.5 Ma).

From model 3, a laterally variable heat production was added, resulting in a more complex 110 °C isotherm shape. Interestingly, the pronounced topography of the Gotthard Massif, forming an ENE–WSW-trending ridge in the southern tunnel section, is not mirrored by the 110 °C isotherm. Evidently, differences in heat production values are capable of blurring the effect of topography. As demonstrated by model run 1, the net topographic effect is small. Thus, for model 3, the perturbation of the 110 °C can be nearly completely attributed to the spatial variation of heat production, which here was assumed to persist to the depth of 10 km. Owing to the lateral decrease in total heat production rate, the shape of the 110 °C isotherm along the tunnel profile shows a steady decrease to the south, with a perturbation of 550 m. The 60 °C isotherm runs more or less horizontally up to tunnel kilometre 12 and then decreases towards the south with a total perturbation of 450 m. For this model run, variations of AFT ages in the range of 1.2 Ma (modelled ages range from 7.6 to 6.4 Ma) along the tunnel are predicted. The perturbation effect of topography and heat production at that depth becomes less pronounced in the northern part of the transect. Thus, differences in heat production are capable of compensating (northern section) and amplifying (southern section) the perturbation of isotherms owing to topography (Fig. 6).

A palaeotopographic evolution with increasing relief after 5 Ma, assuming isotropic thermal parameters, would result in flat isotherms at the time the AFT samples cross the 110 °C isotherm, following the observation of model 1. In the extreme case that the palaeotopography is flat at that time, resulting AFT ages along the tunnel are all the same. Incorporating spatial variable heat production rates (model 4) results in increased isotherm perturbation with depth, with perturbation values of the 60 and 110 °C isotherms of 250 and 500 m, respectively (Fig. 6).

To investigate the effect of a possible palaeotopographic evolution with higher relief and peak elevations in Late Miocene times (6 Ma), which decrease to the present, model 5 was carried out assuming initial maximum elevations of 5 km and a relief of 3 km. The shape of resulting isotherms is comparable to that of model 1, however, with more pronounced isotherm perturbations of 450 m (110 °C) and 750 m (60 °C). The doubling of the vertical isotherm perturbation with respect to model 1 corresponds to a doubling in the FT age difference along the tunnel.

Influence of advective heat transport on low-temperature isotherms

To investigate the impact of heat advection due to faulting on isotherms and AFT ages simple 2D numerical modelling was carried out. A flat topography was assumed with two crustal blocks exhumed with rates of 0.5 and 1 km Ma^{-1} (Fig. 7a).

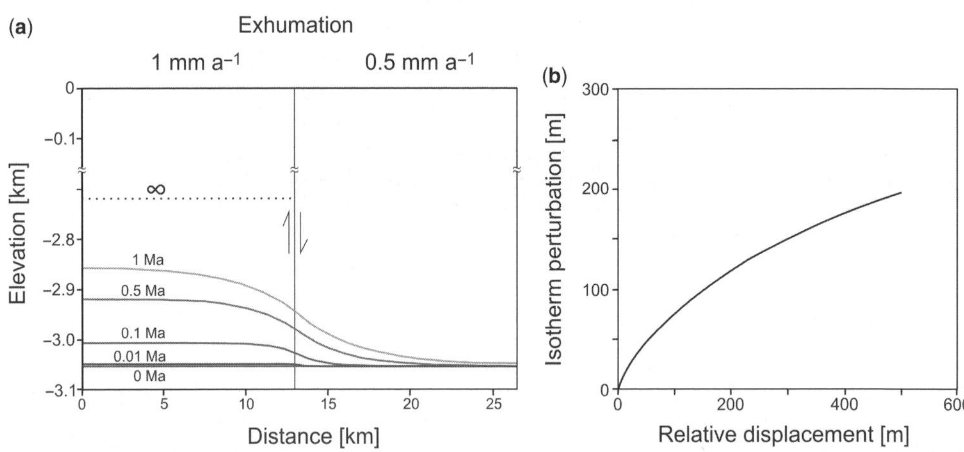

Fig. 7. (**a**) Two-dimensional numerical model with two crustal blocks separated by a perpendicular fault. Crustal blocks differ in their rock uplift rates and resulting exhumation rate assuming steady-state topography. Calculated 110 °C isotherms for different time steps (0, 0.01, 0.1, 0.5 and 1 Ma) are displayed. The total shift of the 110 °C isotherm after infinite time is indicated by a dashed line. (**b**) Relationship between perturbation of the 110 °C isotherm and relative displacement along the fault.

Assuming steady-state conditions and an exhumation rate of 0.5 km Ma^{-1}, the 110 °C isotherm is at a depth of *c.* 3 km. Initiation of a perpendicular fault and increased exhumation (1 km Ma^{-1}) on one side of the fault results in a perturbation of isotherms depending on the duration of displacement. Short-term movements (<0.1 Ma) result in a less pronounced perturbation of the isotherm across the fault (<50 m) and a displacement of the crustal blocks of <50 m. For small displacements heat advection approximates relative rock uplift (Fig. 7b); hence advective heat flow dominates the temperature field. For long-term movements, however, the role of diffusion increases. Thus the perturbation of the 110 °C isotherm is small as compared to the displacement (e.g. for a duration of 1 Ma, a perturbation of 200 m corresponds to a relative displacement of 500 m). Infinite displacement results in a maximum possible isotherm perturbation of *c.* 320 m.

Discussion

In the following some essential requirements for the investigation of isotherm perturbation due to topography are discussed. In addition, we discuss the measured AFT age pattern in light of the modelled predictions of isotherm perturbation.

Advective and convective heat transport

In addition to the discussed thermally relevant parameters (e.g. heat production), near-surface temperatures are affected by advective and convective heat transport. Therefore, an important prerequisite for investigations on isotherm perturbations is that advective and convective heat transport must be negligible or quantifiable over geological timescales.

Advective heat transport may be caused by deformation and spatial differences in rock uplift rates. Studies on brittle deformation structures and hydrothermal mineral precipitation along the Gotthard road tunnel showed that the main phase of brittle faulting had ceased before the structural level of the tunnel samples cooled below 190 °C (Luetzenkirchen 2002). First-order precise levelling of the Swiss national levelling network revealed only a slight gradient of present-day vertical movements within the area of the Gotthard Massif (Fig. 1b), with rates of 0.8–1.0 mm $year^{-1}$ (Kahle 1997). On a smaller scale, however, remote sensing analysis, field work and numerical modelling demonstrate the activity of uphill-facing scarps in the Urseren valley as a response to isostatic vertical movements after the last Ice Age (Dahinden 2001; Persaud & Pfiffner 2004; Hampel & Hetzel 2006; Ustaszewski *et al.* 2008). AFT ages along the Gotthard road tunnel show no jumps in ages (Fig. 2b), thus a differential vertical displacement of individual crustal blocks has been small and is hidden within the 1σ-error of the AFT ages. Assuming an exhumation rate of 0.5 km Ma^{-1} and an average AFT error of 0.6 Ma, only a vertical displacement of more than 300 m along a fault would be significant. We conclude that vertical displacements along fault zones were small and therefore can be neglected for the last 6 Ma. Simple 2D thermal modelling of movements along a single vertical fault show that a displacement of 300 m is capable of generating a perturbation of the 110 °C isotherm of approximately 150 m (Fig. 7). We cannot exclude the fact that mass advection along the Gotthard transect possibly led to a perturbation of isotherms of less than 150 m, which in magnitude is close to the net topography-induced isotherm perturbation (see models 1 and 2, Fig. 6).

Convective heat transport may play an important role in the total heat transport in high mountainous areas (e.g. Whipp & Ehlers 2007). During construction of the Gotthard road tunnel water inflow measurements were carried out, with maximum *in situ* measured rock temperatures of 32 °C (Fig. 2a) (Keller *et al.* 1987; Luetzenkirchen 2002). Except for an area in the central Gamsboden granite-gneiss, which is assumed to be affected by a near-surface convective hydrothermal circuit (Pastorelli *et al.* 2001; Luetzenkirchen 2002), measured rock temperatures are correlated to topography.

We conclude that advective and convective heat transport are negligible, and that conductive heat transport dominates and controls the temperature distribution in the upper crust of the study area.

Predicted v. observed ages

A striking characteristic when comparing the modelled shape of isotherms and measured AFT ages along the Gotthard transect is the incompatibility of the more complex model 3, including spatially variable thermal conductivities and heat production rates. A possible explanation could be that measured radioactivities in the rocks of the Gotthard road tunnel and the resulting heat production rates are not representative of the entire lithological unit, and, thus, cannot be extrapolated straightforwardly. Alternatively, perturbation of isotherms may be compensated for because of the spatially variable exhumation rates increasing to the south, which is likely when the deduced exhumation rates are considered. Mean exhumation rates of between 10 and 6 Ma are higher on the southern flank of the Gotthard Massif compared to the northern flank (Fig. 3). The present-day uplift rates show the same trend (Kahle 1997). Repeating model run 3

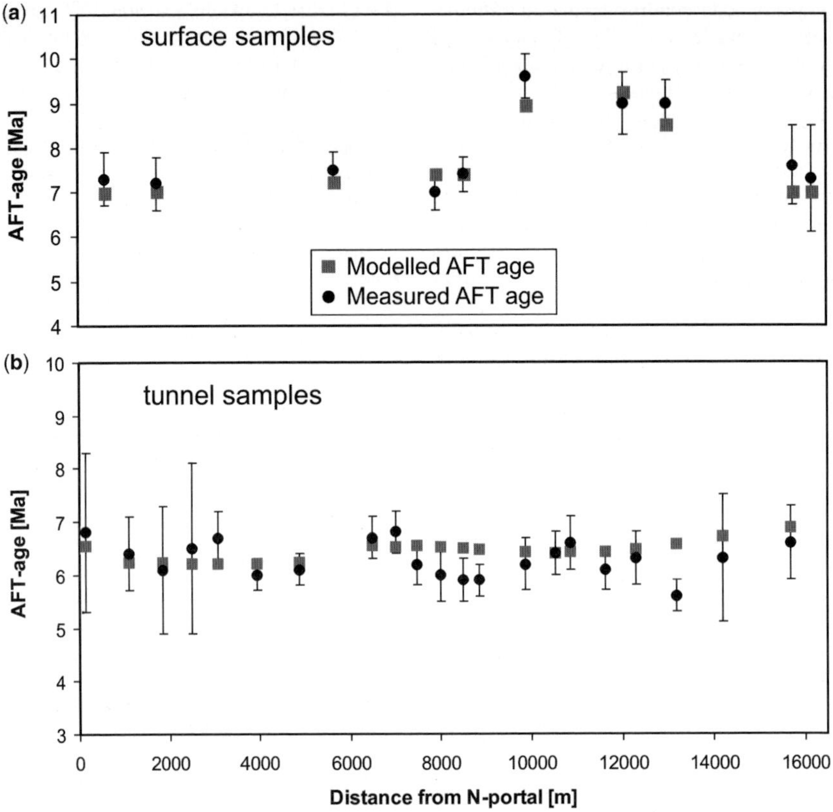

Fig. 8. Measured AFT ages (black dots, with 1σ age errors) and modelled AFT ages (grey squares) of (**a**) surface samples and (**b**) tunnel samples. Model parameters are that of model 3, but with the assumption that exhumation rates linearly increase from north (0.45 km Ma^{-1}) to south (0.55 km Ma^{-1}).

with exhumation rates increasing from north (0.45 km Ma^{-1}) to south (0.55 km Ma^{-1}) results in an AFT age pattern similar to that measured for the surface and tunnel samples (Fig. 8). These exhumation rates are optimized to match the measured AFT ages and the AER-derived exhumation rates. Thus, small spatial differences in exhumation rates are able to compensate for and mask the perturbation of isotherms resulting from other thermal parameters. Therefore, we conclude that measured AFT age variations along the Gotthard road tunnel (6.2 ± 0.6 Ma) do not preclude any presented model or palaeotopographic evolution. The Gotthard transect, however, is characterized by steady exhumation rates (Schaer *et al.* 1975; Wagner *et al.* 1977), and more or less vertically oriented tectonic structures and boundaries (Fig. 1b), facilitating the generation of a steady-state topography with spatially invariant ridges and valleys (Willett & Brandon 2002). But steady-state conditions are rarely achieved, especially relating to the Late Neogene climate change and the increased frequency in climatic oscillations (Whipple & Meade 2006). Most probably, positive feedback between climate change, weathering, erosion and isostatic rebound led to an increase in relief over the last million years (e.g. England & Molnar 1990; Molnar & England 1990). However, resolving the latest Neogene (<5 Ma) exhumation history of the Gotthard Massif is beyond the scope of this paper and will require the use of apatite (U–Th)/He data.

Conclusions

Apatite fission-track data along the Gotthard road tunnel and its corresponding surface line are compared with 3D numerical thermal modelling results. Modelling allows the interplay and relative importance of 3D topography, spatially variable thermal conductivities, heat production rates and exhumation rates on the shape of low-temperature isotherms to be investigated. In addition, the possible impact of advective and convective heat transport was examined.

The following conclusions were obtained.

- The AER of AFT ages yielded exhumation rates of 0.35 ± 0.1 km Ma^{-1} (northern flank of the Gotthard transect) and 0.5 ± 0.2 km Ma^{-1} (southern flank), similar to previously published data (Schaer *et al.* 1975; Michalski & Soom 1990).
- Measured AFT ages from tunnel samples were around 6.2 Ma (5.6–6.8 Ma), with no significant trend. A perturbation of AFT ages due to topography, with young ages correlating with maximum overburden, is not visible along the Gotthard road tunnel. Because AFT age errors range from 5 to 25% (0.3–1.6 Ma), a perturbation of approximately 0.6 Ma will mostly remain hidden behind the data scatter.
- Modelling reveals that spatially variable heat production and exhumation rates strongly influence the shape of near-surface isotherms, and need to be considered for the interpretation of low-temperature thermochronological data. In the specific case of the Gotthard transect, the modelled perturbation of the 110 °C isotherm is in the range of 150–550 m (the corresponding difference in AFT ages along the tunnel is between 0.3 and 1.2 Ma), depending on the input parameters and boundary conditions.
- This study illustrates that topography-induced perturbation of isotherms can be neglected for the interpretation of AFT ages under the given boundary conditions (topographic wavelength 12 km; relief 1.5 km; exhumation rate 0.45 km Ma^{-1}) and petrophysical parameters of the Gotthard transect, such as thermal conductivities.

This study was funded by the German Science Foundation (DFG), project SP 673/2. G. Höckh, D. Mühlbayer-Renner and D. Kost (Universität Tübingen) are gratefully acknowledged for the mineral separation. Thanks also to T. Wenzel (Universität Tübingen) for his support during electron microprobe analysis. M. Zattin and P. van der Beek are thanked for their constructive reviews of this manuscript.

References

BRAUN, J. 2002. Quantifying the effect of recent relief changes on age–elevation relationships. *Earth and Planetary Science Letters*, **200**, 331–343.

BRAUN, J. 2003. Pecube: a new finite-element code to solve the 3D heat equation including the effects of a time-varying, finite amplitude surface topography. *Computers & Geosciences*, **29**, 787–794.

BURTNER, R. L., NIGRINI, A. & DONELICK, R. A. 1994. Thermochronology of Lower Cretaceous source rocks in the Idaho–Wyoming thrust belt. *AAPG Bulletin*, **78**, 1613–1636.

BUSSLINGER, A. & RYBACH, L. 1999. Felstemperaturprognose für tiefliegende Tunnel. *Tunnel*, **1**, 24–32.

DAHINDEN, T. 2001. *Verschiebungsmessungen im Gotthardgebiet*. Diploma thesis, ETH Zürich.

DUNKL, I. 2000. Trackkey: a windows program for calculation and graphical presentation of fission track data. *Computers & Geosciences*, **28**, 3–12.

EHLERS, T. A. & FARLEY, K. A. 2003. Apatite (U–Th)/He thermochronometry: methods and applications to problems in tectonic and surface processes. *Earth and Planetary Science Letters*, **206**, 1–14.

EHLERS, T. A., CHAUDHRI, T. & KUMAR, S. 2005. Computational tools for low-temperature thermochronometer interpretation. *Reviews in Mineralogy & Geochemistry*, **58**, 589–622.

ENGLAND, P. & MOLNAR, P. 1990. Surface uplift, uplift of rocks, and exhumation of rocks. *Geology*, **18**, 1173–1177.

FOEKEN, J. P. T., PERSANO, C., STUART, F. M. & TER VOORDE, M. 2007. Role of topography in isotherm perturbation: apatite (U–Th)/He and fission track results from the Malta tunnel, Tauern Window, Austria. *Tectonics*, **26**, doi:10.1029/2006TC002049.

FREY, M. & FERREIRO MÄHLMANN, R. 1999. Alpine metamorphism of the Central Alps. *Schweizerische Mineralogische und Petrographische Mitteilungen*, **79**, 135–154.

GALBRAITH, R. F. & LASLETT, G. M. 1993. Statistical models for mixed fission track ages. *Nuclear Tracks and Radiation measurements*, **21**, 459–470.

GLEADOW, A. J. W. 1981. Fission-track dating methods: what are the real alternatives? *Nuclear Tracks and Radiation Measurements*, **5**, 3–14.

HAMPEL, A. & HETZEL, R. 2006. Response of normal faults to glacial–interglacial fluctuations of ice and water masses on Earth's surface. *Journal of Geophysical Research*, **111**, doi:10.1029/2005JB004124.

HOUSE, M. A., WERNICKE, B. P. & FARLEY, K. A. 1998. Dating topography of the Sierra Nevada, California, using apatite (U–Th)/He ages. *Nature*, **396**, 66–69.

HOUSE, M. A., WERNICKE, B. P. & FARLEY, K. A. 2001. Paleo-geomorphology of the Sierra Nevada, California, from (U–Th)/He ages in apatite. *American Journal of Science*, **301**, 77–102.

HURFORD, A. J. & GREEN, P. F. 1982. A users' guide to fission track dating calibration. *Earth and Planetary Science Letters*, **59**, 343–354.

HURFORD, A. J. & GREEN, P. F. 1983. The Zeta age calibration of fission-track dating. *Chemical Geology (Isotope Geoscience Section)*, **41**, 285–317.

KAHLE, H.-G. 1997. Recent crustal movements, geoid and density distribution; contribution from integrated satellite and terrestrial measurements. *In*: PFIFFNER, O. A., LEHNER, P., HEITZMANN, P., MÜLLER, S. & STECK, A. (eds) *Results of NRP 20; Deep Structure of the Swiss Alps*. Birkhaeuser, Basel, 251–259.

KAPPELMEYER, O. & HAENEL, R. 1974. *Geothermics with Special Reference to Application. Geoexploration Monographs*. Gebrüder Bornträger, Stuttgart.

KELLER, F., WANNER, H. & SCHNEIDER, T. R. 1987. Geologischer Schlussbericht Gotthard-Strassentunnel. *Beiträge zur Geologie der Schweiz: Geotechnische Serie*, **70**.

KETCHAM, R. A., CARTER, A., DONELICK, R. A., BARBARAND, J. & HURFORD, A. J. 2007. Improved modeling of fission-track annealing in apatite. *American Mineralogist*, **92**, 799–810.

KOHL, T., SIGNORELLI, S. & RYBACH, L. 2001. Three-dimensional (3-D) thermal investigation below high Alpine topography. *Physics of the Earth and Planetary Interiors*, **126**, 195–210.

LABHART, T. P. 1977. *Aarmassiv und Gotthardmassiv. Sammlung geologischer Führer, 63*. Gebr. Bornträger, Berlin.

LABHART, T. P. 1999. Aarmassiv, Gotthardmassiv und Tavetscher Zwischenmassiv: Aufbau und Entstehungsgeschichte. *In*: LÖW, S. & WYSS, R. (eds) *Vorerkundung und Prognose der Basistunnels am Gotthard und am Lötschberg*. A. A. Balkema, Rotterdam, 31–43.

LUETZENKIRCHEN, V. H. 2002. *Structural Geology and Hydrogeology of Brittle Fault Zones in the Central Eastern Gotthard Massif, Switzerland*. PhD thesis, ETH Zürich.

MANCKTELOW, N. S. & GRASEMANN, B. 1997. Time-dependent effects of heat advection and topography on cooling histories during erosion. *Tectonophysics*, **270**, 167–195.

MICHALSKI, I. & SOOM, M. 1990. The Alpine thermo-tectonic evolution of the Aar and Gotthard Massifs, Central Switzerland: fission track ages on zircon and apatite and K–Ar mica ages. *Schweizerische Mineralogische und Petrographische Mitteilungen*, **70**, 373–387.

MOLNAR, P. & ENGLAND, P. 1990. Late Cenozoic uplift of mountain ranges and global climate change: chicken or egg? *Nature*, **346**, 29–34.

NAESER, C. W. 1978. *Fission Track Dating*. US Geological Survey Open-File Report, 76–190.

NIETHAMMER, G. 1910. Die Wärmeverteilung im Simplon. *Ecologae Geologicae Helvetiae*, **11**, 96–120.

PARRISH, R. R. 1983. Cenozoic thermal evolution and tectonics of the Coast Mountains of British Columbia, 1. Fission track dating, apparent uplift rates, and patterns of uplift. *Tectonics*, **2**, 601–631.

PASTORELLI, S., MARINI, L. & HUNZIKER, J.-C. 2001. Chemistry, isotope values (δD, δ^{18}O, δ^{34}SSO^4) and temperatures of the water inflows in two Gotthard tunnels, Swiss Alps. *Applied Geochemistry*, **16**, 633–649.

PERSAUD, M. & PFIFFNER, O. A. 2004. Active deformation in the eastern Swiss Alps: post-glacial faults, seismicity and surface uplift. *Tectonophysics*, **385**, 59–84.

PFIFFNER, O. A. & HEITZMANN, P. 1997. Geological interpretation of the seismic profiles of the Central Traverse (lines C1, C2 and C3-north). *In*: PFIFFNER, O. A., LEHNER, P., HEITZMANN, P., MUELLER, ST. & STECK, A. (eds) *Deep Structure of the Swiss Alps: Results of NRP 20*. Birkhauser, Basel, 115–122.

RAHN, M. K. & GRASEMANN, B. 1999. Fission track and numerical thermal modeling of differential exhumation of the Glarus thrust plane (Switzerland). *Earth and Planetary Science Letters*, **169**, 245–259.

REINERS, P. W., ZHOU, Z., EHLERS, T. A., XU, C., BRANDON, M. T., DONELICK, R. A. & NICOLESCU, S. 2003. Post-orogenic evolution of the Dabie Shan, eastern China, from (U–Th)/He and fission-track thermochronology. *American Journal of Science*, **303**, 489–518.

ŠAFANDA, J. 1999. Ground surface temperature as a function of slope angle and slope orientation and its effect on the subsurface temperature field. *Tectonophysics*, **306**, 367–375.

SCHAER, J. P., REIMER, G. M. & WAGNER, G. A. 1975. Actual and ancient uplift rate in the Gotthard region, Swiss Alps: a comparison between precise levelling and fission-track apatite age. *Tectonophysics*, **29**, 293–300.

SCHALTEGGER, U. 1994. Unravelling the pre-Mesozoic history of Aar and Gotthard Massifs (Central Alps) by isotopic dating – a review. *Schweizerische Mineralogische und Petrographische Mitteilungen*, **74**, 41–51.

SCHMID, S. M., PFIFFNER, O. A., FROITZHEIM, N., SCHÖNBRON, G. & KISSLING, E. 1996. Geophysical–geological transect and tectonic evolution of the Swiss–Italian Alps. *Tectonics*, **15**, 1036–1064.

STÜWE, K. 2007. *Geodynamics of the Lithosphere – Quantitative Description of Geological Problems*. Springer, Berlin.

STÜWE, K. & HINTERMÜLLER, M. 2000. Topography and isotherms revisited: the influence of laterally migrating drainage divides. *Earth and Planetary Science Letters*, **184**, 287–303.

STÜWE, K., WHITE, L. & BROWN, R. 1994. The influence of eroding topography on steady-state isotherms. Application to fission track analysis. *Earth and Planetary Science Letters*, **124**, 63–74.

TURCOTTE, D. L. & SCHUBERT, G. 1982. *Geodynamics – Applications of Continuum Physics to Geological Problems*. Wiley, New York.

USTASZEWSKI, M., HAMPEL, A. & PFIFFNER, O. A. 2008. Composite faults in the Swiss Alps formed by the interplay of tectonics, gravitation and postglacial rebound: an integrated field and modelling study. *Swiss Journal of Geoscience*, **101**, 223–235, doi: 10.1007/s00015-008-1249-1.

WAGNER, G. A. & REIMER, G. M. 1972. Fission track tectonics: The tectonic interpretation of fission track apatite ages. *Earth and Planetary Science Letters*, **14**, 263–268.

WAGNER, G. A., REIMER, G. M. & JÄGER, E. 1977. Cooling ages derived by apatite fission track, mica Rb–Sr and K–Ar dating: the uplift and cooling history of the Central Alps. *Memorie Instituti Geologia ed Mineralogia Università Padova*, **30**, 1–27.

WHIPP, D. M. & EHLERS, T. A. 2007. Influence of Groundwater flow on thermochronometer-derived exhumation rates in the central Nepalese Himalaya. *Geology*, **35**, 851–854.

WHIPPLE, K. X. & MEADE, B. J. 2006. Orogen response to changes in climatic and tectonic forcing. *Earth and Planetary Science Letters*, **243**, 218–228.

WILLETT, S. D. & BRANDON, M. T. 2002. On steady states in mountain belts. *Geology*, **30**, 175–178.

WYDER, R. F. & MULLIS, J. 1998. Fluid impregnation and development of fault breccias in the Tavetsch basement rocks (Sedrun, Central Swiss Alps). *Tectonophysics*, **194**, 89–107.

Pebble population dating as an additional tool for provenance studies – examples from the Eastern Alps

I. DUNKL[1]*, W. FRISCH[2], J. KUHLEMANN[2] & A. BRÜGEL[2]

[1]*Sedimentology, Geoscience Centre of University of Göttingen, Goldschmidtstrasse 3, Göttingen, D-37077, Germany*

[2]*Institute of Geology, University of Tübingen, Sigwartstrasse 10, Tübingen, D-72076, Germany*

**Corresponding author (e-mail: istvan.dunkl@geo.uni-goettingen.de)*

Abstract: Detrital fission-track (FT) dating can be successfully used in provenance studies of siliciclastic sediments to define the characteristic cooling ages of the source regions during erosion and sedimentation. In order to obtain more specific information about potential source regions we have developed the pebble population dating (PPD) method in which pebbles of specific lithotype are merged and dated. Dating of both zircon and apatite crystals from such pebble populations yields age distributions, which reflect the cooling ages of the given lithotype in the source area at the time of sedimentation. By this technique it is possible to define 'FT litho-terrains' in the source regions and thus outline palaeogeological maps.

Two examples are presented from the Eastern Alps. (i) Comparison of FT ages from a sandstone sample and a gneiss PPD sample from an Oligocene conglomerate of the Molasse Basin shows that the youngest age cluster is present only in the sand fraction and derived from the Oligocene volcanic activity along the Periadriatic zone. The lack of the youngest ages in the gneiss pebble assemblage excludes the Oligocene exhumation of the crystalline basement from mid-crustal level. (ii) Pebble assemblages of red Bunter sandstone, gneiss and quartzite were collected from an Upper Miocene conglomerate of the Molasse Foreland Basin and merged as PPD samples. Apatite and zircon FT grain age distributions of these PPD samples, representing the largest ancient East Alpine catchment, allow generating a new combination of palaeogeological and palaeo-FT-age maps of the Eastern Alps for the Late Miocene.

The reconstruction of the palaeogeological and palaeogeographical setting of Alpine-type mountain chains is among the biggest challenges in the research of orogenic belts. Investigations of active, mountainous areas provide limited information, as erosion removed former higher tectonic units and only the most recently exhumed units can be studied. For this reason, interest has turned to peri-orogenic basins that may preserve a longer denudation history of the orogen. The petrography of pebbles, the mineralogical composition of the arenitic sediments and the geochemical signatures of the detritus reflect the former composition of the drainage areas of the orogen (e.g. von Eynatten 2007; Ruiz *et al.* 2007). Detrital fission-track dating, moreover, provides information on exhumation processes (e.g. McGoldrick & Gleadow 1978; Hurford *et al.* 1984; Hurford & Carter 1991; Garver & Brandon 1994; Thomson 1994; Carter 1999; Ruiz *et al.* 2004).

Fission-track (FT) analysis of pooled pebbles of a distinct lithotype combines the methodology of the widely used single-grain dating of sands and the petrographical analysis of single pebbles. Two examples from the Eastern Alps will be presented to illustrate applicability and potential of the pebble population dating method.

Pebble population dating

Besides structural and metamorphic studies there are three basic methods that are used to determine and quantify the exhumation history of a mountain range: (1) thermochronology of rocks exposed in the mountainous area itself; (2) evaluation of the sediment facies, petrography and accumulation rates in the basins adjacent to the mountain range; and (3) detrital FT dating of siliciclastic sediments.

The clusters of individual grain FT ages from a siliciclastic sediment reflect different erosional provinces with different cooling ages at the time of sedimentation. Brandon (1992) introduced the term '*FT source terrain*' to relate each age group to various eroding areas in the hinterland. We have developed a new technique for conglomeratic sediments, which implies FT dating of populations of pebbles with distinct lithologies. By selecting and merging 50–100 pebbles of the same distinct lithotype from the same outcrop we have obtained representative samples (pebble populations) for different lithologies (e.g. gneiss, quartzite, sandstone, granite). Fission-track analysis of individual apatite or zircon grains from these populations produces age distributions characteristic of the various

From: LISKER, F., VENTURA, B. & GLASMACHER, U. A. (eds) *Thermochronological Methods: From Palaeotemperature Constraints to Landscape Evolution Models*. Geological Society, London, Special Publications, **324**, 125–140.
DOI: 10.1144/SP324.10 0305-8719/09/$15.00

lithological units (Fig. 1). The age clusters of the single-grain age distributions represent areas in the hinterland '*FT litho-terrains*' in which the given lithotype at the time of sedimentation was exposed. Several pebble populations can be studied from one outcrop, and the source area of the sediment may be reconstructed in terms of different lithologies with different exhumation histories. We have summarized in Appendix 1 some hints on the collection of PPD samples.

The (i) estimation of amounts of a given lithologies in the pebbles, (ii) the areal distribution of these lithologies in the present-day geological map, (iii) the present-day FT age distribution on the catchment area of the sediment and (iv) the identified apatite and zircon FT age components in the PPD samples allow the compilation of palaeogeological maps of the eroding areas.

Geological setting of the Eastern Alps

Crystalline units

The crystalline part of the Eastern Alps can be divided into two tectonic units distinguished by contrasting cooling histories (Fig. 2) (Frank *et al.* 1987). (a) The Austroalpine nappe complex forms the upper plate. It is composed of medium- to high-grade as well as low-grade Variscan basement slices, and weakly to non-metamorphosed Permo-Mesozoic volcano-sedimentary cover sequences. $^{40}Ar/^{39}Ar$ and Rb/Sr mica cooling ages between 140 and 70 Ma indicate a predominant Eoalpine metamorphic event, but there are 'patches' in which only the older Variscan metamorphism (*c.* 300 Ma) can be detected. (b) The lower unit is the Penninic domain, exposed in tectonic windows along the central axis of the orogen and in the narrow tectonic wedge of the Rhenodanubian flysch zone in front of the orogen. This unit contains both Variscan basement blocks with post-Variscan cover and Mesozoic ophiolites and oceanic metasediments. It has experienced high pressure (in part) and greenschist- to lower-amphibolite-facies metamorphism in Tertiary time. In completely reset areas mica $^{40}Ar/^{39}Ar$ cooling ages are usually around 20 Ma (see compilations of Frank *et al.* 1987; Thöni 1999).

During Palaeogene collision the Austroalpine nappe complex was thrust over the Penninic units. Nappe stacking was followed by lateral tectonic extrusion in Miocene times (Ratschbacher *et al.* 1991). The extrusion process created numerous brittle structures in the Austroalpine realm and

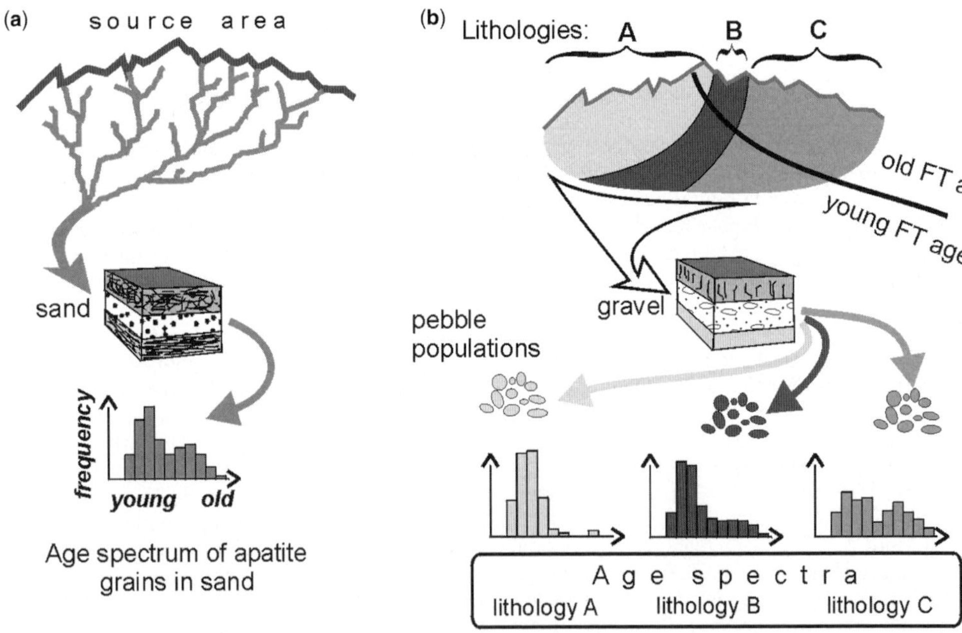

Fig. 1. Schematic cartoons presenting: (**a**) the capability of the detrital geochronology applied to a sandstone sample; and (**b**) the scheme of pebble population method applied to a gravel horizon. In the first case the age distribution characterizes the entire source area, while the PPD samples give litho-specific age information. For example: at the time of sedimentation 'lithology A' delivers pebbles mainly with young FT age, while 'lithology C' provides pebbles with old FT ages.

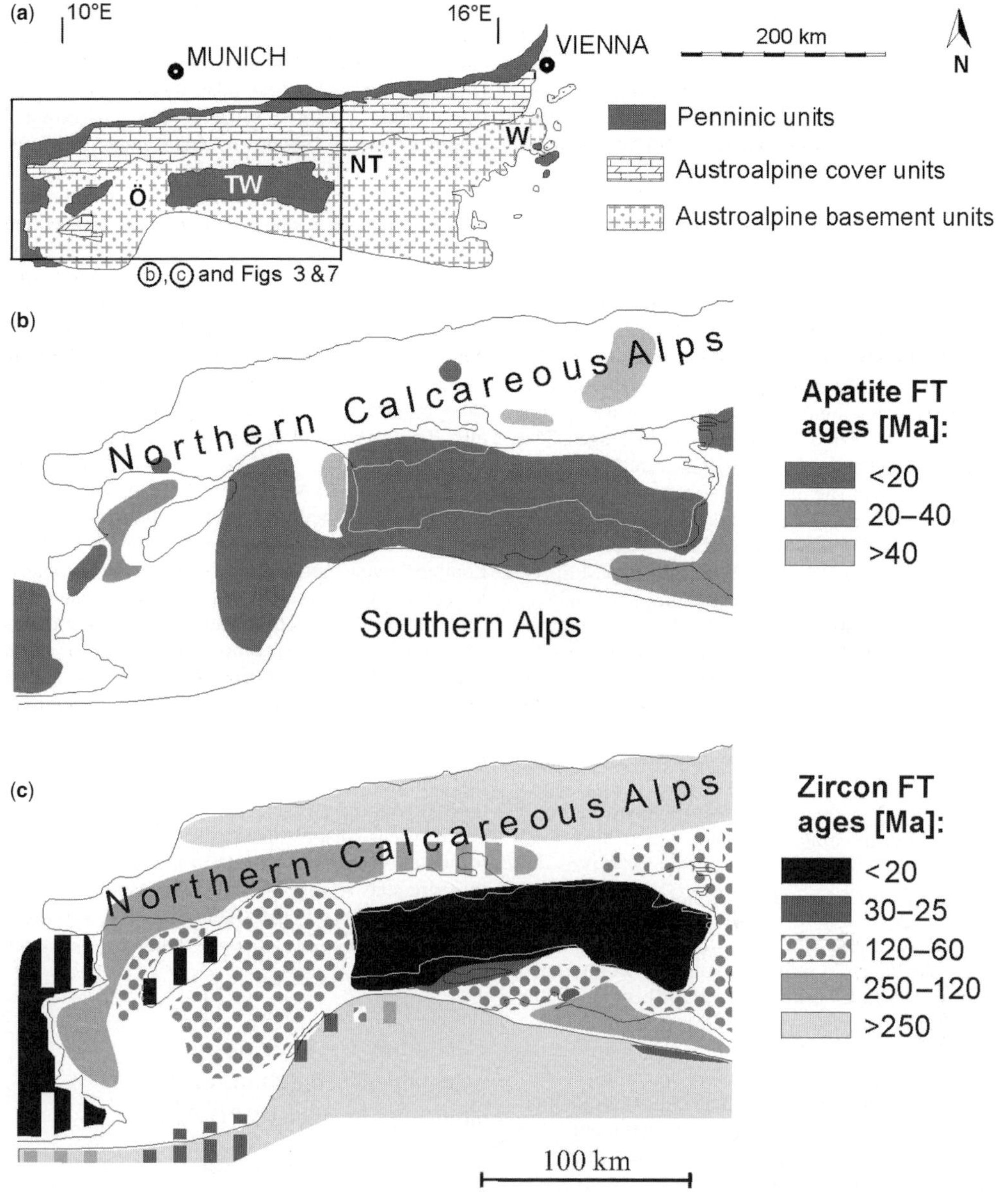

Fig. 2. (**a**) Simplified sketch showing Austroalpine and Penninic units of the Eastern Alps (Ö, Ötztal Alps; TW, Tauern Window; NT, Niedere Tauern; W, Wechsel). Map of fission track ages of the Eastern Alps: (**b**): apatite ages; and (**c**): zircon ages. The contours mark the envelope of the actual FT data coverage; vertical stripes indicate areas with sporadic data points or only an assumed characteristic FT age; white areas, no data. The maps were compiled from 311 apatite and 163 zircon FT ages of: Grundmann & Morteani (1985), Flisch (1986), Hurford (1986), Staufenberg (1987), Hejl & Grundmann (1989), Neubauer *et al.* (1995), Fügenschuh *et al.* (1997), Hejl (1997), Dunkl *et al.* (1998), Elias (1998), Sachsenhofer *et al.* (1998), Viola *et al.* (1999), Reinecker (2000), Stöckhert *et al.* (1999) and I. Dunkl (unpublished data).

exhumed the Penninic units in windows by orogen-parallel tectonic unroofing (Frisch *et al.* 1998, 2000). The margins of the windows are marked by highly contrasting mica ages (Late Tertiary $^{40}Ar/^{39}Ar$ and Rb/Sr ages inside, while mostly Mesozoic ages outside the windows) (Frank *et al.* 1987), which underlines the tectonic nature of unroofing of the lower plate terrains.

Present-day FT age distribution in the Eastern Alps

Zircon fission-track ages clearly show a sharp boundary between the Penninic (ages <20 Ma) and Austroalpine (ages usually >45 Ma) terrains (see compilation of FT ages in Fig. 2 and the citations in the figure caption). Only a slice of Austroalpine basement along the southern border of the Tauern Window shows Tertiary reset FT ages, and this slice is interpreted as the lower part of the upper unit. Contrary to the zircon ages, the tectonic boundary between the upper and lower plate units makes no recognizable change in the apatite FT age map of the Eastern Alps (Fig. 2b). Beyond the Penninic areas a significant part of the Austroalpine basement units yield Neogene apatite FT ages. The areas of Palaeogene apatite FT ages in the Austroalpine unit indicate minor Neogene erosion, and also mark the areas that could have supplied sand or pebbles with Cretaceous and Palaeogene apatite FT ages into the Molasse Basin during Miocene time. However, there are areas in the Austroalpine unit that experienced deep erosion in Neogene time, and the apatite FT chronometer shows Miocene ages (e.g. Ötztal Alps, Niedere Tauern, see Fig. 2). These intensely eroded tectonic blocks were the source regions for the large amount of siliciclastic sediments of the Neogene Molasse basins. As a consequence, the current apatite FT age map of the Eastern Alps does not reflect the distribution of ages in Miocene time.

Denudation history of the Eastern Alps

Beyond the above-listed thermochronometers, the temporal change of sediment volumes accumulated in the peri-Alpine basins provides further constraints on the erosional history of the Alps (see compilation of Kuhlemann 2000). The sudden increase in the amount of coarse conglomerates in molasse sediments in the late Early Oligocene indicates exhumation, surface uplift and relief enhancement as a consequence of collision between the European and Apulian continents and slab breakoff (von Blanckenburg & Davies 1995). In Oligocene and Miocene times the foreland basin was filled with alternating marine and fluviatile sediments (Lemcke 1988). Heavy mineral, petrographical and geochronological studies on sediments indicate that until the Middle Miocene only the Austroalpine upper plate supplied detrital material into the Molasse Basin (see the compilation in Brügel 1998). The first pebbles from the exhumed gneiss domes of the Tauern Window appeared at approximately 13 Ma (Frisch *et al.* 1999). By that time, all the currently exposed parts of the Eastern Alps were already subjected to erosion (Frisch *et al.* 1998).

The first study on the erosion of the Alps using detrital thermochronology was performed on the Gonfolite Lombarda Conglomerate by Gieger & Hurford (1989). The denudation of Eastern Alps was studied by the detrital thermochronological record of the sediments in the Veneto Basin by Zattin *et al.* (2003), and in the northern Molasse Basin and in the intramontane basin remnants by Brügel *et al.* (2003), Dunkl *et al.* (2005) and Kuhlemann *et al.* (2006). The age components identified in the siliciclastic sediments refer more to cooling–exhumation periods of the Eastern Alps. The oldest group of zircon FT ages indicate that a part of the Austroalpine unit was thermally overprinted in Jurassic time, but a Late Cretaceous, Eoalpine event is also documented in many samples. The metamorphic Penninic material – supposedly derived from the Tauern Window–appeared approximately 13 Ma ago and demonstrates the exhumation of the lower plate to the surface. The detrital AFT ages also record the cooling after the Eoalpine event and show a distinct, but probably less important, exhumation phase in Eocene time. The Miocene age components in the sediments show an increase in lag time (difference between the age of the youngest component and the age of sedimentation: Cerveny *et al.* 1988; Brandon & Vance 1992) from 3 to more than 6 Ma during the Late Miocene. This indicates a decrease in the exhumation rate after the Early–Middle Miocene thermo-tectonic climax.

In order to test the capability of the PPD method for detailed provenance studies, we have chosen two siliciclastic formations of the Alpine Molasse for case studies: (1) the 'Inntal Tertiary' of Oligocene age; and (2) the Late Miocene Hausruck Conglomerate.

Case study I – filtering out the volcanogenic contribution of a sandstone by gneiss PPD dating

Geology, samples and results

The sequence studied is part of the 'Inntal Tertiary', one of the oldest conglomerate members of the Molasse Basin. Distal turbidites of Rupelian age and polymict conglomerates of Chattian age (*c.* 28–26 Ma) were deposited both in the foreland and on top of the Northern Calcareous Alps (Fig. 3). The Inntal Tertiary is a basin remnant preserved in the Inntal strike-slip fault zone within the Northern Calcareous Alps (Ortner & Sachsenhofer 1996). The pebble composition is dominated by metamorphic material with some sedimentary lithologies (sandstone and limestone). Volcanic pebbles (andesites–dacites) occur rarely ($\ll$1%)

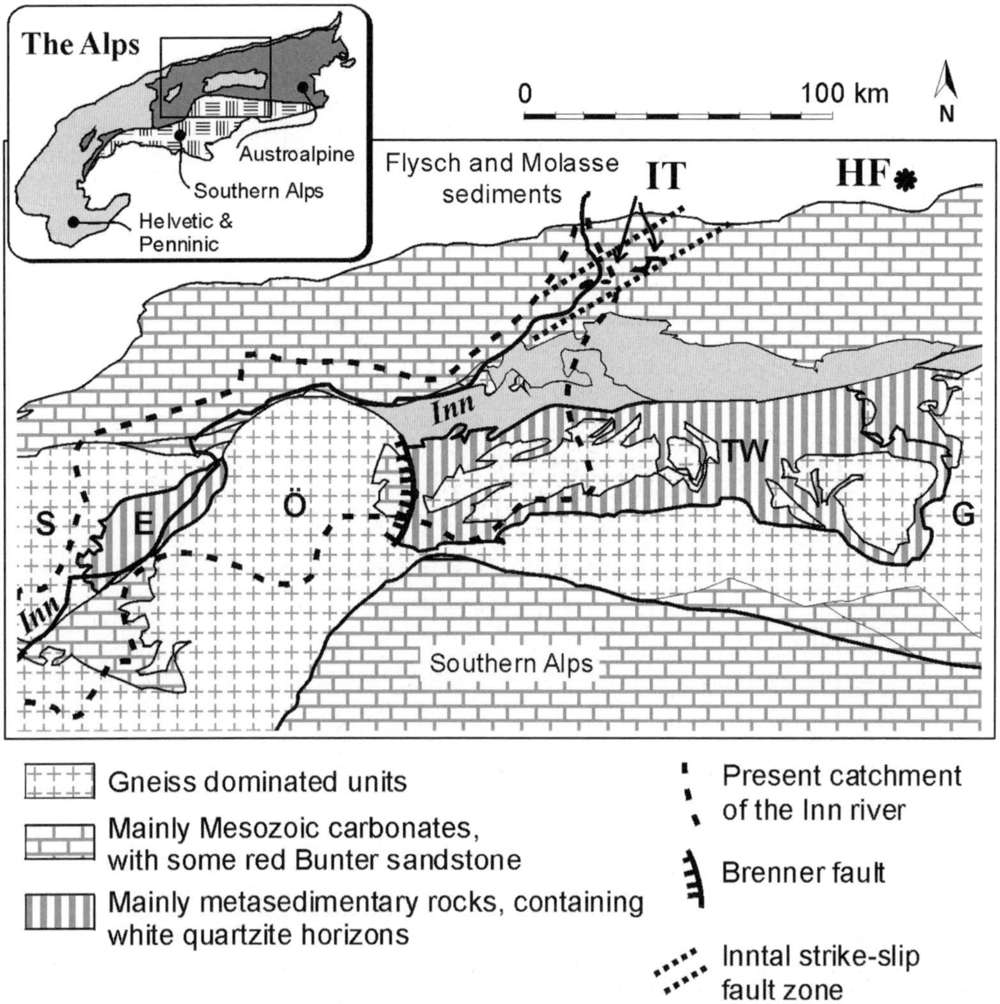

Fig. 3. Geological sketch map of the Eastern Alps with the sample localities and the present-day distribution of the dated lithologies. Ö, Ötztal Alps; S, Silvretta Alps; IT, Inntal Tertiary (within the Northern Calcareous Alps); HF, Hausruck fan; TW, Tauern Window; E, Engadin Window; G, Gurktal Alps. Units marked by a pale grey colour are composed of carbonate rocks and low-grade metapelites, and they do not produce apatite- or zircon-containing sediments during erosion.

(Mair *et al.* 1996; Brügel *et al.* 2000). During the Late Oligocene, the Penninic formations were still buried (Frisch *et al.* 2000), thus the sandy and pebbly material was derived from the uplifted western part of the Austroalpine realm and was transported by the fault-bounded palaeo-Inn River system to the NE (Brügel *et al.* 2003).

Samples were collected in an approximately 100 m-long, 2–3 m-high artificial outcrop of gravel, 400 m SSW of Mosen, 200 m east of Angerberg (Inn Valley, Austria; N47°27′40″/E11°54′38″). We have compiled a gneiss PPD sample, and the apatite and zircon crystals of the sandy matrix were also dated. The laboratory procedure is described in Appendix 2, and the results are in Table 1. The ranges and the means of single-grain apatite FT ages in the sandstone and conglomerate PPD samples are similar, but the fission-track age distributions show differences (Fig. 4). The values of dispersions and the results of Chi-square test indicate that the single-grain ages of the samples have composite distributions. The apatite and zircon FT ages in the sandstone sample form more clusters. The core of the apatite distribution (between 40 and 80 Ma) is identical to the range of the majority of the apatite ages in the gneiss

Table 1. *Apatite and zircon fission-track results obtained from sandstone and pebble population samples from the Alpine Molasse (left). Component analysis of the samples (right). The age components were determined by the 'Binomfit' computer program of Brandon (1996)*

Code	Petrography	Cryst.	Spont. ρ_s (N_s)	Ind. ρ_i (N_i)	Det. ρ_d (N_d)	Disp.	$P(\chi^2)$ (%)	Central age (Ma ± 1σ)	Component I			Component II			Component III		
									Mean (Ma)	CI (Ma)	*W* (%)	**Mean** (Ma)	CI (Ma)	*W* (%)	**Mean** (Ma)	CI (Ma)	*W* (%)
Inntal Tertiary (Late Oligocene, *c*. 27 Ma)																	
Apatite																	
NA-28/D	sand	60	4.03 (1309)	5.64 (1834)	4.08 (12045)	0.26	<1	55.9 ± 3.2	**33**	8	*15*	**59**	7	*79*	**105**	80	*6*
NA-28/B	gneiss PPD	80	6.27 (2912)	7.52 (3492)	4.08 (12045)	0.20	<1	62.7 ± 2.6	**54**	3	*64*	**80**	7	*36*			
Zircon																	
NA-28/D	sand	60	78.4 (5107)	58 (3777)	5.26 (10350)	0.53	<1	47.0 ± 3.5	**33**	2	*67*	**97**	9	*33*			
Hausruck Conglomerate (Late Miocene, *c*. 10 Ma)																	
Apatite																	
NA-40/C	w. qzte PPD	100	2.24 (1505)	12.3 (8259)	4.08 (12045)	0.09	5	13.9 ± 0.5									
NA-40/A	gneiss PPD	100	2.78 (1607)	10.5 (6084)	4.08 (12045)	0.49	<1	21.0 ± 1.3	**15**	1	*71*	**40**	5	*29*			
NA-40/B	red ss. PPD	100	9.61 (3419)	12.6 (4474)	4.08 (12045)	0.30	<1	57.4 ± 2.6	**39**	4	*40*	**74**	6	*60*			
Zircon																	
NA-40/C	w. Qzte PPD	100	33.3 (4701)	49.5 (6983)	5.26 (10350)	0.52	<1	23.2 ± 1.3	**16**	1	*70*	**52**	4	*30*			
NA-40/A	gneiss PPD	100	78 (6634)	61.6 (5230)	5.26 (10350)	0.51	<1	43.5 ± 2.5	**17**	2	*23*	**57**	3	*77*			
NA-40/B	red ss. PPD	100	168 (12005)	29.0 (2066)	5.26 (10350)	0.21	<1	193 ± 6.9	**127**	23	*14*	**209**	14	*86*			

Cryst: number of dated crystals.
Spont., Ind. and Det.: spontaneous, induced and detector track densities and track counts.
Track densities (ρ) are as measured ($\times 10^5$ tr cm^{-2}); number of tracks counted (N) shown in brackets.
Disp.: dispersion, according to Galbraith & Laslett (1993).
$P(\chi^2)$: probability obtaining Chi-square value for ν degree of freedom (where ν = number of crystals − 1).
Central ages calculated using dosimeter glass CN5 with $\zeta_{CN5} = 373.3 \pm 7.1$ (1 SE) for apatite and CN2 with $\zeta_{CN2} = 127.8 \pm 1.6$ (1 SE) for zircon.
CI, 95% confidence interval.
W, weight of the component.

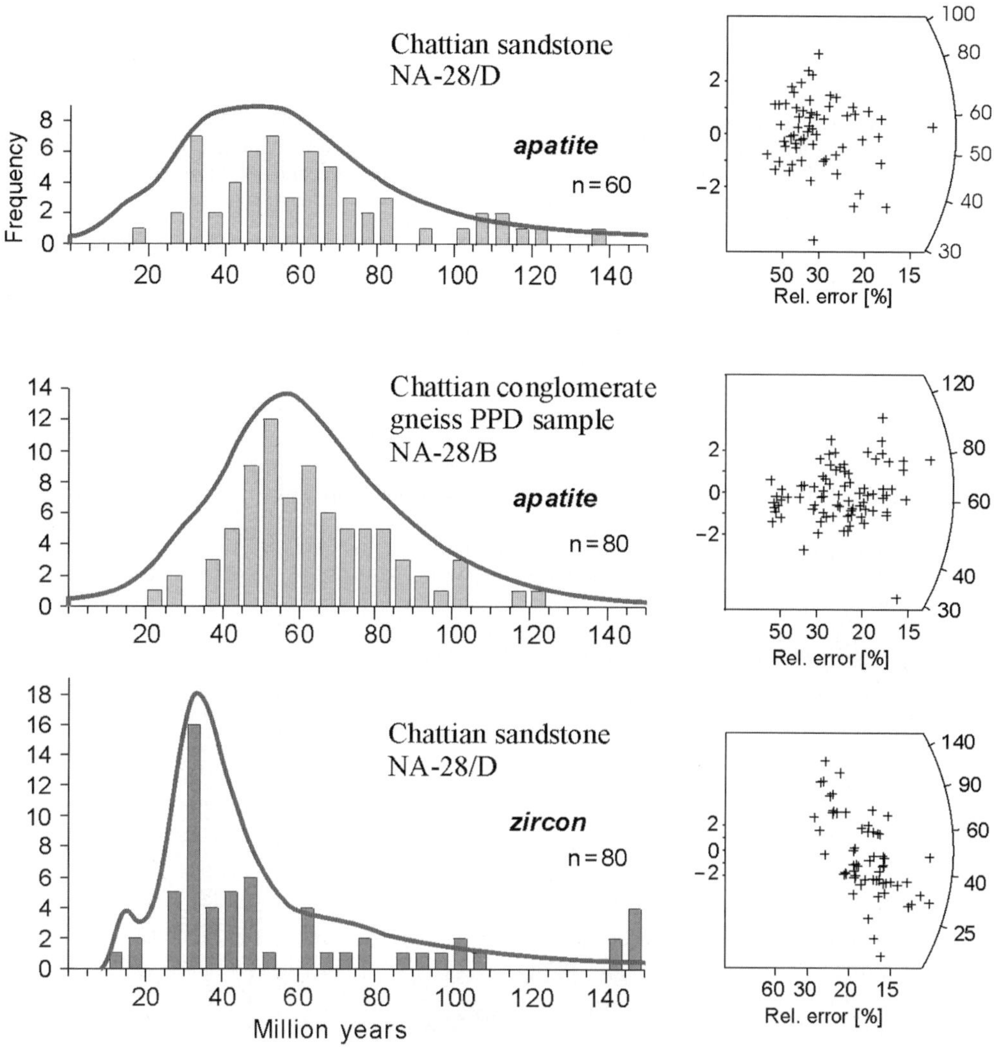

Fig. 4. Single-grain fission-track age distributions of the Inntal Tertiary samples. Age spectra (probability density plots) are computed according to Hurford *et al.* (1984).

PPD sample, but a younger and an older component are also present. Modelling of the age components with the program Binomfit (Galbraith & Green 1990; Brandon 1996) resulted in the following means for these age components: 33 ± 8, 59 ± 7 and 105 ± 80 Ma (Table 1). The poorly constrained oldest component is most probably derived from the 'Upper Austroalpine unit' (*sensu* Tollmann 1977), which is mainly composed of non-metamorphic and greenschist-facies rocks (without gneisses) with approximately 140–120 Ma mica K/Ar ages. This would explain why this age cluster does not appear in the gneiss PPD sample.

Discussion

The appearance of grains with a short lag time is especially important for the reconstruction of Alpine denudation history and palaeogeography in the Palaeogene time. The presence of zircon grains with an approximately 33 Ma FT age in a sediment of 27 ± 2 Ma depositional age can usually indicate a rapid exhumation in the catchment area. However, the youngest zircon age component identified in the sandstone is in apparent contradiction to the older apatite FT ages in the gneiss PPD sample (Fig. 4). Zircon should yield older ages than apatite due to its higher closing temperature. Even within the sandstone sample, the zircon FT ages tend to be younger than the apatite FT ages. Furthermore, apatite grains with ages of less than 40 Ma are rare in the gneiss PPD sample.

This apparent contradiction requires an alternative interpretation. The young zircon age cluster of the sandstone sample must be derived from a source other than the metamorphic basement that delivered the gneiss pebble association. The conglomerate sequence contains a small number of volcanic pebbles derived from the Periadriatic volcanic chain (Mair *et al.* 1996; Brügel *et al.* 2000). This mainly calc-alkaline volcanic–intrusive association (e.g. von Blanckenburg & Davies 1995) was active mainly around 33–28 Ma and formed an approximately 800 km-long belt along the Periadriatic lineament from the Western Alps to the Pannonian Basin. The euhedral shape of the crystals in the young zircon group also supports an origin from the Periadriatic suite (Fig. 5). These magmatic rocks were situated along the southern margin of the catchment of the palaeo-Inn River, some 150–250 km away today from the area of deposition (Brügel *et al.* 2000). Palaeogeographic reconstructions (Frisch *et al.* 1998) show that the transport distance of the volcanogenic pebbles was greater than for the gneissic material. The plagioclase-rich intrusive and volcanic pebbles and the syngenetic ashes were extremely sensitive to weathering, thus the major part of these rocks was decomposed during transport, enriching the accessory minerals, particularly the very durable zircons in the sediment. The low resistance to weathering and the large transport distance explain the scarcity of the volcanic pebbles in the conglomerate. Thus, we suppose that this source lithology supplied into the sandstone the zircon and apatite crystals with young FT ages. The contribution from the Periadriatic magmatic chain to the 'Inntal Tertiary' was much more significant than would be expected from the paucity of volcanogenic pebbles. The young apatite age cluster is very probably also volcanogenic in origin, although this is not evident from the shape

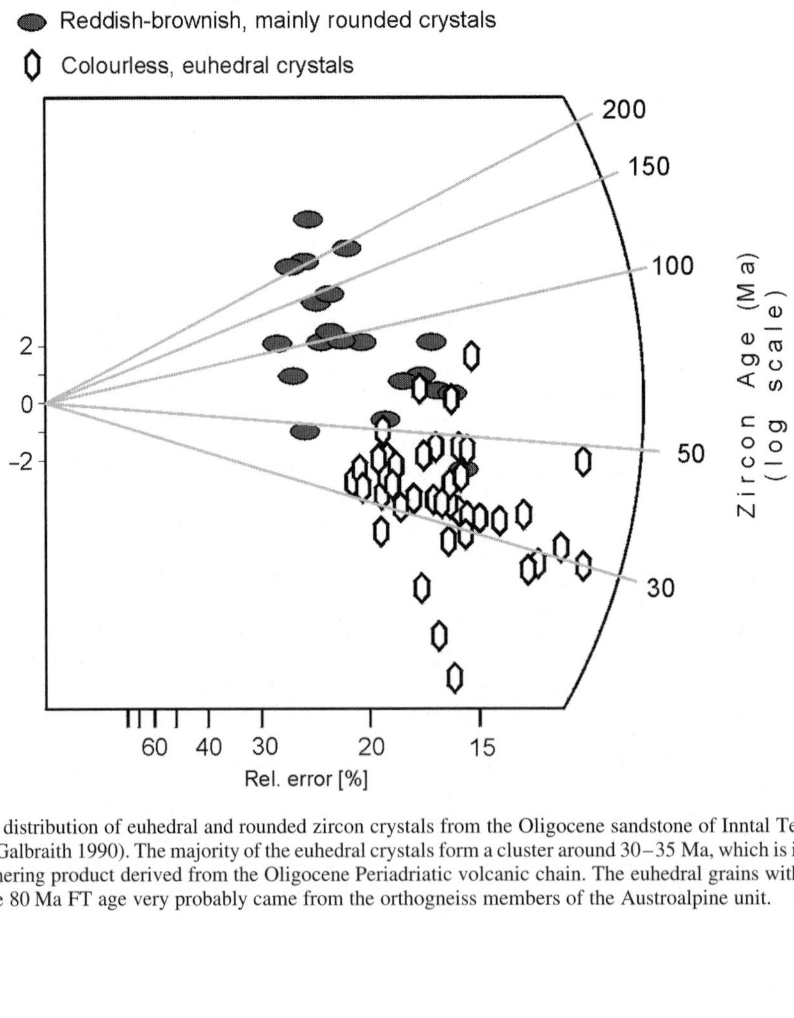

Fig. 5. The distribution of euhedral and rounded zircon crystals from the Oligocene sandstone of Inntal Tertiary on radial plot (Galbraith 1990). The majority of the euhedral crystals form a cluster around 30–35 Ma, which is interpreted as the weathering product derived from the Oligocene Periadriatic volcanic chain. The euhedral grains with an approximate 80 Ma FT age very probably came from the orthogneiss members of the Austroalpine unit.

of the grains, because apatite is much more sensitive to mechanical and chemical corrosion than zircon.

This case study shows that the apatite and zircon crystals were derived mainly from two source lithologies. Different lithologies can have significantly different yield and ratio of datable mineral species. FT analysis of just sandstone samples may provide zircon age distributions with components of short lag time that could be misinterpreted as active, contemporary exhumation of the basement, where in fact the zircons were derived from a volcanic source (see similar scenarios in Ruiz *et al.* 2004). The FT age distributions of PPD samples made from crystalline rocks show the typical cooling ages of the basement, and these cooling age distributions reflect the intensity and rate of the exhumation processes in the hinterland unbiased from the effect of volcanogenic contribution.

Case study II – FT litho-terrains of the Eastern Alps in the Late Miocene

Samples and present-day distribution of the dated lithotypes

The poorly cemented Hausruck Conglomerate is the youngest proximal fluviatile fan preserved in the foreland basin of the Eastern Alps. It was deposited during the Pannonian time (*c.* 10–8.5 Ma: Mackenbach 1984). The transport was directed NE, towards the modern course of the Danube (for the locality see the star in Fig. 3). Owing to the large catchment size and long transport, the pebble spectrum is polymict with a large proportion of polycrystalline quartz (55%). The other lithologies include various metamorphic rocks (32%), white quartzite (7%), sandstone (4%), and a small percentage of granite and limestone (Brügel 1998). Pebble composition, sedimentological data and palaeogeographic considerations suggest that this fan was deposited by the palaeo-Inn River where it entered the Molasse Basin (Brügel *et al.* 2003). We obtained pebble population samples from the most abundant zircon- and apatite-containing lithologies: gneisses, white quartzites and red Bunter sandstones. The studied PPD samples are derived from the lowermost part of the Hausruck Conglomerate sequence, from a gravel pit 300 m SW of Ditting (next to Haag am Hausruck, Austria; N48°10′44″/E13°37′29″).

Figure 3 shows the actual bedrock distribution of these lithologies in the Eastern Alps. Gneisses are the most widespread lithologies of the central zone of the Eastern Alps, and they occur both in the Austroalpine upper plate and Penninic lower plate units. White–light-greenish quartzite partly containing phengitic mica forms marker horizons in both the Austroalpine and the Penninic metamorphic cover sequences. Locally, slightly metamorphosed Lower Triassic (Bunter) sandstones are typical for the base of the western part of the Northern Calcareous Alps and occur only in a structurally high level of the Austroalpine unit.

Results

The fission-track data obtained from the various pebble populations are listed in Table 1 and are shown in Figure 6. Only the apatite age distribution of the white quartzite PPD sample passes the Chi-square test and this indicates the derivation from a single source (central age: 13.9 $\pm$ 0.5 Ma). The FT age distribution was decomposed using the Binomfit program for the component estimation (Galbraith & Green 1990; Brandon 1996). The apatite age distributions have two major components with mean ages of 15$\pm$1 and 40 $\pm$ 5 Ma in the gneiss PPD sample, and with means of 39$\pm$4 and 74 $\pm$ 6 Ma in the Bunter sandstone PPD sample (Table 1). The zircon probability density plots also show complex age distributions. The components are: 16$\pm$1 and 52 $\pm$ 4 Ma in the white quartzite PPD sample (which does not show this bimodality for apatite ages); and 17$\pm$2 and 57 $\pm$ 3 Ma in the gneiss PPD sample. The Bunter sandstone sample contains much older zircon grains than the gneiss PPD sample; it is noticeable that the age distributions have practically no overlapping (Fig. 6).

Discussion

Miocene age components. During the interpretation of the FT results the youngest age component has a prominent, sometimes diagnostic, function (Brandon 1992; Ruiz *et al.* 2004). The lag times of the detected youngest age components are rather short; they range from 4 to 6 Ma for apatite and from 6 to 7 Ma for zircon in both the white quartzite and gneiss PPD samples. The presence of young apatite and zircon ages in these lithologies, and the small difference between the means of the apatite and zircon age components, suggest rapid exhumation in a part of the hinterland.

We believe that the gneiss and white quartzite pebbles have a short lag time derived from the central and eastern Tauern Window, where the present-day zircon FT ages of the basement cluster around 18 Ma (Dunkl *et al.* 2003) and the apatite cooling ages are between 20 and 6 Ma (Staufenberg 1987). The western Tauern Window could not have had a role in the supply of gneiss and white quartzite pebbles because Fügenschuh *et al.* (1997) presented Late Miocene zircon FT ages (*c.* 10 Ma) from the western margin of the Tauern Window, close to the Brenner detachment fault. Hence, this westernmost part of the window was only exhumed later;

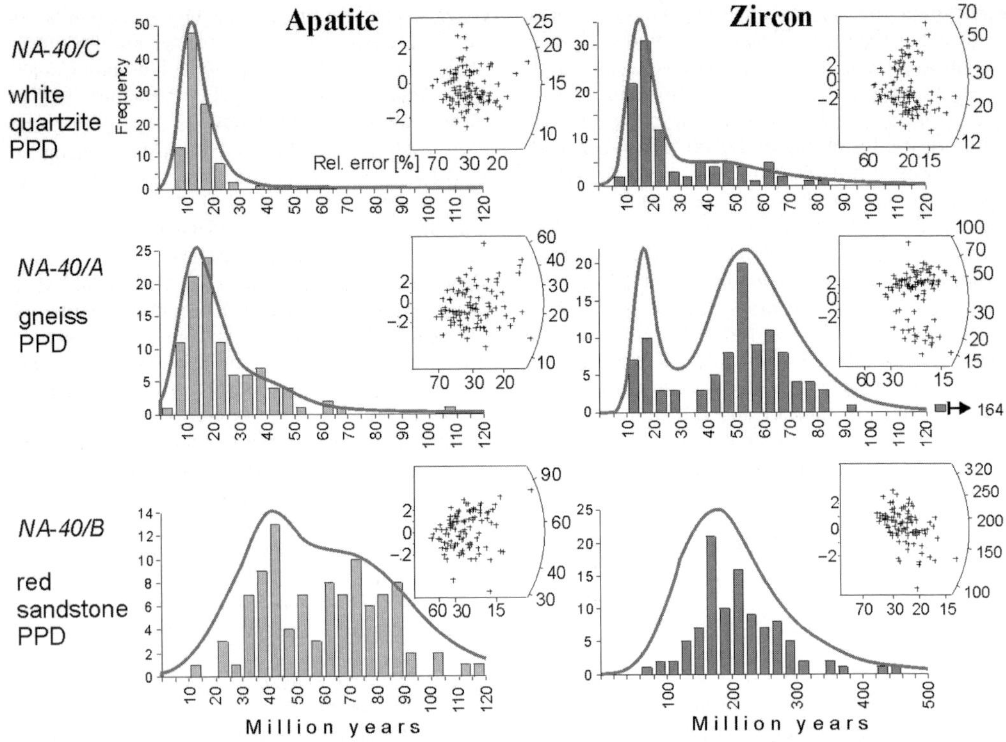

Fig. 6. Age distributions of the pebble population samples derived from the Late Miocene Hausruck Conglomerate (all samples contain 100 data; note the different age scale in the lower right-hand side diagram).

this segment of the lower plate was still covered by the Austroalpine hanging wall during sedimentation of the Hausruck Conglomerate at approximately 10 Ma.

Pre-Miocene age components. The mean values of the zircon age components of the gneiss PPD and the white quartzite samples are very similar, but the gneiss PPD sample contains a significantly higher amount of older grains (Table 1). The presence of Palaeogene apatite FT ages and the higher proportion and slightly older mean of the Palaeogene zircon age group in the gneisses indicate that:

- the gneiss pebbles were derived from both the Penninic and Austroalpine units;
- the Austroalpine source was dominant (a certain ambiguity arises from possible variations of the zircon/apatite ratio in the gneisses);
- high levels of the Austroalpine crystalline basement were not extensively eroded in the catchment area of the palaeo-Inn River during the sedimentation of the Hausruck Conglomerate because the proportion of >50 Ma old apatite ages is very small. This indicates that in the Late Miocene the Ötztal Alps and the Silvretta Alps had already lost the uppermost levels of the Austroalpine gneiss complexes.

The age distributions of the Lower Triassic Bunter sandstone PPD sample are fundamentally different from those of the other lithologies. The zircon ages show a wide scatter, partly due to the large uncertainties of old grains. Some of the zircon ages in the Bunter sandstone are older than the sedimentation age (250–240 Ma) of this rock and thus represent source areas that have not, or have only partly been, thermally overprinted during Alpine metamorphism. The FT age distribution in the red sandstone consists of two age components: a Triassic one (*c.* 209 Ma) and an Early Cretaceous one (*c.* 127 Ma). The older age component is related to cooling after the Permo-Triassic rifting processes (Bertotti *et al.* 1999), or the poorly defined Jurassic tectonothermal event in the Alps that caused partial resetting in certain tectonic blocks (Dunkl *et al.* 1999; Vance 1999). The younger age component of this lithology probably reflects the partial resetting generated by Eoalpine metamorphism in Cretaceous time (Thöni 1981). These Mesozoic thermal events have mainly affected the western part of the Austroalpine nappe pile (Elias 1998). In the eastern part of the presently exposed belt of red sandstones, these rocks generally do not show ductile deformation and the resetting of zircon FT ages.

The apatite age distribution in the Bunter sandstone PPD sample reflects two major source areas, one with Palaeogene and another with Cretaceous ages, which can be related to a western and an eastern source terrain. We suggest that Bunter sandstone pebbles with Palaeogene apatite FT ages originated from the west, where exhumation had been faster since the onset of Oligocene uplift. In this area, the deeper part of the pre-exhumation partial track annealing zone was exposed during the Late Miocene. However, in the east slower erosion had not yet incised into the Early Tertiary partial annealing zone within the Upper Austroalpine nappes. The pebbles of the Hausruck Conglomerate with Cretaceous FT apatite ages were derived from this eastern source terrain.

Palaeogeological implications

The reconstruction of the palaeo-drainage situation of the Inn River during the Late Miocene is based on the FT ages of the pebble population method, and on the recent distribution of the dated lithologies and of the FT ages in the drainage area. The following are the major palaeogeographical and palaeogeological features.

- The Inntal fault is an old and persistent strike-slip fault, repeatedly active since Oligocene times (Ortner & Sachsenhofer 1996). It is responsible for the formation of a longitudinal depression or valley system, which had canalized the runoff of the major part of the western Eastern Alps since Oligocene time (Frisch *et al.* 1998).
- The major rearrangement of tectonic blocks as a result of Early–Middle Miocene large-scale crustal extension and tectonic escape in the course of lateral extrusion was completed before the sedimentation of the Hausruck Conglomerate (Frisch *et al.* 1998, 2000). Thus, we consider the Austroalpine blocks in their present-day positions to be similar to their positions 10 Ma ago.
- The Tauern Window was only partly exhumed and was considerably smaller than today. Figure 7a integrates the lithological and geochronological data, and schematically presents the proportion of different lithologies of the pebbles and their FT ages at time of sedimentation (10 Ma). Figure 7b is an interpretation of the FT ages considering other observations (geology, pebble statistics, tectonics, etc.). The combination of the shading keys for the ages, the black patterns for the lithologies and tectonic units mark the FT litho-terrains.
- We infer that by the Late Miocene time on the exposed surface of the Silvretta Alps and Ötztal Alps the apatite FT ages were already different. Gneiss clasts with Palaeogene apatite FT age were probably derived from the Silvretta Alps (Flisch 1986), whereas Neogene ages were derived probably from the Ötztal Alps (Elias 1998).
- Gneiss pebbles with Miocene apatite and Palaeogene–Late Cretaceous zircon FT ages were mainly derived from the Ötztal Alps. Based on the lack of Cretaceous apatite FT ages in the gneiss PPD sample, we assume that the higher levels of the Ötztal Alps – dominated by the pre-Tertiary – was already eroded by about 10 Ma.
- At 10 Ma the orthogneiss cores within the embryonic Tauern Window were less extended than at present, and in the major part of the window the metasedimentary cover sequences were dominant (Brügel 1998). The western part of the Tauern Window was covered by an extensional allochton, composed of Austroalpine crystalline rocks. The apatite and zircon FT ages of around 10 Ma (Fügenschuh *et al.* 1997) along the western margin of the Tauern Window indicate a later exhumation of that part.
- The exposure of Miocene zircon FT ages during late Miocene time is mainly limited to the area of Penninic formations, but parts of the upper plate south of the Tauern Window and above the western Tauern Window also probably had Miocene zircon FT ages at the time of sedimentation of the Hausruck Conglomerate.

Conclusions

The PPD method is a valuable tool for discriminating the provenance of coarse clastic material. We draw the following conclusions from the two case studies of the pebble population dating method presented in this paper.

- In the case a of synsedimentary volcanic contribution, such a source appears as a young age cluster in the FT single-grain ages in the sandy matrix of the sediment. By dating PPD samples composed of metamorphic lithologies, a volcanogenic population in the sand fraction can be filtered out. Thus, a more reliable geodynamic reconstruction of the source area can be drawn from fission-track dating of pebble population samples than from the detrital thermochronology of sandstones. It is clear from these FT data that during the deposition of the Oligocene sediments of the Inntal Tertiary there was no rapidly cooling crystalline material exposed in the hinterland. The relatively long

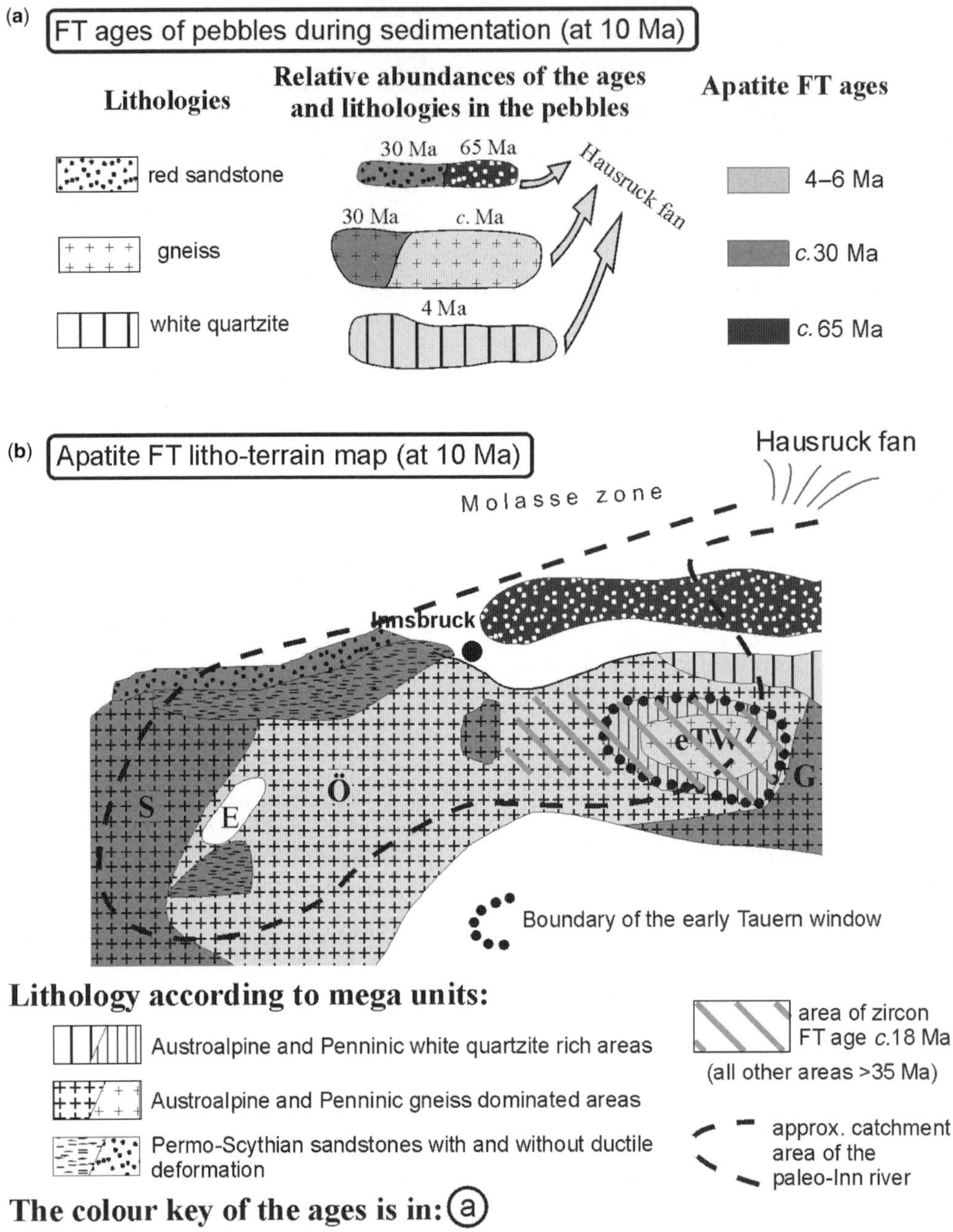

Fig. 7. (**a**) Simplified presentation of the total mass of pebbles in the sediments with their FT age during Hausruck fan sedimentation (average FT age of main clusters minus 10 Ma). (**b**) Palaeogeological map of the western Eastern Alps with the characteristic fission-track ages (10 Ma ago). The area corresponds approximately to the map of Figure 3. eTW, early stage of the Tauern Window; Ö, Ötztal Alps; S, Silvretta Alps; G, Gurktal Alps. The Engadin Window (E) in this early stage of exhumation supplies minor amounts of detrital apatites and zircons due to its dominantly metapelitic composition. Note the retro-deformed shape and position of the main units (cf. Fig. 3).

lag times (several tens of millions of years for both apatite and zircon FT ages) indicate moderate relief and slow exhumation in the western Eastern Alps during the Oligocene.

- PPD samples from different lithologies enable detailed palaeogeological reconstructions. The use of the PPD method on three lithologies from the same outcrop in the Late Miocene Alpine Molasse zone allows these rock types to be connected to distinct source units. The majority of the white quartzite was derived from the rapidly exhumed lower-plate units of the orogen, while the majority of the gneiss pebbles were derived from the upper plate. The gneisses and quartzites, however, show that the Penninic formations in the central and eastern parts of the Tauern Window had already been exhumed to the surface. The Lower Triassic weakly metamorphosed red Bunter sandstones have old apatite and zircon cooling ages, and were, therefore, derived exclusively from a high level of the upper-plate (Austroalpine) unit.
- The PPD method is a powerful tool for revealing the palaeogeodynamics in the catchment areas of river systems feeding syntectonic sedimentary basins. Because it is lithology-based, this technique is capable of providing detailed information about the exhumation history of different units of the source terrain. A combination of PPD results from different lithotypes reveals a detailed reconstruction of FT litho-terrain evolution, which serves as a basis for palaeogeological and palaeogeographical maps.

The German Science Foundation financed this study in the frame of the Collaborative Research Centre 275. The help of M. Kázmér (Budapest) during fieldwork and the computer programs of M. Brandon (New Haven) are gratefully acknowledged. Age standards were made available as a favour from C. W. Naeser (Reston) and A. J. Hurford (London). Thanks are given to M. Brix (Bochum) for the communications by letter on his unpublished FT data. The careful and constructive reviews of J. I. Garver (Schenectady), M. Rahn (Freiburg), M. Brandon (New Haven), P. J. J. Kamp (Hamilton), A. Carter (London), D. Seward (Zurich), E. Hejl (Salzburg), M. G. Fellin (Bologna), G. M. H. Ruiz (Neuchâtel) and B. Ventura (Bremen) were much appreciated, and improved both content and style. Many thanks to D. Patterson for the careful corrections of the English text and improvement of the structure of the manuscript.

Appendix 1: Sampling strategy for pebble population dating

- The pebbles or pebble fragments should be approximately similar in weight to avoid the dominance of apatite and zircon crystals from one or very few pebbles. In practice, we collected pebble (and pebble fragment) samples with a volume of 2–4 l. When the typical pebble size was small (2–3 cm), we split fragments of similar weights from the much larger pebbles. When the sediment was a fanglomerate and dominated by cobbles or boulders, fragments the size of a walnut were collected. Remains of the sandy matrix were carefully removed from the surface of the pebbles. Except for two–five typical rocks slabs (reserved for documentation and thin sections), all of the selected pebbles were crushed together. After homogenization, approximately one-third of the bulk amount was processed using standard mineral concentration techniques.
- An increasing number of pebbles decreases the biasing effect of few atypical pebbles, which may be extremely rich in datable apatite or zircon crystals. We suggest collecting a minimum of 50 pieces for a representative pebble population.
- FT dating of the sandy matrix can supply essential additional information. We recommend this technique, therefore, as a routine process accompanying PPD dating.

Appendix 2: Experimental procedure of fission-track chronology

The samples were treated using heavy liquid and magnetic separation processes; the apatite crystals were embedded in epoxy resin and the zircon crystals in PFA Teflon. For apatite, 1% nitric acid was used with 2.5–3 min etching time (Burchart 1972). In the case of zircon mounts, the eutectic melt of NaOH–KOH was used at a temperature of 210 °C (Gleadow *et al.* 1976). Couples or triplets of mounts were produced to gain enough, properly etched crystals. The etching time varied from 23 to 54 h. Neutron irradiations were made at the Risø reactor (Denmark). The external detector method was used (Gleadow 1981); after irradiation the induced fission tracks in the mica detectors were revealed by etching in 40% HF for 30 min. Track counts were made with a Zeiss-Axioskop microscope–computer-controlled stage system (Dumitru 1993), with magnification of 1000×. The FT ages were determined by the zeta method (Hurford & Green 1983) using age standards listed in Hurford (1998). The error was calculated using the classical procedure, that is by Poisson dispersion (Green 1981). Calculations and plots were made with the TRACKKEY program (Dunkl 2002).

References

BERTOTTI, G., SEWARD, D., WIJBRANS, J., TER VOORDE, M. & HURFORD, A. J. 1999. Crustal thermal regime prior to, during, and after rifting: a geochronological and modeling study of the Mesozoic South Alpine rifted margin. *Tectonics*, **18**, 185–200.

BRANDON, M. T. 1992. Decomposition of fission-track grain-age distributions. *American Journal of Science*, **292**, 535–564.

BRANDON, M. T. 1996. Probability density plot for fission-track grain-age samples. *Radiation Measurements*, **26**, 663–676.

BRANDON, M. T. & VANCE, J. A. 1992. Tectonic evolution of the Cenozoic Olympic Subduction complex, Washington state, as deduced from fission track ages for detrital zircons. *American Journal of Science*, **292**, 565–636.

BRÜGEL, A. 1998. Provenances of alluvial conglomerates from the Eastalpine foreland: Oligo-Miocene denudation history and drainage evolution of the Eastern Alps. *Tübinger Geowissenschaftliche Arbeiten, A*, **40**, 1–168.

BRÜGEL, A., DUNKL, I., FRISCH, W., KUHLEMANN, J. & BALOGH, K. 2000. The record of Periadriatic volcanism in the Eastern Alpine Molasse zone and its paleogeographic implications. *Terra Nova*, **12**, 42–47.

BRÜGEL, A., DUNKL, I., FRISCH, W., KUHLEMANN, J. & BALOGH, K. 2003. Geochemistry and geochronology of gneiss pebbles from foreland molasse conglomerates: geodynamic and paleogeographic implications for the Oligo-Miocene evolution of the Eastern Alps. *Journal of Geology*, **111**, 543–563.

BURCHART, J. 1972. Fission track age determinations of accessory apatite from the Tatra Mountains, Poland. *Earth and Planetary Science Letters*, **15**, 418–422.

CARTER, A. 1999. Present status and future avenues of source region discrimination and characterization using fission track analysis. *Sedimentary Geology*, **124**, 13–45.

CERVENY, P. F., NAESER, N. D., ZEITLER, P. K., NAESER, C. W. & JOHNSON, N. M. 1988. History of uplift and relief of the Himalaya during the past 18 million years – evidence from fission-track ages of detrital zircons from sandstones of the Siwalik Group. *In*: KLEINSPEHN, K. L. & PAOLA, C. (eds) *New Perspectives in Basin Analysis*. Springer, New York, 43–61.

DUMITRU, T. 1993. A new computer-automated microscope stage system for fission-track analysis. *Nuclear Tracks and Radiation Measurement*, **21**, 575–580.

DUNKL, I. 2002. TRACKKEY: a Windows program for calculation and graphical presentation of fission track data. *Computers & Geosciences*, **28**, 3–12.

DUNKL, I., FRISCH, W. & GRUNDMANN, G. 2003. Zircon fission track thermochronology of the southeastern part of the Tauern Window and the adjacent Austroalpine margin. *Eclogae Geologicae Helvetiae*, **96**, 209–217.

DUNKL, I., FRISCH, W. & KUHLEMANN, J. 1999. Fission track record of the thermal evolution of the eastern Alps – review of the main zircon age clusters and the significance of the 160 Ma event. Abstract 4th Workshop on Alpine Geological Studies, Tübingen 21.24.9.99. *Tübinger Geowissenschaftliche Arbeiten, A*, **52**, 77–78.

DUNKL, I., GRASEMANN, B. & FRISCH, W. 1998. Thermal effects of exhumation of a metamorphic core complex on hanging wall syn-rift sediments – an example from the Rechnitz Window, Eastern Alps. *Tectonophysics*, **297**, 31–50.

DUNKL, I., KUHLEMANN, J., REINECKER, J. & FRISCH, W. 2005. Cenozoic relief evolution of the Eastern Alps – constraints from apatite fission track age-provenance of Neogene intramontane sediments. *Austrian Journal of Earth Sciences*, **98**, 92–105.

ELIAS, J. 1998. The thermal history of the Ötztal-Stubai complex (Tyrol, Austria/Italy) in the light of the lateral extrusion model. *Tübinger Geowissenschaftliche Arbeiten, A*, **36**.

FLISCH, M. 1986. Die Hebungsgeschichte der oberostalpinen Silvretta-Decke seit der mittleren Kreide. *Bulletin der Vereinigung schweizer Petroleum-Geologen und Ingenieure*, **53**, 23–49.

FRANK, W., KRALIK, M., SCHARBERT, S. & THÖNI, M. 1987. Geochronological data from the Eastern Alps. *In*: FLÜGEL, H. W. & FAUPL, P. (eds) *Geodynamics of the Eastern Alps*. F. Deuticke, Wien, 272–281.

FRISCH, W., BRÜGEL, A., DUNKL, I., KUHLEMANN, J. & SATIR, M. 1999. Post-collisional large-scale extension and mountain uplift in the Eastern Alps. *Memorie di Scienze Geologiche, Padova*, **51**, 3–23.

FRISCH, W., DUNKL, I. & KUHLEMANN, J. 2000. Postcollisional orogen-parallel large-scale extension in the Eastern Alps. *Tectonophysics*, **327**, 239–265.

FRISCH, W., KUHLEMANN, J., DUNKL, I. & BRÜGEL, A. 1998. Palinspastic reconstruction and topographic evolution of the Eastern Alps during late Tertiary extrusion. *Tectonophysics*, **297**, 1–15.

FÜGENSCHUH, B., SEWARD, D. & MANCKTELOW, N. 1997. Exhumation in a convergent orogen: the western Tauern window. *Terra Nova*, **9**, 213–217.

GALBRAITH, R. F. 1990. The radial plot; graphical assessment of spread in ages. *Nuclear Tracks and Radiation Measurements*, **17**, 207–214.

GALBRAITH, R. F. & GREEN, P. F. 1990. Estimating the component ages in a finite mixture. *Nuclear Tracks and Radiation Measurements*, **17**, 197–206.

GALBRAITH, R. F. & LASLETT, G. M. 1993. Statistical models for mixed fission track ages. *Nuclear Tracks and Radiation Measurements*, **21**, 459–470.

GARVER, J. I. & BRANDON, M. T. 1994. Erosional denudation of the British Columbia Coast Ranges as determined from fission-track ages of detrital zircon from Tofino basin, Olympic Peninsula, Washington. *Bulletin of Geological Society of America*, **106**, 1398–1412.

GIEGER, M. & HURFORD, A. J. 1989. Tertiary intrusives of the Central Alps: their Tertiary uplift, erosion, redeposition and burial in the south-alpine foreland. *Eclogae Geologicae Helvetiae*, **82**, 857–866.

GLEADOW, A. J. W. 1981. Fission-track dating methods: what are the real alternatives? *Nuclear Tracks*, **5**, 3–14.

GLEADOW, A. J. W., HURFORD, A. J. & QUAIFE, R. D. 1976. Fission track dating of zircon: improved etching techniques. *Earth and Planetary Research Letters*, **33**, 273–276.

GREEN, P. F. 1981. A new look at statistics in fission track dating. *Nuclear Tracks*, **5**, 77–86.

GRUNDMANN, G. & MORTEANI, G. 1985. The young uplift and thermal history of the Central Eastern Alps (Austria/Italy), evidence from apatite fission track ages. *Jahrbuch der Geologischen Bundesanstalt, Vienna*, **128**, 197–216.

HEJL, E. 1997. 'Cold spots' during the Cenozoic evolution of the Eastern Alps. *Tectonophysics*, **272**, 159–173.

HEJL, E. & GRUNDMANN, G. 1989. Apatit-Spaltspurendaten zur thermischen Geschichte der Nördlichen Kalkalpen, der Flysch- und Molassezone. *Jahrbuch der Geologischen Bundesanstalt, Vienna*, **132**, 191–212.

HURFORD, A. J. 1986. Cooling and uplift patterns in the Lepontine Alps South Central Switzerland and an age of vertical movement on the Insubric fault line. *Contribution to Mineralogy and Petrology*, **93**, 413–427.

HURFORD, A. J. 1998. Zeta: the ultimate solution to fission-track analysis calibration or just an interim measure? *In*: VAN DEN HAUTE, P. & DE CORTE, F. (eds) *Advances in Fission-track Geochronology*. Kluwer, Dordrecht, 19–32.

HURFORD, A. J. & CARTER, A. 1991. The role of fission track dating in discrimination of provenance. *In*: MORTON, A. C., TODD, S. P. & HAUGHTON, P. D. W. (eds) *Developments in Sedimentary Provenance Studies*. Geolological Society, London, Special Publications, **57**, 67–78.

HURFORD, A. J. & GREEN, P. F. 1983. The zeta age calibration of fission-track dating. *Chemical Geology*, **41**, 285–312.

HURFORD, A. J., FITCH, F. J. & CLARKE, A. 1984. Resolution of the age structure of the detrital zircon populations of two Lower Cretaceous sandstones from the Weald of England by fission-track dating. *Geological Magazine*, **121**, 269–277.

KUHLEMANN, J. 2000. Post-collisional sediment budget of circum-Alpine basins (Central Europe). *Memorie di Scienze Geologiche, Padova*, **52**, 1–91.

KUHLEMANN, J., DUNKL, I., BRÜGEL, A., SPIEGEL, C. & FRISCH, W. 2006. From source terrains of the Eastern Alps to the Molasse basin: detrital record of non-steady-state exhumation. *Tectonophysics*, **413**, 301–316.

LEMCKE, K. 1988. *Geologie von Bayern I.; Das Bayerische Alpenvorland vor der Eiszeit. Erdgeschichte-Bau-Bodenschätze*. E. Schweizer-bart'sche Verlagsbuchhandlung (Nägele u. Obermiller), Stuttgart, 1–175.

MACKENBACH, R. 1984. Jungtertiäre Entwässerungsrichtungen zwischen Passau und Hausruck (O. Österreich). *Sonderveröffentlichung des Geologischen Instituts der Universität zu Köln*, **55**.

MAIR, V., STINGL, V., KROIS, P. & KEIM, L. 1996. Die Bedeutung andesitischer und dazitischer Gerölle im Unterinntal-Tertiär (Tirol, Österreich) und im Tertiär des Mte. Parei (Dolomiten, Italien). *Neues Jahrbuch Geologische und Paläontologische Abhandlungen*, **199**, 369–394.

MCGOLDRICK, P. J. & GLEADOW, A. J. W. 1978. Fission-track dating of lower Palaeozoic sandstones at Tatong, North central Victoria. *Journal of the Geological Society of Australia*, **24**, 461–464.

NEUBAUER, F., DALLMEYER, R. D., DUNKL, I. & SCHIRNIK, D. 1995. Late Cretaceous exhumation of the metamorphic Gleinalm Dome, Eastern Alps: kinematics, cooling history, and sedimentary response in a sinistral wrench corridor. *Tectonophysics*, **242**, 79–98.

ORTNER, H. & SACHSENHOFER, R. F. 1996. Evolution of the Lower Inn Valley Tertiary and constraints on the development of the source area. *In*: WESSELY, G. & LIEBL, W. (eds) *Oil and Gas in Alpidic Thrustbelts and Basins of Central and Eastern Eorope*. European Association of Geoscientists & Engineers (EAGE), Special Publications, **5**, 237–247.

RATSCHBACHER, L., FRISCH, W., LINZER, H.-G. & MERLE, O. 1991. Lateral extrusion in the Eastern Alps, part 2: structural analysis. *Tectonics*, **10**, 257–271.

REINECKER, J. 2000. Stress and deformation: miocene to present-day tectonics in the Eastern Alps. *Tübinger Geowissenschaftliche Arbeiten, A*, **55**, 1–78.

RUIZ, G. M. H., SEWARD, D. & WINKLER, W. 2004. Detrital thermochronology – a new perspective on hinterland tectonics, an example from the Andean Amazon Basin, Ecuador. *Basin Research*, **16**, 413–430.

RUIZ, G. M. H., SEWARD, D. & WINKLER, W. 2007. Evolution of the Amazon Basin in Ecuador with special reference to hinterland tectonics: data from zircon fission-track and heavy mineral analysis. *In*: MANGE, M. A. & WRIGHT, D. T. (eds) *Heavy Minerals in Use*. Developments in Sedimentology, **58**, 907–934.

SACHSENHOFER, R. F., DUNKL, I., HASENHÜTTL, CH. & JELEN, B. 1998. Miocene thermal history of the southwestern margin of the Styrian Basin: coalification and fission track data from the Pohorje/Kozjak area (Slovenia). *Tectonophysics*, **297**, 17–29.

STAUFENBERG, H. 1987. Apatite fission-track evidence for postmetamorphic uplift and cooling history of the Eastern Tauern Window and the surrounding Austroalpine (Central Eastern Alps, Austria). *Jahrbuch der Geologischen Bundesanstalt, Vienna*, **130**, 571–586.

STÖCKHERT, B., BRIX, M. R., KLEINSCHRODT, R., HURFORD, A. J. & WIRTH, R. 1999. Thermochronometry and microstructures of quartz – a comparison with experimental flow laws and predictions on the temperature of the brittle–plastic transition. *Journal of Structural Geology*, **21**, 351–369.

THOMSON, S. N. 1994. Fission-track analysis and provenance studies in Calabrian Arc sedimentary rocks, southern Italy. *Journal of the Geological Society, London*, **151**, 463–471.

THÖNI, M. 1981. Degree and evolution of the Alpine metamorphism in the Austroalpine unit W of the Hohe Tauern in the light of K/Ar and Rb/Sr age determinations on micas. *Jahrbuch der Geologischen Bundesanstalt, Vienna*, **124**, 111–174.

THÖNI, M. 1999. A review of geochronological data from the Eastern Alps. *In*: FREY, M., DESMONS, J. & NEUBAUER, F. (eds) *The New Metamorphic Map of the Alps*; 1:500 000; 1:1 000 000. *Schweizerische Mineralogische und Petrographische Mitteilungen*, **79**, 209–230.

TOLLMANN, A. 1977. *Geologie von Österreich. Vol. 1. Die Zentralalpen*. Deuticke, Wien.

VANCE, J. A. 1999. Zircon fission-track evidence for a Jurassic (Tethyan) thermal event in the Western Alps. *Memorie di Scienze Geologiche, Padova*, **51**, 473–476.

VIOLA, G., MANCKTELOW, N. & SEWARD, D. 1999. The Giudicarie fault system: inherited or primary structural feature? *Terra Nova*, **10**, (Abstr. Suppl. 1), 68.

von Blanckenburg, F. & Davies, J. H. 1995. Slab breakoff: a model for syncollisional magmatism and tectonics in the Alps. *Tectonics*, **14**, 120–131.

von Eynatten, H. 2007. Heavy minerals in the Swiss Molasse Basin: occurrence, frequency, chemistry and thermochronology. *In*: Mange, M. A. & Wright, D. T. (eds) *Heavy Minerals in Use*. Developments in Sedimentology, **58**, 887–905.

Wagner, G., Miller, D. S. & Jaeger, E. 1979. Fission track ages on apatite of Bergell rocks from central Alps and Bergell boulders in Oligocene sediments. *Earth and Planetary Science Letters*, **45**, 355–360.

Yamada, R., Tagami, T., Nishimura, S. & Ito, H. 1995. Annealing kinetics of fission-track in zircon: an experimental study. *Chemical Geology*, **122**, 249–258.

Zattin, M., Stefani, C. & Martin, S. 2003. Detrital fission-track analysis and petrography as keys of alpine exhumation: the example of the Veneto foreland (Southern Alps, Italy). *Journal of Sedimentary Research*, **73**, 1051–1061.

Focused erosion in the Alps constrained by fission-track ages on detrital apatites

MARCO G. MALUSÀ[1*], MASSIMILIANO ZATTIN[2], SERGIO ANDÒ[1], EDUARDO GARZANTI[1] & GIOVANNI VEZZOLI[1]

[1]*Laboratorio di Petrografia del Sedimentario, Dipartimento di Scienze Geologiche e Geotecnologie, Università di Milano-Bicocca, Piazza della Scienza, 4 – 20126 Milano, Italy*

[2]*Dipartimento di Scienze della Terra e Geologico-Ambientali, Università di Bologna, Via Zamboni, 67 – 40127 Bologna, Italy*

**Corresponding author (e-mail: marco.malusa@unimib.it)*

Abstract: Fission-track dating on detrital apatites from modern sands of the Po Delta is used for a provenance study of sediments in the Po River basin. Analysed samples show a fission-track grain-age distribution characterized by two prominent peaks at 7.7 Ma and 17 Ma. The youngest peak accounts for 46% of the total population of dated grains. This young component in the grain-age distribution is consistent with bedrock cooling ages observed in the Western Alps between the External Massifs and the Houiller unit, as well as in the Lepontine dome of the Central Alps and in the Miocene foredeep units of the Apennines, that overall represent only 12% of the orogenic source area. Results suggest that most of the sediment load in the last 10^2–10^5 years was supplied by focused erosion of relatively small areas that experienced short-term erosion rates one order of magnitude higher than in the rest of the belt.

Erosion is a key process in the evolution of orogenic belts. Recent investigations show that erosion patterns in collisional settings are far from homogeneous (e.g. Zeitler *et al.* 2001; Finlayson *et al.* 2002; Garzanti *et al.* 2004*a*; Thiede *et al.* 2005), and focused erosion may occur in relatively small areas due to the interplay between tectonics and climate (Burbank 2002; Reiners & Brandon 2006). Provenance studies based on detrital geochronological techniques are a powerful tool to investigate the present-day pattern of erosion in mountain ranges (e.g. Hurford & Carter 1991; Garver *et al.* 1999; Bernet & Spiegel 2004; Brewer *et al.* 2006). In the Alpine belt, systematic analysis of bulk composition, ranking of metamorphic lithic grains, heavy-mineral assemblages and detrital zircon fission-track (FT) ages suggest that modern river sediments are characterized by specific end-member detrital modes, and they derive from small portions of the drainage where erosion rates are one order of magnitude higher than in surrounding areas (Bernet *et al.* 2001, 2004*a*; Vezzoli 2004; Vezzoli *et al.* 2004; Malusà & Vezzoli 2006).

This paper deals with the analysis of the detrital apatite FT-age signal in the sediment of the Po River, which drains the southern side of the Alps and the northern Apennines and flows eastward across the Po Plain and into the Adriatic Sea. The FT age distributions of detrital apatites from the Po Delta, when compared with the cooling ages of bedrock exposed in the catchment, provide new constraints to the erosional pattern and the sediment supply from the Alps, which represent the largest orogenic source area in the Po basin.

Tectonic setting and bedrock apatite FT ages

The Alpine belt consists of slices of continental crust, slivers of oceanic lithosphere and cover sequences, stacked and deformed during continental collision between the African and European plates (e.g. Coward *et al.* 1989; Roure *et al.* 1990). Rocks exposed in the axial sector of the belt, bounded by the Frontal Pennine Fault and the Periadriatic Fault (Fig. 1), underwent pervasive multistage metamorphism during Cretaceous–Early Oligocene times (Desmons *et al.* 1999*a*, *b*; Frey & Ferreiro Mählmann 1999). Since the pioneering works of Wagner & Reimer (1972) and Schaer *et al.* (1975), several thermochronological techniques have been used to study the exhumation history of the these rocks (e.g. Hunziker *et al.* 1992). After decades of thermochronological analyses a numbers of apatite FT ages has been accummulated all along the chain, and a few major fault-bound blocks that experienced different exhumation histories at shallow-crustal levels can thus be distinguished on the southern side of the Alps (Fig. 2).

From: LISKER, F., VENTURA, B. & GLASMACHER, U. A. (eds) *Thermochronological Methods: From Palaeotemperature Constraints to Landscape Evolution Models*. Geological Society, London, Special Publications, **324**, 141–152.
DOI: 10.1144/SP324.11 0305-8719/09/$15.00

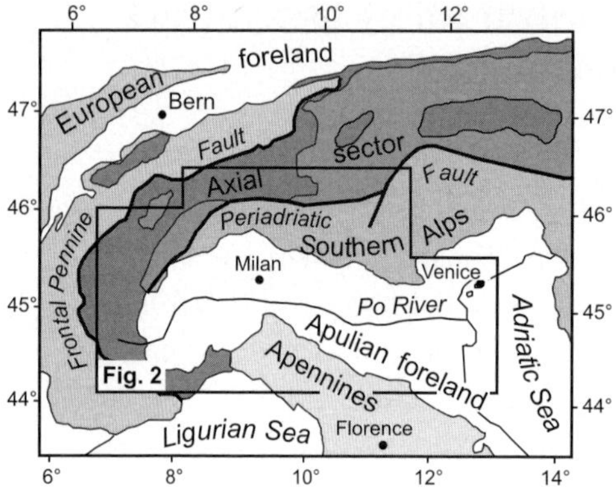

Fig. 1. Sketch map of the Po Plain and adjacent orogenic belts (Alps and Apennines).

In the Western Alps, the Internal Houiller Fault (Malusà 2004) and the Longitudinal Fault System (Barfety & Gidon 1975) split the belt in two major blocks (western and eastern) characterized by different apatite and zircon FT ages (Malusà *et al.* 2005). Within the catchment area of the Po basin, the western block includes the Mt Blanc–Aiguilles Rouges External Massifs and the Houiller and Valaisan units, which are all characterized by apatite ages younger than 10 Ma (Carpéna 1992; Seward & Mancktelow 1994; Fügenschuh *et al.* 1999; Seward *et al.* 1999; Sabil & Menard 2000; Fügenschuh & Schmid 2003; Leloup *et al.* 2005). The eastern block, including the Gran San Bernardo basement, the Internal Penninic units, the Dent Blanche and Sesia-Lanzo Austroalpine units, and the Piedmont ophiolitic units, instead, shows

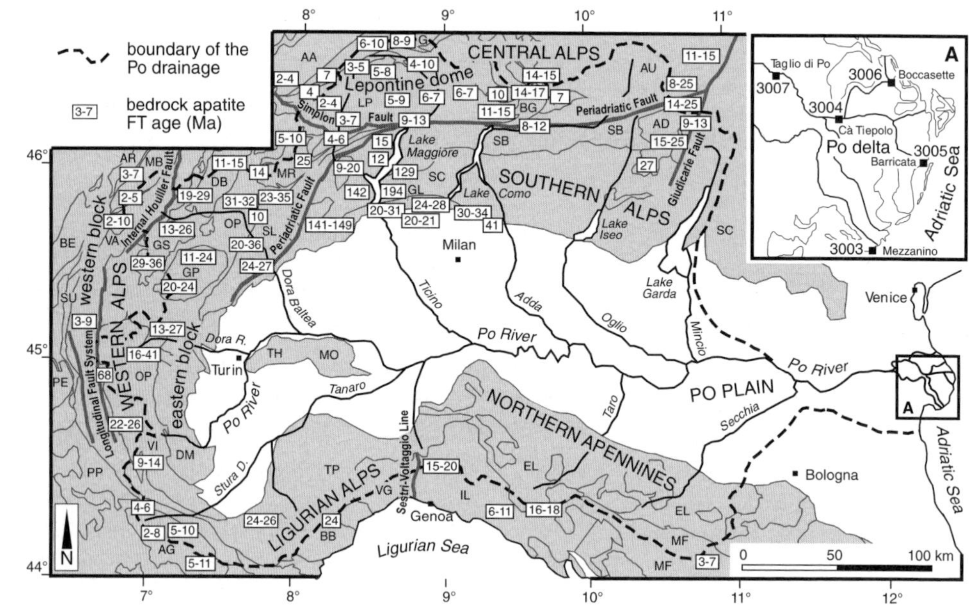

Fig. 2. Compilation of bedrock apatite fission-track ages (in Ma) in the orogenic source areas of the Po basin. Major faults controlling the distribution of apatite fission-track ages are also indicated. Inset: location of the analysed samples from the Po Delta. AA, Aar massif; AD, Adamello pluton; AG, Argentera massif; AU, Austroalpine units (Central Alps); AR, Aiguilles Rouges massif; BB, Briançonnais basement (Ligurian Alps); BE, Belledonne massif; BG, Bergell pluton; DB, Dent Blanche system; DM, Dora–Maira unit; EL, External Ligurid units; G, Gotthard massif; GL, Southalpine Oligo-Miocene clastic wedge; GP, Gran Paradiso unit; GS, Gran San Bernardo basement units; IL, Internal Ligurid units; LP, Lower Penninic units; MB, Mont Blanc massif; MF, Miocene foredeep units; MO, Monferrato; MR, Monte Rosa unit; OP, Piedmont ophiolitic units; PE, Pelvoux massif; PP, Parpaillon nappe; SB, Southalpine basement; SC, Southalpine cover sequences; SL, Sesia–Lanzo unit; SU, Subbriançonnais units; TH, Torino Hill; TP, Tertiary Piedmont Basin; VA, Valaisan units; VG, Voltri Group; VI, Viso massif. Bedrock fission-track data on apatite after Wagner & Reimer (1972), Schaer *et al.* (1975), Wagner *et al.* (1977, 1979), Steiner (1984), Carpéna (1985, 1992), Hurford & Hunziker (1985, 1989), Hurford (1986), Giger & Hurford (1989), Hurford *et al.* (1989, 1991), Munardi (1989), Bürgi & Klötzli (1990), Michalski & Soom (1990), Soom (1990), Giger (1991), Hunziker *et al.* (1992), Bernoulli *et al.* (1993), Seward & Mancktelow (1994), Balestrieri *et al.* (1996, 2004), Martin *et al.* (1998), Fügenschuh *et al.* (1999), Seward *et al.* (1999), Ventura & Pini (1999), Bigot-Cormier *et al.* (2000, 2006), Bogdanoff *et al.* (2000), Sabil & Menard (2000), Viola (2000), Bistacchi *et al.* (2001), Tricart *et al.* (2001, 2007), Ventura *et al.* (2001), Viola *et al.* (2001, 2003), Carrapa (2002), Fügenschuh & Schmid (2003), Timar-Geng *et al.* (2004), Ciancaleoni (2005), Fellin *et al.* (2005), Keller *et al.* (2005), Leloup *et al.* (2005), Malusà *et al.* (2005), Rahn (2005), and Schwartz *et al.* (2007).

apatite ages between 9 and 68 Ma, with a cluster at approximately 20 Ma (Carpéna 1985; Hurford & Hunziker 1985, 1989; Hurford *et al.* 1989, 1991; Bistacchi *et al.* 2001, 2007; Tricart *et al.* 2001, 2007; Balestrieri *et al.* 2004; Malusà *et al.* 2005; Schwartz *et al.* 2007). The same pattern can be observed further to the south, where the Argentera External Massif shows much younger apatite ages (2–11 Ma) (Bigot-Cormier *et al.* 2000, 2006; Bogdanoff *et al.* 2000) than the Briançonnais basement rocks of the Ligurian Alps (24–26 Ma) (Carrapa 2002).

In the Central Alps, on the footwall of the Simplon Fault, apatite ages from the Lepontine dome range from 2 to 10 Ma (Wagner *et al.* 1977; Steiner 1984; Hurford 1986; Timar-Geng *et al.* 2004; Ciancaleoni 2005; Keller *et al.* 2005; Rahn 2005). The youngest ages (2–7 Ma) characterize the western part of the dome, also referred to as Toce subdome (Merle *et al.* 1989). Young apatite ages characterize also the nearby Gotthard area (6–10 Ma) (Schaer *et al.* 1975; Wagner *et al.* 1977; Michalski & Soom 1990; Soom 1990) and the NE part of the Monte Rosa unit (4–10 Ma) (Wagner *et al.* 1977; Keller *et al.* 2005). Conversely, the Austroalpine units of the Central Alps mostly show older apatites ages, ranging from 10 to 25 Ma (Wagner *et al.* 1977, 1979; Martin *et al.* 1998; Viola *et al.* 2003). The same FT age range is observed in the Bergell pluton (Wagner *et al.* 1977, 1979; Munardi 1989).

South of the Periadriatic Fault, most of the Southern Alps were never buried below the apatite partial annealing zone (60–120 °C: Gleadow & Brown 1999) during the Alpine orogeny. Apatites were annealed only in the northernmost basement exposures, where FT ages range between 9 and 20 Ma (Wagner *et al.* 1977; Hurford 1986; Bürgi & Klötzli 1990; Hurford *et al.* 1991). Apatites from the Adamello pluton yield FT ages within the same range (14–27 Ma) (Martin *et al.* 1998; Viola 2000; Viola *et al.* 2003). The younger ages observed along the Periadriatic Fault (8–12 Ma) (Wagner *et al.* 1979; Viola 2000; Viola *et al.* 2001, 2003) are probably related to the activity of this major tectonic discontinuity. Conversely, apatites in the southern part of the Southalpine retrowedge are generally thermally unaffected by the Alpine orogeny. They yielded FT ages between 129 and 194 Ma in the Variscan basement (Giger 1991), and between 20 and 41 Ma in the Oligo-Miocene Gonfolite turbidites (Wagner *et al.* 1979; Giger & Hurford 1989; Bernoulli *et al.* 1993; Fellin *et al.* 2005). A similar trend with annealed apatites in the north and unannealed apatites in the south characterizes the Southern Alps east of the Lake Garda, outside the Po drainage (Zattin *et al.* 2003, 2006).

Another important sediment source in the Po basin is the Northern Apennines fold and thrust belt, made of structural units accreted onto the Adriatic foreland mainly during the Neogene and bounded to the west by the Sestri–Voltaggio Line (Fig. 2). In the uppermost structural unit, that is the ophiolite-bearing Ligurid Nappe, FT in apatites were either locally annealed (6–11 Ma: Balestrieri *et al.* 1996) or document a much older thermal history (ages up to 350 Ma: Zattin unpublished data). Lower units, consisting of thick foredeep turbidites accreted in the evolving thrust belt, are characterized by young FT ages (3–7 Ma) (Ventura & Pini 1999; Ventura *et al.* 2001).

Why dating detrital apatites?

Bedrock versus detrital FT data

Bedrock apatite FT data provide an estimate of the average exhumation rate from the partial annealing zone to the surface, which is equivalent to the erosion rate when exhumation is mainly due to erosion. The timescale the rate is referred to depends on the distribution of FT age in the study area. If apatite ages range between 10 and 100 Ma, exhumation rates are averaged over 10^7–10^8 years timescales. To describe rapid tectonic processes in orogenic settings, which happen within timescales of 10^6 Ma (e.g. Burg *et al.* 1998; Zeitler *et al.* 2001), the time-window covered by 10^7–10^8 FT ages is therefore too large. More adequate constraints may be provided by the analysis of detrital systems. Detrital FT data (e.g. Gallagher *et al.* 1998; Carter 1999; Gleadow & Brown 1999) offer two advantages over bedrock approaches. First, rivers provide an efficient means for statistic sampling over large areas. Therefore, according to Bernet *et al.* (2004*a*), peaks in FT grain-age distributions are a truly representative mirror of bedrock FT ages in the source regions. Second, and this is the novelty of this work, grain-age distributions provide semi-quantitative constraints to the erosional pattern on a 10^2–10^5 years timescale (see discussion in the next paragraph), which represents a more adequate time-window to investigate orogenic processes. This latter evaluation is based on the comparison between the size of the peaks in grain-age distributions and the area of the potential source regions. Therefore, a single detrital FT dataset bears the potential to constrain the pattern of erosion both at long-term (10^6–10^8 years) and short-term (10^2–10^5 years) timescales. We chose to focus our attention on apatite FT, rather than on zircon FT, because apatite ages chiefly reflect the very last steps of the exhumation path, from 3–4 km depth to the surface. Zircon ages, which refer to cooling from higher temperature, may reflect instead longer and more complex thermal paths. The major source of bias in detrital thermochronology are the effects of sedimentary processes such as storage and reworking (see discussion below), but in foreland basins these processes operate on timescales of between 10^2 and 10^5 years, which are far below the resolution of the apatite FT method.

The effect of sedimentary processes

The interpretation of grain-age distributions in modern sands is based on the recognition of different age populations and on their correlation with specific source areas characterized by specific bedrock-age fingerprints. In order to perform reliable correlations, any possible delay of the cooling-age signal in the delta with respect to the source areas must be carefully evaluated. Apatite grains are transported both as bedload and suspended load along the river trunk. To reach the delta, bedload takes an average transportation time that is geologically negligible and far below the resolution of the FT dating method. However, grains may be stored in valley deposits, lakes, alluvial plains and/or in the delta itself, and reworking of stored sediment potentially introduces apatites that have 'aged' while in storage, leading to a shift of the peaks in the FT grain-age distribution. Because of the analytical sensitivity of the FT method, the effects of storage and reworking on the distribution are undetectable when sediments are reworked on timescales of less than 10^6 years.

In valley deposits, cycles of aggradation and incision alternate on timescales of 10^3–10^4 years (e.g. Bull 1991; Bernet *et al.* 2004*a*) (Fig. 3). This is confirmed by the occurrence of several orders of fluvial terraces formed after the Pleistocene Last Glacial Maximum, that are commonly preserved in Alpine valleys (e.g. Polino *et al.* 2002). Therefore, reworking of sediment stored adjacent to mountain rivers does not affect FT grain-age distributions. In foothill regions on the southern side of the Alps, cycles of scouring and filling of major lakes during glacial and interglacial periods occurred within timescales of 10^4–10^5 years. As a consequence these lakes can be considered as ephemeral sediment traps. Conversely, even if recycling occurs extensively in alluvial plains, rapid burial in strongly subsiding foreland basins such as the Po Plain prevents reworking on timescales of more than 10^5 years. Sediments 1 Ma old in the Po Plain are, in fact, buried to depths ranging from approximately 100 to 150 m, near the Southalpine frontal thrusts (Scardia *et al.* 2006), to more than 400 m in the eastern Po Plain (Regione Emilia-Romagna & ENI-AGIP 1998), far beyond the reach of fluvial reworking. Finally, storage and

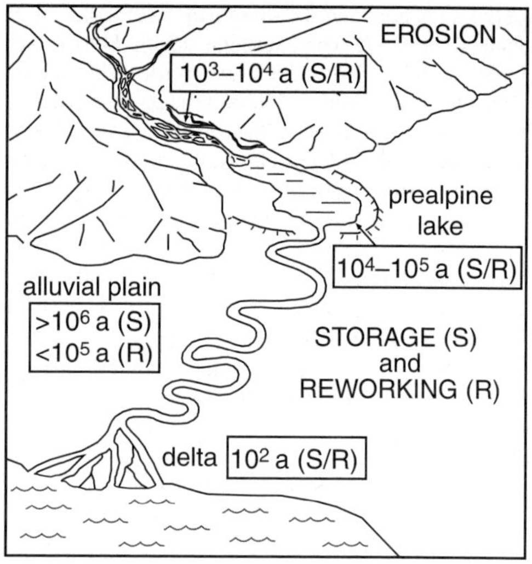

Fig. 3. Conceptual path of the apatite grains from the source areas to the delta. Boxes indicate the timescales for storage (S) and reworking (R) in different settings.

reworking also occurs extensively in the delta, which is currently undergoing erosion and sediment redistribution. Reworking of Po Delta sediments, however, occurred on 10^2 years timescales, because the 'modern' Po Delta formed only at the beginning of the seventeenth century (see maps in Mikhailova 2002) when the Venetians – fearing that discharged sediments might close the mouth of the Venice lagoon – diverted the Po southeastwards, excavating a 5 km-long channel (Taglio di Po: Cencini 1998).

We can reasonably conclude that no significant shift in the FT grain-age distribution is expected between source mountain areas and the Po Delta.

Detrital fission-track ages

Sampling and dating method

Bulk samples of modern river and beach sands, fine to very fine, were collected at five different sites of the Po Delta (Fig. 2), in order to check possible variations within the full population of detrital apatites. Apatite grains were separated in the FT laboratory of the University of Bologna using standard hydrodynamic, magnetic and heavy-liquid separation techniques, and they were prepared for irradiation according to the external-detector method (Hurford 1990; Wagner & Van den Haute 1992). After irradiation at the Oregon State University reactor, the mica detectors were etched in 40% hydrofluoric acid (HF) for 45 min.

One of the advantages of using apatite is that only one mount with a single etching step is needed, because apatite does not show the same differential etching behaviour as zircon (Bernet *et al.* 2004*b*). It is also worth noting that the low temperature of track annealing in apatites ($T > 60$ °C according to Gleadow & Brown 1999) is not a problem for the analysis of modern sands, because apatites did not experience deep burial after erosion from their bedrocks.

Fission tracks were counted using a Zeiss microscope at a magnification of 1250×. As detrital samples usually contain a mixture of apatites with a large range of cooling ages, all suitable grains in each sample were dated independently of their shape, colour or other attributes to ensure that all major components of the grain-age distribution were faithfully represented. The zeta calibration approach (Hurford & Green 1983) was used for age determinations, with $\zeta_{CN5} = 340.05 \pm 4.29$.

The resulting FT grain-age distributions were treated as a mixture of populations, also referred to as peaks. Populations were estimated: (a) according to the binomial peak-fitting method and a log-normal distribution (Galbraith & Green 1990; Brandon 1992, 1996; Stewart & Brandon 2004) using the program 'Binomfit' (Brandon, as summarized in Ehlers *et al.* 2005); and (b) according to the simplex method for the best-fit search and a Gaussian distribution type using the program 'Popshare' (Dunkl & Székely 2003).

It is worth noting that Gaussian approximations of the Poisson-distributed spontaneous and induced track counts tend to break down for young cooling ages. This is especially true for apatites, which commonly have a lower U content than zircons. In these cases, decomposition methods based on the Gaussian distribution will perform poorly, while the binomial peak-fitting method provides unbiased estimates of peak ages for track counts of any size (Brandon 2002). These peaks have to be supported by geological arguments before being treated as true representatives in order to recognize artificial and geologically meaningless solutions.

Results

Samples from the Po Delta show a large percentage of young ages (Fig. 4) and a large span in single-grain ages, ranging from 3 to 261 Ma (Fig. 5). The variance in grain ages is greater than expected from the analytical uncertainty alone, as the observed distributions are mixtures of different grain-age components that appear as discrete and well-defined peaks in the probability density plots (Fig. 6). The clusters in the distribution, when formally analysed using the binomial peak-fitting method, show distinct populations defined by an estimated age and size (Table 1). The youngest peak, which can be loosely viewed as the pooled age of the youngest concordant grain ages in the FT grain-age distribution (Stewart & Brandon 2004), are very similar in all analysed samples. They range from 6.7 Ma (sample 3004) to 8.4 Ma (sample 3003) and overlap within their two stardard deviations (2 SD) (Figs 5, 6 and Table 1). A second

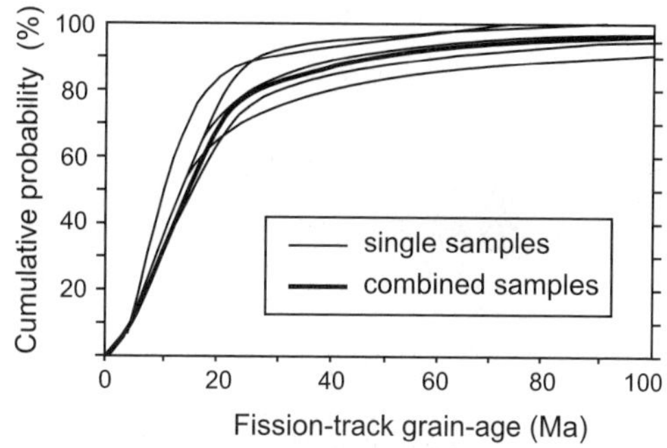

Fig. 4. Cumulative probability plots for the observed FT single-grain age distributions, showing a great percentage of ages greater than 20 Ma.

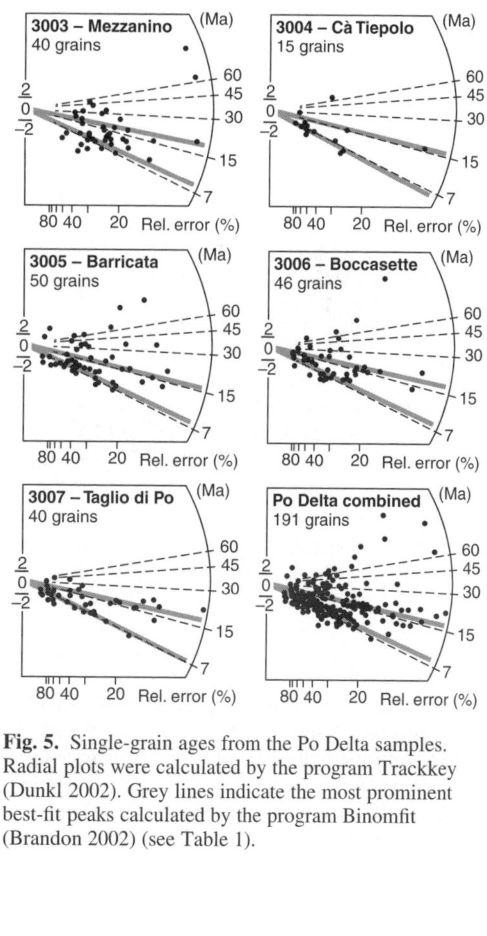

Fig. 5. Single-grain ages from the Po Delta samples. Radial plots were calculated by the program Trackkey (Dunkl 2002). Grey lines indicate the most prominent best-fit peaks calculated by the program Binomfit (Brandon 2002) (see Table 1).

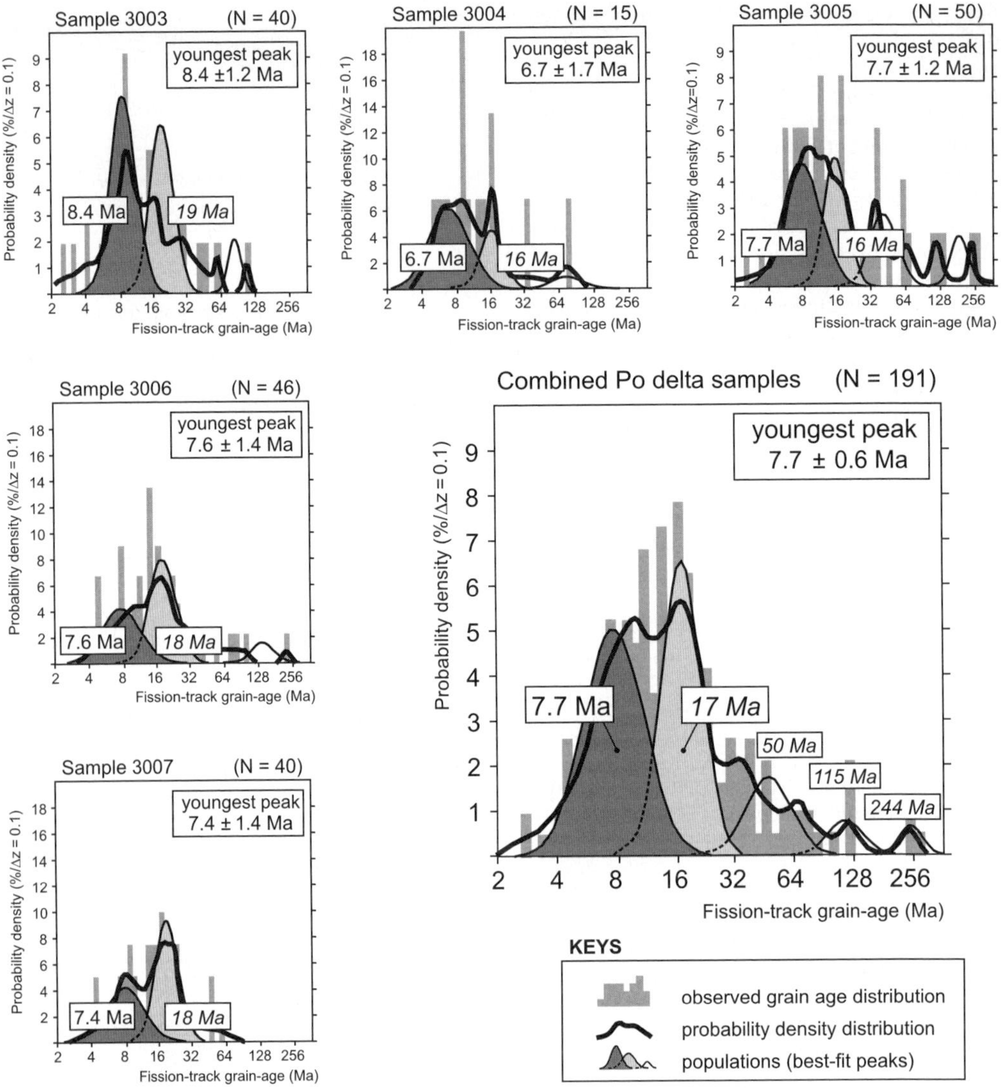

Fig. 6. Probability density plots generated by manual fit, and histograms for FT grain-age distributions from the Po Delta. Plots were constructed according to Brandon (1996). Age is plotted on a logarithmic axis. The probability density scale is the same for both the density plots and the histograms.

older population can also be clearly distinguished in all samples, ranging from 16 Ma (samples 3004 and 3005) to 19 Ma (sample 3003), again with overlap of the mean ages within 2 SD. The occurrence of very similar peaks in different samples attests that sediment mixing averages out variations in sediment yield resulting from episodic erosion and transport events.

Because the old grains only account for a small percentage of detrital apatites, the older components are not well resolved in single samples and range in mean age from Eocene to Jurassic. When data from all five samples are merged, prominent peaks at 7.7 Ma (46% of the distribution) and 17 Ma (40% of the distribution) are much better defined (Figs 5 and 6). Older and smaller peaks, not apparent in single samples, are shown at 50 Ma (10%), 115 Ma (3%) and 244 Ma (1%) (Fig. 6). The same merged distribution was analysed by the simplex method according to a Gaussian distribution type, and very similar results were obtained, including a prominent peak at approximately 8 Ma (43% of the distribution) and an older peak at about 17 Ma (33%). The size of the youngest peak is probably

Table 1. *FT grain-age components of detrital apatites from the Po Delta*

Sample	Location	No. of grains	Range (Ma)	Best-fit procedure*	Youngest peak	Old peaks		
3003	Mezzanino (B)	40	3–122	(a)	8.4 ± 1.2 53%	19.0 ± 3.2 39%	82.2 ± 13.2 8%	–
3004	Cà Tiepolo (R)	15	5–74	(a)	6.7 ± 1.7 67%	16.1 ± 3.6 26%	71.3 ± 45.0 7%	–
3005	Barricata (B)	50	3–261	(a)	7.7 ± 1.2 44%	15.7 ± 2.1 29%	40.4 ± 6.9 19%	186.1 ± 27.3 8%
3006	Boccasette (B)	46	4–240	(a)	7.6 ± 1.5 39%	18.0 ± 2.1 53%	142.2 ± 34.0 8%	–
3007	Taglio di Po (R)	40	4–56	(a)	7.4 ± 1.2 43%	18.0 ± 1.8 57%	–	–
Po Delta combined	–	191	3–261	(a)	7.7 ± 0.6 46%	17.0 ± 1.1 40%	older peaks 14%	
Po Delta combined	–	191	3–261	(b)	8.2 ± 2.9 43%	17.3 ± 3.5 33%	older peaks 24%	

*Best-fit peaks, derived from statistical best-fit procedure after (a) Brandon (1996) and (b) Dunkl & Székely 2003, are indicated by their mean age ± 2 SD and by their relative size (%) in the FT grain-age distribution.
Samples from beaches (B) and rivers (R) were counted by M. Malusà using a zeta calibration (CN-5) of 340.05 ± 4.29.

underestimated, as we did not take into account the low-U grains yielding zero age.

Age of the peaks and potential source areas

Since no significant shift in the FT grain-age distribution is expected between source mountain areas to the Po Delta, the best-fit peaks in the delta sediment are directly compared with bedrock FT ages. The young peak at 7.7 Ma is consistent, in the Alps, with cooling ages observed in the Lepontine dome (LD in Fig. 7), in the western block of the Western Alps (WB) and in the Argentera massif (AG). In the Northern Appennines, similar ages can be found in the Internal Ligurid or Miocene foredeep units (MF in Fig. 7). The contribution from the Ligurid units to the minimum-age peak is probably negligible, as young apatite ages were detected in only on a few exposures outside the Po catchment. The second peak at 17 Ma is consistent

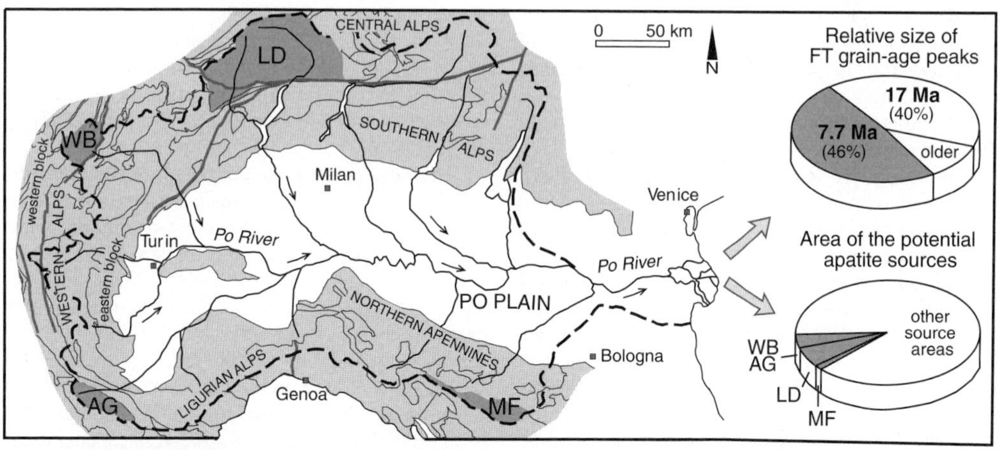

Fig. 7. Potential source areas for the youngest peak (in dark grey). AG, Argentera massif; LD, Lepontine dome; MF, Miocene foredeep units of the Apennines; WB, Western Block of the Western Alps. The diagrams on the right show the relative size of the apatite FT grain-age peaks (above) compared with the area of their potential sources estimated using GIS techniques (below). The peak at 7.7 Ma, which accounts for 46% of the distribution, is consistent with the bedrock FT ages observed in a small portion of the drainage (*c.* 12% of the orogenic source area).

with bedrock FT ages reported from the Briançonnais basement units of the Ligurian Alps, from the eastern block of the Western Alps or from the Austroalpine units of the Central Alps. The Bergell and Adamello plutons, as well as small portions of the Variscan basement of Southern Alps close to the Periadriatic and Giudicarie faults, may contribute to this peak. Finally, the oldest peaks in the distributions are consistent with cooling ages observed in wide areas of the Southern Alps, and in parts of the Ligurid units of the Apennines.

Cooling ages mirrored by the youngest peak attest that the western block of the Western Alps, the Argentera basement, and the Lepontine dome experienced high average exhumation rates on 10^6–10^7 years timescales. Miocene–present exhumation of the western block is probably due to erosion promoted by transpressional tectonic activity and by westward migration of crustal thrusts (Leloup *et al.* 2005; Malusà *et al.* 2005; Malusà & Vezzoli 2006). The same mechanism, followed by syncollisional extension, is also suggested for the exhumation of the Argentera basement (Bogdanoff *et al.* 2000; Bigot-Cormier *et al.* 2006). In the Lepontine dome the exhumation pattern is rather complex, and both erosion of uplifted blocks and tectonic denudation in the footwall of major low-angle faults, such as the Simplon Fault, are considered to have contributed to long-term exhumation (e.g. Schlunegger & Willett 1999).

Size of the peaks and short-term erosion pattern

Besides the age of the peaks, another important parameter of the grain-age distributions is the size of the peaks. The size of the peaks is a function of: (a) the erosion rates in the source areas; and (b) the distribution of apatite-bearing rocks in the catchment.

Rivers sample their drainage areas by yield, which means that fast eroding areas deliver more material to nearby basins. With respect to the area that they occupy in the basin, fastest and slowest eroding sources will thus be, respectively, over- and under-represented in the FT grain-age distribution. This is only valuable if apatite abundance is independent of rock type, which is obviously not the case. Carbonates, mudrocks, metapelites and mafic igneous rocks have negligible apatite concentrations. Large parts of the Po basin, such as the Southern Alps cover sequences, are dominated by carbonates that yield little or no apatite (Garzanti *et al.* 2006). In the Western Alps, Piedmont calcschists and meta-ophiolites also have a low apatite content (Garzanti *et al.* 2004*b*). Conversely, plutonic rocks of intermediate–acid composition and their metamorphic derivatives contain significant amounts of apatite. Orthogneissic basement rocks are extensively exposed all along the Alpine belt, including the Southern Alps (*c.* 40% of the Southalpine rocks drained by the Po River), the eastern and the western blocks of the Western Alps, the Lepontine dome and the Austroalpine units of the Central Alps. Even though the distribution of apatite-bearing rocks in the basin influences the size of the peaks, this is thus unlikely to represent the dominant controlling factor.

We thus believe that the relative size of the peaks in the probability density plots provides important provenance information that allows us to constrain sediment contribution and short-term erosion patterns within the orogenic source areas. The comparison between the size of the peaks and the potential source areas provides a first semi-quantitative indication of the fastest and slowest eroding parts of the basin on 10^2–10^5 years timescales. Detrital apatites from the Po Delta show a well-resolved peak in the distribution at 7.7 Ma that contains nearly half of the apatites (46%), but the potential source areas for these young apatites are quite small (Fig. 7). The Lepontine dome in the Central Alps (less than 3000 km^2), the Argentera basement and the western block in the Western Alps (*c.* 1500 km^2), and the Miocene foredeep units in the Apennines (*c.* 590 km^2) represent, in fact, only 12% of the total mountain area of the Po basin. This fraction of total drainage area produces half of the apatite grains that reach the Po Delta, with an apatite production and delivery rate one order of magnitude greater than in the rest of the belt. These results, even if perhaps influenced to some extent by rock composition, strongly suggest that most of the sediment load delivered to the Po Delta is supplied by small parts of the orogenic source area where short-term erosion rates are one order of magnitude greater than in the adjoining areas (Fig. 7).

This interpretation is consistent with the results of other studies based on different analytical techniques. In the Western Alps, high-resolution petrography and heavy mineral analyses of sediments carried out in the Aosta Valley (Vezzoli 2004) show that the sediment yield in the Dora Baltea catchment varies by an order of magnitude between the upper part of the basin, draining the western block, and its lower part. In the Central Alps, zircon FT grain-age distributions reported by Bernet *et al.* (2004*b*) in the Ticino and Adda rivers are characterized by prominent peaks at about 25–30 Ma that can be referred to bedrock FT cooling ages observed in the eastern part of the Lepontine dome. The size of these peaks are consistent with major sediment supply from the Lepontine units.

This qualitative evaluation of short-term erosion rates based on the size of detrital apatite populations bears the potential to become quantitative, when the apatite concentration in source rocks is evaluated by high-resolution petrography and heavy mineral analysis (e.g. Garzanti & Andò 2007).

Conclusions

FT grain-age distributions, even at river deltas far away from the eroded rocks, are a mirror of the cooling-age pattern in tectonic units exposed in the source areas. So they give a measure of the average exhumation rates experienced by source rocks from the partial annealing zone to the surface. Age distributions of detrital apatites from the Po Delta can be compared with that of their potential sources. The well-resolved youngest peak at 7.7 Ma observed in the Po Delta sediment is consistent with bedrock FT ages observed in the Lepontine dome, in the western block of the Western Alps and in the Argentera basement. These units experienced high average exhumation rates on 10^6-10^7 years timescales.

The comparison between (a) the size of the peaks in the grain-age distribution and (b) the size of potential source areas constrains instead the short-term erosional pattern on 10^2-10^5 timescales. In the Po Delta sediments, the youngest peak contains 46% of the apatites of the distribution, even if the potential sources for these young apatites represent only 12% of the total orogenic source area of the Po drainage. These results, even though influenced to an undetermined extent by rock composition, suggest that most of the sediment load in the last 10^2-10^5 years was supplied by small areas of the belt, where short-term erosion rates were one order of magnitude higher than in adjoining areas because of the complex interplay between tectonic processes, climate and erosion.

This paper benefitted from constructive and stimulating reviews by C. Cederbom, M. Rahn and F. Lisker. We also thank M. Bernet, I. Dunkl and M. L. Balestrieri for their useful advice and manuscript reviews at an early phase of the work, and G. Fellin for sample preparation.

References

BALESTRIERI, M. L., ABBATE, E. & BIGAZZI, G. 1996. Insights on the thermal evolution of the Ligurian Apennines (Italy) through fission-track analysis. *Journal of the Geological Society, London*, **153**, 419–425.

BALESTRIERI, M. L., BIGAZZI, G., CADOPPI, P. & SACCHI, R. 2004. Studio preliminare di alcuni campioni di ortoderivati provenienti dalla Val di Susa mediante il metodo delle tracce di fissione su apatite. *In*: SACCHI, R. (ed.) *Studi geologici in Val di Susa finalizzati ad un nuovo collegamento ferroviario Torino-Lione*. Regione Piemonte, Museo Regionale di Scienze Naturali, Monografie, **41**, 103–115.

BARFETY, J. C. & GIDON, M. 1975. La place des failles longitudinales dans la structure du Briançonnais oriental (Alpes occidentals, France). *Comptes Rendus de L'Académie des Sciences*, **267**, 394–397.

BERNET, M. & SPIEGEL, C. (eds) 2004. *Detrital Thermochronology–Provenance Analysis, Exhumation, and Landscape Evolution of Mountain Belts*. Geological Society of America, Special Paper, **378**.

BERNET, M., BRANDON, M. T., GARVER, J. I. & MOLITOR, B. 2004*a*. Downstream changes of Alpine zircon fission-track ages in the Rhône and Rhine Rivers. *Journal of Sedimentary Research*, **74**, 82–94.

BERNET, M., BRANDON, M. T., GARVER, J. I. & MOLITOR, B. 2004*b*. Fundamentals of detrital zircon fission-track analysis for provenance and exhumation studies with examples from the European Alps. *In*: BERNET, M. & SPIEGEL, C. (eds) *Detrital Thermochronology–Provenance Analysis, Exhumation, and Landscape Evolution of Mountain Belts*. Geological Society of America, Special Paper, **378**, 25–94.

BERNET, M., ZATTIN, M., GARVER, J. I., BRANDON, M. T. & VANCE, J. A. 2001. Steady-state exhumation of the European Alps. *Geology*, **29**, 35–38.

BERNOULLI, D., GIGER, M., MÜLLER, D. W. & ZIEGLER, U. R. F. 1993. Sr-isotope-stratigraphy of the Gonfolite Lombarda Group ('South-Alpine Molasse', northern Italy) and radiometric constraints for its age of deposition. *Eclogae Geologicae Helvetiae*, **86**, 751–767.

BIGOT-CORMIER, F., POUPEAU, G. & SOSSON, M. 2000. Dénudations différentielles du massif cristallin externe alpin de l'Argentera (Sud-Est de la France) révélées par thermochronologie traces de fission (apatites, zircons). *Comptes Rendus de l'Académie des Sciences, Series IIA, Earth and Planetary Science*, **330**, 363–370.

BIGOT-CORMIER, F., SOSSON, M., POUPEAU, G., STÉPHAN, J. F. & LABRIN, E. 2006. The denudation history of the Argentera Alpine External Crystalline Massif (Western Alps, France–Italy): an overview from the analysis of fission tracks in apatites and zircons. *Geodinamica Acta*, **19**, 455–473.

BISTACCHI, A., DAL PIAZ, G. V., MASSIRONI, M., ZATTIN, M. & BALESTRIERI, M. L. 2001. The Aosta–Ranzola extensional fault system and Oligocene–Present evolution of the Austroalpine-Penninic wedge in the northwestern Alps. *International Journal of Earth Sciences*, **90**, 654–667.

BOGDANOFF, S., MICHARD, A., MANSOUR, M. & POUPEAU, G. 2000. Apatite fission track analysis in the Argentera massif: evidence of contrasting denudation rates in the External Crystalline Massifs of the Western Alps. *Terra Nova*, **12**, 117–125.

BRANDON, M. T. 1992. Decomposition of fission-track grain-age distributions. *American Journal of Science*, **292**, 535–564.

BRANDON, M. T. 1996. Probability density plot for fission-track grain-age samples. *Radiation Measurements*, **26**, 663–676.

BRANDON, M. T. 2002. Decomposition of mixed grain age distributions using Binomfit. *On Track*, **24**, 13–18.

BREWER, I. D., BURBANK, D. W. & HODGES, K. V. 2006. Downsteram development of a detrital cooling-age signal: Insights from $^{40}Ar/^{39}Ar$ muscovite thermochronology in the Nepalese Himalaya. *In*: WILLET, S. D., HOVIOUS, N., BRANDON, M. T. & FISHER, D. (eds) *Tectonics, Climate, and Landscape Evolution*. Geological Society of America, Special Paper, **398**, 321–338.

BULL, W. B. 1991. *Geomorphic Responses to Climatic Changes*. Oxford University Press, New York.

BURBANK, D. W. 2002. Rates of erosion and their implications for exhumation. *Mineralogical Magazine*, **66**, 25–52.

BURG, J. P., NIEVERGELT, P. *ET AL*. 1998. The Namche Barwa syntaxis: evidence for exhumation related to compressional crustal folding. *Journal of Asian Earth Sciences*, **16**, 239–252.

BÜRGI, A. & KLÖTZLI, U. 1990. New data on the evolutionary history of the Ivrea Zone (Northern Italy). *Bulletin of the Swiss Association of Petroleum Geology and Engineering*, **56**, 49–70.

CARPÉNA, J. 1985. Tectonic interpretation of an inverse gradient of zircon fission track ages with respect to altitude: alpine thermal history of the Gran Paradiso basement. *Contributions to Mineralogy and Petrology*, **90**, 74–82.

CARPÉNA, J. 1992. Fission track dating of zircon: Zircons from Mont Blanc granite (French–Italian Alps). *Journal of Geology*, **100**, 411–421.

CARRAPA, B. 2002. *Tectonic Evolution of an Active Orogen as Reflected by its Sedimentary Record. An Integrated Study of the Tertiary Piedmont Basin (Internal Western Alps, NW Italy)*. PhD thesis, Vrije University, Amsterdam.

CARTER, A. 1999. Present status and future avenues of source region discrimination and characterization using fission-track analysis. *Sedimentary Geology*, **124**, 31–45.

CENCINI, C. 1998. Physical processes and human activities in the evolution of the Po Delta, Italy. *Journal of Coastal Research*, **14**, 774–793.

CIANCALEONI, L. 2005. *Deformation Process During the Last Stages of the Continental Collision: The Brittle–Ductile Fault Systems in the Bergell and Insubric Areas (Eastern Central Alps, Switzerland*. PhD thesis, University of Neuchatel.

COWARD, M. P., DIETRICH, D. & PARK, R. G. (eds) 1989. *Alpine Tectonics*. Geological Society, London, Special Publications, **45**.

DESMONS, J., APRAHAMIAN, J., COMPAGNONI, R., CORTESOGNO, L. & FREY, M. 1999*a*. Alpine metamorphism of the Western Alps. I. Middle to high T/P metamorphism. *Schweizerische Mineralogische und Petrographische Mitteilungen*, **79**, 89–110.

DESMONS, J., COMPAGNONI, R. & CORTESOGNO, L. 1999*b*. Alpine metamorphism of the Western Alps. I. High P/T and related pre-greenschist metamorphism. *Schweizerische Mineralogische und Petrographische Mitteilungen*, **79**, 111–134.

DUNKL, I. 2002. TRACKKEY: a Windows program for calculation and graphical presentation of fission track data. *Computers & Geosciences*, **28**, 3–12.

DUNKL, I. & SZÉKELY, B. 2003. Component analysis with visualization of fitting – Popshare, a freeware program for evaluation of mixed geochronological data. *Geophysical Research Abstracts*, **5**, 02657.

EHLERS, T. A., CHAUDHRI, T. *ET AL*. 2005. Computational tools for low-temperature thermochronometer interpretation. *Reviews in Mineralogy & Geochemistry*, **58**, 589–622.

FELLIN, M. G., SCIUNNACH, D., TUNESI, A., ANDÒ, S., GARZANTI, E. & VEZZOLI, G. 2005. Provenance of detrital apatites from the Upper Gonfolite Lombarda Group (Miocene, NW, Italy). *GeoActa*, **4**, 43–56.

FINLAYSON, D. P., MONTGOMERY, D. R. & HALLET, B. 2002. Spatial coincidence of rapid inferred erosion with young metamorphic massifs in the Himalayas. *Geology*, **30**, 219–222.

FREY, M. & FERREIRO MÄHLMANN, R. 1999. Alpine metamorphism of the Central Alps. *Schweizerische Mineralogische und Petrographische Mitteilungen*, **79**, 135–154.

FÜGENSCHUH, B. & SCHMID, S. M. 2003. Late stages of deformation and exhumation of an orogen constrained by fission-track data: A case study in the Western Alps. *Geological Society of America Bulletin*, **115**, 1425–1440.

FÜGENSCHUH, B., LOPRIENO, A., CERIANI, S. & SCHMID, S. M. 1999. Structural analysis of the Subbriançonnais and Valais units in the area of Moutiers (Savoy, Western Alps): Paloegeographic and tectonic consequences, *International Journal of Earth Sciences*, **88**, 201–218.

GALBRAITH, R. F. & GREEN, P. F. 1990. Estimating the component ages in a finite mixture. *Nuclear Tracks and Radiation Measurements*, **17**, 197–206.

GALLAGHER, K., BROWN, R. & JOHNSON, C. 1998. Fission track analysis and its applications to geological problems. *Annual Reviews Earth and Planetary Science*, **26**, 519–572.

GARVER, J. I., BRANDON, M. T., RODEN-TICE, M. K. & KAMP, P. J. J. 1999. Exhumation history of orogenic highlands determined by detrital fission track thermochronology. *In*: RING, U., BRANDON, M. T., WILLET, S. D. & LISTER, G. (eds) *Exhumation Processes: Normal Faulting, Ductile Flow, and Erosion*. Geological Society, London, Special Publications, **154**, 283–304.

GARZANTI, E. & ANDÒ, S. 2007. Heavy-mineral concentration in modern sands: implications for provenance interpretation. *In*: MANGE, M. & WRIGHT, D. (eds) *Heavy Minerals in Use*. Developments in Sedimentology, **58**, 517–546.

GARZANTI, E., ANDÒ, S. & VEZZOLI, G. 2006. The Continental crust as source of sand (Southern Alps cross-section, Northern Italy). *Journal of Geology*, **114**, 533–554.

GARZANTI, E., VEZZOLI, G., ANDÒ, S., FRANCE-LANORD, C., SINGH, S. K. & FOSTER, G. 2004*a*. Sand petrology and focused erosion in collision orogens: the Brahmaputra case. *Earth and Planetary Science Letters*, **220**, 157–174.

GARZANTI, E., VEZZOLI, G., LOMBARDO, B., ANDÒ, S., RUSSO, M., MAURI, E. & MONGUZZI, S. 2004*b*. Collision-orogen provenance (Western Alps). Detrital signatures and unroofing trends. *Journal of Geology*, **112**, 145–164.

GIGER, M. 1991. *Geochronologische und Petrographische Studien an Gerollen und Sedimenten der Gonfolite Lombarda Gruppe (Sudschweiz und Norditalien) und ihr Vergleich mit dem Alpinen Hinterland*. PhD thesis, University of Bern, Switzerland.

GIGER, M. & HURFORD, A. J. 1989. Tertiary intrusives of the Central Alps: Their Tertiary uplift, erosion, redeposition and burial in the south-alpine foreland. *Eclogae Geologicae Helvetiae*, **82**, 857–866.

GLEADOW, A. J. W. & BROWN, R. W. 1999. Fission track thermochronology and long-term denudational response to tectonics. *In*: SUMMERFIELD, M. A. (ed.) *Geomorphology and Global Tectonics*. Wiley, Chichester, 57–75.

HUNZIKER, J. C., DESMONS, J. & HURFORD, A. J. 1992. Thirty-two years of geochronological work in the Central and Western Alps: a review on seven maps. *Mémoires de Géologie, Lausanne*, **13**, 1–59.

HURFORD, A. J. 1986. Cooling and uplift patterns in the Lepontine Alps South Central Switzerland and an age of vertical movement on the Insubric fault line. *Contributions to Mineralogy and Petrology*, **92**, 413–427.

HURFORD, A. J. 1990. Standardization of fission-track dating calibration: reccomendation by the Fission Track Working Group of the I.U.G.S. Subcommission on Geochronology. *Chemical Geology*, **80**, 171–178.

HURFORD, A. J. & CARTER, A. 1991. The role of fission track dating in discrimination of provenance. *In*: MORTON, A. C., TODD, S. P. & HAUGHTON, P. D. W. (eds) *Developments in Sedimentary Provenance Studies*. Geological Society, London, Special Publications, **57**, 67–78.

HURFORD, A. J. & GREEN, P. F. 1983. The zeta age calibration of fission-track dating. *Chemical Geology Isotope Geosciences*, **1**, 285–317.

HURFORD, A. J. & HUNZIKER, J. C. 1985. Alpine cooling history of the Monte Mucrone eclogites (Sesia–Lanzo zone): fission track evidence. *Schweizerische Mineralogische und Petrographische Mitteilungen*, **65**, 325–334.

HURFORD, A. J. & HUNZIKER, J. C. 1989. A revised thermal history for the Gran Paradiso massif. *Schweizerische Mineralogische und Petrographische Mitteilungen*, **69**, 319–329.

HURFORD, A. J., FLISCH, M. & JAGER, E. 1989. Unravelling the thermo-tectonic evolution of the Alps: a contribution from fission track analysis and mica dating. *In*: COWARD, M. P., DIETRICH, D. & PARK, R. G. (eds) *Alpine Tectonics*. Geological Society, London Special Publications, **45**, 369–398.

HURFORD, A. J., HUNZIKER, J. C. & STOCKHERT, B. 1991. Constraints on the late thermotectonic evolution of the Western Alps: evidence for episodic rapid uplift. *Tectonics*, **10**, 758–769.

KELLER, L. M., HESS, M., FÜGENSCHUH, B. & SCHMID, S. M. 2005. Structural and metamorphic evolution of the Camughera–Mocucco and the Monte Rosa units southwest of the Simplon line. *Eclogae Geologicae Helvetiae*, **98**, 19–49.

LELOUP, P. H., ARNAUD, N., SOBEL, E. R. & LACASSIN, R. 2005. Alpine thermal and structural evolution of the highest external crystalline massif: the Mont Blanc. *Tectonics*, **24**, 1–26.

MALUSÀ, M. G. 2004. *Post-metamorphic Evolution of the Western Alps: Kinematic Constraints From a Multidisciplinary Approach (Geological Mapping, Mesostructural Analysis, Fission-track Dating, Fluid Inclusion Analysis)*. PhD thesis, CNR-IGG, University of Torino.

MALUSÀ, M. G. & VEZZOLI, G. 2006. Interplay between erosion and tectonics in the Western Alps. *Terra Nova*, **18**, 104–108.

MALUSÀ, M. G., POLINO, R., ZATTIN, M., BIGAZZI, G., MARTIN, S. & PIANA, F. 2005. Miocene to Present differential exhumation in the Western Alps: Insights from fission track thermochronology. *Tectonics*, **24**, 1–23.

MARTIN, S., BIGAZZI, G., ZATTIN, M., VIOLA, G. & BALESTRIERI, M. L. 1998. Neogene kinematics of the Giudicarie fault (Central–Eastern Alps Italy): New apatite fission-track data. *Terra Nova*, **10**, 217–221.

MERLE, O., COBBOLD, P. R. & SCHMID, S. 1989. Tertiary kinematics in the Lepontine dome. *In*: COWARD, M. P., DIETRICH, D. & PARK, R. G. (eds) *Alpine Tectonics*. Geological Society, London, Special Publications, **45**, 113–134.

MICHALSKI, I. & SOOM, M. 1990. The Alpine thermo-tectonic evolution of the Aar and Gotthard massifs, Central Switzerland: fission track ages on zircon and apatite and K/Ar mica ages. *Schweizerische Mineralogische und Petrographische Mitteilungen*, **70**, 373–388.

MIKHAILOVA, M. V. 2002. The hydrological regime and the peculiarities of formation of the Po River Delta. *Water Resources*, **29**, 370–380.

MUNARDI, W. 1989. *Der Osthang der Valle della Mera Zwischen Samolaco und Dubino: Petrographische, Geochemische und Geochronologische Bearbeitung der Gesteine unter Besonderer Berucksichtigung des Novate-Granits*. Diploma thesis, University of Bern.

POLINO, R. (coord) 2002. *Geological Map of Italy (scale 1 : 50 000), sheet 132-152-153 Bardonecchia*. Servizio Geologico d'Italia, Litografia GEDA (Nichelino).

RAHN, M. K. 2005. Apatite fission track ages from the Adula nappe: late-stage exhumation and relief evolution. *Schweizerische Mineralogische und Petrographische Mitteilungen*, **85**, 233–245.

REGIONE EMILIA-ROMAGNA & ENI-AGIP. 1998. *Riserve idriche sotterranee della Regione Emilia-Romagna*. A cura di G. Di Dio. S.EL.CA, Firenze.

REINERS, P. W. & BRANDON, M. T. 2006. Using thermochronology to understand orogenic erosion. *Annual Review of Earth and Planetary Science*, **34**, 419–466.

ROURE, F., HEITZMANN, P. & POLINO, R. 1990. Deep structures of the Alps. *Mémoires de la Société géologique de France*, **156**, 1–367.

SABIL, N. & MENARD, G. 2000. Apport des traces de fission sur apatite aux reconnaissances du tunnel d'Ambin. *In*: *Proceedings of the 18th Réunion des sciences de la terre 2000*. Société Géologique de France, Paris, 228–229.

SCARDIA, G., MUTTONI, G. & SCIUNNACH, D. 2006. Subsurface magnetostratigraphy of Pleistocene sediments from the Po Plain (Italy): Constraints on rates of sedimentation and rock uplift. *Geological Society of America Bulletin*, **118**, 1299–1312.

SCHAER, J. P., REIMER, G. M. & WAGNER, G. A. 1975. Actual and ancient uplift rate in the Gotthard region, Swiss Alps: a comparison between precise levelling and fission-track apatite age. *Tectonophysics*, **29**, 293–300.

SCHLUNEGGER, F. & WILLET, S. 1999. Spatial and temporal variations in exhumation of the central Swiss Alps and implications for exhumation mechanisms. *In*: RING, U., BRANDON, M. T., LISTER, G. S. & WILLETT, S. D. (eds) *Exhumation Processes: Normal Faulting, Ductile Flow and Erosion*. Geological Society, London, Special Publications, **154**, 157–179.

SCHWARTZ, S., LARDEAUX, J. M., TRICART, P., GUILLOT, S. & LABRIN, E. 2007. Diachronous exhumation of HP-LT metamorphic rocks from southwestern Alps: evidence from fission-track analysis. *Terra Nova*, **19**, 133–140.

SEWARD, D. & MANCKTELOW, N. S. 1994. Neogene kinematics of the central and western Alps: Evidence from fission-track dating. *Geology*, **22**, 803–806.

SEWARD, D., FORD, M., BURGISSER, J., LICKORISH, H., WILLIAMS, E. A. & MECKEL, L. D., III. 1999. Preliminary results of fission-track analyses in the Southern Pelvoux area, SE France. *Memorie di Scienze Geologiche, Padova*, **51**, 25–31.

SOOM, M. A. 1990. *Abkühlungs- und Hebungsgeschichte der Externmassive und der penninischen Decken beidseits der Simplon-Rhone-Linie seit dem Oligozän: Spaltspurdatierungen an Apatite/Zirkon und K-Ar-Datierungen an Biotit/Muskowit (Westliche Zentralalpen)*. PhD thesis, University of Bern.

STEINER, H. 1984. Radiometrische Altersbestimmungen an Gasteinen der Maggia-Decke (Penninikum der Zentralapen). *Schweizerische Mineralogische und Petrographische Mitteilungen*, **64**, 227–259.

STEWART, J. S. & BRANDON, M. T. 2004. Detrital-zircon fission-track ages for the 'Hoh Formation': Implications for late Cenozoic evolution of the Cascadia Subduction wedge. *Geological Society of America Bulletin*, **116**, 60–75.

THIEDE, R. C., ARROWSMITH, J. R., BOOKHAGEN, B., MCWILLIAMS, M. O., SOBEL, E. R. & STRECKER, M. R. 2005. From tectonically to erosionally controlled development of the Himalayan orogen. *Geology*, **33**, 689–692.

TIMAR-GENG, Z., GRUJIC, D. & RAHN, M. 2004. Deformation at the Leventina-Simano nappe boundary, Central Alps, Switzerland. *Eclogae Geologicae Helvetiae*, **97**, 265–278.

TRICART, P., SCHWARTZ, S., SUE, C., POUPEAU, G. & LARDEAUX, J. M. 2001. La dénudation tectonique de la zone ultradauphinoise et l'inversion du front briançonnais au sud-est du Pelvoux (Alpes occidentales): une dynamique miocène à actuelle. *Bulletin de Société Géologique de France*, **172**, 49–58.

TRICART, P., VAN DER BEEK, P., SCHWARTZ, S. & LABRIN, E. 2007. Diachronous late-stage exhumation across the western Alpine arc: constraints from apatite fission-track thermochronology between the Pelvoux and Dora-Maira Massifs. *Journal of the Geological Society, London*, **164**, 163–174.

VENTURA, B. & PINI, G. A. 1999. Constraining the time of displacement along faults with apatite fission tracks. An example from the Mt. Cervarola Sandstones (northern Apennines, Italy). *Memorie di Scienze Geologiche, Padova*, **51**, (2), 476–477.

VENTURA, B., PINI, G. A. & ZUFFA, G. G. 2001. Thermal history and exhumation of the Northern Apennines (Italy): evidence from combined apatite fission track and vitrinite reflectance data from foreland basin sediments. *Basin Research*, **13**, 435–448.

VEZZOLI, G. 2004. Erosion in the Western Alps (Dora Baltea Basin): 2. Quantifying sediment yield. *Sedimentary Geology*, **171**, 247–259.

VEZZOLI, G., GARZANTI, G. & MONGUZZI, S. 2004. Erosion in the Western Alps (Dora Baltea Basin): 1. Quantifying sediment provenance. *Sedimentary Geology*, **171**, 227–246.

VIOLA, G. 2000. *Kinematics and Timing of the Periadriatic Fault System in the Giudicarie Region (Central–Eastern Alps)*. PhD thesis, University of Zurich.

VIOLA, G., MANCKTELOW, N. & SEWARD, D. 2001. Late Oligocene–Neogene evolution of Europe–Adria collision: New structural and geochronological evidence from the Giudicarie fault system (Italian eastern Alps). *Tectonics*, **20**, 999–1020.

VIOLA, G., MANCKTELOW, N., SEWARD, D., MEIER, A. & MARTIN, S. 2003. The Pejo fault system: an example of multiple tectonic activity in the Italian Eastern Alps. *Geological Society of America Bulletin*, **115**, 515–532.

WAGNER, G. & VAN DEN HAUTE, P. 1992. *Fission-Track Dating*. Kluwer, Dordrecht.

WAGNER, G. A. & REIMER, M. 1972. Fission-track tectonics: the tectonic interpretation of fission track apatite ages. *Earth and Planetary Science Letters*, **14**, 263–268.

WAGNER, G. A., REIMER, G. M. & JAGER, E. 1977. Cooling ages derived by apatite fission track, mica Rb-Sr, and K-Ar dating: the uplift and cooling history in the central Alps. *Memorie dell'Istituto di Geologia e Mineralogia dell'Università di Padova*, **30**, 1–27.

WAGNER, G. A., MILLER, D. S. & JAGER, E. 1979. Fission track ages on apatite of Bergell rocks from Central Alps and Bergell boulders in Oligocene sediments. *Earth and Planetary Science Letters*, **45**, 355–360.

ZATTIN, M., CUMAN, A., FANTONI, R., MARTIN, S., SCOTTI, P. & STEFANI, C. 2006. Cooling during burial: the thermal record in the sediments of the Dolomite region (Central Alps, Italy). *Tectonophysics*, **414**, 191–202.

ZATTIN, M., STEFANI, C. & MARTIN, S. 2003. Detrital fission-track analysis and petrography as keys of alpine exhumation: the example of the Veneto foreland (Southern Alps, Italy). *Journal of Sedimentary Research*, **7**, 1051–1061.

ZEITLER, P. K. & MELTZER, A. S. 2001. Erosion, Himalayan geodynamics, and the geomorphology of metamorphism. *GSA Today*, **11**, 4–9.

Exhumation of the Sierra de Cameros (Iberian Range, Spain): constraints from low-temperature thermochronology

P. DEL RÍO[1]*, L. BARBERO[1] & F. M. STUART[2]

[1]*Departamento de Geología, Universidad de Cádiz, 11510 Puerto Real, Cádiz, Spain*

[2]*Isotope Geosciences Unit, Scottish Universities Environmental Research Centre, Rankine Avenue, Technology Park, East Kilbride G75 0QF, UK*

**Corresponding author (e-mail: pedro.bermejo@uca.es)*

Abstract: We present new fission-track and (U–Th)/He data from apatite and zircon in order to reconstruct the exhumation of the Sierra de Cameros, in the northwestern part of Iberian Range, Spain. Zircon fission-track ages from samples from the depocentre of the basin were reset during the metamorphic peak at approximately 100 Ma. Detrital apatites from the uppermost sediments retain fission-track age information that is older than the sediment deposition age, indicating that these rocks have not exceeded 110 °C. Apatites from deeper in the stratigraphic sequence of the central part of the basin have fission-track ages of around 40 Ma, significantly younger than the stratigraphic age, recording the time of cooling after peak metamorphic conditions. Apatite (U–Th)/He ages in samples from these sediments are 31–40 Ma and record the last period of cooling during Alpine compression. The modelled thermal history derived from the uppermost sediments indicates that the thermal pulse associated with peak metamorphism was rapid, and that the region has cooled continuously to the present. The estimated palaeogeothermal gradient is around 86 °C km^{-1} and supports a tectonic model with a thick sedimentary fill (*c.* 8 km) and explains the origin of the low-grade metamorphism observed in the oldest sediments.

Convergence of the Eurasian and Iberian plates during the Palaeogene–early Miocene led to the formation of the Pyrenees in the northern Iberian margin. The Iberian Range, located to the south of the Pyrenees, is an intraplate mountain chain that formed as a result of migration of compressional deformation towards the inner part of the Iberian microplate during Alpine collision. The Cameros Basin, located in the northwestern most part of the Iberian Chain (Fig. 1), represents a synrift sequence formed prior to the Alpine convergence during the Upper Jurassic–Mid Cretaceous extensional period. It was inverted during the subsequent Alpine shortening, either by the reactivation of the synrift normal faults or on newly formed faults and thrust (Casas-Sainz & Gil-Imaz 1998).

The stratigraphy, mineralogy, metamorphism and structure of the Sierra de Cameros have been studied for several decades (e.g. Tischer 1966; Mas *et al.* 1993; Casas-Sainz & Gil-Imaz 1998; Mata *et al.* 2001). However, the timing and mechanism of the basin inversion, the origin of the greenschist-facies metamorphism of sediments in the deepest part of the basin and the geometry of the sedimentary filling remain unclear. Three tectonic models have been proposed to explain the formation of the Cameros Basin and the geometry of the synrift deposits.

In the first model Guiraud (1983) and Guiraud & Séguret (1984) proposed an east–west-striking, south-dipping normal fault in the basement as the structure responsible for the formation of a syncline basin. The sedimentary infilling of this basin progressively onlapped the structure as the depocentre migrated northwards (Fig. 2a). In this model the total thickness of the sedimentary fill would not exceed 5 km. During the Tertiary the basin was inverted, with the normal faults being reactivated as reverse faults.

In a second model Mas *et al.* (1993) and Guimerà *et al.* (1995) (Fig. 2b) interpreted the Cameros Basin as a sag basin that formed on a ramp of an extensional, subhorizontal fault located several kilometres deep in the Variscan basement. The sedimentary units that fill the basin decrease in thickness towards the southern border and onlap to the north on the Jurassic, prerift deposits. The depocentres are located on the normal fault ramp and migrated progressively towards the north. Again, the maximum vertical thickness reached by the sedimentary fill would be about 5 km. Tectonic inversion of the basin was caused by a newly formed thrust developed along the northern border, and a thrust system to the south. In these two models the origin of low-grade metamorphism remains unclear and some authors claim for a hydrothermal origin (e.g. Casquet *et al.* 1992).

From: LISKER, F., VENTURA, B. & GLASMACHER, U. A. (eds) *Thermochronological Methods: From Palaeotemperature Constraints to Landscape Evolution Models*. Geological Society, London, Special Publications, **324**, 153–166.
DOI: 10.1144/SP324.12 0305-8719/09/$15.00

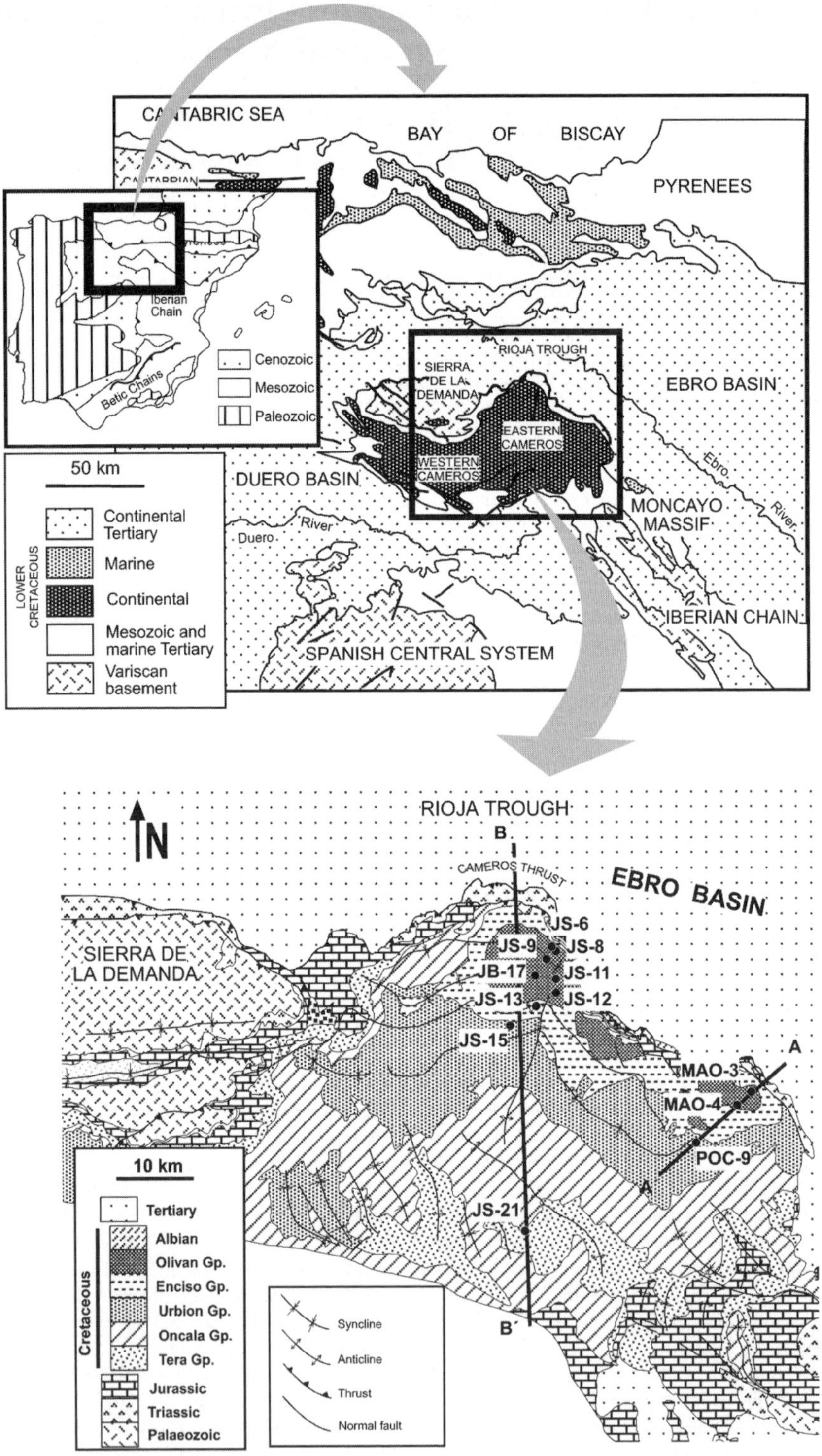

Fig. 1. Schematic geological map of the Sierra de Cameros showing the location of the study area within the context of the Iberian plate and the locations of cross-sections and samples for this study.

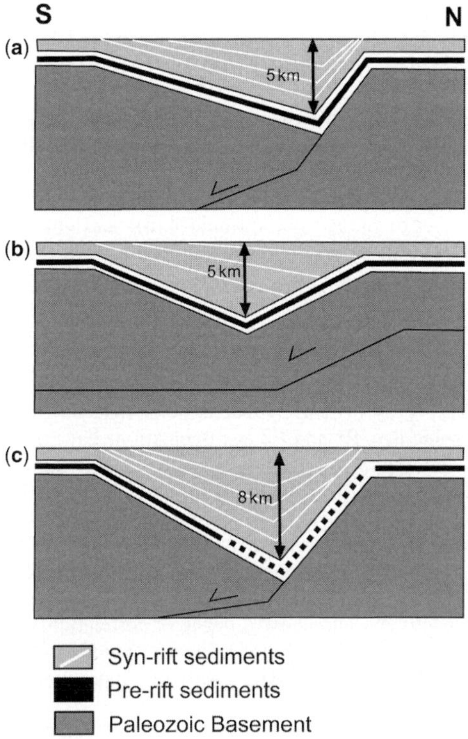

Fig. 2. Summary of the tectonic models proposed to explain the Cameros Basin evolution (modified after Gil-Imaz *et al.* 2002). (**a**) Synsedimentary synclinal model proposed by Guiraud (1983) and Guiraud & Séguret (1984), with a maximum vertical thickness of about 5 km. (**b**) Sag-basin model after Mas *et al.* (1993) and Guimerá *et al.* (1995) in which the units onlap onto the prerift sequence; a maximum vertical thickness of about 5 km is reached in this model. (**c**) Synsedimentary synclinal model proposed by Casas-Sainz & Gil-Imaz (1998), with vertical stacking of the units reaching a maximum vertical thickness of about 8 km.

Casas-Sainz & Gil-Imaz (1998) proposed an alternative model in which the basin exhibits a half-graben geometry controlled by a listric fault developed along previous Variscan structures. This fault does not extend beyond the limits of the basin, and the units stack vertically and decrease in thickness towards the basin margins (Fig. 2c). The maximum vertical thickness reached by the sedimentary fill according to this model was 8 km. During the Tertiary basin inversion the Upper Triassic sediments (Keuper facies) were the detachment level for thrusts involving the Mesozoic cover. This model can satisfactorily explain the origin of low-grade metamorphism in the deep basin mainly as a consequence of burial. The interpretation of the magnetic fabric of rocks (ASM: Gil-Imaz *et al.* 2000), and fluid inclusions in quartz and calcite filling extensional cracks (Mata *et al.* 2001) also support this hypothesis.

The Sierra de Cameros shows several features that make it an interesting area for the study of the evolution of intraplate rifting process and far-field effect of the plate boundaries. These include: (1) a well-preserved, thick synrift sedimentary sequence; (2) a unique low-grade metamorphic event and several folding stages that did not appear in other parts of the Iberian Chain; and (3) the presence of thick, well-preserved foreland basin deposits (Ebro Basin) in which it is possible to study the evolution of the eroded part of the Sierra de Cameros.

The application of low-temperature thermochronometers such as fission track and (U–Th)/He to detrital minerals provides a unique tool to better understand the thermal history of the Sierra de Cameros and, consequently, its geological evolution. This is facilitated by the presence of detrital apatite and zircon crystals in the detrital rocks of the sedimentary sequence of the Cameros Basin. The combination of two low-temperature thermochronological methods allows for new constraints on the palaeogeothermal gradients to be placed, which in turn can be compared to previous data from other independent techniques.

Geological framework

The Sierra de Cameros forms the main part of the northwestern sector of the Iberian Range. It is bordered by two continental basins, the Ebro and Duero basins to the north and south respectively, and by two Paleozoic massifs, the Sierra de la Demanda to the west and the Moncayo Massif to the east (Fig. 1). Sediments of the Sierra de Cameros were deposited in a Mesozoic basin that formed as a consequence of northward propagation of rifting related to the progressive opening of the North Atlantic Ocean and the Bay of Biscay. Transtensional rifting of the Bay of Biscay resulted in the progressive destruction of the Middle–Late Jurassic carbonate platforms and the development of a new extensional system in the Iberian Range (Mas *et al.* 2002). The basin exhibited a half-graben-like geometry that was controlled by a system of extensional listric faults with NW–SE strikes that were developed on earlier Variscan structures. The synrift stratigraphic sequence spans from Upper Jurassic (Tithonian–Berriasian) to the upper–lower Cretaceous boundary (Albian–Cenomanian) (Mas *et al.* 1994; Muñoz *et al.* 1997). Depositional sequences are classically divided into five lithostratigraphic groups: the Tera, Oncala, Urbión, Enciso and Oliván groups (Tischer 1966). Sediments are largely fluvial sandstones, lutites and conglomerates (Tera, Urbión and Oliván groups), and lacustrine carbonates with siliciclastic influence (Oncala and Enciso groups). The clastic sequences are characterized by compositional immaturity (arkoses are abundant) characteristic of proximal facies. The geochemistry of the detrital sediments reveals an acid-igneous or metamorphic source rock (Mata *et al.* 2000). The sediments may have been sourced in areas close to the Spanish Central System, which is dominated by acid granites and orthogneisses. This is consistent with the palaeo-current directions (Tischer 1966) and the compositional immaturity of the sediments. Depositional ages used in the present work are based on *Charophyceae* biozones (Schudack 1987), pollen analysis (Martín i Closas 1989) and correlations made by Mas *et al.* (1993), which give depositional times for the Cameros Basin as follows: lower Albian for the Oliván group; Aptian for the Enciso group; Barremian–Valanginian for the Urbión group; Berriasian for the Oncala group; and Tithonian for the Tera group.

The presence of chloritoid in the deepest sediments cropping out at present indicates that the region underwent a low-grade metamorphism that peaked of approximately at 326–350 °C and 1 kb (Casquet *et al.* 1992; Mata *et al.* 2001). Ar–Ar and K–Ar ages for peak temperatures are approximately 100 Ma (Goldberg *et al.* 1988; Casquet *et al.* 1992). During the Tertiary the basin underwent tectonic inversion related to Alpine compression that produced folding and exhumation of the whole Mesozoic basin The Cameros Basin sediments were thrusted 25–30 km north along the Cameros thrust over the Tertiary molasse of the Ebro Basin (Guimerà *et al.* 1995; Casas-Sainz & Gil-Imaz 1998). During the Palaeocene and Eocene, the Sierra de Cameros was the source area for the detrital sediments that were deposited in the Ebro Basin, in particular for its westernmost part (the Rioja Trough) where they reach a maximum thickness of 5 km (Muñoz-Jiménez & Casas-Sainz 1997).

Results

Fission-track results are shown in Table 1. Radial plots of zircon fission-track (ZFT) ages of the two samples from profile A–A′ are displayed in Figure 3. Zircons from sample MAO-3 from the Oliván group have two age populations that are considerably older than the sediment stratigraphic age (Lower Albian). Zircons from POC-9 (Urbión group) yield a single fission-track age population of about 85 Ma. This is considerably younger than the sediment deposition age (Valanginian–Barremian).

Apatite fission-track (AFT) data have been obtained in three samples along the profile A–A′ (Fig. 3). Sample MAO-3 (Oliván) shows three age populations: 60.9 ± 2.9 Ma (P1), 80.0 ± 4.1 Ma (P2) and 122.9 ± 15.5 Ma (P3). Mean track length associated with the P1 and P2 age populations are 12.8 ± 2.2 and 11.7 ± 2.4 μm, respectively, showing in both cases a negatively skewed distribution (Fig. 4). D_{par} of the P1 and P2 populations have average values of 1.38 ± 0.14 and 1.38 ± 0.13 μm, respectively. Fluorine (1.32–1.49 apfu (atoms per formula unit)) and chlorine (0.02–0.03 apfu) contents show a negative correlation with the D_{par} values (Fig. 5). Sample MAO-4 (lower part of the Oliván Group) shows two age populations of 54.3 ± 2.4 Ma (P1) and 69.8 ± 5.6 Ma (P2) that are younger than its stratigraphic age, and a third population of 127.3 ± 24.7 Ma (P3) that is older than its stratigraphic age. P1 and P2 age populations have mean track lengths of 11.3 ± 2.3 and 11.6 ± 2.2 μm, respectively, showing in both cases a negatively skewed distribution (Fig. 4). Average D_{par} measurements are 2.2 ± 0.3 μm for P1, 2.1 ± 0.3 μm for P2 and 2.1 ± 0.3 μm for P3. Fluorine and chlorine contents are 1.26–1.41 and 0.02–0.03 apfu, respectively, and show a negative correlation with the measured D_{par} (Fig. 5). Finally, sample POC-9 exhibits a single AFT age of approximately 48.3 ± 4.1 Ma.

AFT data from nine sandstone samples along the profile B–B′ are shown in Figure 1. Most samples have several age populations, most of them younger than the depositional age (see radial plots in Fig. 6). Sample JS-6 (Oliván) has an age population of 123.2 ± 14.9 Ma, which is older than the depositional age (Lower Albian), but the other two components are younger. Apatites from samples JS-9 and JS-11, both from the middle part of the Oliván group, yield a single age population of 55.6 ± 3.2 and 34.5 ± 1.9 Ma, respectively. The number of confined tracks in samples from the B–B′ profile is far too low to allow a reasonable statistical evaluation of the data. Mean track length in samples JS-9 and JS-11 are 10.61 ± 2.53 ($n = 19$) and 11.32 ± 2.34 ($n = 9$) μm, respectively. D_{par} values for these samples are in the range of 2.4 and 6.0 μm.

(U–Th)/He ages (AHe) have been determined on apatites from two sandstone samples from profile B–B′ (see Fig. 6 and Table 2). Sample JS-21 (Tera) yields AHe age of 31.1 ± 1.5 Ma, and two splits of apatites from JS-12 (Oliván) of 39.8 ± 2.8 and 37.6 ± 1.8 Ma.

Thermal modelling

Thermal modelling has been performed on samples MAO-3 and MAO-4 from the Oliván group in which track lengths associated with the two youngest age populations have been clearly separated. From the

Table 1. *Apatite and zircon fission-track age data from the Sierra de Cameros*

Sample No. (mineral type)*	Stratigraphic age *Group*	No. of grains	$\rho_D^{\dagger}$(10^6 tr cm^{-2}) (N_D)	ρ_s (10^6 tr cm^{-2}) (N_s)	ρ_i (10^6 tr cm^{-2}) (N_i)	Central age $\pm 1\sigma$ (Ma) $P(\chi^2)$	D_{par} (μm)/MTL (μm) $\pm 1\sigma$	Age populations (Ma $\pm 1\sigma$), N_f, W (%) P1	P2	P3	Fit statistics
MAO-3 (Ap)	Lower Albian *Oliván*	42	1.132 *(5249)*	1.955 *(4995)*	5.269 *(13462)*	71.7 ± 3.0 *0.00*	P1 = 1.4 ± 0.1/12.8 ± 2.2 P2 = 1.4 ± 0.1/11.7 ± 2.4 P3 = 1.3 ± 0.4	60.9 ± 2.9 N_f = 20.5 W = 13	80.0 ± 4.1 N_f = 19.2 W = 12	122.9 ± 15.5 N_f = 4.1 W = 18	$\chi^2 = 39$ $l = 37$ $P(F) = 0\%$
MAO-3 (Zr)	Lower Albian *Oliván*	24	0.394 *(1826)*	13.524 *(3511)*	1.7256 *(448)*	211.4 ± 18.7 *0.04*		184.9 ± 13.4 N_f = 19.4 W = 27	419.6 ± 82.7 N_f = 4.6 W = 39		$\chi^2 = 24$ $l = 21$ $P(F) = 0\%$
JS-6 (Ap)	Lower Albian *Oliván*	74	1.007 *(4669)*	1.547 *(5583)*	4.696 *(16948)*	58.0 ± 3.4 *0.00*	P1 = 3.8 ± 1 P2 = 3.1 ± 1 P3 = 3.4 ± 2	45.2 ± 2.8 N_f = 45.1 W = 15	57.8 ± 6.0 N_f = 24.4 W = 15	123.2 ± 14.9 N_f = 3.5 W = 23	$\chi^2 = 74$ $l = 67$ $P(F) = 0\%$
JS-8 (Ap)	Lower Albian *Oliván*	20	1.124 *(5212)*	1.499 *(1041)*	5.053 *(3509)*	55.7 ± 3.4 *0.07*	P1 = 4.5 ±3 P2 = 2.6 ± 1	49.2 ± 3.4 N_f = 13.0 W = 18	70.7 ± 7.0 N_f = 7.0 W = 19		$\chi^2 = 23$ $l = 17$ $P(F) = 0\%$
MAO-4 (Ap)	Lower Albian *Oliván*	46	1.141 *(5289)*	1.503 *(3848)*	4.943 *(12655)*	59.1 ± 2.4 *0.00*	P1 = 2.2 ± 0.3/11.3 ± 2.3 P2 = 2.1 ± 0.3/11.6 ± 2.2 P3 = 2.1 ± 0.3	54.3 ± 2.4 N_f = 32.5 W = 15	69.8 ± 5.6 N_f = 12.5 W = 15	127.3 ± 24.7 N_f = 1.0 W = 20	$\chi^2 = 42$ $l = 41$ $P(F) = 0\%$
JS-9 (Ap)	Lower Albian *Oliván*	15	0.929 *(4307)*	1.452 *(878)*	3.973 *(2402)*	55.6 ± 3.2 *2.55*	P1 = 2.4 ± 0.4				
JB-17 (Ap)	Lower Albian *Oliván*	23	1.066 *(4942)*	1.220 *(663)*	3.532 *(1919)*	64.4 ± 6.2 *0.00*		21.4 ± 5.1 N_f = 2.4 W = 36	60.0 ± 4.1 N_f = 15.1 W = 24	105.6 ± 12.6 N_f = 5.5 W = 26	$\chi^2 = 25$ $l = 18$ $P(F) = 0\%$
JS-11 (Ap)	Lower Albian *Oliván*	15	0.883 *(4093)*	0.722 *(556)*	2.830 *(2179)*	34.5 ± 1.9 *90.44*	P1 = 6.0 ± 1.4				
JS-12 (Ap)	Lower Albian *Oliván*	61	1.065 *(4941)*	1.163 *(4598)*	4.193 *(16578)*	48.5 ± 2.3 *0.00*	P1 = 2.3 ± 0.7 P2 = 2.2 ± 0.5	41.9 ± 1.6 N_f = 43.3 W = 15	62.7 ± 3.6 N_f = 16.7 W = 15		$\chi^2 = 56$ $l = 56$ $P(F) = 0\%$
JS-13 (Ap)	Apitan *Enciso*	35	0.831 *(3855)*	1.065 *(1985)*	3.226 *(6015)*	46.5 ± 2.7 *0.00*	P1 = 4.8 ± 1.0 P2 = 4.3 ± 1.1	40.8 ± 1.7 N_f = 30.6 W = 18	88.6 ± 12.7 N_f = 4.4 W = 20		$\chi^2 = 32$ $l = 32$ $P(F) = 0\%$
JS-15 (Ap)	Hauterivian–Barremian *Urbión*	22	1.101 *(5102)*	0.931 *(747)*	3.486 *(2796)*	50.6 ± 4.3 *0.00*	P1 = 4.7 ± 1.1 P2 = 4.7 ± 1.5 P3 = 4.0 ± 0.1	39.0 ± 3.8 N_f = 11.8 W = 24	58.2 ± 8.2 N_f = 8.0 W = 21	100.7 ± 19.0 N_f = 2.2 W = 23	$\chi^2 = 21$ $l = 17$ $P(F) = 2\%$

(Continued)

Table 1. *Continued*

Sample No. (mineral type)*	Stratigraphic age *Group*	No. of grains	$\rho_D^{\dagger}$ (10^6 tr cm^{-2}) (N_D)	ρ_s (10^6 tr cm^{-2}) (N_s)	ρ_i (10^6 tr cm^{-2}) (N_i)	Central age $\pm 1\sigma$ (Ma) $P(\chi^2)$	D_{par} (μm)/MTL (μm) $\pm 1\sigma$	Age populations (Ma $\pm 1\sigma$), N_f, W (%)			Fit statistics
								P1	P2	P3	
POC-9 (Ap)	Hauterivian–Barremian *Urbión*	25	1.168 *(5417)*	0.270 *(202)*	1.099 *(822)*	48.3 ± 4.1 *82.61*					
POC-9 (Zr)	Hauterivian–Barremian *Urbión*	20	0.3854 *(1787)*	9.3645 *(1615)*	2.9108 *(502)*	84.7 ± 5.5 *65.73*					
JS-21 (Ap)	Tithonian *Tera*	21	0.863 *(4001)*	1.322 *(1347)*	6.139 *(6256)*	28.7 ± 1.5 *0.13*	P1 = 3.2 ± 0.3 P2 = 3.2 ± 0.3	25.5 ± 1.8 $N_f = 13.1$ $W = 17$	34.6 ± 2.9 $N_f = 7.9$ $W = 15$		$\chi^2 = 19$ $l = 18$ $P(F) = 0\%$

*Ap, apatite; Zr, zircon.

†CN-5 for apatites and CN-2 for zircons.

Samples MAO-3, MAO-4 and POC-9 measured by Luis Barbero ζ (CN-5) = 337.8 ± 5.2 and ζ (CN-2) = 137.5 ± 4.5.

All samples JS measured by Pedro del Rio ζ (CN-5) = 339.2 ± 8.2.

ρ_D, glass monitor track density; ρ_s, spontaneous track density; ρ_i, induced track density; $N_{D,s,i}$, number of tracks counted to determinate the reported track density.

N_f, number of grains in age population; W, relative standard deviation; χ^2, goodness of fit parameter; l, degrees of fredoom; $P(F)$, probability that the random variation alone could produce the observed F value for peak 2 or 3.

MTL, mean track length.

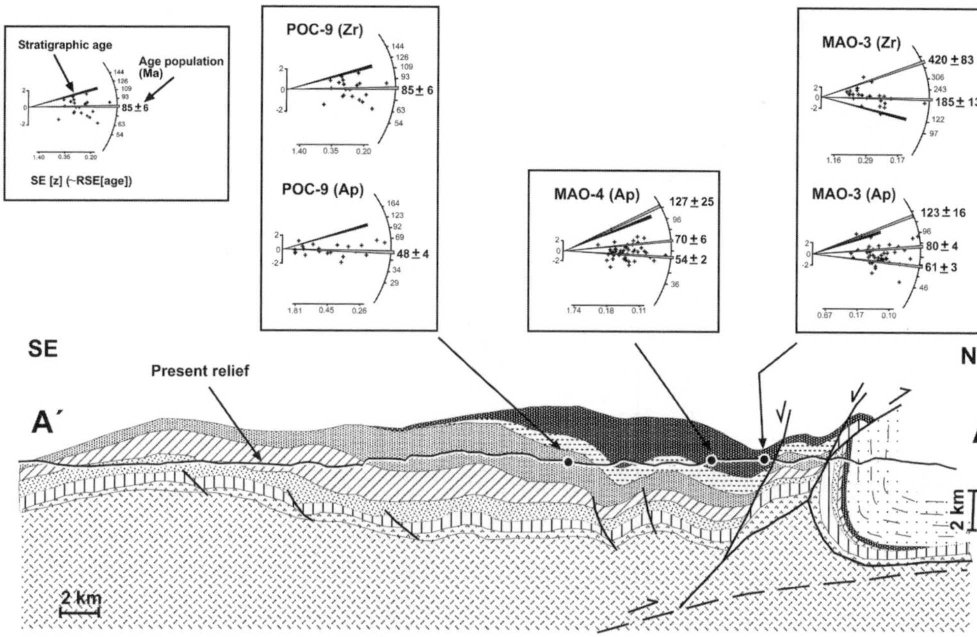

Fig. 3. Cross-section along A–A′ (see Fig. 1 for the location and legend) modified after Gil-Imaz (2001), showing the location of the samples and AFT and ZFT radial plots.

analysis of the age population distribution in these two samples, it is clear that both have undergone partial resetting after deposition. Both populations retain an inherited thermal signal from the source areas. Constraints used for modelling include: (1) the depositional age of the sediment (Lower Albian); (2) the fact that the AFT system is not fully reset indicates that maximum temperatures reached after deposition should be less than approximately 120 °C (in agreement with the stratigraphic position of these two samples from the Oliván group); and (3) for the predepositional history a free temperature constraint has been used for times in the 280–120 Ma interval. The chlorine content of these two samples is similar to the Durango apatite, therefore the Laslett *et al.* (1987) annealing model has been used. Modelling was performed using the AFTSolve program (Ketcham *et al.* 2000). Results from modelling indicate that: (1) the post-depositional history is similar in both samples, showing that a fast heating period soon after deposition was followed by continuous cooling from maximum palaeotemperatures of 110 °C at 108–80 Ma to the present day (which is consistent with the timing of the low-grade metamorphic event of the area); and (2) a pre-Albian inherited history is poorly constrained by modelling.

Discussion

The single ZFT age component that is younger than the depositional age in sample POC-9 (top of the Urbión group) indicates that, at least in the eastern part of the basin (profile A–A′), temperatures above 300–350 °C were reached (Yamada *et al.* 1995; Tagami *et al.* 1998), probably during the low-grade metamorphic event (108–86 Ma: Casquet *et al.* 1992). However, the two old age populations of samples MAO-3 and MAO-4 indicate that the Oliván group sediments were only partially reset. The vertical distance between samples POC-9 and MAO-3, according to the vertical stacking model of Casas-Sainz & Gil-Imaz (1998), can be easily deduced by adding the sedimentary thickness for the Oliván and Enciso groups, and is about 2.8 km. The difference in palaeotemperature between these two samples, on the basis of the apatite and zircon fission-track data (ZFT closure temperature of 350 °C: Tagami *et al.* 1998) and an AFT closure temperature of 110 °C (Green *et al.* 1989), is estimated to be around 240 °C (ZFT ages are clearly reset in sample POC-9 but not in sample MAO-3). The palaeogeothermal gradient calculated with these data is close to 86 °C km^{-1}. Based on low-grade metamorphic parageneses, Mata *et al.* (2001)

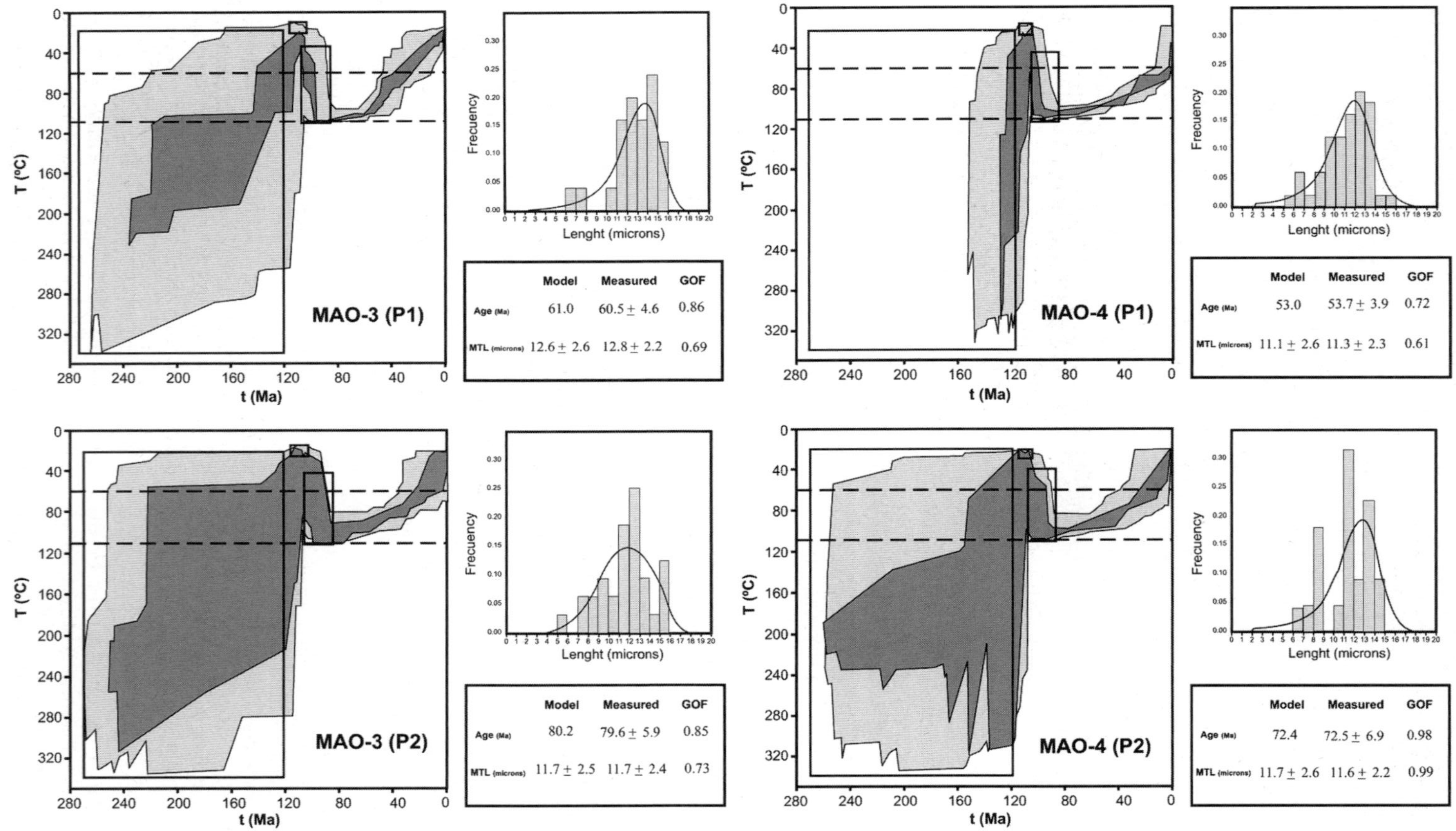

Fig. 4. Thermal models and mean track-length distribution for samples MAO-3 (left) and MAO-4 (right) for the AFT age populations P1 (up) and P2 (down). The dashed lines correspond to partial annealing zone boundaries for AFT. The T–t path envelopes are plotted. The light grey area includes the acceptable paths, and the dark grey area includes the good paths. The constraints are shown as boxes (explanation is in the text).

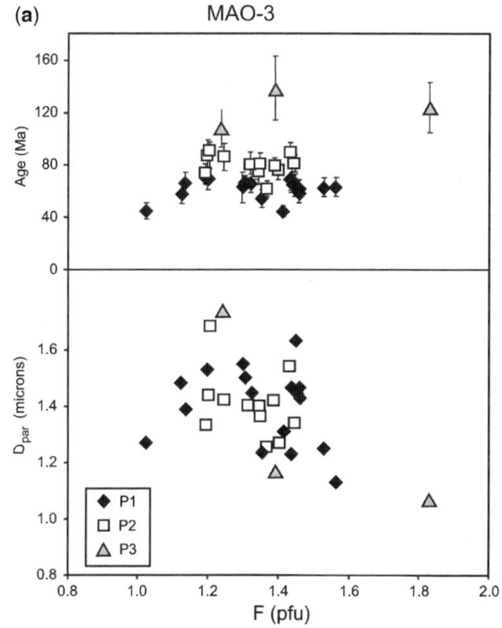

Fig. 5. Fluorine v. AFT ages and D_{par} values for samples (**a**) MAO-3 (upper part of the Oliván group) and (**b**) MAO-4 (lower part of the Oliván group).

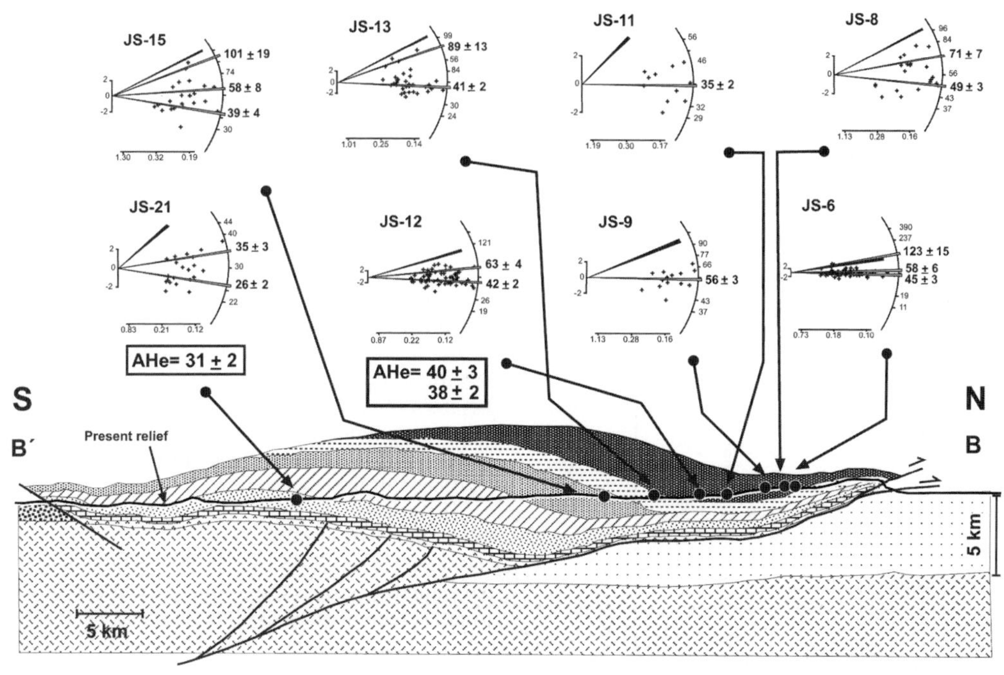

Fig. 6. Cross-section along B′–B (see Fig. 2 for the location and legend) (structural interpretation from Casas-Sainz & Gil-Imaz 1998, p. 812, fig. 8A) showing the location of the samples and the radial plots for AFT and AHe ages.

Table 2. *(U–Th)/He age results in apatites from the B′–B profile in the Sierra de Cameros*

Sample No.	Stratigraphic age *Group*	No. of grains	U (ng)	Th (ng)	4He (cm^3)	Mean radius (μm)	Raw age (Ma)	FT (μm)	Corrected age (Ma)	Error $\pm 2\sigma$ (Ma)
JS-12-I	Lower Albian *Oliván*	2	0.104	0.035	3.3×10^{-10}	35	24.4	0.61	39.8	2.8
JS-12-III	Lower Albian *Oliván*	2	0.034	0.122	1.8×10^{-10}	30	22.8	0.53	31.1	1.5
JS-21-III	Tithonian *Tera*	2	0.246	0.106	8.1×10^{-10}	40	24.5	0.70	35.2	1.7

FT, alpha-ejection correction factor.

calculated a similarly high maximum palaeo-geothermal gradient of approximately 70 °C km^{-1} at the end of the extensional stage. Temperatures obtained in samples located within the Urbión group can be attained in the basin depocentre if we consider a tectonic model of vertical superposition of the sedimentary units with a high geothermal gradient. When considering the alternative model of lateral stacking of sedimentary units (Mas *et al.* 2002), the deepest part of the basin cropping out at present (the base of the Urbión group) would only be covered by a maximum sedimentary thickness of 2 km (Mata *et al.* 2001). This implies that a geothermal gradient of more than 150 °C km^{-1} is necessary to explain the high temperatures reached in the Urbión group. Therefore, the sag basin model of Mas *et al.* (1993) and Guimerà *et al.* (1995) requires a geothermal gradient of at least 150 °C km^{-1} that, on the basis of the FT and fluid inclusion data and in the absence of magmatism or a hydrothermal fluid activity, seems unrealistic (Casquet *et al.* 1992). Furthermore, other arguments against this hypothesis come from the source–sediment balance, because if we compare the volume of sediments deposited in the Ebro foreland basin eroded from the Sierra de Cameros it is possible to estimate that the amount of removed material from the Cameros Basin cannot be balanced with a tectonic model which proposes a maximum thickness of eroded sediments of 2 km (see e.g. Muñoz-Jiménez & Casas-Sainz 1997).

AFT data from all samples from profile A–A′ and B–B′ (except JS-6) show cooling ages younger than their depositional age and record the time of cooling from more than 110 °C. However, AFT age populations range from 21 to 124 Ma. The presence of several age components in samples with evidence of total resetting of the fission tracks in apatite (for example, the presence of chloritoid in the mineral paragenesis) could be explained by differences in chemical composition of the apatites, which produces different kinetic behaviour during annealing (Green *et al.* 1986; Ketcham *et al.* 1999). A study of D_{par} and the chemical composition of apatites in these samples is underway to shed more light on the origin of the dispersion of the results.

The youngest age components of the AFT data from the B–B′ profile (Fig. 7a) record a first cooling event at 45–55 Ma in the northern part of the Cameros Basin, close to the present-day thrust front (as seen in samples JS-6, JS-8 and JS-9). The second, more important, cooling event is recorded by samples JS-11–JS-15 (see Fig. 7) at 30–40 Ma. This coincides with an important period of tectonic activity, with a horizontal displacement

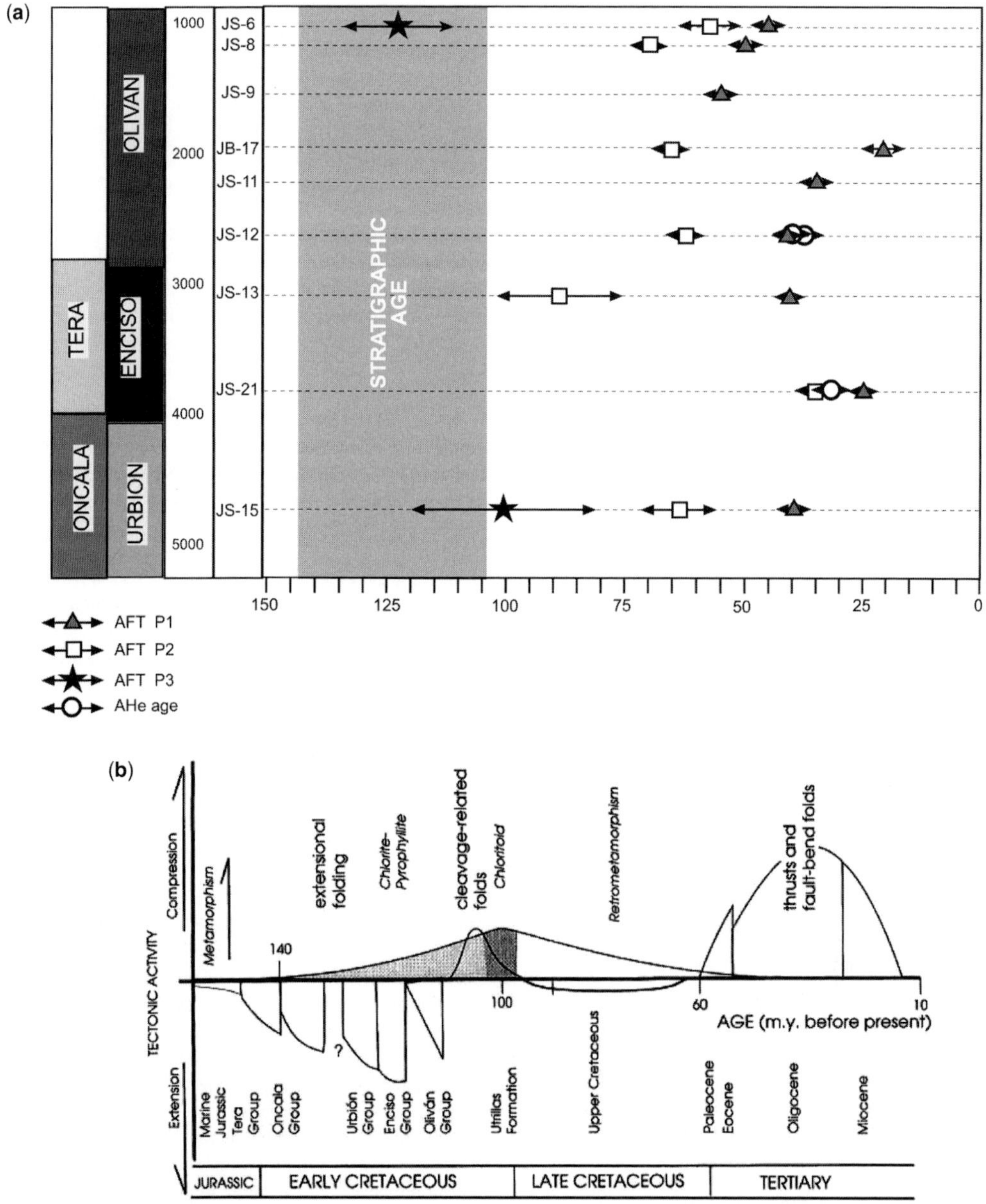

Fig. 7. (**a**) AFT age populations and AHe cooling ages v. the estimated depth of the samples at the time of deposition from samples along the B′–B profile. (**b**) The main thermal events that occurred in the Cameros Basin (Casas-Sainz & Gil-Imaz 1998, p. 814, fig. 9B).

velocity of the thrusts of 0.8 and 1.25 mm year^{-1} for the central and eastern part of the basin, respectively (Muñoz-Jiménez & Casas-Sainz 1997), and so we consider that this age can constrain the age of the main pulse of tectonic inversion in the area. The AHe ages of JS-12 and JS-21 are similar to the youngest apatite fission-track age component in these samples (Fig. 7a), consistent with rapid cooling through the apatite partial annealing zone (PAZ; from 110 to 60 °C) to the AHe closure temperature (60–70 °C: Farley 2000).

The cooling rate can be determined by combining the AFT and ZFT data. For sample POC-9, in the eastern part of the Urbion group, is around

5.4–8.9 °C Ma^{-1}. These rates indicate that the basin cooled rapidly between 85 and 48 Ma. The modelled thermal histories for samples MAO-3 and MAO-4 from the upper and lower parts of the Olivan group show a fast heating during the metamorphic event, up to 100 °C, of about 5–8 °C Ma^{-1}. A continuous regional cooling of around 0.6–1.6 °C Ma^{-1} is observed from the thermal peak (*c*. 100 Ma) to present. This is consistent with the progressive exhumation of the Cameros Basin during the Tertiary that resulted in the accumulation of approximately 5 km of Tertiary sediments in the western part of Ebro Basin, the Rioja Trough, which represents the foreland basin of the Sierra de Cameros (Muñoz-Jiménez & Casas Sainz 1997).

Conclusions

The Sierra de Cameros was an intracratonic sedimentary basin filled by continental synrift sediments during the lower Cretaceous. During the extensional stage, the sedimentary fill attained temperatures of more than 110 °C, as demonstrated by the young AFT ages of all samples except those of the uppermost Olivan group sediments. Temperatures above 300 °C were reached in the Urbión group, as revealed by the young ZFT age of sample POC-9. This is consistent with the low-grade metamorphism observed in the central part of the basin. From an independent estimate of the palaeogeothermal gradient (86 °C km^{-1}), a sediment thickness of 8 km is necessary to explain these temperatures and the sediment volume found in the Ebro foreland basin. These data suggest that a model of vertical sediment accumulation is more feasible than the lateral stacking model. Alpine compression produced the tectonic inversion of the basin during the Tertiary. Combining ZFT and AFT data, we calculated that between 85 and 48 Ma rocks from the Urbión group were cooled at a rate of 5.4–8.9 °C Ma^{-1}. An increase in the cooling rate is identified by the coincidence of the youngest age component of AFT results and the AHe cooling ages, at about 40 Ma. This represents the time of maximum uplift and cooling, which produced the exhumation of the main part of the basin related to the Alpine orogenesis.

We thank B. Ventura and two anonymous reviewers for the helpful comments made to a previous version of this manuscript. This work has been supported by MCYT project BTE2002-04168-C03-02 and by the Consolider-Ingenio 2010 Programme under project CSD2006-0041 'Topo-Iberia'. This research is part of the objectives of the Junta de Andalucía PAI group RNM-160. We thank D. Vilbert and J. Foeken for assistance at SUERC. The SUERC He laboratory is supported by the Scottish Universities.

Appendix

Analytical methods

Thirteen samples of sandstones were collected along two cross-sections (north–south and NE–SW) perpendicular to the strike of regional structures. Mineral separations were carried out using standard methods. In this work we use the ζ approach for fission-track analyses of individual apatite and zircon crystals (Hurford & Green 1983). Apatites and zircons were mounted in U-free epoxy resin and in a piece of Teflon, respectively, and polished with 1 and 3 μm diamond and 0.25 μm aluminium oxide. To reveal the spontaneous tracks, the apatite crystals were etched in 5 M HNO_3 for 20 ± 1 s. The zircons were prepared according to the procedure described by Naeser *et al.* (1987). Three replicates for each sample were etched with a low-melting-point binary eutectic of NaOH and KOH at 239 °C for a period of between 6 and 100 h to ensure all of the zircon was adequately etched.

We applied the external detector method using thin U-free muscovite sheets as detectors. We used two dosimeters at the top and bottom of the irradiation capsule to determine the flux of thermal neutrons' irradiation; CN-5 and CN-2 for apatite and zircon, respectively. The irradiation was performed at HIFAR (Australia). The neutron fluence was 9×10^{15} and 1×10^{15} n cm^{-2} for apatites and zircons, respectively. All of the irradiated mica sheets were etched in 5 M HF at room temperature for 40 min. For track counting and length measurements, a magnification of 1250× was used.

The best-fit binomial peak-fitting routine of Galbraith & Green (1990) was used to deconvolve the fission-track grain-age spectra in samples containing more than one age population. Calculations were performed using the Windows version of the BINOMFIT program of Brandon (1992, 1996). The procedures were based on the maximum-likelihood method, where the best-fit solution is determined by comparing the age distribution of the data to a predicted mixed binomial distribution.

Apatite grains for (U–Th)/He age determinations were hand-picked using a stereographic binocular microscope (220× magnification). Only grains with good crystal shape and no mineral inclusions were selected. Two crystals per sample were packed in Pt foil tubes and loaded in batches of 40 into an ultra-high vacuum extraction cell. Helium was extracted by heating the foil with an 808 nm diode laser to approximately 1000 °C. After a short clean-up step, helium was measured using a Hiden HAL3F gas source quadrupole mass spectrometer following the procedures of Foeken *et al.* (2006). Samples were then diluted in 5% HNO_3 and spiked with ^{230}Th and ^{235}U, and the U and Th concentrations measured using a VG PQ2.5 inductively coupled plasma mass spectrometer (Balestrieri *et al.* 2005). Ages were corrected for alpha-recoil using the procedure of Farley *et al.* (1996).

Apatite compositions were measured by electron microprobe (Jeol Superprobe JXA-8900-M) using operating conditions of 15 kV, 20 nA and a beam diameter of

2–5 μm, and a ZAF (atomic number (Z), absorption (A) and fluorescence (F)) correction procedure.

References

BALESTRIERI, M. L., STUART, F. M., PERSANO, C., ABBATE, E. & BIGAZZI, G. 2005. Geomorphic development of the scarpment of the Eritrean margin, southern Red Sea from combined apatite fission track and (U-Th)/He thermochronometry. *Earth and Planetary Science Letters*, **231**, 97–110.

BRANDON, M. T. 1992. Decomposition of fission-track grain-age distributions. *American Journal of Science*, **292**, 535–564.

BRANDON, M. T. 1996. Probability density plot for fission-track grain-age samples. *Radiation Measurements*, **26**, 663–676.

CASAS-SAINZ, A. M. & GIL-IMAZ, A. 1998. Extensional subsidence, contractional folding and thrust inversion of the Estern Cameros Massif, northern Spain. *Geological Rundschau*, **86**, 802–818.

CASQUET, A., GALINDO, C. *ET AL.* 1992. El metamorfismo en la cuenca de los Cameros, Geocronología e implicaciones tectónicas. *Geogaceta*, **11**, 22–25.

FARLEY, K. A. 2000. Helium diffusion from apatite: general behavior as illustrated by Durango fluorapatite. *Journal of Geophysical Research*, **105**, 2903–2914.

FARLEY, K. A., WOLF, R. A. & SILVER, L. T. 1996. The effects of long alpha-stopping distances on (U–Th)/He ages. *Geochimica et Cosmochimica Acta*, **60**, 4223–4229.

FOEKEN, J. P. T., STUART, F. M., DOBSON, K. J., PERSANO, C. & VILBERT, D. 2006. A diode laser system for heating minerals for (U–Th)/He chronometry. *Geochemistry, Geophysics, Geosystems*, **7**, 1–9, Q04015, doi:10.1029/2005GC001190.

GALBRAITH, R. F. & GREEN, P. F. 1990. Estimating the component ages in a finite mixture. *Nuclear Tracks and Radiation Measurements*, **17,** 197–206.

GIL-IMAZ, A. 2001. *La Estructura de la Sierra de Cameros: Deformación dúctil y su Significado a Escala Cortical*. PhD thesis. *Instituto de Estudios Riojanos, Colección Ciencias de la Tierra*, **23**, 305.

GIL-IMAZ, A., POCOVÍ, A., LAGO, M. & PARÉS, J. M. 2000. Effect of lithostatic pressure and tectonic deformation on the magnetic fabric (anisotropy of magnetic susceptibility) in low-grade metamorphic rocks. *Journal of Geophysical Research*, **105**, 305–317.

GIL-IMAZ, A., VILLALAIN, J. J., BARBERO, L., GONZALEZ, G., MATA, P. & CASAS, A. M. 2002. Aplicación de técnicas geoquímicas, geofísicas y mineralógicas al estudio de la cuenca de Cameros. Implicaciones geométricas y evolutivas. *Zubía, Instituto de Estudios Riojanos*, **14**, 65–98.

GOLDBERG, J. M., GUIRAUD, M., MALUSKI, H. & SÉGURET, M. 1988. Caractères pétrologiques et âge du métamorphisme en contexte distensif du basin sur décrochement de Soria (Crétacéinférieur, Nord Espagne). *Les Comptes Rendus de ĺ Académie des Sciences, Paris*, **307**, 521–527.

GREEN, P. F., DUDDY, I. R., GLEADOW, A. J. W., TINGATE, P. R. & LASLETT, G. M. 1986. Thermal annealing of fission tracks in apatite: 1. A qualitative description. *Chemical Geology (Isotope Geosciences Section)*, **59**, 237–253.

GREEN, P. F., DUDDY, I.R, LASLETT, G. M., HEGARTY, K. A., GLEADOW, A. J. W. & LOVERING, J. F. 1989. Thermal annealing of fission tracks in apatite: 4. A quantitative modelling techniques and extension to geological timescales. *Chemical Geology (Isotope Geosciences Section)*, **79**, 155–182.

GUIMERÀ, J., ALONSO, A. & MAS, J. R. 1995. Inversion of an extensional-ramp basin by a newly formed thrust: the Cameros basin (N Spain). *In*: BUCHANAN, J. G. & BUCHANAN, P. G. (eds) *Basin Inversion*. Geological Society, London, Special Publications, **88**, 433–453.

GUIRAUD, M. 1983. *Evolution Tectono-sedimentarie du Basin Wealdien (Crétacé inferieur) en relais de Décrochements de Logroño-Soria (N-W Espagne)*. PhD thesis, Languedoc, 1–186.

GUIRAUD, M. & SÉGURET, M. , 1984. Releasing solitary overstep model for the Late Jurassic–Early Createceous (Wealdian) Soria strike-slip basin (North Spain). *In*: BIDDLE, K. T. & CHRISTIE-BLICK, N. (eds) *Strike-slip Deformation, Basin Formation and Sedimentation*. SEPM, Special Publications, **37**, 159–175.

HURFORD, A. & GREEN, P. F. 1983. The zeta calibration of fission-track dating. *Chemical Geology*, **1**, 285–317.

KETCHAM, R. A., DONELICK, R. A. & CARLSON, W. D. 1999. Variability of apatite fission track annealing kinetics III: extrapolation to geological time scale. *American Mineralogist*, **84**, 1235–1255.

KETCHAM, R. A., DONELICK, R. A. & DONELICK, M. B. 2000. AFTSolve: a program for multi-kinetic modeling of apatite fission-track data. *Geological Materials Research*, **2**, 32.

LASLETT, G. M., GREEN, P. F., DUDDY, I. R. & GLEADOW, A. J. W. 1987. Thermal annealing of fission tracks in apatite: 2. A quantitative analysis. *Chemical Geology (Isotope Geosciences Section)*, **65**, 1–13.

MARTÍN I CLOSAS, C. 1989. *Els caròfits del Cretacé inferior de les Conques perifériques del Bloc de Ĺ Ebre*. PhD thesis, Universidad de Barcelona.

MAS, J. R., ALONSO, A. & GUIMERÀ, J. 1993. Evolución tectonosedimentaria de una cuenca extensional intraplaca: la cuenca finijurásica-eocretácica de los Cameros (La Rioja-Soria). *Revista de la Sociedad Geológica de España*, **6**, 129–144.

MAS, J. R., ALONSO, A. & GUIMERÀ, J. 1994. Réplica del artículo: Evolución tectonosedimentaria de una cuenca extensional intraplaca: la cuenca finijurásica-eocretácica de los Cameros (La Rioja-Soria). *Revista de la Sociedad Geológica de España*, **7**, 347–355.

MAS, R. & SEGURA, M. *ET AL.* 2002. The Iberian Basin. *In*: GIBBONS, W. & MORENO, T. (eds) *The Geology of Spain*. Geological Society, London, 284–292.

MATA, M. P., OSÁCAR, M. C. & LÓPEZ-AGUAYO, F. 2000. Una aproximación al área fuente del Weald de Cameros: datos geoquímicos. *Geotemas*, **1**, (3) 263–265.

MATA, M. P., CASAS, A. M., CANALS, A., GIL, A. & POCOVI, A. 2001. Thermal history during Mesozoic extension and Tertiary uplift in the Cameros Basin, northern Spain. *Basin Research*, **13**, 91–111.

MUÑOZ, A., SORIA, A., CANUDO, J. I., CASAS, A., GIL-IMAZ, A. & MATA, M. P. 1997. Caracterización estratigráfica y sedimentológica del Albiense marino del borde Norte de la Sierra de Cameros. Implicaciones paleogeográficas. *Cuadernos de Geología Hispánica*, **21–22**, 427–435.

MUÑOZ-JIMÉNEZ, A. & CASAS-SAINZ, A. M. 1997. The Rioja trough: tectono-sedimentary evolution of a foreland symmetric basin. *Basin Research*, **9**, 65–85.

NAESER, N. D., ZEITLER, P. K., NAESER, C. W. & CERVENY, P. F. 1987. Provenience studies by fission track dating of zircon – Etching and counting procedures. *Nuclear Tracks and Radiation Measurements*, **13**, 121–126.

SCHUDACK, M. 1987. Charophytenflora und fazielle Entwicklung der Grenzschichten mariner Jura/Wealden in der Nordwestlichen Iberischen Ketten (mit Vergleichen zu Asturien und Kantabrien). *Paleontographica Abteilung B*, **204**, 108.

TAGAMI, T., GALBRAITH, R. F., YAMADA, R. & LASLETT, G. M. 1998. Revised annealing kinetics of fission tracks in zircon and geological implications. *In*: VAN DEN HAUTE, P. & DE CORTE, F. (eds) *Advances in Fission-Track Geochronology*, Volume 10. Kluwer, Dordrecht, 99–112.

TISCHER, G. 1966. El delta Weáldico de las montañas Ibéricas Occidentales y sus enlaces tectónicos. *Notas y Comunicaciones Instituto Geológico y Minero de España*, **81**, 53–78.

YAMADA, R., TAGAMI, T., NISHIMURA, S. & ITO, H. 1995. Annealing kinetics of fission track in zircon: an experimental study. *Chemical Geology*, **122**, 249–258.

Late- and post-Variscan evolution of the Ardennes in France and Belgium: constraints from apatite fission-track data

CHANGHAI XU[1,2]*, JEAN LOUIS MANSY[1], PETER VAN DEN HAUTE[3], FRANCOIS GUILLOT[1], ZUYI ZHOU[2], JUN CHEN[2] & JOHAN DE GRAVE[3]

[1]*UMR PBDS (CNRS) Université Lille 1, F-59655 Villeneuve d'Ascq Cédex, France*

[2]*State Key Laboratory of Marine Geology, Tongji University, Shanghai 200092, China*

[3]*Geological Institute, Ghent University, Krijgslaan 281, B9000, Ghent, Belgium*

**Corresponding author (e-mail: xchxch@tongji.edu.cn)*

Abstract: Apatite fission-track (AFT) analyses were performed on 13 Late Palaeozoic samples in order to unravel the late- to post-Variscan evolution of the Ardennes. The dated AFT ages cover a range from 290 ± 33 Ma to 168 ± 12 Ma, and the mean confined track lengths correspond to a unimodal distribution, with means varying between 13.1 ± 0.1 μm and 11.7 ± 0.3 μm. These ages for the sedimentary rocks are clearly younger than the respective stratigraphic ages, indicative of a cooling through the apatite partial annealing zone after post-depositional complete annealing. All available AFT data (290–146 Ma) from this region might be classified as three groups, that is 290–229 Ma, 218–198 Ma and 190–146 Ma, at least in correlation with three exhumation events. Using an inverse model, four major cooling episodes are identified from the modelled temperature–time (T–t) paths. The first rapid cooling (4.2–5.4 °C Ma^{-1}, 320–300 Ma) corresponds to the late-Variscan rapid thrusting that ceased at about 300 Ma. The second cooling episode (0.2–4.0 °C Ma^{-1}, up to 230 Ma) activated differentially, and was probably controlled by the post-Variscan transtension. The third cooling regime (0.1–0.3 °C Ma^{-1}, 230–45 Ma) in the Ardennes Allochthon is slow, and represents a long-term and slow exhumation. In the Brabant Parautochthon, however, it is subdivided into 0.7 °C Ma^{-1} (225–110 Ma) and 0.2 °C Ma^{-1} (110–45 Ma). The last accelerated cooling (0.7–1.1 °C Ma^{-1}, since 45 Ma) that affected the whole Ardennes is associated with a south–north compression during the Pyrenean phase.

The Rhenohercynian zone of the central European Variscides developed from a passive continental margin and then underwent Variscan collision with deformation propagation from south (335–330 Ma) to north (325–300 Ma) (Oncken *et al.* 1999). For the Ardennes part of it, focused by this study, the Palaeozoic history has been well constrained (e.g. Helsen 1995; Mansy *et al.* 1999; Fielitz & Mansy 1999; Franke *et al.* 2000). The post-Variscan development of the Ardennes, however, is comparatively poorly known. Apatite fission-track (AFT) analysis is a useful tool for evaluating the low-temperature evolution of rocks over a temperature range from the surface down to approximately 110 °C, equivalent to about 3.7 km of exhumation if a thermal gradient of 30 °C km^{-1} is assumed (e.g. Wagner & Van den Haute 1992; Gleadow & Brown 2000). Vercoutere & Van den Haute (1993) reported a first batch of AFT ages of igneous rocks from the southern Brabant Massif ranging between 209 ± 13 Ma and 146 ± 15 Ma, along with confined track lengths of 14.2–13.0 μm. They correlated them with a relatively rapid cooling phase under the control of an important uplift in Jurassic times. Another study of Palaeozoic rocks from the NE Ardennes (Glasmacher *et al.* 1998) provides an AFT age range from 130 ± 22 Ma to 239 ± 26 Ma, with mean track lengths varying between 12.4 and 13.2 μm. In addition, zircon fission-track ages obtained by Brix (2002) consist of two groups: one of 322–218 Ma from the High Ardennes is attributed, at least partly, to a Variscan doming; and the other, of 442–395 Ma, from the Dinant and Namur basins records much of the thermal information of sediment provenance. In this study, an attempt is made to evaluate the late- to post-Variscan tectonothermal evolution on the basis of new AFT data from the central and the eastern Ardennes, combined with other published results.

Regional geological background

The doubly-vergent Variscan orogen developed from subduction and collision of three major zones: Rhenohercynian, Saxothuringian and Moldanubian (Schäfer *et al.* 2000). The Variscan Ardennes, as part of the Rhenohercynian foreland fold–thrust

From: LISKER, F., VENTURA, B. & GLASMACHER, U. A. (eds) *Thermochronological Methods: From Palaeotemperature Constraints to Landscape Evolution Models*. Geological Society, London, Special Publications, **324**, 167–179.
DOI: 10.1144/SP324.13 0305-8719/09/$15.00

zone, comprises the Ardennes Allochthon and the Brabant Parautochthon (Fig. 1). The Ardennes Allochthon is an area of thin-skinned northwards-tapering tectonics on the roof of the Brabant Massif. The Midi fault, as a low-angle frontal thrust between them, brings Lower Devonian rocks immediately over Upper Carboniferous (Márton *et al.* 2000). The Ardennes Allochthon falls into two units: the Dinant Synclinorium and the High Ardennes. For the Dinant Synclinorium, the Lower–Late Devonian rocks are continental to marine strata, and the Late Carboniferous coal-bearing sequence represents the filling of a foreland basin (Mansy *et al.* 1999). The metamorphism for this synclinorium reaches uppermost anchizonal conditions (245–310 °C) in the central and northern parts, and diagenetic conditions (120 °C) in the southern to southeastern parts (Helsen 1995). The High Ardennes consists of the Lower Devonian Ardennes Anticlinorium, the Lower Devonian Neufchâteau Syclinorium and several Lower Palaeozoic inliers. The metamorphism reaches anchizonal conditions (245–310 °C) for the Ardennes Anticlinorium, epizonal levels (300–340 °C) for the Neufchâteau Syclinorium (Helsen 1995) and 300–500 °C for those inliers (Fielitz & Mansy 1999). Across the Stavelot Inlier, the NE–SW-trending Malmédy Graben is filled with the Permian conglomerate deposits (e.g. Glasmacher *et al.* 1998; Bultynck *et al.* 2001) and attributed to a strike-slip control (Geukens 1995).

The Brabant Massif has a Cambrian core bordered by Ordovician–Silurian strata. The Border fault between the Brabant and Ardennes areas displays a clear gravity gradient on a Bouguer anomaly map (Mansy *et al.* 1999), which was reactivated as a strike-slip element probably in late- to post-Variscan times (Sintubin 1999). The Brabant Parautochthon, as part of the southern Brabant Massif, retained a thin para-autochthonous cover of Mid-Devonian clastic and Early Carboniferous carbonate sediments uncomfortably overlying the Caledonian basement. A thick, intensely deformed Namurian–Westphalian coal-bearing clastic succession developed from Boulonnais to Aachen areas, representing a foreland basin at the northwards-migrating deformational front of the Rhenohercynian zone (Mansy *et al.* 1999). This parautochthon has the Mid-Devonian–Carboniferous sediments reaching a diagenetic–lower anchizonal metamorphism (Fielitz & Mansy 1999). The southern Brabant Massif is covered by Upper Cretaceous–Cenozoic strata. The thickness of the Upper Cretaceous has a range of 0–300 m from SW to NE (Legrand 1968). Palaeocene–Early Eocene strata distributed widely, but their depositional centres from the middle Eocene to the Pliocene became more and more restricted, and moved northeastwards (De Batist & Versteeg 1999).

In addition, a weathering cover widely developed in the Ardennes as a result of a long-term exhumation of the basement. In the Haute–Lesse, a 65 m-thick kaolinitic weathering sequence demonstrates an age succession of 135–18 Ma from top to bottom, dated by various dating methods, whereas the Pb/Pb dating on phosphates from the fractures reveals a Late Permian–Early Triassic weathering process (Yans & Dupuis 2005). In the western Ardennes Allochthon, it is commonly found that the Wealden continental deposits (Berriasian–Barremian: Hutt *et al.* 1996) filled the hollows of Palaeozoic basements. Even in the Sarre, south of the Ardennes, dated weathering ages of volcanic Permian rocks vary between 142 and 120 Ma (Lippolt *et al.* 1998). It is thus suggested that the Ardennes, Brabant and Rhenish areas were exposed in the Early Cretaceous and remained as large plateaus (Thiry *et al.* 2006). If these eroded weathering products are considered, the restored weathering process (at least in the Ardennes) would have initiated earlier than in the Early Cretaceous.

Sampling and analytical methods

Thirteen outcrop samples were collected from the Ardennes for AFT analyses. These Late Palaeozoic samples comprised one igneous and 12 sedimentary rocks, covering a variety of stratigraphic units from Lochkovian to Namurian and Permian. These samples scatter, in general, perpendicular to the main Variscan structural trend, with elevations from 350 to 95 m. The sample descriptions are listed in Table 1.

Bulk rocks were crushed, pulverized, sieved and dried at temperatures of less than 50 °C. The apatites for each sample were extracted from a 1.5–2.0 kg, 74–297 μm fraction using heavy liquid (bromoform, 2.81 g cm^{-3}; diiodomethane, 3.3 g cm^{-3}) and magnetic separation techniques. After having been cleaned up by handpicking, the apatite concentrates were mounted in Araldite epoxy resin, and polished using, in succession, 6.0 μm, 3.0 μm and 1.0 μm diamond pastes in order to expose internal surfaces. Etching to reveal spontaneous tracks was performed using 2.5% HNO_3 at room temperature for 70 s. All etched mounts, U-doped glass dosimeter (IRMM540: De Corte *et al.* 1998) and age standards (Durango and FCT: Hurford 1990) were covered with low-U mica external detectors and sent for neutron irradiation. The irradiation for most samples were performed in the Triga Mark II Reactor at Pavia University (Italy) (thermal flux $= 0.9997 \times 10^{12}$ n cm^{-2} s^{-1}), with a total thermal neutron fluence of $\phi_{TH} = 7.198 \times 10^{15}$ n cm^{-2}. Samples AD11 and

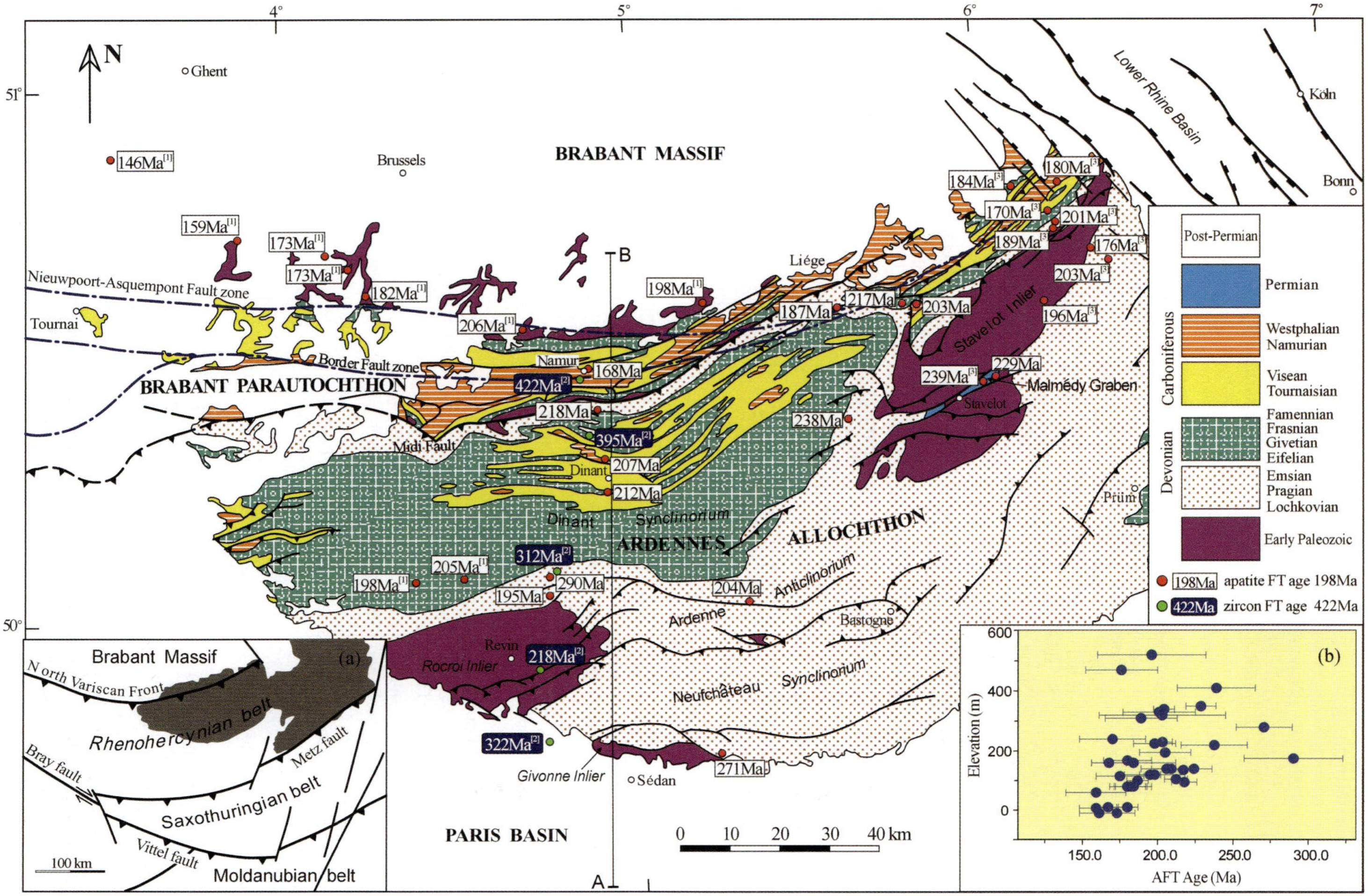

Fig. 1. Structural units and available fission-track data of the Ardennes and Brabant areas. The locations of the Border fault and the Nieuwpoort–Asquempont fault zones are obtained through the downward continuation (*c.* 4 km) of gravity Bouguer data, for the removal of the effects overlapped by the Midi fault; Fission-track data sources: [1] from Vercoutere & Van den Haute 1993; [2] from Brix 2002; [3] Glasmacher *et al.* 1998; others from this study.

Table 1. *Description of Late Palaeozoic samples from the Ardennes used for AFT analyses*

Sample	Elevation (m)	Location	Stratigraphic age	Lithology
AD18	160	Namur	Namurian	Medium sandstone
AD15	95	2.5 km SW of Dave	Emsian	Coarse sandstone
AD14	140	Anhée	Namurian	Medium sandstone
AD13	105	2.0 km SW of Neffe	Famennian	Medium sandstone
AD11	175	1.5 km West of Vireux	Emsian	Medium sandstone
AD10	120	1.0 km South of Montigny	Lochkovian	Coarse sandstone
AD108	136	Goffontaine	Emsian	Medium–coarse sandstone
AD107	230	Pépinster	Lochkovian–Pragian	Coarse sandstone
AD106	350	Malmédy	Permian	Conglomerate
AD103	100	Tilff	Emsian	Coarse sandstone
AD102	220	Heyd	Emsian	Conglomerate
AD101	340	St Hubert	Lochkovian	Fine-grained sandstone
AD100	280	Muno	Mid–late Devonian	Muno Kersantite

AD18 were irradiated in the Thetis Reactor at Ghent University (Belgium) (thermal flux = $8.735479 \times 10^{10}\ n\ cm^{-2}\ s^{-1}$) using a total thermal neutron fluence of $\phi_{TH} = 1.56917 \times 10^{15}\ n\ cm^{-2}$. All mica detectors after irradiation were removed and etched in 40% HF at room temperature for 40 min in order to reveal induced fission tracks.

Etched tracks were counted in transmitted light at a total magnification of 1000× (100× dry objective, 10× eyepieces) using a FT Autoscan system. Confined track-length measurements were carried out through a drawing tube connected to a computer-controlled digital tablet. The AFT ages for the samples were calculated using ζ calibration factors for two irradiations with varying neutron fluence, that is $\zeta_{IRMM05} = 227.2 \pm 3.3$ for the first author, and $\zeta_{IRMM03} = 179.5 \pm 4.2$, which were averaged from eight age standard mounts for each irradiation. Radial plots were used for sedimentary samples to evaluate the distribution of AFT grain ages and the degree of homogeneity. The radial plotting was run using the BINOMFIT program (Brandon 2002), which provides the optimal solution through an iterative search of peak ages and number of peaks. Based on AFT grain age, confined track length and other constraints, the temperature–time (T–t) paths for the samples were modelled using the software of AFTSolve 1.3.1 (Ketcham *et al.* 2000) and an inverse procedure (Gallagher 1995).

AFT results and interpretation

All analytical data for the samples are listed in Table 2, and spatially represented in Figure 1. Most of them have good measurements, with more than 20 grain ages and in excess of 50 confined track lengths for each. The radial plots for sedimentary rocks are shown in Figure 2. All dated samples, except for AD14, yielded pooled AFT ages varying between 290 ± 33 Ma and 168 ± 12 Ma, and the confined track lengths for them correspond to a unimodal distribution, with means ranging from 12.8 ± 0.1 μm to 11.7 ± 0.3 μm. Sample AD14 has a low $P(\chi^2)$ test (0.08), and therefore a weighted mean age (207 ± 3 Ma) is used for it.

The two oldest AFT ages were obtained from samples AD11 and AD100 in the High Ardennes. Sample AD11, an Emsian sandstone from the northern flank of the Ardennes Anticlinorium, has an age of 290 ± 33 Ma pooled from 20 single-grain ages ($P\chi^2 = 1.00$) that are all significantly younger than the stratigraphic age (Fig. 2). Its mean track length is 12.3 ± 0.2 μm, averaged from 82 confined track lengths, indicative of a cooling through the apatite partial annealing zone (APAZ) after post-depositional complete annealing. Sample AD100 close to the Givonne Inlier is a Muno kersantite that intruded in the Mid–Late Devonian. This sample yielded an AFT pooled age of 271 ± 19 Ma, averaged from 15 single-grain ages ($P\chi^2 = 0.90$), with a mean track length of 12.2 ± 0.2 μm from 35 confined track lengths, reflecting a cooling after the emplacement. The youngest AFT age in this study, 168 ± 12 Ma, was obtained from AD18, a medium-grained Namurian sandstone from the Namur foreland basin. Its 34 single-grain ages belonging to the same population ($P\chi^2 = 0.95$) are clearly younger than the sample depositional age (Fig. 2), with the implication that the AFT system was reset after sedimentation. Its 73 confined track lengths correspond to a broad unimodal distribution with a mean of 12.6 ± 0.3 μm.

Most samples from the Ardennes have AFT ages varying between 187 ± 7 Ma and 238 ± 22 Ma. Except for sample AD106, the others are with the AFT data illustrating a thermal history after post-depositional complete annealing. In the High Ardennes, samples AD10, AD101 and AD102 from the Ardennes Anticlinorium are Emsian–Lochkovian sandstones and a conglomerate, and

Table 2. *Apatite fission-track ζ-age data for Palaeozoic rocks in the Ardennes**

Sample	Grains	Spontaneous tracks		Induced tracks		$P(\chi^2)$	N_S/N_I	ρ_S/ρ_I	Glass dosimeter		Age (Ma ± 1σ)	Confined track length		
		ρ_S ($\times 10^5$ tracks cm^{-2})	N_S	ρ_I ($\times 10^5$ tracks cm^{-2})	N_I				ρ_D ($\times 10^5$ tracks cm^{-2})	N_D		Tracks	SD	Mean length (μm ± 1σ)
AD18	34	2.12	1777	3.20	270	0.95	6.582	7.237	2.88	11 503	168 ± 12	73	2.3	12.6 ± 0.3
AD15	43	23.95	2146	14.74	1368	0.49	1.569	1.624	12.43	12 425	218 ± 8	52	1.3	12.4 ± 0.2
AD14	14	24.20	1021	15.25	635	0.08	1.608	1.642	12.48	12 484	207 ± 3	78	1.2	13.1 ± 0.1
AD13	38	28.83	2539	19.22	1673	0.37	1.518	1.597	12.51	12 513	212 ± 8	90	1.3	12.8 ± 0.1
AD11	20	1.61	1022	1.40	90	1.00	11.356	11.705	2.91	11 653	290 ± 33	82	2.2	12.3 ± 0.2
AD10	38	23.62	2351	16.66	1693	0.17	1.389	1.455	12.55	12 547	195 ± 7	88	1.2	12.7 ± 0.1
AD108	45	31.45	3517	20.87	2323	0.70	1.514	1.571	12.83	12 832	217 ± 7	77	1.2	12.6 ± 0.1
AD107	48	28.73	3320	20.25	2358	0.52	1.408	1.464	12.90	12 904	203 ± 7	125	1.3	12.8 ± 0.1
AD106	25	29.21	1592	18.61	1007	0.50	1.581	1.601	12.96	12 960	229 ± 10	65	1.5	12.0 ± 0.2
AD103	33	23.45	1971	18.15	1536	0.19	1.283	1.438	12.99	12 994	187 ± 7	74	1.6	12.4 ± 0.2
AD102	11	9.65	315	5.56	193	0.99	1.632	1.671	13.06	13 054	238 ± 22	13	1.2	11.7 ± 0.3
AD101	43	32.29	2823	23.46	2025	0.30	1.394	1.436	13.10	13 101	204 ± 7	72	1.3	12.2 ± 0.2
AD100	15	10.50	635	6.07	343	0.90	1.851	1.926	13.15	13 149	271 ± 19	35	1.4	12.2 ± 0.2

*ρ_S, ρ_I and ρ_D are, respectively, the areal density of spontaneous, induced tracks and induced tracks of irradiated mica-detector against a glass dosimeter, N_S, N_I and N_D are respectively, the number of counted spontaneous, induced tracks and induced tracks in the mica detector. $P(\chi^2)$ is the chi-squared probability for the dated grains. The weighted mean age is used for AD14, and the pooled ages are used for other samples.

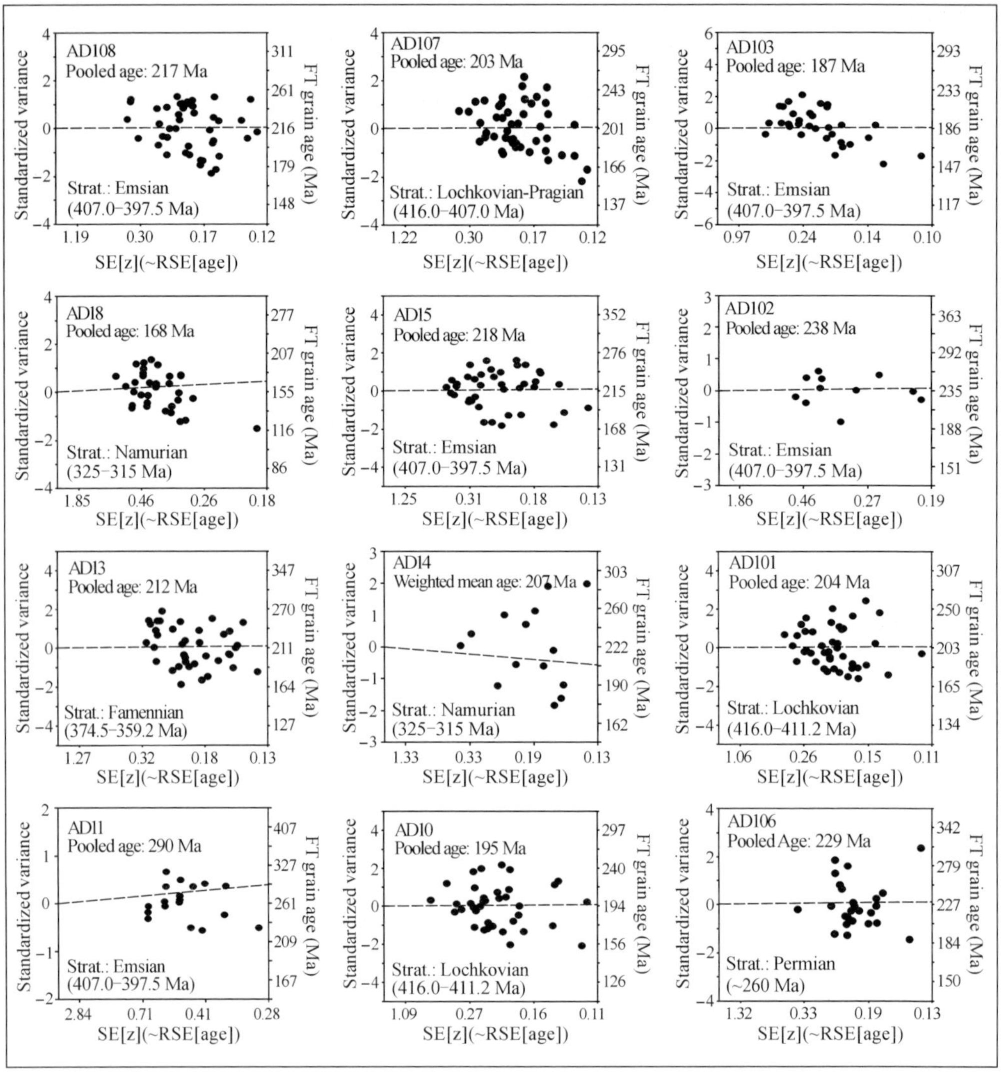

Fig. 2. Radial plots of AFT ages for Late Palaeozoic samples, using the BINOMFIT program (Brandon 2002).

their pooled ages averaged from 11 to 43 single-grain ages ($P\chi^2 = 0.17$–0.99) are 195 ± 7 Ma, 204 ± 7 Ma and 238 ± 22 Ma, respectively. These pooled ages, together with single-grain ages, are without exception much younger than the respective depositional ages (Fig. 2). Track-length distributions for AD101 ($N = 72$) and AD10 ($N = 88$) are unimodal, with means of between 12.7 ± 0.1 μm and 12.2 ± 0.2 μm. AD102 yields the shortest track length of 11.7 ± 0.3 μm, averaged from 13 confined track lengths. Sample AD106 from the Malmédy Graben is a conglomerate of the Permian Malmédy Formation. (Bultynck *et al.* 2001). It yields a pooled age of 229 ± 10 Ma, slightly younger than the depositional age. Its 65 confined track lengths display a unimodal distribution, with a mean of 12.0 ± 0.2 μm. Twenty-five single-grain ages calculated for sample AD106, despite them all passing a χ^2-test with 50% probability, mostly plot below and partly above the sample stratigraphic age (Fig. 2), implying a partial annealing after sedimentation rather than a complete annealing. Samples AD13, AD15, AD103, AD107 and AD108 from the Dinant Synclinorium are Lochkovian, Emsian and Famennian medium- to coarse-grained sandstones. These samples have pooled AFT ages ranging from 218 ± 8 Ma to 187 ± 7 Ma, respectively averaged from 33–48 single-grain ages ($P\chi^2 = 0.19$–0.70) that are all younger than the sample depositional ages (Fig. 2). Their confined track lengths ($N = 52$–125) correspond to a unimodal distribution, with means varying between 12.4 ± 0.2 μm and 12.8 ± 0.1 μm. Sample AD14, a Namurian sandstone from the Dinant Synclinorium, has a weighted mean age of 207 ± 3 Ma averaged from 14 grain ages ($P\chi^2 = 0.08$), coupled with a mean confined track length of 13.1 ± 0.1 μm ($N = 78$).

These AFT data, together with other published results (Vercoutere & Van den Haute 1993; Glasmacher *et al.* 1998), are shown in Figure 1. Four points can be noted. (1) Three age groups, that is 290–229, 218–198, and 190–146 Ma, might be classified for the Ardennes that, in general, become younger from SSE to NNW (partly overlapped), implying three exhumation events developed successively. (2) For the High Ardennes, its southern part had an exhumation through the APAZ in 290–271 Ma, earlier than that of the northeastern part (238–229 Ma). The southern part also demonstrates an asymmetric domical architecture. Its core concentrates on the Rocroi Inlier and is marked by the younger ages of 204–195 Ma, whereas the domical flanks are characterized by the older ages of 290–271 Ma. An abrupt change in age from the core (195 Ma) to the northern flank (290 Ma) indicates a large differential exhumation, therefore, in correlation with the control of a fault zone between them. (3) The AFT ages of 190–146 Ma concentrate on both the southern Brabant Massif and the NE fringe of the Ardennes Allochthon, and again lies on the Border fault zone and to the north of it. That is to say, the differential exhumation between Ardennes and Brabant areas is probably controlled by the post-Variscan reactivation of the Border fault zone. (4) The AFT samples from the Ardennes, owing to the controls of multiple exhumation events, do not show any systematic correlation between age and elevation (Fig. 1).

Thermal history modelling

Apatite track annealing occurs during heating over 60–125 °C, mainly as a function of time, temperature and composition. Thus, track-annealing models obtained from annealing experiments allow unknown temperature–time (T–t) histories of rocks to be evaluated (e.g. Laslett *et al.* 1987; Crowley *et al.* 1991; Donelick *et al.* 1999). In this study, the F-apatite annealing equation (Crowley *et al.* 1991) was used because fluorapatite grains are major components of Upper Palaeozoic sediments in the Ardennes (Glasmacher *et al.* 1998). Using an inverse procedure (Gallagher 1995), the thermal history was modelled for the sample according to apatite FT age and confined track length using AFTSolve 1.3.1 software (Ketcham *et al.* 2000), in which 10 000 candidate T–t paths were generated using a Monte Carlo algorithm with no maximum heating or cooling rates set. Other constraints for the modelling are stratigraphic ages, illite crystallinity, vitrinite reflectance, conodont alteration index data, illite K–Ar ages (reviewed by Fielitz & Mansy 1999; Han *et al.* 2000) and zircon FT data (Brix 2002), with approximately 20 °C set for the surface temperature. The modelled T–t paths for 12 samples in Figure 3 have an obvious variation both temporally and spatially throughout the Ardennes, and are thereby classified as four major groups.

Those for samples AD100 and AD11 from the High Ardennes make the first T–t group. Using a zircon FT age of 322 Ma (Brix 2002) with the PAZ of approximately 160–330 °C (Tagami 2005), and an illite K–Ar age of 326 Ma with 245–310 °C from conodont alteration index and illite crystallinity data (Fielitz & Mansy 1999) as constraints, the T–t paths for this group consist of three segments, that is a rapid cooling of 4.2–5.4 °C Ma^{-1} down to 105–102 °C in 320–300 Ma, a slow cooling of 0.2 °C Ma^{-1} down to 54–65 °C in 300–45 Ma and an accelerated cooling of 0.7–1.0 °C Ma^{-1} up to the surface (since 45 Ma). The second T–t group is derived from samples AD10, AD101, AD13, AD14, AD15, AD103, AD107 and AD108 of the Ardennes. The lower limits for T–t modelling, based on conodont alteration index, illite crystallinity, vitrinite reflectance (Fielitz & Mansy 1999; Han *et al.* 2000) and zircon FT data (Brix 2002), use an illite K–Ar age of 326 Ma with 245–310 °C for AD10 and AD101, an illite K–Ar age of 297 Ma with 140–195 °C for AD13, an age of the diagenetic metamorphism of approximately 300 Ma with 160–210 °C for AD14, an illite K–Ar age of 280 Ma with 240–300 °C for AD15, and an age of illite K–Ar of 273 Ma with 200–250 °C for AD103, AD107 and AD108. For this group, a rapid cooling of 1.3–4.0 °C Ma^{-1} down to 120–85 °C occurred in 280–230 Ma, followed by a slow cooling of 0.1–0.3 °C Ma^{-1} down to 55–70 °C in 230–45 Ma and an accelerated cooling of 0.8–1.1 °C Ma^{-1} since 45 Ma. The third T–t group modelled from sample AD106 of the Malmédy Graben, using a sample stratigraphic age of approximately 260 Ma (Glasmacher *et al.* 1998) and a diagenetic temperature (close to the APAZ) as constraints, displays three thermal stages, that is a rapid heating of 2.8 °C Ma^{-1}, reaching 105 °C in 260–230 Ma, a slow cooling of 0.2 °C Ma^{-1} down to 75 °C in 230–45 Ma and a rapid cooling of 1.2 °C Ma^{-1} up to the surface (since 45 Ma). The fourth T–t group from sample AD18 of the Brabant Parautochthon, using a sample stratigraphic age (325–310 Ma) and a diagenetic metamorphism (180–220 °C) as

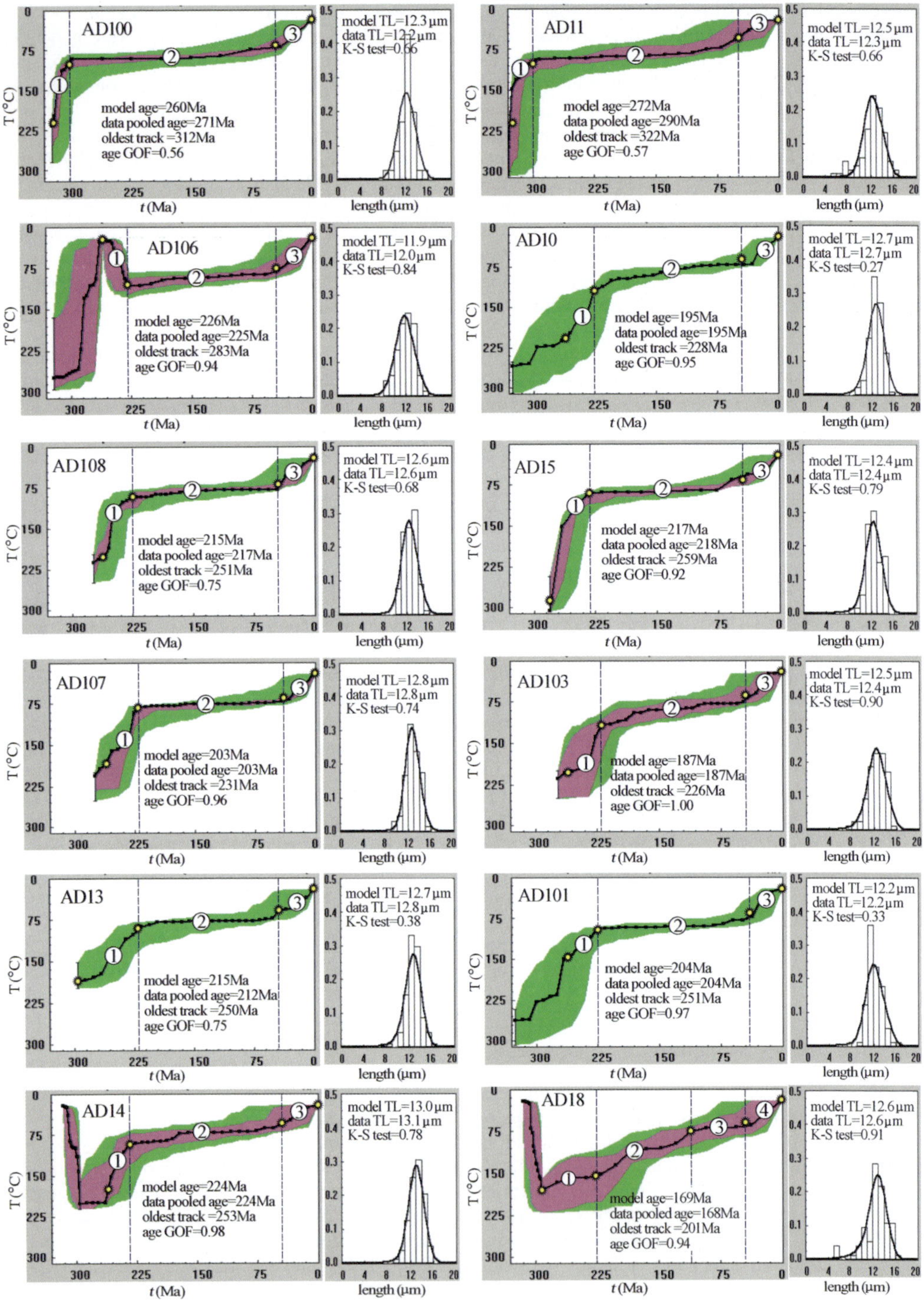

Fig. 3. Modelled T–t paths according to apatite FT data from the Ardennes, using AFTSolve 1.3.1 (Ketcham *et al.* 2000). In T–t space the green models are 'acceptable' fits, and the magenta models are 'good' fits. A 'good' fit is with K–S and GOF values of more than 0.50, and for an 'acceptable' fit values of more than 0.05 are required. The K–S test is the Kolmogorov–Smirnov test; the GOF is a Goodness of Fit.

constraints, records four major cooling stages, that is a slow cooling of 0.4 °C Ma^{-1} down to 155 °C in 292–225 Ma, a fast cooling of 0.7 °C Ma^{-1} down to 75 °C in 225–110 Ma a slow cooling of 0.2 °C Ma^{-1} down to 60 °C in 110–45 Ma and a rapid cooling up to the surface (0.9 °C Ma^{-1}, since 45 Ma).

These modelled T–t paths indicate that the Ardennes had suffered from four major thermal regimes in the late- to post-Variscan period, which can be in correlation with four types of tectonic controls. The rapid exhumation caused by the late-Variscan thrust is proposed as being responsible for the first rapid cooling of 4.2–5.4 °C Ma^{-1} (320–300 Ma), corresponding to the thick intensely formed Namurian–Westphalian fillings in the northern foreland basins. It is at approximately 300 Ma that the late-Variscan thrust ceased in the Ardennes. Between 280 Ma and 230 Ma the differential thermal regimes dominated regionally, that is a slow cooling of 0.4 °C Ma^{-1} in the Brabant Parautochthon, a very slow cooling of 0.2 °C Ma^{-1} along two flanks around the Rocroi core, and a fast cooling of 1.2–4.0 °C Ma^{-1} both in the Dinant Synclinorium and in most of the High Ardennes coexistent with a rapid heating of 2.8 °C Ma^{-1} in the Malmédy Graben. This thermal geometry is preferably correlated with the controls of post-Variscan transtensional tetonics, which is also a key mechanism to interpret the subsidence patterns of the Permian Saar–Nahe Basin (Korsch & Schäfer 1991) and the Late Permian–Mid-Triassic Paris Basin (Guillocheau *et al.* 2000). In contrast, most areas of the Ardennes from 230 to 45 Ma underwent a long and slow cooling (0.1–0.3 °C Ma^{-1}), much less differential than before, corresponding to a long-term and very slow exhumation that resulted in a weathering progression in the Ardennes (Thiry *et al.* 2006). In this period the Brabant Parautochthon underwent cooling with a subdivision of two stages, that is 0.7 °C Ma^{-1} in 225–110 Ma and 0.2 °C Ma^{-1} in 110–45 Ma, indicating that an obvious differential exhumation occurred in 225–110 Ma between the Ardennes and the southern Brabant areas. Although the T–t path from AFT modelling is not sufficient to offer a good assessment on the thermal evolution of the low-temperature segment, it is worth noting that the final accelerated cooling of 0.7–1.18 °C Ma^{-1} (since 45 Ma) simultaneously affected the whole Ardennes, in time according with the Pyrenean compression (Schmid *et al.* 1996, 2004).

Discussion

In order to examine the correlation between post-Variscan thermal regimes and structural controls, an integrated south–north profile across the central Ardennes was made (Fig. 4), using gravity Bouguer data (database from the Royal Observatory of Belgium), 1/1 000 000 geological map of France (2003) and a balanced section by Raoult & Meilliez (1987). In this profile the variation in exhumed amounts calculated from the modelled T–t paths is displayed in the three stages of 300–230 Ma, 230–45 Ma and since 45 Ma, assuming a geothermal gradient of approximately 40 °C km^{-1} (Han *et al.* 2000) for the first stage and approximately 35 °C km^{-1} for two other stages. In Figure 4, the most prominent features are the two major sets of crustal fault systems: the Variscan south-dipping thrusts are dominant and imbricate downwards into the Midi fault crossing the middle–lower crust, while most of the nearly vertical faults in the middle–upper crust are interpreted as the reactivated and the post-Variscan faults. Among these faults, two are important, the Border fault and Fault 1, because they controlled post-Variscan differential exhumation among the High Ardennes, the Dinant Synclinorium and the southern Brabant Massif. In the gravity field, the Border fault zone between the Brabant and Ardennes areas is characterized by a sharp gravity gradient trending east–west in the central part, becoming less sharp towards the west and the east (Mansy *et al.* 1999), which is regarded as a strike-slipping reactivation probably of late- to post-Variscan age (Sintubin 1999). Interpreted as a blind Variscan thrust, Fault 1 would create a large fault-propagation fold in the High Ardennes with a considerable elevation difference in comparison with the Dinant Synclinorium. This fault was overlapped by the post-Variscan vertical fault.

It is also seen in Figure 4 that the exhumed amounts in 300–230 Ma along the south–north profile demonstrate a considerable variation, that is 1.9–5.0 km in the Rocroi Inlier and the Dinant Synclinorium, 0.2–0.3 km in two flanks around the Rocroi Inlier, and 0.6 km in the Brabant Parautochthon, in general corresponding to two asymmetric domical architectures. This exhumation geometry, besides the contribution from late-Variscan elevation difference, is preferably related to the controls of two important vertical fault zones among the High Ardennes, the Dinant Synclinorium and the Brabant Massif, and regionally with the transtension that can be further traced in the late- to post-Variscan basins adjacent to the Ardennes. The Saar–Nahe Basin, as a Permo-Carboniferious half-graben within the orogen, is transtensional in character, its geometry being mainly controlled by strike-slip movements on the South Metz fault (Korsch & Schäfer 1991). In the Paris Basin the scattered Late Carboniferous–Permian deposition has two preferential directions along NE–SW and NW–SE, parallel to the Metz fault and the Bray fault zones (Ziegler & Dèzes 2006). The

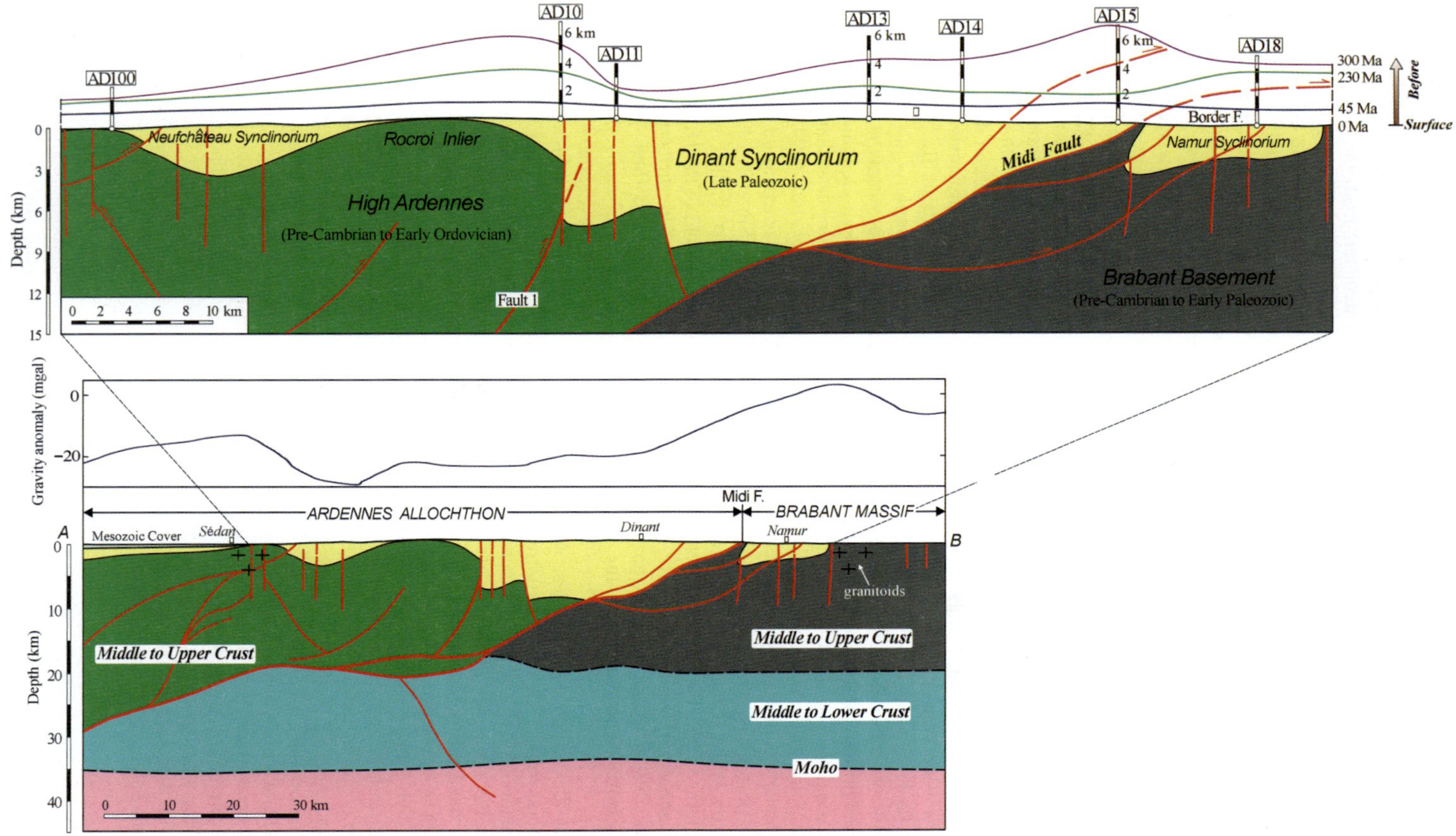

Fig. 4. Relationship between the restored differential exhumation and the crustal structure along an integrated south–north profile crossing the central Ardennes.

Scythian–Carnian sequences are confined to the eastern part of the Paris Basin along NE–SW and east–west subsidence zones, respectively parallel to the South Metz fault and the Vittel fault zones (Guillocheau *et al.* 2000), being in correlation with the controls of NE–SW-trending wrench faults (Goggin *et al.* 1997).

The Ardennes in 230–45 Ma was characterized by a slow differential exhumation, with the exhumed amounts along the south–north profile (Fig. 4) changing from 2.3 km in the Rocroi Inlier to 0.7–1.4 km in the Dinant Synclinorium and the southern High Ardennes, again constituting a domical structure. This slow exhumation accords with a long weathering progression revealed by the dating data (Yans & Dupuis 2005). In contrast, the southern Brabant Massif in 230–110 Ma, clearly different from the Ardennes, retained a fast cooling rate of 0.7 °C Ma^{-1} with an exhumation of 2.3 km, which is presumably related to the influence of the Border fault zone. The last accelerated cooling (since 45 Ma) affected the whole Ardennes, with exhumed amounts of 0.9–1.3 km along the south–north profile (Fig. 4). This accelerated exhumation regionally coincides with the major collision (50–35 Ma) of the Eastern Alps followed by the post-collisional thrust (35–7 Ma) that caused a northwestwards propagation of the deformation front (Schmid *et al.* 1996, 2004). In response, the depositional centres within the Brabant Massif became more and more restricted from the middle Eocene to the Pliocene, and moved northeastwards (De Batist & Versteeg 1999). Similarly, the Tertiary Paris Basin developed with a very low subsidence and basin-scale shrinkage in a compressional framework (Guillocheau *et al.* 2000). This basin remained relatively stable from the Danian to the Ypresian, and a short wavelength folding took place in the Lutetian–Lower Oligocene. The compression in the Paris Basin continued until the Miocene, and caused overall uplift and erosion.

Conclusions

- Thirteen Late Palaeozoic samples from the Ardennes, including 12 Lochkovian–Namurian sedimentary rocks and one kersantite, yielded AFT ages ranging from 290 ± 33 Ma to 168 ± 12 Ma. The confined track lengths correspond to a unimodal distribution, with means varying between 13.1 ± 0.1 μm and 11.7 ± 0.3 μm. These AFT ages for the sedimentary rocks are obviously younger than the respective stratigraphic age, indicative of a cooling through the APAZ after post-depositional complete annealing.
- All available AFT data from the Ardennes range from 290 ± 33 Ma to 146 ± 15 Ma, which might be further classified as three groups, that is 290–229 Ma, 218–198 Ma and 190–146 Ma, with the implication that at least three exhumation events took place. The age group of 290–229 Ma concentrates on the High Ardennes, the group of 218–198 Ma covers the whole Ardennes, and the group of 190–146 Ma lies on the Border fault zone and to the north of it. There is no systematic correlation shown between these AFT ages and the elevations.
- The modelled *T–t* paths register four major cooling episodes in the Ardennes. The early fast cooling episode (4.2–5.4 °C Ma^{-1}, 320–300 Ma) corresponds to a late-Variscan rapid thrusting that ceased at approximately 300 Ma. The second episode (up to 230 Ma) is marked by a differential cooling (0.2–4.0 °C Ma^{-1}), which might be attributed to the controls of major vertical fault zones and regionally related with the transtension. The third cooling episode of the Ardennes Allochthon (0.1–0.3 °C Ma^{-1}, 230–45 Ma) is slow, and represents a long-term and slow exhumation. But in the Brabant Parautochthon, it is subdivided into 0.7 °C Ma^{-1} in 225–110 Ma and 0.2 °C Ma^{-1} in 110–45 Ma. The last accelerated cooling episode (0.7–1.1 °C Ma^{-1}, since 45 Ma) affected the whole Ardennes, structurally in correlation with the controls of the compression during the Pyrenean phase.

This work is part of a post-doctoral programme financially supported by the Université Lille 1. Authors are grateful to Mrs N. Selen for providing technical support for mineral separation. Dr M. Patin helped to improve the earlier English manuscript. We are indebted to O. Averbuch, M. R. Brix and A. Trentesaux for their valuable suggestions in the preparation of this paper. Constructive comments from M. R. Brix and one anonymous reviewer are gratefully acknowledged.

References

BRANDON, M. T. 2002. Decomposition of mixed grain age distributions using BINOMFIT. *On Track*, **24**, 13–18.

BRIX, M. R. 2002. Thermal history of Palaeozoic rocks in the Meuse Valley between Charleville–Mézières and Namur assessed from zircon fission track data. *Aardkundige Mededelingen*, **12**, 93–95.

BULTYNCK, P., GEUKENS, F. & SMOLDEREN, A. 2001. Permian lithostratigraphic units, Malmédy graben (Belgium). *Geologica Belgica*, **4**, 105–106.

CROWLEY, K. D., CAMERON, M. & SCHAEFER, R. L. 1991. Experimental studies of annealing etched fission tracks in fluorapatite. *Geochimica et Cosmochimica Acta*, **55**, 1449–1465.

DE BATIST, M. & VERSTEEG, W. H. 1999. Seismic stratigraphy of the Mesozoic and Cenozoic in northern Belgium: main results of a high-resolution reflection seismic survey along rivers and canals. *Geologie en Mijnbouw*, **77**, 17–37.

DE CORTE, F., BELLEMANS, F., VAN DEN HAUTE, P., INGELBRECHT, C. & NICHOLL, C. 1998. A new U doped glass certified by the European commission for the calibration of fission-track dating. *In*: VAN DEN HAUTE, P. & DE CORTE, F. (eds) *Advances in Fission-track Geochronology*. Kluwer Academic, Dordrecht, 67–78.

DONELICK, R. A., KETCHAM, R. A. & CARLSON, W. D. 1999. Variability of apatite fission track annealing kinetics II: crystallographic orientation effects. *American Mineralogist*, **84**, 1224–1234.

FIELITZ, W. & MANSY, J. L. 1999. Pre- and synorogenic burial metamorphism in the Ardenne and neighbouring areas (Rhenohercynian zone, central European Variscides). *Tectonophysics*, **309**, 227–256.

FRANKE, W., HAAK, V., ONCKEN, O. & TANNER, D. (eds). 2000. *Orogenic Process: Quantification and Modelling in the Variscan Belt*. Geological Society, London, Special Publications, **179**.

GALLAGHER, K. 1995. Evolving temperature histories from apatite fission-track data. *Earth and Planetary Science Letters*, **136**, 421–435.

GEUKENS, F. 1995. Strike slip deformation des deux côtés du Graben de Malmedy. *Annales de la Société Géologique de Belgique*, **118**, 139–146.

GLASMACHER, U., ZENTILLI, M. & GRIST, A. M. 1998. Apatite fission track thermochronology of Paleozoic sandstones and the Hill-intrusion, northern Linksrheinisches Schiefergebirge, Germany. *In*: VAN DEN HAUTE, P. & DE CORTE, F. (eds) *Advances in Fission-track Geochronology*. Kluwer Academic, Dordrecht, 151–172.

GLEADOW, A. J. W. & BROWN, R. W. 2000. Fission track thermochronology and the long-term denudational response to tectonics. *In*: SUMMERFIELD, M. A. (ed.) *Geomorphology and Global Tectonics*. Wiley, Chichester, 57–75.

GOGGIN, V., JACQUIN, T. & GAULIER, J. M. 1997. Three-dimensional accommodation analysis of the Triassic in the Paris Basin: a new approach in unraveling the basin evolution with time. *Tectonophysics*, **282**, 205–222.

GUILLOCHEAU, F., ROBIN, C. *ET AL.* 2000. Meso-Cenozoic geodynamic evolution of the Paris Basin: 3D stratigraphic constraints. *Geodinamica Acta*, **13**, 189–246.

HAN, G., PREAT, A., CHAMLEY, H., DECONINCK, J. F. & MANSY, J. L. 2000. Paleozoic clay mineral sedimentation and diagenesis in the Dinant and Avesnes Basins (Belgium, France): relationships with Variscan tectonism. *Sedimentary Geology*, **136**, 217–238.

HELSEN, S. 1995. Burial history of Palaeozoic strata in Belgium as revealed by conodont colour alteration data and thickness distributions. *International Journal of Earth Sciences*, **84**, 738–747.

HURFORD, A. J. 1990. Standardization of fission track dating calibration: recommendation by the Fission Track Working Group of the IUGS Subcommission on Geochronology. *Chemical Geology* (Isotope Geoscience Section), **80**, 171–178.

HUTT, S., MARTILL, D. M. & BARKER, M. J. 1996. The first European allosaurid dinosaur (Lower Cretaceous, Wealden Group, England). *Neues Jahrbuch für Geologie und Paläontologie, Monatshefte*, **10**, 635–644.

KETCHAM, R. A., DONELICK, R. A. & DONELICK, M. B. 2000. AFTSolve: a program for multi-kinetic modeling of apatite fission-track data. Geological Material Research, **2**, 1–32.

KORSCH, R. J. & SCHÄFER, A. 1991. Geological interpretation of DEKORP deep seismic reflection profiles 1C and 9 N across the variscan Saar–Nahe Basin southwest Germany. *Tectonophysics*, **191**, 127–146.

LASLETT, G. M., GREEN, P. F., DUDDY, I. R. & GLEADOW, A. J. W. 1987. Thermal annealing of fission tracks in apatite 2. A quantitative analysis. *Chemical Geology* (Isotope Geoscience Section), **65**, 1–13.

LEGRAND, R. 1968. *Le massif du Brabant*. Mémoires pour servir à l'Explication des Cartes Géologiques et Minières de la Belgique, **9**, 148.

LIPPOLT, H. J., BRANDER, T. & MANKOPE, N. R. 1998. An attempt to determine formation ages of goethites and limonites by $(U + Th)-^{4}He$ dating. *Neues Jahrbuch für Mineralogie Monatshefte*, **11**, 505–528.

MANSY, J. L., EVERAERTS, M. & DE VOS, W. 1999. Structural analysis of the adjacent Acadian and Variscan fold belts in Belgium and northern France from geophysical and geological evidence. *Tectonophysics*, **309**, 99–116.

MÁRTON, E., MANSY, J. L., AVERBUCH, O. & CSONTOS, L. 2000. The Variscan belt of the northern France-southern Belgium: geodynamic implications of new palaeomagnetic data. *Tectonophysics*, **324**, 57–80.

ONCKEN, O., VON WINTERFELD, C. & DITTMAR, U. 1999. Accretion and inversion of a rifted passive margin – the Late Palaeozoic Rhenohercynian fold and thrust belt. *Tectonics*, **18**, 75–91.

RAOULT, J. F. & MEILLIEZ, F. 1987. The Variscan Front and the Midi Fault between the Channel and the Meuse River. *Journal of Structural Geology*, **9**, 473–479.

SCHÄFER, F., ONCKEN, O., KEMNITZ, H. & ROMER, R. L. 2000. Upper-plate deformation during collisional orogeny: a case study from the German Variscides (Saxo-Thuringian Zone). *In*: FRANKE, W., HAAK, V., ONCKEN, O. & TANNER, D. (eds) *Orogenic Process: Quantification and Modelling in the Variscan Belt*. Geological Society, London, Special Publications, **179**, 218–302.

SCHMID, S. M., FÜGENSCHUH, B., KISSLING, E. & SCHUSTER, R. 2004. Tectonic map and overall architecture of the Alpine orogen. *Eclogae Geologica Helvetica*, **97**, 93–117.

SCHMID, S. M., PFIFFNER, O. A., FROITZHEIM, N., SCHÖNBORN, G. & KISSLING, E. 1996. Geophysical–geological transect and tectonic evolution of the Swiss–Italian Alps. *Tectonics*, **15**, 1036–1064.

SINTUBIN, M. 1999. Arcuate fold and cleavage patterns in the southeastern part of the Anglo-Brabant Fold belt (Belgium): tectonic implications. *Tectonophysics*, **309**, 81–97.

TAGAMI, T. 2005. Zircon fission-track thermochronology and application to fault studies. *Review in Mineralogy and Geochemistry*, **58**, 95–122.

THIRY, M., QUESNEL, F. *ET AL.* 2006. Continental France and Belgium during the early Cretaceous paleo-weatherings and paleolandforms. *Bulletin de la Société Géologique de France*, **177**, 155–175.

VERCOUTERE, C. & VAN DEN HAUTE, P. 1993. Post-Palaeozoic cooling and uplift of the Brabant Massif as revealed by apatite fission track analysis. *Geological Magazine*, **130**, 639–646.

WAGNER, G. & VAN DEN HAUTE, P. 1992. *Fission Track Dating*. Kluwer Academic, Dordrecht.

YANS, J. & DUPUIS, C. 2005. Timing of saprolitisation in the Haute–Lesse area (Belgium). *Geophysical Research Abstracts*, **7**, 07064. European Geosciences Union.

ZIEGLER, P. A. & DÈZES, P. 2006. Crustal evolution of Western and Central Europe. *In*: GEE, D. & STEPHENSON, R. (eds) *European Lithosphere Dynamics*. Geological Society, London, Memoirs, **32**, 43–56.

Thermal and denudation history of the Lusatian Block (NE Bohemian Massif, Germany) as indicated by apatite fission-track data

BARBARA VENTURA[1]*, FRANK LISKER[1] & JÜRGEN KOPP[2]

[1]*Universität Bremen, Fachbereich Geowissenschaften, Postfach 330 440, 28334 Bremen, Germany*

[2]*Landesamt für Geowissenschaften und Rohstoffe Brandenburg, Stahnsdorfer Damm 77, 14532 Kleinmachnow, Germany*

**Corresponding author (e-mail: bventura@uni-bremen.de)*

Abstract: The Lusatian Block in eastern Germany is part of the Variscan Bohemian Massif. It is located at the intersection of two regional fault–thrust systems that have been controlling the regional geological and landscape evolution since at least late Mesozoic times, the Elbe Fault System and the Eger Graben. Although the Lusatian Block has traditionally been described as a morphological high throughout the Mesozoic, timing, style and amount of denudation have not yet been quantified. Apatite fission-track (AFT) analysis of basement rocks yields ages varying between approximately 70 and 95 Ma, with mean track-lengths ranging from 13.6 to 14.1 μm. Thermal history modelling of the AFT data points to a minimum denudation of 3 km of the Lusatian Block in the Late Cretaceous. Onset of denudation can be related to transpression along the Lusatian Thrust as a main element of the Elbe Fault System. Moreover, the thermal history models provide tentative evidence for a second cooling–denudation event, which affected Lusatia in the late Palaeogene. This final denudation of at least 1 km probably occurred as a far-field response to the uplift of the northern shoulder of the Cenozoic Eger Graben.

The Lusatia region covers an area of approximately 10 000 km^2 in eastern Germany at the border with the Czech Republic and Poland. Its geographical limits roughly coincides with the rivers Elbe and Oder/Neisse to the west and east, and with the Fleming moraine tract and the Eger Graben to the north and south, respectively (Fig. 1). The geomorphology of Lusatia shows the transition from the flat plain of the NE German Basin at almost sea level in the north towards the rugged hills adjacent to the northern shoulder of the Eger Graben.

Tectonically, the Lusatian Block is bounded by two regional fault–thrust systems that have been controlling the geological and landscape evolution of Germany since at least the late Mesozoic times, the Elbe Fault System and the Eger Graben (Fig. 1). The WNW–ESE-striking, up to 100 km-wide, Elbe Fault System consists of several major tectonic dislocations and extends for more than 800 km from the SE North Sea to SW Poland (Scheck *et al.* 2002). During Late Carboniferous–early Permian times the Elbe Fault System was part of a post-Variscan wrench fault system and probably acted as the southern boundary fault during the formation of the Permian basins along the Trans-European Suture Zone (Scheck *et al.* 2002). Geophysical and structural data as well as stratigraphic evidence suggest that the northern segment of the Elbe Fault System was particularly active during Cretaceous – early Cenozoic times, when it responded to regional compression with up to 4 km of rock uplift and the formation of internal flexural highs (e.g. Scheck *et al.* 2002). Compressional deformation is supposed to have continued throughout the early Cenozoic and may be ongoing. At present, the northern segment of the Elbe Fault System defines the inverted southern margin of the NE German Basin as a remnant of the much larger Permian Central European Basin. The southern segment of the Elbe Fault System intersects the WSW–ENE-striking Eger Graben in the study area. This Cenozoic rift has formed since the Oligocene, paralleled by the reactivation of fault systems delimiting the Bohemian Massif to the SW and NE (Schröder 1987).

While the overall geological evolution of the Elbe Fault System and the Eger Graben is reasonably constrained by geophysical and structural data, poor or missing sedimentary records – particularly the absence of post-Cretaceous sediments – does not allow the timing and magnitude of tectonic activity to be resolved satisfactorily, and the long-term denudation history due to inversion within the Central European Basin to be derived. Here we present and interpret apatite fission-track (AFT) data from three boreholes from Lusatia in order to

From: LISKER, F., VENTURA, B. & GLASMACHER, U. A. (eds) *Thermochronological Methods: From Palaeotemperature Constraints to Landscape Evolution Models*. Geological Society, London, Special Publications, **324**, 181–192.
DOI: 10.1144/SP324.14 0305-8719/09/$15.00

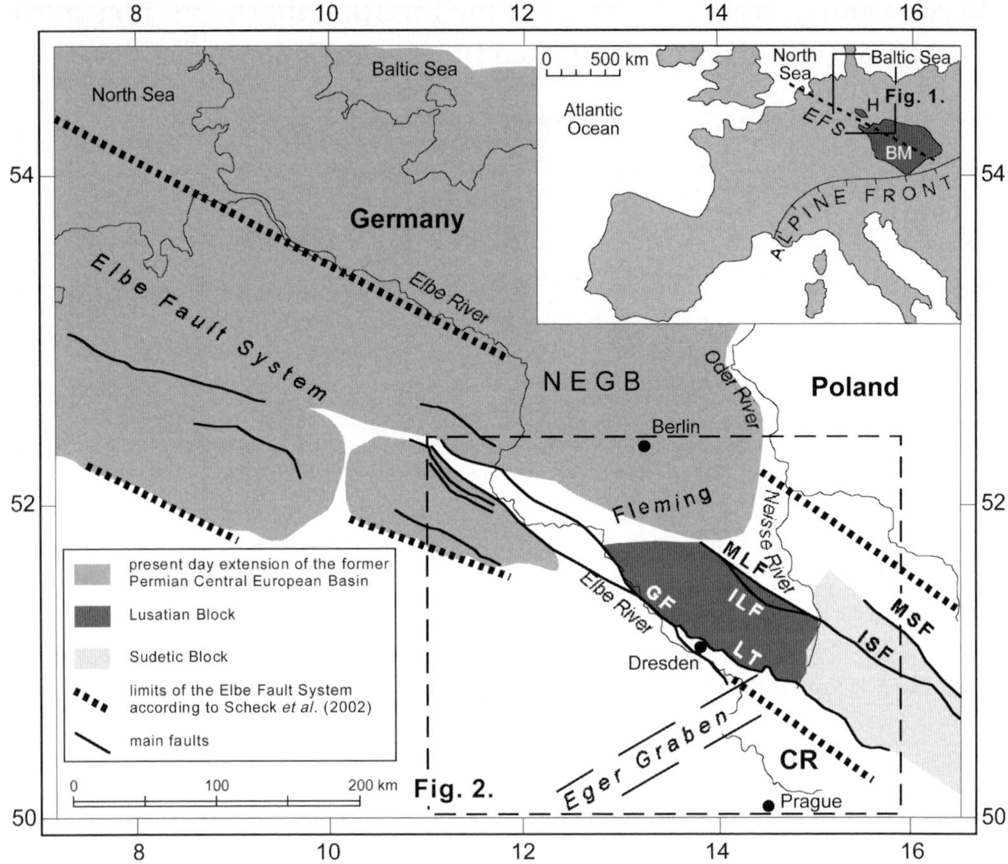

Fig. 1. Location of the study area (outlined) in a regional tectonic framework (modified after Scheck *et al.* 2002). BM, Bohemian Massif; CR, Czech Republic; EFS, Elbe Fault System; GF, Grossenhein Fault; H, Harz Mountains; ISF, Intra Sudetic Fault; ILF, Intra Lusatian Fault; LT, Lusatian Thrust; NEGB, NE German Basin.

evaluate the potential burial of the presently exposed basement within the Central European Basin. Thermal history modelling is applied to constrain onset, timing, course and amount of regional post-Variscan denudation. Moreover, the time–temperature pattern provides insight into the relationship between uplift–denudation processes and sedimentation within Tertiary lignite basins and the Eger Graben. The crucial position of the Lusatian Block at the intersection of the two regional fault systems enables differentiation to be made between the style and magnitude of the uplift–denudation of the two structures, and any possible denudational interferences to be recognized.

Geological setting

The Lusatian Block, together with the adjacent Torgau–Doberlug Synclinorium and Görlitz Synclinorium, forms the northernmost unit of the Bohemian Massif (Fig. 2). It comprises a variety of rocks of late Proterozoic–early Palaeozoic age that were partially deformed during the Variscan Orogeny and are locally intruded by Variscan granitoids. To the north, the basement of the Lusatian Block also includes late Proterozoic–early Cambrian granitoids, an anchimetamorphic Proterozoic turbiditic succession, and a series of partially metamorphosed volcanic and sedimentary rocks (e.g. Linnemann 1995; Hammer *et al.* 1999; Hammerschmidt *et al.* 2003).

Post-Variscan erosion of Lusatia resulted in regional peneplanation and provided detritus to the originally adjacent extensive Permian Central European Basin (Fig. 1), which occupied a large part of central and northern Europe (e.g. Van Wees *et al.* 2000). However, the presence of a small patch of Late Permian strata in northern Lusatia and within the Intra-Sudetic Basin along the NE rim of the Lusatian Block indicate that Lusatia was not

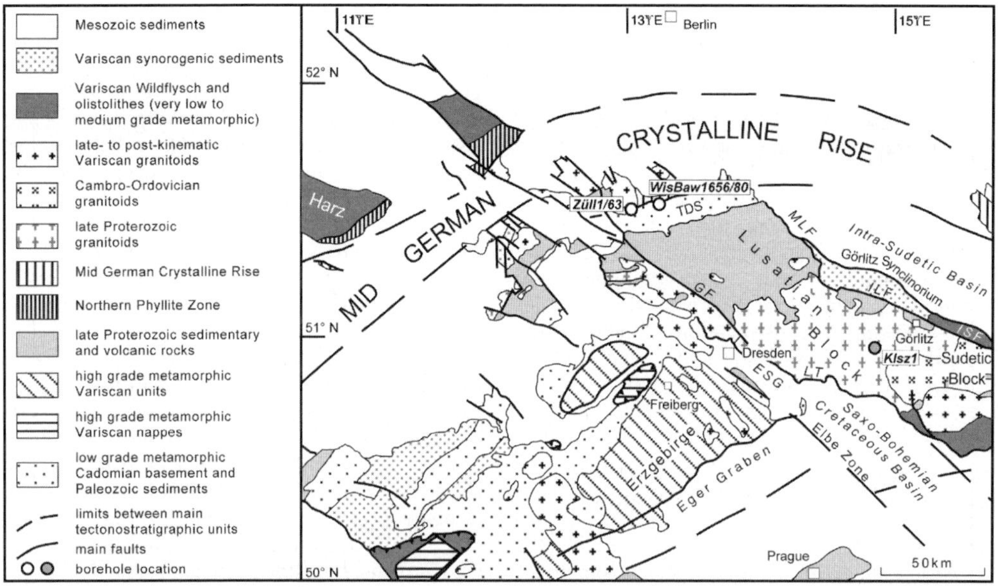

Fig. 2. Schematic geotectonic map of the northern Bohemian Massif, without Cenozoic deposits (modified after Tait *et al.* 1997; Cajz *et al.* 2001). EG, Elbsandsteingebirge; GF, Grossenhein Fault; ISF, Intra Sudetic Fault; ILF, Intra Lusatian Fault; LT, Lusatian Thrust. Jurassic–Cretaceous sedimentary rocks are not present in the study area, but only SW of the Lusatian Block (Elbsandsteingebirge). The locations of the boreholes Klsz1 (solid circle), WisBaw1656/80 and Züll1/63 (empty circles) are indicated. The area represented corresponds to the study area outlined in Figure 1.

uplifted permanently during Mesozoic times (e.g. Cajz *et al.* 2001). While Triassic and Jurassic sedimentary rocks are known only from drill cores and subsurface mining from adjacent areas (Cajz *et al.* 2001), superficial Cretaceous siliciclastic rocks with a thickness of up to about 1000 m are preserved within the Elbe River valley, immediately SW of the Lusatian Block. This sedimentary sequence was deposited at the northern rim of the Saxo-Bohemian Basin (Fig. 2), which connected the Tethys with the NE German Basin in mid–late Mesozoic times (e.g. Malkovsky 1987).

The oldest Cretaceous strata of the Saxo-Bohemian Basin in the study area consist of mid–late Cenomanian fluvial sediments (Kossmat 1925) cropping out between Freiberg and Dresden (Fig. 2). Mineralogical and petrographical analyses indicate that they were derived mainly from the western Erzgebirge (Voigt 1995 and references therein). Between the mid-Cenomanian and the mid-Coniacian, the palaeocurrent pattern changed and deposition occurred mainly from the Lusatian Block to the west. Thrusting of the Lusatian Block onto the basin in the Late Cretaceous interrupted sedimentation and induced regional denudation of Lusatia. This was accompanied by sub-aerial weathering and erosion, and by the formation of extensive weathering mantles (Störr 1983).

The Lusatian Block is bounded by two continental-scale tectonic features of almost perpendicular strike: to the west and to the east by a series of NNW–SSE-trending parallel dislocations belonging to the Elbe Fault System (Scheck *et al.* 2002); and to the south by the WSW–ENE-trending Eger Graben (Figs 1 and 2). Both fault systems represent long-lasting structures that were reactivated in the wider context of significant changes in relative motion between the European and African plates since the Late Cretaceous (Kley & Voigt 2008).

The SW border of the Lusatian Block is constituted by the Grossenhain Fault–Lusatian Thrust (Fig. 2), a strike-slip fault with a vertical displacement of up to 550 m (Cajz *et al.* 2001). To the north, the Main Lusatian Fault separates the basement of the Lusatian Block from Permo-Carboniferous–Cretaceous sedimentary rocks of the Intra-Sudetic Basin, with a vertical displacement exceeding 1000 m. These two transpressional fault segments of the Elbe Fault System are recognized down to crustal depths of 7–8 km (DEKORP Group 1994). Although direct evidence is missing, stratigraphic correlation and relative age constraints suggest two stages of tectonic activity that are probably related to the inversion of the Central European Basin during the Early–Mid Jurassic and the Late Cretaceous.

Starting in the Oligocene, extension of the northern Bohemian Massif triggered the formation of the NE–SW-trending Eger Graben (Fig. 2) and associated volcanic activity on the southern margin of Lusatia (Malkovsky 1987). Effusive rocks of the Eger volcanism are still preserved on the southern margin of the Lusatian Block, whereas remnants of Neogene magmatism within central Lusatia are restricted to intrusive dykes (e.g. Cajz *et al.* 2001). Evolution of the Eger Graben was paralleled by the reactivation of fault systems delimiting the Bohemian Massif to the SW and NE (Schröder 1987). Contemporaneously, a series of lignite basins evolved in northern Lusatia and in the adjacent portion of the NE German Basin so that the basement is accessible only through boreholes.

Apatite fission-track analysis results

Ten samples from magmatic rocks were collected for AFT analysis from three different boreholes located east of the River Elbe (Fig. 2) between surface level and a depth of approximately 500 m. Sample details, AFT data and counting parameters, along with preparation and experimental details, are presented in Table 1. The four samples from the borehole Klsz1 (Kleinschweidnitz) in the south of the study area consist of Variscan (late Carboniferous) granodiorite (Table 1). The six samples from the boreholes WisBaw1656/80 (Herzberg) and Züll1/63 (Züllsdorf) in the north of the study area comprise Cambro-Ordovician and Variscan (Carboniferous–Lower Permian) volcanic rocks and granitoids (Table 1).

The analysed samples yielded sufficient apatite crystals to obtain 10 AFT ages, whereas $\geq$75 confined track-lengths could be measured only for seven samples (Table 1). The apatite ages (central ages $\pm 1\sigma$) vary from 72$\pm$6 to 94 $\pm$ 8 Ma, and therefore considerably post-date the Cadomian and Variscan magmatic–metamorphic events. They are also significantly younger than the AFT ages from more internal parts of the Bohemian Massif reported by Hejl *et al.* (1997), Glasmacher *et al.* (2002) and Ventura & Lisker (2003). The ages show a steep correlation with sample depths (Fig. 3). Samples show low dispersion, $\leq$3%, of the individual grain ages and pass the χ^2-test with high probabilities, of $\geq$50% (Table 1 and Fig. 4). This implies that all grains dated in these samples are consistent with one homogeneous grain-age population (Galbraith 1981; Galbraith & Laslett 1993).

The track-length distributions of the seven samples with $\geq$75 measured confined track-lengths are very similar (Fig. 4). Their mean track-lengths (MTL; $\pm 1\sigma$ errors) vary in a narrow range between 13.15$\pm$0.16 and 14.12 $\pm$ 0.12 μm, and are correlated with standard deviations of between 1.66 and 1.12 μm (Table 1). The MTL do not show any significant variation with depth (Fig. 3). The track-length histograms show unimodal distributions (Fig. 4). However, while the track-length distributions of the four Kleinschweidnitz samples from the south of the study area are narrow with small standard deviations, the distributions of the Herzberg and Züllsdorf samples from the north of the study area (six samples) are broader and are characterized by a negative skew of shorter track-lengths with standard deviations of approximately 1.6 μm.

To verify variations of track annealing in response to apatite chemical composition, and in particular to the F/Cl ratio (Green *et al.* 1986; Barbarand *et al.* 2003), the mean fission-track etch-pit diameter, D_{par} (Burtner *et al.* 1994), was measured for all modelled samples (Table 1). The obtained D_{par} values between 1.01 and 1.74 μm are characteristic for apatites less resistant to annealing (Carlson *et al.* 1999; Ketcham *et al.* 1999).

Interpretation and thermal history modelling

Despite the large spatial scatter of the samples, AFT ages are remarkably consistent and suggest a common period of cooling commencing at least in the Late Cretaceous. The fact that the AFT ages at the surface coincide with those at depth suggests that they represent a time close to the onset of a rapid cooling event. Moreover, the broader track-length distributions of the Herzberg and Züllsdorf samples in northern Lusatia (Fig. 4) combined with the relatively old AFT ages point to a more complex cooling history, referring to a relatively long residence in the upper part of the apatite partial annealing zone (PAZ: 60–110 °C) or to minor reheating to lower PAZ temperatures before final cooling (for interpretation of cooling patterns cf. Gleadow *et al.* 1986; Gleadow & Brown 2000).

To verify the course of cooling, thermal history modelling of the AFT data was performed using the program HeFTy (Ketcham *et al.* 1999; Ketcham 2005) for samples with $\geq$20 single-grain ages and $\geq$75 measured confined tracks. Based on a Monte Carlo algorithm and independent geological (thermal) constraints, the program generates random t–T paths. The thermal histories rely on the initial assumption of four time–temperature (t–T) envelopes. The first two t–T envelopes were defined to allow initial cooling from temperatures above the upper limit of the PAZ (>125 °C) to (near-) surface temperatures, as indicated by the occurrence of Oligocene strata in the north of the study area (Eissmann 2002). However, the second

Table 1. *Apatite fission-track analytical results*

Sample (Rock type) Age	Location RW; HW	Elev. (m) Depth (m)	N	ρ_D ($10^6 cm^{-2}$) (N_D)	ρ_s ($10^6 cm^{-2}$) (N_s)	ρ_I ($10^6 cm^{-2}$) (N_I)	U (ppm) CC	$P(\chi^2)$ (%) Disp. (%)	AFTA (Ma)	MTL (μm) SD (μm) (n)	D_{par} (μm) T_{Max} (°C)
Klsz1-a* (Granodiorite) Carboniferous	Kleinschweidnitz 5475300; 5659600	267 11	20	1.257 (6397)	0.422 (163)	1.122 (433)	12 0.31	93 0	87 ± 8	13.84 ± 0.16 1.38 (75)	1.51 *c.* 65
Klsz1-b* (Granodiorite) Carboniferous	Kleinschweidnitz 5475300; 5659600	267 68	20	1.261 (6397)	0.554 (211)	1.601 (610)	15 0.27	88 0	81 ± 7	13.84 ± 0.15 1.39 (84)	1.39 *c.* 80
Klsz1-c* (Granodiorite) Carboniferous	Kleinschweidnitz 5475300; 5659600	267 130	16	1.267 (6397)	0.360 (117)	1.100 (358)	12 0.43	90 0	76 ± 8	13.56 ± 0.14 1.18 (76)	–
Klsz1-d* (Granodiorite) Carboniferous	Kleinschweidnitz 5475300; 5659600	267 196	20	1.270 (6397)	0.449 (188)	1.466 (614)	14 0.30	85 1	72 ± 6	14.12 ± 0.12 1.12 (83)	1.35 *c.* 65
WisBaw1656/80-c* (Volcanic breccia) Carboniferous–Lower Permian	Herzberg 97885; 34340	79 388	20	1.315 (6397)	1.592 (811)	4.601 (2344)	50 0.58	56 3	84 ± 4	13.15 ± 0.16 1.66 (100)	1.01 *c.* 65
WisBaw1656/80-d* (Granodiorite) Carboniferous	Herzberg 97885; 34340	79 439	20	1.319 (6397)	0.436 (203)	1.424 (663)	14 0.32	77 1	75 ± 6	13.89 ± 0.22 1.10 (25)	–
WisBaw1656/80-e† (Granodiorite) Carboniferous	Herzberg 97885; 34340	79 480	9	1.465 (7188)	0.370 (96)	1.132 (294)	12 0.93	66 0	88 ± 11	–	–

(Continued)

Table 1. *Continued*

Sample (Rock type) Age	Location RW; HW	Elev. (m) Depth (m)	N	ρ_D (10^6cm^{-2}) (N_D)	ρ_s (10^6 cm^{-2}) (N_s)	ρ_I (10^6 cm^{-2}) (N_I)	U (ppm) CC	$P(\chi^2)$ (%) Disp. (%)	AFTA (Ma)	MTL (μm) SD (μm) (n)	D_{par} (μm) T_{Max} (°C)
Züll1/63-a*[a] (Gabbro) Cambro-Ordovician	Züllsdorf 4569653; 5721938	79 113	25	1.247 (6397)	0.487 (266)	1.254 (685)	12 0.23	88 0	89 ± 7	13.64 ± 0.16 1.57 (100)	1.74 *c.* 65
Züll1/63-b*[a] (Gabbro) Cambro-Ordovician	Züllsdorf 4569653; 5721938	79 165	7	1.250 (6397)	0.306 (67)	0.845 (185)	8 0.24	96 0	84 ± 12	–	–
Züll1/63-c*[a] (Gabbro) Cambro-Ordovician	Züllsdorf 4569653; 5721938	79 245	18	1.254 (6397)	0.468 (208)	1.153 (513)	12 0.28	99 0	94 ± 8	13.98 ± 0.17 1.66 (100)	–

Mineral concentrates for apatite FT analysis were obtained using standard crushing, heavy liquids and magnetic techniques. Apatites were mounted in epoxy resin on glass slides and polished to expose internal grain surfaces. Apatite mounts were etched for 20 s in 5 M HNO_3 at 20 °C to reveal the spontaneous fission tracks and prepared for analysis with the external detector method (Gleadow 1981). Neutron irradiation was carried out in the well-thermalized Thetis reactor at Ghent. Thermal neutron fluence was monitored using standard reference glass CN5* (Bellemans *et al.* 1995) and IRMM† (De Corte *et al.* 1998). After irradiation, the mica detectors were etched in 48% HF for 20 min at 20 °C to reveal the induced fission tracks. Spontaneous and induced fission tracks were counted, and confined track lengths and etch-pit diameters were measured using a Zeiss Axioplan microscope at 1250× and 2000× magnification, respectively, under dry objectives. Track length and etch-pit diameter measurements were carried out using a digitizing tablet calibrated against a grid following the recommendation of Laslett *et al.* (1982), with a precision to ±0.2 μm. Whenever possible, at least 20 grains were counted, and 100 confined track lengths were measured. Central ages (Galbraith & Laslett 1993) were calculated using the ζ-calibration approach of Hurford & Green (1983) with $*\zeta = 371.7 \pm 10.7$ and $^{\dagger}\zeta = 325.4 \pm 9.4$ determined with apatite age standards Fish Canyon, Durango and Mt Dromedary. Errors quoted as $\pm 1\sigma$, according to the conventional method of Green (1981).

RW, Rechtswert; HW, Hochwert; Elev., elevation of the drill site; Depth, position of the sample in the drill cores; N, number of grains analysed; $(\rho, N)_{D,S,I}$, track density and number of tracks counted in dosimeter glass, sample and external detector; U, uranium; CC, correlation coefficient, $P(\chi^2)$, chi-square probability; Disp., dispersion; AFTA, apatite fission-track central age; MTL, mean track length; n, number of measured confined tracks; SD, standard deviation; D_{par}, mean etch-pit diameter (five measurements per crystal); T_{Max}, maximum palaeotemperature modelled with HeFTy (Ketcham 2005; only for samples with ≥20 counted grains and ≥75 measured track lengths) according to the annealing algorithm of Ketcham *et al.* (1999); [a], apatite fission-track age from Ventura *et al.* (2003). Modelling parameters are described in the caption to Figure 5. Analyst: B. Ventura.

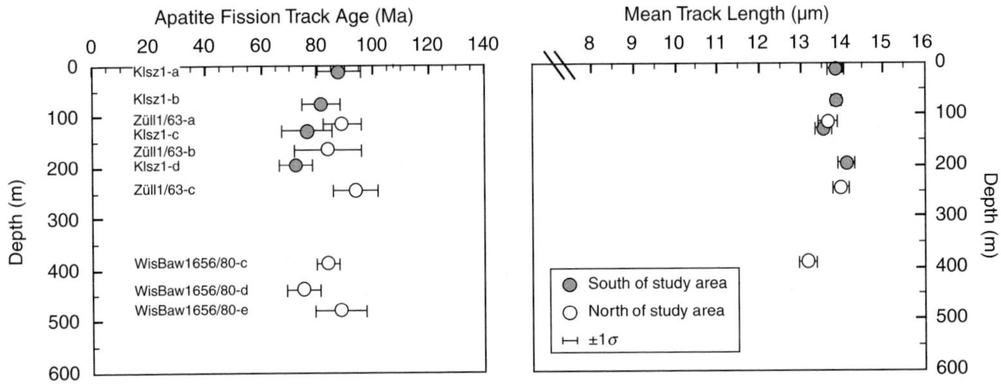

Fig. 3. Diagrams showing apatite fission-track age and mean track length v. depth ($\pm 1\sigma$) for samples in the south (solid circles) and in the north (empty circles) of the Lusatia region. Mean track lengths are shown only for samples with $\geq$75 confined tracks measured.

temperature frame remains open to avoid forcing of cooling to surface conditions. The third t–T envelope was specified to allow post-Oligocene reheating in order to account for the observed track-length distributions, and the last envelope accounts for the final cooling from maximum palaeotemperatures (Fig. 5).

Combining the quantitative constraints of all five modelled FT samples suggests that at least one cooling event affected the Lusatia Block in the Late Cretaceous (Fig. 5). A second heating–cooling episode in the late Palaeogene is allowed but not required as the modelled maximum palaeotemperatures of most samples did not significantly exceed approximately 60 °C during the Oligocene, and are therefore beyond the sensitivity of the AFT method. Samples underwent sudden cooling of at least *c.* 30 °C (from >110 °C to the modelled maximum palaeotemperatures of *c.* 80 °C, see Table 1) in the Late Cretaceous. Subsequently, they resided at temperatures $\leq$80 °C until the late Palaeogene. Whether they resided at PAZ temperatures or were subjected to further cooling and Cenozoic reheating cannot be constrained from the AFT data alone (see the Discussion). However, the constraint point pattern of the inverse modelling and the GOF (goodness of fit) values of direct models favour early Palaeogene (near-) surface cooling rather than a single cooling stage scenario. The highest calculated maximum palaeotemperature of approximately 80 °C constrains maximum cooling of *c.* 75 °C through the upper part of the apatite PAZ to surface/borehole temperatures of about 5 °C (Bramer 1991) since the late Palaeogene. Final cooling most probably commenced around 30 Ma, although any time within 40 and 20 Ma would be compatible with modelling and regional stratigraphy.

Discussion

Late Cretaceous denudation

Thermal history modelling of the AFT data presented here consistently reveals an initial cooling stage for all samples from Lusatia during the Late Cretaceous. As no evidence of increased hydrothermal activity is known from the literature nor is obvious from the sample appearance, this cooling stage is interpreted in terms of denudation.

Late Cretaceous cooling–denudation coincides with a series of geological indications for Cretaceous tectonic activity in easternmost Germany and Bohemia. Denudation of the Lusatian Block commences coeval with or subsequent to the inversion of southern part of the NE German Basin (Scheck & Bayer 1999; Ventura unpublished data), and is consistent with the contemporaneous sedimentation in the adjacent Saxo-Bohemian Basin. This cooling–denudation event is also supported by the timing of the tectonic activity along the Lusatian Thrust (Figs 1 and 2). Structural data, stratigraphic observations and palaeocurrent directions suggest the Lusatian Block as being the main source for the sedimentation within the Saxo-Bohemian Basin (Voigt 1995) and within its southern continuation in Bohemia (Uličný 2001) in the Late Cretaceous.

Late Cretaceous denudation refers to cooling of at least 30–50°C (higher temperature limit of the PAZ–modelled maximum palaeotemperatures). More realistic estimates can be derived from geological constraints. Remnants of Late Cretaceous–Eocene extensive weathering mantles, in the form of up to 100 m-thick kaolin saprolites lying directly on the granitic basement and partially covered by Tertiary deposits in the central part of the Lusatian Block and of the Erzgebirge (Störr 1983), indicate

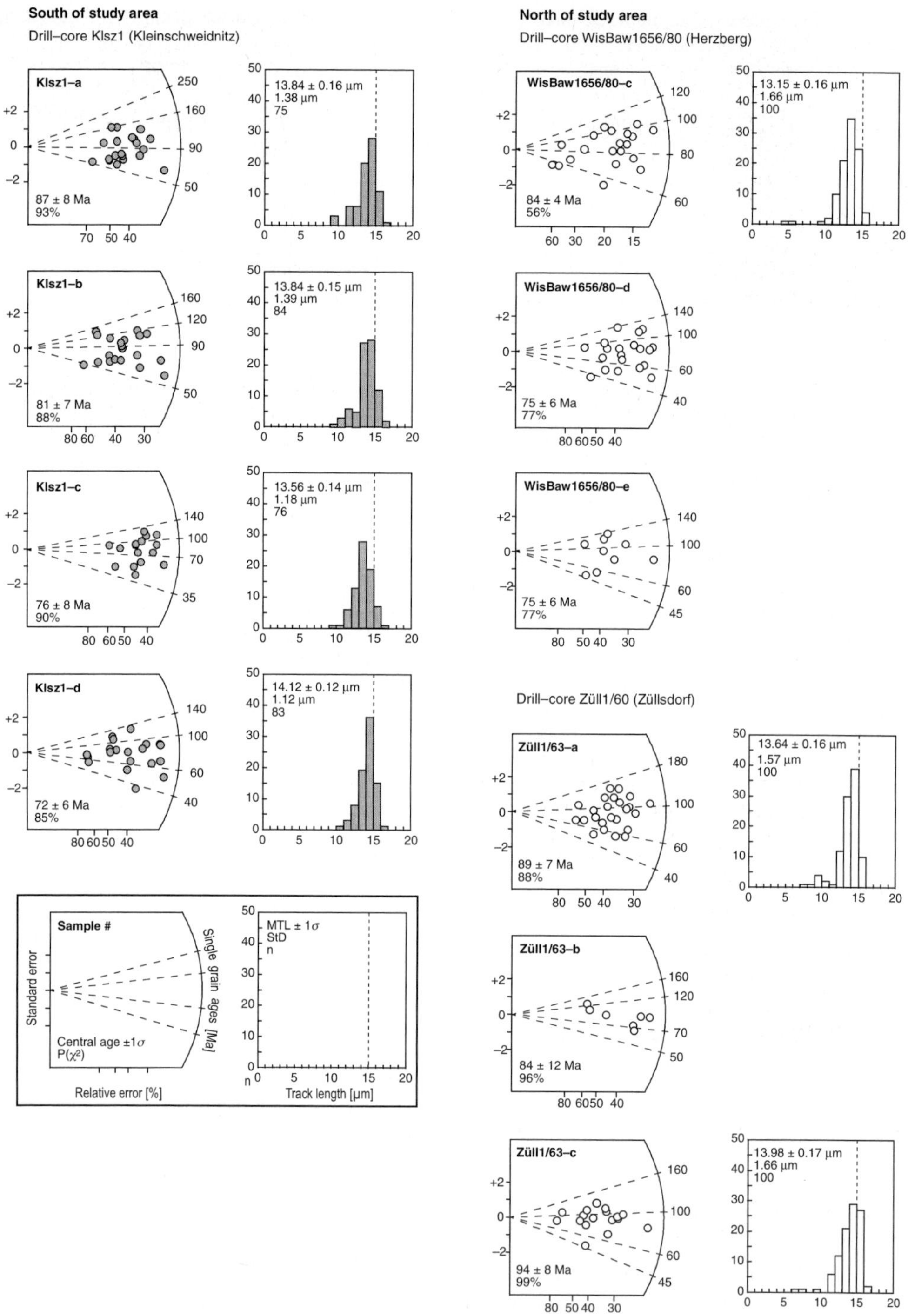

Fig. 4. Radial plots (left) of apatite fission-track ages calculated with TrackKey 4.2 (Dunkl 2002) and track-length histograms (right) for all analysed samples. MTL, mean track length; StD, standard deviation; *n*, number of tracks measured. Track-length distribution for samples WisBaw1656/80-d, WisBaw1656/80-e and Züll 1/63-b are not presented because the number of tracks measured is less than 75 (see also Table 1).

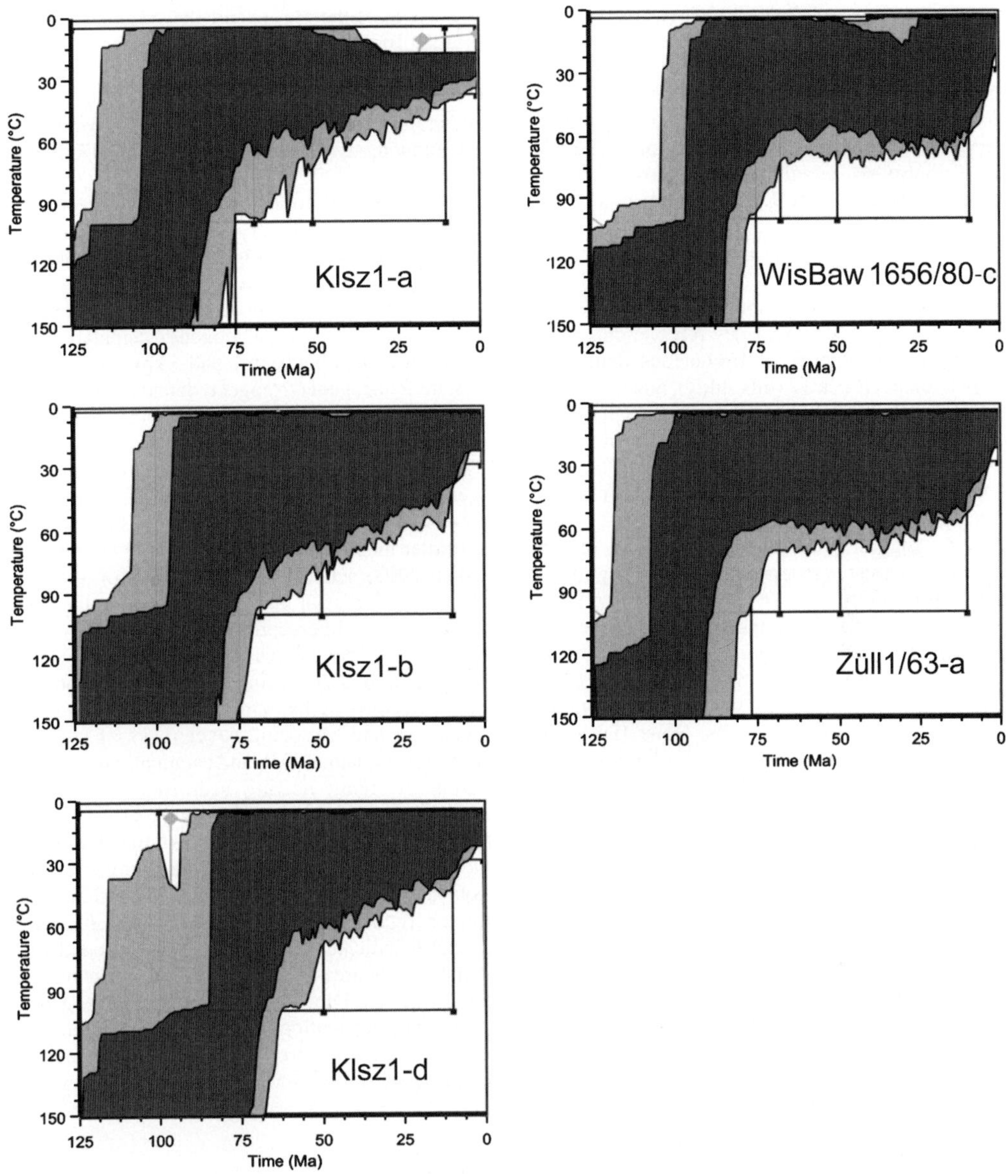

Fig. 5. Thermal histories of the samples with at least 20 counted crystals and 75 measured confined tracks obtained by inverse modelling with HeFTy (Ketcham 2005), and using the Ketcham *et al.* (1999) annealing model. Input parameters include track density and length, and D_{par} as composition proxy. Dark path envelopes represent 1000 successful t–T paths with default 'good' goodness of fit (GOF) >0.5 (light: 'acceptable', GOF > 0.05). Direct-control modelling produced model data that coincide with the observed data within 1σ. It favours early Palaeogene (near-) surface cooling rather than a single cooling stage scenario.

intense prolonged weathering under warm and humid climatic conditions (Migon & Lidmar-Bergström 2001; Gilg *et al.* 2003). The presence of this erosion surface and the sedimentation within the lignite basins of northern Lusatia commencing in the Oligocene suggest cooling of the basement up to surface or near-surface temperatures was already occuring in the Late Cretaceous–early Cenozoic.

Taking into account a Late Cretaceous surface temperature of approximately 20 °C (e.g. Voigt *et al.* 2006), a minimum Late Cretaceous–Early

Eocene cooling of about 90 °C (from >110 °C down to surface temperature) can be estimated. For a conventional geothermal gradient of <30 °C km^{-1}, the minimum amount of Late Cretaceous–Early Eocene exhumation was then 3 km. Such substantial denudation occurring in the vicinity of the structurally controlled margin of the Saxo-Bohemian Basin characterizes the Lusatian Thrust as a main regional tectonic feature during Cretaceous times. The Lusatian Thrust persists towards the SE as the structural limit of the Bohemian Cretaceous Basin, while its northwestern continuation defines the southern margin of present-day NE German Basin. Therefore, the amount of Cretaceous denudation close to the fault–thrust (this study), and the character and stratigraphic age of the preserved Cretaceous deposits (Voigt 1995), require a sedimentary sequence within the Cretaceous Saxo-Bohemian Basin that was originally much thicker than the preserved sandstone succession.

We suggest initiation/reactivation of the Lusatian Thrust associated with a significant strike-slip component in response to regional NNE–SSW-directed compression subsequent to mid-Jurassic crustal separation in the Central Tethys (e.g. Malkowsky 1987; Bayer *et al.* 1999). This model is in accordance with: (a) the prevailing dextral shearing along the Elbe Zone (DEKORP Group 1994); (b) the formation of a large regional flower structure below the Lusatian Thrust/Fault; (c) the preservation of small pull-apart structures along the same fault; (d) the characteristics of the Saxo-Bohemian Cretaceous sedimentary rocks (Voigt 1995; Uličný 2001); and (e) the absence of tectonic subsidence throughout the Cretaceous in the adjacent NE German Basin (Scheck & Bayer 1999).

Late Palaeogene denudation

A late Palaeogene cooling stage from modelled maximum palaeotemperatures between <60 and *c.* 80 °C (Fig. 5) is constrained only tentatively by AFT data. However, Palaeogene burial and subsequent denudation of the Lusatia Block appears plausible from diagenetic constraints, geomorphological indications and palaeogeographical considerations. Based on field studies and petrographical analyses of thin sections, Voigt (1995) estimated the Cenomanian–Coniacian strata of the Elbe valley to be compacted to approximately 30% of their original thickness. Although not yet quantified, such a degree of diagenesis requires a considerable overburden. This overburden probably preserved the weathering-sensitive (kaolinitic) Cretaceous erosion surface that is now exposed on several locations in Saxony (Migon & Lidmar-Bergström 2001), and mantled the palaeorelief between the Elbe Basin and the basement of the Lusatian Block.

For the maximum palaeotemperatures modelled in this study and a present day borehole/surface temperature of approximately 5 °C (Bramer 1991), and allowing *c.* 15–25 °C for combined climatic cooling since the climatic optimum and cooling effects subsequent to the heat-flow increase associated with the late Tertiary volcanism in southernmost Lusatia and the neighbouring Eger Graben, denudational cooling may be between <30 and 60 °C. Applying a conventional geothermal gradient of 20–30 °C km^{-1}, Palaeogene denudation of Lusatia was than between 1 and 3 km.

Late Palaeogene–Neogene denudation of Lusatia is probably related to its position in the hinterland of the Erzgebirge and to the uplift of this morphological feature as the northern rift shoulder of the Eger Graben since the Oligocene. This interpretation is supported by contemporaneous denudation of similar magnitude of the Erzgebirge (Ventura & Lisker 2003) and by the formation of the late Oligocene–Pliocene main depocentres of the Eger Graben at its interception with the Elbe Fault System (e.g. Špičáková *et al.* 2000). Far-field response to the Eger rifting is probably responsible for subrosion and halokinetic isostasy within the Palaeozoic and Mesozoic cover rocks of northern Lusatia (Eissmann 2002), and eventually for the formation of lignite-bearing sedimentary basins in central East Germany over about the last *c.* 30 Ma. Typically, the respective basins are dominated by thick Mid-Eocene–Mid-Oligocene continental sequences unconformably overlain by significantly thinner, lignite-bearing Quaternary deposits. Such a pattern indicates that the associated denudation of the source areas would have been most intense during the late Palaeogene. In contrast, Neogene–Quaternary denudation may not exceed an amount of some hundred metres in Lusatia, as indicated by the presence of Oligo-Miocene volcanic rocks.

Conclusions

The thermal history of the Lusatia region verifies that the long-term denudation of eastern Germany is primarily controlled by the tectonic activity of regional crustal structures, particularly the Elbe Fault System and the Eger Graben. Post-Variscan regional denudation is likely to have occurred in two impulses during the Cretaceous and late Palaeogene times.

Whilst inversion of the southern margin of the Central European Basin (Fig. 1) SW of the Elbe Fault System commenced in the Late Jurassic and ceased in the Late Cretaceous (Wagner 1989; Ventura & Lisker 2003), denudation of the Lusatia region east of the Elbe has been detected so far only for the Late Cretaceous (this study). This denudation pattern satisfyingly agrees with the prevailing model of several phases of transpressional tectonics of the Elbe Fault within a varying stress field triggered by the early Mesozoic Pangea break-up (e.g. Schröder 1987; Ventura & Lisker 2003). Alternatively, it also could be explained by the Cretaceous inversion of the whole southern part of the Central European Basin (including the Lusatian Block), and the subsequent uplift of a Lusatian Block indenter owing to transpression. Such a scenario would place the inversion hinge at the southern rim of the present Mid-German Crystalline Rise (Fig. 1). The erosion products of initial basin inversion were probably deposited within the NE German Basin, which contains up to more than 2200 m of Jurassic–Lower Cretaceous clastic sedimentary rocks and up to more than 1500 m of Late Cretaceous sediments, and where a marked acceleration in subsidence has been recognized in the Late Cretaceous (e.g. Scheck & Bayer 1999).

Late Palaeogene–Neogene rifting of the Eger Graben initiated denudation of considerably lower magnitude probably because the deposits from the southern part of the Central European Basin were already consumed, and the exposed basement is much more resistant to erosion. Accordingly, a significant component of denudation is compensated by uplift while the Eger rifting is still in progress.

B. Ventura gratefully acknowledges the financial support provided to this project through a grant from the Deutsche Forschungsgemeinschaft (VE 244/3). A. Friebe and D. Netscholta from the Geological Survey of Saxony kindly provided the samples from drill-core Klsz1. The paper benefited from critical comments of L. Barbero, U. Glasmacher and M. Wipf.

References

BARBARAND, J., CARTER, A., WOOD, I. & HURFORD, T. 2003. Compositional and structural control of fission-track annealing in apatite. *Chemical Geology*, **198**, 107–137.

BAYER, U., SCHECK, M. *ET AL.* 1999. An integrated study of the NE German Basin. *Tectonophysics*, **314**, 285–307.

BELLEMANS, F., DE CORTE, F. & VAN DEN HAUTE, P. 1995. Composition of SRM and CN U-doped glasses: Significance for their use as thermal neutron fluence monitors in fission track dating. *Radiation Measurements*, **24**, 153–160.

BRAMER, H. 1991. *Physische Geographie: Mecklenburg-Vorpommern, Brandenburg, Sachsen-Anhalt, Sachsen, Thüringen*. Haak, Gotha.

BURTNER, R. L., NIGRINI, A. & DONELICK, R. A. 1994. Thermochronology of Lower Cretaceous source rocks in the Idaho–Wyoming thrust belt. *AAPG Bulletin*, **78**, 1613–1636.

CAJZ, V., CYMERAMAN, Z. & KOZDRÓJ, W. 2001. Tertiary volcanic rocks. *In*: KOZDRÓJ, W., KRENTZ, O. & OPLETAL, M. (eds) *Comments on the Geological Map Lausitz–Jizera–Karkonosze (Without Cenozoic Sediments), 1:100 000*. Polish Geological Institute, Warsaw.

CARLSON, W. D., DONELICK, R. A. & KETCHAM, R. A. 1999. Variability of apatite fission-track annealing kinetics: I. Experimental results. *American Mineralogist*, **84**, 1213–1223.

DE CORTE, F., VAN DEN HAUTE, P., BELLEMANS, F., INGELBRECHT, C. & NICHOLL, C. 1998. The uranium doped glass IRMM-540 distributed by the European Commission: A new tool in fission-track age calibration. Paper presented at the 19th International Conference on Nuclear Tracks in Solids, Besancon, France.

DEKORP GROUP. 1994. Crustal structure of the Saxothuringian Zone: Results of the deep seismic profile MVE-90(East). *Zeitschrift für Geologische Wissenschaften*, **22**, 647–769.

DUNKL, I. 2002. TRACKKEY: a Windows program for calculation and graphical presentation of fission track data. *Computers and Geosciences*, **28**, 3–12.

EISSMANN, L. 2002. Tertiary geology of the Saale–Elbe Region. *Quaternary Science Reviews*, **21**, 1245–1274.

GALBRAITH, R. F. 1981. On statistical models for fission track counts. *Mathematical Geology*, **13**, 471–478.

GALBRAITH, R. F. & LASLETT, G. M. 1993. Statistical models for mixed fission track ages. *Nuclear Tracks and Radiation Measurements*, **21**, 459–470.

GILG, H. A., WEBER, B., KASBOHM, J. & FREI, R. 2003. Isotope geochemistry and origin of illite–smectite and kaolinite from the Seilitz and Kemmlitz kaolin deposits, Saxony, Germany. *Clay Minerals*, **38**, 95–112.

GLASMACHER, U. A., MANN, U. & WAGNER, G. N. A. 2002. Thermotectonic evolution of the Barrandian, Czech Republic, as revealed by apatite fission-track analysis. *Tectonophysics*, **359**, 381–402.

GLEADOW, A. J. W. 1981. Fission-track dating methods: What are the real alternatives? *Nuclear Tracks*, **5**, 3–14.

GLEADOW, A. J. W. & BROWN, R. 2000. Fission-track thermochronology and the long-term denudational response to tectonics. *In*: SUMMERFIELD, M. A. (ed.) *Geomorphology and Global Tectonics*. Wiley, Chichester, 57–75.

GLEADOW, A. J. W., DUDDY, I. R., GREEN, P. F. & HEGARTY, K. A. 1986. Fission track-lengths in the apatite annealing zone and the interpretation of mixed ages. *Earth and Planetary Science Letters*, **78**, 245–254.

GREEN, P. F. 1981. A new look at statistics in fission-track dating. *Nuclear Tracks*, **5**, 77–86.

GREEN, P. F., DUDDY, I. R., GLEADOW, A. J. W., TINGATE, P. R. & LASLETT, G. M. 1986. Thermal annealing of fission tracks in apatite, 1. A Qualitative Description. *Chemical Geology (Isotope Geoscience Section)*, **59**, 237–253.

HAMMER, J., EIDAM, J., RÖBER, B. & EHLING, B.-C. 1999. Prävariskischer und variskischer granitoider

Magmatismus am NE-Rand des Böhmischen Massivs – Geochemie und Petrogenese. *Zeitschrift für Geologische Wissenschaften*, **27**, 401–415.

HAMMERSCHMIDT, K., KOPP, J. & HAUCKE, L. 2003. Sm-Nd Dating on Minerals of a Silica Undersaturated, Ca-rich, Crustally Derived Gabbro of the Mid-German Crystalline Zone, drill hole Züllsdorf, Southern Brandenberg (Germany). *Zeitschrift der Gessellschaft für Geowissenschaften*, **31**, 225–238.

HEJL, E., COYLE, D., LAL, N., VAN DEN HAUTE, P. & WAGNER, G. A. 1997. Fission-track dating of the western border of the Bohemian massif: thermochronology and tectonic implications. *Geologische Rundschau*, **86**, 210–219.

HURFORD, A. J. & GREEN, P. F. 1983. The zeta age calibration of fission-track dating. *Chemical Geology*, **1**, 285–317.

KETCHAM, R. A. 2005. Forward and inverse modeling of low temperature thermochronometry data. *Reviews in Mineralogy and Geochemistry*, **58**, 275–314.

KETCHAM, R. A., DONELICK, R. A. & CARLSON, W. A. 1999. Variability of apatite fission-track annealing kinetics: III. Extrapolation to geological time scales. *American Mineralogist*, **84**, 1235–1255.

KLEY, J. & VOIGT, T. 2008. Late Cretaceous intraplate thrusting in central Europe: Effect of Africa–Iberia–Europe convergence, not Alpine collision. *Geology*, **36**, 839–842.

KOSSMAT, F. 1925. *Übersicht der Geologie von Sachsen*. Geologisches Landesamt, Leipzig.

LAI, L. N. 1985. Die geologische Struktur und die tektonische Analyse des nordwestlichen Teiles der Elbtalzone. *Freiberger Forschungshefte*, C, **390**, 68–85.

LASLETT, G. M., KENDALL, W. S., GLEADOW, A. J. W. & DUDDY, I. R. 1982. Bias in measurement of fission-track length distributions. *Nuclear Tracks*, **6**, 79–85.

LINNEMANN, U. G. 1995. The Neoproterozoic terranes of Saxony (Germany). *Precambrian Research*, **73**, 235–250.

MALKOVSKY, M. 1987. The Mesozoic and Tertiary basins of the Bohemian Massif and their evolution. *Tectonophysics*, **137**, 31–42.

MIGON, P. & LIDMAR-BERGSTRÖM, K. 2001. Weathering mantles and their significance for geomorphological evolution of central and northern Europe since the Mesozoic. *Earth-Science Reviews*, **56**, 285–324.

SCHECK, M. & BAYER, U. 1999. Evolution of the Northeast German Basin – inferences from a 3D structural model and subsidence analysis. *Tectonophysics*, **313**, 145–169.

SCHECK, M., BAYER, U., OTTO, V., LAMARCHE, J., BANKA, D. & PHARAOH, T. 2002. The Elbe Fault System in North Central Europe – a basement controlled zone of crustal weakness. *Tectonophysics*, **360**, 281–299.

SCHRÖDER, B. 1987. Inversion tectonics along the Western margin of the Bohemian Massif. *Tectonophysics*, **137**, 93–100.

ŠPIČÁKOVÁ, L., ULIČNÝ, D. & KOUDELKOVÁ, G. 2000. Tectonosedimentary Evolution of the Cheb Basin (NW Bohemia, Czech Republic) between Late Oligocene and Pliocene: A Preliminary Note. *Studia Geophysica et Geodaetica*, **44**, 556–580.

STÖRR, M. 1983. *Die Kaolinlagerstätten der Deutschen Demokratischen Republik. Schriftenreihe für geologische Wissenschaften*, **18**.

TAIT, J. A., BACHTADSE, V., FRANKE, W. & SOFFEL, H. C. 1997. Geodynamic evolution of the European Variscan fold belt: palaeomagnetic and geological constraints. *Geologische Rundschau*, **86**, 585–598.

THOMSON, S. N. & ZEH, A. 2000. Fission-track thermochronology of the Ruhla Crystalline Complex: New constraints on the post-Variscan thermal evolution of the NW Saxo-Bohemian Massif. *Tectonophysics*, **324**, 17–35.

ULIČNÝ, D. 2001. Depositional systems and sequence stratigraphy of coarse-grained deltas in a shallow-marine, strike-slip setting: the Bohemian Cretaceous Basin, Czech Republic. *Sedimentology*, **48**, 599–628.

VAN WEES, J.-D., STEPHENSON, R. A. *ET AL.* 2000. On the origin of the Southern Permian Basin, Central Europe. *Marine and Petroleum Geology*, **17**, 43–59.

VENTURA, B. & LISKER, F. 2003. Long-term landscape evolution of the north-eastern margin of the Bohemian Massif: apatite fission track data from the Erzgebirge (Germany). *International Journal of Earth Sciences*, **92**, 691–700.

VENTURA, B., LISKER, F. & KOPP, K. 2003. Apatite fission track data from the drill-core Züllsdorf 1/63: implications for the reconstruction of the Mid German Crystalline High. *Zeitschrift der Deutschen Gesellschaft für Geowissenschaften*, **31**, 251–261.

VOIGT, S., GALE, A. S. & VOIGT, T. 2006. Sea-level change, carbon cycling and palaeoclimate during the Late Cenomanian of northwest Europe; an integrated palaeoenvironmental analysis. *Cretaceous Research*, **27**, 836–858.

VOIGT, T. 1995. Faziesentwicklung und Ablagerungssequenzen am Rand eines Epikontinentalmeeres – Die Sedimentationsgeschichte der Sächsischen Kreide. *In*: *Fakultät für Geowissenschaften, Geo technik und Bergbau*. TU Bergakademie, Freiberg, 130.

WAGNER, S. 1989. Geothermische Krustenmodellierung im Suedteil der DDR. *Zeitschrift für Angewandte Geologie*, **35**, 1–3.

WETZEL, H. U. 1985. Interpretation von Bruchzonen im Osterzgebirge und ihre Fortsetzung in Deckgebirgseinheiten der Lausitz. *Zeitschrift für Geologische Wissenschaften*, **13**, 111–121.

A reappraisal of low-temperature thermochronology of the eastern Fennoscandia Shield and radiation-enhanced apatite fission-track annealing

BARRY P. KOHN[1]*, MATEVZ LORENCAK[1], ANDREW J. W. GLEADOW[1], FABIAN KOHLMANN[1], ASAF RAZA[1], KIRK G. OSADETZ[2] & PETER SORJONEN-WARD[3]

[1]*School of Earth Sciences, University of Melbourne, Melbourne, Victoria 3010, Australia*

[2]*Geological Survey of Canada, 3303 33rd Street N.W., Calgary, Alberta, Canada T2L 2A7*

[3]*Geological Survey of Finland, P.O. Box 1237, FI-70211, Kuopio, Finland*

**Corresponding author (e-mail: b.kohn@unimelb.edu.au)*

Abstract: We assess the proposal of Hendriks & Redfield (*Earth and Planetary Science Letters*, **236**, 443–458, 2005) that cross-over of the predicted apatite fission track (AFT) > (U–Th–Sm)/He (AHe) age relationship in the southeastern Fennoscandian shield in southern Finland reflects α-radiation-enhanced annealing (REA) of fission tracks at low temperatures and that more robust estimates of the denudation history are recorded through reproducible AHe data. New AHe results from southern Finland showing variable dispersion of single-grain ages may be biased by different factors operating within grains, which tend to give a greater weighting towards older age outliers. AHe ages from mafic rocks show the least dispersion and tend to be consistently lower than their coexisting AFT ages. In general, it is at the younger end of the single-grain variation range from such lithologies where most meaningful AHe ages can be found. AHe data from multigrain aliquots are, therefore, of limited value for evaluating thermal histories in southern Finland, especially when compared against coexisting AFT data as supporting evidence for REA.

New, large datasets from the southern Canadian and Western Australian shields show the relationship between AFT age, single-grain age or mean track length as a function of U content (determined by the external detector method). These do not display the moderately strong inverse correlations previously reported from southern Finland in support of REA. Rather, the trends are inconsistent and generally exhibit weak positive or negative correlations. This is also the case for plots from both shields, as well as those from southern Finland, where AFT parameters are plotted against effective U concentration [eU] [based on U and Th content determined by inductively coupled plasma-mass spectroscopy (ICP-MS)], which weights decay of the parents more accurately in terms of their α-productivity. Further, samples from southern Finland yield values of chi-square χ^2 >5%, indicating that there is no significant effect of the range of uranium content between grains within samples on the AFT ages, and that they are all consistent with a single population.

The oldest AFT ages in southern Finland apatites (amongst the oldest recorded from anywhere) are found in gabbros, which also have the highest Cl content of all samples studied. We suggest, that it is Cl content rather than REA that has influenced the annealing history of the apatites, which have experienced a history including reburial into the partial annealing zone by Caledonian Foreland basin sedimentation. The study of apatite from low U and Th rocks, with relatively low levels of α-radiation damage may provide the most practical approach for producing reliable results for AFT and AHe thermochronometry studies in cratonic environments.

Reconstructed thermal histories based on patterns of relatively young apatite fission-track (AFT) ages (mostly <450 Ma) suggest that some cratons have undergone discrete episodes of regional-scale Phanerozoic cooling from modestly elevated palaeotemperatures (in some places >110 °C) (Brown *et al.* 1994; Harman *et al.* 1998). This cooling has commonly been attributed to transient or episodic burial and denudation of extensive, relatively thin – although in places up to km-scale – low-conductivity platform sediments, rather than to the erosion of large sections of crystalline shield (e.g. Cederbom *et al.* 2000; Kohn *et al.* 2002*b*, 2005; Belton *et al.* 2004; Lorencak *et al.* 2004; Weber *et al.* 2005).

The advent of modern apatite (U–Th–Sm)/He (AHe) thermochronometry (e.g. Zeitler *et al.* 1987; Farley 2002) has extended the sensitivity of

From: LISKER, F., VENTURA, B. & GLASMACHER, U. A. (eds) *Thermochronological Methods: From Palaeotemperature Constraints to Landscape Evolution Models*. Geological Society, London, Special Publications, **324**, 193–216.
DOI: 10.1144/SP324.15 0305-8719/09/$15.00

low-temperature thermochronometers to significantly lower temperatures (*c.* 40–80 °C) than those accessible to the AFT system. Ideally, therefore, in cratonic environments, information provided by the AHe system should be strongly complementary to AFT data, potentially constraining very low magnitudes of heating and cooling, and thereby contributing further to resolving the low-temperature history of cratonic lithosphere (e.g. Lorencak *et al.* 2004; Söderlund *et al.* 2005; Flowers *et al.* 2006).

A significant problem has emerged, however, in that the sequence of AFT > AHe age (predicted from their relative decay-product retention properties), as often reported from cratonic environments (e.g. Lorencak *et al.* 2004; Flowers 2009), may also be reversed. The AHe analyses often yield erratic and unreproducible results, frequently older than the corresponding consistent and reproducible AFT data (e.g. Lorencak 2003; Murrell 2003; Belton *et al.* 2004). Speculation for this observation initially centred on the possibility that it might be related to a cross-over in the closure temperatures of the two systems at very low cooling rates, but it has become apparent that many of the results may be due to problems with one or other of the two dating systems (e.g. Hendriks & Redfield 2005; Green *et al.* 2006; Hansen & Reiners 2006). Several different reasons for the anomalous AHe ages are now apparent (see Fitzgerald *et al.* 2006), including, for example, the presence of U- and Th-enriched micro-inclusions, crystal size variations and U–Th zoning (Farley 2002). Further, Shuster *et al.* (2006), Flowers *et al.* 2007 and Flowers (2009) reported that accumulating radiation damage progressively increases the ^{4}He retentivity of apatites (the trapping model), thereby increasing the effective closure temperature and apparent ages obtained in slowly cooled rocks. This phenomenon results in a range of apparent ages correlating with both ^{4}He concentration and effective U concentration [eU] expressed as U ppm + 0.235 Th ppm (see also Flowers *et al.* 2007). The finding of enhanced ^{4}He retention in response to the accumulation of radiation damage is also consistent with the empirical observations of Green *et al.* (2006).

Others have expressed an alternative view, regarding the AFT ages as 'too young', and preferring to accept at face value apparently 'too old' coexisting AHe ages that replicate well (Hendriks & Redfield 2005; Söderlund *et al.* 2005). Hendriks & Redfield (2005) suggested that an inverse correlation between AFT age, or mean track length, and uranium content in the eastern Fennoscandian Shield indicates α-radiation-enhanced annealing (REA) of fission tracks at low temperatures. The effect of this postulated REA, as a previously unaccounted factor in addition to temperature sensitivity, implies that modelled AFT data in cratonic settings (and particularly in U-and Th-rich apatites in general) and using conventional thermal annealing models only may lead to an overestimation of palaeotemperatures. Hendriks & Redfield (2005) suggested that where there is a clear departure from the predicted AFT > AHe age relationship accompanied by reproducible AHe ages, then it is those ages rather than the AFT data that provide information on the true cratonic cooling history, i.e. the AFT system fails 'to properly record long-term stability' (p. 453). On the basis of this conclusion and other geological evidence, Hendriks & Redfield (2005) discarded the notion of burial of the Fennoscandian Shield under a regionally extensive, deep Caledonian foreland basin, as previously proposed by several workers (e.g. Zeck *et al.* 1988; Larson *et al.* 1999; Cederbom *et al.* 2000; Lorencak 2003; Murrell & Andriessen 2004). The papers by Hendriks & Redfield and Söderlund *et al.* have caused considerable controversy and debate (e.g. Green & Duddy 2006; Larson *et al.* 2006). Although new developments with respect to the effects of radiation damage in the understanding of the AHe system (as described earlier) have appeared since publication of the work of Hendriks & Redfield, an evaluation of REA is still of the utmost importance for fission-track interpretation because it questions a fundamental premise on which AFT thermochronology is based.

In this work we further investigate the question of REA in apatite from cratonic environments by: (1) assessing new AHe data, mainly based on single-grain analyses, from some of the same aliquots previously studied by Lehtovaara (1976) and Lorencak (2003) from the eastern domain of the Fennoscandian shield in southern Finland, as well as on two new samples; (2) evaluating the relationship of the AHe results to AFT data from aliquots of the same sample; and (3) comparing the eastern Fennoscandia data plots (including effective U concentration) using the same criteria for sample selection as used by Hendriks & Redfield (2005) with new plots constructed for two other classic cratons [Western Australia (Yilgarn and Pilbara) and southern Canada (Trans-Hudson–Superior)], where geological and isotopic evidence also support the removal of sedimentary cover (e.g. Kohn *et al.* 2002*b*, 2005; Patchett *et al.* 2004).

Samples and analytical methods

Apatite from 15 samples covering a range of lithologies (see Table 1) from southern Finland were analysed for AFT and AHe thermochronology by Lorencak (2003). This work has not previously been published and is presented here in addition to

Table 1. *Apatite fission-track data from southern Finland*

Sample No.	Lithology	Locality	Longitude (°E)	Latitude (°N)	No. of grains	Standard track density ($\times 10^6$ cm^{-2})	Fossil track density ($\times 10^6$ cm^{-2})	Induced track density ($\times 10^6$ cm^{-2})	U (ppm)	Chi-square probability (%)	Age dispersion (%)	*Fission track age (Ma) ($\pm 1\sigma$)	Mean track length $\pm$ std error (μm)	Std. dev. (μm)	‡Range of Cl (wt. %)
9-73	Px-Hb Gabbro	Hyvinkaa	23.78	60.63	20	1.174 (4234)	7.289 (2617)	1.947 (699)	20.7	95.0	0.0	**756 ± 34**	12.05 ± 0.14 (100)	1.41	0.35–0.63
9-73					12	1.106 (4022)	8.372 (1697)	1.850 (375)	20.9	94.8	0.0	**896 ± 53**†	12.10 ± 0.18 (100)	1.77	–
15-73	Hb Gabbro	Parikkala	29.68	61.63	20	1.197 (4234)	3.912 (2028)	0.901 (467)	9.4	3.1	19.6	**899 ± 64**	12.05 ± 0.16 (99)†	1.62	0.21–0.35
17-73	Hb Gabbro	Parikkala	29.73	61.61	6	1.220 (4234)	3.949 (230)	0.789 (46)	8.1	98.2	0.0	**1027 ± 167**	11.78 ± 0.28 (42)†	1.83	0.15–0.32
19-73	Rapakivi Granite	Viipuri	27.75	60.87	16	1.266 (4234)	4.229 (885)	1.854 (388)	18.3	17.6	11.5	**510 ± 36**	10.75 ± 0.24 (70)	2.01	0.00–0.01
30-73	Gabbronorite	Ylivieska	24.46	64.08	20	1.289 (4234)	3.347 (3025)	0.909 (822)	8.8	79.2	0.3	**812 ± 34**	11.60 ± 0.15 (100)	1.41	0.10–0.21
30-73					11	1.222 (4022)	5.541 (1043)	1.519 (286)	15.5	36.8	0.2	**803 ± 55**†	11.88 ± 0.17 (96)	1.62	–
35-73	Hb Gabbro	Ylivieska	24.51	64.06	10	1.312 (4234)	4.900 (831)	1.739 (295)	16.6	12.4	14.2	**659 ± 57**	11.67 ± 0.30 (41)	1.91	0.13–0.45
35-73					12	1.245 (4022)	4.684 (871)	1.403 (261)	14.1	46.2	8.4	**755 ± 58**†	11.71 ± 0.17 (100)	1.67	–
A82	Granite	Hankavesi	24.03	62.50	20	1.146 (3795)	2.421 (2352)	0.921 (895)	10.0	43.6	1.6	**527 ± 23**	11.38 ± 0.18 (100)	1.77	0.00–0.03
A121	Granite	Loimaa	23.10	60.78	16	1.166 (3795)	3.934 (1727)	1.533 (673)	16.4	64.7	1.7	**524 ± 25**	11.49 ± 0.16 (100)	1.59	0.00–0.04
A220	Granodiorite	Imatra	28.69	61.16	20	1.186 (3795)	5.098 (1961)	2.355 (906)	24.8	9.6	10.5	**449 ± 23**	10.97 ± 0.20 (100)	2.05	0.00–0.05
A223	Granodiorite	Ignonvaara	31.05	62.65	9	1.206 (3795)	2.683 (960)	0.869 (311)	9.0	84.1	0.0	**646 ± 43**	11.35 ± 0.18 (91)	1.71	0.02–0.06
A339	Granodiorite	Silvevaara	31.26	62.84	20	1.226 (3795)	3.447 (1871)	1.211 (657)	12.3	54.8	3.5	**607 ± 30**	11.28 ± 0.19 (101)	1.90	0.00–0.01
A596	Diorite	Tuusmäki	28.00	62.01	8	1.246 (3795)	6.352 (496)	3.279 (256)	32.9	69.4	0.7	**426 ± 34**	10.76 ± 0.31 (44)	2.06	0.00–0.03
A602	Granite	Obbnäs	24.46	60.02	7	1.265 (3795)	3.610 (670)	2.241 (416)	22.1	0.8	21.4	**367 ± 39**	11.37 ± 0.23 (65)	1.85	–
A1028	Gneiss	Sahinperä	26.64	63.49	4	1.285 (3795)	4.881 (353)	1.867 (135)	18.2	58.4	0.0	**586 ± 60**	11.35 ± 0.42 (23)	2.00	0.04–0.18
A1033	Augen gneiss	Petosenmäki	27.62	62.84	2	1.305 (3795)	5.615 (115)	1.465 (30)	14.0	77.7	0.0	**854 ± 176**	10.22 ± 0.55 (10)	1.74	0.00–0.01

Brackets show number of tracks counted or measured.
Standard and induced track densities measured on external mica detectors ($g = 0.5$) and fossil track densities on internal mica surfaces.
*Central ages (Galbraith & Laslett 1993) determined by ML (Matevz Lorencak) using zeta (ζ) = 365 ± 6 for dosimeter glass Corning-5.
†Track lengths and central ages determined by AR (Asaf Raza) using $\zeta = 384 \pm 5$ for dosimeter glass Corning-5.
‡Chlorine was determined on a JEOL JXA-5A electron microprobe running at an accelerating voltage of 15 kV, defocused beam size of 15–20 μm and beam current of 29 nA; samples were calibrated using Durango apatite. For values <0.02% Cl errors are about ±100%; for values of *c.* 0.10%, Cl errors are about ±25%; and for values *c.* 1%, Cl errors are about ±15%. Analyses were performed at Geotrack International Pty Ltd.

new analytical data from the same samples. The geological setting of the study area has been described elsewhere (Lehtovaara 1976; Lorencak 2003; Murrell 2003; Murrell & Andriessen 2004). Six of the apatites (those labelled -73) are from samples described by Lehtovaara (1976). The remaining nine (prefixed by A-) are from samples provided by Peter Sorjonen-Ward and were used to refine the areal distribution of our samples (Fig. 1). In addition, we also use AFT and AHe data reported from previous studies in the southern Canadian (Lorencak 2003; Lorencak *et al.* 2004) and Western Australia shields (Weber 2002), as well as unpublished data from both areas analysed by Kohn and co-workers – in all 41 samples from Canada and 35 samples from Western Australia.

The methodology for AFT analysis of all samples reported here including the southern Canadian Superior–Trans-Hudson shield of central Ontario and eastern Manitoba, and the Western Australian Shield, follows that described by Lorencak *et al.* (2004). Analyses for AHe thermochronometry were carried out in two stages. The early AHe analyses (those shown in Table 2 comprising multi-grains with no determination of Sm content) were carried out on nine samples, chosen on the basis of apparent AFT age, location, lithology and availability of suitable grains, following the protocol

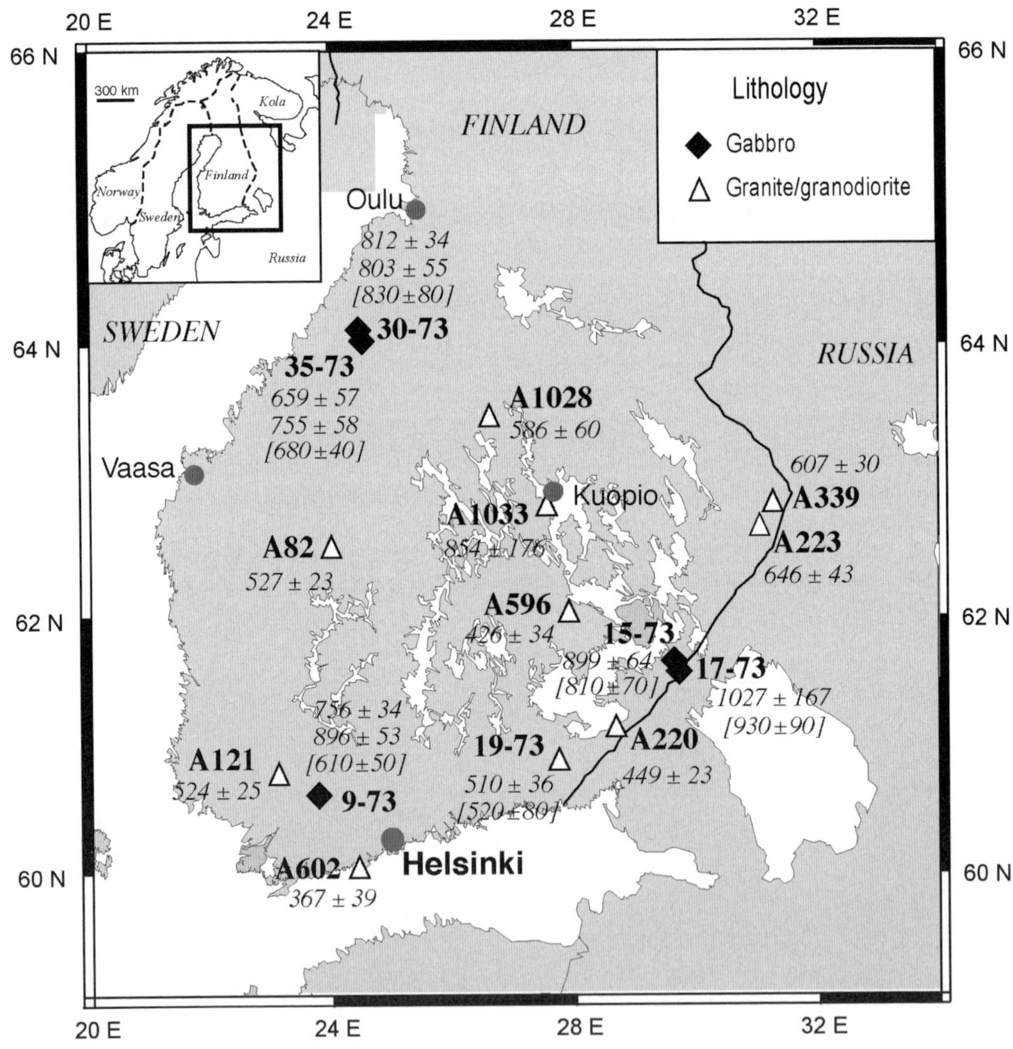

Fig. 1. Map of the study area in southern Finland showing sample locations, lithologies and AFT ages in italics (see Table 1 for analytical details). Numbers in square brackets are AFT ages reported for the samples by Lehtovaara (1976).

Table 2. *(U–Th–Sm)/He apatite data from southern Finland*

Sample No.	Lab. No.	No. of grains	Gas ncc	Mass (mg)	U ppm	Th ppm	Sm ppm	Th/U ratio	MWAR (μm)[1]	SD[2]	Grain length (μm)	R_s (μm)[3]	Raw Age (Ma)	F_T^4	Corrected Age (Ma)	Error (±1σ)	Weighted mean age[5]
9-73	1084	10	8.536	0.0167	16.6	23.1	–	1.40	41.7	4.4	113.6	45.8	188	0.64	**293**	18	
9-73	1156	10	10.395	0.0228	13.0	22.5	–	1.78	50.0	10.1	92.3	48.6	201	0.68	**293**	18	**[293]**
9-73	2872	1	7.188	0.0091	25.5	32.1	190.1	1.26	74.2	–	248.5	85.7	192	0.80	**239**	15	
9-73	2873	1	5.012	0.0123	11.8	15.7	78.8	1.32	75.3	–	216.5	83.8	211	0.81	**261**	16	
9-73	2874	1	16.007	0.0243	8.5	10.5	59.1	1.24	85.9	–	327.0	102.0	471	0.84	**558**	35	
9-73	2875	1	3.106	0.0135	8.3	11.0	82.7	1.32	74.4	–	242.1	85.4	170	0.81	**209**	13	
9-73	2876	1	4.465	0.0133	10.4	12.5	76.7	1.20	77.7	–	220.0	86.1	202	0.81	**248**	15	
																	(240 ± 12)[6]
15-73	1085	10	6.124	0.0215	6.4	10.1	–	1.62	44.1	6.5	119.5	48.3	262	0.68	**387**	24	
15-73	1157	10	7.767	0.0225	5.7	16.6	–	2.97	43.9	6.9	127.9	49.0	290	0.67	**433**	27	
15-73	1270	9	3.941	0.0192	5.7	9.2		1.67	43.7	6.3	110.4	47.0	269	0.66	**404**	25	
15-73	1262	8	2.408	0.0106	4.3	7.5	–	1.80	39.1	5.0	106.8	42.9	236	0.62	**382**	24	**[400 ± 11]**
15-73	2877	1	1.121	0.0025	7.0	11.0	115.0	1.67	38.1	–	212.5	48.5	389	0.64	**603**	37	
15-73	2878	1	0.667	0.0026	8.0	8.0	93.0	1.06	45.1	–	186.4	54.5	214	0.69	**310**	19	
15-73	2879	1	2.025	0.0039	7.0	13.0	59.0	1.83	49.3	–	226.9	60.8	417	0.71	**584**	36	
15-73	2880	1	2.100	0.0034	7.0	13.0	68.0	1.96	53.2	–	179.4	61.5	505	0.72	**698**	43	
15-73	2881	1	1.503	0.0054	6.0	13.0	34.0	2.11	50.5	–	209.1	61.0	247	0.73	**337**	21	
																	(390 ± 65)
Double dissolution																	
15-73	1290	5	1.529	0.0067	5.4	9.0	–	1.67	38.0	3.9	102.1	41.5	245	0.61	**402**	25	
15-73	1291	5	2.646	0.0071	7.1	12.3	–	1.73	37.7	6.4	110.5	42.2	302	0.62	**489**	30	
																	438 ± 43
17-73	1088	9	5.885	0.0116	18.6	10.2	–	0.50	41.0	9.3	97.6	43.3	195	0.62	**313**	19	
17-73	1158	8	4.795	0.0095	23.2	20.9	–	0.93	39.9	7.4	98.6	42.6	158	0.60	**261**	16	
																	[283 ± 26]
19-73	1086	10	25.534	0.0104	22.8	128.3	–	5.60	37.3	5.0	96.7	40.4	373	0.57	**655**	41	
19-73	2893	1	19.551	0.0034	40.3	238.6	789.1	5.92	40.9	–	204.8	51.1	467	0.68	**690**	43	
19-73	2894	1	0.840	0.0031	4.8	29.0	125.2	6.07	41.9	–	173.9	50.6	189	0.68	**280**	17	
19-73	2895	1	8.681	0.0014	39.8	290.0	1395.6	7.28	33.1	–	173.0	41.7	462	0.58	**803**	50	
19-73	2896	1	6.876	0.0017	30.0	200.8	1168.0	6.68	37.7	–	167.6	46.2	417	0.62	**677**	42	

(Continued)

Table 2. *Continued*

Sample No.	Lab. No.	No. of grains	Gas ncc	Mass (mg)	U ppm	Th ppm	Sm ppm	Th/U ratio	MWAR (μm)[1]	SD[2]	Grain length (μm)	R_s (μm)[3]	Raw Age (Ma)	F_T^4	Corrected Age (Ma)	Error (±1σ)	Weighted mean age[5]
19-73	2897	1	1.945	0.0014	13.7	63.9	1091.8	4.66	38.1	–	137.5	44.8	384	0.61	**631**	39	
19-73	3049	1	8.232	0.0033	15.8	89.0	1038.9	5.64	48.8	–	138.2	54.1	523	0.70	**752**	47	
19-73	3050	1	5.677	0.0021	25.5	162.7	287.3	6.39	40.2	–	127.0	45.8	347	0.64	**544**	34	**(661 ± 38)**[7]
30-73	1087	9	7.158	0.0210	9.2	14.0	–	1.50	50.9	9.3	98.4	50.3	220	0.68	**326**	20	
30-73	1160	9	12.229	0.0192	13.1	26.6	–	2.10	47.9	7.4	93.8	47.6	268	0.66	**405**	25	**[357 ± 39]**
35-73	1161	9	66.590	0.0181	62.4	89.7	–	1.48	42.3	4.2	107.9	45.6	366	0.65	**558**	35	
35-73	1263	8	46.352	0.0170	57.0	91.1	–	1.64	42.8	3.5	125.2	47.8	300	0.67	**452**	28	
35-73	1264	8	56.994	0.0186	65.9	101.1	–	1.58	48.5	4.6	101.9	49.3	317	0.68	**470**	29	**[485 ± 30]**
35-73	2898	1	25.408	0.0060	50.1	82.4	86.7	1.64	61.6	–	157.1	66.4	481	0.76	**632**	39	
35-73	2899	1	42.134	0.0132	44.4	84.4	47.6	1.90	77.2	–	220.3	85.8	396	0.81	**488**	30	
35-73	2900	1	57.330	0.0120	58.5	74.2	48.9	1.27	72.0	–	230.7	82.3	494	0.81	**611**	38	
35-73	2901	1	24.089	0.0082	41.3	77.3	115.8	1.87	62.4	–	254.2	75.2	391	0.76	**514**	32	
35-73	2902	1	16.762	0.0184	15.1	25.0	39.7	1.66	87.9	–	236.6	96.1	347	0.83	**417**	26	**(508 ± 39)**
Double dissolution																	
35-73	1294	5	38.754	0.0090	73.4	126.0	–	1.72	40.5	5.1	114.0	44.8	338	0.64	**529**	33	
35-73	1296	4	18.235	0.0056	61.3	98.4	–	1.61	35.0	2.2	118.1	40.5	302	0.60	**505**	31	*516 ± 12*
A82	1070	10	16.249	0.0147	17.8	41.9	–	2.40	41.0	3.9	118.2	45.7	321	0.63	**509**	32	
A82	3282	1	10.058	0.0152	8.7	21.3	82.6	2.44	76.2	–	260.6	88.4	382	0.82	**468**	29	
A82	3283	1	19.513	0.0135	16.5	40.0	100.1	2.43	72.1	–	259.1	84.6	442	0.81	**547**	34	
A82	3284	1	25.900	0.0245	9.4	23.2	68.9	2.46	98.5	–	251.0	106.1	557	0.85	**659**	41	
A82	3285	1	24.256	0.0140	18.3	52.5	102.2	2.87	80.0	–	316.3	95.8	447	0.82	**547**	34	
A82	3286	1	5.412	0.0091	8.5	16.4	64.6	1.94	72.9	–	170.2	76.6	383	0.79	**484**	30	**(526 ± 31)**

A121	3292	1	3.538	0.0092	2.3	3.0	39.6	1.31	74.9	–	163.0	77.0	949	0.79	**1197**	74	
A121	3293	1	7.905	0.0065	18.2	24.7	67.1	1.35	59.6	–	259.4	72.7	401	0.76	**526**	33	
A121	3295	1	14.693	0.0072	21.5	30.1	102.7	1.40	54.1	–	243.4	66.4	560	0.76	**735**	46	
A121	3296	1	1.581	0.0049	8.9	11.8	100.6	1.33	55.2	–	160.5	61.6	220	0.74	**296**	18	
																	(390 ± 106)[8]
A220	1082	10	24.947	0.0182	24.9	5.4	–	0.22	40.7	4.6	137.5	47.1	416	0.67	**620**	38	
A220	3277	1	5.726	0.0044	22.0	3.0	104.0	0.15	48.4	–	187.5	57.7	437	0.74	**593**	37	
A220	3278	1	8.248	0.0044	24.0	3.0	130.0	0.13	52.3	–	159.2	59.1	588	0.75	**789**	49	
A220	3279	1	11.080	0.0046	33.0	6.0	125.0	0.17	49.5	–	186.8	58.7	544	0.74	**731**	45	
A220	3280	1	5.364	0.0038	16.0	3.0	116.0	0.20	44.9	–	188.4	54.4	648	0.72	**896**	56	
A220	3281	1	4.126	0.0042	18.0	3.0	95.0	0.15	53.2	–	149.1	58.8	413	0.74	**555**	34	
																	(666 ± 60)
Double dissolution																	
A220	1297	5	5.878	0.0049	26.3	4.2	–	0.16	33.5	3.3	119.9	39.3	356	0.58	**613**	38	
A220	1298	5	18.181	0.0055	53.5	32.2	–	0.60	34.4	1.8	125.6	40.5	420	0.58	**723**	45	
																	659 ± 54
A223	3297	1	12.568	0.0039	18.0	14.0	67.0	0.78	56.3	–	123.4	58.0	1088	0.73	**1497**	93	
A223	3298	1	25.615	0.0124	14.0	7.0	29.0	0.51	73.3	–	229.9	83.4	946	0.81	**1164**	72	
A223	3299	1	3.275	0.0044	15.0	3.0	13.0	0.20	56.5	–	138.2	60.2	373	0.75	**500**	31	
A223	3300	1	17.549	0.0100	22.0	3.0	21.0	0.13	73.7	–	182.5	78.7	610	0.81	**755**	47	
A223	3301	1	27.045	0.0149	30.0	4.0	29.0	0.12	74.6	–	266.4	87.4	463	0.83	**559**	35	
																	(571 ± 67)[9]
A339	1083	10	11.696	0.0150	10.8	16.2	–	1.50	40.8	5.4	130.2	46.6	430	0.63	**674**	42	
A339	3287	1	14.225	0.0157	13.0	33.0	30.0	2.49	83.1	–	226.8	91.2	344	0.83	**417**	26	
A339	3288	1	3.170	0.0099	5.0	7.0	13.0	1.35	70.9	–	195.2	78.0	392	0.79	**494**	31	
A339	3289	1	11.417	0.0127	11.0	14.0	46.0	1.31	67.9	–	273.6	81.6	504	0.80	**627**	39	
A339	3290	1	4.561	0.0087	8.0	9.0	20.0	1.15	64.2	–	210.6	73.8	403	0.78	**514**	32	
A339	3291	1	4.685	0.0055	10.0	16.0	33.0	1.52	55.4	–	178.5	63.4	476	0.75	**639**	40	
																	(511 ± 41)

[1]MWAR is the mass-weighted average radius of apatite crystals analysed.
[2]Standard deviation of the MWAR is used as a guide for the 'tightness' of the range of single crystal radii picked within a sample.
[3]R_s is the equivalent sphere radius of the crystal/s (see text).
[4]F_T is the α-ejection correction after Farley *et al.* (1996).
[5]Multi-grain mean ages indicated by square brackets, single grain mean ages indicated by round brackets and double dissolution mean ages on multi-grains indicated in italics.
[6]Analysis #2874 omitted from calculation.
[7]Analysis #2894 omitted from calculation.
[8]Analysis #3292 omitted from calculation.
[9]Analyses #3297 and #3298 omitted from calculation.

described by Lorencak *et al.* (2004). Analytical methods for later AHe analyses (single-grain analyses showing Sm content in Table 2) followed an established laboratory routine for laser He extraction (House *et al.* 2000), but outgassed using a solid-state diode laser with 820 nm wavelength and fibre-optic coupling. U–Th–Sm data used in the age calculations were acquired via total dissolution of outgassed apatite in HNO_3 and analysed using a second-generation Varian quadrupole ICP-MS housed in the School of Earth Sciences, University of Melbourne.

All AHe ages were calculated and corrected for α-emission following the approach of Farley *et al.* (1996). Analytical uncertainties for the University of Melbourne (U–Th–Sm)/He facility are assessed to be approximately 6.2% ($\pm 1\sigma$), which incorporates the α-correction-related constituent and takes into account an estimated 5 μm uncertainty in grain-size measurements, gas analysis and ICP-MS uncertainties. Durango apatite standard was run as an internal standard with each batch of samples analysed and served as a check on analytical accuracy. The uncertainty on U, Th and Sm content is estimated to be <2% (at $\pm 2\sigma$).

Results

Our AFT data from southern Finland show a wide scatter of central ages, ranging from 367 ± 39 to 1027 ± 167 Ma (Fig. 1 and Table 1). Mean horizontal confined track lengths (HCTL) range between 10.22 and 12.10 μm, with unimodal distributions and a relatively small number of long lengths. Standard deviations range from 1.41 to 2.06 μm. Ages and track lengths were determined on three of the gabbros (9-73, 30-73 and 35-73) by two different analysts (Table 1). In each case ages are concordant at the $\pm 2\sigma$ level and mean track lengths are identical within the error levels shown. In Figure 1 our AFT ages are also compared with those published from the same aliquots by Lehtovaara (1976), who used an absolute neutron dose determination approach. Overall, our data are in broad agreement with those reported from previous AFT work in the southern Fennoscandian Shield in Finland (e.g. Lehtovaara 1976; Larson *et al.* 1999; Murrell 2003; Murrell & Andriessen 2004).

While an areal relationship between the ages is not always straightforward, a relationship between age and lithology is apparent. Samples from granite, granodiorite and diorite show a broad trend of apparent younging towards the southern and western coast, and where located in relatively close proximity, e.g. samples A223 and A339, yield concordant ages (within error limits). With the exception of sample A1033 (an augen gneiss), all older AFT ages are found in gabbros, which also show longest mean HCTL. However, owing to the low apatite yield from A1033, only two grains were analysed from this sample; its data are therefore not considered robust and are not discussed further.

For fourteen of the samples, the same apatite grains previously used for AFT age determination were probed for chlorine content. All non-gabbro samples show a chlorine content of <0.06 wt% Cl, while the five gabbros analysed consistently show a higher Cl content ranging from 0.1 to 0.6 wt% Cl (results are summarized in Table 1 and single-grain data shown in Fig. 2).

Despite the relatively low absolute Cl content of all apatites studied, the 'fluorapatite' end members (<0.06 wt% Cl) typically yield younger single-grain ages (ranging from *c.* 330 to 800 Ma) compared to those from gabbros (ranging from *c.* 550 to 1190 Ma), even though there is some age overlap between the two broad groupings (Fig. 2).

As indicated previously, AHe analyses were carried out in two stages. Multigrain analyses were first determined on nine samples and these were later supplemented by single-grain analyses on seven of the samples previously analysed, as well as on two further samples A121 and A223 (Table 2). However, no further single-grain analyses were attempted on samples 17-73 and 30-73, as no grains of suitable quality were found in the available concentrates.

All samples are characterized by a relatively wide degree of scatter of AHe ages and low reproducibility. For the purposes of comparison, weighted mean ages for samples are shown in Table 2, where they are presented separately for multigrain analyses, for single-grain analyses and for multigrain analyses that have been subjected to a double dissolution protocol (see later). In general, as is the case for the AFT data, a broad correlation with sample lithology is suggested (Table 2). Samples from non-gabbro lithologies yield older mean weighted multigrain ages (509–674 Ma) and mean single-grain ages (390–666 Ma) compared to gabbros, which yield mean weighted multigrain ages of 283–516 Ma and mean single-grain ages of 240–508 Ma, although there is some age overlap between individual analyses. Non-gabbro lithologies also tend to yield AHe ages older than or close to their coexisting AFT ages, whereas those for gabbros are consistently lower and often less dispersed (Fig. 3). In samples where both multi- and single-grain AHe analyses have been carried out, multigrain ages usually fall in the middle of or at the higher end of the range of ages covered by single-grain ages (Table 2 and Fig. 3).

In some samples the intra-sample single-grain age dispersion is so broad (e.g. samples 9-73,

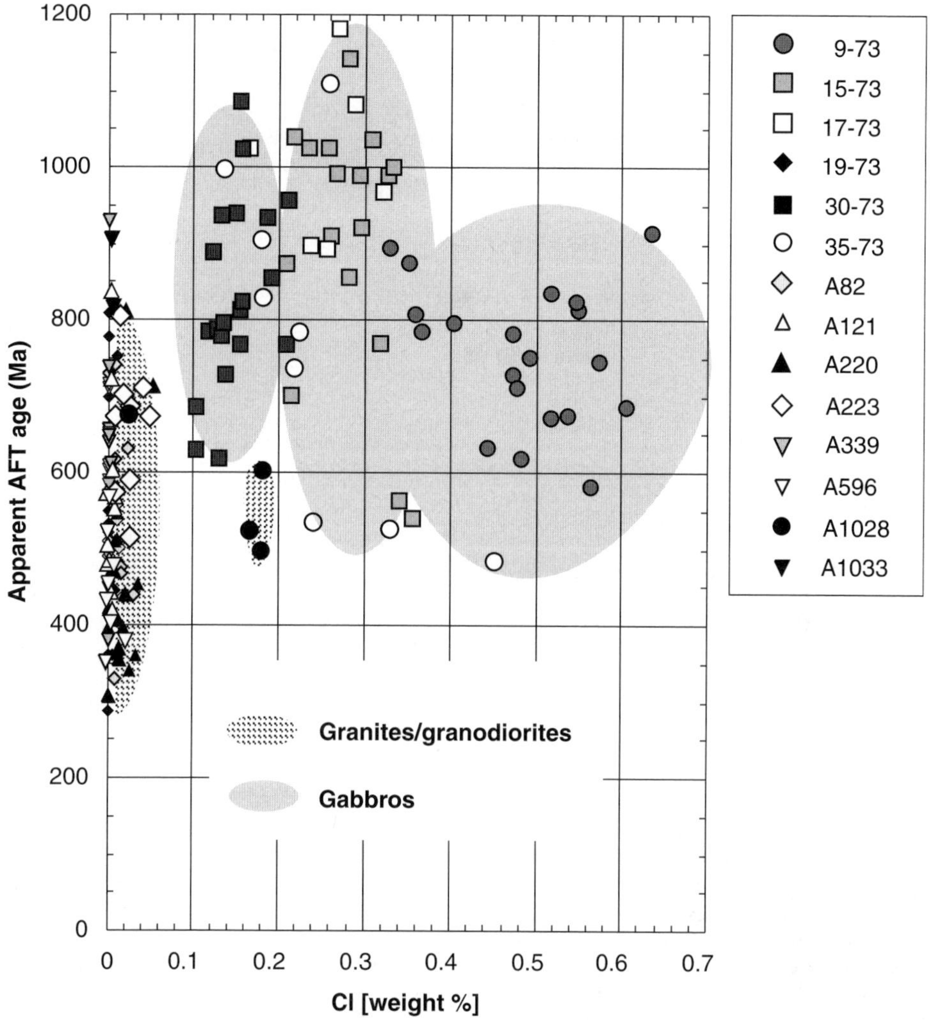

Fig. 2. Plot of apatite fission-track age v. chlorine content for grains from samples studied from southern Finland.

19-73 and A223) that some individual analyses were omitted from calculation of the mean ages (Table 2). The criterion used in rejecting these outliers follows that adopted for the removal of single crystal $^{40}Ar/^{39}Ar$ ages, where any grain outside $\pm 2\sigma$ of the mean of the population is excluded and a revised weighted mean calculated (e.g. Deino & Potts 1990). For other samples (e.g. A121) the age range between the single-grain analyses is so broad that the weighted mean age is meaningless and this data will not be discussed further.

The gabbros, all from the Lehtovaara (1976) study, are located in three different areas (Fig. 1). Sample 9-73, excluding one age outlier, yields a weighted mean single-grain AHe age of 240 ± 12 Ma and is the most reproducible of all the samples studied, even though there is a relatively large variation in radii between grains analysed. Relatively good age replication was also obtained for multigrain analyses of sample 15-73 from the Parikkala intrusion in the eastern part of the study area (Fig. 1), although single-grain analyses show a wider dispersion despite the limited range of grain sizes analysed. The U–Th content of this sample is uniformly low across all aliquots analysed. If only the four multigrain analyses are considered for this sample, then the weighted mean age is 400 ± 11 Ma, which is concordant with the weighted mean age of 390 ± 65 Ma from the single-grain ages. However, the range of single-grain ages is very broad, so the interpretation of any weighted mean ages in this sample needs to be treated with

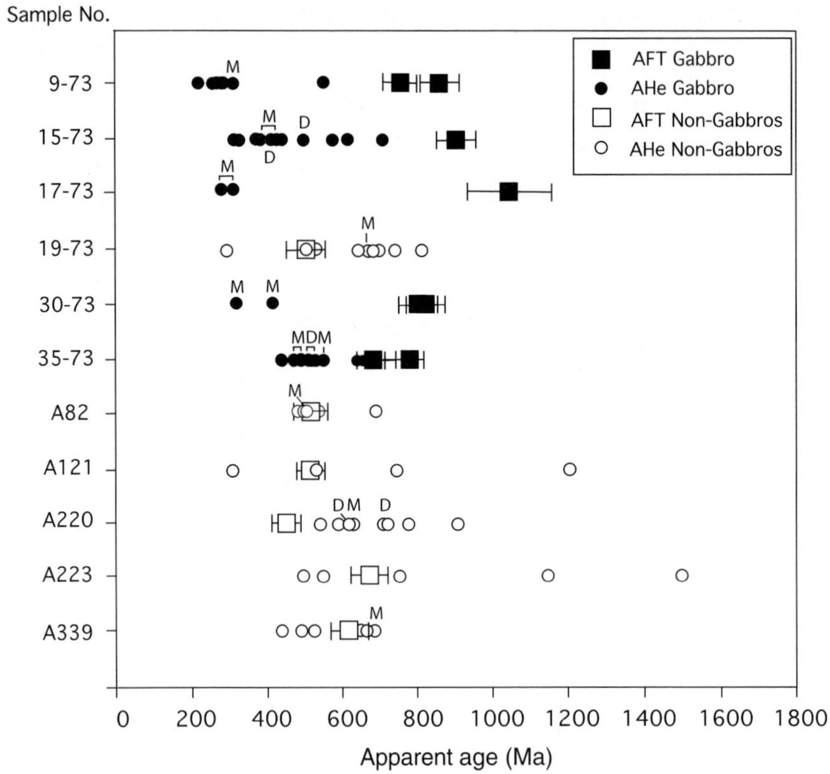

Fig. 3. Relationship between AFT and AHe ages from the same aliquot for all samples studied. Unless specified otherwise AHe ages are for single grains; M, mean weighted age for multigrains; D, mean weighted age for multiple grains dissolved using a double dissolution protocol (see the text for further details).

caution. Comparison of these ages with sample 17-73, located only about 4 km to the SE in the same gabbroic body and for which two multigrain analyses (with similar grain radii) yield a weighted mean of 283 ± 26 Ma, are markedly discordant. In the northwestern part of the study area, sample 35-73, which contains the highest U and Th content measured in any of the gabbroic apatites studied, also yields the oldest weighted mean ages of 485 ± 30 Ma for multigrain analyses and 508 ± 39 Ma for single-grain analyses. Again, the apparent concordance of the weighted mean ages does not, in reality, reflect the scatter in single-grain age data. Lehtovaara (1976) described this sample from the Ylivieska gabbro as being located at the contact between the gabbro and quartz diorite and the granodiorite country rock, and intimated that this could account for the higher apatite uranium content. Two multigrain AHe ages from sample 30-73, located approximately 4 km to the NE of sample 35-73 within the same intrusive body, yield a significantly younger weighted mean age of 357 ± 39 Ma, and have a considerably lower U and Th content.

Interpretation and discussion

Apatite fission-track data

Apparent AFT ages plotted against their mean HCTL are shown in Figure 4. A correlation with sample lithology is suggested: all gabbroic samples are characterized by older apparent ages and longer mean HCTL. This boomerang plot can be interpreted following Green (1986), in showing a general decrease of mean HCTL with decreasing age, displaying a partial 'boomerang' only, with one sample (A-602) showing an increase in mean HCTL compared to the general trend. These results suggest that most samples experienced partial, but not total, track annealing during Palaeozoic time. This is consistent with sampling at the present erosion level, being close (<*c.* 500 m) to a sub-Cambrian peneplain (Lorencak 2003; Murrell & Andriessen 2004), and with $^{40}Ar/^{39}Ar$ K-feldspar thermochronometry, which shows that outcrops of the study area were last at temperatures of >150 °C during the Mesoproterozoic (Murrell 2003).

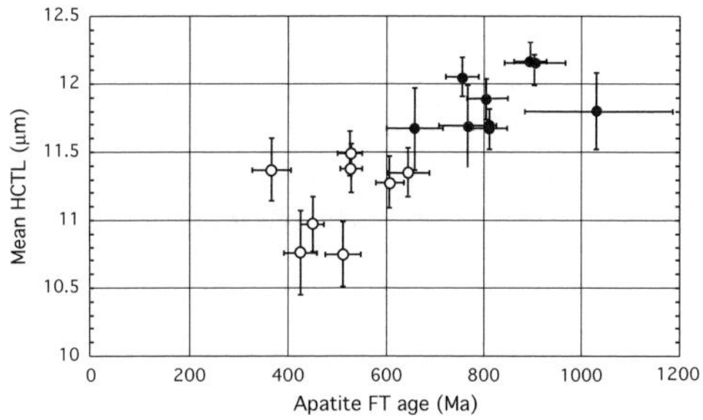

Fig. 4. Boomerang plot for samples from southern Finland showing the relationship between apparent apatite fission-track age and their mean horizontal confined track length (TINTs). Only samples with more than six grains were counted and 40 lengths measured are considered in this plot. Open circles represent non-gabbros; solid circles represent gabbros.

The wide range of AFT ages can be explained by a combination of factors. The granitic lithologies show a younging age trend towards the south and west. Given the absence of significant structures that might have experienced Phanerozoic reactivation and the unlikely possibility of large variations in sedimentary overburden over the study area, Murrell & Andriessen (2004) attributed this pattern to lateral variations in heat flow within the crystalline basement (based on data of Kukkonen 1989, 1993), influencing the degree of partial annealing experienced by these samples. Other samples yielding apparent Mesoproterozoic AFT ages (some of the oldest reported from anywhere in the world) are invariably from gabbros, an observation also evident in the data of Murrell & Andriessen (2004). Although variations in thermal conductivity or the relatively low heat production in gabbros may also have influenced annealing behaviour, it is well established that the rate of fission-track annealing in apatite varies with chemical composition (particularly chlorine content) (Gleadow & Duddy 1981; Green *et al.* 1985; Barbarand *et al.* 2003; Green & Duddy 2006) and related mineralogical properties (Carlson *et al.* 1999). Although chlorine substitution probably exerts the most important effect, the possible influence of other trace elements (including rare earths) has also been reported (Barbarand *et al.* 2003). Differences in chlorine content will produce variations in the degree of age reduction of apatites and we attribute this effect to the older ages observed in the gabbros, which also display a higher chlorine content (Fig. 2), and also to increased apatite solubility, expressed by the track etching rate and etch pit size (D_{par}) (e.g. Donelick 1993; Barbarand *et al.* 2003) as observed by Murrell (2003). Although the variation in chlorine content is relatively low, it has a strong effect on the annealing history of our samples, which are thought to have experienced some reburial into the partial annealing zone by Caledonian Foreland basin sediments.

Time–temperature models presented by Lorencak (2003), Murrell (2003) and Murrell & Andriessen (2004) relate this residence to heating (*c.* 60–80 °C), attributed to a transient Early Palaeozoic episode of burial by sediments deposited in a broad foreland basin formed as a flexural response to topographic loading adjacent to the Caledonian orogeny. This explanation is in accord with studies reported by several workers from elsewhere in the Fennoscandia shield (e.g. Zeck *et al.* 1988; Nikishin *et al.* 1996; Samuelsson & Middleton 1998; Larson & Tullborg 1998; Larson *et al.* 1999, 2006; Cederbom *et al.* 2000; Cederbom 2001; Lorencak 2003; Huigen & Andriessen 2004; Murrell & Andriessen 2004; Kohonen & Rämö 2005). Furthermore, the AFT data show a low proportion of long tracks in their HCTL distributions, and this is in accordance with published thermal history models for eastern Fennoscandia, which suggest that final cooling (from *c.* 60 to 65 °C) occurred relatively late (Late Cretaceous–Early Tertiary) in the cooling history (Lorencak 2003; Murrell & Andriessen 2004). A geological history involving burial under a 'deep, extensive foreland basin' cover has been rejected by Hendriks & Redfield (2005, 2006) in favour of an interpretation incorporating postulated long-term REA. They also suggest that the AHe data available from the eastern Fennoscandia shield, all based on multigrain analyses reported by Lorencak (2003) and Murrell (2003), appear to be more compatible with the absence of a thick sedimentary cover and that samples have resided close to the present surface for a period of several hundreds of million years.

Apatite (U–Th–Sm)/He data

As indicated previously, much of the AHe data show a wide age dispersion and poor reproducibility. Low reliability is observed particularly in the non-gabbro samples, where mean weighted multigrain and many single-grain ages are older than or concordant (within error limits) with their coexisting AFT ages (Table 2 and Fig. 3). Further, many AHe ages predate the Early Palaeozoic Caledonian orogeny and the postulated foreland basin sedimentation, which is presumed from previously published AFT models to be associated with the timing of maximum Phanerozoic palaeotemperatures. There are several factors that may contribute to low reproducibility of AHe or 'too old' ages (particularly when compared to their coexisting AFT ages). This is especially the case in apatites that have experienced slow cooling, through either reheating or lengthy residence at temperatures within the AHe partial retention zone (*c.* 40–80 °C) during which He retentivity changes rapidly as a function of temperature (e.g. Reiners & Farley 2001; Farley & Stockli 2002; Fitzgerald *et al.* 2006; Hansen & Reiners 2006; Shuster *et al.* 2006).

One possible explanation for erroneously old ages is the presence of U- and Th-rich micro-inclusions, such as zircon or monazite (e.g. Farley *et al.* 1996; Farley 2002; Farley & Stockli 2002), which are not dissolved by the HNO_3 protocol, thereby resulting in 'parentless' He in the analysis. Such inclusions were noted in many granitic apatites, but were avoided during grain selection. An approach to account for all parent isotopes in apatite grains was applied with some success by McInnes *et al.* (1999) using HF to dissolve zircon and xenotime inclusions, and also more recently by Vermeesch *et al.* (2007) to dissolve zircon inclusions. In order to evaluate the possible impact of micro-inclusions not visible under binocular microscope examination in our samples, we dissolved three samples (gabbros 15-73 and 35-73, and granite A-220), using a procedure that we here term 'double dissolution' (DD). Following routine He degassing, DD was carried out at the former CSIRO, North Ryde, Sydney laboratory essentially following a protocol described by Evans *et al.* (2005, method 1, p. 1162) (see also Table 2), which dissolves apatite as well as the more refractory actinide-bearing minerals (including monazite). The results for duplicate DD analyses on multigrains are summarized in Table 2 and Figure 3. For sample 15-73 one of the analyses falls in the same age range as the weighted mean multigrain age, while a second yields a higher age. Samples 35-73 and A-220 replicate within error limits of their weighted mean multigrain ages despite their relatively small grain radii, and the significantly higher U and Th concentrations for the second DD analysis of A-220. Overall, the DD results do not show a significant reduction or improved replication of ages compared to the weighted mean multigrain ages for aliquots dissolved using the HNO_3 procedure. This suggests that apparently old AHe ages in our samples are not due to the presence of parentless He originating from micro-inclusions. It is acknowledged, however, that even if problematic micro-inclusions were present, the DD protocol may not universally explain excess He problems because their presence and distribution could complicate α-ejection corrections (Lorencak 2003; Vermeesch *et al.* 2007).

Older AHe ages for the multigrain analyses, particularly in samples where U and Th concentrations are low, could also be attributed to the presence of otherwise unaccounted ^{4}He produced from ^{147}Sm. However, Sm was analysed in the single-grain runs, and its maximum contribution to the ^{4}He content is estimated at 1.6%, although values are typically $<1\%$. This is also reflected in the general concordance between multigrain (no Sm analysis) and single-grain (with Sm analysis) weighted mean AHe ages from most samples (Table 2 and Fig. 3).

Several authors have described the effects of U and Th zonation on AHe ages (e.g. Farley 2002; Farley & Stockli 2002; Meesters & Dunai 2002*b*). In our samples a first-order assessment of the distribution of U and possible zoning was made from the mica detectors used for AFT dating. Bearing in mind that this does not cover the specific grains chosen for AHe analysis, it nevertheless provides a general indication, except perhaps in cases where U content is very low. Overall, the U distribution is homogeneous in all of the gabbros studied, whereas varying degrees of U zonation are apparent in many of the granitic–granodioritic apatites, some of which show extreme variability between grains within the same sample, e.g. A121 and A220 (Lorencak 2003). We examined the homogeneity of U, Th and Sm within aliquots of the samples used for the DD experiment by laser ablation-inductively coupled plasma mass spectroscopy (LA-ICPMS) analysis with a 20 μm-diameter laser beam on core and rim areas of grains (Fig. 5). Gabbro 15-73 shows a relatively low and uniform U and Th distribution. Gabbro 35-73 is noteworthy for its high U and, in particular, Th content, but nevertheless the distribution is relatively uniform, with the exception of a U- and Sm-rich rim in grain 2. In granodiorite A220, most grains exhibit higher U and Sm content at their rims, but U zonation can vary by a factor of 2, whereas Th content is uniformly low. Depending on the style and degree of U and Th zonation, the maximum error introduced for an extreme case is $\pm 33\%$, but in most cases is likely to be considerably less (Farley 2002). In our samples any U and Th zonation influencing the observed AHe age dispersion is likely to be more significant in the non-gabbro apatites.

A further consideration for explaining AHe age dispersion is whether there is any observable age correlation with grain size (e.g. Reiners & Farley 2001). Following Meesters & Dunai (2002*a*) we calculated the radii of spheres with equivalent surface-area-to-volume ratios (R_s), based on measurements of grain dimensions (see Table 2) following a procedure described by Hansen & Reiners (2006). Such geometries provide a more realistic and sensitive assessment of spherical diffusion volume for He diffusion models in apatite. No clear systematic relationship is apparent between (R_s) and (U–Th–Sm)/He age for our samples, not even in sample 9-73, which shows the most reproducible AHe ages. It is noted, however, for this sample

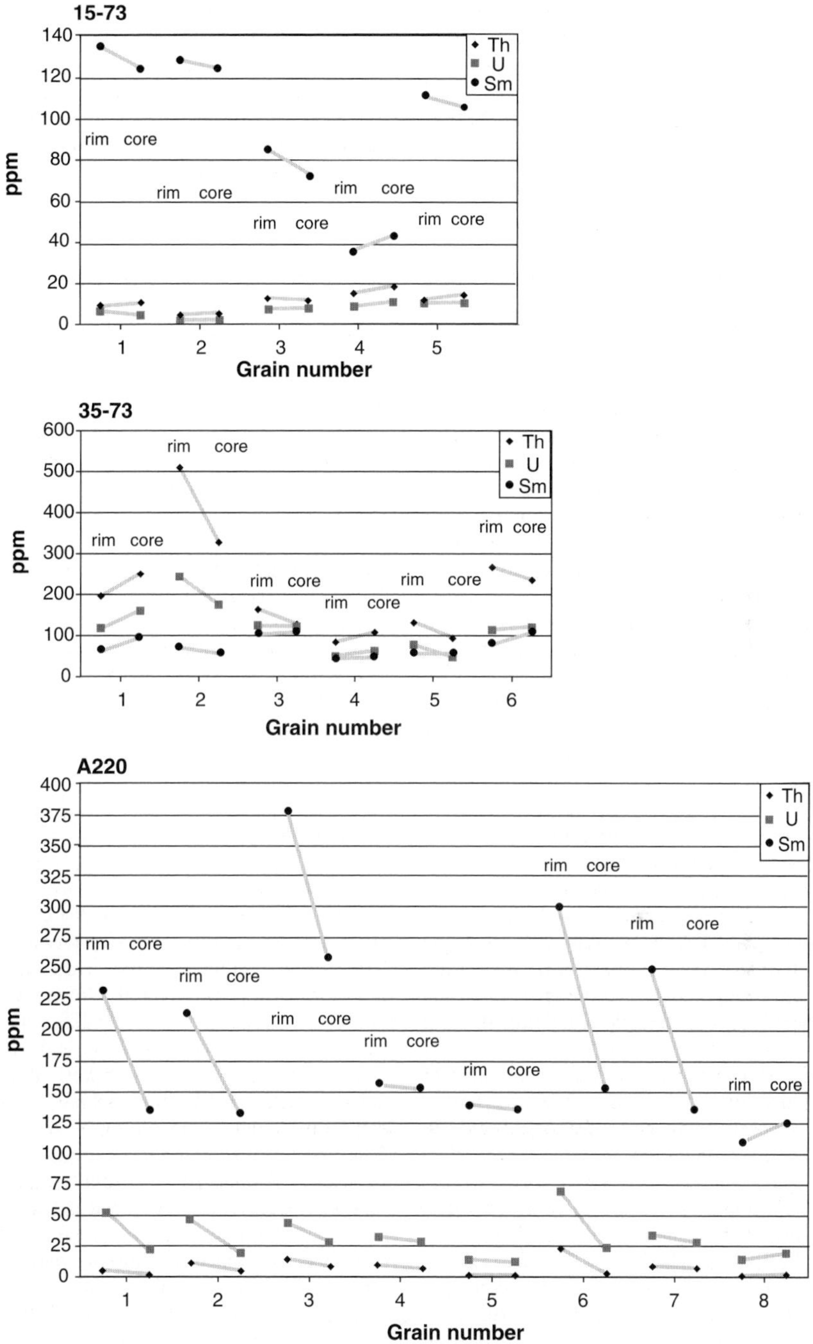

Fig. 5. Summary plot showing the distribution of U, Th and Sm for each of the apatite grains from southern Finland targeted for LA-ICPMS analysis. Each grain shown is represented by an analysis at the rim and core. For easier visualization of trends, each grain pair is joined with a tie line. See the text for further discussion.

and also for A82 and A121 that the highest R_s values coincide with the oldest AHe ages, but this relationship does not hold for other samples, and in some cases samples with large differences in R_s yield similar AHe ages. The above observations appear to rule out a significant grain-size effect on our samples.

Dunai (2005) observed that where the thermal history of samples involves lengthy residence in the He partial retention zone, the effect of α-ejection on He diffusion can result in ages that are too old by 10–20% compared to ages without α-ejection. In the case of our samples even a severe overcorrection by up to 20% would still be unable to account for the poor reproducibility of much of our AHe data. Furthermore, many AHe ages, particularly in granitic samples, would still remain older than their coexisting AFT ages, which would be inconsistent with predictions based on the conventional understanding of the relevant diffusion and annealing kinetics.

Farley (2000) reported step heating He diffusion experiments on Durango apatite, and observed that an apparent increase in He retentivity (and closure temperature) at higher temperatures may be related to defects induced by α-radiation damage. Experimental data recently reported by Shuster *et al.* (2006) test Farley's observation in further detail and strongly suggest that the accumulation of α-radiation damage in apatite, results in the formation of 'traps' that impede He diffusion. These results imply that He retentivity and the effective He diffusion kinetics are functions that evolve with time. Modelling shows that this effect will be particularly evident as a positive correlation between AHe ages with ^{4}He concentration and effective U concentration [eU] (which weights decay of the parents in terms of their He productivity) within samples with a range of grain [eU]. Provided that such samples have experienced a lengthy thermal history, thus allowing their He diffusion kinetics (and closure temperatures) to diverge between different grains prior to cooling or partial resetting of the He clocks, a range of AHe ages between grains will result even though they have all experienced the same thermal history (see also Flowers *et al.* 2007). Samples subjected to reheating after accumulation of extensive radiation damage as suggested by the thermal history of our samples would be predicted to be more retentive than previously thought. Shuster *et al.* (2006) emphasized, however, that the trapping model will not necessarily result in such AHe ages being significantly older than their coexisting AFT ages and does not necessarily provide a ubiquitous solution to cases of the AFT $<$ AHe age relationship.

Plots of [eU] v. single-grain AHe ages for our southern Finland samples are shown in Figure 6a (gabbros) and b (granites and granodiorites).

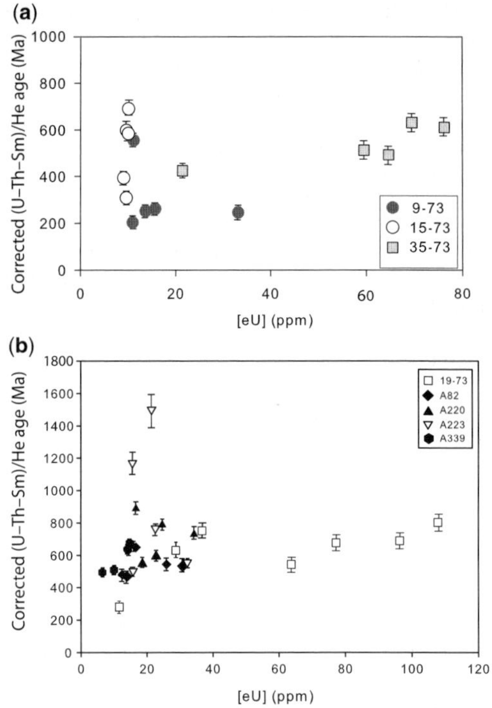

Fig. 6. AHe age ($\pm 1\sigma$) v. effective U concentration [eU], which takes into account the decay of the two parents for their α-productivity and is calculated as $^{238}U_{ICP\text{-}MS}$ (ppm) + 0.235 $Th_{ICP\text{-}MS}$ (ppm) from southern Finland for: (**a**) gabbros and (**b**) non-gabbros.

Samples 19-73 and 35-73, which have the greatest range of [eU], show a positive correlation with AHe age, as predicted by the 'trapping' model. No such correlation is apparent, however, for other samples, e.g. 15-73, A220, A223 and A339, for which [eU] is relatively low and restricted in range. These samples, however, show a wide range of ages, suggesting that factors other than radiation damage are required to explain their AHe age dispersion. This is especially apparent in samples 15-73 and 17-73, which are both from the Parikkala gabbro intrusion, and as they are separated by only about 4 km they would be expected, therefore, to share a similar thermal history; sample 15-73, nevertheless, yields older AHe ages despite having lower [eU] apatite (Table 2).

Whereas AHe age dispersion for two of our samples with a broad range of [eU] can be explained by α-radiation damage causing trapping, further analyses of these samples would be required before attempting to extract their thermal histories using the trapping model. Other samples appear to require different explanations to account for their age dispersion, including U and Th zonation and possibly ^{4}He implantation from external sources (not discussed here in any detail, but see Fitzgerald *et al.* 2006, table 3).

We therefore conclude that AHe data from our samples are of limited value in evaluating thermal histories in the eastern Fennoscandia shield, and need to be treated with the utmost caution when comparing them with coexisting AFT data in support of arguments for REA. Vermeesch *et al.* (2007) proposed the use of the 'pooled age' as the best way to compare multi- and single-grain ages or to compare two sets of multigrain ages with one another. However, the broad dispersion of single-grain ages for most of our samples calls into question the value of using weighted mean AHe ages on multigrain aliquots. Such ages only average the spread of single-grain ages apparent in most samples and will tend to be biased by high U, Th, Sm or He content, which may reflect different factors (e.g. radiation damage, zonation or He implantation) operating within anomalous grains and between them, thus biasing the result towards the 'too old' ages. In such cases, it is often at the younger end of the single-grain variation where the meaningful ages are likely to found, a finding also suggested by Fitzgerald *et al.* (2006). This is particularly the case for our gabbro samples, which show the least age dispersion and ages consistently lower than their AFT ages. This finding is also consistent with data from mafic and ultramafic rocks reported by Kohn *et al.* (2002*a*) and Lorencak *et al.* (2004). After excluding one data outlier, gabbro sample 9-73 with a restricted range of low [eU] apatite yields the most reproducible AHe data for any of our samples and below we evaluate its thermal history together with its coexisting AFT data.

Thermal history modelling

The AFT data presented here have been modelled using the AFTSolve multi-kinetic fission-track annealing model (Ketcham *et al.* 2000) (Fig. 7). The models presented differ from those in Lorencak (2003), which were based on the MonteTrax software (Gallagher 1995) in that they also incorporate the fission-track kinetic annealing indicator Cl (wt%) data (see Table 1 and Fig. 2), and are also based on new AFT data for samples 15-73 and 35-73. Potential models were run using 15 000 Monte Carlo iterations, and the main constraints imposed were to account for the presence of the sub-Cambrian peneplain across the study area and the present-day exposure at a mean annual surface temperature of 5 ± 5 °C (Murrell 2003). Shaded regions mark envelopes of statistically acceptable fits between observed and modelled fission-track parameters – the wider envelope contains paths with a merit function of at least 0.05 ('acceptable' fit) and a narrower envelope containing paths with a merit function of at least 0.5 ('good' fit), which can be considered as the limit of statistical precision (Ketcham *et al.* 2000). Only sample 9-73 yielded suitable data for both AHe and AFT that were modelled together using HeFTy (Ketcham 2005), whereas all other models are based on AFT data only. Best-fit models, including those for sample 9-73, suggest a phase of mid-Paleozoic heating to temperatures of approximately 60–80 °C, as described by previous AFT models. Further, the models serve to emphasize that southern Finland samples experienced a lengthy history of slow cooling and residence in the partial annealing zone. This is not consistent with a scenario in which samples reside close to the present surface over hundreds of millions of years as invoked by REA.

Radiation enhanced annealing

The case for REA being factored into apatite fission-track studies due to α-damage resulting principally from the decay of ^{238}U and, to a lesser extent, ^{232}Th, ^{235}U and ^{147}Sm has been outlined by Hendriks & Redfield (2005, 2006). This has been partly argued on the basis of experiments related to nuclear waste disposal showing α-particle emission lattice defect recovery (annealing) in apatite, with an efficiency based on dose rate and time (e.g. Chaumont *et al.* 2002; Ewing & Wang 2002). Hendriks & Redfield further argued that α-annealing is probably the dominant process of lattice recovery in apatite at temperatures of less than about 60 °C, where thermal annealing is less significant. Hence, for apatites from southern Finland, which they suggest have resided at near-surface temperatures over a timescale of several 10^8 years (and lack a dominant population of long horizontal confined track lengths), they proposed REA as an alternative to thermal history models invoking sedimentary burial or variable heat flow as advocated by Lorencak (2003), Murrell (2003) and Murrell & Andriessen (2004).

Central to the argument for REA of fission tracks in apatite are data plots showing an inverse correlation between uranium content and either AFT age or mean track length. To test this proposal further we constructed plots with polynomial linear regression lines and correlation coefficients calculated on new and old AFT data from southern Finland (Fig. 8), and new data from the southern Canadian and Western Australian shields (Figs 9 & 10, respectively). In selecting acceptable AFT data for our plots we used the same criteria as applied by Hendriks & Redfield (2005). In our plots we distinguish between gabbro and

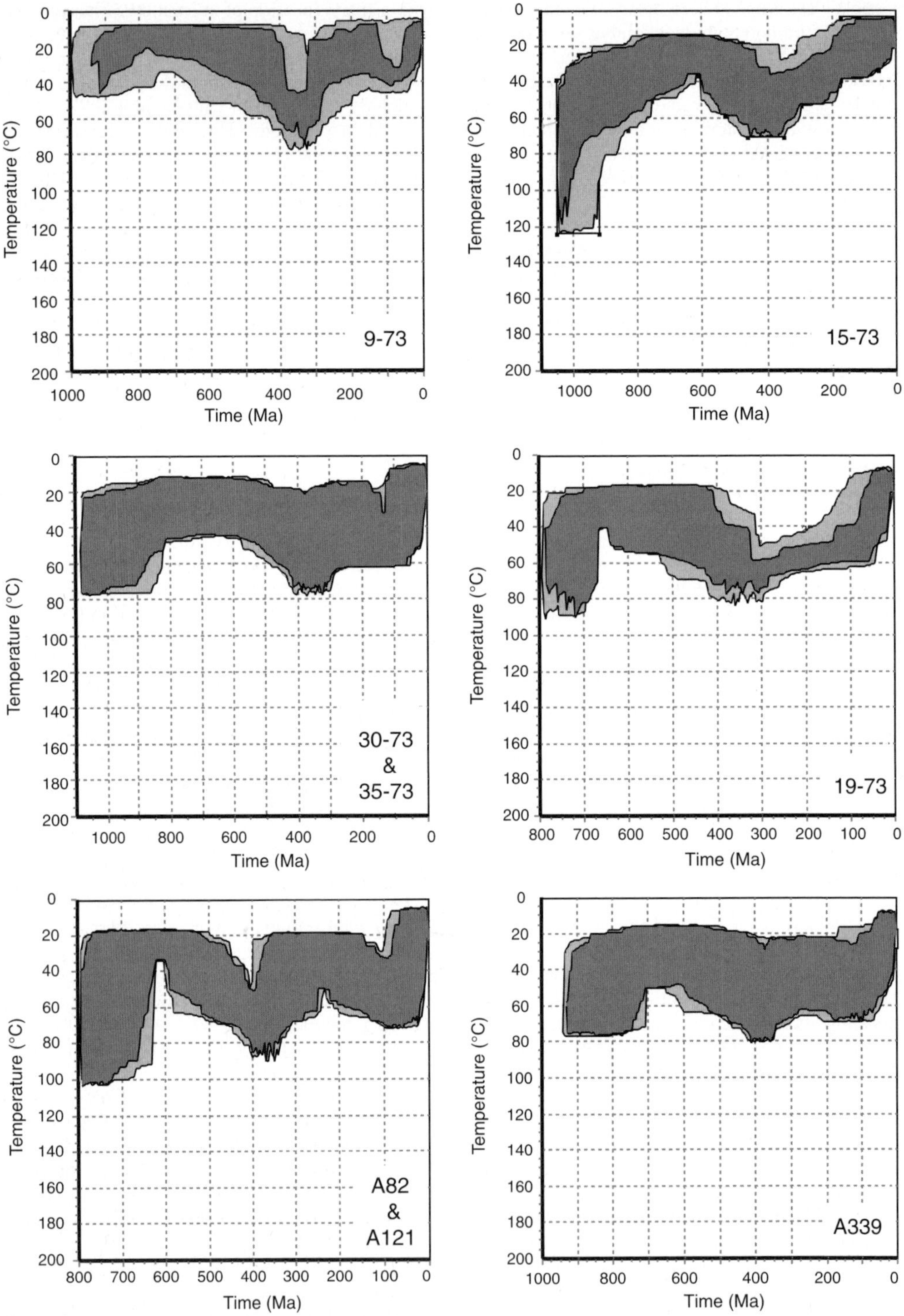

Fig. 7. Summary of time–temperature inversion models for representative southern Finland samples (all with 12 or more grains counted and 70 lengths measured) using the multi-kinetic fission-track annealing model HeFTy (Ketcham 2005). See the text for further discussion.

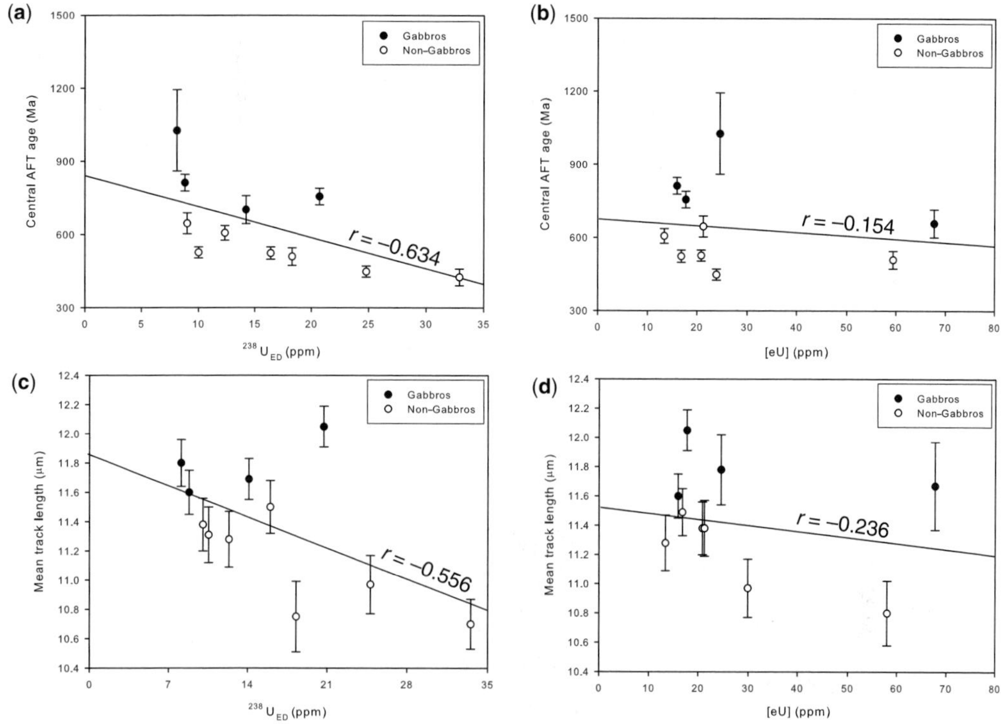

Fig. 8. Central apatite fission-track ages (**a**, **b**) and mean horizontal track lengths (TINTs) (**c**, **d**) from southern Finland plotted against ^{238}U concentration determined by the external detector method (using a dosimeter Corning-5 glass standard with a value of 12.5 ppm) and [eU] (ppm) (determined by ICP-MS) with polynomial linear regression lines and correlation coefficients shown for each plot. Gabbro and non-gabbro data are plotted separately.

non-gabbro samples, following Hendriks & Redfield (2005), but in some cases there are too few gabbro samples to justify separate calculations, so the regression lines shown are derived for all samples shown in each plot. We also include separate plots using the effective U concentration [eU], which weights more accurately the decay of the parents in terms of their α-productivity. The U and Th data used in the plots arise from AHe analyses on apatite aliquots from the same concentrates as used for AFT determinations, but clearly not on exactly the same grains.

Our plots show the negative trends with moderate correlation for uranium content v. AFT age or mean track length (Fig. 8a, c) reported from the eastern Fennoscandia shield (southern Finland) by Hendriks & Redfield (2005). When plotted against the effective U concentration [eU], however, only a weak negative correlation is apparent (Fig. 8b, d). Similar plots for samples from the southern Canadian and Western Australian shields, using more data than available for southern Finland, however, do not show any consistent trends and correlations are weak, suggesting that REA does not need to be factored into interpretations of their thermal history (Figs 9 & 10).

For some samples from southern Finland (35-73, 17-73 and A223), a bias is introduced when U determined by the external detector method is compared to that factored in to the value for [eU]. This is most evident in gabbro 35-73 where most of the apatites have a high density of spontaneous tracks that are not countable, hence only low U grains were chosen for the AFT determination. Such grains constitute less than 5% of the apatite sample, whereas for ICP-MS determinations (and picking of grains for AHe measurement) from the same concentrate no such bias in grain selection was introduced, and the U and Th measurements more closely reflect the chemistry of the entire apatite population (e.g. compare U content for sample 35-75 in Tables 1 & 2). Hence, the [eU] determined is not directly related to the accompanying AFT parameters in Figure 8b and d. If sample 35-73 is removed from these figures, then the correlation remains weak for Figure 8b ($r = -0.241$), but moderate for Figure 8d ($r = -0.678$) and is strongly influenced by one sample (19-73) with high [eU]. In most of

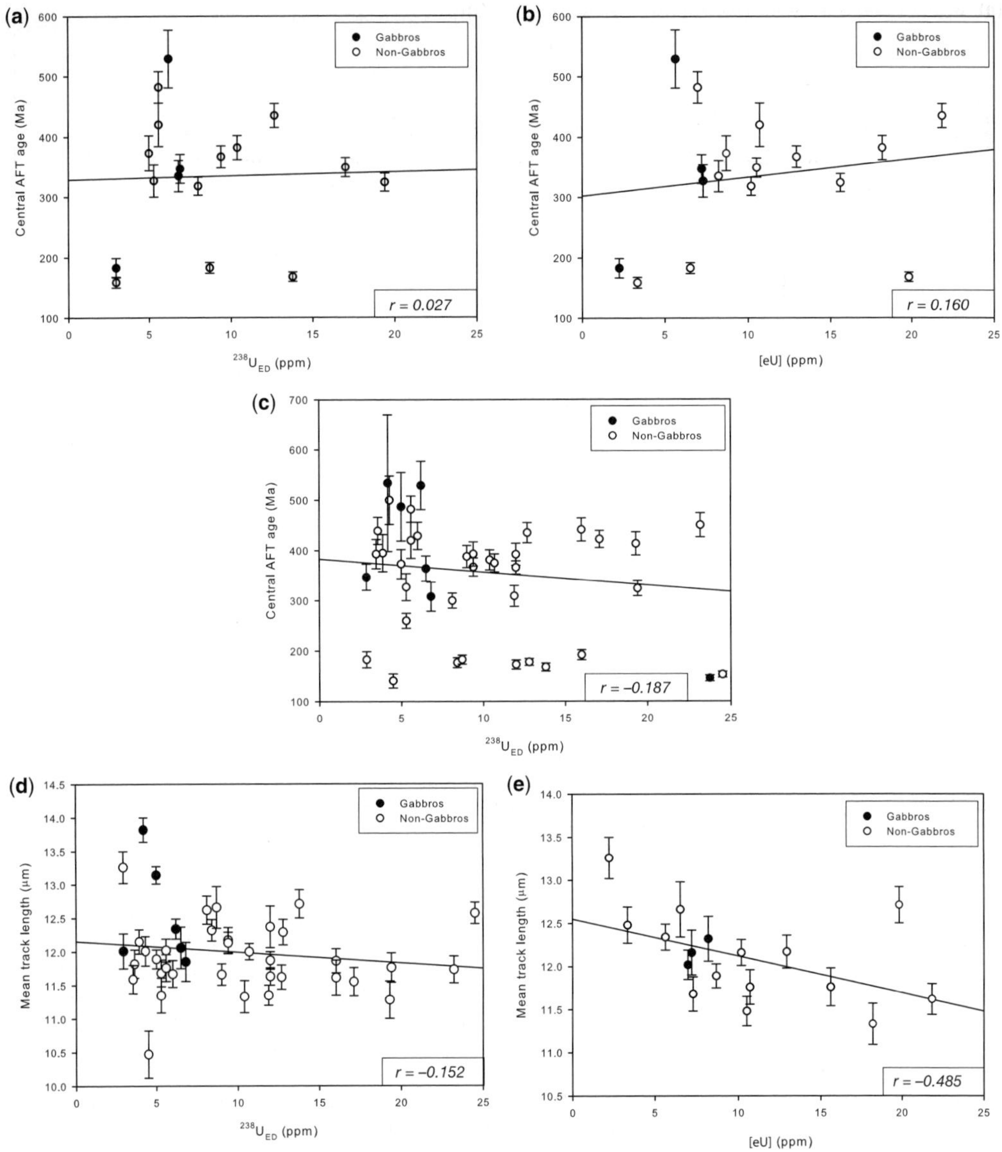

Fig. 9. Central apatite fission-track ages (**a–c**) and mean horizontal track lengths (TINTs) (**d, e**) from the southern Canada Superior shield in Ontario and eastern Manitoba plotted against ^{238}U concentration determined by the external detector method (using dosimeter Corning-5 glass standard with a value of 12.5 ppm) and [eU] (ppm) (determined by ICP-MS). Polynomial linear regression lines and correlation coefficients (at bottom right) are shown for each plot. Plots (a), (b) and (e) compare data from the same samples ($n = 17$), whereas plots (c) and (d) show data for a further 24 samples for which ICP-MS U and Th data from the same shield area are not available.

our samples, though, the grain-selection bias is not significant, as can be seen in the relatively strong correlation of ^{238}U content determined by ICP-MS compared to ^{238}U content determined by the external detector method in aliquots of the same apatite concentrate, particularly in the case of samples from the southern Canadian and Western Australian shields (Fig. 11). The relationship is more scattered for some of the southern Finland samples (for which there are fewer data) due to the non-random selection of some grains for AFT determination as outlined earlier, with ICP-MS U values being higher, particularly for those samples yielding old AFT ages. Of further interest in Figure 8b is that

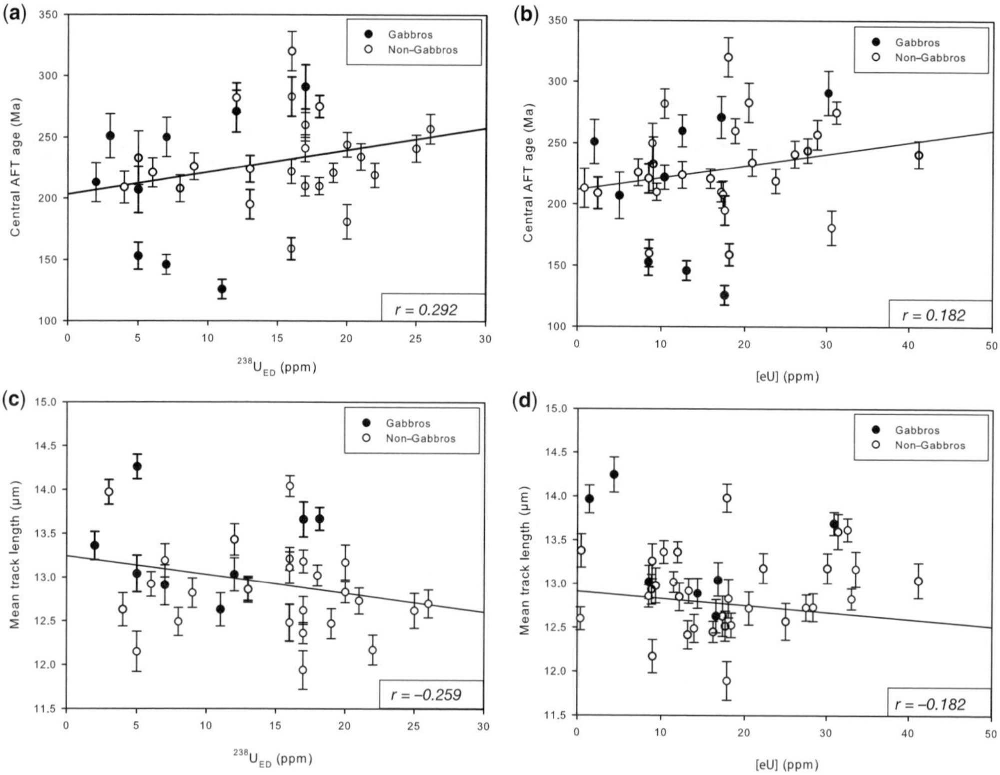

Fig. 10. Central apatite fission-track ages (**a**, **b**) and mean horizontal track lengths (TINTs) (**c**, **d**) from the Western Australia shield plotted against ^{238}U concentration determined by the external detector method (using dosimeter Corning-5 glass standard with a value of 12.5 ppm) and [eU] (ppm) (determined by ICP-MS). Polynomial linear regression lines and correlation coefficients (bottom right) are shown for each plot. See the text for further discussion.

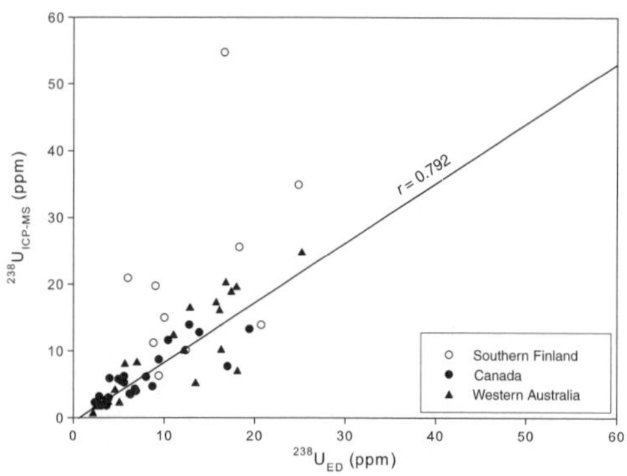

Fig. 11. Plot of ^{238}U content determined by ICP-MS v. ^{238}U content determined by the external detector method on apatites from samples studied in the southern Finland, southern Canada and Western Australia shields. Determinations were carried out on apatite from the same aliquots for each sample, but not on the same grains. Correlation is moderate to very good for most samples, but tends to be weaker for the southern Finland samples. See text for further discussion.

sample 19-73 (Rapakivi granite), which shows high [eU] (due to a particularly high Th content), yields an older AFT age (510 ± 36 Ma) than a proximal low [eU] granodiorite sample (e.g. A220, 449 ± 23 Ma). This is contrary to the relationships predicted from REA.

In Figure 12 we present plots of the spontaneous/induced track ratio (ρ_s/ρ_i) from which fission-track ages are calculated v. ^{238}U content determined by the external detector method for single apatite grains in representative southern Finland samples. Hendriks & Redfield (2005, 2006) noted that these correlations might not be rigorous because of the dependency between the two variables. However, because of the reasonably good agreement between U content derived from both the external detector and ICP-MS methods (Fig. 11) (except for sample 35-73 as outlined earlier), we believe most of the correlations are independent. This is particularly evident for samples showing the strongest negative correlations, which are found in gabbroic apatites (samples 9-73, 15-73 and 30-73; c.f. U data in

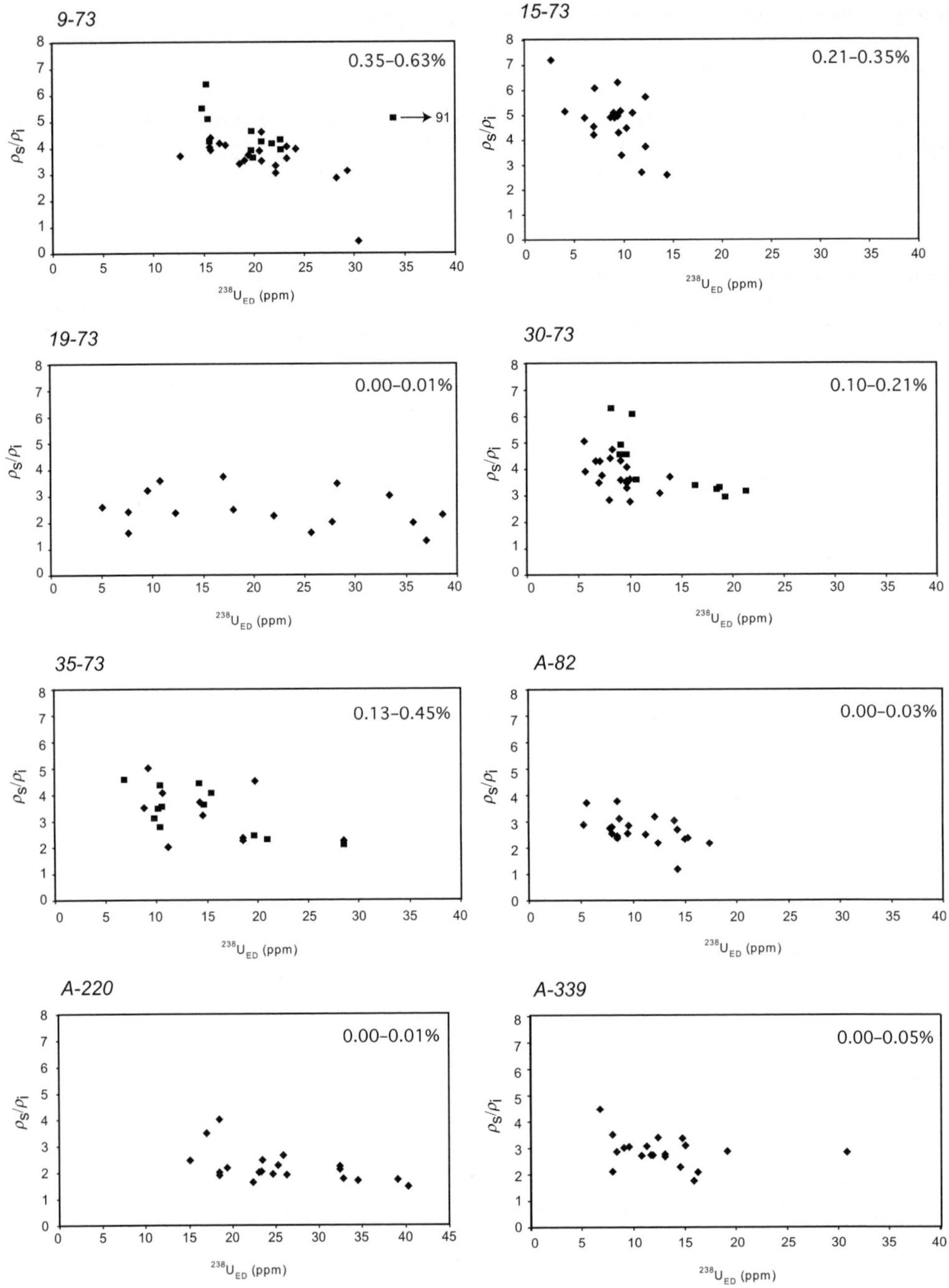

Fig. 12. Plots of the spontaneous/induced track ratio (ρ_s/ρ_i) from which fission-track ages are calculated v. ^{238}U content determined by the external detector method for single apatite grains in representative samples studied in southern Finland. Range of Cl content in apatite for each sample is shown in the top right of each panel.

Tables 1 & 2), and which also have the highest Cl content (also shown). Conversely, correlations are either weak or absent for Cl-poor apatites from non-gabbroic rocks. Note also that for sample 9-73 one grain has an approximately three–sixfold higher U content than other grains, but does not show a markedly different age, contrary to expectations from REA. This can also be seen in other samples from southern Finland, such as 15-73, 30-73, A-82, A-121, A-220 and A-339, which all have at least a threefold range in U content between different grains within a sample (as well as relatively low Th content, see Table 2) and a limited range of Cl content (Table 1), yet all yield values of chi-square (χ^2) >5% (Table 1). This again indicates that there cannot be any significant effect of uranium content on the AFT ages.

AFT ages of more than 450 Ma, which are widespread in the eastern Fennoscandian shield, generally exceed those found on other cratons studied to date, suggesting that samples in this region record an exceptionally lengthy low-temperature thermal history and prolonged residence in the partial annealing zone. Those few samples, in this study and previously published (Murrell 2003), that have AFT ages of <450 Ma are confined to southernmost Finland, in an area where some Palaeozoic transgression may be inferred, on the basis of preserved sedimentary sequences flanking the southern margin of the shield. Moreover, the crystalline basement in this southernmost part of Finland includes high-heat production granites (Kukkonen 1989, 1993), which might lead to enhanced heat flow if buried beneath low-conductivity sediments, as also proposed by Murrell & Andriessen (2004). Calcite–galena–fluorite veins in this area, and in the Palaeozoic basin sequences on the flanking basins, have also yielded Palaeozoic Pb/Pb and Sm–Nd model ages (Sundblad *et al.* 2004). As is evident from Figure 1, Palaeozoic AFT ages are recorded by samples A602, A220 and A596. Within the area circumscribed by these samples, gabbroic samples 9-73, 15-73, 17-73, nevertheless, record considerably older ages. We suggest in this context that despite a thermal overprint during the early to mid-Palaeozoic time, Cl content has had a strong affect on the annealing history of these apatites, particularly on those from gabbros.

Although we find no strong evidence for systematic REA in the cratons investigated, our work invites further experimental study to further evaluate the existence of such a possible effect, particularly in cratonic rocks or in high U- and Th-bearing apatites from different geological environments. This can be achieved through AFT determinations of single grains using the ICP-MS approach of Hasebe *et al.* (2004) in samples showing a wide range of [eU] located within close geographical proximity and sharing a similar thermal history. Preliminary AFT results reported by Kohn *et al.* (2008) using this approach on crystalline rocks in the northeastern Brazil and eastern Fennoscandia shields show no clear correlation between AFT age and [eU] or REE, and do not support an observable REA effect.

Conclusions

New AHe data from the southern Finland shield show variable dispersion of single-grain ages. The range is biased on different factors operating within grains, which generally tend to give a greater weighting towards older age outliers. AHe ages from mafic rock types show the least dispersion and tend to be consistently lower than their coexisting AFT ages. In general, it is at the younger end of the single-grain variation range of such rock types where meaningful AHe ages can be found. AHe data from multigrain aliquots are, therefore, of limited value for evaluating thermal histories in the southern Finland shield, and need to be treated with caution when used as a yardstick for comparison with coexisting AFT data in support of arguments for REA.

The relationship between U content v. AFT central age and mean track length for samples from two classic cratons (southern Canada and Western Australia) show weakly positive or negative correlations, unlike the consistent moderately strong negative trends of the kind previously reported from the southern Finland shield and used to support arguments for REA. Such weak correlation is also the case for plots from both shields, as well as those from southern Finland, where AFT parameters are plotted against effective U concentration [eU] (based on U and Th content determined by ICP-MS), which weights decay of the parents more accurately in terms of their α-productivity. Further, samples from southern Finland yield values of χ^2 >5%, indicating that there is no significant effect of the range of uranium content between grains within a sample on the AFT ages, and that they are all consistent with a single population.

The eastern Fennoscandia shield, where AFT ages are typically in excess of 450 Ma and usually exceed those found on most other cratons, is an example of a region where samples record a lengthy low-temperature thermal history including reb-urial by foreland basin sediments and subsequent cooling. We suggest that in this context a moderate range of Cl content has a strong affect on the annealing history of these apatites, particularly on gabbros, which yield the oldest ages. Our data, therefore, do not show any discernible effects that could be attributed to REA in the cratonic environments studied.

In the eastern Fennoscandia shield and elsewhere in cratonic environments, the study of apatite from low U and Th rocks, with relatively low levels of α-radiation damage, may provide the most practical approach for producing reliable results for both the AFT and AHe thermochronometry methods.

This work was supported by grants from the Australian Institute of Nuclear Science and Engineering, and the Australian Research Council. We are grateful to J. Lehtovaara for supplying apatite splits from some of his samples; P. Crowhurst, N. Evans and K. Blacklock (CSIRO) for guidance on the dissolution experiments and other aspects of chemistry for the early (U–Th)/He analyses; J. Woodhead (University of Melbourne) for help in analysing U and Th for some of the earliest (U–Th)/He determinations; and M. Norman (ANU) for his invaluable assistance with the LA-ICPMS analyses. A. Blythe, P. Green and an anonymous reviewer provided constructive criticism and comment on an earlier draft of the manuscript.

References

BARBARAND, J., CARTER, A., WOOD, I. & HURFORD, A. 2003. Compositional and structural control of fission-track annealing in apatite. *Chemical Geology*, **198**, 107–137.

BELTON, D. X., BROWN, R. W., KOHN, B. P., FINK, D. & FARLEY, K. A. 2004. Quantitative resolution of the debate over antiquity of central Australian landscapes and implications for the tectonic and geomorphic stability of cratonic interiors. *Earth and Planetary Science Letters*, **219**, 21–34.

BROWN, R. W., SUMMERFIELD, M. A. & GLEADOW, A. J. W. 1994. Apatite fission track analysis: its potential for the estimation of denudation rates and implications for models of long term landscape development. *In*: KIRKBY, M. J. (ed.) *Process Models and Theoretical Geomorphology*. Wiley, Chichester, 23–53.

CARLSON, W. D., DONELICK, R. A. & KETCHAM, R. A. 1999. Variability of apatite fission-track annealing kinetics: I. Experimental results. *American Mineralogist*, **84**, 1213–1223.

CEDERBOM, C. 2001. Phanerozoic, pre-Cretaceous thermotectonic events in southern Sweden revealed by fission track thermochronology. *Earth and Planetary Science Letters*, **188**, 199–209.

CEDERBOM, C., LARSON, S. Å., TULLBORG, E.-L. & STIBERG, J.-P. 2000. Fission track thermochronology applied to Phanerozoic thermotectonic events in central and southern Sweden. *Tectonophysics*, **316**, 153–167.

CHAUMONT, J., SOULET, S., KRUPA, J. C. & CARPENA, J. 2002. Competition between disorder creation and annealing in fluorapatite waste forms. *Journal of Nuclear Materials*, **301**, 122–128.

DEINO, A. & POTTS, R. 1990. Single crystal $^{40}Ar/^{39}Ar$ dating of the Ologesilie Formation, southern Kenya Rift. *Journal of Geophysical Research*, **95**, 8453–8470.

DONELICK, R. A. 1993. Apatite etching characteristics versus chemical composition. *Nuclear Tracks and Radiation Measurements*, **21**, 604.

DUNAI, T. 2005. Forward modeling and interpretation of (U–Th)/He ages. *In*: REINERS, P. & EHLERS, T. (eds) *Low Temperature Thermochronology: Techniques, Interpretations, and Applications. Reviews in Mineralogy and Geochemistry*, **58**, 259–274.

EVANS, N. J., BYRNE, J. P., KEEGAN, J. T. & DOTTER, L. E. 2005. Determination of uranium and thorium in zircon, apatite and fluorite: application to laser (U–Th)/He thermochronology. *Journal of Analytical Chemistry*, **60**, 1159–1165.

EWING, R. C. & WANG, L. 2002. Phosphates as nuclear waste forms. *In*: KOHN, M., RAKOVAN, J. & HUGHES, J. M. (eds) *Phosphates – Geochemical, Geobiological and Material Importance. Reviews in Mineralogy and Geochemistry*, **48**, 673–699.

FARLEY, K. A. 2000, Helium diffusion from apatite: General behavior as illustrated by Durango apatite. *Journal of Geophysical Research*, **105**, 2903–2914.

FARLEY, K. A. 2002. (U–Th)/He dating: techniques, calibrations and applications. *In*: PORCELLI, C., BALLENTINE, C. & WIELER, R. (eds) *Noble Gases in Geochemistry and Cosmochemistry. Reviews in Mineralogy and Geochemistry*, **47**, 819–843.

FARLEY, K. A. & STOCKLI, D. F. 2002. (U–Th)/He dating of phosphates: apatite, monazite and xenotime. *In*: KOHN, M., RAKOVAN, J. & HUGHES, J. M. (eds) *Phosphates – Geochemical, Geobiological and Material Importance. Reviews in Mineralogy and Geochemistry*, **48**, 557–577.

FARLEY, K. A., WOLF, R. A. & SILVER, L. T. 1996. The effects of long alpha-stopping distances on (U–Th)/He ages. *Geochimica et Cosmochimica Acta*, **60**, 4223–4229.

FITZGERALD, P. G., BALDWIN, S. L., WEBB, L. E. & O'SULLIVAN, P. B. 2006. Interpretation of (U–Th)/He single grain ages from slowly cooled crustal terranes: a case study from the transantarctic mountains of southern victoria land. *Chemical Geology*, **225**, 91–120.

FLOWERS, R. M. 2009. Exploiting radiation damage control on apatite (U–Th)/He dates in cratonic regions. *Earth and Planetary Science Letters*, **277**, 148–155.

FLOWERS, R. M., BOWRING, S. A. & REINERS, P. W. 2006. Low long-term erosion rates and extreme continental stability documented by ancient (U–Th)/He dates. *Geology*, **34**, 925–928.

FLOWERS, R. M., SHUSTER, D. L., WERNICKE, B. P. & FARLEY, K. A. 2007. Radiation damage control on apatite (U–Th)/He dates from the Grand Canyon region, Colorado Plateau. *Geology*, **35**, 447–450.

GALBRAITH, R. F. & LASLETT, G. M. 1993. Statistical models for mixed fission track ages. *Nuclear Tracks*, **21**, 459–470.

GALLAGHER, K. 1995. Evolving temperature histories from apatite fission-track data. *Earth and Planetary Science Letters*, **136**, 421–435.

GLEADOW, A. J. W. & DUDDY, I. R. 1981. Early Cretaceous volcanism and early breakup history of southeastern Australia: Evidence from fission track dating of volcaniclastic sediments. *In*: CRESSWELL, M. M. & VELLA, P. (eds) *Gondwana Five – Proceedings of the Fifth International Gondwana Symposium*. Balkema, Rotterdam, 295–300.

GLEADOW, A. J. W., BELTON, D. X., KOHN, B. P. & BROWN, R. W. 2002. Fission track dating of phosphate minerals and the thermochronology of apatite. *In*: KOHN, M., RAKOVAN, J. & HUGHES, J. M. (eds) *Phosphates – Geochemical, Geobiological and*

Material Importance. Reviews in Mineralogy and Geochemistry, **48**, 579–630.

GREEN, P. F. 1986. On the thermo-tectonic evolution of Northern England: evidence from fission track analysis. *Geological Magazine*, **153**, 493–506.

GREEN, P. F. & DUDDY, I. R. 2006. Interpretation of apatite (U–Th)/He ages and fission track ages from cratons. *Earth and Planetary Science Letters*, **244**, 541–547.

GREEN, P. F., CROWHURST, P. V., DUDDY, I. R., JAPSEN, P. & HOLFORD, S. P. 2006. Conflicting (U–Th)/He and fission track ages in apatite: enhanced He retention, not anomalous annealing behaviour. *Earth and Planetary Science Letters*, **250**, 407–427.

GREEN, P. F., DUDDY, I. R., GLEADOW, A. J. W., TINGATE, P. R. & LASLETT, G. M. 1985. Fission track annealing in apatite: Track length measurements and the forms of the Arrhenius plot. *Nuclear Tracks*, **10**, 323–328.

HANSEN, K. & REINERS, P. W. 2006. Low temperature thermochronology of the southern East Greenland continental margin: evidence from apatite (U–Th)/He and fission track analysis and implications for intermethod calibration. *Lithos*, **92**, 117–136.

HARMAN, R., GALLAGHER, K., BROWN, R. W., RAZA, A. & BIZZI, L. 1998. Accelerated denudation and tectonic/geomorphic reactivation of the cratons of northeastern Brazil during the Late Cretaceous. *Journal of Geophysical Research*, **103**, 27 091–27 105.

HASEBE, N., BARBARAND, J., JARVIS, K., CARTER, A. & HURFORD, A. 2004. Apatite fission-track thermochronometry using laser ablation ICP-MS. *Chemical Geology*, **207**, 135–145.

HENDRIKS, B. W. H. & REDFIELD, T. F. 2005. Apatite fission track and (U–Th)/He data from Fennoscandia: an example of underestimation of fission track annealing in apatite. *Earth and Planetary Science Letters*, **236**, 443–458.

HENDRIKS, B. W. H. & REDFIELD, T. F. 2006. Reply to: Comment on 'Apatite Fission Track and (U–Th)/He data from Fennoscandia: an example of underestimation of fission track annealing in apatite' by B. W. H. Hendriks and T. F. Redfield. *Earth and Planetary Science Letters*, **248**, 568–576.

HOUSE, M., FARLEY, K. A. & STOCKLI, D. 2000, Helium chronometry of apatite and titanite using Nd-YAG laser heating. *Earth and Planetary Science Letters*, **183**, 365–368.

HUIGEN, Y. & ANDRIESSEN, P. A. M. 2004. Thermal effects of Caledonian foreland basin formation, based on fission track analyses applied on basement rocks in central Sweden. *Physics and Chemistry of the Earth*, **29**, 683–694.

KETCHAM, R. A. 2005. Forward and inverse modeling of low-temperature thermochronometry data. *In*: REINERS, P. & EHLERS, T. (eds) *Low Temperature Thermochronology: Techniques, Interpretations, and Applications. Reviews in Mineralogy and Geochemistry*, **58**, 275–314.

KETCHAM, R. A., DONELICK, R. A. & DONELICK, M. B. 2000. AFTSolve: a program for multi-kinetic modelling of apatite fission-track data. *Geological Material Research*, **2**, 1(electronic version).

KOHN, B. P., GLEADOW, A. J. W., BROWN, R. W., GALLAGHER, K., O'SULLIVAN, P. B. & FOSTER, D. A. 2002*b*. Shaping the Australian crust over the last 300 million years: insights from fission track thermotectonic and denudation studies of key terranes. *Australian Journal of Earth Sciences*, **49**, 697–717.

KOHN, B. P., GLEADOW, A. J. W., BROWN, R. W., GALLAGHER, K., LORENCAK, M. & NOBLE, W. P. 2005. Visualising thermotectonic and denudation histories using apatite fission track thermochronology. *In*: REINERS, P. & EHLERS, T. (eds) *Low Temperature Thermochronology: Techniques, Interpretations, and Applications. Reviews in Mineralogy and Geochemistry*, **58**, 527–565.

KOHN, B. P., FOSTER, D. A. & FARLEY, K. A. 2002*a*. Low temperature thermochronology of apatites with exceptional compositional variations: the Stillwater Complex, Montana revisited. *Sociedad Geologica de España, Geotemas*, **4**, 103–105.

KOHN, B. P., GLEADOW, A. J. W., FREI, S. & KOHLMANN, F. 2008. Low-temperature thermochronology in ancient terranes: A progress report. *In*: GARVER, J. I. & MONTARIO, M. J. (eds) *Proceedings from the 11th International Conference on Thermochronometry, Anchorage, Alaska, September 2008*, Union College, Schenectady, NY, 142–143.

KOHONEN, J. & RÄMÖ, O. T. 2005. Sedimentary rocks, diabases, and late cratonic evolution. *In*: LEHTINEN, M., NURMI, P. A. & RÄMÖ, O. T. (eds) *Precambrian Geology of Finland – Key to the Evolution of the Fennoscandian Shield*. Elsevier, Amsterdam, 563–604.

KUKKONEN, I. T. 1989. Terrestrial heat flow and radiogenic heat production in Finland, the central Baltic Shield. *Tectonophysics*, **164**, 219–230.

KUKKONEN, I. T. 1993. Heat flow map of northern and central parts of the Fennoscandian Shield based on geochemical surveys of heat producing elements. *Tectonophysics*, **225**, 3–13.

LARSON, S. Å. & TULLBORG, E.-L. 1998. Why Baltic Shield zircons yield late Paleozoic lower intercept ages on U–Pb Concordia. *Geology*, **26**, 919–922.

LARSON, S. Å., CEDERBOM, C. E., TULLBORG, E.-L. & STIBERG, J.-P. 1999. Sveconorwegian and Caledonian foreland basins in the Baltic Shield revealed by fission track thermochronology. *Terra Nova*, **11**, 210–215.

LARSON, S. Å., CEDERBOM, C. E., TULLBORG, E.-L. & STIBERG, J.-P. 2006. Comment on paper by Hendriks and Redfield in *Earth and Planetary Science Letters* **236**, 443–458. *Earth and Planetary Science Letters*, **248**, 560–567.

LEHTOVAARA, J. 1976. Apatite fission track dating of Finnish Precambrian intrusives. *Annales Academioae Scientarium Fennicae*, **117**, 1–94.

LORENCAK, M. 2003. *Low Temperature Thermochronology of the Canadian and Fennoscandian Shields: Integration of Apatite Fission Track and (U–Th)/He Methods*. PhD thesis, University of Melbourne.

LORENCAK, M., KOHN, B. P., OSADETZ, K. G. & GLEADOW, A. J. W. 2004. Combined apatite fission track and (U–Th)/He thermochronometry in a slowly cooled terrane: results from a 3440 m deep drill hole in the southern Canadian Shield. *Earth and Planetary Science Letters*, **227**, 87–104.

McInnes, B. I. A., Farley, K. A., Sillitoe, R. H. & Kohn, B. P. 1999. Application of apatite (U–Th)/He dating to the determination of the sense and amount of vertical fault displacement at the Chuquicamata Mine, Chile. *Economic Geology*, **94**, 937–948.

Meesters, A. G. C. A. & Dunai, T. J. 2002*a*. Solving the production–diffusion equation for finite diffusion domains of various shapes (part I): implications for low-temperature (U–Th)/He thermochronometry. *Chemical Geology*, **186**, 333–344.

Meesters, A. G. C. A. & Dunai, T. J. 2002*b*. Solving the production–diffusion equation for finite diffusion domains of various shapes (part II): Application to cases with α-ejection and non-homogeneous distribution of the source. *Chemical Geology*, **186**, 347–363.

Murrell, G. R. 2003. *The Long-term Thermal Evolution of Central Fennoscandia*. PhD thesis, Vrjie Universiteit, Amsterdam.

Murrell, G. R. & Andriessen, P. A. M. 2004. Unravelling a long-term multi-event thermal record in the cratonic interior of southern Finland through apatite fission track thermochronology. *Physics and Chemistry of the Earth*, **29**, 695–706.

Nikishin, A. M., Ziegler, P. A. *et al.* 1996. Late Precambrian to Triassic history of the East European Craton: dynamics of sedimentary basin evolution. *Tectonophysics*, **268**, 23–63.

Patchett, P. J., Embry, A. F. *et al.* 2004. Sedimentary cover of the Canadian Shield through Mesozoic time reflected by Nd isotopic and geochemical results for the Sverdrup Basin, Arctic Canada. *Journal of Geology*, **112**, 39–57.

Reiners, P. W. & Farley, K. A. 2001. Influence of crystal size on apatite (U–Th)/He thermochronology: an example from the Bighorn Mountains, Wyoming. *Earth and Planetary Science Letters*, **188**, 413–420.

Samuelsson, J. & Middleton, M. F. 1998. The Caledonian foreland basin in Scandinavia: constrained by the thermal maturation of the Alum shale. *GFF*, **120**, 307–314.

Shuster, D. L., Flowers, R. M. & Farley, K. A. 2006. The influence of natural radiation damage on helium diffusion kinetics in apatite. *Earth and Planetary Science Letters*, **249**, 148–161.

Söderlund, P., Juez-Larre, J., Page, L. M. & Dunai, T. 2005. Extending the time range of apatite (U–Th)/He thermochronometry in slowly cooled terranes: Palaeozoic to Cenozoic exhumation history of southeast Sweden. *Earth and Planetary Science Letters*, **239**, 266–275.

Sundblad, K., Alm, E., Huhma, H., Vaasjoki, M. & Sollien, D. B. 2004. Early Devonian tectonic and hydrothermal activity in the Fennoscandian Shield; evidence from calcite-fluorite-galena mineralization. *In*: Mertanen, S. (ed.) *5th Nordic Paleomagnetic Workshop, 'Supercontinents, remagnetizations and geomagnetic modelling'. Extended Abstract Volume*. Geological Survey of Finland Report, **Q29.1/2004/*1***, 67–71.

Vermeesch, P., Seward, D., Latkoczy, C., Wipf, M., Günther, D. & Baur, H. 2007. α-emitting mineral inclusions in apatite, their effect on (U–Th)/He ages, and how to reduce it. *Geochimica et Cosmochimica Acta*, **71**, 1737–1746.

Weber, U. D. 2002. *Thermotectonic Evolution of the Precambrian Shield area of Northern West Australia*. PhD thesis, University of Melbourne.

Weber, U. D., Kohn, B. P., Gleadow, A. J. W. & Nelson, D. R. 2005. Low temperature Phanerozoic history of the northern Yilgarn Craton, Western Australia. *Tectonophysics*, **400**, 127–151.

Zeck, H. P., Andriessen, P. A. M., Hansen, K., Jensen, P. K. & Rasmussen, B. L. 1988. Paleozoic paleo-cover of the southern part of the Fennoscandian shield – fission track constraints. *Tectonophysics*, **149**, 61–66.

Zeitler, P. K., Herczig, A. L., McDougall, I. & Honda, M. 1987. U–Th–He dating of apatite: a potential thermochronometer. *Geochimica et Cosmochimica Acta*, **51**, 2865–2868.

Weathering of granite and granitic regolith in Corsica: short-term ^{10}Be versus long-term thermochronological constraints

JOACHIM KUHLEMANN[1]*, INGRID KRUMREI[1], MARTIN DANIŠÍK[1] & KLAAS VAN DER BORG[2]

[1]*University of Tübingen, Institute of Geosciences, Sigwartstrasse 10, D-72076 Tübingen, Germany*

[2]*Subatomic Physics Department, Van de Graaff Laboratory, Utrecht University, P.O. Box 80.000, 3508 TA Utrecht, The Netherlands*

**Corresponding author (e-mail: kuhlemann@uni-tuebingen.de)*

Abstract: *In situ* ^{10}Be concentrations in granites on Corsica (Mediterranean), exposed to subalpine climate, yield weathering rates of between 9 and 20 mm ka^{-1}, when averaged over the last 30–100 ka. Weathering rates rise with increasing precipitation and brittle deformation. Thermal history modelling of apatite fission-track (AFT) data confirms that average denudation rates of 5–20 mm ka^{-1} were typical during the Neogene. Short- and long-term denudation rates are in the same range and indicate that a late Neogene global increase in denudation rates has probably not affected geomorphically stable uplifted palaeosurfaces of hilly to low mountainous local relief. A Southwards decrease in short- and long-term denudation rates indicate a linear relationship with decreasing precipitation. Post-glacial downwearing of moraine matrix, as constrained by soil-derived humic etching of dated glacial boulders, yields 40–140 mm ka^{-1}, which indicates a poor preservation potential of moraines older than the last glaciation. This weathering rate is of the order of 3 times faster than that found for long-term regolith formation from granites based on of geomorphic evidence.

Weathering rates of crystalline rocks in passive tectonic settings are known to be very low under arid conditions in timescales of several million years (e.g. Bierman & Turner 1995; Cockburn *et al.* 2000), but under humid conditions the role of climate is disputed (e.g. Riebe *et al.* 2000). Natural denudation rates ranging between 5 and 11 mm ka^{-1} found in the partly rugged humid mountains of Sri Lanka (Von Blanckenburg 2005) indicate that thick tropical soils slow down weathering, and may, therefore, unvalidate a positive correlation between precipitation and denudation rates. Erosion rates exceeding 10 mm ka^{-1}, as found by Riebe *et al.* (2000) in granites, appear neither to be related to precipitation nor to temperature, but instead to tectonic activity. Since ongoing tectonic deformation can increase erosion rates by an order of magnitude (Sklar & Dietrich 2001), a discrimination of other factors controlling weathering rates hinges on a comparison of tectonically passive sites (very low stress, no strain) and uniform common lithologies, such as granites (Kuhlemann *et al.* 2007). Corsica, as a tectonically inactive crystalline block, provides a great opportunity to test which factors, besides ongoing deformation, govern the rates of denudation at variable temporal scales. Moreover, Corsica is characterized by remnants of ancient landscapes, including hilly to mountainous palaeorelief. Palaeo-landscapes are common phenomena that provide an opportunity to compare surface process rates both of both modern and ancient relief.

Preservation of palaeo-landscapes and, even small-scale geomorphic features, for millions to tens of millions of years is self-evident if weathering rates and long-term denudation rates are very low. Although denudation rates in subalpine–alpine climate conditions are one or two magnitudes higher, weathering of solid rock and regolith appears not to destroy the shape of ancient landscapes in several cases in North America, reflecting geomorphic stability of palaeosurfaces (Small *et al.* 1997; Small & Anderson 1998; Anderson 2002). For a better assessment of geomorphic stability in subalpine–alpine climate on timescales varying between 10^4 and 10^7 years, a combination of cosmogenic nuclide analysis, low-temperature thermochronology and geomorphic evidence is useful.

In most geomorphically 'stable' mountainous landscapes, regolith is a common surface cover, whereas solid rock and rock castles are less frequently exposed. For long-term denudation rates in Corsica, geomorphic evidence (Kuhlemann *et al.* 2005*b*, 2007) and thermochronologic data

From: LISKER, F., VENTURA, B. & GLASMACHER, U. A. (eds) *Thermochronological Methods: From Palaeotemperature Constraints to Landscape Evolution Models*. Geological Society, London, Special Publications, **324**, 217–235.
DOI: 10.1144/SP324.16 0305-8719/09/$15.00

(Zarki-Jakni *et al.* 2004; Danišik 2005; Danišik *et al.* 2007) are available. Long-term regolith formation by chemical weathering in tectonically passive granite landscapes has rarely been assessed, whereas short-term regolith formation and denudation has been studied in tectonically active settings such as the Alps (Wittmann *et al.* 2007). For an approximation of short-term maximum regolith weathering rates in Corsica, we have studied regolith downwearing on granitic moraine deposits.

In this paper, we compare late Pleistocene weathering rates of granite rock castles on subalpine palaeosurfaces and the downwearing of regolith with long-term denudation rates in Corsica (western Mediterranean). For short-term weathering rates of granite basement we use ^{10}Be concentrations, assuming steady-state grusification. For downwearing rates of regolith we use ^{10}Be exposure ages of glacial boulders with weathering girdles and geomorphic evidence of long-term landscape change. Apatite fission-track ages, length measurements and modelling of cooling paths are used to constrain Neogene average denudation rates. The aim is to quantify rates of geomorphic change and to study the stability of landforms on timescales of 10^4-10^7 years. Since the role of climate, namely precipitation, for denudation is a matter of controversy, we intend to validate this in a natural laboratory where the effective sealing of crystalline rocks by tropical soil is excluded.

Regional setting

The island of Corsica is located in the northwestern Mediterranean basin, between 41° and 43° latitude (Fig. 1). Together with Sardinia, it was part of the southern margin of Europe–Eurasia for much of late Palaeozoic and Mesozoic times. During the Variscan orogeny in the Carboniferous, both islands were partly situated in the internal zone that experienced high-grade metamorphic overprint and the formation of large granite intrusions (Rossi & Cocherie 1991). The larger western part of Corsica is formed by a Variscan crystalline basement with dominant calc-alkaline granites. Some alkaline granites and their volcanic equivalents formed in Permian times in a phase of transtension and high heat flow (Fig. 2b) (Rossi & Cocherie 1991). In the Cretaceous, both islands were part of the Iberian microplate during counterclockwise rotation that ended with a local orogeny and the building of the Pyrenees (Durand-Delga 1978; Egal 1992). In the early Cenozoic, the southern European margin, including Corsica but not Sardinia, was subducted below the Adriatic microplate. Large parts of Variscan Corsica became covered by a flysch layer several kilometres thick (Danišík *et al.* 2007). Subduction culminated with the Alpine collision, which caused thrusting and nappe stacking in eastern Corsica (Alpine Corsica: Egal 1992). Alpine collision was followed by rapid uplift and erosion of the flysch deposits in the Oligocene. Following this exhumation slowed down, except for the eastern part of Corsica where strong extension triggered basin formation (Cavazza *et al.* 2001; Fellin *et al.* 2005; Danišík *et al.* 2007). Extension had followed after the establishment of the NW-dipping Apennine subduction zone in the Oligocene, which started to roll back in an easterly direction (Egal 1992). Both Corsica and Sardinia were detached from the European margin, and were counterclockwise rotated into their present position up until Early Miocene times (Vigliotti & Langenheim 1995). The termination of rotation was followed by an uplift event by about 17 Ma, a stagnation period and another uplift event by around 10 Ma (Kuhlemann *et al.* 2005*b*). As a result of young differential uplift, palaeosurfaces are preserved both in Corsica and in Sardinia. Rugged recent relief and elevations up to 2706 m in the NW of Corsica contrast with high-uplifted palaeosurface relicts up to 30 km^2 in size (Figs 2a & 3), which display minor to moderate relief.

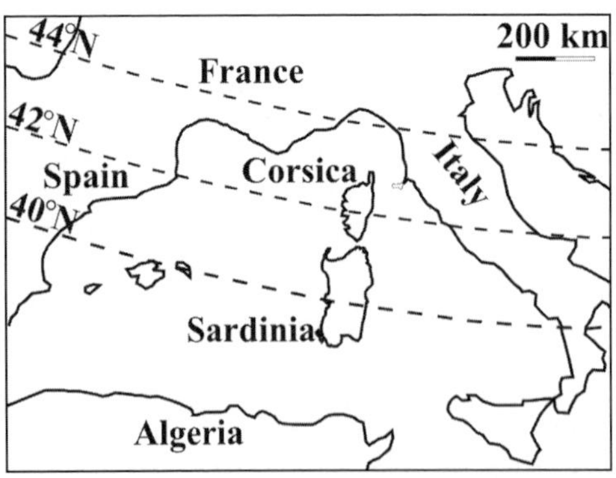

Fig. 1. Geographical location of Corsica.

The role of climate is of special importance for rates of denudation. The recent climate in Corsica at sea level is characterized by subtropical Mediterranean-type conditions (dry warm summer and temperate wet winter), whereas temperate conditions are found in the montane zone between 1000 and 1800 m. Average annual precipitation increases from about 600 mm year^{-1} on the coast to more than 1500 mm year^{-1} inland (Fig. 2b) (Bruno *et al.* 2001). The subalpine sampling sites at elevations up to 2350 m experience frost for half of the year. During stadials of the late Pleistocene, average temperatures in subalpine and alpine environments were lowered by about 8 °C (Kuhlemann *et al.* 2005*a*, 2008). Glaciations strongly modified mesoscale relief and are the major cause for

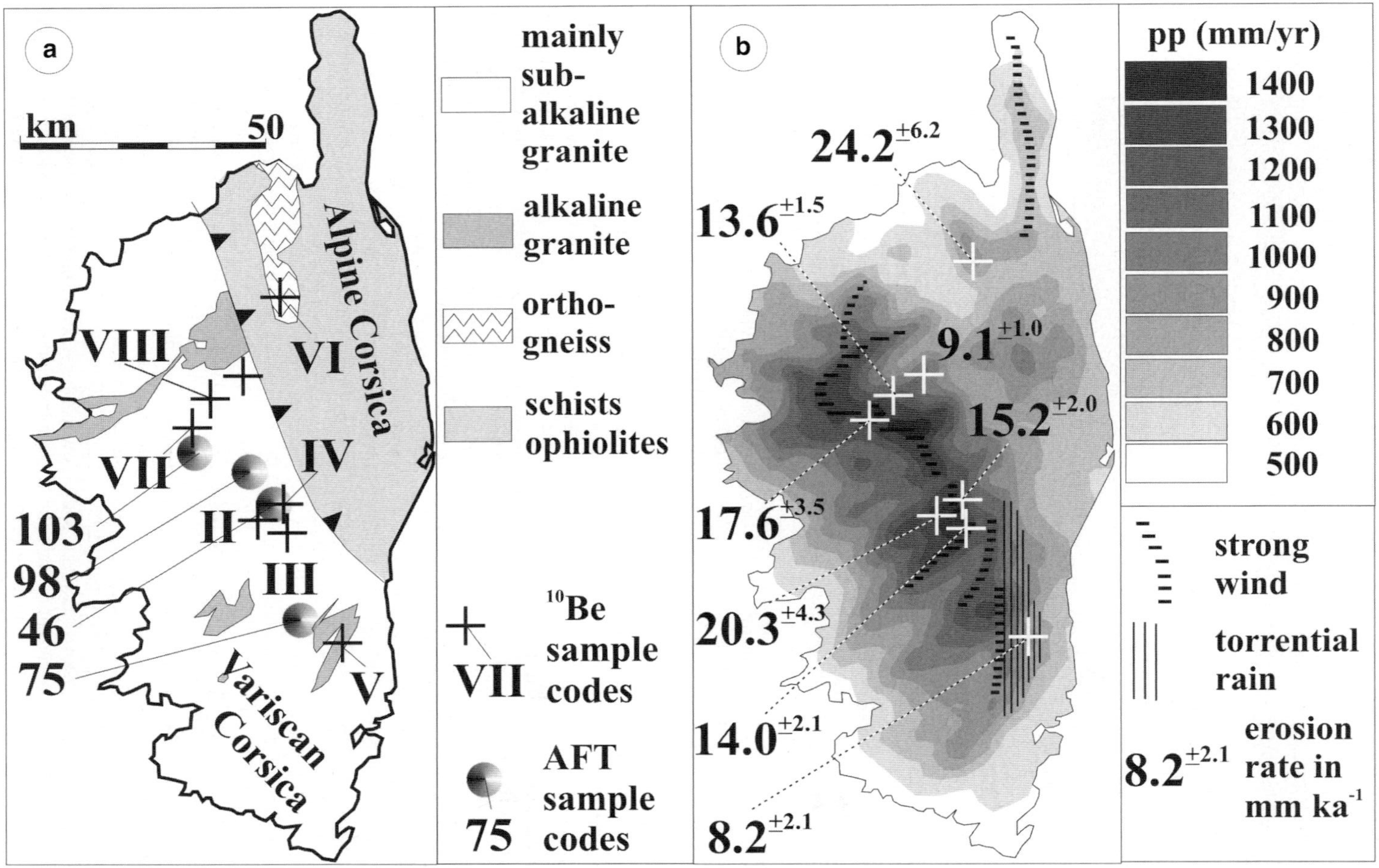

Fig. 2. (**a**) Simplified sketch map of lithologies in Corsica with sampling locations. (**b**) Distribution of annual precipitation (pp) in Corsica according to Bruno *et al.* (2001), and zones of torrential rain and strong wind.

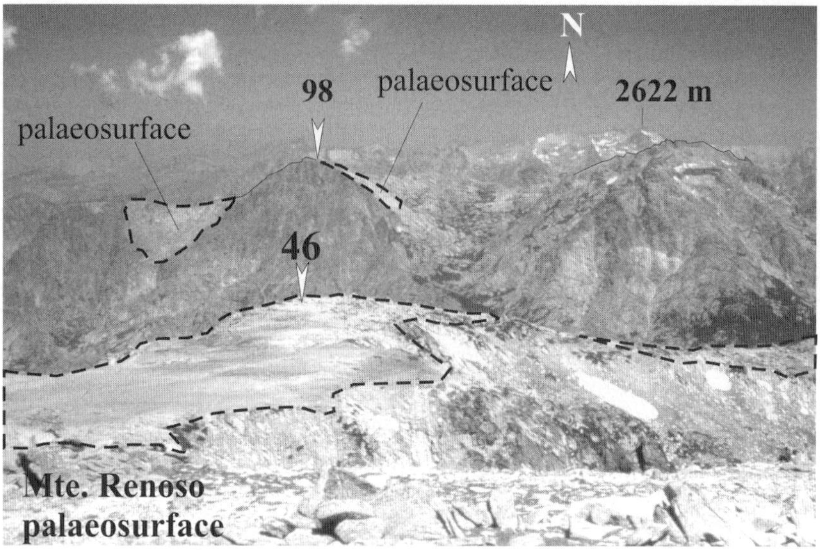

Fig. 3. View from Mte. Renoso to the north on the high-elevation palaeosurface rising up to the top of this mountain (2352 m) and to isolated relicts of this palaeosurface in the mid-distance that are too small to be mapped in Figure 2c. The latter relicts are limited by rock walls towards the north and south. All palaeosurfaces are indicated by hatched lines.

palaeosurface destruction (Kuhlemann *et al.* 2005*a*, *b*, 2007). In Pliocene (Fauquette *et al.* 1999) and Miocene times (Bruch *et al.* 2007) precipitation rates in the northwestern Mediterranean realm were about twice as high as today; however, decreasing southwards to a modern level at the latitude of southern Spain (Jiménez-Moreno & Suc 2007).

Field evidence of weathering on palaeosurfaces and of regolith downwearing

Typical high-elevation palaeosurface relicts in southern Corsica display largely vegetated flats with isolated tors, steep hills built of calc-alkaline granite boulders and up to 200 m-high rock castles with a boulder mantle. The rock castles are surrounded by rarely exposed grusy bedrock with a regolith cover, from 0.5 m to a few metres thick, of granite grains with some blocks. In central and northern Corsica, rock castles are smaller and rise up to 30 m, surrounded by regolith, which locally forms a periglacial flat. Although periglacial climate conditions in general favour spallation of blocks (see Small *et al.* 1997), platy block fragments are subordinate on top of the regolith of palaeosurfaces below 2200 m a.s.l. (m above sea level) and typically consist of aplitic or basaltic–andesitic dyke material. Above 2200 m, the number of platy block fragments gradually increases depending of the granite type as climate conditions change from subalpine to alpine (Fig. 3). However, even the highest sampling site IV at 2352 m is dominated by grusification.

Evidence of chemical (rain) and biochemical (lichen) etching of bare bedrock is found in various granite types, but especially in alkali-feldspar granites, where features such as karren, typical of massive limestone, have been described as 'pseudokarst' and 'silica karstification' by Klaer (1956). On flat summit surfaces, fissure-controlled networks of decimetre- (dm-) to metre-wide, dm-deep potholes and channels are typical (Kuhlemann *et al.* 2007).

Weathering and downwearing of regolith developed on moraine material is deduced from exposure-dated glacial boulders in the upper montane zone. We selected three representative sites in the northern (dry), central (wet) and southern (moderatly wet) parts of Corsica where glacial boulders are located on a flat substratum (Fig. 4). We considered post-glacial *in situ* soil weathering only in flat terrain, away from running water.

It was observed that the lower part of boulders, which had been buried in the soil, can be differentiated from those parts always rising above the soil by distinct surface features. Parts below the surface experienced concave dissolution by organic acids within the soil, which has subsequently been lowered by weathering. Effectiveness of etching by organic acids is highest at depths of several decimetres where water fills the pore spaces for most of the year and where organic acids are most abundant. The concave surfaces had been etched in mm- to

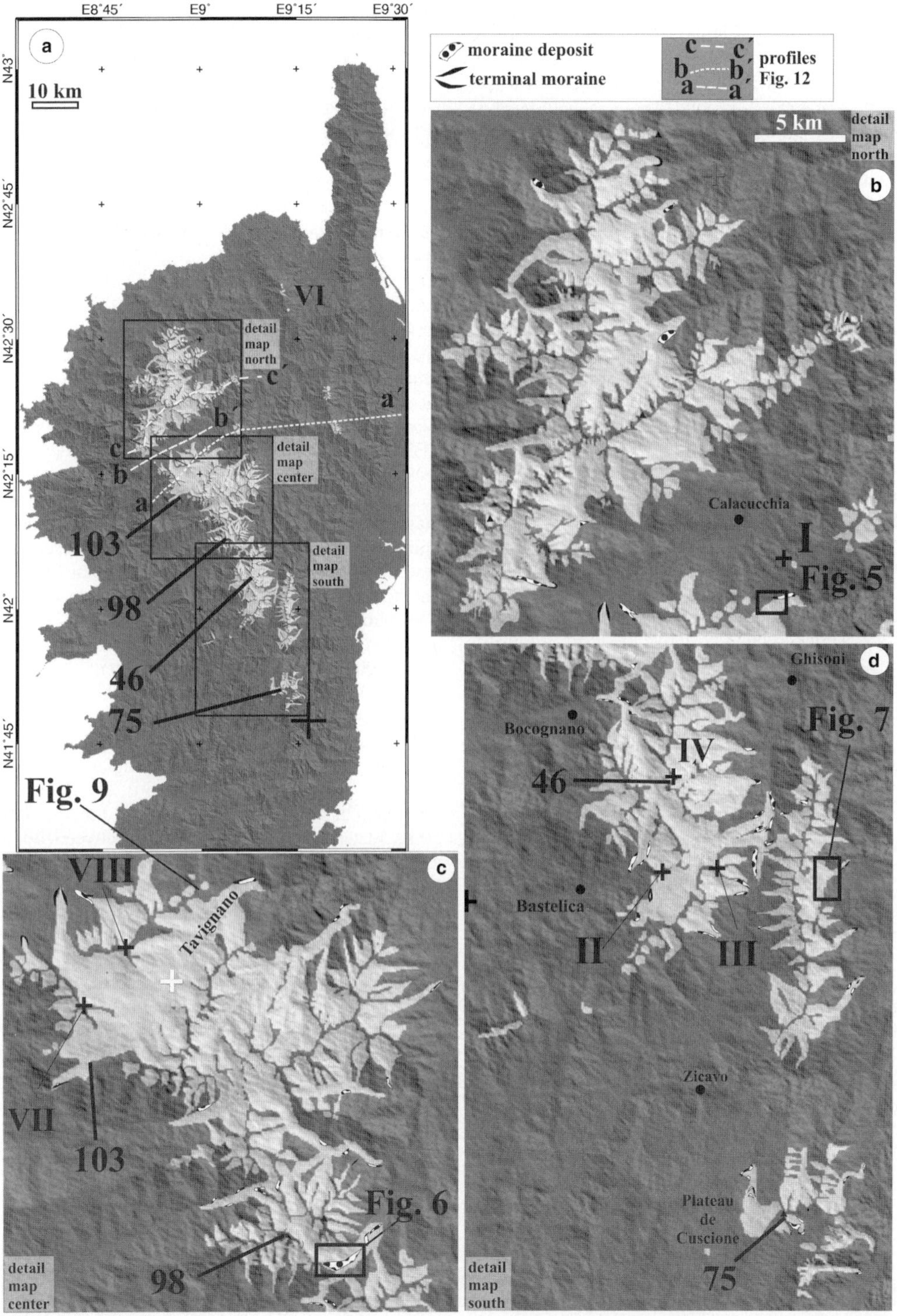

Fig. 4. Map of the Wuermian maximum glacier extension in Corsica, probably during Marine Isotope Stage (MIS) 4 for most of Corsica (modified from Kuhlemann *et al.* 2005*a*, 2007). The location of local maps and AFT sampling sites are indicated.

cm-scale, with quartz crystals rising above the surrounding feldspar grains. Older concave parts of such boulders often expose limonitic crusts. In several cases soil-parallel smooth concave brownish weathering girdles surround the glacial boulders just below the top level of soil formation. Often there is more than just one weathering girdle found. Vertical extent and number of weathering girdles varies within close distances, possibly reflecting meaningless self-organization of etching. However, the constantly exposed upper parts typically expose a convex, densely lichen-covered surface, often with cm-scale silica karstification. Above the topmost former soil level, spallation of several cm-thick granite plates is common. It is, however, also found occasionally below this level.

For measuring the rates of weathering and erosion, the focus of the sampling strategy both for cosmogenic nuclide analysis and low-temperature thermochronology was dedicated to calc-alkaline granites of rock castles on palaeosurfaces. An alkaline granite and a granite gneiss were taken as reference samples for the lithology-dependent variation of erosion rates. The regional cover included transects with a west–east and north–south precipitation gradient. Sampling sites on palaeosurface relicts were chosen to be above the level of Pleistocene glaciers, and to be exposed to wind and solar radiation to minimize shielding by snow and ice during stadials. In the case of regolith downwearing, the selected sites also show different annual precipitation rates.

Analytical methods

For the introduction of the analytical methods, we start with the analysis of ^{10}Be for short-term weathering rates.

Effects of topographic shielding for our rock castle sampling sites, with slope angles ranging between 2° and 8°, are negligible. Samples were taken from planar tops of granite tors rising to at least several metres above the regolith cover of the palaeosurface. Exposure ages of glacial boulders with geomorphic evidence of former soil-cover levels (see below) had to be corrected for the effects of topographic shielding between 7° and 12°. This correction was made by GIS software on a high-resolution digital elevation model (DEM). Elevation and latitude were determined from French IGN 1:25 000 topo-graphic maps.

Chemical treatment generally followed that of Kohl & Nishiizumi (1992). After optical evaluation and x-ray fluorescence (XRF) control measurements of Al and trace-element content, 15–50 g purified quartz samples were prepared for accelerator mass spectrometry (AMS) measurements at the van de Graaf Laboratory in Utrecht. Samples taken from glacial boulders were prepared as targets in Tuebingen and were measured at ETH Zurich.

^{10}Be forms when oxygen nuclei within quartz are exposed to high-energy cosmic ray particles (Nishiizumi *et al.* 1989). The production of ^{10}Be (P) decreases with depth (z) according to the equation:

$$P(z) = P_0 e^{-z/z*}$$

where the scale length, $z* = \lambda/\rho$, is the ratio of the absorption mean free path (λ) of 157 g cm^{-2} (Lal 1991) and the density of the solid (ρ), here granite and gneiss have a density of 2.7 g cm^{-3}. For local surface production rates (P_0) and latitude-elevation coefficients at the rock castles we used a calculation scheet provided by Dunai (2001). For the calculation of exposure ages of glacial boulders we used an open-access MS-Excel^{TM} calculation sheet TEBESEA (Abramowski 2004). This includes corrections for shielding by topography, surface cover (sediment and/or snow), sample thickness, for geomagnetic variations and, if necessary, for tectonic uplift. Exposure ages are calculated for the scaling system according to the method by Dunai (2001), but the production is scaled by capture of slow negative muons and fast muon reactions, with attenuation lengths of 1510 and 4320 g cm^{-2} (Heisinger *et al.* 2002*a*, *b*). For the scaling system of Dunai (2001), we used the marine average sea-level palaeo-temperature record of Cacho *et al.* (2002) and the nominal exposure age to recalculate the time-integral of higher altitude temperature. We assumed an average air pressure of 1014 hPa (hectopascal = 100 pa) at sea level for the last 50 ka instead of the recent range between 1015 and 1016 hPa, due to higher cyclon frequency (Kuhlemann *et al.* 2005*a*). The sea-level high-latitude (SLHL) production rates of negative muon capture and fast muon reactions are also adopted from Heisinger *et al.* (2002*a*, *b*). For the SLHL ^{10}Be-production rate, Abramowski (2004) rescaled published calibration values (Kubik *et al.* 1998; Stone *et al.* 1998; Dunai 2000; Klein & Gosse 2002; Kubik & Ivy-Ochs 2004) with the different scaling systems, and used the resulting mean (5.47 atoms g^{-1} year^{-1}: Dunai 2001, modified). The used attenuation length for neutron spallations in rock is 155 g cm^{-2} (Gosse & Phillips 2001). The potential error of the production rate is estimated at 6%. To correct for slightly reduced local surface production rates according to the shape of glacial boulders, we adopted the modelling results of Masarik & Wieler (2003). The potential error of temperature and pressure estimated for the calculation of ^{10}Be-production is about 2%, and that resulting from snow shielding of the rock castles is about 2%.

We now calculate the total 1σ error of our data. For a calculation of maximum erosion rates (ε) in surface samples we assume a dynamic equilibrium of erosion rates and nuclide production rates on the timescale of 100 ka (see Clark *et al.* 1995), which ideally represents steady state:

$$N_{(0)} = \frac{P_0}{(\varepsilon/z^*) + \lambda} \Longrightarrow \varepsilon = z^* \left(\frac{P_0}{N_{(0)}} - \lambda \right).$$

Here, λ (year^{-1}) is the decay constant of ^{10}Be. According to Small *et al.* (1997), the assumption of steady-state erosion is not always valid in alpine settings. Field evidence, however, confirms that this assumption is reasonable for the temperate sites sampled in Corsica.

For apatite fission track (AFT) dating, 5–10 g granite samples were taken from the same palaeosurfaces, as those sampled for ^{10}Be analysis. The AFT samples selected for the present study did not strictly originate from the same rock castles as those chosen for ^{10}Be measurements, except for one sample, because of the lack of apatite crystals in the sampled rocks. Instead, samples of sufficient quality from the close vicinity were considered. AFT dating is sensitive to the cooling of source rocks in the temperature range 60–110 °C (Gleadow & Duddy 1981). FT dating was performed using the external detector method and the zeta calibration approach (Hurford & Green 1983). Thermal histories based on apatite fission-track age and length data were modelled using the HeFTy modelling program (Ketcham 2005).

Dating of glacial boulders and soil downwearing

The first study site in northern Corsica (for the location see Fig. 4) is located in a relatively dry and largely deforested area. A local map of the site (Fig. 5a) shows a succession of four lateral moraines, which mark the upper level of the largest ice field that existed in Corsica (Kuhlemann *et al.* 2005*a*).

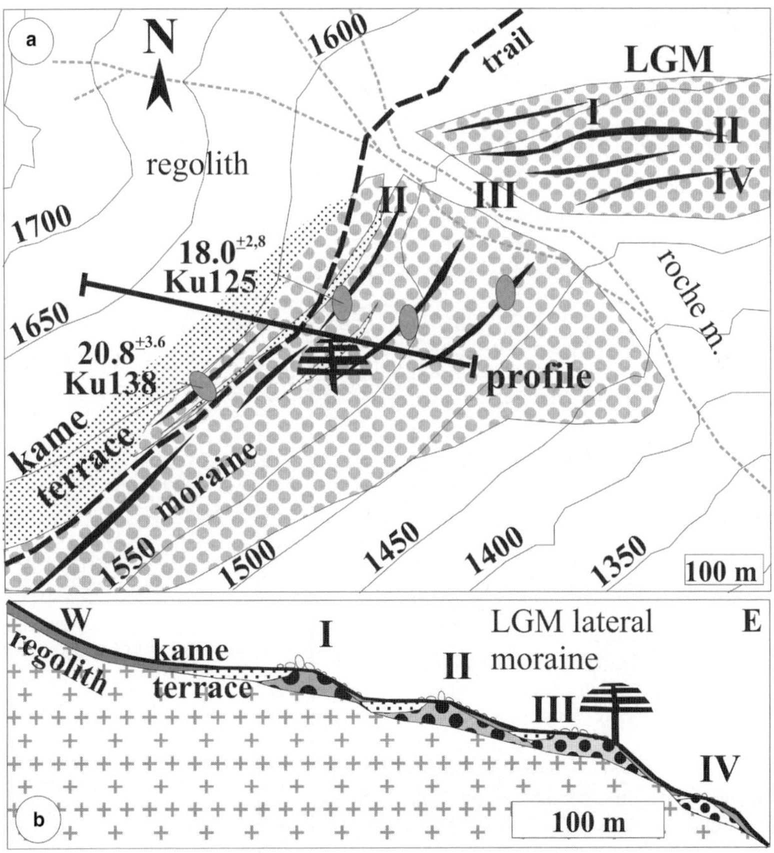

Fig. 5. Sketch map of lateral moraines in northern Corsica, at the northern slope of the Tavignano valley with (**a**) exposure ages of glacial boulders (Table 1) and (**b**) a schematic profile across the set of moraines. LGM stadial moraines are marked with Roman numbers according to their relative depositional age.

Glacial boulders with a top rising more than 1.5 m above the substrate were considered for exposure dating. The results of exposure dating, noted in Table 1 with detailed information, confirm a last glacial maximum (LGM) depositional age. A schematic profile across the lateral moraine succession shows the depositional setting with a stepwise repetition of lateral moraines, which have been superficially reworked by meltwater, each accompanied by its own kame terrace. Concerning ancient levels of soil cover, two dated boulders in this site show two or three different weathering girdles, respectively (Fig. 6). The boulder shown in Figure 6b exposes three tilted subparallel weathering girdles, which indicate that this boulder has been tilted recently. Presumably, this happened during the late Holocene, when its obelisk-like upright position in the substrate became unstable. The highest weathering girdle of the first boulder (Fig. 5a) is found approximately 1 m above the present top of the soil. Downslope, the vertical distance is much larger. In the case of the second boulder, this vertical distance is in the range of 1.2–1.3 m. A shortcoming of the latter site is the existence of a nearby trail that is a modern equivalent of a shepherd's trails, which has been used at least since Roman times. Thus, the obtained downwearing rate is at a maximum value.

The second study site is located at the Vizzavona Pass in central Corsica, within a wet local setting (Kuhlemann *et al.* 2005*a*; for the location see Fig. 4). Similar to the first and numerous other sites, a set of four latero-terminal moraines is exposed at the Vizzavona Pass (Fig. 7 and Table 1). Boulders are generally covered by an approximately 5 cm-thick layer of moss. Intense soil weathering on moderate slopes in most parts of the composite moraine caused the formation of a sieve-type top layer, unsuitable for the formation and detection of a concave soil formation level. Nevertheless, on top of the outmost lateral moraine ridge, close to the pass, a flat has been formed, most probably by meltwater erosion. A gneiss boulder on this flat shows a less obvious weathering girdle compared to granite boulders, owing to spallation of the fractured gneiss. It is placed on smaller boulders that are still largely buried in the moraine. The etched top surface of the gneiss boulder indicates that it has not been significantly tilted during weathering and removal of moraine matrix. The weathering girdle in the southern half of the boulder (Fig. 8, left) is still horizontal, whereas in the northern part of the boulder the weathering girdle, representing the former topmost soil level, is dipping towards the north, subparallel to the present slope. The setting indicates that the removal of about 1.5 m of moraine matrix on the flat is a maximum estimate of natural weathering if the site was unaffected by human activity, which is not the case. The former

Table 1. *Exposure ages of glacial boulders assuming no erosion (column 8) or 15 mm ka^{-1} erosion (column 10; Kuhlemann* et al. *2007), with 1σ error of the measurements (columns 9 and 11, respectively)*

Column 1 Sample code	2 Lat. (°)	3 Long. (°)	4 Altitude (m)	5 Qz (g)	6 $^{10}_{4}Be$ (atoms g^{-1})	7 Local production rate (atoms g^{-1} year^{-1})	8 No erosion exposure age (ka)	9 Error (ka)	10 15 mm ka^{-1} erosion exposure age (ka)	11 Error (ka)	12 Measurement error (%)
Ku 117	42.01	9.23	1405	17.17	25.14	15.99	15.7	1.8	18.9	3.1	4.5
Ku 103	42.11	9.12	1190	16.56	20.30	13.49	15.1	2.0	17.9	3.2	6.9
Ku 110	42.12	9.12	1110	23.07	21.34	12.60	16.9	2.4	20.9	4.1	8.4
Ku 111	42.12	9.12	1130	17.87	18.74	12.81	14.7	2.2	17.4	3.5	9.9
Ku 112	42.12	9.12	1115	24.15	20.02	12.65	15.8	2.5	19.2	4.1	11.2
Ku 138				20.33	31.53		17.0		20.7		5.1
Ku 125	42.28	9.01	1553	20.13	27.54	18.08	15.3	1.8	18.0	2.8	4.2

Neutron attenuation length 155 g cm^{-2} (Gosse & Phillips 2001).
Muon attenuation lengths 1510 g cm^{-2} (negative) and 4320 g cm^{-2} (fast) (Heisinger *et al.* 2001*a*, *b*).
SLHL production rate 5.44 atom g^{-1} year^{-1} (Kubik & Ivy-Ochs 2004).
Muon production rates of Heisinger *et al.* (2002*a*, *b*).
Local production rate with correction for topographic shielding, and shape factor (Masarik & Wieler 2003).
No correction for vegetation, snow or sediment cover.
Blank-corrected, NIST SRM 4325 ^{10}Be-standard, but 1.51 Ma half-life accoridng to Goose & Phillips (2001).
Local production rate after Dunai (2000), corrected for topographic shielding and average temperature during exposure time.

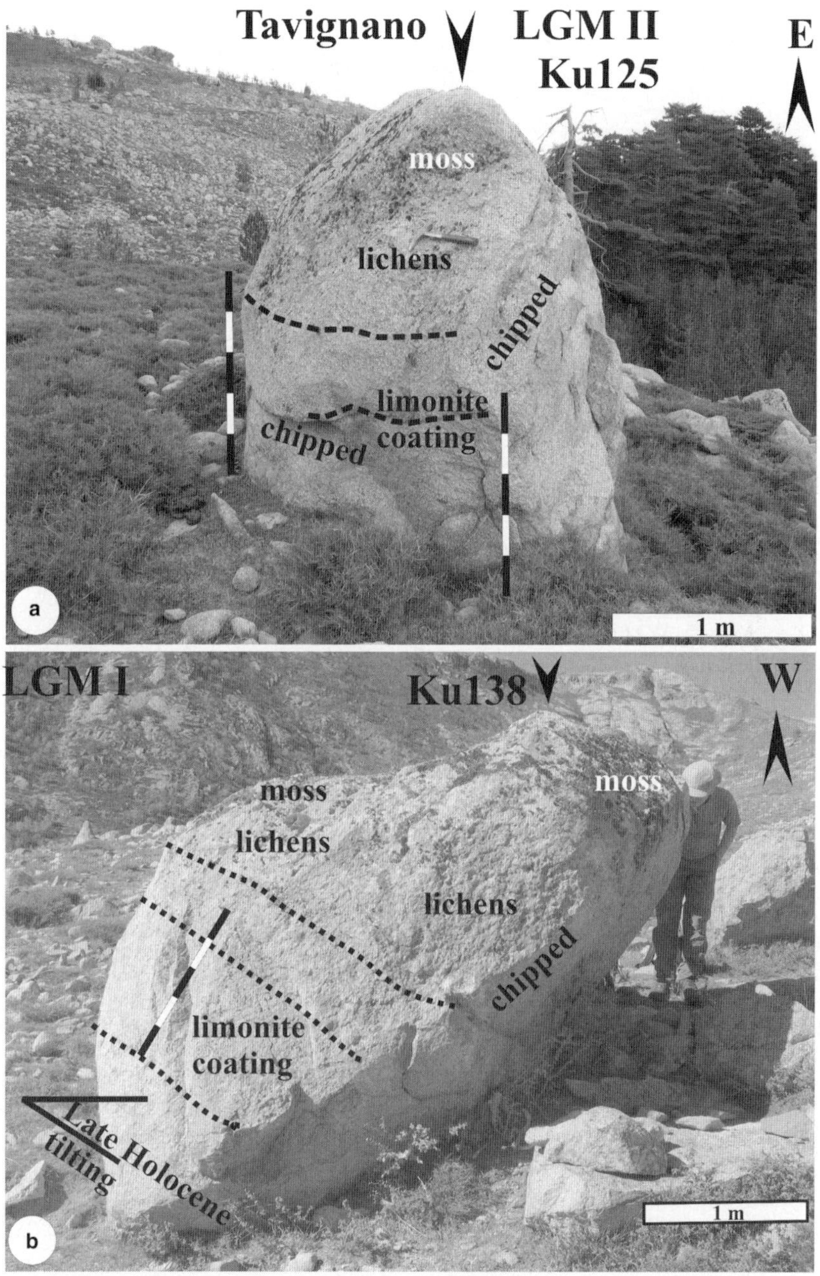

Fig. 6. Glacial boulders marked in Figure 5, exposing weathering girdles (see the text).

beech forest cover of the moraine was most probably removed during the construction of the castle built only 200 m from this site in the fourteenth century. The blind-end 10 cm-deep trail passing the boulder has frequently been used by tourists since the late nineteenth century. Prior to tourism, soldiers may have established the trail. As people were likely to have bypassed the glacial boulder (obstacle) during these times, local downwearing of approximately 1.5 m may, indeed, reflect the natural rate.

The third study site is located within the generally steep easterly slopes of the Verde range in southeastern Corsica, where good preservation of

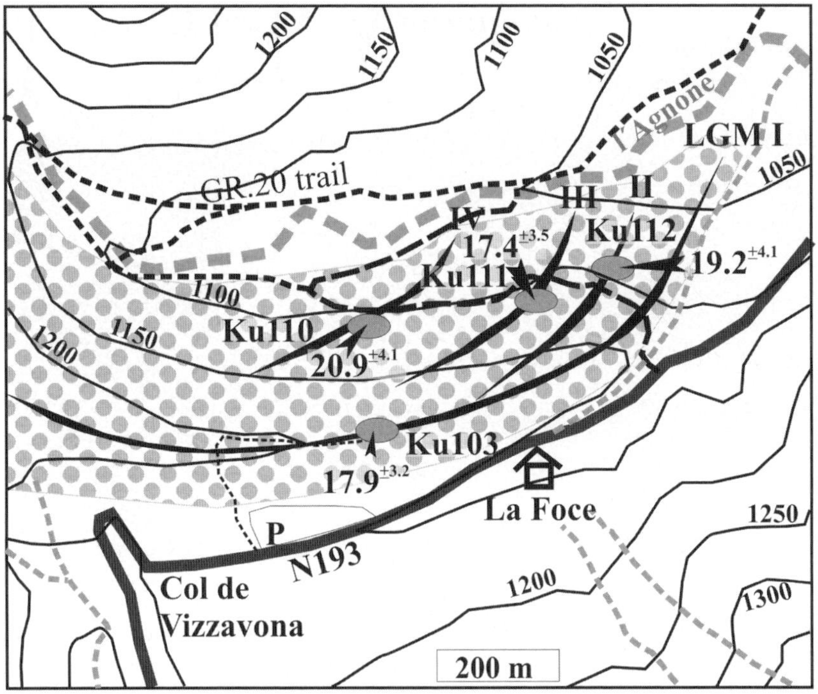

Fig. 7. Sketch map of lateral moraines in central Corsica at the Vizzavona Pass with exposure ages of glacial boulders (Table 1). LGM stadial moraines are marked with Roman numbers according to their relative depositional age.

moraines is generally not favourable. The ridge-parallel strike of normal faults and local listric tilt of a downthrown palaeosurface, however, has facilitated excellent preservation of LGM moraines in one valley (Fig. 9). Mapping of this moraine complex again shows a set of each of four latero-terminal moraines from two joining tributary valleys. In the lower steep valley, sparse relicts of moraines testify to a stronger glacier advance prior to LGM. Large parts of the LGM moraine complex were deposited in the structurally controlled flat and expose numerous examples of boulders with concave weathering features (Fig. 10). Among the best of these boulders is one located close to the trail (Ku115), which displays three weathering girdles on an almost flat substrate (Fig. 10a). The local soil downwearing since deposition is about 0.7–1.0 m. A second example is situated in a dense beech forest in a flat part of the valley floor. A large boulder with a flat top represents an ideal site for exposure dating (Table 1) and also serves as a detection site for a single weathering girdle, 1.2–1.3 m above present substratum (Fig. 10b). Anthropogenic impact in this site is unlikely. Inferred precipitation in this site (Table 2) is almost as high as at the previous site at Vizzavona, due to its higher elevation.

Fig. 8. Glacial boulder marked in Figure 7, exposing a weathering girdle (see the text).

To summarize the observations on weathering girdles formed by soil etching, the selected sites exemplified the local downwearing of granitic moraine matrix since the LGM, which was between 0.7 and 1.5 m. A natural soil downwearing of about 1 m was found to be a reasonable average for the upper montane zone of Corsica (Table 2). The maximum of *c.* 1.5 m was found at the wettest site. According to the local precipitation rates, the

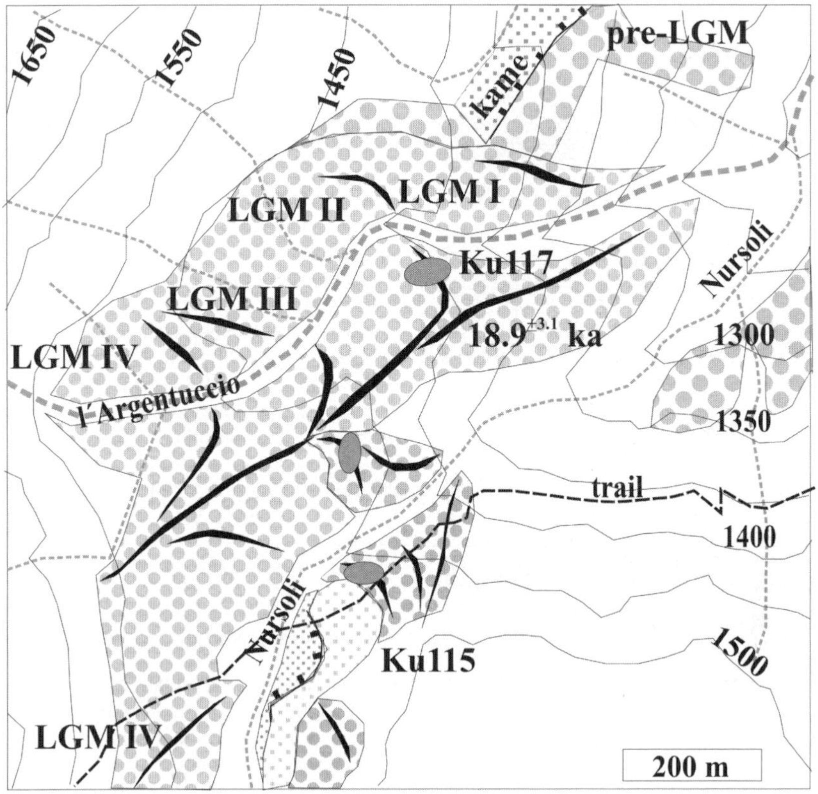

Fig. 9. Sketch map of lateral moraines, named 'Bibus moraine', in southern Corsica with an exposure age of a glacial boulder (Table 1). LGM stadial moraines are marked with Roman numbers according to their relative depositional age.

lowest values should be expected in the northern dry site, but this was not observed. This site is likely to have been subject to anthropogenic impact for millenia. The lowest value is observed in a moderately wet setting, which indicates a high natural variability of soil downwearing.

The timespan of etching activity by humid acids in soil is a matter of discussion. The upper time limit is the termination of the LGM (*c.* 18 ka), whereas the onset of the Holocene (11.7 ka) is the lower age limit. Some soil formation and thus etching could be expected during the Bølling–Allerød interstadial. We assume that about 13 ka is a reasonable estimate of the time taken for soil etching to have an effect on the glacial boulders. Downwearing of moraine matrix and soil ranges between 130 and 40 mm ka^{-1}, with an average of about 77 mm ka^{-1}.

In situ ^{10}Be weathering rates

We observed ^{10}Be concentrations between 0.37×10^6 and 1.25×10^6 atoms g^{-1} (Table 3). The acquisition and interpretation of the data is described in more detail by Kuhlemann *et al.* (2007). We interpret these data in terms of maximum weathering rates, assuming dynamic equilibrium of cosmogenic nuclide production, decay and erosion. This assumption is based on field evidence, which shows that episodic spallation of larger rock pieces from tors on the palaeosurface is rare, whereas fairly regular weathering processes such as dissolution and grusification is dominant. Maximum weathering rates of the palaeosurface samples range between 8.2 and 24 mm ka^{-1} (Table 3 and Fig. 2c). The average weathering rate of the calc-alkaline samples is about 15 mm ka^{-1}.

The weathering rates show a combined lithological and climatological effect. The lowest rate of 8.2 mm ka^{-1} is found in the alkaline granite reference sample, despite the frequent torrential rain at this site. (sample V; Fig. 2). A low weathering rate of 9.1 mm ka^{-1} was found in a calc-alkaline granite in a zone of reduced precipitation (I). It seems that the higher weathering resistance of the alkaline granite (sample V), indicated by the

Fig. 10. Glacial boulders marked in Figure 7, exposing levels of weathering girdles (see the text).

geomorphic setting in Corsica, is counterbalanced by precipitation. Medium weathering rates ranging from 13.6 to 17.6 mm ka^{-1} are found on, or close to, the main drainage divide in windy and rainy sites of weakly sheared calc-alkaline granite (samples III, IV, VII and VIII; Fig. 2). Highest rates of 20 and 24 mm ka^{-1} are found in calc-alkaline granite with intense shearing (II) and a reference sample of sheared granite gneiss (VI), respectively. Our interpretation of the effects of lithology on weathering rates, however, cannot be quantified on the basis of the presented data, but rather provides a semi-quantitative hint to the degree of variation. Moreover, the degree of shearing of granites is difficult to quantify.

A focus on calc-alkaline granite samples exposes a tendency for westwards-increasing erosion rates in northwestern Corsica, which matches the regional precipitation trend. With increasing elevation, precipitation isolines generally rise above the northern interior as a result of foehn effects (Fig. 11). The foehn effect of temperature rise and precipitation drop matches with the change in vegetation from wet and cool maritime forest on the western windward side and dry pine forest on the interior leeward side (Fig. 11). Moreover, the tree line rises from the western slope of Corsica to the interior by about 200 m, indicating a rise in annual temperature of *c.* 1.2 °C. The weathering of bare rock surfaces probably depends not only on the amount of precipitation, but also on the presence of moisture and frequency of wetness, as indicated by vegetation. Weathering rates obtained close to the drainage divide in the NE match with those found in the SW, where precipitation rates are similar.

Long-term erosion rates (AFT dating)

For this study, we evaluate a representative set of four samples of a north–south section of Corsica with respect to pre-Quaternary Neogene cooling, in order to compare these data with weathering and erosion rates obtained from cosmogenic nuclides (Table 4). Evaluation of a long-term west–east gradient of erosion in the west–east cross-section covered by the precipitation profile (Fig. 11) failed as all AFT sampled were devoid of suitable apatite crystals. Thus, a sample in the close vicinity was selected. The time–temperature paths modelled from the AFT age and the track-length data show envelopes of acceptable (light grey) and good (dark grey) and a 'best-fit' (black) line. To assess average cooling during the stagnation period, we chose the time between 17 and *c.* 2 Ma, after a thermo-tectonic event with the removal of 1 km-thick layer of Eocene flysch sediments (Danišík *et al.* 2007), and prior to Quaternary cooling. Apparent slight acceleration of cooling in

Table 2. *Regolith and moraine matrix downwearing since the LGM, according to mapping and exposure dating*

Sample code	Location	Inferred precipitation (mm year^{-1})	15 mm ka^{-1} erosion exposure age (ka)	Minimum post-LGM weathering (m)
Ku 138	Tavignano I	1000	20.8 ± 3.6	1.2–1.3
Ku 125	Tavignano II	1000	18.0 ± 2.8	0.9–1.0
Ku 103	Vizzavona I	1550	17.9 ± 3.2	1.4–1.6
Ku 115	Nursoli III	1450		0.7–1.0
Ku 117	Argentuccio II	1450	18.9 ± 3.1	1.2–1.3

Table 3. *Measured concentration of ^{10}Be in the samples (column 6). For the calculation of erosion rates (column 10) we assume that ^{10}Be production equals erosion ('Steady state'). For comparison, the nominal but meaningless exposure age (zero erosion) is given (column 12). The related 1σ errors are shown in columns 11 and 12, respectively. For parameters and references see Table 1*

Column 1 Sample site	2 Code	3 Altitude (m)	4 N. lat. ° ′ ″	5 E. long. ° ′ ″	6 ^{10}Be ($\times 10^6$ atoms g^{-1})	7 Error ($\times 10^6$ atoms g^{-1})	8 Analyt. error (%)	9 Local ^{10}Be production rate (atoms g^{-1} $year^{-1}$)	10 Erosion rate (mm ka^{-1})	11 1σ error (mm ka^{-1})	12 Nominal exposure age (years)	13 1σ error age (years)
Pinerole/Niolu	I	1759	42 18 47	09 02 55	1.23	0.11	8.94	19.79	9.1	1.0	63161	7311
Plateau d'Ese	II	1759	42 00 00	09 07 06	0.55	0.11	20	19.54	20.3	4.3	28429	6058
Mte Giovanni	III	1950	41 00 00	09 09 14	0.92	0.12	13	22.59	14.0	2.1	41146	6160
Mte Renoso	IV	2352	42 03 36	09 08 04	1.14	0.12	10.5	30.32	15.2	2.0	37932	4870
Bavella	V	1420	41 46 36	09 14 15	1.03	0.25	24.3	15.13	8.2	2.1	69790	17699
Tenda	VI	1453	42 30 58	09 15 02	0.37	0.09	24.3	15.64	24.2	6.2	23871	6066
Cimatella	VII	2098	42 14 13	08 56 28	0.82	0.15	18.3	25.24	17.6	3.5	32764	6459
Punta Artica	VIII	2327	42 15 51	08 58 18	1.25	0.1	8	29.88	13.6	1.5	42244	4589

the Quaternary indicated by the thermal modelling includes an artefact, which only partly exposes atmospheric cooling of 3–4 °C. To estimate a reasonable average cooling rate and its error range, we extracted the upper and lower limits of the 'good' model envelope and the 'best-fit' line to amalgamate an average cooling rate. Atmospheric cooling at sea level from 17 Ma (approximating to a climate optimum) until the Quaternary (1.8 Ma) has to be considered (subtracted). In Corsica, the amount of cooling is not well known, but data from the circum-Alpine realm suggest 3–4 °C of cooling, driven by a drop in winter temperatures (Mosbrugger *et al.* 2005). Atmospheric cooling as forced by surface uplift also has to be considered (subtracted). Geomorphic evidence from southern and eastern Corsica suggests that at 10 Ma a major uplift event occurred that pushed up NE (Alpine) Corsica by more than 1 km. In southern and northwestern Corsica uplift was in the range of 500–900 m. When applying an atmospheric lapse rate of 0.6 °C per 100 m, cooling between 3 and 5.4 °C would result from surface uplift. The combination of cooling by surface uplift, together with regional atmospheric cooling, accounts for *c.* 6–9 °C of total cooling than is not related to the exhumation of rock by erosion which we are trying to detect. The average of 7 °C is of importance if, like in the present case study, stagnation or slow cooling of rocks is found. For the calculation of erosion rates, we used recent heat flow data (Della Vedova *et al.* 1995), which indicates increasing heat flow from the SE (25 °C km^{-1}) to the NW (30 °C km^{-1}) of Corsica.

The modelled time–temperature paths show fast cooling through the apatite partial annealing zone in the Oligocene–early Miocene times followed by a period of modest decrease in temperature or thermal stagnation until present. The fast cooling period was extensively interpreted by Danišík *et al.* (2007) and is not a topic of this paper. In brief, the cooling is related to tectonic and erosional denudation of Variscan basement from below the Alpine nappes and weakly consolidated Eocene flysch cover, which was induced both by the collapse of the Alpine wedge and the opening of Ligurian–Provençal rift system. After the flysch cover was removed by erosion, cooling and erosion rates were strongly reduced when the basement became exposed. Here, a focus is dedicated to a period of stagnation after the Oligocene–early Miocene cooling event.

The lowest average erosion rate of approximately 5 mm ka^{-1} during Neogene stagnation of cooling is found in the southernmost sample, which is from a relatively dry region (2134 m). Towards the north, erosion rates increase to 14 mm ka^{-1} in the Renoso massif (2350 m), and

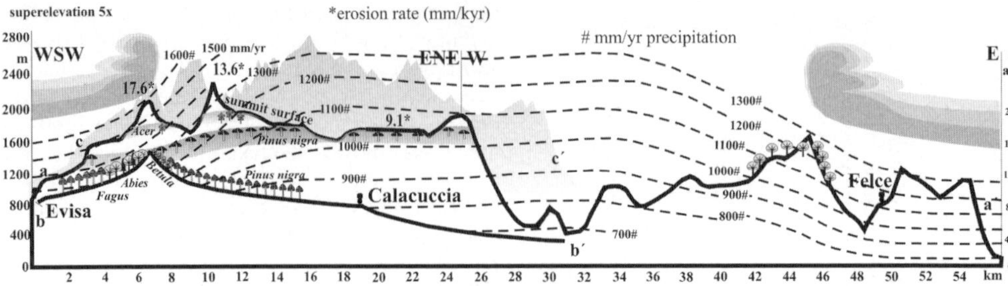

Fig. 11. WSW–ENE cross-sections in northern Corsica to illustrate potential precipitation and vegetation trends (for the location of profiles see Fig. 4a). The lower profile a–a′ connects the climate stations Evisa (1250 mm year^{-1}) and Calacuccia (850 mm year^{-1}) across the mountain pass. The upper profile b–b′ follows the crest of the sampling sites VII, VIII and I, and continues towards the east coast. The precipitation isolines are hand-fitted to the relief and the foehn effect is indicated by vegetation. The uppermost tree line is represented by isolated stands of *Acer pseudoplatanus* (and *Sorbus aucuparia*). A continuous tree line is formed by *Pinus nigra ssp laricio* and *Fagus sylvatica*.

to 20 mm ka^{-1} in northern-central Corsica in the regions of Mte d'Oro and Niolo, both in grusy granite at altitudes of 2250 and 1350 m, respectively, exposed to roughly similar precipitation conditions. Summarizing these data, we note that erosion rates in most of the Miocene–late Pliocene range between 5 and 20 mm ka^{-1} in calc-alkaline granite. In average at all four sites, a Neogene long-term erosion rate of about 13 mm ka^{-1} is found on the basis of AFT data and track-length measurements. In the granite gneiss of NE Corsica (Tenda massif), long-term average erosion ranges between 15 and 20 mm ka^{-1} over the last 12 Ma (Fellin *et al.* 2005), matching the aforementioned erosion rates of calc-alkaline granites of northern Corsica. The latter sampling site at an elevation of 1509 m is located 1 km north of our ^{10}Be sample IV (1453 m).

Comparison of short- and long-term erosion rates

Average denudation rates of approximately 15 mm ka^{-1}, obtained by cosmogenic nuclide dating of calc-alkaline granite rock castles on palaeosurfaces, for the last 10^5 years are in the range of those calculated from modelled cooling paths of apatite fission-track data for large parts of the Neogene (*c.* 20–3 Ma), despite the broad range of time frames studied (Fig. 12). The data do not support a late Neogene increase in erosion rates on palaeosurfaces. In fact, glacial relief is progressively destroying the palaeosurfaces and average regional erosion rates must have increased as a result of cyclic glaciation.

For an independent assessment of the effect of lithology on weathering rates, geomorphic evidence is available for long-term perspectives. Mountains of alkaline granite in Corsica stand about 150 m above surrounding calc-alkaline granite, into which they intruded during Permian times. If average weathering rates of *c.* 15 mm ka^{-1} in calc-alkaline granites were twice as high as those measured for alkaline granite (8.2 $\pm$ 2.1 mm ka^{-1}; 7.5 mm ka^{-1} calculated for simplicity), it would take 20 Ma of downwearing for 300 m of calc-alkaline granite and 150 m for alkaline granite, if weathering started at a plane surface and vertical tectonic offset between these two granites is excluded. This is just the amount of time available since the removal of the flysch cover (Fig. 12).

Comparing average soil downwearing rates of approximately 77 mm ka^{-1} in the last 10^4 years with average erosion rates of solid calc-alkaline granites of 13 mm ka^{-1}, a 6 times faster weathering of regolith on moraines as a result of etching humic acids and high internal surface of the sediment appears reasonable. This number implies that weathering of a 10 m-thick moraine deposit from the penultimate glaciation would leave nothing but some loose glacial boulders behind that could only survive weathering if their original diameter was larger than 4 m.

A weathering rate of about 77 mm ka^{-1} of granitic moraine matrix, however, is not applicable to the rate of regolith formation from solid granite in the subsurface, although the uppermost level of both unconsolidated material may look similar. For an estimate of long-term regolith formation rate, geomorphic evidence in southernmost Corsica provides some hints. This region has never been affected by glaciation, just by periglacial processes. Calc-alkaline rock castles and bell-shaped peaks (Fig. 13) rise up to 300 m above flat or hilly regolith-filled valleys (Kuhlemann *et al.* 2005*b*). If within the last 20 Ma (starting between 25 and 17 Ma; Fig. 11) 300 m of calc-alkaline granite were removed by

Table 4. *Apatite fission-track data with analytical results, parameters and references**†

Sample code	UTM32/WGS84 X	UTM32/WGS84 Y	Altitude (m)	N	ρ_s	N_s	ρ_i	N_i	ρ_d	N_d	$P(\chi^2)$ (%)	Age (Ma)	$\pm 1\sigma$ (Ma)	MTL (µm)	SD (µm)	SE (µm)	N (L)	D_{par} (µm)
XC-46	510622	4656974	2247	36	1.006	112	4.022	448	5.517	4936	100	23.4	2.6	14.2	1.2	0.1	65	1.63
XC-75	517225	4633070	2133	25	2.912	230	10.013	791	5.902	5647	100	28.5	2.3	14.3	1.0	0.1	46	1.79
XC-98	505899	4664212	2254	25	1.666	143	8.052	691	6.679	5647	100	22.9	2.2	14.3	1.2	0.1	68	1.74
XC-103	495552	4672943	1375	25	1.713	150	8.601	753	6.029	5647	100	19.9	1.9	13.9	1.2	0.1	71	1.51

*Note please that all fission track data are published in Danišík *et al.* (2007). All samples are Variscan granites.

†N, number of dated apatite crystals; ρ_s (ρ_i), spontaneous (induced) track densities ($\times 10^5$ tracks cm^{-2}); N_s (N_i), Number of counted spontaneous (induced) tracks; ρ_d, dosimeter track density ($\times 10^5$ tracks cm^{-2}); N_d, number of tracks counted on dosimeter; $P(\chi^2)$, probability obtaining chi-square value (χ^2) for n degrees of freedom (where n = no. of crystals − 1); Age ± 1σ, central age ± 1 standard error (Galibraith & Laslett 1993); MTL, mean track length; SE, standard error of mean track length; SD, standard deviation of track length distribution; N(L), number of horizontal confined tracks measured; D_{par}, average etch-pit diameter of fission tracks. Ages were calculated using zeta calibration method (Hurford & Green 1983), glass dosimeter CN-5, and zeta value of 324.93 ± 6.46 year cm^{-2}.

weathering (see earlier), a local relief of 300 m implies that regolith formation operates about twice as fast as bare rock weathering (*c.* 25 mm ka^{-1}), in order to produce 300 m of local relief. Regolith formation would operate 3 times slower than moraine weathering.

Discussion

The compilation of erosion rates by Cockburn & Summerfield (2004) noted quite similar magnitudes for seasonally wet temperate–alpine settings of western America and SE Australia. The impact of precipitation, degree of deformation and petrography on erosion rates, as tentatively indicated by our data, is apparently in conflict with conclusions drawn from studies of ^{10}Be concentrations in small catchments of various climatic setting (Riebe *et al.* 2000; Von Blanckenburg 2005). These studies, however, show a fairly wide scatter of data, particularly in mid-latitudes, whereas our study cannot provide enough data for a statistically robust assessment of the quantitative importance of the controlling factors. The data of Riebe *et al.* (2000), however, confirm that ongoing tectonic activity strongly increases erosion rates, possibly by an order of magnitude. Thus, second-order factors such as climate, degree of deformation and petrography can only be detected if study areas, such as Corsica, are tectonically inactive, including those that are passively uplifted.

An important control of precipitation for erosion rates in tectonically active mountain ranges, such as southern New Zealand, has been found as a major reason for asymmetric exhumation, confirmed by thermochronological data (Batt *et al.* 2000) and surface process modelling (Batt & Braun 1999). In the southern New Zealand, eroded rocks are strongly deformed and there appears to be a more or less linear relationship between erosion and precipitation rates. In this tectonically active setting, regional-scale erosion depends on the transport capacity of rivers, which depends on precipitation. In the passive setting of Corsica, the weathering rate during the etching of bare granite may also increase roughly linearly with precipitation. In support of this hypothesis, Brady *et al.* (1999) report a linear relationship for plagioclase weathering and rainfall in Hawaii under recent conditions. If such a linear relationship for precipitation and weathering rates existed for humid–semi-humid conditions, and moderate precipitation rates existed as a general rule, Miocene average precipitation in Corsica should have been roughly similar to that of the present.

Reconstructions of Mediterranean palaeoprecipitation in the Miocene indicate a strong

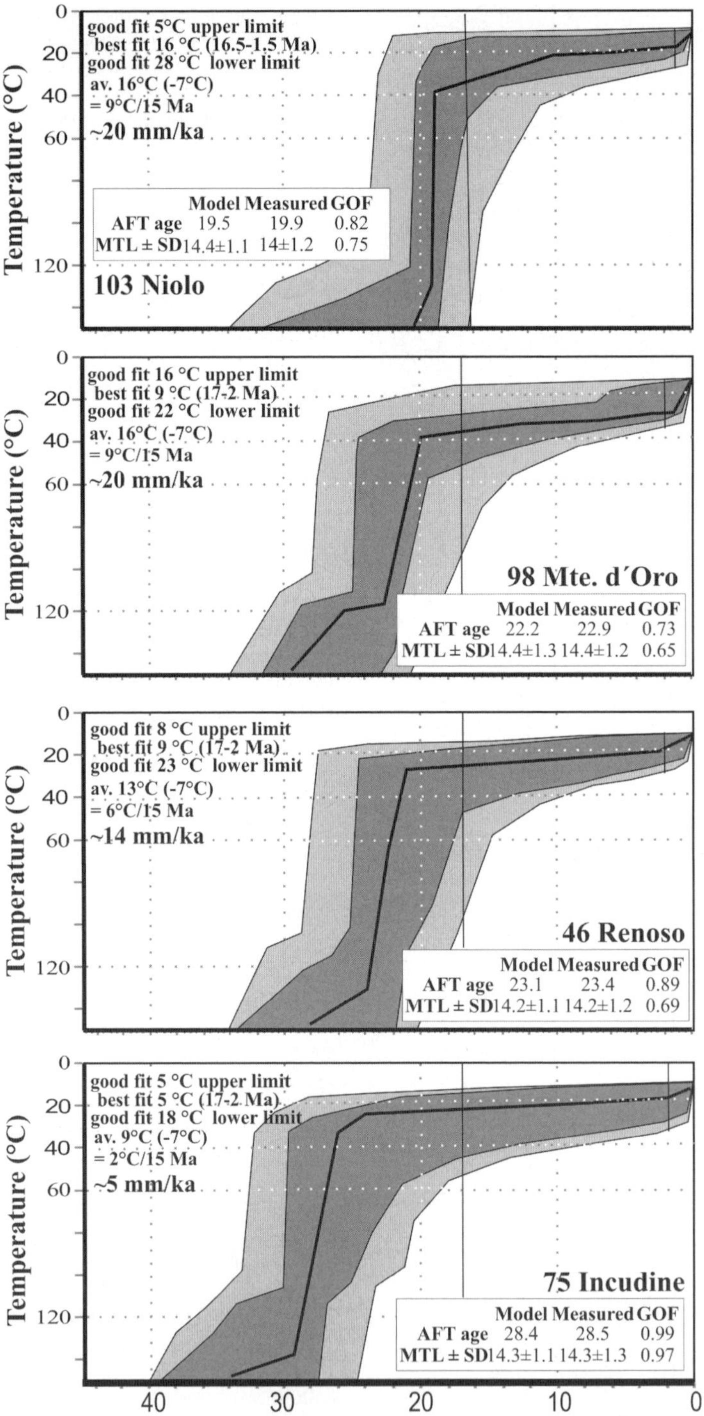

Fig. 12. Cooling paths of samples taken from the palaeosurface relicts based on the modelling of fission-track ages and track lengths (Ketcham 2005). Results are displayed in time–temperature diagrams (left-hand diagrams). Light grey paths, acceptable fit; dark grey paths, good fit; thick black line, best fit; MTL, mean track length in μm; SD, standard deviation in μm; GOF, goodness of fit (statistical comparison of the measured input data and modelled output data, where a 'good' result corresponds to a value of 0.5 or higher; 'the best' result corresponds to a value of 1). Note that the modelled cooling paths are valid only inside the temperature range of 60–120 °C, and outside this temperature range they must not necessarily represent the real thermal trajectory of a sample.

Fig. 13. Rock castle in southern Corsica, rising 300 m above the ancient valley floor on the palaeosurface.

gradient from a dry southern Iberia to a humid central Europe (Jiménez-Moreno & Suc 2007) and a dry southern Greece to the western Paratethys margin (Bruch *et al.* 2007). Temporal variation in time is quite low throughout the Miocene (Mosbrugger *et al.* 2005), although punctuated by a wet event at about 10 Ma (Boehme *et al.* 2006). Southern lowlands of France were characterized by precipitation rates of between 1200 and 1300 mm year^{-1} in the Middle Miocene (Bruch *et al.* 2007), and, hence, Miocene precipitation rates in near-coast sites were much higher than at present. This also applies to the NW Mediterranean realm in Pliocene times (Fauquette *et al.* 1999). In Corsica semi-humid conditions are confirmed by flora remains (Ferrandini & Loÿe-Pilot 1992), such as *Magnolia* and *Laurus*, which are not typical of very dense vegetation cover. Charcoals are frequently found in Middle Miocene near-shore sediments, indicating that forest fires occurred often. However, relief-controlled landward increase in precipitation in the mountains may have been less strong in the Miocene, owing to convective-type rather than advective-type precipitation compared to the present. To our knowledge, the relative role of these precipitation types has not yet been quantitatively assessed for the Miocene climate. The Miocene elevation of Corsica's palaeosurface above sea level between 16 and 10 Ma is not well constrained, but was probably several hundreds of metres lower than that after 10 Ma, by which time a major phase of uplift had occurred (Kuhlemann *et al.* 2005*b*). Before 17 Ma, southernmost Corsica was hilly. Lower elevation would probably have caused less precipitation. Therefore, we speculate that the decrease in precipitation in Corsica at sea level between the Middle Miocene optimum climate and the late Pliocene has been compensated for by surface uplift, which triggered enhanced precipitation. At present, precipitation rates increase by 60–70 mm year^{-1} for 100 m elevation, deduced from data of Bruno *et al.* (2001). This would increase precipitation at 700 m a.s.l., as for the abovementioned estimated uplift around 10 Ma, up to around 1050 mm year^{-1}, compared to 600 mm year^{-1} at sea level. Other factors controlling weathering rates, such as the protection of the uplifted palaeosurface by dense vegetation cover and sealing soil, are difficult to assess for geological timescales. Preserved Middle Miocene reddish palaeosoils in the eastern plain (Durand-Delga 1978), however, indicate that fair infiltration properties are not compatible with the sealing of the substrate.

The comparison of short- and long-term erosion rates, as revealed from ^{10}Be concentrations and AFT ages and modelled cooling paths from track-length measurements, remains controversial as during stagnation periods our approach of combining the most likely cooling path with the upper and lower limits of the 'good-fit' model scenario still includes an error range that is hard to assess. The apparent long-term decrease in erosion rates towards the less high and drier south, as a result of a long-term precipitation gradient, is of striking simplicity.

Conclusions

Geomorphic evidence, combined with thermochronological data and cosmogenic exposure ages, is shown to be a valuable tool for assessing long-term rates of surface processes. Comparing process rates obtained by different methods, the effects of climate, lithology and fracturing can better discriminated.

On average, calc-alkaline granites are weathering at average rates of approximately 15 ± 6 mm ka^{-1} and 13 ± 8 m Ma^{-1}, as reflected by short-term ^{10}Be weathering rates and long-term thermochronological data, respectively. The generation of local relief by granite etching in the last *c.* 18 Ma indicates that regolith formation from solid granite operates at rates about twice as fast. Regolith formation would operate 3 times slowes than moraine weathering. As a consequence, relatively fast weathering of granitic moraine matrix in a temperate climate with approximately 1300 mm year^{-1} of annual precipitation would degrade a moraine to an accumulation of rotated glacial boulders within tens of thousands of years, without any fluvial erosion. Granite weathering of glacial boulders would destroy all remnants of the penultimate glaciation, except for a few boulders with diameters in excess of 4 m at the time of deposition.

This study has been funded by the German Science Foundation (DFG Projects KU1298/2, 1298/3 and 1298/7). We are grateful to G. Höckh, D. Kost and D. Mühlbayer-Renner (Tübingen) for quartz purification, and to Dr P. Kubik (Zürich) for measuring the exposure ages of glacial boulders. We would like to thank P. Bierman and C. Schlüchter for constructive reviews. We are also grateful to the editors for their support and constructive comments.

References

ABRAMOWSKI, U. 2004. *The Use of ^{10}Be Exposure Dating of Erratic Boulders in the Reconstruction of the Late Pleistocene Glaciation History of Mountainous Regions, With Examples From Nepal and Central Asia.* PhD thesis, http://opus.ub.uni-bayreuth.de/volltexte/2005/150/pdf/dissua04.pdf, 167 pp. TEBESEA: http://opus.ub.uni-bayreuth.de.

ANDERSON, R. S. 2002. Modeling of tor-dotted crests, bedrock edges and parabolic profiles of the high alpine surfaces of the Wind River Range, Wyoming. *Geomorphology*, **46**, 35–58.

BATT, G. E. & BRAUN, J. 1999. The tectonic evolution of the Southern Alps, New Zealand; insights from fully thermally coupled dynamical modelling. *Geophysical Journal International*, **136**, 403–420.

BATT, G. E., BRAUN, J., KOHN, B. P. & MCDOUGALL, I. 2000. Thermochronological analysis of the dynamics of the Southern Alps, New Zealand. *GSA Bulletin*, **112**, 250–266.

BIERMAN, P. & TURNER, J. 1995. ^{10}Be and ^{26}Al evidence for exceptionally low rates of Australian bedrock incision and the likely existance of pre-Pleistocene landscapes. *Quaternary Research*, **44**, 378–382.

BOEHME, M., ILG, A., OSSIG, A. & KÜCHENHOFF, H. 2006. A new method to estimate paleoprecipitation using fossil amphibians and reptiles and the Middle and Late Miocene precipitation gradients in Europe. *Geology*, **34**, 425–428.

BRADY, P. V., DORN, R. I., BRAZEL, A. J., CLARK, J., MOORE, R. B. & GLIDEWELL, T. 1999. Direct measurement of the combined effects of lichen, rainfall, and temperature on silicate weathering. *Geochemica et Cosmochemica Acta*, **63**, 3293–3300.

BRUCH, A. A., UHL, D. & MOSBRUGGER, V. 2007. Miocene climate in Europe – Patterns and evolution. A first synthesis of NECLIME. *Palaeogeography, Palaeoclimatology, Palaeoecology*, **253**, 1–7.

BRUNO, C., DUPRÉ, G., GIORGETTI, G., GIORGETTI, J. P. & ALESANDRI, J. 2001. *Chì tempu face? Méteorologie, climat et microclimates de la Corse.* CNDP-CRDP de Corse/Méteo France. Ajaccio, France.

CACHO, I., GRIMALT, J. O. & CANALS, M. 2002. Response of the western Mediterranean Sea to rapid climatic variability during the last 50,000 years: a molecular biomarker approach. *Journal of Marine Systems*, **33–34**, 253–272.

CAVAZZA, W., ZATTIN, M., VENTURA, B. & ZUFFA, G. G. 2001. Apatite fission-track analysis of Neogene exhumation in northern Corsica (France). *Terra Nova*, **13**, 51–57.

CLARK, D. H., BIERMAN, P. R. & LARSEN, P. 1995. Improving in situ cosmogenic chronometers. *Quaternary Research*, **44**, 367–377.

COCKBURN, H. A. P. & SUMMERFIELD, M. A. 2004. Geomorphological applications of cosmogenic isotope analysis. *Progress in Physical Geography*, **28**, 1–42.

COCKBURN, H. A. P., BROWN, R. W., SUMMERFIELD, M. A. & SEIDL, M. A. 2000. Quantifying passive margin denudation and landscape development using a combined fission-track thermochronology and cosmogenic isotope analysis approach. *Earth and Planetary Science Letters*, **179**, 429–435.

DANIŠÍK, M. 2005. *Cooling History and Relief Evolution of Corsica (France) as Constrained by Fission Track and (U–Th)/He Thermochronology.* Tübinger Geowissenschaftliche Arbeiten, Reihe A, **72**.

DANIŠÍK, M., KUHLEMANN, J., DUNKL, I., SZÉKELY, B. & FRISCH, W. 2007. Burial and exhumation of Corsica (France) in the light of fission track data. *Tectonics*, **26**, TC1001, doi: 10.1029/2005TC001938.

DELLA VEDOVA, B., LUCAZEAU, F., PASQUALE, V., PELLIS, G. & VERDOYA, M. 1995. Heat flow in the tectonic provinces crossed by the southern segment of the European Geotraverse. *Tectonophysics*, **244**, 57–74.

DUNAI, T. J. 2000. Scaling factors for production rates of in-situ produced cosmogenic nuclides: a critical reevaluation. *Earth and Planetary Science Letters*, **176**, 157–169.

DUNAI, T. J. 2001. Reply to comment on 'Scaling factors for production rates of in situ produced cosmogenic nuclides: a critical reevaluation' by Darin Desilets, Marek Zreda and Nathaniel Lifton. *Earth and Planetary Science Letters*, **188**, 289–298.

DURAND-DELGA, M. 1978. *Corse – Guides géologiques régionaux.* Masson, Paris.

EGAL, E. 1992. Structures and tectonic evolution of the external zone of Alpine Corsica. *Journal of Structural Geology*, **14**, 1215–1228.

FAUQUETTE, S., SUC, J.-P. *ET AL.* 1999. Climate and biomes in the West Mediterranean area during the Pliocene. *Palaeogeography, Palaeoclimatology, Palaeoecology*, **152**, 15–36.

FELLIN, M. G., ZATTIN, M., PICOTTI, V., REINERS, P. W. & NICOLESCU, S. 2005. Relief evolution in northern Corsica (western Mediterranean): constraints on uplift and erosion at the long-term and short-term time scales. *Journal of Geophysical Research*, **110**, F01016, doi: 10.1029/2004JF000167.

FERRANDINI, A. & LOŸE-PILOT, M. D. 1992. Téctonique en distension et décrochement au Burdigalian–Tortonian en Corse: l'example du bassin Francardo-Ponte-Leccia (Corse Centrale). *Géologie Alpine, Série Spéciale, Résumés de colloques*, **1**, 30–31.

GALBRAITH, R. F. & LASLETT, G. M. 1993. Statistical models for mixed fission track ages. *Nuclear Tracks Radiation and Measurements*, **21**, 459–470.

GLEADOW, A. J. W. & DUDDY, I. R. 1981. A natural long-term annealing experiments for apatite. *Nuclear Tracks and Radiation Measurements*, **5**, 169–174.

GOSSE, J. C. & PHILLIPS, F. M. 2001. Terrestrial in situ cosmogenic nuclides: theory and application. *Quaternary Science Reviews*, **20**, 1475–1560.

HEISINGER, B., LAL, D. *ET AL*. 2002*a*. Production of selected cosmogenic radionuclides by muons 1. Fast muons. *Earth and Planetary Science Letters*, **200**, 345–355.

HEISINGER, B., LAL, D., JULL, A. J. T., KUBIK, P., IVY-OCHS, S., KNIE, K. & NOLTE, E. 2002*b*. Production of selected cosmogenic radionuclides by muons 2. Capture of negative muons. *Earth and Planetary Science Letters*, **200**, 357–369.

HURFORD, A. J. & GREEN, P. F. 1983. The zeta age calibration of fission-track dating. *Chemical Geology*, **1**, 285–317.

JIMÉNEZ-MORENO, G. & SUC, J.-P. 2007. Middle Miocene latitudinal climatic gradient in Western Europe: evidence from pollen records. *Palaeogeography, Palaeoclimatology, Palaeoecology*, **253**, 208–225.

KETCHAM, R. A. 2005. Forward and inverse modeling of low-temperature thermochronometry data. *In*: REINERS, P. W. & EHLERS, T. A. (eds), *Low-Temperature Thermochronology: Techniques, Interpretations, and Applications. Reviews in Mineralogy and Geochemistry*, **58**, 275–314.

KLAER, W. 1956. Verwitterungsformen im Granit auf Korsika. *Petermanns Geographische Mitteilungen/Ergänzungsheft*, **261**, 1–146.

KLEIN, J. & GOSSE, J. C. 2002. *Production rates of ^{10}Be and ^{26}Al in mid-latitudes and high altitudes*. Goldschmidt Conference Abstracts A403.

KOHL, C. P. & NISHIIZUMI, K. 1992. Chemical isolation of quartz for measurement of in-situ-produced cosmogenic nuclides. *Geochimica et Cosmochimica Acta*, **56**, 3583–3587.

KUBIK, P. W. & IVY-OCHS, S. 2004. A re-evaluation of the 0–10 ka 10Be production rate for exposure dating obtained from the Köfels (Austria) landslide. *Nuclear Instruments and Methods in Physics Research Section B*, **223–224**, 618–622.

KUBIK, P. W., IVY-OCHS, S., MASARIK, J., FRANK, M. & SCHLÜCHTER, C. 1998. ^{10}Be and ^{26}Al production rates deduced from an instantaneous event within the dendro-calibration curve, the landslide of Köfels, Ötz valley, Austria. *Earth and Planetary Science Letters*, **161**, 231–241.

KUHLEMANN, J., FRISCH, W., SZÉKELY, B., DUNKL, I., DANIŠÍK, M. & KRUMREI, I. 2005*a*. Würmian maximum glaciation in Corsica. *Austrian Journal of Earth Sciences*, **97**, 68–81.

KUHLEMANN, J., SZÉKELY, B., FRISCH, W., DANIŠÍK, M., DUNKL, I., MOLNÁR, G. & TIMÁR, G. 2005*b*. DEM analysis of mountainous relief in a crystalline basement block, Cenozoic relief generations in Corsica (France). *Zeitschrift für Geomorphologie, N.F.*, **49**, 1–21.

KUHLEMANN, J., VAN DER BORG, K., DANIŠÍK, M. & FRISCH, W. 2007. In situ ^{10}Be-rosion rates in granites of subalpine Miocene paleosurfaces in the western Mediterranean (Corsica, France). *International Journal of Earth Sciences*, **97**, 549–564, doi: 10.1007/s00531-007-0169-z.

KUHLEMANN, J., KRUMREI, I., ROHLING, E., KUBIK, P., IVY-OCHS, S. & KUCERA, M. 2008. Regional synthesis of Mediterranean atmospheric circulation during the Last Glacial Maximum. *Science*, **321**, 1338–1340, doi: 10.1126/science.1157638.

LAL, D. 1991. Cosmic ray labeling of erosion surfaces, in situ nuclide production rates and erosion models. *Earth and Planetary Science Letters*, **104**, 424–439.

MASARIK, J. & WIELER, R. 2003. Production rates of cosmogenic nuclides in boulders. *Earth and Planetary Science Letters*, **216**, 201–208.

MOSBRUGGER, V., UTESCHER, T. & DILCHER, D. L. 2005. Cenozoic continental climatic evolution of Central Europe. *Proceedings of the National Academy of Sciences*, **102**, 14,964–14,969, doi: 10.1073pnas.0505267102.

NISHIIZUMI, K., WINTERER, E. L., KOHL, C. P., KLEIN, J., MIDDLETON, R., LAL, D. & ARNOLD, J. R. 1989. Cosmic ray production rates of ^{10}Be and ^{26}Al in quartz from glacially polished rocks. *Journal of Geophysical Research*, **94**, 17,907–17,915.

RIEBE, C. S., KIRCHNER, J. W., GRANGER, D. E. & FINKEL, R. C. 2000. Erosional equilibrium and disequilibrium in the Sierra Nevada, inferred from cosmogenic ^{26}Al and ^{10}Be in alluvial sediment. *Geology*, **28**, 803–806.

ROSSI, P. & COCHERIE, A. 1991. Genesis of a Variscan batholith; field, petrological and mineralogical evidence from the Corsica–Sardinia Batholith. *Tectonophysics*, **195**, 319–346.

SKLAR, L. S. & DIETRICH, W. E. 2001. Sediment and rock strength controls on river incision into bedrock. *Geology*, **29**, 1087–1090.

SMALL, E. E. & ANDERSON, R. S. 1998. Pleistocene relief production in Laramide mountain ranges, western United States. *Geology*, **26**, 123–126.

SMALL, E. E., ANDERSON, R. S., REPKA, J. L. & FINKEL, R. 1997. Erosion rates of alpine bedrock summit surfaces deduced from in situ ^{10}Be and ^{26}Al. *Earth and Planetary Science Letters*, **150**, 413–425.

STONE, J. O., BALLANTYNE, C. K. & FIFIELD, L. K. 1998. Exposure dating and validation of periglacial weathering limits, northwest Scotland. *Geology*, **26**, 587–590.

VIGLIOTTI, L. & LANGENHEIM, V. E. 1995. When did Sardinia stop rotating? New paleomagnetic results. *Terra Nova*, **7**, 424–435.

VON BLANCKENBURG, F. 2005. The control mechanisms of erosion and weathering at basin scale from cosmogenic nuclides in river sediment, *Earth and Planetary Science Letters* (Frontiers Article), **237**, 462–479.

WITTMANN, H., VON BLANCKENBURG, F., KRUESMANN, T., NORTON, K. P. & KUBIK, P. W. 2007. The relation between rock uplift and denudation from cosmogenic nuclides in river sediment in the Central Alps of Switzerland. *Journal of Geophyical Research – Earth Surface*, **112**, 1–20, F04010, doi:10.1029/2006JF000729.

ZARKI-JAKNI, B., VAN DER BEEK, P., POUPEAU, G., SOSSON, M., LABRIN, E., ROSSI, P. & FERRANDINI, J. 2004. Cenozoic denudation of Corsica in responce to Ligurian and Tyrrhenian extension: results from apatite fission-track thermochronology. *Tectonics* **23**, TC1003, doi: 10.1029/2003TC001535.

Multi-method chronometry of the Teletskoye graben and its basement, Siberian Altai Mountains: new insights on its thermo-tectonic evolution

JOHAN DE GRAVE[1]*, MIKHAIL M. BUSLOV[2], PETER VAN DEN HAUTE[1], JAMES METCALF[3,4], BORIS DEHANDSCHUTTER[5] & MICHAEL O. MCWILLIAMS[3]

[1]*Laboratory of Mineralogy and Petrology, Ghent University, Krijgslaan 281/S8, 9000 Gent, Belgium*

[2]*Institute of Geology, Siberian Branch of the Russian Academy of Science, Koptyuga Avenue 3, 630090 Novosibirsk, Russia*

[3]*Department of Geological and Environmental Sciences, 450 Serra Mall, Braun Hall, Stanford University, Stanford, CA 94305-2115, USA*

[4]*Present address: Department of Earth Sciences, Heroy Geology Laboratory, Syracuse University, Syracuse, NY 13244-1070, USA*

[5]*Federal Agency for Nuclear Control, Ravensteinstraat 36, 1000 Brussels, Belgium*

**Corresponding author (e-mail: johan.degrave@ugent.be)*

Abstract: The Altai Mountains form an intracontinental, transpressive deformation belt in the NW Central Asian orogenic system. Using a multi-method chronometric approach, the thermo-tectonic history of the basement underlying the Teletskoye graben area is constrained in more detail. The results provide new insights into the Siberian Altai basement evolution from the Early Palaeozoic to the present. Zircon SHRIMP (sensitive high-resolution ion microprobe) U–Pb ages (Late Ordovician–Early Silurian, 460–420 Ma) indicate an earlier crystallization age for the basement granitoids than previously thought (Late Devonian–Early Carboniferous, 370–350 Ma), while new multi-mineral $^{40}Ar/^{39}Ar$ age spectra suggest continuous basement cooling throughout the Devonian–Carboniferous. Reactivation of long-lived Palaeozoic structures controls the Teletskoye graben formation since the Plio-Pleistocene as a distant effect of India–Eurasian convergence. Deformation is propagated through Central Asia and Siberia along an inherited structural network closely associated with its basement fabric. A similar reactivation affected the Altai during the Mesozoic. Modelled apatite fission-track data suggest Late Jurassic–Cretaceous (150–80 Ma) cooling, interpreted to be related to denudation and the tectonic reactivation that we link to the coeval Mongol–Okhotsk orogeny. From the Late Cretaceous until the Pliocene, the thermal history models indicate a period of stability. Roughly around 5 Ma ago renewed cooling is observed that possibly represents the denudation and growth of the present-day Altai, and provides the context for the Teletskoye graben formation. A modelled Late Cenozoic cooling can be a result of, or overemphasized by, a modelling artefact. Some caution should be taken not to overinterpret this cooling phase.

The Altai Mountains represent an active intracontinental mountain belt located at the border zone of Siberia (Russia–Gorny Altai Republic), Kazakhstan, Mongolia and China (Fig. 1). Our study deals with the Siberian Altai Mountains, in particular with the area around Lake Teletskoye (NE Siberian Altai; Fig. 1). Lake Teletskoye occupies a young, narrow graben basin that formed less than 2 Ma ago (Dehandschutter *et al.* 2002) (Fig. 2) as an extensional feature between the Gorny Altai and West Sayan tectonic units (Fig. 3). The basin exhibits a typical graben morphology and is composed of two distinct parts: the north–south-oriented main southern graben; and a smaller northern section in an east–west direction, separated by a branch of the West Sayan fault zone, expressed as the submerged Lepnova ridge (Fig. 2). The southern graben is long (77 km), narrow (4 km) and bordered by steep fault-controlled slopes rising 500 m (north) to over 2000 m (south) above the lake level. The lake is 325 m deep and accumulated 800 m of sediments since its formation about 2 Ma ago (Seleznev *et al.* 2001).

From: LISKER, F., VENTURA, B. & GLASMACHER, U. A. (eds) *Thermochronological Methods: From Palaeotemperature Constraints to Landscape Evolution Models*. Geological Society, London, Special Publications, **324**, 237–259.
DOI: 10.1144/SP324.17 0305-8719/09/$15.00

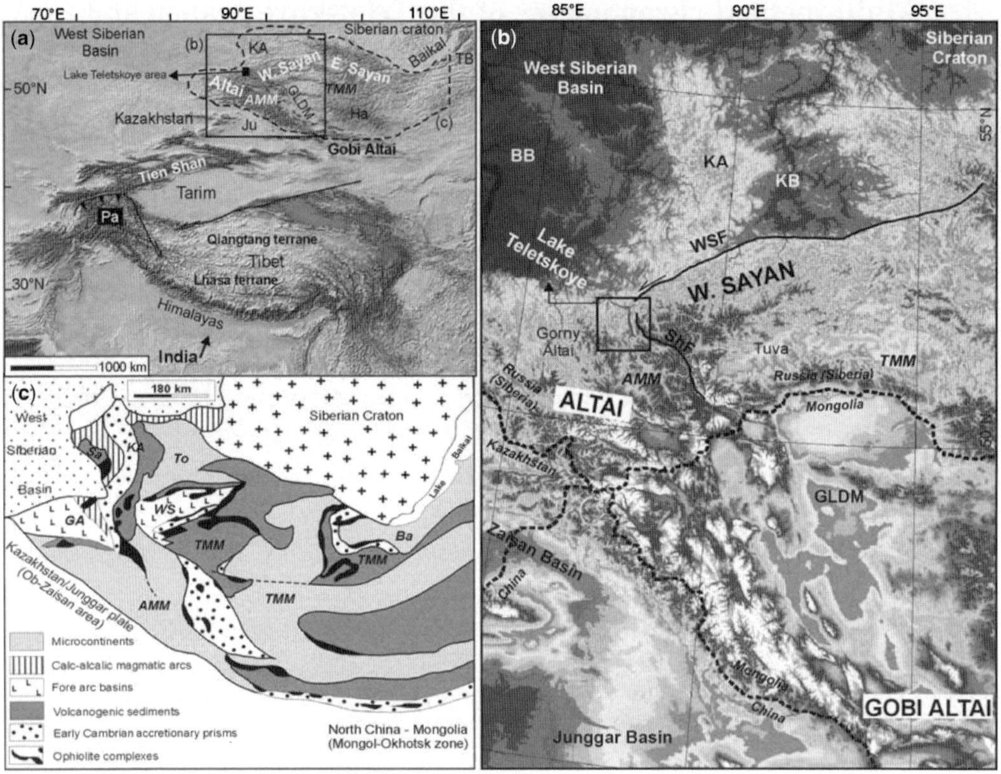

Fig. 1. (**a**) General location of the Altai Mountains in the Central Asian Orogenic System. (**b**) Detailed digital terrain model of the Altai Mountains. (**c**) Sketch map of the Palaeozoic terrane subdivision in the Siberian Altai–Sayan area (after Dobretsov *et al.* 1996). AMM, Altai–Mongolia microcontinent; Ba, Barguzin microcontinent; BB, Biya–Barnaul basin; GA, Gorny Altai; Ha, Hangai; Ju, Junggar Basin; KA, Kuznetsk–Alatau (ridge); KB, Kuznetsk Basin; GLDM, Great Lakes' Depression of Mongolia; Pa, Pamirs, Sa, Salair terrane; ShF, Shapshal fault; TB, Transbaikal Mountains; TMM, Tuva–Mongolia microcontinent; To, Tomsk microcontinent; WS, West Sayan; WSF, West Sayan fault.

The (Siberian) Altai Mountains are part of a vast area in Central Asia and Siberia that is being subjected to ongoing Late Cenozoic, mainly transpressive, intracontinental deformation between the active Himalaya–Tibet–Pamir orogenic zone and the rigid backstop of the Siberian craton. The Tien Shan, Altai, Sayan, Hangai and Transbaikal mountains are some of the major composing units (Fig. 1). These mountain belts alternate with large, relatively undeformed basins, such as Tarim, Junggar and the Great Lakes' Depression of Mongolia (GLDM) (Fig. 1). In Russian literature this Central Asian–Siberian intracontinental deformation zone is often called the Ural–Mongolian fold belt, and corresponds to 'the Altaids' as defined by Şengör *et al.* (1993). We refer to this region as the Central Asian Orogenic System (CAOS: Briggs *et al.* 2007). The driving forces for CAOS deformation are generated by ongoing India–Eurasia convergence and the indentation of India into Eurasia (Molnar & Tapponnier 1975; Tapponnier & Molnar 1979; Avouac & Tapponnier 1993).

The Indian–Eurasian convergence has a significant impact on Asian Cenozoic tectonics. Large-scale effects include the extrusion or lateral lithospheric escape of Tibet and SE Asia (Peltzer & Tapponnier 1988; Le Pichon *et al.* 1992) along vast shear zones, mainly directed eastwards to the less constraining Pacific plate boundary. In addition, crustal thickening constructed the Himalayan orogen and the Tibetan Plateau (e.g. Yin & Harrison 2000 and references therein). The widespread reactivation of ancient CAOS mobile belts represents a third major effect and has resulted in the world's largest, most active, intracontinental orogenic system. Deformation propagates to Eurasia's continental interior along an intricate pre-existing structural network associated with its Palaeozoic basement and, hence, reactivation is basement controlled (Dobretsov *et al.* 1995; Dricker *et al.* 2002).

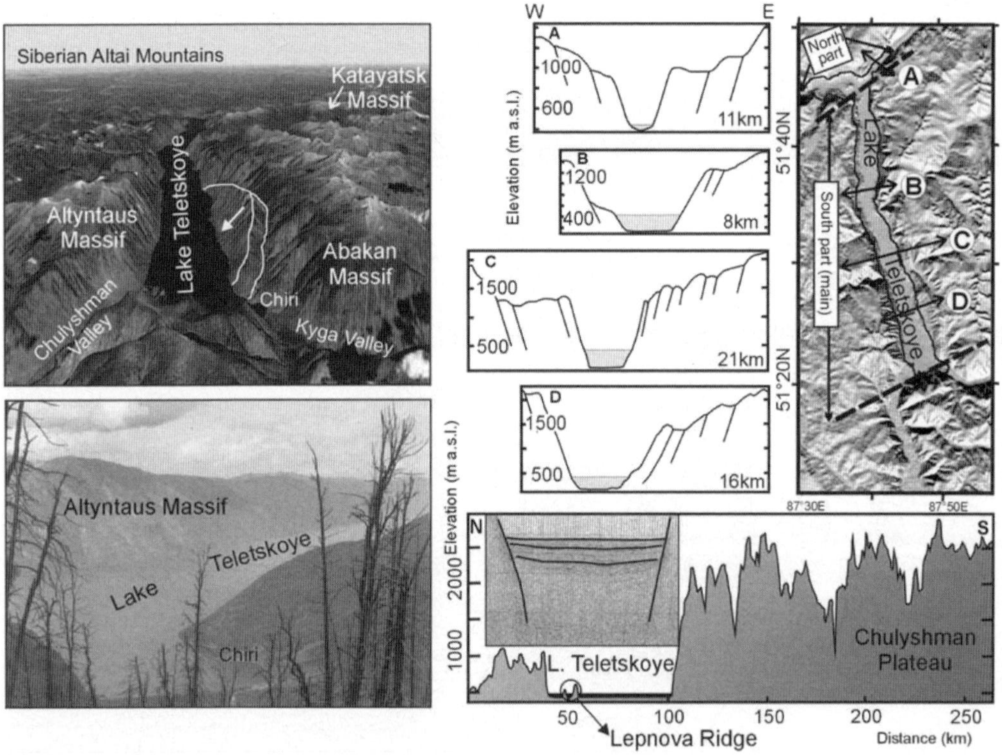

Fig. 2. Graben features of the Teletskoye basin.

This involves primarily the more easily deformable sutures and mobile belts extending between the more rigid, often microcontinental, fragments of the blocky tectonic collage of the CAOS. The Late Cenozoic formation and growth of the modern Altai between the rigid Palaeozoic Altai–Mongolia (AMM) and Tuva–Mongolia (TMM) microplates are an example of this deformation process (Fig. 1) (Buslov *et al.* 2004).

Despite recent advances in our understanding of the evolution of the CAOS, geochronological and thermochronological data are sparse. This is especially true for the Siberian Altai region. Therefore, an important rationale for undertaking this study was to obtain more data in this field and to constrain the thermo-tectonic history of the area. We focus on the key area of the Teletskoye graben (Fig. 1), where we sampled crystalline basement rocks (mainly granitoids and gneisses). A range of dating techniques such as sensitive high-resolution ion microprobe (SHRIMP) U–Pb zircon dating, $^{40}Ar/^{39}Ar$ dating of muscovite and biotite, and apatite fission-track (AFT) thermochronology was applied in order to shed new light upon Palaeozoic basement formation and the evolution in the Teletskoye area, on its poorly understood Mesozoic reactivation, and on the timing of Late Cenozoic India–Eurasia far-field tectonics and the building of the modern Siberian Altai.

Palaeozoic geodynamics and basement structure

The geology of the Palaeozoic basement of the Teletskoye graben area has been discussed extensively by Buslov & Sintubin (1995) and Smirnova *et al.* (2002). The graben developed at the junction zone of two major Palaeozoic tectonic terranes, Gorny Altai and West Sayan (Fig. 3). These developed and evolved as independent geodynamic terranes that eventually became accreted to the Siberian craton during Palaeo-Asian Ocean evolution in the Neoproterozoic and Early Palaeozoic (*c.* 650–510 Ma) (Buslov *et al.* 2001; Khain *et al.* 2003). Gorny Altai and West Sayan are separated by the West Sayan fault, that branches off in a set of Palaeozoic thrusts and oblique strike-slip faults, which form the units' boundary in the western Teletskoye basement (Fig. 3). During the Palaeo-Asian Ocean closure, peri-Gondwanan fragments or microcontinents,

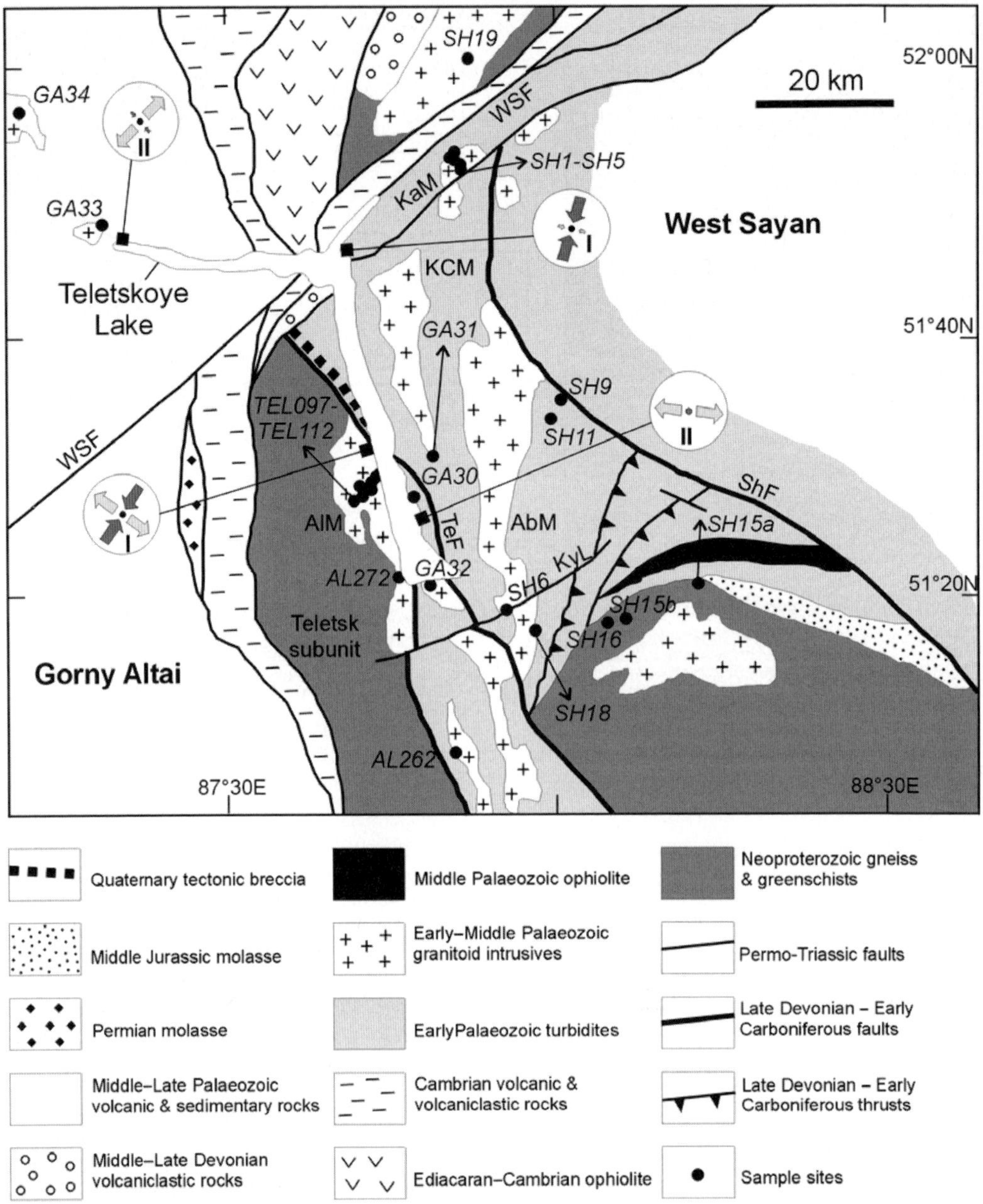

Fig. 3. Simplified geological map of the Teletskoye region with indication of the sample sites. AbM, Abakan massif; AlM, Altyntaus massif; KaM, Katayatsk massif; KCM, Kokshi–Chelyush massif; KyL, Kyga Lineament; ShF, Shapshal fault; TeF, Teletsk fault; WSF, West Sayan fault. Late Cenozoic formation of the Teletskoye graben: Pliocene–Pleistocene horizontal paleo-stress tensors are shown (after Dehandschutter *et al.* 2002); 'I' indicates a first Late Pliocene–Pleistocene phase of formation of the Teletskoye graben (transpressive phase of initial opening); and 'II' indicates a second Pleistocene–Holocene phase of formation of the Teletskoye graben (recent extensive phase).

such as the AMM, converged and collided with Siberia, trapping 'oceanic terranes' such as island arcs, accretionary prisms and seamounts. Gorny Altai and West Sayan are examples of such terranes. Along with AMM, they form the bulk of the present-day Siberian Altai basement.

The *Gorny Altai unit* forms the northern and NW basement that underlies the current Teletskoye basin (Fig. 3). It is composed of Ediacaran–Early Cambrian volcanic and volcaniclastic rocks of the Kuznetsk–Altai–Khantaishirin island arc that formed during Palaeo-Asian Ocean subduction under Siberia. It also includes seamount rocks that were accreted to the island arc in the Early Cambrian (Buslov *et al.* 2002). In the Late Cambrian–Early Ordovician, the amalgamated island arc terrane was attached to Siberia and underwent major deformation.

Ordovician–Silurian passive margin and shelf sediments overlie the island arc rocks, reflecting a period of subduction cessation. Up-sequence, Early–Middle Devonian volcanic rocks bare witness to an active margin that typified Siberia and its accreted rim during the progressive closure of a remnant Palaeo-Asian Ocean tract (Ob-Zaysan, Kazakhstan, Fig. 1) (Buslov *et al.* 2001), resulting finally in the Late Devonian–Early Carboniferous collision of AMM with Siberia, and trapping Gorny Altai in between. During the Permo-Triassic, as a consequence of the collision between Kazakhstan and Siberia, large-scale strike-slip motions reshaped the Siberian orogenic rim (Şengör *et al.* 1993; Buslov *et al.* 2003, 2004). During this collision, major terrane reorganization and sinistral shearing along the West Sayan fault juxtaposed the Gorny Altai and West Sayan units as they appear today.

The *West Sayan unit* can be subdivided in two subunits, separated by the Late Devonian–Early Carboniferous sinistral Teletsk fault zone (Fig. 3). Late Devonian mylonites are associated with the fault. The Teletsk subunit is located SW of the fault zone and consists of Neoproterozoic (formerly Riphean) greenschists and gneisses, intruded by the pre-Devonian Altyntaus Massif. The Altyntaus Massif granitoids and gabbros are exposed along the SW flank of the modern Teletskoye graben (Fig. 2). Altyntaus granites constitute a sizeable portion of our sample collection. East of the Teletsk fault zone, the second (unnamed) subunit of West Sayan (or West Sayan proper) composes most of the basement east of the graben. It is formed by Cambrian–Ordovician (meta)turbidites and zonal (cordierite–biotite) granite-gneiss domes presumed to be of Late Devonian–Early Carboniferous age (Buslov & Sintubin 1995; Smirnova *et al.* 2002). Several of these rocks were sampled for this study (Fig. 3 and Table 1). By the Cambrian–Ordovician periods, West Sayan was

Table 1. *Sample location and lithology details*

Sample	Latitude	Longitude	Altitude (m)	Locality	Lithology
TEL 097	51°24′07″N	87°39′59″E	2300	Altyntaus massif	granite
TEL 101	51°23′49″N	87°41′31″E	2210	Altyntaus massif	granite
TEL 105	51°26′10″N	87°41′22″E	2000	Altyntaus massif	granite
TEL 107	51°27′37″N	87°41′23″E	1585	Altyntaus massif	granite
TEL 108	51°27′39″N	87°41′23″E	1390	Altyntaus massif	granite
TEL 109	51°28′19″N	87°41′29″E	1155	Altyntaus massif	granite
TEL 110	51°28′22″N	87°41′41″E	940	Altyntaus massif	granite
TEL 111	51°28′27″N	87°41′52″E	820	Altyntaus massif	granite
TEL 112	51°28′39″N	87°42′00″E	610	Altyntaus massif	granite
AL 262	51°07′43″N	87°47′55″E	480	Chulyshman valley	granite
AL 272	51°21′56″N	87°44′50″E	435	Altyntaus massif	granite
SH 1	51°45′13″N	87°55′40″E	1850	Katayatsk massif	diorite
SH 2	51°44′42″N	87°55′44″E	1950	Katayatsk massif	diorite
SH 3	51°44′32″N	87°55′44″E	2010	Katayatsk massif	diorite
SH 4	51°44′58″N	87°55′55″E	2100	Katayatsk massif	diorite
SH 5	51°45′31″N	87°55′11″E	1720	Katayatsk massif	granodiorite
SH 6	51°20′56″N	87°56′56″E	650	Kyga valley/Abakan massif	granite
SH 9	51°24′11″N	88°08′21″E	2250	Kojildukir massif	sandstone
SH 11	51°23′07″N	88°11′38″E	1700	Kosbashi massif	sandstone
SH 15a	51°20′30″N	88°15′30″E	2150	Karakol Creek	gneiss
SH 15b	51°18′16″N	88°20′09″E	1950	Tugunluarchikkil lake	gneiss
SH 16	51°17′37″N	88°17′45″E	1900	Tugunluarchikkil lake	granite
SH 18	51°16′51″N	87°58′46″E	2050	Ljukjil lake	gneiss
SH 19	52°01′08″N	87°51′51″E	470	Blyka village	granite
GA 30	51°27′00″N	87°46′32″E	440	Bele terrace	gneiss
GA 31	51°23′22″N	87°55′24″E	2300	Kokshi–Chelyush massif	granodiorite
GA 32	51°21′00″N	87°49′30″E	440	Kyga valley/Chiri outpost	diorite
GA 33	51°48′53″N	87°10′36″E	420	Artybash village	diorite
GA 34	51°55′13″N	87°05′42″E	400	Kebezen village	diabase

an integral part of the accreted Siberian shelf where the turbidites accumulated (Torsvik *et al.* 1995; Bachtadse *et al.* 2000; Sennikov *et al.* 2000). In the SE part of the Teletskoye basement, the Kyga lineament (Chiri, Kyga valley) is a structure that branches off the Shapshal fault zone to the east (Fig. 3) and where this major fault zone relays to the basin (Dehandschutter *et al.* 2002). The West Sayan, Teletsk and Shapshal fault zones are important, long-lived structures, representing a continental wrench system that is not only responsible for major Palaeozoic tectonic activity, but is also the site of post-Palaeozoic reactivation episodes.

Meso-Cenozoic reactivation and formation of the Teletskoye graben

Mesozoic reactivation

Although the Mesozoic is perceived as a period of tectonic quiescence for the Teletskoye region, it has previously been suggested that Mesozoic tectonic activity in adjacent regions affected the Siberian orogenic rim, including at least some parts of the Siberian Altai (Delvaux *et al.* 1995*b*; Dobretsov *et al.* 1996; Buslov *et al.* 2003). After the Permo-Triassic, tectonic activity in the Siberian Altai region subdued. From the Late Triassic and Jurassic on, basins formed in the Altai basement (Fig. 4) in an active continental regime (e.g. Dobretsov *et al.* 1996). At that time the Altai were adjacent to several large basins where active subsidence and extension occurred. To the north and NW the Siberian Altai Mountains give way to the Kuznetsk and Biya–Barnaul basins that are, in fact, parts of the extensive West Siberian Basin (Fig. 4). Jurassic–Cretaceous extension and rifting in the West Siberian Basin resulted in the deposition of terrigenous (molasse) and in some parts marine sediments (e.g. Vyssotski *et al.* 2006). To the SW of the Altai, the Zaisan and Junggar Basin (Fig. 4) accumulated km-scale thick Mesozoic sediments, while its basement underwent Jurassic reactivation (Allen & Vincent 1997; Hendrix 2000). East of the Altai several basins formed in the GLDM and in the Gobi region (Fig. 4) where Mesozoic extension is documented by Jurassic–Cretaceous terrigeneous sediments (Graham *et al.* 2001; Johnson 2004). During the Jurassic, further north and east of the GLDM, the Mongol–Okhotsk Ocean that extended between Siberia and the composite North China–Mongolia (Amurian) continent closed (Fig. 1), inducing the Mongol–Okhotsk orogeny that affected a broad region in Mongolia, North China and Siberia (Van der Voo *et al.* 1999; Zorin 1999; Kravchinsky *et al.* 2002; Tomurtogoo *et al.* 2005).

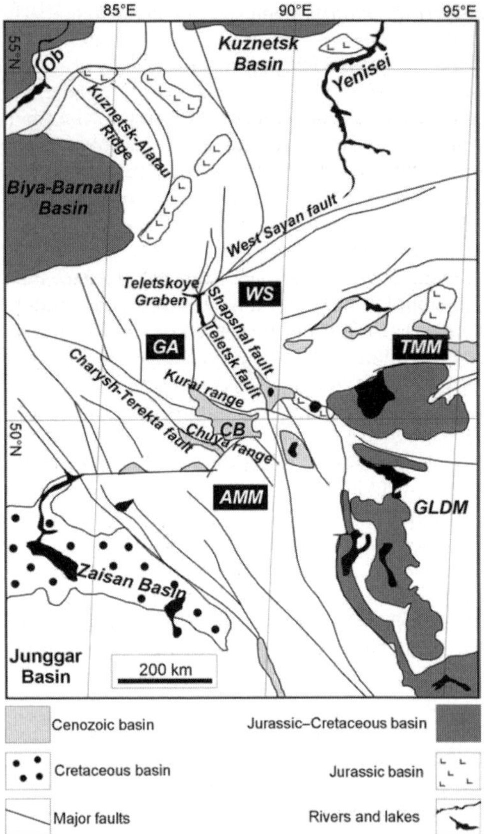

Fig. 4. Sketch map of intramontane and orogen-adjacent Mesozoic and Cenozoic basins in the Altai orogen (after Dobretsov *et al.* 1996). The Biya–Barnaul and Kuznetsk basins are part of the West Siberian Basin. Basins: CB, Chuya Basin; GLDM, Great Lakes' Depression of Mongolia. Tectonic units: AMM, Altai–Mongolia microcontinent; GA, Gorny Altai unit; TMM, Tuva–Mongolia microcontinent; WS, West Sayan unit.

Mesozoic tectonic forces also affected the Altai basement, in particular during the Late Jurassic and Cretaceous (Dobretsov *et al.* 1996; Buslov *et al.* 2003). In our study area, Jurassic fault-controlled basins with coarse continental sediments developed along the Shapshal fault (Fig. 3). In the Late Cretaceous and throughout much of the Palaeogene, tectonic activity subdued again and the region was subjected to widespread peneplanation and red-bed formation (Nikolaeva & Shuvalov 1995; Dobretsov *et al.* 1996). Similar observations of Mesozoic tectonics were made in the Mongolian Altai (e.g. Howard *et al.* 2003 and references therein).

Cenozoic reactivation

After an Early Cenozoic period of quiescence, Late Cenozoic tectonic movements related to India–Eurasia convergence occurred, creating a series of active intracontinental mountain belts (e.g. Tapponnier & Molnar 1979; Peltzer & Tapponnier 1988; Le Pichon *et al.* 1992; Avouac & Tapponnier 1993). These distant tectonic effects have reactivated major Palaeozoic structures in the Siberian Altai basement, and are responsible for the Late Cenozoic building and morphology of the modern Altai orogen and the formation of the Teletskoye graben.

The Cenozoic evolution of the Siberian Altai can best be illustrated by the evolution of the Chuya Basin, south of the Teletskoye graben area (Fig. 4). A Late Cretaceous–recent stratigraphic section, 1.5 km thick, that includes several hiatuses from episodes of orogenesis and erosion is observed (Buslov *et al.* 1999). The basin is situated in Gorny Altai, and its basement is composed of Ediacaran–Cambrian (650–500 Ma: Buslov *et al.* 2002) volcanic arc and accretionary prism rocks and Devonian active margin deposits. Remnants of the Late Cretaceous–Palaeogene peneplain are preserved on top of some basement blocks adjacent to the basin, while small pockets of Late Cretaceous marine sediments are also found. These basal units are overlain by the oldest Chuya sediments of Late Paleocene–Eocene to, perhaps, Early Oligocene age according to pollen data (Buslov *et al.* 1999). Oligocene lacustrine sediments are mostly fine grained. In the Mongolian Altai similar observations were made (Howard *et al.* 2003). This situation persisted into the Miocene and is encountered in other locations of the Siberian Altai (Dobretsov *et al.* 1996). Towards the Late Miocene–Early Pliocene intercalations of coarser material appear, and by the Late Pliocene (3–2.6 Ma according to Russian literature cited in Buslov *et al.* 1999), coarse sands, breccias and conglomerates dominate and demonstrate the rise of the bordering Chuya and Kurai mountain ranges (Fig. 4). The Late Pliocene, therefore, is thought to be the time of onset of large-scale regional orogenic reactivation of Gorny Altai (e.g. Dobretsov *et al.* 1996). The reactivation in the Altai region is mainly of transpressive nature, involving a counter-clockwise rotation (Thomas *et al.* 2002).

The young and ongoing tectonic activity in the Siberian Altai is mainly manifested along reactivated Palaeozoic structures, and recorded by structural, geomorphological, paleoseismic and geodetic techniques (e.g. Deev *et al.* 1995; Delvaux *et al.* 1995*a*; Dehandschutter *et al.* 2002; Dricker *et al.* 2002; Kolmogorova & Kolmogorov 2002). Late Cenozoic and active, mainly transpressive, movements associated with the development of the Mongolian, Chinese and Gobi Altai ranges have also been extensively documented (e.g. Cunningham *et al.* 1996; Philip & Ritz 1999; Howard *et al.* 2003; Pollitz *et al.* 2003; Vergnolle *et al.* 2003). However, in contrast to the Siberian Altai, the Mongolian Altai to the SE show earlier signs of erosion and uplift in the Miocene and even Late Oligocene, while the Pliocene is also characterized by the deposition of molasse-type sediments in intramontane basins. This indicates that since the Pliocene an orogen-wide tectonic pulse affected and shaped the Altai region as a whole, while earlier (Miocene) tectonics seem to have affected the Mongolian Altai.

Formation of the Teletskoye graben

The Late Pliocene tectonic pulse is still active, as shown by strong earthquakes (e.g. the recent, 27 September 2003, $M = 7.3$ quake induced by strike-slip along the Charysh–Terekta fault zone near the Chuya Basin (Fig. 4): Vysotsky *et al.* 2006; Nissen *et al.* 2007). The Teletskoye graben is a major feature in the neotectonic evolution of the Siberian Altai. Dehandschutter *et al.* (2002) identified two basin formation phases of Quaternary age, each with a characteristic kinematic nature (Fig. 3). A first, transpressive phase, related to movements along the West Sayan and Shapshal related faults resulted in initial basin opening as a series of pull-apart structures that formed in the weaker mylonitic zone of the Teletsk fault. A second phase of pure (east–west) extensional movements occurred along normal faults at angle to the strike-slip faults, producing two main leading, north–south-oriented, fault zones along which the basin experienced extensional collapse. This led to the typical graben morphology of the current basin (Fig. 2). The faults controlling the Teletskoye basin evolution clearly cross-cut Late Pleistocene glacial deposits and structures, and give evidence that the basin is of Late Pleistocene–Holocene age (Dehandschutter *et al.* 2002). Pseudotachylytes found in the southern border fault of the basin might indicate the rapid subsidence of the Teletsk graben (Theunissen *et al.* 2002).

Remnants of the Cretaceous–Palaeogene peneplanation surface can be found on the summit areas of the faulted blocks around the Teletskoye basement. The neotectonic disruption of this surface is evident, it has been vertically offset by several hundreds of meters to over 3 km in the southern part of the area. It is clear from this discussion that the Teletskoye basement underwent significant Late Cenozoic deformation, with a considerable vertical component.

Samples and analytical techniques

After separation using conventional density and magnetic methods, and handpicking of the relevant mineral fractions (grain size 80–250 μm), 29 samples from the Teletskoye crystalline basement were retained (Table 1 and Fig. 3). Three zircon samples covering the immediate graben area (TEL105, SH1 and SH6) were mounted for SHRIMP U–Pb geochronology. Muscovite (five samples: TEL097, TEL101, TEL105, TEL109 and TEL112) and biotite (three samples: TEL097, TEL101 and TEL105) from the Altyntaus granites were used for $^{40}Ar/^{39}Ar$ dating. Apatite was mounted for fission-track thermochronology. The apatite samples cover a broader area around the graben and involve 28 samples. All but one sample (TEL097) yielded good-quality apatites for AFT dating. TEL097, however, did produce nice muscovites and biotites that were used for $^{40}Ar/^{39}Ar$-dating, as reported later in this paper.

Zircon SHRIMP U–Pb dating

Around 50 grains of each zircon sample were handpicked based on clarity, colour (originally pale yellow and pale orange crystals were chosen for later thermochronology studies, darker crystals were excluded), absence of inclusions and defects, and on euhedral crystal morphology (for the same reason). These grains were then mounted in epoxy, polished and gold-coated for U–Pb dating. The zircon mounts were imaged using reflected-light digital photography and cathodoluminescence (CL) to illuminate internal zoning associated with variations in rare earth element (REE) (e.g. U, Th) content. Based on these images, spot locations for the ion-probe were selected and targeted on the basis of surface smoothness, and U and Th content (dark zones). For each zircon mount up to 15 spots (30 μm) were analysed. After every fourth analysis a calibration measurement was carried out, using the Stanford–USGS (US Geological Survey) internal R33 zircon standard. We used the SHRIMP-RG (sensitive high-resolution ion microprobe with reverse geometry) at the joint Stanford–USGS (Menlo Park) facility (California, USA). The intensity of the primary $^{16}O_2^-$ beam was 4–6 nA. Mass discrimination exceeded 5000. Each analysis entailed five scans on secondary ion peak intensities of: $^{90}Zr_2^{16}O$, ^{204}Pb, background, ^{206}Pb, ^{207}Pb, ^{208}Pb, ^{238}U, $^{248}(ThO)$ and $^{254}(UO)$, with counting times ranging between 2 and 16 s at each peak. $^{206}Pb/^{238}U$ and $^{207}Pb/^{235}U$ ratios were determined, and the corresponding ages were calculated using the data reduction program SQUID (Ludwig 2001) and corrected for common Pb (Cumming & Richards 1975).

Muscovite and biotite $^{40}Ar/^{39}Ar$ dating

Muscovites and biotites from the Altyntaus granite samples were carefully handpicked to obtain pure grain separates. The separates (2–5 mg) were subsequently wrapped in pure Cu foil. As a standard for neutron fluence monitoring, Taylor Creek Rhyolite sanidine (1–2 mg) with an assigned age of 27.92 ± 0.17 (USGS standard 85G003: Duffield & Dalrymple 1990) was wrapped in Al foil for total-fusion analysis. A standard packet was inserted between every six mica packets. The packets were baked at 120 °C for 1 h and then sealed in a quartz-glass vial. The vial was irradiated with fast neutrons at the TRIGA reactor facility of the Oregon State University (USA) for 9 MWh. After irradiation the Al packages were unwrapped and about five sanidine grains were fused with the laser, and the Ar-release analysed to determine the irradiation parameter, *J*. An uncertainty of 0.5% was assumed (at 1σ) for all *J*-values (Hacker *et al.* 1996); these were calculated for each sample individually based on the relative position of the packages in the irradiated vial. The stepwise heating analyses were conducted in a double-vacuum Staudacher-type resistance furnace programmed with a specific stepwise heating program between temperatures of 700 and 1400 °C. At each step the gas was extracted and purified by SAES ST-172 and ST101 getters before being led to the MAP-216 spectrometer equipped with a Baur–Signer ion source and a Johnston MM1 multiplier. The collected gas was analysed for ^{40}Ar, ^{39}Ar, ^{38}Ar, ^{37}Ar and ^{36}Ar. Blank and baseline corrections were made. After each analysis the furnace was 'burnt out' at 1600 °C for two 30 min-cycles. A blank was measured before the next sample was dropped into the furnace.

Apatite fission-track (AFT) thermochronology

Apatite from the Teletskoye area rock samples was embedded in epoxy, polished and analysed with the external detector (ED) method. Spontaneous tracks were etched with a 2.5% HNO_3 solution for 70 s at 22 °C and induced tracks in the muscovite ED with 40% HF for 40 min at 22 °C. Apatite-ED wafers were irradiated in several batches in the well-thermalized channels of the Thetis reactor at Ghent University. Thermal neutron fluences of around 2×10^{15} cm^{-2} were administered and monitored using metal activation monitors (diluted Au–Al and Co–Al alloys: Van den haute *et al.* 1998). Where possible, at least 1000 spontaneous tracks were counted per sample, spread over minimally 20 grains. AFT ages are conventional ζ-ages (Hurford 1990) with an overall weighted mean zeta of 253.1 ± 2.4 a cm^2, calculated based on analyses of 32 Durango and Fish Canyon Tuff apatite age standards and the IRMM-540 dosimeter glass (De Corte *et al.* 1998). Standards co-embedded with dosimeter glass and activation monitors were regularly spaced in the sample package to detect and correct for any axial thermal neutron fluence gradient, and to allow the interpolation of induced glass dosimeter track densities (ρ_d for ζ calibration) for individual samples.

Where possible, 100 natural confined tracks were measured to construct an AFT length–frequency distribution. Because no chemical composition data from our apatite samples is available, AFT age and length data were modelled using the Laslett *et al.* (1987) annealing model (with initial track length parameter $l_o = 16.3$ μm) and the AFT-Solve thermal history modelling software (Ketcham *et al.* 2000). In a first run the thermal history model was only constrained by two time–temperature (t–T) constraints: a low-T (10 °C) benchmark reflecting current ambient temperatures, and a high-T bar-constraint well above 120 °C (AFT accumulation threshold: Gleadow & Duddy 1981; Wagner & Van den haute 1992 and references therein) significantly predating apparent AFT ages. An age of 260 Ma was chosen based on amphibole $^{40}Ar/^{39}Ar$ ages from the Teletskoye area (Dehandschutter *et al.* 1997). After an initial run, additional constraints (as few as possible) were placed reiteratively along the general trend to refine the model. Care was taken to let the T-interval be wider than a statistically acceptable t–T path envelope (Ketcham *et al.* 2000). More details on the procedures can be found in De Grave & Van den haute (2002). A note of caution, however, should be conveyed here. It has been shown that in some cases there exists a certain tendency in the aforementioned AFT thermal history modelling strategy to produce a Late Cenozoic cooling merely as an artefact (e.g. Ketcham *et al.* 2000). So it can be hazardous to interpret Late Cenozoic cooling simply on the basis of AFT thermal history models. It should be meticulously tested against independent geological evidence.

Results and discussion

SHRIMP data

The three analysed zircon separates were all taken from granitoid massifs close to the Teletskoye graben, and in particular from the West Sayan basement unit (Fig. 3): TEL105, Altyntaus massif; SH1, Katayatsk massif; and SH6, Abakan massif. All SHRIMP U–Pb ages of these zircons are, to a large extent, internally concordant within each sample (Fig. 5) and yield ages of: 419.0 ± 11.0 Ma for TEL105, 420.7 ± 6.6 Ma for SH1 and 460.1 ± 6.5 Ma for SH6. These ages are interpreted as the crystallization ages of the zircons and, hence, of the emplacement of the granitoid massifs. These results clearly show a temporal (and probably a genetic) relationship between the Altyntaus and Katayatsk massifs, both having an intrusion age of 420 Ma (Ludlow, Late Silurian), while the Abakan massif with an age of 460 Ma (Late Ordovician) is obviously older. The Altyntaus and Abakan

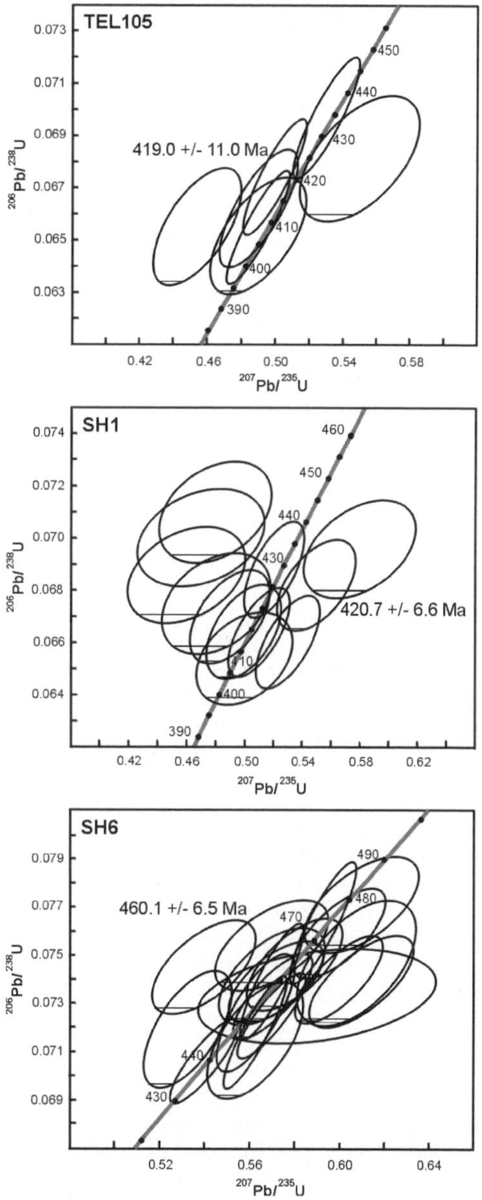

Fig. 5. Tera–Wasserburg concordia diagrams and U–Pb ages of the Teletskoye zircons. Error ellipses are 2σ.

massifs are from two different (sub)units, separated by the Teletsk fault zone: the Teletsk and West Sayan (sub)units, respectively (Fig. 3). The Katayatsk massif crops out within the West Sayan strike-slip fault zone that separates the West Sayan and Gorny Altai units. It might, therefore, represent an allochthonous massif that is not linked to the West Sayan basement and that is more closely related to the intrusion of the Altyntaus massif. Whatever the case, the ages obtained are significantly older than the emplacement ages of the bulk of the plutonic massifs in the Gorny Altai unit. The majority of the Gorny Altai plutons exhibit Devonian–Early Carboniferous zircon U–Pb ages, roughly in the range of 400–320 Ma (Vladimirov *et al.* 1997, 2001 and references therein; Dobretsov & Vladimirov 2001 and references therein).

The Silurian and especially the Ordovician ages obtained here correspond to ages of many intrusive complexes in the eastern part of West Sayan and Tuva (Fig. 1). An important phase of intrusive magmatism in this area yields ages of 440–490 Ma (Lebedev *et al.* 1993), 457 Ma (Kostitsyn *et al.* 1998), 413–483 Ma (Distanova 2000 and references therein) and 432–505 Ma (Rudnev *et al.* 2004). These authors relate their findings to either post-collisional magmatism or post-collisional deformation events during the Palaeozoic accretionary history and assembly of the Siberian–Central Asian continent (Distanova 2000). The collision involved here is that of the TMM with the Siberian–Central Asian mobile belt (Fig. 1). Salnikova *et al.* (2001) demonstrated that the TMM collision occurred around 500 Ma, and that the TMM was already inactive by 450–480 Ma, when several intrusions occurred in the Tuva region. Buchan *et al.* (2002) dated several ophiolites (zircon Pb–Pb) in northern and central Mongolia, and concluded that a major collision event occurred between 540 and 450 Ma. The West Sayan granitoid massifs from the Teletskoye basement could be related to the TMM event and Late Palaeozoic strike-slip deformation (Buslov *et al.* 2003) would then have positioned the West Sayan unit in contact with Gorny Altai. The ages from the Teletskoye granitoids can further underscore this hypothesis.

Similar ages have also been reported for several areas of northern and central Mongolia (south of the Baikal area, Fig. 1), closer to where the TMM and other microcontinents collided with the Siberian–Central Asian mobile belt (Buslov *et al.* 2001; Salnikova *et al.* 2001). Kröner *et al.* (2005) gave a review of zircon U–Pb ages from this area and report an age range of felsic igneous rocks of 417–460 Ma, corresponding with the ages obtained for the Teletskoye basement rocks. These authors described the formation of their samples in an island arc environment. Windley *et al.* (2002) provide evidence for a Cambrian–Ordovician continental magmatic arc in the Chinese part of the Altai orogen, with zircon U–Pb ages of around 505 Ma, followed by a later collision of an island arc with the Chinese Altai block in the Late Silurian–Early Devonian, giving U–Pb ages for younger batholiths in the Chinese Altai of about 415–380 Ma. Recently, Yuan *et al.* (2007) have dated additional batholith samples from this area to find ages in the range of about 412 to 359 Ma. In addition, an even younger, Late Carboniferous–Permian, phase of intrusion was found by these authors and dated at between 318 and 267 Ma. Briggs *et al.* (2007) have also reported Permian ages of 278–286 Ma for granitoids from the Chinese Altai region, as well as Ordovician ages (448–451 Ma) for deformed granitoids and orthogneiss.

Muscovite $^{40}Ar/^{39}Ar$ dating

The results of muscovite $^{40}Ar/^{39}Ar$ dating of five undeformed Altyntaus granites are depicted in Figure 6. Stepwise heating of the muscovites in the resistance furnace yielded well-defined plateaus in the age spectra. Except sample TEL112, that exhibits a younger plateau age of 361 Ma (Late Devonian), the samples constrain a tight age range of 390–395 Ma (Middle Devonian). Because all muscovites were retrieved from undeformed granites with no indication of shearing, these Devonian $^{40}Ar/^{39}Ar$ ages are interpreted as cooling ages of the Altyntaus massif after its emplacement in the Teletskoye basement 420 Ma ago. These results confirm that the Altyntaus granites were in a phase of post-magmatic cooling by the Middle Devonian. This cooling would have reached $350 \pm 50\ ^{\circ}C$ given the closure temperature (T_C) of muscovite for Ar-diffusion (McDougall & Harrison 1999).

Earlier results of muscovite $^{40}Ar/^{39}Ar$ dating in the Teletskoye region yielded somewhat younger dates between about 370 and 380 Ma (Buslov *et al.* 2001; data by Travin reported in Smirnova *et al.* 2002). These dates were obtained from an Altyntaus gabbro sample, and from schists and mylonites from the Teletsk fault zone (Fig. 3), and were interpreted as recording Late Devonian initiation of movements along the Teletsk fault zone. Amphibole $^{40}Ar/^{39}Ar$ dating of these schists and mylonites also gives ages of 370–380 Ma (Dehandschutter *et al.* 1997; Smirnova *et al.* 2002). They correspond with ages from the Kurai–Chuya area of Gorny Altai (Fig. 4). Buslov *et al.* (2001) interpret them as reflecting a period of large-scale shearing in the Gorny Altai region owing to collision of the AMM with Siberia.

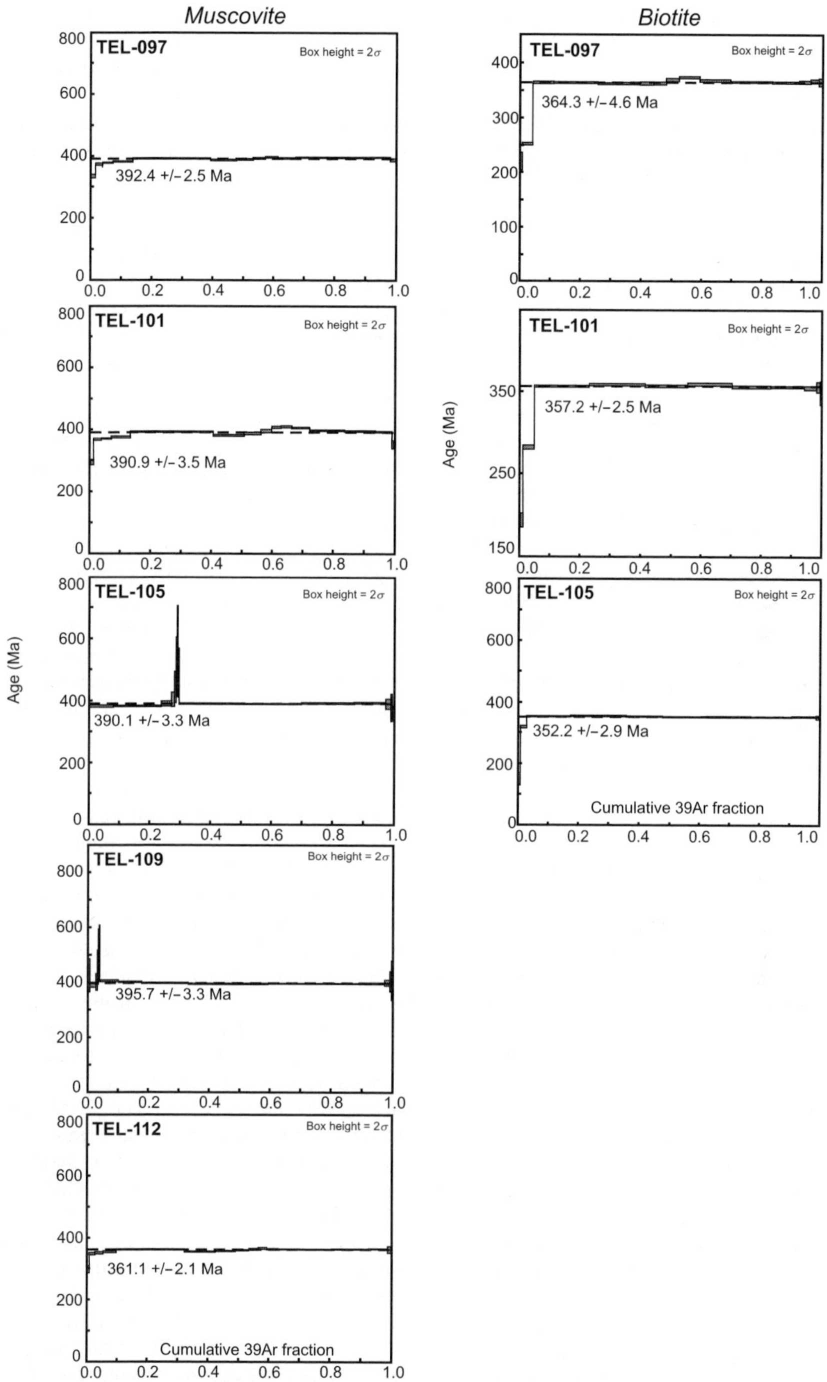

Fig. 6. Muscovite $^{40}Ar/^{39}Ar$ age spectra and plateau ages of the Altyntaus granite samples; and biotite $^{40}Ar/^{39}Ar$ age spectra and plateau ages of the Altyntaus granite samples.

Biotite $^{40}Ar/^{39}Ar$ dating

Three biotite separates from the same Altyntaus granites used for muscovite Ar-dating were also dated, and yielded age spectra with well-defined plateaus (Fig. 6) and ages between 353 and 364 Ma (Devonian–Carboniferous transition). These biotites were separated from the same undeformed granite samples, they therefore represent a further stage in the continuous cooling of the Altyntaus massif. The biotite ages are obviously younger than the muscovite $^{40}Ar/^{39}Ar$ ages due to a lower T_C for Ar-diffusion in biotite (300 ± 20 °C: McDougall & Harrison 1999). The region at that time experienced large-scale fault movements (mainly shearing and strike-slip) as a result of accretionary tectonics (Buslov *et al.* 2001). Perhaps activity of the Teletsk fault zone resulted in uplift and denudation of the rocks above the Altyntaus massif, bringing it closer to the surface and facilitating its cooling. However, the cooling rate drops significantly after about 390 Ma. Late Devonian–Early Carboniferous biotite Ar-ages between 329 and 365 Ma for the Teletsk fault zone schists are reported by Smirnova *et al.* (2002). Dehandschutter *et al.* (1997) dated two biotite samples from mylonitic gneisses from the Teletsk fault zone and obtained ages of 380 and 375 Ma.

AFT data and modelling

AFT age and length data are presented in Table 2 and Figure 7. Preliminary results for 13 samples have already been reported (De Grave & Van den haute 2002). No totally reset (Cenozoic) AFT ages were found. Apparent AFT ages for most samples are Late Cretaceous ranging between 65 and 105 Ma with a single Early Cretaceous outlier of 123 Ma (SH15a; Table 2). Sample SH6 gives the youngest apparent age of about 60 Ma (Palaeocene). This sample, together with other samples from lower elevations (for example TEL112 and AL272, Table 1), give younger ages that probably reflect mixed ages. They retain part of their 'Mesozoic signature' as annealing of the older tracks was not completed. These samples were most probably brought at temperatures just within the apatite partial annealing zone (APAZ) during a Mesozoic denudation episode. They hence started accumulating tracks that were shortened significantly at these higher APAZ temperatures with respect to samples higher up in the rock column. This is clearly shown by the Altyntaus vertical profile samples where the lowermost samples (TEL111 and TEL112) also have the lowest mean track lengths. As a consequence of a low spontaneous track density, the youngest sample in the Teletskoye region, SH6, unfortunately did not contain a sufficient amount of confined track lengths in order to yield a meaningful length distribution.

AFT mean track lengths (MTL) and length distributions show clear signs of thermal track shortening. Given the fact that we are dealing with crystalline basement apatites and that no geological evidence is present to imply any heating event (magmatic, frictional or burial) post-dating the apparent Late Cretaceous AFT ages, these length distributions suggest that most of the samples stayed relatively long at APAZ temperatures (60–120 °C: Gleadow & Duddy 1981; Wagner & Van den haute 1992 and references therein). MTL values vary between 11.2 and 14.4 μm; distributions are relatively broad ($1.3 < \sigma < 2.0$ μm), asymmetric and negatively skewed (Fig. 7). Both in shape and in MTL (12–14 μm) and σ (1.0–2.0 μm) values, the distributions appear typical basement AFT length distributions as defined by Gleadow *et al.* (1986). It is now widely accepted that AFT length (and annealing kinetics) can vary with chemical composition of the analysed apatite grain (e.g. Green *et al.* 1986; Barbarand *et al.* 2003*a*, *b*) and that this has a clear effect on the exact characteristics of the AFT length distributions. Unfortunately, for our apatite grains no chemical data are available and it is, hence, not possible to assess the exact effects here.

The AFT age and length data were modelled according to the procedures mentioned earlier and within general trends, a three-stage thermal history model was revealed: (1) Late Jurassic–Cretaceous rapid cooling; (2) Late Cretaceous–Palaeogene stability with only slow cooling; and (3) Late Neogene–Quaternary cooling (Fig. 7). Not all modelled envelopes show these stages as clearly as for example sample TEL112, but the broad envelopes for some (e.g. sample GA 31) do not exclude this general subdivision either. These results confirm the earlier data reported by De Grave & Van den haute (2002), and on a broader regional scale can also be compared to findings by Yuan *et al.* (2006) in the Chinese Altai Mountains and by Jolivet *et al.* (2007) and Vassallo *et al.* (2007) in the Mongolian Altai territory.

The Late Jurassic–Cretaceous cooling brought the apatites below the minimal temperature of total annealing around 150 Ma. Here we consider this temperature threshold to be of the order of 120 °C (Gleadow & Duddy 1981; Wagner & Van den haute 1992 and references therein). As mentioned, this value can vary somewhat according to the exact chemical composition (Cl/F ratio) of the apatite (Ketcham *et al.* 1999). Considering the thickened Phanerozoic crust underneath the Altai–Sayan area of some 50–60 km (Zorin *et al.* 1993), with no evidence of hot mantle upwelling (Dricker

Table 2. *AFT age and length data: ρ_s, ρ_i, and ρ_d are the density of spontaneous, induced tracks and induced tracks in an external detector (ED) irradiated against a dosimeter glass. The ρ_d-values are interpolated values from regularly spaced glass dosimeters (IRMM-540), expressed as 10^5 tracks cm^{-2}; ρ_s and ρ_i are expressed as 10^6 tracks cm^{-2}. N_s, N_i and N_d are the number of counted spontaneous, induced tracks and induced tracks in the ED. N_d is also an interpolated value. $P(\chi^2)$ is the chi-squared probability that the dated grains have a constant ρ_s/ρ_i ratio. An ζ-value of 253.1 ± 2.4 years cm^2 was used for the calculation of $t(\zeta)$. AFT length data are reported as a mean track length (l_m) with standard deviation (σ), obtained from the measurement of a number (n_l) of natural, horizontal confined tracks. For samples indicated by (*), preliminary results were reported earlier by De Grave & Van den haute (2002)*

Sample	n	ρ_s (±1σ)	N_s	ρ_i (±1σ)	N_i	ρ_d (±1σ)	N_d	ρ_s/ρ_i	$P(\chi^2)$	$t(\zeta)$	l_m	n_l	σ
TEL 101*	49	3.047 (0.073)	1732	2.199 (0.060)	1332	5.789 (0.134)	1879	1.377 ± 0.050	0.01	100.1 ± 4.4	12.2	70	1.9
TEL 105*	50	2.799 (0.054)	2676	2.167 (0.048)	2039	5.794 (0.134)	1881	1.312 ± 0.039	0.31	95.5 ± 3.7	12.4	100	1.7
TEL 107*	50	4.336 (0.088)	2444	2.999 (0.072)	1721	5.800 (0.134)	1882	1.507 ± 0.047	0.26	109.7 ± 4.4	12.5	100	1.7
TEL 108*	19	4.319 (0.184)	549	3.901 (0.171)	520	5.804 (0.134)	1884	1.104 ± 0.068	0.88	80.6 ± 5.4	–	–	–
TEL 109*	50	2.663 (0.057)	2189	2.599 (0.056)	2130	5.809 (0.134)	1885	1.066 ± 0.032	0.09	77.9 ± 3.0	12.3	100	1.7
TEL 110*	50	6.084 (0.112)	2956	5.958 (0.111)	2905	5.814 (0.134)	1887	1.027 ± 0.027	0.75	75.1 ± 2.7	12.3	100	1.8
TEL 111*	50	5.296 (0.099)	2879	4.978 (0.095)	2758	5.819 (0.134)	1889	1.067 ± 0.028	0.97	78.1 ± 2.8	12.4	100	1.5
TEL 112*	49	4.115 (0.083)	2480	4.942 (0.092)	2899	5.826 (0.134)	1891	0.889 ± 0.024	0.01	65.2 ± 2.4	11.2	100	1.6
AL 272	41	3.306 (0.058)	3305	2.171 (0.047)	2170	3.304 (0.082)	1611	1.567 ± 0.043	0.43	65.2 ± 2.5	11.9	100	2.0
AL 262	40	0.699 (0.035)	1150	0.350 (0.014)	591	3.944 (0.111)	1262	2.144 ± 0.108	0.94	106.1 ± 6.2	–	–	–
SH 1*	50	2.551 (0.049)	2718	2.033 (0.044)	2177	5.747 (0.133)	1865	1.269 ± 0.036	0.95	91.6 ± 3.5	12.8	100	1.8
SH 2*	50	4.013 (0.062)	4148	3.089 (0.055)	3167	5.751 (0.133)	1867	1.324 ± 0.031	0.39	95.7 ± 3.3	13.3	100	1.5
SH 3*	52	1.618 (0.030)	2961	1.293 (0.027)	2350	5.756 (0.133)	1868	1.287 ± 0.036	0.7	93.1 ± 3.5	13.2	100	1.3
SH 4*	50	6.291 (0.109)	3304	4.542 (0.094)	2324	5.760 (0.133)	1870	1.429 ± 0.039	0.80	103.4 ± 3.8	12.8	100	1.8
SH 5*	51	1.442 (0.028)	2746	1.160 (0.025)	2200	5.765 (0.133)	1871	1.296 ± 0.037	0.81	93.9 ± 3.6	13.1	62	1.6
SH 6	40	1.299 (0.036)	1322	0.993 (0.031)	1011	3.320 (0.083)	1616	1.429 ± 0.060	>0.99	59.8 ± 3.0	–	–	–
SH 9	25	1.728 (0.057)	929	1.133 (0.046)	609	3.374 (0.083)	1642	1.704 ± 0.089	>0.99	72.4 ± 4.2	11.3	24	1.9
SH 11	28	1.409 (0.060)	546	0.703 (0.043)	269	3.379 (0.083)	1645	2.239 ± 0.167	0.31	95.0 ± 7.5	–	–	–
SH 15a	50	0.837 (0.026)	1020	0.314 (0.016)	383	3.383 (0.083)	1647	2.904 ± 0.174	0.80	123.1 ± 8.1	12.8	95	1.4
SH 15b	30	1.670 (0.074)	513	1.081 (0.059)	332	3.387 (0.083)	1649	1.613 ± 0.114	>0.99	68.8 ± 5.2	–	–	–
SH 16	30	5.454 (0.144)	1430	2.871 (0.104)	759	4.083 (0.080)	2613	1.947 ± 0.087	0.98	99.8 ± 5.0	14.4	100	1.4
SH 18	30	0.913 (0.044)	427	0.492 (0.033)	224	3.399 (0.084)	1654	1.926 ± 0.159	0.99	82.3 ± 7.1	–	–	–
SH 19	40	0.511 (0.023)	501	0.227 (0.015)	222	3.405 (0.084)	1657	2.452 ± 0.198	>0.99	104.8 ± 8.9	11.2	80	2.0
GA 30	50	1.185 (0.039)	924	0.844 (0.033)	658	3.822 (0.083)	2153	1.488 ± 0.076	>0.99	71.6 ± 4.0	13.8	100	1.4
GA 31	40	0.989 (0.033)	890	0.532 (0.024)	479	3.818 (0.082)	2150	1.910 ± 0.108	0.98	91.6 ± 5.6	13.8	73	1.6
GA 32	60	0.916 (0.026)	1237	0.618 (0.021)	834	3.815 (0.082)	2149	1.587 ± 0.071	0.82	76.2 ± 3.9	13.6	100	1.3
GA 33	60	1.119 (0.031)	1320	0.620 (0.023)	705	3.807 (0.082)	2144	1.811 ± 0.084	0.97	86.7 ± 4.5	–	–	–
GA 34	30	4.650 (0.153)	926	2.170 (0.106)	420	3.804 (0.082)	2142	2.210 ± 0.130	>0.99	105.5 ± 6.7	–	–	–

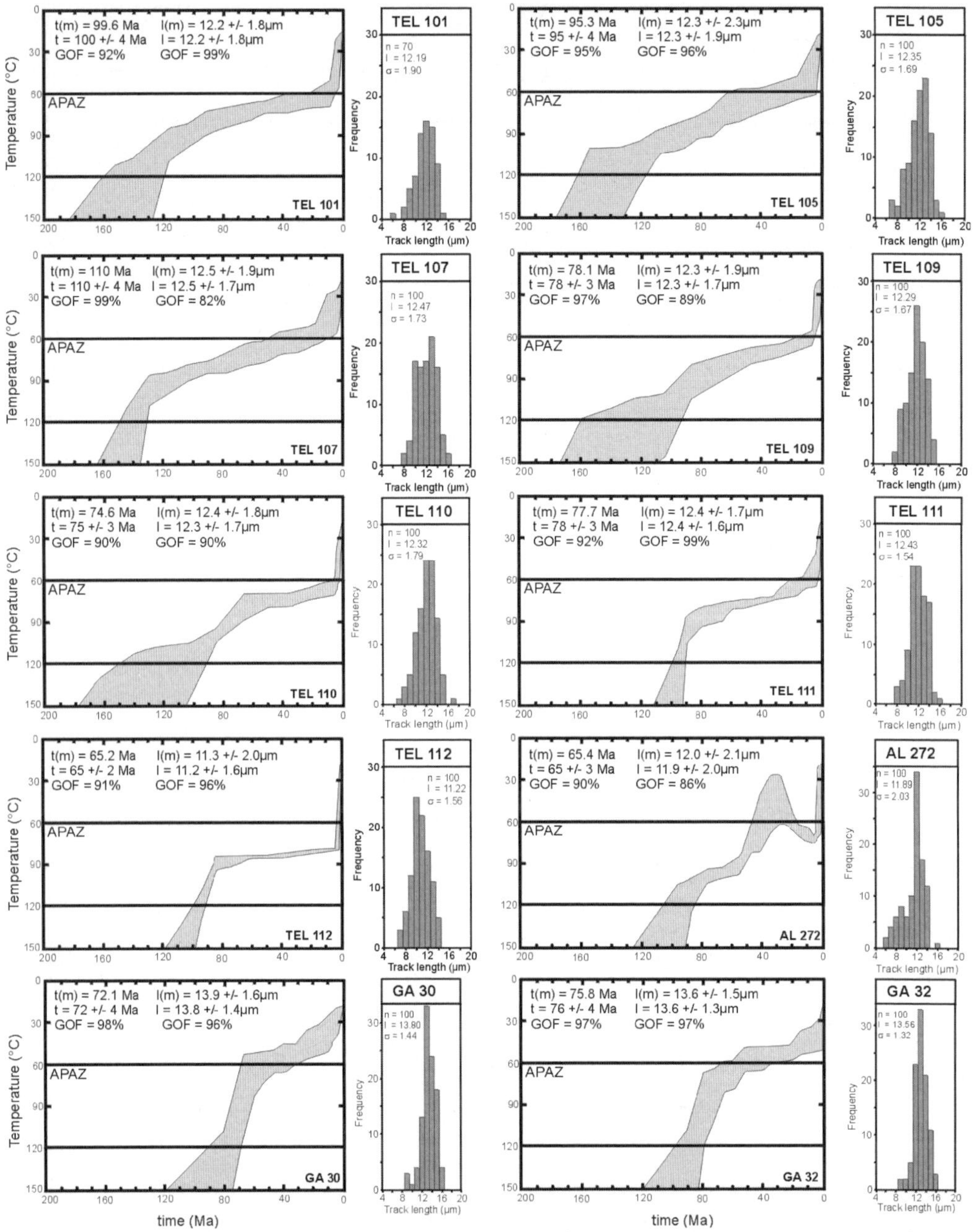

Fig. 7. AFTSolve (Ketcham *et al.* 2000) thermal history models of the Teletskoye apatites: APAZ, Apatite Partial Annealing Zone; *t* (m), modelled AFT age; *t*, observed AFT age; *l* (m), modelled mean track length; *l*, observed mean track length; GOF, goodness of fit. The *t*–*T* paths are represented by a statistical good-fit envelope (grey shading). See the text for the description and interpretation. AFT ages and length distributions (histograms) of these samples: *n*, number of measured confined tracks; *l*, mean track length (in µm); σ, standard deviation of track-length distribution (in µm).

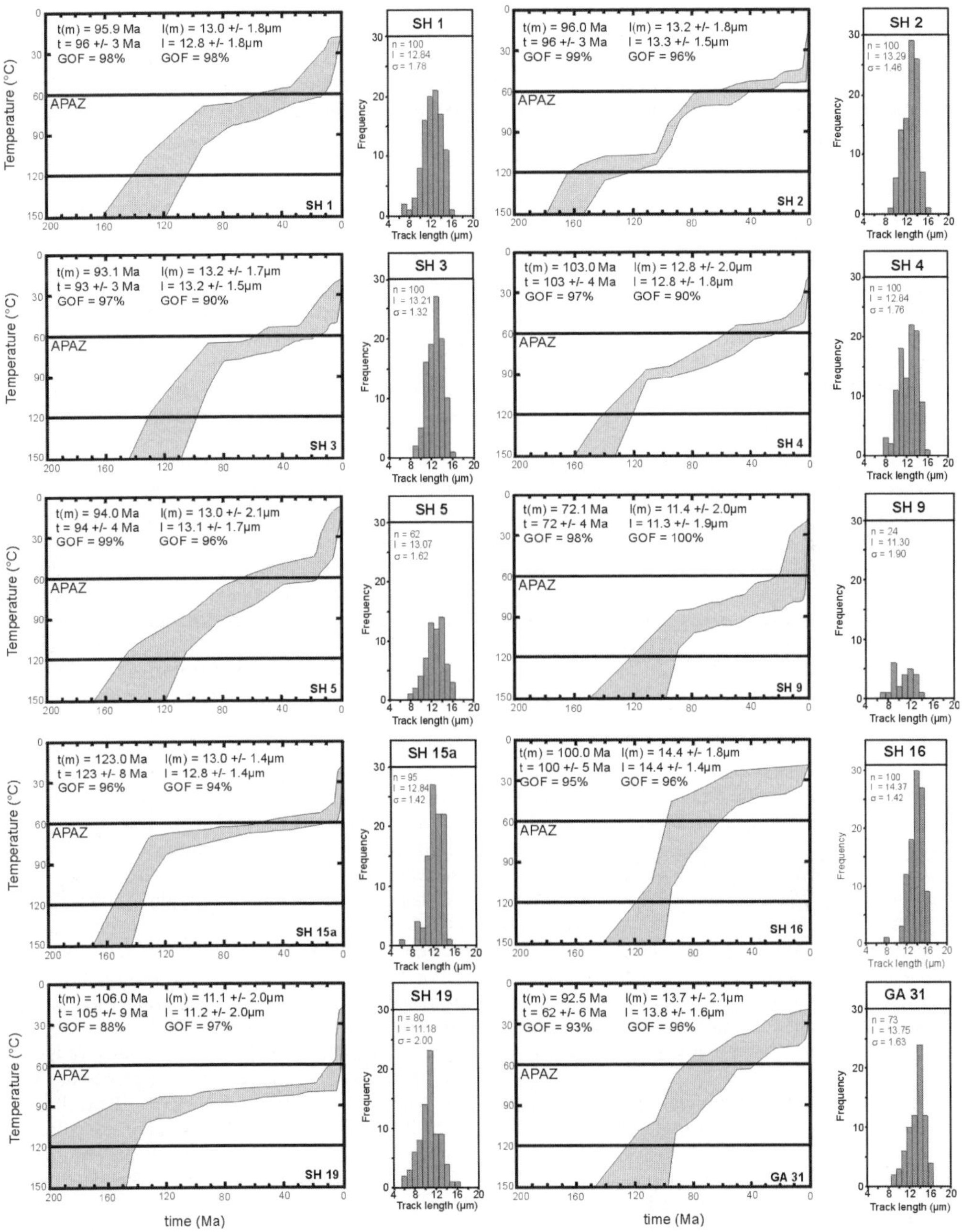

Fig. 7. (*Continued*).

et al. 2002), it is not surprising to find relatively low to moderate heat flows and geothermal gradients under the Altai–Sayan topography. Approximate values in the region range from more or less 15–20 °C km^{-1} (Duchkov *et al.* 1995; Hu *et al.* 2000) up to 25 °C km^{-1} (Cermak 1993). So, based on this, when we consider a geothermal gradient of 20–25 °C km^{-1}, this would imply that most of the apatite-bearing rocks were exhumed to depths shallower than about 5 km.

According to the general trend of the models, the Mesozoic cooling lasted on average until 80–90 Ma ago. This approximation is based on the models where this episode is clearly distinctive and encompasses the majority of the samples. At the cessation of Mesozoic cooling, the rocks still resided in the APAZ, reaching upper APAZ temperatures (of the order of 80–60 °C) and, thus, corresponding to a depth of 2.5–4 km. At these conditions partial AFT annealing persisted and pre-existing tracks were shortened. Variation in annealing degree is predominantly due to the present outcrop altitude (400–2300 m) of the particular sample, and hence the associated palaeo-depth. As mentioned above, additional scatter related to differences in chemical composition between apatite samples can, however, not be ruled out (Green *et al.* 1986; Barbarand *et al.* 2003*a*, *b*).

After the Late Jurassic–Cretaceous cooling, by and large, near-horizontal *t–T* paths endured from the Late Cretaceous, throughout the Palaeogene, until the Late Miocene. They mark a period of stability, with the rocks staying at upper APAZ temperatures, while slow cooling brought some down to lower temperatures (50–60 °C) of AFT retention. The near-horizontal *t–T* paths are modestly disturbed by rapid Late Miocene–Pliocene to Recent cooling as early as 15–20 Ma, but with most cooling confined to the last 5–10 Ma. One sample (AL272) results in a somewhat deviant model, with a moderate reheating episode (upper APAZ temperatures at about 60–70 Ma) before final cooling sets in. It is not clear at this time if this feature is a modelling artefact or if it actually relates to a real event. In any case, viewing the regional geological setting, it is unlikely that considerable burial affected this sample site, nor is there any sign of magmatic or hydrothermal activity in this period in the wide surroundings of the study area. Hence, if a real event transpired, frictional heating along the reactivated Teletsk fault (Fig. 3) is the most plausible. But this is only applicable, if at all, to the latter sample that was collected in the fault zone itself. The recent rapid cooling eventually brought the samples to ambient surface temperatures and is associated with their exhumation to present outcrop positions. The long APAZ residence time (Late Mesozoic–Late Cenozoic) and associated track shortening explains the low MTL values and the relatively broad and negatively skewed track-length distributions.

Again, we would like to caution the reader concerning possible Late Cenozoic cooling as a modeling artifact. It can be perilous to interpret the Late Cenozoic cooling simply on the basis of AFT thermal history models without assessing additional geological information from the study area or adjacent areas. Below we attempt to make a case in favour of keeping the Late Cenozoic cooling in the overall thermal history.

The Late Jurassic–Cretaceous (160–65 Ma) cooling of the Teletskoye basement is contemporaneous with significant denudation as recorded in sediments of intramontane basins in the Altai and Sayan mountains and the larger orogen-adjacent basins (Fig. 4) (Dobretsov *et al.* 1996; Novikov 2002). We therefore interpret the Mesozoic apparent AFT ages and the modelled Late Jurassic–Cretaceous cooling as being largely the result of denudation of the Teletskoye basement. Taking into account different sample elevations (between 400 and 2300 m) and assuming a geothermal gradient of 20–25 °C km^{-1} (Cermak 1993; Duchkov *et al.* 1995; Hu *et al.* 2000), the area might have experienced 1–3 km of Late Jurassic–Cretaceous denudation recorded by the AFT system.

This interpretation would seem to be reasonable, as during the Jurassic and Cretaceous period the Altai area, including the Teletskoye basement blocks, was subjected to tectonic reactivation and sediments derived from the Mesozoic orogen were transported to large adjacent basins (for instance the West Siberian Basin, Junggar, GLDM; Fig. 4), and to smaller fault-controlled, intramontane depressions (Dobretsov *et al.* 1996; Novikov 2002; Howard *et al.* 2003). A remnant of such a Jurassic intramontane basin is preserved close to the Teletskoye graben (Fig. 3). Jurassic–Early Cretaceous molasse deposits unconformably overlie the Palaeozoic basement of many of these basins. Mesozoic sediment thickness reaches several kilometres in the large adjacent basins (e.g. Vyssotski *et al.* 2006), and in the smaller intramontane basins it can reach over 2 km (Dobretsov *et al.* 1996).

Mesozoic tectonic reactivation of the Altai–Sayan is coeval with the final closure of the Mongol–Okhotsk Ocean, and the ensuing Mongol–Okhotsk orogeny. Convergence and ultimate collision of the Siberian and North China–Mongolian (Amur) continents resulted in the development of the Mongol–Okhotsk orogenic belt in the Late Jurassic–Early Cretaceous (Van der Voo *et al.* 1999; Zorin 1999; Kravchinsky *et al.* 2002; Cogné *et al.* 2005). Incipient collision occurred in the western part of the Mongol–Okhotsk belt (South Baikal–East Sayan–Central Mongolia; Fig. 1) around 170–180 Ma ago (Kravchinsky *et al.* 2002; Cogné *et al.* 2005). Oblique collision lasted until final closure in the east (current Sea of Okhotsk region) in the Early Cretaceous (110–140 Ma ago). Deformation was not solely confined to the collision zone, but migrated through the hinterland, affecting the Mongolian Altai (Dobretsov *et al.* 1996; Novikov 2002), the Baikal area (Van der Beek *et al.* 1996) and the Siberian cratonic rim (Ermikov 1994; Zorin 1999).

Our interpretation of our AFT data suggests that the Siberian Altai, including the Teletskoye area, experienced far-field effects of this orogeny, leading to an important phase of denudation. Van der Beek *et al.* (1996) interpreted their AFT results of Mesozoic denudation in the Baikal area also in terms of distant Mongol–Okhotsk effects. Their AFT ages and thermal history models show Late Cretaceous cooling, which is somewhat younger than observed in our results. This does agree, however, with the oblique collision model as outlined above. A Mesozoic compressive tectonic regime in the Siberian Altai is demonstrated by the Jurassic Kuznetsk–Alatau thrust system and its related basins at the northern edge of the present-day Altai Mountains (Fig. 4) (Novikov 2002; Buslov *et al.* 2003). Hitherto, the kinematics and tectonic–geodynamic implications of this thrust are poorly understood, but it does clearly indicate Jurassic tectonic activity.

Accumulation of thick Mesozoic deposits and evidence from geophysical exploration in large adjacent basins also indicate active extension, and subsidence in the West Siberian Basin (Vyssotski *et al.* 2006), the Mongolian–Baikal basins (Ermikov 1994; Graham *et al.* 2001; Johnson 2004) and Junggar (Hendrix 2000) (Figs 1 and 4). Therefore, denudation and cooling of the Altai basement was probably an interaction of extension in large adjoining basins creating the accommodation space for Altai–Sayan-derived sediments, on one hand, and tectonic reactivation with denudation of the Altai–Sayan orogen, on the other.

The near-horizontal t–T paths at lower APAZ–upper AFT retention temperatures (about 50–80 °C) in the Late Cretaceous–Palaeogene reflect a period of prolonged stability, tectonic quiescence and peneplanation. Rocks consequently remained approximately at the position and depth they reached after Mesozoic denudation. Relaxation of the isotherms occurred and the rocks stayed at more or less constant temperatures. Throughout Siberia a lateritic peneplain with red beds developed (Ermikov 1994; Delvaux *et al.* 1995*a*; Dobretsov *et al.* 1996). Remnants of this surface can be found in the Siberian Altai Mountains (Deev *et al.* 1995; Dehandschutter *et al.* 2002; Novikov 2002). At present this surface is extensively deformed and broken as a result of Late Cenozoic tectonic activity. The identification of this phase of stability and its interpretation agrees with cooling curves based on AFT data by Van der Beek *et al.* (1996) obtained in the adjoining Baikal region.

The Late Cenozoic cooling phase in our thermal history models is interpreted as being associated with the tectonic reactivation of the region and the building and denudation of the modern Siberian Altai orogen. The reactivation is seen as a distant effect of ongoing India–Eurasia convergence. According to our models, cooling is mainly confined to the last 5–10 Ma (De Grave *et al.* 2007). This corresponds with other indicators that reveal an incipient reactivation in the Late Miocene and a clear intensification of transpressive tectonic movements in the Plio-Pleistocene (Delvaux *et al.* 1995*b*; Buslov *et al.* 1999; Dehandschutter *et al.* 2002; Novikov 2002).

Sediments produced by the denudation of the modern Altai are deposited in the large adjacent basins in a similar setting as that during the Late Jurassic–Cretaceous. In addition, reminiscent of the Mesozoic history, formation of Cenozoic, fault-controlled basins in the orogen itself also accommodate part of the Late Cenozoic sediments. In this respect, the Teletskoye graben is a prime example. In the graben and other Siberian Altai basins (the Chuya Basin for example, see Fig. 4), coarse Late Neogene to Holocene sediments are observed (Dobretsov *et al.* 1996; Buslov *et al.* 1999; Dehandschutter *et al.* 2002).

Although, to some extent, AFT thermal history modelling may overestimate the amount of denudation associated with this event as a consequence of model-induced Late Cenozoic cooling as mentioned earlier, the Late Neogene–Quaternary cooling of the Siberian Altai basement and its timing are, in our opinion, accurately reflected in the modelled t–T paths. Ample independent geological and tectonic evidence summarized in previous sections, and studies in adjoining areas, support our observations. We hence interpret the young rapid cooling to be related to Late Neogene–Quaternary denudation in a transpressive tectonic regime that is still active today. We conclude that reactivation and building of the modern Siberian Altai orogenic edifice is largely constrained to the last 5 Ma. However, as briefly stated, in contrast to the Siberian Altai, the Mongolian Altai and Gobi Altai to the SE (Fig. 1) show earlier signs of uplift, erosion and deposition of coarse clastic sediments in the Miocene and even Late Oligocene (Cunningham *et al.* 1996; Howard *et al.* 2003).

Other thermochronological investigations of the Altai–Sayan area report very similar AFT data. This is, for example, the case in Jolivet *et al.* (2007) and Vassallo *et al.* (2007). Their AFT results from the Mongolian Altai area show ages roughly between 65 and 200 Ma, MTL values between 11.1 and 13.8 μm, with standard deviations of the length distributions between 1.3 and 2.1 μm. Modelling using the Ketcham *et al.* (1999) annealing kinetics shows a Jurassic–Early Cretaceous cooling, followed by stability at upper APAZ–lower AFT retention temperatures up until about 5 $\pm$ 3 Ma, when rapid Pliocene cooling sets in (so a Late Cenozoic cooling is clearly observed even when the AFT data are modelled according to the Ketcham *et al.* 1999 multi-kinetic approach). In contrast to the sedimentological data, the AFT results from the Mongolian (and Gobi) Altai show great similarities with the AFT data reported in this paper. The apparent AFT ages are, in general, older (116–199 Ma) than those reported here for the Teletskoye region in the Siberian Altai (65–123 Ma). This offset could be due to the more southern position of the Mongolian and Gobi Altai. The latter part of the Altai orogen is located close to the Tien Shan orogen (Fig. 1). The Mesozoic ages obtained by Jolivet *et al.* (2007) and Vassallo *et al.* (2007) seem to be more closely associated with AFT ages from the Tien Shan area (e.g. De Grave *et al.* 2007). The Tien Shan area was reactivated earlier on in the Mesozoic, namely in the Early Jurassic–Early Cretaceous (around 125–200 Ma: Kapp *et al.* 2007), by the collision of the Tibetan blocks (Qiangtang and Lhasa; Cimmerian orogeny) in distinction to the Late Jurassic–Early Cretaceous (approximately at 100–150 Ma: Cogné *et al.* 2005) Mongol–Okhotsk orogeny that seems to have affected the Siberian Altai positioned more to the north (Fig. 1).

A long-lived episode of Late Mesozoic–Early Neogene stability is reflected in the near-horizontal t–T paths found in the modelling of our data, as well as in the Van der Beek *et al.* (1996), Jolivet *et al.* (2007) and Vassallo *et al.* (2007) data, and can therefore also be construed as a regional feature, that is tectonic stability over the entire Altai–Gobi Altai–Sayan–Baikal region.

Both in the Siberian Altai (our data) and the Mongolian and Gobi Altai (Jolivet *et al.* 2007; Vassallo *et al.* 2007) this stability is disturbed by Late Cenozoic tectonic activity, responsible for the current reactivation of the Altai orogen and the building of the modern Altai–Sayan–Gobi Altai mountain belt. In both cases it appears that the peak activity is constrained to the Plio-Pleistocene (largely during the last 5 Ma).

Another AFT study, this time from the Chinese Altai (Fig. 1) (Yuan *et al.* 2006), also supports the findings reported above. These authors find apparent AFT ages for basement rocks between about 50 and 160 Ma, although the vast majority of the ages fall in the 60–100 Ma range. MTL values vary between 11.3 and 14.5 μm (most values 12–14 μm) with standard deviations of 1.4–2.8 μm (most values 1.5–2.0 μm) are reported. Modelling using the Ketcham *et al.* (1999) annealing equations results in a three-stage cooling history, mimicking in general terms the ones found for the Siberian (our data) and the Mongolian Altai (Jolivet *et al.* 2007; Vassallo *et al.* 2007): (1) Cretaceous cooling to temperatures in the upper APAZ–AFT retention window; (2) long-lived Late Cretaceous–Palaeogene stability; and (3) Miocene onset of recent cooling.

Conclusions

In this paper we present a multi-method chronometric study of the Teletskoye graben basement in the Siberian Altai Mountains. The Altai mountain belt is part of the extensive intracontinental Central Asian Orogenic System. Based on several analytical techniques (zircon U–Pb SHRIMP dating, muscovite and biotite $^{40}Ar/^{39}Ar$ dating, and AFT thermochronology) a general thermal history of the Teletskoye basement is constructed (Fig. 8). This basement was largely formed in the Palaeozoic during accretion and growth of the Siberian–Central Asian continent. Several Palaeozoic collision–accretion events and Late Palaeozoic strike-slip deformation are responsible for the architecture of the Teletskoye basement. Some of these events are reflected in the geochronological data we present here.

The Late Ordovician–Early Silurian granitoid intrusion ages in the Teletskoye basement can be related to the collision of the Tuva–Mongolia microplate with the West Sayan tectonic unit of the Siberian mobile belt. After emplacement of these granitoids, they experienced cooling through the muscovite and biotite Ar-closure temperatures (about 350 °C) in the Middle Devonian (muscovite) to the Late Devonian–Early Carboniferous (biotite). This is also the time during which new collision–accretion affected the Siberian Altai region and involves, for instance, the docking of the Altai–Mongolia microcontinent with the Gorny Altai tectonic unit of the Siberian mobile belt. This event induced large-scale strike-slip and wrench deformation in Gorny Altai and West Sayan, and might have facilitated the exhumation and cooling of the Teletskoye basement.

The Late Palaeozoic–Early Mesozoic appears to be characterized by a period of relative stability and thermal equilibrium. Although an important deformation event is widely recognized to have been active in the Permo-Triassic, it is not reflected in our data. This is probably due to the fact that this deformation was of pure strike-slip, displacing blocks laterally along reactivated fault zones. Owing to the limited vertical offsets produced by this event, it would not be revealed in the AFT data.

After the Late Palaeozoic–Early Mesozoic period of apparent quiescence, Late Mesozoic cooling is registered in the AFT data. It is of Late Jurassic–Cretaceous age and predominantly related to the denudation of the Teletskoye basement during an episode of tectonic reactivation. The reactivation is mainly constrained to the pre-existing structural fabric of the region and is possibly a distant effect of the Mongol–Okhotsk orogeny. AFT modelling reveals a phase of Late Cretaceous–Palaeogene stability, supported by

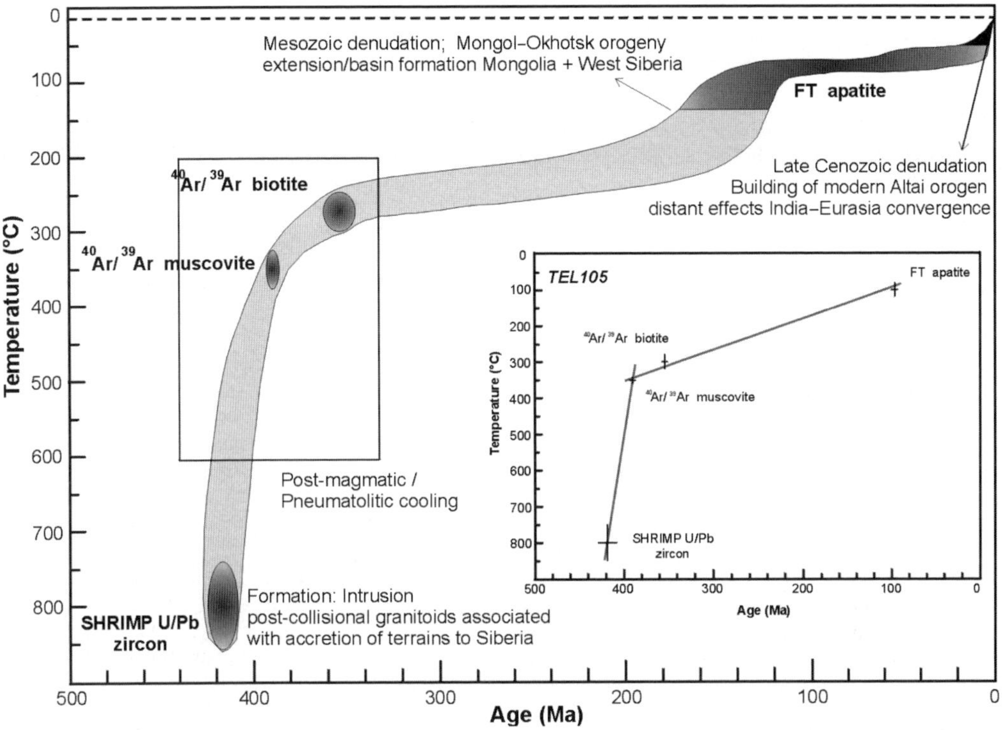

Fig. 8. General thermal history model of the Teletskoye graben basement and thermo-tectonic interpretation based on the different geochronometers used in this study. Zircon U–Pb closure temperatures based on Hanchar & Watson (2003), Ar–Ar closure temperatures on McDougall & Harrison (1999) and AFT closure temperature on Wagner & Van den haute (1992). Inset: example of all data grouped for a single sample (TEL105).

field evidence in the form of a regional lateritic peneplanation surface. Renewed cooling of the basement in the Plio-Pleistocene eventually brought the sampled rocks to their current outcrop positions. This renewed cooling is interpreted as the manifestation of the building and denudation of the modern Altai orogen and is supported by ample field evidence. This evidence is important in order to rule out possible AFT modelling artefacts. Rise and denudation of the Altai mountain ranges, and in particular in the Teletskoye region, is the response to a young (5–10 Ma) episode of tectonic reactivation. This reactivation is, again, most pronounced along the ancestral basement architecture and is seen as a result of far-field effects of the ongoing indentation of India into the Eurasian continent.

We thank F. De Corte, R. Jonckheere and A. De Wispelaere for support at various stages in the AFT research. D. Delvaux, R. Hus, J. Klerkx, L. Smirnova, K. Theunissen and others are acknowledged for assistance during fieldwork and sampling. P. Vermeesch and K. Stuebner provided help during Ar-analyses for which we are very grateful. We are indebted to J. Wooden and F. Mazdab for their valuable assistance during SHRIMP analyses. Comments by an anonymous reviewer and by P. O'Sullivan greatly enhanced this paper. This research was supported by the Fund for Scientific Research – Flanders (Belgium), the Flemish Institute of Innovation by Science and Technology (Belgium) and the Fulbright Scholar Exchange Program (Belgium–USA). J. De Grave is currently post-doctoral fellow of the Fund for Scientific Research – Flanders (Belgium).

References

ALLEN, M. B. & VINCENT, S. J. 1997. Fault reactivation in the Junggar region, northwest China: the role of basement structures during Mesozoic–Cenozoic compression. *Journal of the Geological Society, London*, **154**, 151–155.

AVOUAC, J. P. & TAPPONNIER, P. 1993. Kinematic model of active deformation in Central Asia. *Geophysical Research Letters*, **20**, 895–898.

BACHTADSE, V., PAVLOV, V. E., KAZANSKY, A. Y. & TAIT, J. A. 2000. Siluro-Devonian paleomagnetic results from the Tuva terrane (southern Siberia, Russia): implications for the paleogeography of Siberia. *Journal of Geophysical Research*, **105**, 13,509–13,518.

BARBARAND, J., CARTER, A., HURFORD, A. J. & WOOD, I. G. 2003*a*. Compositional and structural control of fission-track annealing in apatite. *Chemical Geology*, **198**, 107–137.

BARBARAND, J., HURFORD, A. J. & CARTER, A. 2003*b*. Variation in apatite fission-track length measurement: implications for thermal history modelling. *Chemical Geology*, **198**, 77–106.

BRIGGS, S. M., YIN, A., MANNING, C. E., CHEN, Z.-L., WANG, X.-F. & GROVE, M. 2007. Late Paleozoic tectonic history of the Ertix Fault in the Chinese Altai and its implications for the development of the Central Asian Orogenic System. *Geological Society of America Bulletin*, **119**, 944–960.

BUCHAN, C., PFÄNDER, J. *ET AL.* 2002. Timing of accretion and collisional deformation in the Central Asian Orogenic Belt: implications of granite geochronology in the Bayankhongor ophiolite zone. *Chemical Geology*, **192**, 23–45.

BUSLOV, M. M. & SINTUBIN, M. 1995. Structural evolution of the Lake Teletskoe zone, Altai–Sayan folded area. *Russian Geology and Geophysics*, **36**, 81–87.

BUSLOV, M., SAPHONOVA, I. *ET AL.* 2001. Evolution of the Paleo-Asian Ocean (Altai–Sayan region, Central Asia) and collision of possible Gondwana-derived terranes with the southern marginal part of the Siberian continent. *Geosciences Journal*, **5**, 203–224.

BUSLOV, M. M., WATANABE, T. *ET AL.* 2003. Role of strike-slip faults in Late Palaeozoic-Early Mesozoic tectonics and geodynamics of the Altai-Sayan and East Kazakhstan folded zone. *Russian Geology and Geophysics*, **44**, 49–75.

BUSLOV, M. M., WATANABE, T. *ET AL.* 2004. Late Palaeozoic faults of the Altai region, Central Asia: tectonic pattern and model of formation. *Journal of Asian Earth Sciences*, **23**, 655–671.

BUSLOV, M. M., WATANABE, T., SAPHONOVA, I. YU., IWATA, K., TRAVIN, A. & AKIYAMA, M. 2002. A Vendian-Cambrian island arc system of the Siberian continent in Gorny Altai (Russia, Central Asia). *Gondwana Research*, **5**, 781–800.

BUSLOV, M. M., ZYKIN, V. S., NOVIKOV, I. S. & DELVAUX, D. 1999. Cenozoic history of the Chuya depression (Gorny Altai): structure and geodynamics. *Russian Geology and Geophysics*, **40**, 1687–1701.

CERMAK, V. 1993. Lithospheric thermal regimes in Europe. *Physics of the Earth and Planetary Interiors*, **79**, 179–193.

COGNÉ, J. P., KRAVCHINSKY, V. A., HALIM, N. & HANKARD, F. 2005. Late Jurassic–Early Cretaceous closure of the Mongol–Okhotsk Ocean demonstrated by new Mesozoic paleomagnetic results from the Trans-Baikal area (SE Siberia). *Geophysical Journal International*, **163**, 813–832.

CUMMING, G. L. & RICHARDS, J. R. 1975. Ore lead isotope ratios in a continuously changing Earth. *Earth and Planetary Science Letters*, **28**, 155–171.

CUNNINGHAM, W. D., WINDLEY, B. F., DORJNAMJAA, D., BADAMGAROV, J. & SAANDAR, M. 1996. A structural transect across the Mongolian western Altai: active transpressional mountain building in Central Asia. *Tectonics*, **15**, 142–156.

DE CORTE, F., BELLEMANS, F., VAN DEN HAUTE, P., INGELBRECHT, C. & NICHOLL, C. 1998. A new U doped glass certified by the European Commission for the calibration of fission-track dating. *In*: VAN DEN HAUTE, P. & DE CORTE, F. (eds) *Advances in Fission-track Geochronology*. Kluwer, Dordrecht, 67–78.

DEEV, E. V., VYSOTSKY, E. M., NOVIKOV, I. S. & MISTRYUKOV, A. A. 1995. Geomorphological analysis of the Lake Teletskoe Region. *Russian Geology and Geophysics*, **36**, 122–131.

DE GRAVE, J. & VAN DEN HAUTE, P. 2002. Denudation and cooling of the Lake Teletskoye region in the Altai Mountains (South Siberia) as revealed by apatite fission-track thermochronology. *Tectonophysics*, **349**, 145–159.

DE GRAVE, J., BUSLOV, M. M. & VAN DEN HAUTE, P. 2007. Distant effects of India–Eurasia convergence and intracontinental deformation in Central Asia: constraints from AFT thermochronology. *Journal of Asian Earth Science*, **29**, 188–204.

DEHANDSCHUTTER, B., DELVAUX, D. & BOVEN, A. 1997. The Lake Teletsk tectonic depression (Altai): new kinematic data and chronological relations. *Annual Report of the Department of Geology and Mineralogy, Royal Museum for Central Africa, Tervuren, Belgium*, **1995–1996**, 147–167.

DEHANDSCHUTTER, B., VYSOTSKY, E., DELVAUX, D., KLERKX, J., BUSLOV, M. M., SELEZNEV, V. S. & DE BATIST, M. 2002. Structural evolution of the Teletsk graben (Russian Altai). *Tectonophysics*, **351**, 139–167.

DELVAUX, D., FERNANDEZ-ALONSO, M. *ET AL.* 1995*a*. Evidence for active tectonics in Lake Teletskoe (Gorny Altai). *Russian Geology and Geophysics*, **36**, 100–112.

DELVAUX, D., MOEYS, R., STAPEL, G., MELNIKOV, A. & ERMIKOV, V. 1995*b*. Paleostress reconstructions and geodynamics of the Baikal region, Central Asia, Part 1. Palaeozoic and Mesozoic pre-rift evolution. *Tectonophysics*, **252**, 61–101.

DISTANOVA, A. N. 2000. Early Palaeozoic granitoid associations of the Altai–Sayan folded area: their types and role in paleogeodynamic reconstructions. *Russian Geology and Geophysics*, **41**, 1202–1213.

DOBRETSOV, N. L. & VLADIMIROV, A. G. (eds). 2001. *Continental Growth in the Phanerozoic: Evidence from Central Asia. Geology, Magmatism and Metamorphism of the Western part of Altai–Sayan Fold Region. IGCP-420, 3rd Workshop.* UIGGM (United Institute of Geology, Geophysics and Mineralogy), Siberian Branch of the Russian Academy of Science (SB-RAS), Field Excursion Guide, Novosibirsk.

DOBRETSOV, N. L., BERZIN, N. A., BUSLOV, M. M. & ERMIKOV, V. D. 1995. General aspects of the evolution of the Altai region and the interrelationships between its basement pattern and the neotectonic structural development. *Russian Geology and Geophysics*, **36**, 3–15.

DOBRETSOV, N. L., BUSLOV, M. M., DELVAUX, D., BERZIN, N. A. & ERMIKOV, V. D. 1996. Meso- and Cenozoic tectonics of the Central Asian mountain belt: effects of lithospheric plate interaction and mantle plumes. *International Geology Review*, **38**, 430–466.

DRICKER, I. G., ROECKER, S. W., VINNIK, L. P., ROGOZHIN, E. A. & MAKEYEVA, L. I. 2002. Upper-mantle anisotropy beneath the Altai–Sayan region of

Central Asia. *Physics of the Earth and Planetary Interiors*, **131**, 205–223.

DUCHKOV, A. D., KLERKX, J. & KAZANTSEV, S. A. 1995. Heat flow in Lake Teletskoe. *Russian Geology and Geophysics*, **36**, 132–142.

DUFFIELD, W. A. & DALRYMPLE, G. B. 1990. The Taylor Creek Rhyolite of New Mexico: a rapidly emplaced field of lava domes and flows. *Bulletin of Volcanology*, **52**, 475–487.

ERMIKOV, V. D. 1994. Mesozoic precursors of the Cenozoic rift structures of Central Asia. *Bulletin des Centres Recherches, Explorations-Productions, Elf Aquitaine*, **18**, 123–134.

GLEADOW, A. J. W. & DUDDY, I. R. 1981. A natural long-term track annealing experiment for apatite. *Nuclear Tracks*, **5**, 169–174.

GLEADOW, A. J. W., DUDDY, I. R., GREEN, P. F. & LOVERING, J. F. 1986. Confined fission track lengths in apatite: a diagnostic tool for thermal history analysis. *Contributions to Mineralogy and Petrology*, **94**, 405–415.

GRAHAM, S. A., HENDRIX, M. S. *ET AL.* 2001. Sedimentary record and its implications of Mesozoic rifting in southeast Mongolia. *Geological Society of America Bulletin*, **113**, 1560–1579.

GREEN, P. F., DUDDY, I. R., GLEADOW, A. J. W., TINGATE, P. R. & LASLETT, G. M. 1986. Thermal annealing of fission tracks in apatite. 1. A qualitative description. *Chemical Geology (Isotopes Geoscience Section)*, **59**, 237–253.

HACKER, B. R., MOSENFELDER, J. L. & GNOS, E. 1996. Rapid emplacement of the Oman ophiolite: thermal and geochronologic constraints. *Tectonics*, **15**, 1230–1247.

HANCHAR, J. M. & WATSON, E. B. 2003. Zircon saturation thermometry. *In*: HANCHAR, J. M. & HOSKIN, P. W. O. (eds) *Zircon. Reviews in Mineralogy and Geochemistry*, **53**, 89–112.

HENDRIX, M. S. 2000. Evolution of Mesozoic sandstone compositions, southern Junggar, northern Tarim, and western Turpan basins, northwest China: a detrital record of the ancestral Tian Shan. *Journal of Sedimentary Research*, **70**, 520–532.

HU, S., HE, L. & WANG, J. 2000. Heat flow in the continental area of China: a new dataset. *Earth and Planetary Science Letters*, **179**, 407–419.

HURFORD, A. J. 1990. Standardization of fission track dating calibration: recommendation by the Fission Track Working Group of the I.U.G.S. Subcommission on Geochronology. *Chemical Geology (Isotope Geoscience Section)*, **80**, 171–178.

HOWARD, J. P., CUNNINGHAM, W. D., DAVIES, S. J., DIJKSTRA, A. H. & BADARCH, G. 2003. The stratigraphic and structural evolution of the Dzereg Basin, western Mongolia: clastic sedimentation, transpressional faulting and basin destruction in an intraplate, intracontinental setting. *Basin Research*, **15**, 45–72.

JOHNSON, C. L. 2004. Polyphase evolution of the East Gobi basin: sedimentary and structural records of Meso-Cenozoic intraplate deformation in Mongolia. *Basin Research*, **16**, 79–99.

JOLIVET, M., RITZ, J.-F. *ET AL.* 2007. Mongolian summits: an uplifted, flat, old but still preserved erosion surface. *Geology*, **35**, 871–874.

KAPP, P., DECELLES, P. G., GEHRELS, G. E., HEIZLER, M. & DING, L. 2007. Geological records of the Lhasa–Qiangtang and Indo-Asian collisions in the Nima area of central Tibet. *Geological Society of America Bulletin*, **119**, 917–932.

KETCHAM, R. A., DONELICK, R. A. & CARLSON, W. D. 1999. Variability of apatite fission-track annealing kinetics: III. Extrapolation to geologic time scales. *American Mineralogist*, **84**, 1235–1255.

KETCHAM, R. A., DONELICK, R. A. & DONELICK, M. B. 2000. AFTSolve: a program for multi-kinetic modeling of apatite fission-track data. *Geological Materials Research*, **2**, 1–32.

KHAIN, E. V., BIBIKOVA, E. V. *ET AL.* 2003. The Palaeo-Asian ocean in the Neoproterozoic and early Palaeozoic: new geochronological data and palaeotectonic reconstructions. *Precambrian Research*, **122**, 329–358.

KOLMOGOROVA, P. P. & KOLMOGOROV, V. G. 2002. Recent vertical crustal movements in the Altai-Sayan province in relation to neotectonics and seismicity. *Russian Geology and Geophysics*, **43**, 567–578.

KOSTITSYN, Y. A., ALTUKHOV, E. N. & FILINA, N. P. 1998. Rb–Sr dating of alkali granites in north-eastern Tuva. *Russian Geology and Geophysics*, **39**, 925–931.

KRAVCHINSKY, V. A., COGNÉ, J.-P., HARBERT, W. P. & KUZMIN, M. I. 2002. Evolution of the Mongol–Okhotsk ocean as constrained by new paleomagnetic data from the Mongol–Okhotsk suture zone, Siberia. *Geophysical Journal International*, **148**, 34–57.

KRÖNER, A., WINDLEY, B. F. *ET AL.* 2005. Accretionary growth in the Central Asian Orogenic Belt of Mongolia and Kazakhstan during the Neoproterozoic and Palaeozoic and comparison with the Arabian–Nubian shield and the present southwest Pacific. *In*: SKLYAROV, E. V. (ed.) *Structural and Tectonic Correlation Across the Central Asia Orogenic Collage: Northeastern Segment. Guidebook and Abstract Volume of the Siberian Workshop IGCP-480*. Institute of the Earth's Crust, Siberian Branch of the Russian Academy of Science, Irkutsk, 226–229.

LASLETT, G. M., GREEN, P. F., DUDDY, I. R. & GLEADOW, A. J. W. 1987. Thermal annealing of fission tracks in apatite. 2. A quantitative analysis. *Chemical Geology (Isotopes Geoscience Section)*, **65**, 1–13.

LEBEDEV, V. I., VLADIMIROV, A. G. *ET AL.* 1993. U–Pb dating and subdivision of Precambrian–Early Palaeozoic metamorphic and magmatic complexes of West Sangilen (SE Tuva). *Russian Geology and Geophysics*, **34**, 35–41.

LE PICHON, X., FOURNIER, M. & JOLIVET, L. 1992. Kinematics, topography, shortening and extrusion in the India–Eurasia collision. *Tectonics*, **11**, 1085–1098.

LUDWIG, K. R. 2001. *SQUID 1.02 Add-In for Excel*™. Berkeley Geochronology Center, Special Publications, **2**.

MCDOUGALL, I. & HARRISON, T. M. 1999. *Geochronology and Thermochronology by the* $^{40}Ar/^{39}Ar$ *Method*, 2nd edn. Oxford University Press, New York.

MOLNAR, P. & TAPPONNIER, P. 1975. Cenozoic tectonics of Asia: effects of a continental collision. *Science*, **189**, 419–426.

NIKOLAEVA, T. V. & SHUVALOV, V. F. 1995. Mesozoic and Cenozoic evolution of surface topography in Mongolia. *Geomorfologiya*, **2**, 54–65.

NISSEN, E., EMMERSON, B. *ET AL.* 2007. Combining InSAR and seismology to study the 2003 Siberian Altai earthquakes – dextral strike-slip and anticlockwise rotations in the northern India–Eurasia collision zone. *Geophysical Journal International*, **169**, 216–232.

NOVIKOV, I. S. 2002. Late Palaeozoic, Middle Mesozoic, and Late Cenozoic stages of the Altai orogeny. *Russian Geology and Geophysics*, **43**, 432–443.

PELTZER, G. & TAPPONNIER, P. 1988. Formation and evolution of strike-slip faults, rifts, and basins during the India–Asia collision: an experimental approach. *Journal of Geophysical Research*, **93**, 15,085–15,117.

PHILIP, H. & RITZ, J.-F. 1999. Gigantic paleolandslides associated with active faulting along the Bogd fault (Gobi-Altay, Mongolia). *Geology*, **27**, 211–214.

POLLITZ, F., VERGNOLLE, M. & CALAIS, E. 2003. Fault interaction and stress triggering of twentieth century earthquakes in Mongolia. *Journal of Geophysical Research*, **108**, 2503, doi: 10.1029/2002 JB002375.

RUDNEV, S. N., VLADIMIROV, A. G., PONOMARCHUK, V. A., KRUK, N. N., BABIN, G. A. & BORISOV, S. M. 2004. Early Palaeozoic granitoid batholiths of the Altai-Sayan folded region (lateral–temporal zoning and sources). *Doklady Earth Sciences*, **396**, 492–495 (in Russian).

SALNIKOVA, E. B., KOZAKOV, I. K. *ET AL.* 2001. Age of Palaeozoic granites and metamorphism in the Tuvino–Mongolian Massif of the Central Asian mobile belt: loss of a Precambrian microcontinent. *Precambrian Research*, **110**, 143–164.

SELEZNEV, V. S., DEHANDSCHUTTER, B., BUSLOV, M. M., SOLOVIEV, V. M. & DE BATIST, M. 2001. The sediment load of Lake Teletskoye: a seismic study. *Geological Sciences Annals of the Royal Museum for Central Africa, Tervuren, Belgium*, **105**, 239–262.

ŞENGÖR, A. M. C., NATAL'IN, B. A. & BURTMAN, V. S. 1993. Evolution of the Altaid tectonic collage and Palaeozoic crustal growth in Eurasia. *Nature*, **364**, 299–307.

SENNIKOV, N. V., YOLKIN, E. A., IZOKH, N. G. & KLETS, A. G. 2000. The Ordovician molasse complex of the southwestern part of West Sayan (Manchurek Formation). *Russian Geology and Geophysics*, **41**, 30–43.

SMIRNOVA, L. V., THEUNISSEN, K. & BUSLOV, M. M. 2002. Formation of the Late Palaeozoic structure of the Teletsk region: kinematics and dynamics (Gorny Altai–West Sayan junction). *Russian Geology and Geophysics*, **43**, 100–113.

TAPPONNIER, P. & MOLNAR, P. 1979. Active faulting and Cenozoic tectonics of the Tien Shan, Mongolia, and Baykal regions. *Journal of Geophysical Research*, **84**, 3425–3459.

THEUNISSEN, K., SMIRNOVA, L. & DEHANDSCHUTTER, B. 2002. Pseudotachylytes in the southern border fault of the Cenozoic intracontinental Teletsk basin (Altai, Russia). *Tectonophysics*, **351**, 169–180.

THOMAS, J. C., LANZA, R., KAZANSKY, A., ZYKIN, V., SEMAKOV, N., MITROKHIN, D. & DELVAUX, D. 2002. Paleomagnetic study of Cenozoic sediments from the Zaisan basin (SE Kazakhstan) and the Chuya depression (Siberian Altai): tectonic implications for central Asia. *Tectonophysics*, **351**, 119–137.

TOMURTOGOO, O., WINDLEY, B. F., KRÖNER, A., BADARCH, G. & LIU, D. Y. 2005. Zircon age and occurrence of the Adaatsag ophiolite and Muron shear zone, central Mongolia: constraints on the evolution of the Mongol–Okhotsk ocean, suture and orogen. *Journal of the Geological Society, London*, **162**, 125–134.

TORSVIK, T. H., TAIT, J., MORALEV, V. M., MCKERROW, W. S., STURT, B. A. & ROBERTS, D. 1995. Ordovician paleogeography of Siberia and adjacent continents. *Journal of the Geological Society, London*, **152**, 279–287.

VAN DEN HAUTE, P., DE CORTE, F., JONCKHEERE, R. & BELLEMANS, F. 1998. The parameters that govern the accuracy of fission-track age determinations: a re-appraisal. *In*: VAN DEN HAUTE, P. & DE CORTE, F. (eds) *Advances in Fission-track Geochronology*. Kluwer, Dordrecht, 33–46.

VAN DER BEEK, P., DELVAUX, D., ANDRIESSEN, P. A. M. & LEVI, K. G. 1996. Early Cretaceous denudation related to convergent tectonics in the Baikal region, SE Siberia. *Journal of the Geological Society, London*, **153**, 515–523.

VAN DER VOO, R., SPAKMAN, W. & BIJWAARD, H. 1999. Mesozoic subducted slabs under Siberia. *Nature*, **397**, 246–249.

VASSALLO, R., JOLIVET, M. *ET AL.* 2007. Uplift age and rates of the Gurvan Bogd system (Gobi-Altay) by apatite fission track analysis. *Earth and Planetary Science Letters*, **259**, 333–346.

VERGNOLLE, M., POLLITZ, F. & CALAIS, E. 2003. Constraints on the viscosity of the continental crust and mantle from GPS measurements and postseismic deformation models in western Mongolia. *Journal of Geophysical Research*, **108**, 2502, doi: 10.1029/2002JB002374.

VLADIMIROV, A., PONOMAREVA, A. *ET AL.* 1997. Late Palaeozoic–Early Mesozoic granitoid magmatism in Altai. *Russian Geology and Geophysics*, **38**, 755–770.

VLADIMIROV, A. G., KOZLOV, M. S. *ET AL.* 2001. Major epochs of intrusive magmatism of Kuznetsk Alatau, Altai, and Kalba (from U–Pb isotope dates). *Russian Geology and Geophysics*, **42**, 1089–1109.

VYSOTSKY, E. M., NOVIKOV, I. S. & SAGATOVA, A. R. 2006. When firm land collapses: the Chuya earthquake in Gorny Altai. *Science First Hand*, **4**, 32–39.

VYSSOTSKI, A. V., VYSSOTSKI, V. N. & NEZHDANOV, A. A. 2006. Evolution of the West Siberian Basin. *Marine and Petroleum Geology*, **23**, 93–126.

WAGNER, G. A. & VAN DEN HAUTE, P. 1992. *Fission Track-Dating*. Kluwer, Dordrecht.

WINDLEY, B. F., KRÖNER, A., GUO, J., QU, G., LI, Y. & ZHANG, C. 2002. Neoproterozoic to Palaeozoic geology of the Altai orogen, NW China: new zircon age data and tectonic evolution. *Journal of Geology*, **110**, 719–737.

YIN, A. & HARRISON, T. M. 2000. Geological evolution of the Himalayan–Tibetan orogen. *Annual Review in Earth and Planetary Sciences*, **28**, 211–280.

YUAN, W., CARTER, A., DONG, J., BAO, Z., AN, Y. & GUO, Z. 2006. Mesozoic–Tertiary exhumation history of the Altai Mountains, northern Xinjiang, China: new constraints from apatite fission track data. *Tectonophysics*, **412**, 183–193.

YUAN, C., SUN, M. ET AL. 2007. Accretionary orogenesis of the Chinese Altai: insights from Paleozoic granitoids. *Chemical Geology*, **242**, 22–39.

ZORIN, YU. A. 1999. Geodynamics of the western part of the Mongolia–Okhotsk collisional belt, Trans-Baikal region (Russia) and Mongolia. *Tectonophysics*, **306**, 33–56.

ZORIN, YU. A., BELICHENKO, V. G. ET AL. 1993. The South Siberia–Mongolia transect. *Tectonophysics*, **225**, 361–378.

Post Pan-African thermo-tectonic evolution of the north Mozambican basement and its implication for the Gondwana rifting. Inferences from $^{40}Ar/^{39}Ar$ hornblende, biotite and titanite fission-track dating

M. C. DASZINNIES[1,2]*, J. JACOBS[1,3], J.-A. WARTHO[4] & G. H. GRANTHAM[5]

[1]*Department of Geosciences, University of Bremen, P.O. Box 330440, D-28334 Bremen, Germany*

[2]*Present address: SINTEF Petroleum Research, Basin Modelling Group, S. P. Andersens vei 15 B, NO-7031, Trondheim, Norway*

[3]*Department of Earth Sciences, University of Bergen, Allegaten 41, N-5007 Bergen, Norway*

[4]*John de Laeter Centre of Mass Spectrometry, Department of Applied Geology, Curtin University, Perth, WA 6845, Australia*

[5]*Council for Geosciences, P/Bag X112, 0012 Pretoria, South Africa*

**Corresponding author (e-mail: matthias.c.daszinnies@sintef.no)*

Abstract: This research paper investigates the thermo-tectonic history of the north Mozambican basement subsequent to the Pan-African metamorphism. Six $^{40}Ar/^{39}Ar$ hornblende, three $^{40}Ar/^{39}Ar$ biotite and 25 titanite fission-track data place new constraints on the earliest timing of rifting in the central sector of Gondwana, and demonstrate a close linkage between the geometric rift configuration and the ductile metamorphic basement fabrics during the initial dispersal of the supercontinent.

The $^{40}Ar/^{39}Ar$ hornblende and biotite ages range from *c.* 542 to 456 Ma and from *c.* 448 to 428 Ma, respectively. These data record slow basement cooling after the latest Pan-African metamorphism at rates of *c.* 7–11 °C Ma^{-1} between Early and Late Ordovician times. Locally, syn- to post-tectonic granitoid emplacements around 500–450 Ma delayed basement cooling to Late Ordovician–Early Silurian times. The titanite fission-track (TFT) ages fall into two age groups of *c.* 378–327 Ma and *c.* 284–219 Ma. The older TFT ages record very slow cooling from the Late Ordovician–Early Silurian to below 275 $\pm$ 25 °C in the Late Devonian–Early Carboniferous at slow rates of less than 1 °C Ma^{-1}. This slow cooling is related to decreasing denudation in association with the establishment of pre-Karoo peneplains in central Gondwana. The younger TFT ages record denudation due to rift flank uplift in the context of initial Gondwana disintegration in the Mozambican sector. Corresponding Early–Late Permian crustal extension proceeded obliquely to a NW–SE tensional palaeo-stress field and was associated with a brittle reactivation of easterly trending ductile basement fabrics. In total, up to $\leq$9–12 km of denudation is deduced from the TFT results since Permo-Carboniferous times.

The present-day location of the north Mozambican basement at the East African margin (Fig. 1) indicates that this part of central Gondwana has been a favoured site of continental extension during the supercontinent's dissolution (Reeves *et al.* 2002). However, rift basins fringing northern Mozambique exhibit spatially contrasting and temporally discontinuous sedimentary records. Rift basins located to the north (Metangula Basin) and west (Shire Valley) of the basement (Figs 1 & 2) predominantly comprise continental sediments of Late Palaeozoic–Late Triassic/Early Jurassic age (Castaing 1991; Catuneanu *et al.* 2005). The marginal Rovuma and Mozambique rift basins solely exhibit Early–Middle Jurassic–Cenozoic marine and terrestrial strata that also overlap onto the basement (Pinna *et al.* 1993; Salman & Abdula 1995) (Figs 1 & 2). These entirely rift-related sedimentary records only permit a very sparse and insufficient resolution of the Palaeozoic–Mesozoic rift history of central Gondwana (Dingle & Scrutton 1974). Based on them, neither the geological evolution since the latest Pan-African event not the timing of the earliest rifting, i.e. the onset of Gondwana's disintegration, is adequately constrained in the north Mozambican sector.

The current outline of the north Mozambican coast conspicuously coincides with the regional ductile fabric trends of the metamorphic basement. The coastline broadly parallels the northerly ductile trends of the Mozambique Belt at the Tanzanian–Mozambican border and swings into an east to NE

From: LISKER, F., VENTURA, B. & GLASMACHER, U. A. (eds) *Thermochronological Methods: From Palaeotemperature Constraints to Landscape Evolution Models*. Geological Society, London, Special Publications, **324**, 261–286.
DOI: 10.1144/SP324.18 0305-8719/09/$15.00

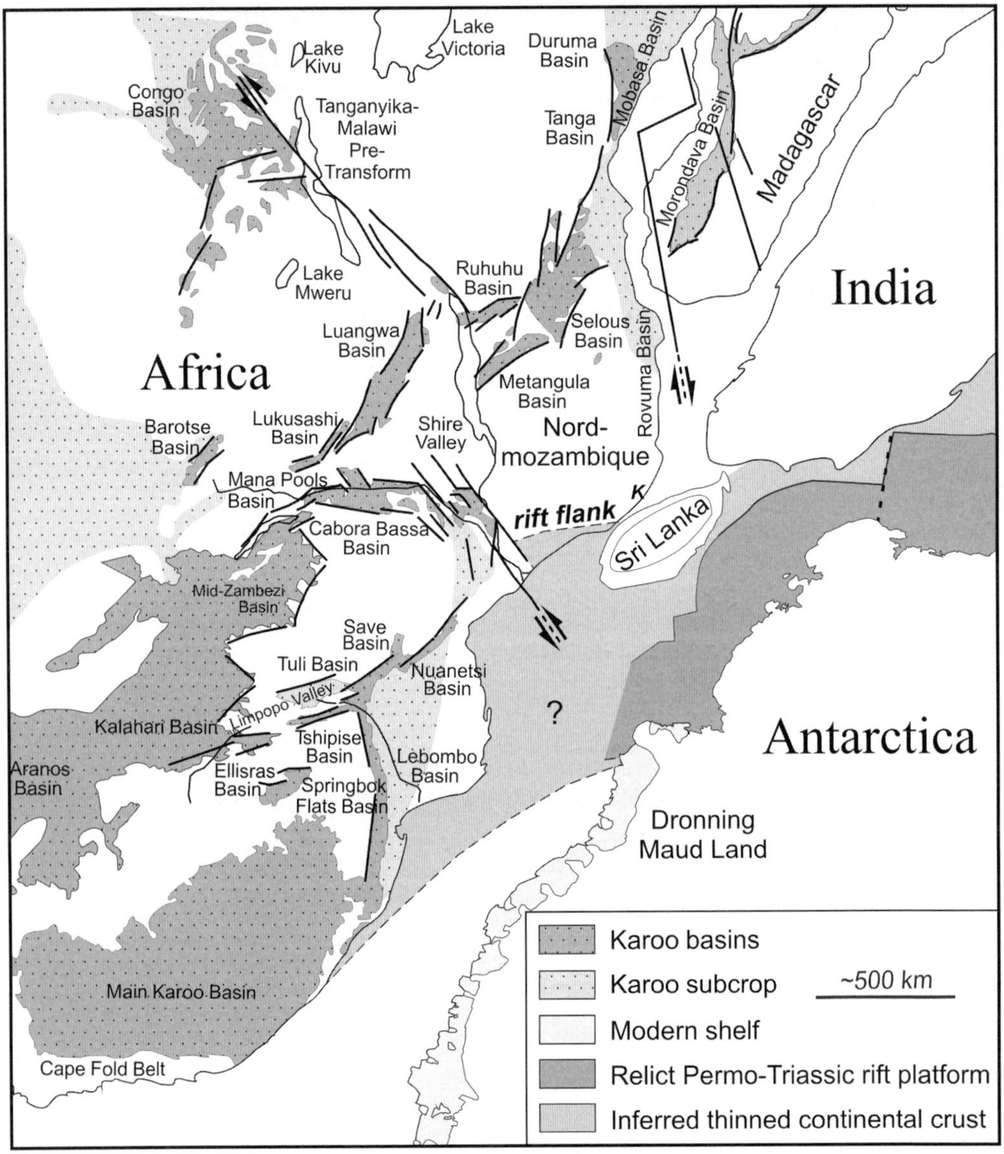

Fig. 1. Illustrations of the distribution of the Karoo basins in central Gondwana. It also shows the north Mozambican rift structure and its westwards extension into the Middle Zambezi Basin, as well as a possible eastward linkage with proposed Permo-Triassic rift basin between India and Antarctic (Harrowfield *et al.* 2005). Dashed lines represent supposed boundaries. Intercontinental white areas are thinned continental crust. *K* represents isolated outcrops of Karoo age volcanics of *c.* 180–160 Ma in age (Jaritz *et al.* 1977; Grantham *et al.* 2005*b*). The map was compiled and modified after Castaing (1991), Catuneanu *et al.* (2005) and Harrowfield *et al.* (2005).

trend to the south of the Lurio Belt similar to the ductile fabrics (Fig. 2). This consistency among the coastline and basement fabric trends might point towards an influence of the Pan-African age structural inventory on the locus and geometry of crustal extension during incipient rifting.

Owing to the absence of comprehensive sedimentary records, thermochronological methods provide an invaluable tool to investigate the cooling history of crystalline basement rocks. If cooling can be related to denudation, the regional thermotectonic history can be interpreted. This study

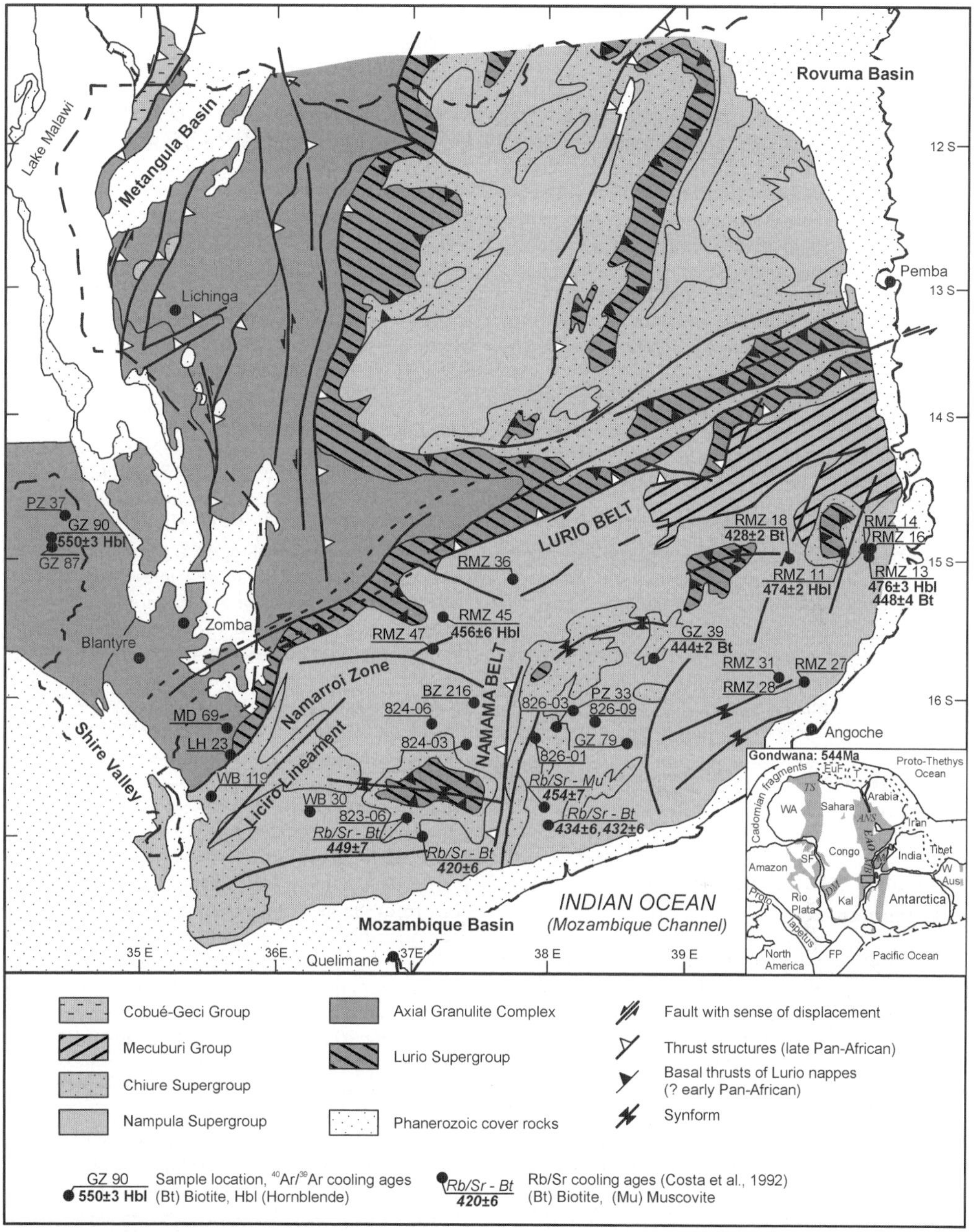

Fig. 2. Generalized map depicting major litho-tectonic units of the crystalline basement in northern Mozambique and southern Malawi. The map was modified after Pinna *et al.* (1993), Andreoli (1984), Kröner *et al.* (2001) and Perits *et al.* (2002) with superimposed locations of samples used for $^{39}Ar/^{40}Ar$ and TFT analyses. The inset depicts Pan-African Mobile Belts (grey) in a Gondwana reconstruction (Kusky *et al.* 2003; Jacobs & Thomas 2004). Abbreviations: ANS, Arabian Nubia Shield; DM, Damara Belt; EAAO, East African–Antartic orogen; EuF, European fragments; FP, Falkland Plateau; Kal, Kalahri craton; M, Madagascar; MB, Mozambique Belt; SF, San Francisco craton; T, Turkey; TS, Trans-Saharan; W Aus, Western Australia; WA, West Africa craton.

combines medium- and low-temperature thermochronological methods involving $^{40}Ar/^{39}Ar$ dating of hornblende and biotite, and titanite fission-track (TFT) analyses. It aims to constrain the thermotectonic evolution subsequent to the latest metamorphic event for southern basement in northern Mozambique, sampled along an approximately 260 km-wide east–west-trending profile (Fig. 2). We attempt to constrain the earliest timing of the Gondwana dispersal in the region. By evaluating the spatial cooling patterns of the basement we try to outline the geometric configuration of the earliest rift setting and to investigate its relation to the Pan-African age structural heritage within northern Mozambique.

$^{40}Ar/^{39}Ar$ dating of minerals yields ages that can record the cooling to below their mineral-specific closure temperatures (Dodson 1973). For cooling rates of 1–100 °C Ma^{-1} and grain radii of *c.* 150 μm, hornblende and biotite yield closure temperatures of 570–483 and 345–281 °C, respectively (McDougall & Harrison 1999). In the case of TFT dating, the thermochronometer partially retains fossil fission tracks over a temperature range of 310–265 ± 10 °C (Coyle & Wagner 1998), and yields ages that can record cooling to below 275 ± 25 °C (Kohn *et al.* 1993). A combination of these three thermochronometers permits inferences on rock cooling paths spanning temperatures from typical lower-amphibolite metamorphic facies conditions down to conditions of about 15–10 km in depth.

Geological setting and previous geochronology

The Mesoproterozoic basement of northern Mozambique was subjected to an intense tectonic and metamorphic overprint under amphibolite–granulite facies conditions during the Pan-African amalgamation of the Gondwana supercontinent (Kröner *et al.* 1997, 2001). In East Africa, this basement represents the southern termination of the north–south-trending Mozambique Belt (MB) (Fig. 2, inset), which constitutes a fundamental suture in the assembly of Gondwana and possibly extends southwards into Dronning Maud Land, Antarctica (Jacobs *et al.* 1998; Jacobs & Thomas 2004).

The study area comprises the Axial Granulite Complex in the west and the southern basement south of the Lurio Belt (Pinna & Marteau 1987; Pinna *et al.* 1993) in the east (Fig. 2).

The Axial Granulite Complex

The Axial Granulite Complex mainly consists of granulitic orthogneisses and charnockites. Its prominent ductile fabrics (D_{M2}) oscillate in trend between NW–SE and NE–SW (Pinna *et al.* 1993). The peak granulite facies metamorphism is dated between 571 and 549 Ma by U–Pb sensitive high-resolution ion micro-probe (SHRIMP) and Pb/Pb evaporation analyses of individual metamorphic zircon grains (Kröner *et al.* 2001). Subsequently, isobaric cooling caused a static retrograde amphibolite facies overprint (Pinna *et al.* 1993; Kröner *et al.* 2001). Amphibolite facies deformations of likely Pan-African age (D_{P1}–D_{P3}), as exemplified in the Geci Group (Lulin 1985), are mainly confined to transpressive thrust zones that trend between ENE–WSW and NW–SE (e.g. Costa *et al.* 1992).

The Lurio Belt foreland

The southern basement region is bound to the north by the ENE–WSE-trending Lurio Belt (Cadoppi *et al.* 1987; Pinna *et al.* 1993; Kröner *et al.* 1997). This belt is a discontinuous high-strain zone of dismembered layers, pods and lenses of granulite facies rocks (Norconsult Consortium 2007), and extends from the Indian Ocean into southern Malawi (Fig. 2). The southern basement dominantly consists of biotite–hornblende and migmatic, leucoratic gneisses (Nampula Supergroup). Supracrustal, mainly meta-sedimentary units of allochthonous (Chiure Supergroup) and autochthonous origin (Mecuburi Group), cover this basement. Remenant, granulite facies ortho- and paragneisses (Granulite Klippen) rest as unrooted klippen on the Nampula basement. Single-grain Pb/Pb evaporation and U–Pb SHRIMP ages of *c.* 615 Ma of metamorphic zircons from the granulites are interpreted to date the peak granulite facies metamorphism in the southern basement and in the Lurio Belt (Kröner *et al.* 1997). This event is post-dated by an unconstrained retrograde amphibolite facies overprint (Pinna *et al.* 1993). One Rb–Sr biotite cooling age of 449 ± 7 Ma (Costa *et al.* 1992) is reported from a meta-granitoid in the Nampula basement (Fig. 2).

In the southern basement the principle ductile fabrics (D_{M2}) display easterly trends (NE–SW to NW–SE) that broadly parallel the Lurio Belt orientation (Pinna *et al.* 1993). The NNE–SSW-trending Namama Thrust Belt is located in the central southern basement (Fig. 2). According to Cadoppi *et al.* (1987) the eastward thrusting altered the basement's ductile fabric (D_{M2} of Pinna *et al.* 1993), at all scales, to strongly aligned east–west trends. The Namama Thrust Belt is considered to be younger than the Lurio Belt (Cadoppi *et al.* 1987) and is probably of Pan-African age (Sacchi *et al.* 2000).

Gabbroic–granitic intrusives of late Pan-African age (*c.* 500 Ma) cross-cut all units in the southern basement. Some intrusions exhibit weak, linear fabrics and elliptic shapes that parallel the Lurio Belt trend. These features are interpreted to indicate their syn- to late-kinematic emplacement (Pinna *et al.* 1993). Rb–Sr biotite cooling ages obtained from these bodies (Fig. 2) range between 420 and 434 Ma (Sacchi *et al.* 1984; Costa *et al.* 1992). Undeformed pegmatite occurrences in the southern basement are largely regarded as Pan-African in age (Afonso 1976; Araujo 1976; Costa *et al.* 1992). For one them a Rb–Sr muscovite age of 454 ± 7 Ma is reported by Costa *et al.* (1992) (Fig. 2).

In this paper we follow the tectonostratigraphic scheme of Pinna *et al.* (1993). We would like to make the reader aware that recent mapping efforts in Mozambique will soon result in a significantly revised tectonostratigraphic framework (Bjerkgård *et al.* 2006; Hollick *et al.* 2006; Thomas *et al.* 2006; Viola *et al.* 2006).

Late Palaeozoic–Early Mesozoic intracontinental rift basin

In the Late Carboniferous–Late Triassic/Early Jurassic, the regional tectonic regime of NE Gondwana was governed by tensional stresses that propagated from the diverging southern Tethyan margin southwards into the supercontinent (Wopfner 2002). Within the East African sector this tectonic regime resulted in the formation of graben and extended intracratonic rift structures. These rifts are filled with thick and mainly terrigenous sedimentary deposits (Karoo Group equivalents) that are of remarkable similarity across eastern Africa (Wopfner 1993, 1994; Visser & Praekelt 1996; Reeves *et al.* 2002). In the entire region, Late Triassic deposits are terminated by a major unconformity that reflects the incipient Gondwana break-up and terminates the intracratonic rift stage (Catuneanu *et al.* 2005 and references therein).

Rift basins fringing the north Mozambican basement to the west (Shire Valley, Cabora Bassa Basin), north (Metangula Basin) and NE (Selous Basin) uniformly preserved Karoo age, terrestrial infillings of Early/Late Permian–Triassic and/or Jurassic age (Fig. 2). Within these basins, synsedimentary faulting observed in the oldest Karoo age strata indicates a common occurrence of incipient sedimentation and tectonic activity in Early–Late Permian times (Habgood 1963; Verniers *et al.* 1989; Castaing 1991; Catuneanu *et al.* 2005 and references therein). The Cabora Bassa Basin and the Shire Valley (Fig. 2) formation was linked to the evolution of the sinistral Zambesi pre-transform system, and was governed by a prominent NW–SE and a subordinate NE–SW tensional stress fields (Castaing 1991). Instead, the evolution of the Metangula and Selous basins was guided by the development of the Malagasy Chasm rift system between Africa and Madagascar (Verniers *et al.* 1989; Wopfner & Kaaya 1991; Wopfner 2002). In SW Madagascar, intracontinental rift basins contain Late Carboniferous/Early Permian–Late Triassic strata (Besairie & Collignon 1972), whereby earliest pull-apart basins initiated along NE–SW-trending sinistral strike-slip zones in the Early–Late Permian (Schandelmeier *et al.* 2004).

Mesozoic marginal rift basins

The Southern Rovuma Basin, located along the NE coast of Mozambique, represents the southern termination of the pericratonic East African marginal basin. In the south the north Mozambican basement is bordered by the Mozambique Basin (Figs 1 & 2). Both basins mainly contain marine sedimentary deposits of Early/Middle Jurassic–Cenozoic age (Salman & Abdula 1995). In Rovuma Basin Karoo age strata are inferred below Jurassic deposits based on seismic correlations with the Tanzanian Selous Basin. In the Mozambique Basin their existence is only supposed (Salman & Abdula 1995).

Analytical procedures

$^{40}Ar/^{39}Ar$ *analysis*

Hornblende and biotite grains were extracted following standard mineral separation procedures. Approximately 300 μm-sized grains were selected from the 150–315 μm sieve fraction. The mineral separates were individually wrapped in aluminium foil packets, and all the samples were inserted into an aluminium irradiation package. Biotite age standard Tinto B (K–Ar age of 409.2 ± 0.7 Ma: Rex & Guise 1995) was placed at 5 mm intervals along the package to monitor the neutron flux gradient. The package was Cd-shielded and irradiated in the 5C position of the McMaster University Nuclear Reactor, Hamilton, Canada, for 90 h. Upon return, the samples were loaded into an ultra-high vacuum laser chamber fitted with a Kovar viewport and baked to 120 °C overnight to remove adsorbed atmospheric argon from the samples and chamber walls.

A 110 W Spectron Laser Systems continuous-wave neodymium–yttrium–aluminium–garnet (CW–Nd–YAG) ($\lambda = 1064$ nm) infra-red laser, fitted with a TEM00 aperture, was used to slowly laser step-heat the mineral samples. The laser was fired through a Merchantek computer-controlled X–Y–Z sample chamber stage and microscope system, fitted with a high-resolution CCD (charged-coupled device) camera, 6× computer-controlled zoom, high-magnification objective lens and two light sources for sample illumination. The gases released by laser heating were 'gettered' using three SAES AP10 getter pumps to remove all active gases. The remaining noble gases were equilibrated into a high-sensitivity mass spectrometer (MAP 215-50), operated at a resolution of 570 and fitted with a Balzers SEV 217 multiplier. The automated extraction and data acquisition system was computer controlled, using a LabView program. The mean 5 min extraction system blank Ar isotope measurements obtained during the experiments were 1.18×10^{-12}, 1.32×10^{-14}, 4.65×10^{-15}, 6.70×10^{-14} and 1.43×10^{-14} cm^3 STP (standard temperature and pressure) for ^{40}Ar, ^{39}Ar, ^{38}Ar, ^{37}Ar and ^{36}Ar, respectively. The Ar isotope analyses were corrected for system blanks, mass discrimination ($^{40}Ar/^{36}Ar = 281.0$), radioactive decay of ^{37}Ar, and minor interference reactions from Ca and K ($^{39}Ar/^{37}Ar_{Ca} = 0.00065$, $^{36}Ar/^{37}Ar_{Ca} = 0.000255$ and $^{40}Ar/^{39}Ar_K = 0.0015$). Errors quoted on the ages (Tables 1 & 2) are 1σ and include the *J* value error (Mitchell 1968). The $^{40}Ar/^{39}Ar$ ages were calculated using the decay constant of Steiger & Jäger (1977). *J* values and errors are noted on the sample $^{40}Ar/^{39}Ar$ data tables (Table 2). The $^{40}Ar/^{39}Ar$ data presented in this study were undertaken at the Western Australian Argon Isotope Facility (Curtin University and the University of Western Australia).

Titanite fission-track analysis

Titanite separates with grain sizes of 150–315 μm were extracted using conventional preparation techniques including crushing, sieving, Wilfley table, heavy liquid and magnetic separation. Batches of titanite grains were embedded in epoxy resin, then ground and polished to expose internal crystal surfaces. The titanite fission tracks were revealed by etching the polished crystal mounts in an acid solution of 1 part concentrated HF, 2 parts concentrated HNO_3, 3 parts concentrated HCl and 6 parts H_2O. Samples were etched individually at room temperature for 17–27 min (Naeser & McKee 1970). Distinctly recognizable terminations of confined tracks were used as an evaluation criterion for sufficient fission-track etching. All the samples were loaded into aluminium capsules and irradiated at the FRM II research reactor facility in Garching, Germany. Corning dosimeter glasses (CN 2) were used to monitor a neutron fluence gradient. An irradiation time of 60 s was applied to obtain a total thermal neutron fluence of 0.5×10^{16} neutrons cm^{-2}. Induced tracks were recorded in white micas following the external detector approach (Gleadow 1981) and revealed after irradiation by etching the micas for 15 min in 40% HF at 21 ± 1 °C. All track density measurements in mounts and micas were performed using a Zeiss® Axiophot microscope. The fission-track ages were calculated according to the zeta calibration method (Hurford 1990; Galbraith & Laslett 1993) using a weighted mean zeta factor (ζ) of 134.5 ± 8.2 obtained from Fish Canyon Tuff and Mt Dromedary TFT age standards. Errors are quoted at the 2σ level and were derived according to conventional methods (Green 1981).

Results

$^{40}Ar/^{39}Ar$ hornblende analysis

The $^{40}Ar/^{39}Ar$ results of six hornblende samples (BZ 216, GZ 90, PZ 37, RMZ 11, RMZ 13 and RMZ 45) are presented in Tables 1 and 2, and Figure 3. All hornblende samples exhibit disturbed age spectra and yield no plateau ages. In all samples, old ages are preserved in the low-temperature steps of the sample's age spectra (Fig. 3a–f). Some of these ages are anomalously old of up to 3.8 Ga (e.g. sample BZ 216, Fig. 3a). All samples contained very high $^{40}Ar^*$ concentrations of 95–99%, with only the occasional presence of 11–36% atmospheric ^{40}Ar in sample PZ 37 (Table 2). No statistically reliable inverse isochron ages were obtained. Pseudo-plateau and weighted mean ages for all hornblende samples range from 462.5 ± 2.7 to 550.7 ± 3.2 Ma, and are statically unreliable (Table 2). They either comprise less than 45% of the cumulative ^{39}Ar (BZ 216 and PZ 37) or yield mean squared weighted deviate (MSWD) and probability values higher than 2.5 and lower than 0.05 (GZ 90, RMZ 11, RMZ 13 and RMZ 45).

$^{40}Ar/^{39}Ar$ biotite analysis

The $^{40}Ar/^{39}Ar$ results of three biotite samples GZ 39, RMZ 13 and RMZ 18 are given in Tables 1 and 2, and Figure 4a–c. The biotite samples contain high concentrations of $^{40}Ar^*$ (92–100%), with only the occasional presence of 11–28% atmospheric ^{40}Ar in samples GZ 39 and RMZ 13 (Table 2). Only sample GZ 39 yielded a statistically reliable inverse isochron age of 440.4 ± 4.1 Ma. Biotite samples GZ 39 and RMZ 18 yielded plateau ages of 443.8 ± 0.4 and 428.4 ± 0.3 Ma. The highest-temperature steps (i.e. > 80% cumulative ^{39}Ar) of both samples revealed discordant $^{40}Ar/^{39}Ar$ age spectra (Fig. 4a, c). Sample RMZ 13 yielded a discordant $^{40}Ar/^{39}Ar$ age spectrum (Fig. 4b) with no plateau age. A calculated weighted mean age of 448.2 ± 3.7 Ma (steps 2–6) is consistent with the sample's total fusion age of 445.9 ± 10.4 Ma (Table 1). The first low-temperature steps of GZ 39 and RMZ 13 yielded ages that are younger than the plateau and weighted mean ages

Table 1. *Results of $^{40}Ar/^{39}Ar$ hornblende and biotite analyses*

Step	$^{36}Ar/^{39}Ar$	$\pm 1\sigma$	$^{37}Ar/^{39}Ar$	$\pm 1\sigma$	$^{40}Ar^{*}/^{39}Ar$	$\pm 1\sigma$	$\%^{40}Ar^{*}$	Cum. % ^{39}Ar	Age (Ma)	$\pm 1\sigma$
Hornblende										
BZ 216										
1	0.031968	0.006445	0.986919	0.028020	349.732503	2.212949	97.37	0.20	3869.41	12.85
2	0.006243	0.001156	2.728420	0.021142	220.124316	0.512934	99.17	1.33	3154.44	8.23
3	0.001416	0.000095	3.399730	0.017253	23.044950	0.031808	98.22	8.18	727.71	3.11
4	0.001222	0.000085	3.543060	0.025649	15.612262	0.030807	97.74	15.84	523.42	2.45
5	0.000720	0.000241	3.791943	0.031466	13.846273	0.071766	98.49	21.24	471.28	2.99
6	0.001581	0.000122	3.161398	0.045993	13.558327	0.036948	96.67	31.88	462.64	2.33
7	0.001013	0.000050	3.641793	0.036380	14.773656	0.015967	98.01	46.47	498.85	2.23
8	0.001071	0.000030	3.812191	0.042203	14.231090	0.014135	97.82	57.47	482.77	2.16
9	0.001079	0.000068	3.762541	0.037817	13.613811	0.022209	97.71	67.05	464.31	2.16
10	0.000603	0.000074	4.005727	0.024829	13.470823	0.025718	98.69	79.48	460.00	2.18
11	0.001267	0.000146	3.625625	0.016354	13.527152	0.048550	97.31	83.94	461.70	2.51
12	0.000653	0.000041	3.957767	0.020543	13.591164	0.016215	98.60	100.00	463.63	2.11
Total fusion age = 1003.4 ± 1133.6 Ma, *J* value = 0.021560 ± 0.000108										
GZ 90										
1	0.002301	0.000146	2.456882	0.016064	22.476271	0.047316	97.06	2.59	712.84	3.20
2	0.000992	0.000064	2.387667	0.015752	18.514820	0.049140	98.44	8.52	605.95	2.92
3	0.000362	0.000110	2.456473	0.026630	16.822836	0.043184	99.37	11.97	558.29	2.70
4	0.000538	0.000045	2.316737	0.022783	16.721306	0.018715	99.06	20.38	555.39	2.45
5	0.000500	0.000021	2.495926	0.010075	16.737354	0.021711	99.13	29.23	555.85	2.48
6	0.000410	0.000014	2.476219	0.006787	16.653821	0.014297	99.28	43.26	553.46	2.42
7	0.000238	0.000031	2.590435	0.007219	16.399811	0.016299	99.57	49.38	546.17	2.41
8	0.000403	0.000000	2.400349	0.005985	16.592527	0.004636	99.29	61.34	551.70	2.38
9	0.000471	0.000033	2.202266	0.020097	16.346616	0.020737	99.16	67.03	544.65	2.43
10	0.000224	0.000018	2.476167	0.017132	16.456251	0.011906	99.60	77.46	547.80	2.39
11	0.000339	0.000020	2.408623	0.009644	16.677319	0.022609	99.40	86.81	554.13	2.48
12	0.000363	0.000014	2.468989	0.020111	16.467898	0.014588	99.35	100.00	548.13	2.40
Total fusion age = 569.5 ± 45.9 Ma, *J* value = 0.021559 ± 0.000108										
PZ 37										
1	0.027605	0.000105	4.795990	0.052506	27.301070	0.104714	76.99	0.56	834.98	4.22
2	0.089966	0.001109	5.335098	0.045883	46.849253	0.335388	63.80	1.18	1259.48	7.92
3	0.012678	0.000028	6.435820	0.032841	29.971019	0.069612	88.89	2.51	899.20	3.91
4	0.003188	0.000004	5.030265	0.049770	17.610746	0.024486	94.92	5.54	580.62	2.58
5	0.001446	0.000357	4.687974	0.026548	16.655755	0.106100	97.50	10.29	553.49	3.86
6	0.001057	0.000154	4.498755	0.018551	16.158381	0.046009	98.10	22.14	539.20	2.69

(*Continued*)

Table 1. *Continued*

Step	$^{36}Ar/^{39}Ar$	$\pm 1\sigma$	$^{37}Ar/^{39}Ar$	$\pm 1\sigma$	$^{40}Ar^*/^{39}Ar$	$\pm 1\sigma$	$\%^{40}Ar^*$	Cum. % ^{39}Ar	Age (Ma)	$\pm 1\sigma$
7	0.001356	0.000226	4.439132	0.031948	16.870235	0.073552	97.68	25.14	559.62	3.20
8	0.000966	0.000064	4.155384	0.034079	16.280563	0.021442	98.28	40.21	542.72	2.43
9	0.000912	0.000025	4.536082	0.015826	16.181961	0.008882	98.36	53.81	539.88	2.35
10	0.001674	0.000077	4.052126	0.016037	16.098877	0.024255	97.02	58.20	537.48	2.43
11	0.000998	0.000078	3.948953	0.030853	16.240262	0.026588	98.22	73.93	541.56	2.47
12	0.000774	0.000060	4.271916	0.035165	16.238207	0.020075	98.61	90.86	541.50	2.41
13	0.000541	0.000000	4.276148	0.034619	16.272831	0.014912	99.03	100.00	542.50	2.39
Total fusion age = 651.7 ± 210.1 Ma, *J* value = 0.021558 ± 0.000108										
RMZ 11										
1	0.002071	0.000001	1.805152	0.011411	16.135771	0.010117	96.35	3.77	538.51	2.35
2	0.000713	0.000000	1.631983	0.013544	13.703791	0.012508	98.49	9.29	466.93	2.10
3	0.000196	0.000000	2.083660	0.006827	13.986771	0.008592	99.59	13.84	475.41	2.11
4	0.000419	0.000000	1.799740	0.028872	13.934258	0.015722	99.12	21.78	473.84	2.14
5	0.000167	0.000054	2.161815	0.020805	14.043136	0.022280	99.65	37.09	477.09	2.20
6	0.000367	0.000055	1.895692	0.008180	14.019303	0.017721	99.23	50.47	476.38	2.16
7	0.000255	0.000000	2.032152	0.006206	13.951377	0.011752	99.46	59.96	474.35	2.12
8	0.000549	0.000073	1.849881	0.007221	13.751567	0.023784	98.84	65.00	468.37	2.19
9	0.000482	0.000001	1.478668	0.049551	13.825448	0.040317	98.98	73.12	470.58	2.40
10	0.000000	0.000000	2.106406	0.013741	14.004789	0.016991	100.00	80.96	475.95	2.16
11	0.000167	0.000000	1.862911	0.010342	13.996620	0.006078	99.65	83.82	475.70	2.10
12	0.000340	0.000000	1.957515	0.008483	14.028065	0.007438	99.29	100.00	476.64	2.11
Total fusion age = 479.2 ± 18.2 Ma, *J* value = 0.021556 ± 0.000108										
RMZ 45										
1	0.001467	0.000168	1.495542	0.064667	16.154574	0.098520	97.39	7.73	539.11	3.68
2	0.000390	0.000032	1.904748	0.008070	13.487405	0.010584	99.15	47.79	460.48	2.60
3	0.000655	0.000083	1.582240	0.027034	13.127612	0.026887	98.55	69.87	449.61	2.15
4	0.000000	0.000000	1.988715	0.018017	13.348557	0.032656	100.00	82.41	456.29	2.25
5	0.000233	0.000000	1.880454	0.007654	13.571388	0.006813	99.50	86.06	463.01	2.60
6	0.000298	0.000000	1.950149	0.004221	13.191889	0.010850	99.34	99.13	451.56	2.30
7	0.002418	0.000012	2.177812	0.100262	13.088641	0.067693	94.82	100.00	448.43	2.86
Total fusion age = 466.9 ± 29.9 Ma, *J* value = 0.021559 ± 0.000108										
Hornblende										
RMZ 13										
1	0.002836	0.000212	2.988312	0.013327	17.243466	0.064602	95.37	3.00	570.16	3.60
2	0.001289	0.000001	3.046942	0.036717	14.266128	0.014036	97.40	9.45	483.72	2.17
3	0.000327	0.000046	3.205138	0.022329	14.047255	0.016597	99.32	16.39	477.20	2.16
4	0.000273	0.000085	3.245300	0.025361	14.006588	0.032799	99.43	20.14	475.98	2.31

5	0.000507	0.000095	3.081394	0.031997	13.959224	0.033506	98.94	30.21	474.57	2.32
6	0.000848	0.000081	2.728519	0.026984	13.908267	0.028004	98.23	44.40	473.04	2.25
7	0.000717	0.000078	2.964814	0.034257	13.879942	0.026517	98.50	52.59	472.20	2.23
8	0.000721	0.000090	2.899146	0.062228	14.204426	0.039599	98.52	62.59	481.88	2.42
9	0.000562	0.000000	3.462113	0.015176	13.977041	0.010603	98.83	73.58	475.10	2.12
10	0.000930	0.000080	2.931609	0.028501	13.830189	0.027456	98.05	81.52	470.71	2.23
11	0.000740	0.000119	3.175439	0.021192	13.874109	0.036536	98.45	86.85	472.02	2.35
12	0.001685	0.000195	2.573737	0.024448	13.558608	0.058774	96.46	90.11	462.55	2.70
13	0.000070	0.000091	3.499486	0.011615	14.071987	0.028740	99.85	100.00	477.94	2.27
Total fusion age = 482.1 ± 25.9 Ma, *J* value = 0.021555 ± 0.000108										
Irradiation standard used = Tinto B biotite (409.24 ± 0.71 Ma)										
Biotite										
GZ 39										
1	0.011680	0.002009	0.000000	0.000000	8.965713	0.594464	72.20	0.99	318.75	19.43
2	0.003772	0.000539	0.002945	0.021168	12.114483	0.159828	91.57	4.27	418.55	5.20
3	0.000682	0.000076	0.007890	0.002870	12.982153	0.029098	98.47	27.59	445.10	2.10
4	0.000324	0.000108	0.000000	0.000000	12.935932	0.034885	99.27	35.79	443.70	2.20
5	0.000204	0.000068	0.000000	0.000000	12.933735	0.022260	99.54	48.82	443.63	2.80
6	0.000255	0.000064	0.002128	0.002101	12.924705	0.019603	99.42	62.67	443.36	2.60
7	0.000190	0.000048	0.000528	0.001563	12.877354	0.014933	99.57	81.31	441.92	2.20
8	0.000117	0.000040	0.012501	0.002493	12.850149	0.013735	99.73	92.32	441.09	2.00
9	0.000081	0.000085	0.015071	0.005261	12.853508	0.027671	99.81	97.55	441.19	2.10
10	0.000000	0.000000	0.030525	0.006027	13.033229	0.007971	100.00	99.16	446.65	2.00
11	0.001047	0.000001	0.035236	0.011596	12.724409	0.015440	97.63	100.00	437.25	2.00
Total fusion age = 429.2 ± 35.7 Ma, *J* value = 0.021554 ± 0.000108										
RMZ 13										
1	0.005050	0.000103	0.000000	0.000000	12.300142	0.037992	89.18	3.23	424.31	2.23
2	0.000920	0.000013	0.001692	0.000003	13.107143	0.020349	97.97	28.88	448.95	2.80
3	0.000743	0.000000	0.006078	0.001491	12.937691	0.008148	98.33	39.60	443.81	1.99
4	0.000119	0.000000	0.004797	0.000963	13.110753	0.007466	99.73	56.21	449.06	2.00
5	0.000102	0.000035	0.005773	0.000554	13.185036	0.012573	99.77	85.09	451.31	2.40
6	0.000628	0.000379	0.013758	0.002717	13.066838	0.114919	98.60	87.73	447.73	4.10
7	0.000000	0.000000	0.044075	0.014505	13.474908	0.400574	100.00	88.22	460.07	12.24
8	0.000087	0.000091	0.016849	0.000979	13.064757	0.029675	99.80	95.54	447.67	2.18
9	0.000330	0.000168	0.027560	0.016723	13.347708	0.050315	99.28	99.49	456.23	2.53
10	0.002616	0.001317	0.071897	0.127766	12.481085	0.390841	94.17	100.00	429.87	12.13
Total fusion age = 445.9 ± 10.4 Ma, *J* value = 0.021557 ± 0.000108										
RMZ 18										
1	0.002598	0.000173	0.015448	0.010784	12.685405	0.055949	94.29	2.56	436.05	2.59
2	0.000683	0.000018	0.001561	0.000771	12.522462	0.008021	98.41	27.91	431.06	1.94

(*Continued*)

Table 1. *Continued*

Step	$^{36}Ar/^{39}Ar$	$\pm 1\sigma$	$^{37}Ar/^{39}Ar$	$\pm 1\sigma$	$^{40}Ar^*/^{39}Ar$	$\pm 1\sigma$	$\%^{40}Ar^*$	Cum. $\%^{39}Ar$	Age (Ma)	$\pm 1\sigma$
3	0.000348	0.000032	0.001412	0.001558	12.428754	0.011991	99.18	55.96	428.19	1.95
4	0.000000	0.000113	0.000000	0.000000	12.462420	0.034782	100.00	63.80	429.22	2.19
5	0.000139	0.000047	0.004156	0.002294	12.442172	0.014372	99.67	82.87	428.60	1.96
6	0.001139	0.000163	0.003685	0.032744	12.230228	0.050871	97.32	85.60	422.09	2.45
7	0.000000	0.000000	0.002661	0.007882	12.435996	0.012731	100.00	96.93	428.41	1.95
8	0.001587	0.000227	0.000000	0.000000	11.963503	0.069610	96.23	98.89	413.87	2.84
9	0.000980	0.000985	0.022325	0.066121	12.234369	0.294736	97.69	99.34	422.22	9.26
10	0.002021	0.000674	0.000000	0.000000	11.628928	0.203067	95.11	100.00	403.49	6.57

Total fusion age = 424.3 ± 9.0 Ma, *J* value = 0.021553 ± 0.000108
Irradiation standard used = Tinto B biotite (409.24 ± 0.71 Ma)

See the section on 'Analytical proceedures' for analytical and data processing details.

(Table 2 and Fig. 4b, c). RMZ 18 exhibits low-temperature step ages that are slightly older than the obtained plateau age (Fig. 4a and Table 2).

Titanite fission-track analysis

The ages and analytical details of 25 TFT analyses are presented in Table 3. Their spatial distribution pattern is depicted in Figures 5 and 6. The TFT ages range from 384 ± 40 to 219 ± 24 Ma and do not show a distinct trend with elevation (Fig. 5a, b). The associated errors are quoted at 2σ confidence levels. The χ^2 probability values of all samples are larger than 5%. All TFT ages are younger than the $^{40}Ar/^{39}Ar$ hornblende and biotite ages of this study.

On the age v. latitude plot (Fig. 5c) two TFT age clusters are apparent, although partly overlapping within their 2σ confidence intervals. In the north of the study area (Figs 5c & 6) eight TFT ages (termed the *O* group) range from 378 ± 40 to 291 ± 30 Ma and group around a weighted mean age of 329 ± 24 Ma. A younger TFT age population of 17 samples with ages spanning between 306 ± 34 and 219 ± 24 Ma (termed the *Y* group) clusters around a weighted mean age of 255 ± 20 Ma. The *Y* group occupies the central–southern part of the study area. The grouping is confirmed by a discriminant analysis (Bahrenberg *et al.* 1992), using the TFT ages, latitudes, longitudes and elevations (Table 3) as class parameters. A Wilks' Λ value of 0.1612 indicates a distinct class separation, and a probability value of $P(\chi^2) > 92\%$ shows that the estimated discriminant functions are significant.

TFT samples from three north–south traverses, A, B and C, delineate the age v. latitude trends across the region from west to east (Fig. 6). Profile B covers the largest section, whereas profile C represents a composite of two locations (Figs 5c, d & 6). Generally, the TFT ages decrease from north to south. The ages remain constant in the NE–SW to ENE–WSW directions between traverses A and B (35.5°E–37°E), and in the east–west direction between traverse B and C (37°E–40.5°E) (Fig. 6). Traverses A and B exhibit an abrupt change in the TFT ages from the *O* to the *Y* group over a range of approximately 0.5° longitude (*c.* 55 km). The location of sudden TFT age change in profile A appears to be offset by approximately 1° towards the south against the one in profile B (Fig. 5c, d).

Interpretation

$^{40}Ar/^{39}Ar$ hornblende data

The observed discordant hornblende age spectra (Fig. 3a–f) are interpreted to result from the degassing of inclusions or impurities incorporated in the analysed crystals (Lee *et al.* 1991; Lee 1993).

Table 2. *Summary of $^{40}Ar/^{39}Ar$ results for hornblende and biotite samples*

Samples ID	Location			Rock type	Age spectrum			Inverse isochron				Total fusion	Weighted mean		
	Longitude	Latitude	Elevation (m)		Age ± 1σ (Ma)	*n* (*N*) % ^{39}Ar	MSWD *P*	Age ± 1σ (Ma)	$^{40}Ar/^{36}Ar$ intercept	*n*	MSWD *P*	Age ± 1σ (Ma)	Age ± 1σ (Ma)	*n* % ^{39}Ar	MSWD *P*
Hornblende															
BZ 216	37.44806	−16.03861	428	Amphibolite	*No plateau to establish*	*(12)*		*No isochron to regress*				1003.4 ± 1133.6	*462.5 ± 1.1*	*9–12* *42.4*	*0.8* *0.49*
GZ 90	34.34167	−14.84944	1150	Amphibolite	*No plateau to establish*	*(12)*		*514 ± 16*	*3484 ± 1300*	*12*	*7.6* *0*	569.5 ± 45.9	*550.7 ± 1.6*	*4–12* *88.03*	*3* *0.003*
PZ 37	34.45778	−14.68889	1300	Amphibolite	*542.05 ± 0.31*	*11–13 (13)* *41.8*	*1.2* *0.3*	*543.6 ± 2.6*	*216 ± 110*	*11–13*	*0.79* *0.37*	651.7 ± 210.1	*541.9 ± 1.35*	*11–13* *41.81*	*0.054* *0.95*
RMZ 11	40.16833	−14.94222	184	Monzonite	*No plateau to establish*	*(12)*		*465.5 ± 3.75*	*1434 ± 250*	*12*	*532* *0*	479.2 ± 18.2	*473.8 ± 1.2*	*2–12* *80.05*	*2.7* *0.03*
RMZ 13	40.37083	−14.91000	117	Orthogneiss	*No plateau to establish*	*(13)*		*467.4 ± 2.85*	*719 ± 150*	*2–9*	*7.6* *0*	482.1 ± 25.9	*475.8 ± 1.4*	*2–13* *90.55*	*3.3* *0*
RMZ 45	37.15000	−15.43194	946	Gneiss	*No plateau to establish*	*(7)*		*No isochron to regress*				466.9 ± 29.9	*455.5 ± 3.1*	*2–7* *92.3*	*7.3* *0*
Biotite															
GZ 39	38.76139	−15.71056	271	Amphibolitic gneiss	443.79 ± 0.37	3–6 (11) 58.4	0.93 0.42	440.4 ± 1.2	553 ± 160	3–10	0.49 0.82	429.2 ± 35.7	443.9 ± 1.05	3–6 58.39	0.13 0.94
RMZ 13	40.37083	−14.91000	117	Orthogneiss	*No plateau to establish*	*(10)*		*449.1 ± 2.65*	*128 ± 50*	*10*	*15* *0*	445.9 ± 10.4	448.2 ± 1.85	2–6 55.3	1.9 0.11
RMZ 18	39.75833	−14.98472	317	Gneiss	428.42 ± 0.27	3–5 (10) 55.0	0.56 0.57	*427.2 ± 2.35*	*456 ± 150*	*1–5*	*2.9* *0.032*	424.3 ± 9.0	428.6 ± 1.15	3–5 54.96	0.062 0.94

N, total number of heating steps; *n*, number of heating steps used to calculate the plateau age (including the % of ^{39}Ar contributing to the plateau age) or inverse isochron age. MSWD values for plateau ages and inverse isochron ages were calculated with $n-1$ and with $n-2$ degrees of freedom, respectively. *Italics:* rejected results due to the plateau age containing $<50\%$ to the cumulative % ^{39}Ar, disturbed spectra or significant differences of individual steps at the 2σ level (MSWD >2.5 and *P* (probability) <0.05).

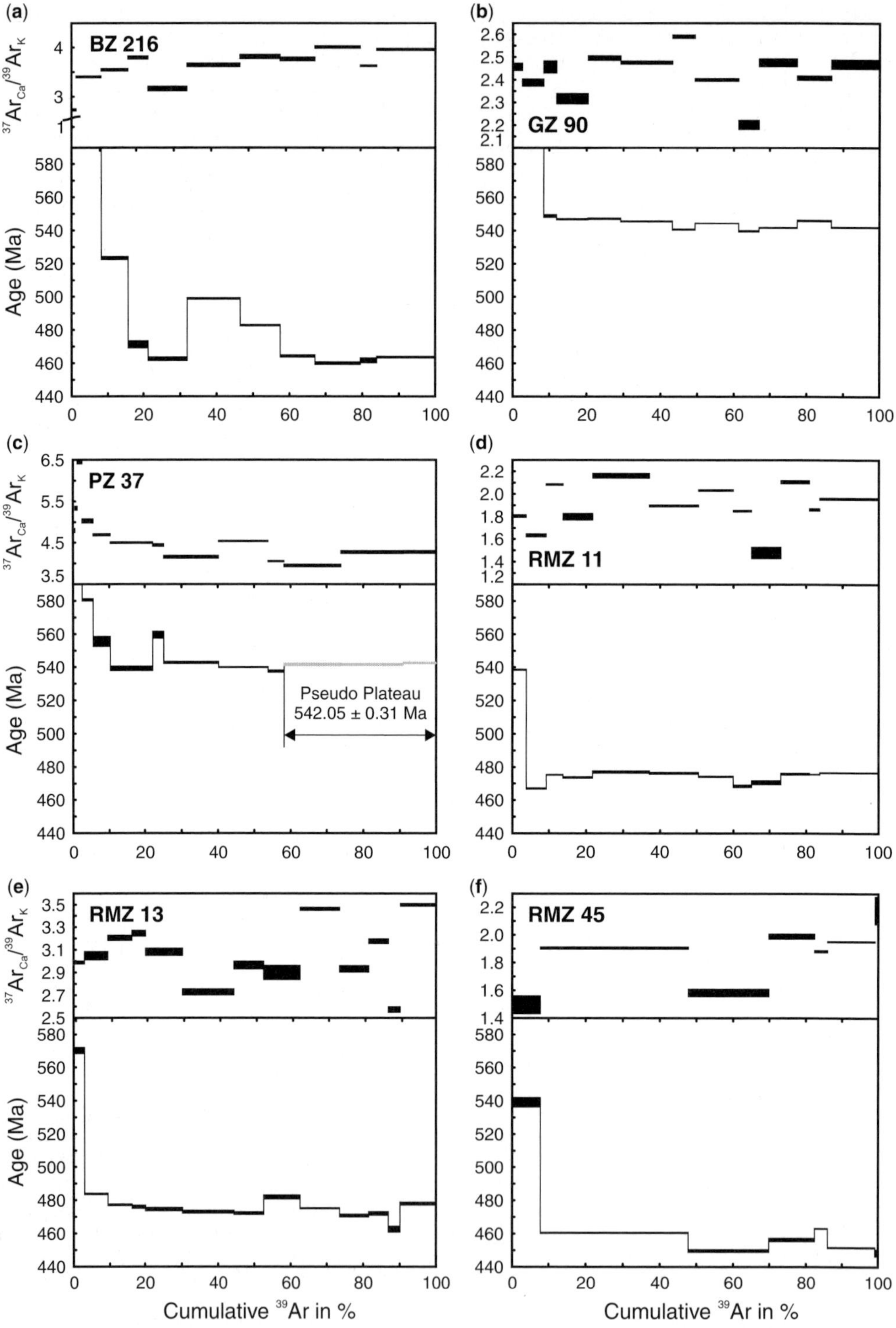

Fig. 3. $^{37}Ar/^{39}Ar$ and apparent age spectra plots of incremental heating steps from hornblende sample analyses. Insets are presented for samples exhibiting strongly discordant spectra and depict entire age range of incremental heating steps. Errors for each step are $\pm 1\sigma$ and do not include the error of the J value.

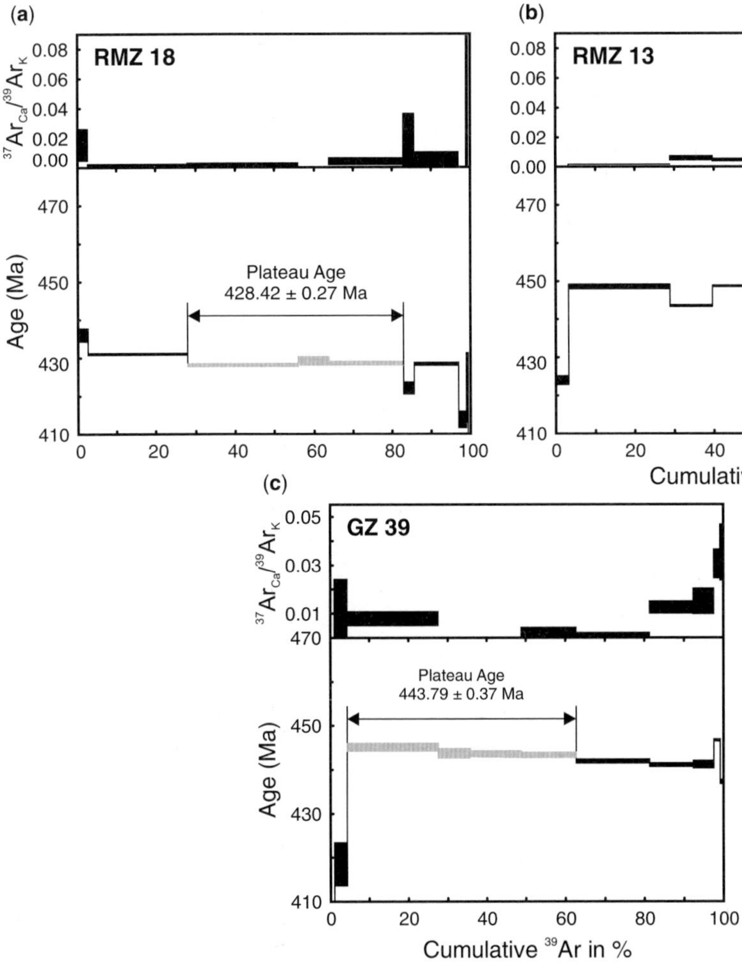

Fig. 4. $^{37}Ar/^{39}Ar$ and apparent age spectra plots of incremental heating steps from biotite samples analyses. Insets are presented for samples exhibiting strongly discordant spectra and depict entire age range of incremental heating steps. Errors for each step are $\pm 1\sigma$ and do not include the error of the J value.

A degassing of non-hornblende phases is corroborated as the discordant hornblende age spectra are usually accompanied by similar variations in their corresponding $^{37}Ar/^{39}Ar$ ratios (Fig. 3a–f) (Di Vincenzo *et al.* 2003).

We interpret the old ages of the first, low-temperature age spectra steps in all hornblende samples (Fig. 3a–f) to reflect the release of excess ^{40}Ar (Damon 1968) from extended crystal defects by short-circuit diffusion (Lee 1993, 1995; Lo *et al.* 2000). The very high radiogenic $^{40}Ar^*$ and very low atmospheric ^{40}Ar concentrations in all hornblende samples (Table 1) suggest that their argon budgets were not significantly altered by mixing with meteoric fluids. As optical thin-section inspection revealed fresh, unaltered mineral phases, a large-scale, external excess ^{40}Ar input by non-meteoric fluids is considered less likely. We suggest that the observed excess ^{40}Ar may either have accumulated during original mineral formation or may be derived from internal build-up above the isotopic system closure (Baxter *et al.* 2002; Baxter 2003). As a consequence, the weighted mean and pseudo-plateau ages of all hornblende samples might also be affected by excess ^{40}Ar. Some of the low-temperature excess ^{40}Ar in PZ37 also appears to be associated with a Ca-rich phase (Fig. 3c).

We infer that the discrepant total fusion and weighted mean ages of BZ 216 and PZ 37 reflect a significant alteration by excess ^{40}Ar and degassing of non-hornblende phases. The $^{40}Ar/^{39}Ar$ ages of these two samples are interpreted as geologically

Table 3. *Results of titanite fission-track analysis*

Sample	Lithology	Longitude	Latitude	Elevation (m)	No. of grains	ρ_s ($\times 10^7$ cm^{-2}) (N_s)	ρ_i ($\times 10^7$ cm^{-2}) (N_i)	ρ_d ($\times 10^7$ cm^{-2}) (N_d)	$P(\chi^2)$ (%)	Central age $\pm$ 2σ (Ma)	U (ppm)
RMZ 13	Orthogneiss	40.37083	−14.91000	142	24	0.695 (2553)	0.134 (492)	0.102 (2049)	36.5	348 ± 40	46.1
RMZ 14	Orthogneiss	40.35556	−14.97000	58	25	0.586 (1527)	0.131 (342)	0.116 (2049)	59.9	338 ± 44	46.6
RMZ 16	Orthogneiss	40.32722	−14.91028	47	21	0.853 (1915)	0.189 (423)	0.111 (2049)	39.7	330 ± 40	72.8
RMZ 31	Gneiss	39.68361	−15.84694	31	25	0.631 (2224)	0.152 (537)	0.113 (2049)	55.4	306 ± 34	51.3
RMZ 36	bt-gneiss	37.72972	−15.14361	657	25	0.171 (614)	0.031 (112)	0.106 (2049)	92.6	378 ± 40	14.0
GZ 87	gt-Amphibolite	34.34556	−14.91611	1150	25	0.697 (2661)	0.148 (565)	0.104 (2049)	30.8	327 ± 38	49.2
GZ 90	Amphibolite	34.34167	−14.84944	1150	30	0.456 (3349)	0.093 (682)	0.102 (2049)	15.5	329 ± 36	31.8
MD 69	Felsic gneiss	35.63489	−16.22693	599	20	0.861 (5667)	0.399 (2625)	0.231 (4492)	71.7	327 ± 20	64.1
BZ 216	Amphibolite	37.44806	−16.03861	500	20	0.941 (1813)	0.294 (566)	0.107 (2049)	22.0	230 ± 28	101.4
GZ 39	Amphibolitic gneiss	38.76139	−15.71056	250	25	0.322 (1957)	0.089 (541)	0.105 (2049)	11.6	254 ± 32	40.3
GZ 79	Amphibolite	38.56639	−16.32750	50	24	0.296 (1176)	0.078 (308)	0.107 (2049)	19.9	271 ± 42	28.0
PZ 33	Amphibolite	38.17444	−16.09139	250	13	0.430 (1541)	0.12 (431)	0.110 (2049)	23.5	260 ± 32	45.3
RMZ 11	Monzonite	40.16833	−14.94222	184	23	1.102 (2368)	0.283 (608)	0.114 (2049)	99.6	291 ± 30	92.4
RMZ 27	Granite	39.87333	−15.87722	91	13	1.015 (1084)	0.272 (291)	0.116 (2049)	95.8	284 ± 40	82.9
RMZ 28	Granite	39.87167	−15.88444	92	13	1.005 (1216)	0.279 (338)	0.111 (2049)	43.7	262 ± 36	90.9
RMZ 47	Granitic gneiss	37.14708	−15.64833	575	14	1.024 (836)	0.287 (234)	0.115 (2049)	9.2	271 ± 48	97.6
823-06	Orthogneiss	36.95478	−16.87181	180	5	1.288 (349)	0.410 (111)	0.120 (2049)	69.5	248 ± 56	129.1
824-03	Migmatite	37.39403	−16.33947	352	32	0.613 (2656)	0.185 (800)	0.119 (2049)	72.7	260 ± 26	63.5
824-06	Migmatite	37.13939	−16.19142	330	25	0.796 (3020)	0.243 (922)	0.116 (2049)	44.4	252 ± 24	79.7
826-01	Gneiss	37.90036	−16.29083	240	13	1.060 (987)	0.351 (327)	0.117 (2049)	16.1	231 ± 32	123.7
826-03	Granite	38.05367	−16.21033	380	24	0.959 (1596)	0.341 (567)	0.118 (2049)	97.9	219 ± 24	116.7
826-09	Granite	38.33333	−16.17339	290	12	1.438 (955)	0.396 (263)	0.119 (2049)	22.2	284 ± 48	127.1
LH 23	Amphibolite	35.66108	−16.41903	454	15	0.605 (2245)	0.375 (1394)	0.242 (4492)	31.2	257 ± 22	60.3
WB 30	hbl-bt-gneiss	36.24128	−16.82791	336	22	0.510 (2708)	0.301 (1597)	0.235 (4492)	12.9	263 ± 22	49.9
WB 119	Metagabbro	35.52135	−16.71980	127	20	0.698 (5169)	0.433 (3207)	0.236 (4492)	17.2	251 ± 16	67.0

N_s, N_i and ρ_s, ρ_i denote counted spontaneous and induced fission tracks and fission-track densities, respectively. N_d and ρ_d are fission tracks and fission track densities of co-irradiated dosimeter glasses IRMM 540. *n*, number of counted grains; *P*, probability of inferred Poissonian distribution of single grain ages. See the text for analytical and data processing details. Samples marked † are *O* group samples, †† are *Y* group samples and * are used for internal cross-validation.

bt-gneiss, biotite gneiss; gt-amphibolite, garnet amphibolite; hbl-bt-gneiss, hornblende-biotite gneiss.

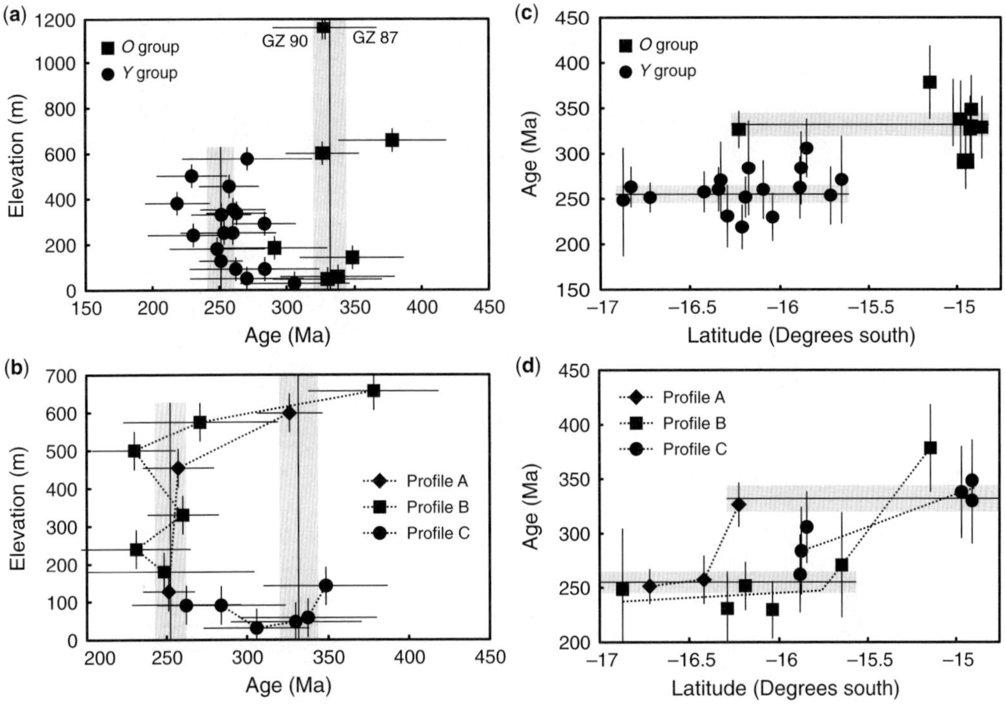

Fig. 5. Diagrams depicting the results of the TFT analyses; weighted mean ages of *Y* and *O* populations are depicted as black horizontal or vertical lines, and the grey bars indicate the associated $\pm 2\sigma$ confidence intervals. (**a**) Latitude v. age plot. (**b**) Age v. elevation plot. (**c**) Age v. elevation plot of traverses A, B and C. (**d**) Latitude v. age plot of traverses A, B and C.

meaningless. Because of the good agreement between the total fusion (466 ± 30–569 ± 46 Ma) and weighted mean ages (455 ± 6–550 ± 3 Ma) in samples RMZ 11, RMZ 13, RMZ 45 and GZ 90 (Fig. 3d–f, and Tables 1 & 2), we suggest that these samples are least affected by excess ^{40}Ar. Their weighted mean ages of approximately 455–550 Ma are considered as the best estimations for hornblende $^{40}Ar/^{39}Ar$ cooling ages in the region.

$^{40}Ar/^{39}Ar$ biotite data

The discordant $^{40}Ar/^{39}Ar$ age spectra pattern of RMZ 13 and the high-temperature age spectra steps of GZ 39 and RMZ 18 (cumulative $^{39}Ar > 80\%$) are accompanied by similar variations in their corresponding $^{37}Ar/^{39}Ar$ ratios (Fig. 4a–c and Table 1). These patterns are interpreted as resulting from the degassing of non-pristine biotite phases, e.g. inclusions (Lo *et al.* 2000; Kuiper 2002; Di Vincenzo *et al.* 2003).

The low ages in the first, low-temperature spectra steps of GZ 39 and RMZ 13 are not accompanied by significantly lowered $^{37}Ar/^{39}Ar$ ratios (Fig. 4b, c and Table 1). Both observations indicate either loss of ^{40}Ar or the degassing of non-biotite phases (Lo *et al.* 2000). The slightly older ages in the first, low-temperature age-spectra step of sample RMZ 18 may reflect the degassing of minor amounts of excess ^{40}Ar from extended defects by short-circuit diffusion (Lo *et al.* 2000).

Because of minor interferences from ^{40}Ar loss in RMZ 13 and excess ^{40}Ar in RMZ 18, we suggest that both their weighted mean and plateau ages of 448.2 ± 3.7 and 428.4 ± 0.3 Ma provide good biotite $^{40}Ar/^{39}Ar$ cooling age estimates in the region. Owing to a possible influence of ^{40}Ar loss, the plateau age of sample GZ 39, 443.8 ± 0.4 Ma, is regarded as a minimum biotite $^{40}Ar/^{39}Ar$ cooling age estimate.

Cooling rates for $^{40}Ar/^{39}Ar$ hornblende and biotite results

The $^{40}Ar/^{39}Ar$ analyses of sample RMZ 13 yield hornblende and biotite cooling ages of 475.8 ± 2.8 Ma and 448.2 ± 3.7 Ma, respectively ($\Delta t =$ 21–34 Ma: Fig. 1 and Table 2). Iteratively derived cooling rates range from *c.* 10.5 to 6.5 °C Ma^{-1} and correspond to closure temperatures

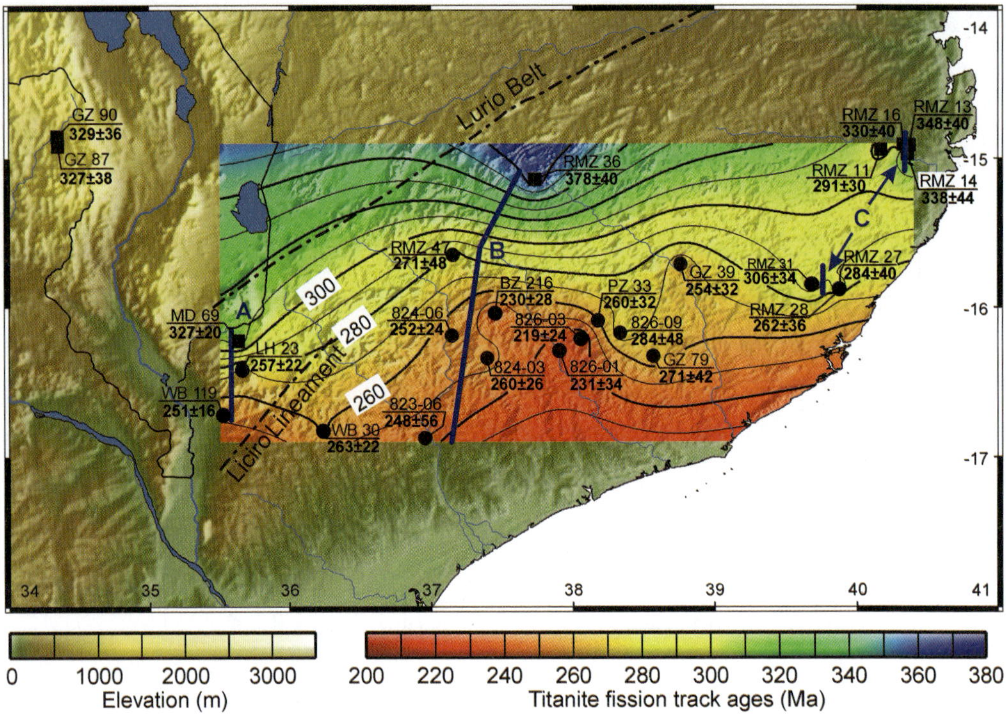

Fig. 6. Topographic map of northern Mozambique with superimposed spatial distribution pattern of TFT ages. In general, the TFT ages remain constant in the east–west direction and young from north to south. Contour lines were calculated using Generic Mapping Tool (GMT) 4.0, module surface. Sample locations are given as black squares and black circles for the *O* and *Y* groups, respectively. The open grey boxes indicate ungrouped samples. TFT age trend traverses A, B and C are plotted in Figure 5 and are drawn here as blue lines.

of *c.* 524–513 and 311–305 °C for hornblende and biotite, respectively. Hornblende sample RMZ 11 has a proximate location to, and yields a similar hornblende $^{40}Ar/^{39}Ar$ cooling age of 473.8 ± 2.4 Ma as, sample RMZ 13 (Table 2 and Fig. 2). Therefore, similar cooling rate and closure temperature values are inferred for RMZ 11. Identical biotite $^{40}Ar/^{39}Ar$ cooling ages of sample RMZ 13 (448.2 ± 3.7 Ma) and sample GZ 39 (443.8 ± 0.4 Ma) imply similar minimum cooling rates of 6.5–10.5 °C Ma^{-1} and closure temperatures of 311–305 °C for GZ 39. Hornblende sample RMZ 45 (455.5 ± 6.2 Ma) and biotite sample RMZ 18 (428.4 ± 0.3 Ma) yielded younger ages than the hornblende and biotite cooling ages of sample RMZ 13. The approximated cooling rates of 5–50 °C Ma^{-1} as well as the associated closure temperatures of 556–511 and 334–302 °C, are also applied to RMZ 45 and RMZ 18, respectively.

Titanite fission-track data

The χ^2 probability values of all TFT samples (>5%) demonstrate that in each sample the dispersion of the single-grain ages is explained by a Poissonian distribution and reflects a single population. As no track-length information was obtained, all TFT ages are regarded as minimum cooling ages, with a TFT closure temperature of 275 ± 25 °C (Kohn *et al.* 1993).

Annealing characteristics of titanite fission-tracks are fairly complex and rather poorly constrained (Jonckheere & Wagner 2000). We therefore interpret the broad TFT age scatters in both age groups (*c.* 90 Ma, Table 3) and the partial age overlaps between them within their 2σ errors (Fig. 5) to result from varying annealing characteristics. Age scatter due to variable annealing behaviour is a known feature in other mineral fission-track chronometers such as apatite (Barbarand *et al.* 2003).

The absence of a TFT age–elevation trend (Fig. 5c) suggests a non-uniform age–elevation relationship in the region, representing diverse and complex cooling path records (e.g. Braun 2002). We interpret this as reflecting two different cooling path records for the *O* and *Y* group samples.

Only samples RMZ 11 and RMZ 31 yield TFT ages of *c.* 300 Ma (Table 3). Hence, we suggest that the spatial separation between the *O* and *Y* group TFT cooling history records is broadly approximated by the interpolated 300 Ma age contour (Fig. 6).

As the present-day altitude (*c.* 1200 m) of *O* group samples GZ 87 and GZ 90 (Figs 5a & 6, and Table 3) is a Neogene feature due to elevation of the western Malawi rift flank over its eastern counterpart by about 500 m (Chorowicz 2005 and references therein), we infer a common cooling history for all *O* group samples.

Relating the $^{40}Ar/^{39}Ar$ biotite results for RMZ 13 (448.2 ± 3.7 Ma) to TFT samples RMZ 13 (348 ± 40 Ma) and RMZ 16 (330 ± 40 Ma) yields cooling rates of less than 0.6 °C Ma^{-1}. A similarly low cooling rate of less than 0.9 °C Ma^{-1} is obtained for $^{40}Ar/^{39}Ar$ biotite sample RMZ 18 (428.4 ± 0.3 Ma) and the closest TFT sample is RMZ 14 (338 ± 44 Ma) (Tables 1, 3, and Figs 2, 3, 6 & 7). These very low cooling rates indicate that the *O* group samples cooled very slowly through the TFT partial annealing zone (PAZ: 310–265 ± 10 °C: Coyle & Wagner 1998) in the Late Carboniferous–Early Permian times (*c.* 320–290 Ma).

One *Y* group cooling rate estimate of less than 0.3 °C Ma^{-1} is derived from the GZ 39 $^{40}Ar/^{39}Ar$ biotite (443.8 ± 0.4 Ma) and TFT (254 ± 32 Ma) age results (Tables 1 & 3, and Fig. 2). Relating the *Y* group TFT ages to Rb–Sr biotite cooling ages (450–420 Ma) of Costa *et al.* (1992) (Table 3, and Figs 2 & 6) yields even lower cooling rate estimates of much less than 0.3 °C Ma^{-1}. Based on the abrupt north–south change in TFT ages (Fig. 5c, d), we argue that these extremely low rates do not represent a southwards decrease in cooling since post-late Ordovician times. We suggest, instead, that these rates result from a complex, multistage cooling history of *Y* group TFT samples. The absence of an age–elevation trend in the *Y* group TFT ages (Fig. 5a, b) may indicate a period of more rapid cooling (cf. Braun 2002), whereby the youngest TFT ages, located furthest south, cooled last.

We suggest that initially all TFT age samples cooled at similar rates of less than 1 °C Ma^{-1} from about 450–420 Ma onwards. The *Y* group samples were located at a deeper crustal level (*c.* 1 km) than the *O* group samples. During, presumably, Late Carboniferous–Early Permian times (*c.* 300–290 Ma) the *O* group samples had

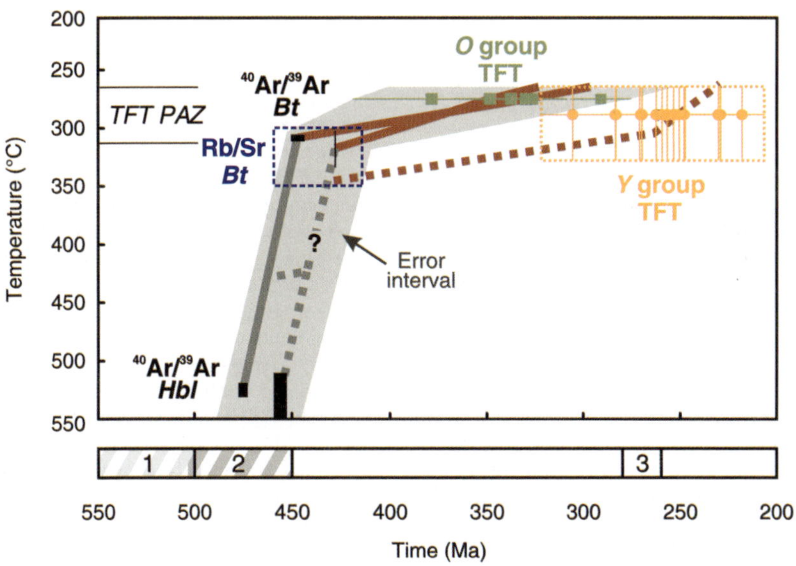

Fig. 7. Diagram illustrating the *t–T* cooling paths of the southern basement in northern Mozambique. Black rectangles represent *t–T* spaces constrained by the $^{40}Ar/^{39}Ar$ hornblende (*Hbl*) and biotite (*Bt*) analyses. Rb–Sr biotite (*Bt*) cooling ages of Costa *et al.* (1992) are represented by the blue dashed outlined box. Older (*O*) and younger (*Y*) TFT age groups are depicted by green squares and orange circles, respectively. Constrained and inferred cooling paths are shown in solid and dashed brown lines, respectively. The light grey area denotes a general *t–T* trend envelope, inferred from error intervals of thermochronological data and cooling rates. An inferred temperature range, where *Y* group TFT samples resided prior to their uplift and cooling is indicated by the orange dotted outlined box. Numbers 1, 2 and 3 indicate periods of the Pan-African metamorphism, granite and pegmatite intrusions and incipient rifting in the Early–Late Permian, respectively. The question mark denotes an exemplary cooling path for the post-metamorphic thermally influenced basement.

cooled through the TFT PAZ, whereas the *Y* group samples still resided within and/or above the high-temperature threshold of the TFT PAZ (*c.* >310 °C). Then, the *Y* group samples cooled more rapidly to below *c.* 275 ± 25 °C at more recent times (*c.* <280 Ma) (Fig. 7). Alternatively, all TFT sample groups shared a common crustal level position and cooling history through the TFT PAZ, similar to the *O* group. At younger times (*c.* <280 Ma) the *Y* group samples experienced a thermal overprint (causing partial to entire TFT age resetting) and subsequently renewed cooling.

Discussion

Proterozoic–Early Palaeozoic cooling in the SW Axial Granulite Complex

The hornblende $^{40}Ar/^{39}Ar$ cooling age of GZ 90 (551 ± 3 Ma) coincides with the youngest age constraint of the granulite facies metamorphism, dated between about 571 and 549 Ma (Kröner *et al.* 2001), in the southern Axial Granulite Complex (Fig. 2). We suggest that hornblende sample GZ 90 records the cooling to below about 550–500 °C at *c.* 550 Ma, subsequent to the peak granulite facies metamorphism. Furthermore, we suggest that $^{40}Ar/^{39}Ar$ cooling age of sample GZ 90 dates the cooling from the static, retrograde amphibolite facies overprint (Pinna *et al.* 1993; Kröner *et al.* 2001) in the SW Axial Granulite Complex.

Early Palaeozoic cooling in the southern basement

$^{40}Ar/^{39}Ar$ hornblende cooling ages of samples RMZ 13 and RMZ 11 date cooling to below about 525 °C at 476–474 Ma in the eastern part of the southern basement. Both cooling ages are younger than the syn-tectonic (= D_{M2}) granulite facies metamorphism, dated at approximately 615 Ma by Kröner *et al.* (1997) and recently at about 551 Ma by Grantham *et al.* (2005*a*) using U–Pb SHRIMP single zircon dating. A time difference of around 30–80 Ma exists between the peak granulite facies metamorphism and the $^{40}Ar/^{39}Ar$ hornblende cooling ages. Both $^{40}Ar/^{39}Ar$ hornblende ages are also younger than the inferred timing (*c.* 500 Ma) for the syn- to late-kinematic granitic intrusions (Pinna *et al.* 1993). The $^{40}Ar/^{39}Ar$ hornblende cooling ages could, therefore, either point towards a period of very slow cooling or a protracted, yet not completely resolved, multiphase evolution of the basement during Pan-African times.

The formation of the Namama Thrust Belt (Fig. 2) is inferred to post-date the peak granulite facies metamorphism (Sacchi *et al.* 2000). The related thrust tectonics account for the distribution of some late-phase, high-grade metamorphic assemblages (Sacchi *et al.* 2000) and, presumably, reworked a large region of the southern basement including some granitoids emplaced at about 500 Ma (Pinna *et al.* 1993). We suggest that the $^{40}Ar/^{39}Ar$ hornblende cooling ages of RMZ 11 and RMZ 13 date the cooling from the Namama Thrust Belt thermo-tectonic event at 476–474 Ma with subsequent cooling rates of 6.5–10.5 °C Ma^{-1} (Fig. 7). These cooling ages also delimit the age of the amphibolite–granulite facies Namama Thrust Belt metamorphism to ≥ *c.* 500 Ma. An unconstrained but widely recognized amphibolite facies overprint in the southern basement, described by Pinna *et al.* (1993), post-dates the peak granulite facies metamorphism. The cooling path traced by the $^{40}Ar/^{39}Ar$ hornblende ages appears to be in line with a cooling from such an overprint. We therefore consider it reasonable that cooling to below about 525 °C at 476–474 Ma was not solely limited to the eastern part, but occurred throughout large parts of the southern basement.

An $^{40}Ar/^{39}Ar$ hornblende cooling age of 455.5 ± 6.2 Ma (RMZ 45: Fig. 2) coincides with a pegmatite Rb–Sr muscovite cooling age of 454 ± 7 Ma (Costa *et al.* 1992) (Fig. 2). Recent monazite U–Pb CHIME (chemical Th–U-total Pb isochrone method) analyses on pegmatites and U–Pb SHRIMP zircon analyses on two undeformed granites from the southern basement yielded intrusion ages of between *c.* 500 and 450 Ma (Grantham *et al.* 2005*a*). These data indicate localized thermal influences by igneous activity in the central southern basement between about 500 and 450 Ma. The preserved younger $^{40}Ar/^{39}Ar$ hornblende age of RMZ 45 is interpreted to record cooling to below 560–510 °C at around 455 Ma (Fig. 7) from this a late- to post-metamorphic reheating.

$^{40}Ar/^{39}Ar$ biotite sample RMZ 13 and GZ 39 recorded the cooling to below about 310–300 °C at 448.2 ± 3.7 and 443.8 ± 0.4 Ma, at rates of 7–11 °C Ma^{-1} in the eastern part of the southern basement, respectively (Figs 2 & 7). Both cooling ages coincide with a Rb–Sr biotite cooling age of 449 ± 7 Ma (Costa *et al.* 1992) from a basement gneiss in the Namama Thrust Belt hinterland (Fig. 2). These results reflect widespread cooling of the southern basement to below *c.* 350–300 °C at around 450 Ma (Fig. 7).

Younger Rb–Sr biotite cooling ages of 434–420 Ma from mainly undeformed granitoids (Sacchi *et al.* 1984; Costa *et al.* 1992) coincide with the $^{40}Ar/^{39}Ar$ biotite cooling age of RMZ 18 (428.4 ± 0.3 Ma). We suggest that biotite sample RMZ 18 records cooling to below 335–300 °C from the basement reheating event between *c.* 500

and 450 (Fig. 7). The pegmatite and granite emplacements locally delayed the final basement cooling to below about 350 °C to around 434–420 Ma (Fig. 7). Conclusively, the $^{40}Ar/^{39}Ar$ biotite and Rb–Sr biotite cooling ages (Table 2 and Fig. 2) indicate that the southern basement experienced ultimate, widespread cooling to below *c.* 350–300 °C after the latest Pan-African thermo-tectonic events, which occurred between >615–551 Ma and ≤450 Ma, at around 450–420 Ma, i.e. in Late Ordovician–Early Silurian times (Fig. 7).

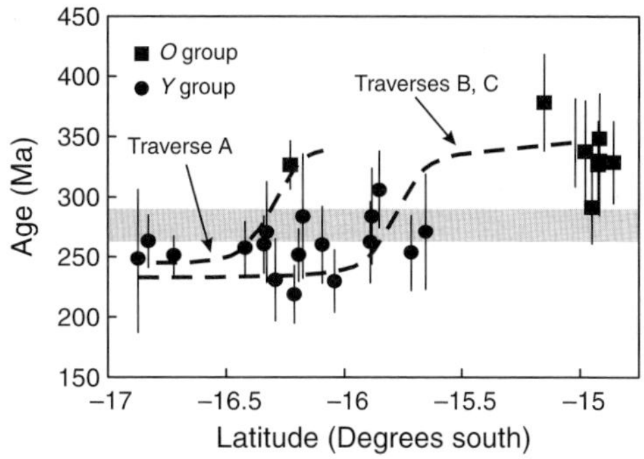

Fig. 8. Diagram depicting Latitude v. TFT age plot of *Y* and *O* populations. Dashed sigmoid lines represent the inferred north–south age decrease trends along the TFT age traverses A, B and C. These sigmoid trends reflect the thermochronological imprint of a denuding, uplifted rift flank. The horizontal grey bar shows the timing of incipient Permian rifting (*c.* 260–280 Ma) in central Gondwana.

Late Palaeozoic cooling and denudation history

Further cooling to below *c.* 275 ± 25 °C is recorded by the *O* group TFT ages (Figs 6 & 7: *c.* 380–290 Ma) and occurred in Late Devonian–Early Carboniferous times. With respect to the $^{40}Ar/^{39}Ar$ analyses (*c.* 7–11 °C Ma^{-1}), the cooling rates dropped by one order of magnitude to values of less than 1 °C Ma^{-1} (Fig. 7). We infer that this cooling rate decrease corresponds to reduced denudation rates within the southern basement. Such a decrease in denudation is in line with the establishment of pre-Karoo peneplains (before Late Carboniferous), subsequent to a prolonged period of erosion in central Gondwana (Wopfner & Kaaya 1991; Catuneanu *et al.* 2005).

Preliminary apatite fission-track results (Daszinnies *et al.* 2004) and remnants of Karoo age lavas (Jaritz *et al.* 1977; Grantham *et al.* 2005*b*) along the continental margin (Figs 1, 6 & 7) indicate that a second, younger cooling event to below *c.* 275 ± 25 °C, recorded by the *Y* group TFT ages, must have occurred prior to Late Triassic–Early Jurassic times (*c.* >200–180 Ma).

Permo-Triassic rift basins adjacent to the north Mozambican basement provide evidence for incipient rifting in Early–Late Permian times at about 280–260 Ma (Castaing 1991; Catuneanu *et al.* 2005). The onset of rifting post-dates the *O* group TFT ages, but pre-dates and overlaps with the *Y* group TFT ages (Fig. 8). We infer that the cooling recorded by the *Y* group TFT samples is related to regional incipient rifting activity. Basement cooling due to rift-related denudation and exhumation could result from:

- crustal extension, i.e. rift basin formation;
- erosional base level lowering due to incipient rifting;
- erosion of an uplifted rift flank adjacent to a rift basin.

In addition, basement cooling could also reflect the fading of a thermal event accompanying the rifting process (e.g. Bott 1995; Ziegler & Cloething 2004).

In the southern basement regional-scale brittle and ductile structures, except the remnant basal granulites klippen and Namama Thrust Belt ductile shear zones (Fig. 2), are absent (Pinna 1995; Norconsult Consortium 2007). Based on the present-day exposure of mid-crustal rocks, the removal of formerly existing upper-crustal structures can not be strictly discarded. However, owing to the apparent lack of regional-scale extensional structures, we regard crustal extension on the southern basement as a less likely cause for the *Y* group cooling pattern (Figs 6 & 7).

TFT dating (PAZ: 310–265 ± 10 °C) is underpinned by similar basic physical principles as FT dating on apatites (PAZ: 110–60 °C ± 10 °C), but is sensitive to recording cooling processes at higher temperatures, i.e. at deeper crustal levels (Wagner & Van den Haute 1992). We infer that thermo-tectonic processes affecting the entire crust would, for both thermo-chronometers, yield FT results that differ in age (although still cooling rate dependent) but bear resemblance within their spatial FT age distribution patterns.

Models of rift-margin evolution involving erosional base-level lowering triggered by incipient rifting produce synthetic patterns of AFT ages that decrease continuously from interior towards rift basin (Van der Beek 1995). As the TFT age pattern in the southern basement displays a prominent drop in age from north to south (Figs 5 & 6), we discard base-level lowering as a suitable explanation for the observed TFT age pattern (Figs 6–8).

All post Pan-African igneous activity in northern Mozambique is younger than approximately 180 Ma (e.g. Eby *et al.* 1995; Grantham *et al.*

2005*b*). We hence consider a rift-related thermal overprinting in the southern basement as a less likely cause for the observed TFT age distribution. Moreover, the absence of a rift-related Permian magmatism argues for passive rifting conditions (Bott 1995; Ziegler & Cloething 2004).

All three TFT age–latitude profiles (Fig. 5d) show a rapid age decrease over a distance of about 50 km from north to south. In the absence of regional-scale faults between both TFT age groups, a sharp age change over a discrete zone is regarded less likely (Figs 2 & 5–7). We suggest that the observed age-decrease pattern between the *O* and *Y* TFT age groups can be approximated by a sigmoid trend line (Fig. 8). Such a sigmoid age-decrease pattern, including the distance over which the TFT ages decrease, strongly resembles sigmoid AFT age patterns that are:

- observed across rifted continental margins displaying prominent escarpments (Gallagher *et al.* 1994);
- predicted by numerical models combining lithosphere extension and surface processes (Van der Beek *et al.* 1995).

The youngest TFT ages (*c.* <250 Ma) are located furthest south and predate the regional timing of incipient rifting (*c.* 280–260 Ma: Fig. 8). This finding is also in good agreement with sigmoid AFT age patterns found across rifted continental margins. There, the youngest AFT ages occur on the uplifted rift flanks close to the basin edge and commonly predate the timing of rifting as they were exhumed from below the AFT PAZ (Van der Beek *et al.* 1995; Gallagher *et al.* 1998).

We conclude that the *Y* group TFT samples were also exhumed from below the TFT PAZ; $T \geq 310$ °C. The *Y* group TFT ages, therefore, directly record the cooling from the denuding uplifted rift flank, whereby cooling results from interwoven processes of tectonic uplift, denudation and local isostatic compensation (Van der Beek *et al.* 1995 and references therein). Consequently, the youngest TFT age samples furthest south (Figs 6–8) represent the zone of highest denudation, and highest tectonic and isostatic uplift on the rift flank, located proximate to the edge of a rift basin (Fig. 9). As passive rifting conditions are considered, rift-flank uplift occurred synchronous with incipient rifting (Braun & Beaumont 1989) in Early–Late Permian times (*c.* 280–260 Ma) and the *Y* group TFT samples recorded the subsequent cooling. We interpret the north–south decreasing TFT age pattern (Figs 5–7) in the southern basement as the thermo-chronological imprint of an east- to ENE-trending rift flank that was uplifted and denuded during Karoo times. Thereby, the *O* group TFT samples represent the unaffected hinterland, inland of the uplifted region. Rift-flank uplift is flexural in nature (e.g. Bott 1995) and this fact is in line with the lack of regional extensional structures on the southern basement.

A corresponding rift basin was probably located in the vicinity of the southern continental margin (Fig. 9), suggestive that the uplifted rift flank could be up to approximately 200 km wide. This is quite a large distance, but is in line with numerical modelling predictions (Van der Beek *et al.* 1994, 1995) that indicate that the extent of the uplifted section scales with the elevation of the pre-existing surface and the presence of a regional uplift. However, the elevation of the Permo-Carboniferous prerift surface is unknown, as is the existence of a synchronous regional uplift.

Between longitudes of approximately 37–40.5°E the regional ductile basement fabrics display easterly trends (Cadoppi *et al.* 1987; Pinna *et al.* 1993; Norconsult Consortium 2007) (Fig. 9a). They broadly parallel the easterly contour trends of the TFT age pattern and, hence, the east to ENE orientation of the uplifted rift flank (Fig. 9). Within this region the NW–SE-orientated Early Permian tensional stress regime (Castaing 1991) (Fig. 9b) encloses acute angles of approximately 40°–85° with the regional basement fabric trends, and an acute angle of about 50° with the general trend of the rift flank and its associated southern rift structure (Fig. 9b). This suggests that pre-existing basement fabrics trending at high angles of 45°–50° to the tensional stress direction were activated by oblique extensional faulting during rifting (Morley *et al.* 2004) (Fig. 9b). West of longitude 37°E, near the Liciro lineament, the TFT age contours approach an NE–SW orientation (Figs 6 & 9) and coincides with the NE-trending ductile basement fabrics (Norconsult 2006). Here also, steeply dipping joints parallel the ductile basement fabrics in trend (Fig. 9b). We infer that this parallelism of brittle and ductile structures indicates brittle, extensional reactivation of ductile basement fabrics. Crustal extension, therefore, could have taken place in a more orthogonal mode to the west of longitude 37°E.

The east–west-orientated extensional fault trends in the Cabora Bassa Basin (Fig. 2) are attributed to a reactivation of easterly-trending crystalline basement fabrics (Castaing 1991). We suppose that the Cabora Bassa Basin and the Permian rift basin to the south of the southern basement were linked via the Zambesi pre-transform system during incipient rifting in the Early–Late Permian (Fig. 2). Harrowfield *et al.* (2005) envisioned the formation of an extensive Permo-Triassic rift basin bordering Australia/Antarctica and India. If this is true, a branch could extend from the Australo-Antarctic rift basin into central Africa via northern Mozambique (Fig. 2).

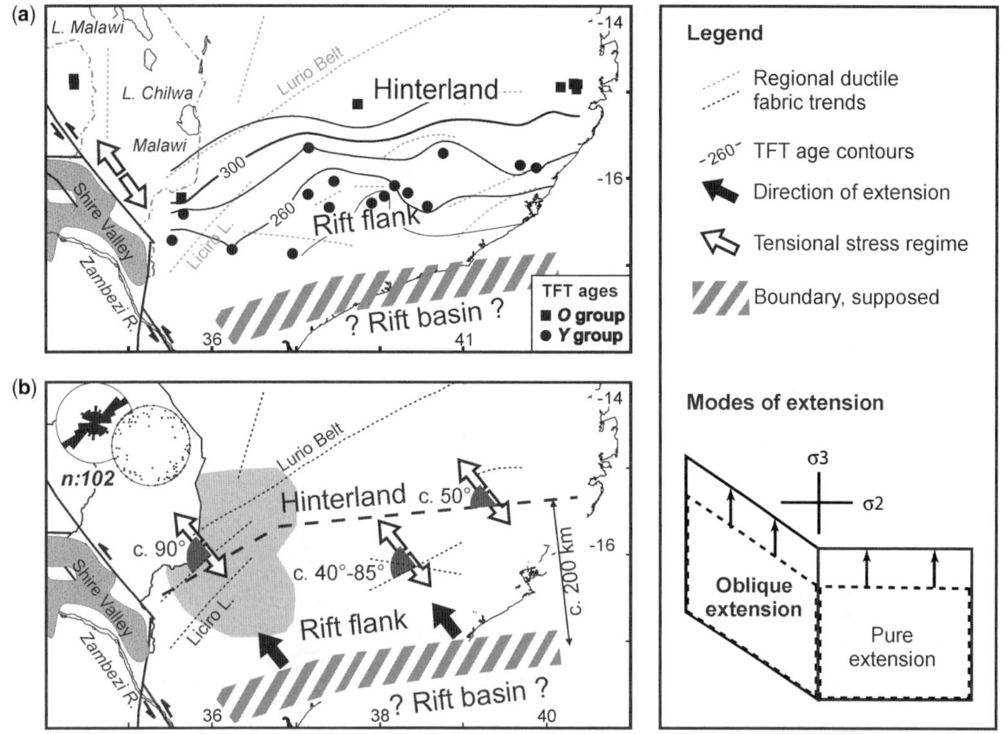

Fig. 9. Illustration (**a**) depicts the spatial TFT age pattern in northern Mozambique and the inferred rift flank. The black squares and dots show the sample locations of the older and younger TFT age populations, respectively. The ductile basement fabrics are derived from Pinna *et al.* (1991) and the orientation of the Permian tensional stress field is obtained from Castaing (1991). Illustration (**b**) shows the inferred rift geometry. Thick dashed lines trace the approximated boundary between the hinterland and the rift flank. The grey circle segments are graphical representations of the acute angles formed by trends of ductile fabrics and the rift structure with the tension stress field, respectively. Orientation measurements of joints obtained from the light grey area shown in the Rose and Schmidt diagram. Modes of extension after Morley *et al.* (2006).

At rifted margins, geothermal gradients are fairly constant as the upper crustal isotherms are not significantly disturbed (Stüwe *et al.* 1994; Brown & Summerfield 1997) by the sufficiently low (<100 m Ma^{-1}) exhumation rates (Leeder 1991; Van der Beek 1995). Estimating the amount of denudation is generally difficult as direct paleao-heat flows are unknown. A geothermal gradient of 25–30 °C km^{-1}, based on present-day heat flow records and vitrinite reflectance data from the Karoo basins of southern Tanzania, has been used in the reconstruction of the Late Palaeozoic–Cenozoic denudation histories on the Malawi and Rukwa rift flanks (Van der Beek *et al.* 1998). These values are adopted here. As the youngest TFT samples of the *Y* population were exhumed from below the TFT PAZ ($T > 310$ °C), approximately 10–12 km of crust have been removed by denudation from the rift flank since its uplift in the Early–Late Permian times. Further inland, the TFT ages of the *O* group indicate an amount of around 9–11 km of denudation since the Early Carboniferous (*c.* 330 Ma, $T_c = 275$ °C). Our denudation estimates are regarded as extreme upper values as they assume a constant geothermal gradient over very long periods. According to Jacobs & Thomas (2004), the thickened continental crust in the central part of East African–Antarctic orogen was thinned by an orogenic collapse at *c.* 530–490 Ma. This could suggest that subsequently average crustal thicknesses (*c.* 30–50 km) prevailed in the area of northern Mozambique. If high denudation of up to 12 km (*c.* one-third of the average thickness of continental crust) has amounted since the Early Carboniferous, the question arises to what extent it can be evidenced by other observations. One could expect that the present-day crustal thickness in the southern basement should be distinctly thinner compared to formerly adjacent regions in Africa and East Antarctica. In relation, the present-day Mohorovičić discontinuity underneath northern Mozambique should be located at a shallower depth.

An alternative interpretation to the rift-margin uplift model (preferred by one of us, G. H. Grantham) is as follows. The data clearly show that the TFT ages young southwards, implying greater degrees of exhumation towards the south. Early workers have interpreted the Lurio Belt as a significant high-strain zone and the overthrust klippen of granulite grade lithologies as derived from the Lurio Belt (e.g. Pinna *et al.* 1993; Sacchi *et al.* 2000). Grantham *et al.* (2003) have suggested that the Lurio Belt is correlatable with the shear zone that separates the Central Highlands Complex from the Vijayan Complex in Sri Lanka (see the tight-fit reconstruction of Gondwana after Lawyer *et al.* 1998 and others including Reeves used in Grantham *et al.* 2003) upon which similar klippen structures of Highland Complex rocks rest (Kröner 1991; Kriegsman 1995). Recent pressure–temperature (P–T) estimates from the granulite structure in the southern basement (Mugeba Klippe) show that the syn-tectonic peak metamorphic assemblages indicate conditions of approximately 10 kb and 900 °C, after which post-tectonic static cooling to *c.* 8 kb and *c.* 700 °C occurred (Roberts *et al.* 2005). U–Pb zircon SHRIMP data indicate that the high-grade metamorphism occurred at about 556 Ma. Pb/Pb zircon data from the same Mugeba Klippe yielded marginally older ages (Kröner *et al.* 1997). However, the evaporation method used by Kröner *et al.* (1997) lacks the high spatial resolution of the SHRIMP. The data of Kröner *et al.* (1997) are considered unreliable because the U–Pb zircon SHRIMP data show complex zircons with concordant ages of about 556 Ma and discordant zircons suggesting an upper intercept of around 1120 Ma.

During the period from *c.* 550 Ma, at about 8 kb and 700 °C, rapid exhumation of the southern basement to *c.* 3.5–2.8 kb and *c.* 280 °C at 245 Ma followed (assuming a geothermal gradient of 25–30 °C km^{-1}). This exhumation implies an uplift rate of about 15 km (4.5 kb) over approximately 300 Ma or 50 m Ma^{-1}. The southern basement contains a significant volume of largely undeformed granites, with most being *c.* 500 Ma in age (*c.* 453–530 Ma: Grantham *et al.* 2005*a*). The hornblende $^{40}Ar/^{39}Ar$ ages are broadly consistent with these conditions, with the granite minimum melt temperatures being around 650–700 °C. The large-scale genesis of the granites from *c.* 550 to 500 Ma is consistent with the southern basement being the footwall of the overthrust granulite klippen remnants. The temperature increase due to burial in the footwall would promote dehydration reactions at depth contributing to an extensive partial melting of the *c.* 1100 Ma gneisses to form the *c.* 450–530 Ma granites. Rapid exhumation and inversion of the southern basement would have followed the Lurian front deformation. On a larger scale, considering a cross-section at approximately 550–600 Ma ago from Cobue-Geci region to the southern Kirwanveggan (Antarctica), the central portion of granulite klippen region experienced depths at 6–8 kb about 500 Ma ago. The orogenic belt between these two areas has been exhumed. A simple cooling at a rate of 2.5 °C Ma^{-1} could be assumed for the cooling history from 550 Ma, 900 °C and 10 kb until Karoo times, i.e. the surface at Gondwana break-up at 180–200 Ma.

Conclusions

This research documents the successful application of a combined $^{40}Ar/^{39}Ar$ and titanite fission-track dating approach to resolve a multistage post-metamorphic basement cooling history. Moreover, we demonstrate the applicability of titanite fission-track dating to the thermally sensitive regime of a rift setting, a realm traditionally investigated by low-temperature thermochronology such as apatite fission-track analyses.

Within the SW Axial Granulite Complex, one $^{40}Ar/^{39}Ar$ hornblende age dates the basement cooling from the peak granulite facies metamorphism to below about 550–500 °C in middle Cambrian times. In the east of the southern basement cooling from a young Pan-African thermo-tectonic imprint, the Namama Thrust Belt development ($\geq$500 Ma), is recorded by $^{40}Ar/^{39}Ar$ hornblende and biotite ages at low rates of 7–11 °C Ma^{-1} from 525 to 300 °C. Within the southern basement, widespread cooling from this thermo-tectonic event to below 350 °C was already achieved by about 450 Ma. Granite and pegmatite emplacements at approximately 500–450 Ma were partly deformed by the Namama Thrust Belt development, and caused localized reheating and, locally, delayed cooling to Late Ordovician–Early Silurian times. Cooling from the latest Pan-African imprint to below *c.* 550–500 °C occurred about 25 Ma earlier in the Axial Granulite Complex than in the southern basement regions. This time difference is attributed to the more complex thermo-tectonic evolution of the southern basement in Early Cambrian times.

Throughout the sample area, similar TFT ages of the *O* group samples (380–290 Ma) indicate a common cooling to below 275 $\pm$ 25 °C during the Late Devonian–Early Carboniferous and up to about 9–11 km of denudation since. A cooling rate decrease to less than 1 °C Ma^{-1} between the Late Ordovician/Early Silurian and the Late Devonian/Early Carboniferous is linked to a reduction in denudation due to the establishment of pre-Karoo peneplains (<Late Carboniferous) as inferred by Wopfner & Kaaya (1991). The younger TFT age group (310–220 Ma) records the cooling of a denuding, broadly east–west-trending, rift flank, uplifted synchronously with incipient rifting in Early–Late Permian times at about 280–260 Ma. This rift event marks the onset of Gondwana's dispersal in the north Mozambican segment. An amount of up to 10–12 km of denudation is estimated on the rift flank since the onset of rifting. A corresponding rift basin was located to the south of the southern basement and linked to the Zambezi rift system via the Zambezi pre-transform system. The rift basin formed in response to a regional NW–SE tensional stress regime by oblique crustal extension that exploited and reactivated the pervasive, easterly-trending Pan-African age ductile basement fabrics. This rift basin formation hightlights the strong influence exerted by the Mozambique Belt as a zone of crustal weakness for the rift localization during Gondwana's disintegration in the Mozambican sector. This rift setting links up to a network of Karoo-age African rift basins that formed by transtensional and extensional fracturing of weak anisotropic lithosphere, which was metamorphosed in late Neoproterozoic–Early Cambrian times (Rogers *et al.* 1995; Visser & Praekelt 1998).

This research is part of the PhD studies of M.Ch. Daszinnies. The funding by the University of Bremen, ZF 05/101/01 is gratefully acknowledged. B. Emmel and F. Lisker are thanked for constructive reviews and valuable comments for improvements of a previous manuscript version. This paper also benefited from reviews of M. Raab and G. Viola. The Norges geologiske underøkelse (NGU) is thanked for the possibility to join the mapping project in northern Mozambique.

Appendix

$^{40}Ar/^{39}Ar$ *analysis*

Optical thin-section inspection (10–100× magnification) revealed that all samples contain fresh, unaltered mineral phases. The euhedral–subhedral hornblende and biotite crystals yield in grain sizes up to 1.5 mm and 1.2 mm, respectively. All samples, with the exception of RMZ 45, contain minor amounts of hornblende (<2–5%) or biotite (<3%) that exhibit partial intergrowth with other mineral phases. Hornblende samples BZ 216, GZ 90, PZ 37, RMZ 11 and RMZ 13 yield opaque inclusions. Inclusions of titanite were observed in hornblende samples BZ 216, GZ 90 and RMZ 13, and inclusions of apatite were found in hornblende samples GZ 90 and PZ 37. Biotite samples GZ 39, RMZ 13 and RMZ 18 contained inclusions of opaques and apatite. Zircon and zircon plus titanite inclusions were detected in biotite samples RMZ 18 and RMZ 13, respectively.

In order to assess the geological significance of $^{40}Ar/^{39}Ar$ ages, a series of age and $^{37}Ar/^{39}Ar$ spectra plots are presented. Isoplot version 2.49 (Ludwig 2001) was used to plot the figures and calculate the ages (Figs 3a–f & 4a–c). Within the $^{40}Ar/^{39}Ar$ age spectra, age plateaus were defined according to the criteria outlined by Ludwig (2001). In assessing whether all the data fit within the estimated error limits, indices of goodness of fit are used, including mean squared weighted deviates (MSWD) (McIntyre *et al.* 1966). MSWD values were calculated for weighted mean ages with $n - 1$ degrees of freedom (Ludwig 2001). MSWD values ranging between 1 and 2.5 are accepted as meaningful goodness of fit indicators (Roddick 1978). MSWD values of >1 generally indicate either underestimated errors or the presence of non-analytical scatter, whereas values of <1 suggest overestimated analytical errors. Weighted mean ages were calculated using both the analytical and J value errors on a series of pseudo-plateau steps, quoted with 95% confidence errors. The total fusion age is an unweighted mean age of all the steps including the analytical and J value errors, quoted with 1σ errors in the $^{40}Ar/^{39}Ar$ data tables (Tables 1 & 2).

For the calculation of closure temperatures and associated cooling rate we adopted the following diffusion parameters. *Hornblende*: a frequency factor (D_0) of $0.024\ cm^2\ s^{-1}$, an activation energy (E) of $64.1\ kcal\ mol^{-1}$ and a diffusion geometry (A) of 55 for spherical diffusion (Harrison 1981). *Biotite*: $D_0 = 0.075\ cm^2\ s^{-1}$, $E = 47.1\ kcal\ mol^{-1}$ (Grove & Harrison 1996) and $A = 27$ for cylindrical diffusion for mica (Onstott *et al.* 1991; Hames & Bowring 1994). As mineral closure temperatures (T_c) are directly linked to the cooling rate (Dodson 1973), the closure temperatures for corresponding $^{40}Ar/^{39}Ar$ hornblende and biotite analyses were calculated using cooling rates ($\Delta T/\Delta t$) that are consistent with the age difference (Δt) between the hornblende and biotite cooling ages.

A closure temperature of T_c approximately 350 °C (Möller *et al.* 2000) is estimated throughout this study for cited Rb–Sr cooling ages (Fig. 2).

References

AFONSO, R. S. 1976. *A geologica de Moçambique Notiçia explicativa da carta de Moçambique*. Imprensa Nacional de Mocambique, Maputo.

ANDREOLI, M. A. G. 1984. Petrochemistry, tectonic evolution and metasomatic mineralisations of Mozambique Belt granulites from S-Malawi and Tete (Mozambique). *Precambrian Research*, **25**, 161–186.

ARAUJO, J. R. 1976. Mozambique Belt: uma interpretaçao geochronologica. *Memoir Notice Publication Museum Mineralogica Geologica Universidad Coimbra Centro Estudidar Geologica*, **81**, 85–102.

BAHRENBERG, G., GIESE, G. & NIPPER, J. 1992. *Statistische Methoden in der Geographie*, Bd 2. Borntraeger, Stuttgart.

BARBARAND, J., CARTER, A., WOOD, I. & HURFURD, T. 2003. Compositional and structural control of fission-track annealing in apatite. *Chemical Geology*, **198**, 107–137.

BAXTER, E. F. 2003. Quantification of the factors controlling the presence of excess ^{40}Ar or ^{4}He. *Earth and Planetary Science Letters*, **216**, 619–634.

BAXTER, E. F., DEPAOLO, D. J. & RENNE, P. R. 2002. Spatially correlated anomalous ^{40}Ar/^{39}Ar 'age' variations in biotites about a lithologic contact near Simplon Pass, Switzerland; a mechanistic explanation for excess Ar. *Geochimica et Cosmochimica Acta*, **66**, 1067–1083.

BESAIRIE, H. & COLLIGNON, M. 1972. Géologie de Madagascar; I. Les terraines sédimentaires. *Annales Géologiques de Madagascar*, **35**, 1–463.

BJERKGÅRD, T., NORDGULEN, Ø. ET AL. 2006. Overview of the Proterozoic Geology in Niassa and Cabo Delgado provinces, N. Mozambique. *In*: SCIENTIFIC COMMITTEE (eds) *CAG 21 Abstract Volume*, Maputo, Mozambique, 15–18.

BOTT, M. H. P. 1995. Mechanisms of rifting: geodynamic modeling of continental rift systems. *In*: OLSEN, K. H. (ed.) *Continental Rifts: Evolution, Structure, Tectonics*. Developments in Geotectonics, **25**, 27–43.

BRAUN, J. 2002. Quantifying the effect of recent relief changes on the age–elevation relationships. *Earth and Planetary Science Letters*, **200**, 331–343.

BRAUN, J. & BEAUMONT, C. 1989. A physical explanation of the relation between flank uplift and the breakup unconformity at rifted continental margins. *Geology*, **17**, 760–764.

BROWN, R. & SUMMERFIELD, M. A. 1997. Some uncertainties in the derivation of rates of denudation from thermochronologic data. *Earth Surface Processes and Landforms*, **22**, 239–248.

CADOPPI, P., COSTA, M. & SACCHI, R. 1987. A cross section of the Namama Thrust Belt (Mozambique). *Journal of African Earth Sciences*, **6**, 493–504.

CASTAING, C. 1991. Post Pan-African tectonic evolution of South Malawi in relation to the Karroo and Recent East African Rift Systems. *Tectonophysics*, **191**, 53–73.

CATUNEANU, O., WOPFNER, H., ERIKSSON, P. G., CAIRNCROSS, B., RUBIGE, B. S., SMITH, R. M. H. & HANCOX, P. J. 2005. The Karoo basins of south-central Africa. *Journal of African Earth Science*, **43**, 211–253.

CHOROWICZ, J. 2005. The East African rift system. *Journal of African Earth Science*, **43**, 79–410.

COSTA, M., FERRARA, G., SACCHI, R. & TONARINI, S. 1992. Rb/Sr dating of the Upper Proterozoic basement of Zambesia, Mozambique. *Geologische Rundschau*, **81**, 487–500.

COYLE, D. A. & WAGNER, G. A. 1998. Positioning the titanite fission-track partial annealing zone. *Chemical Geology*, **149**, 117–125.

DAMON, P. E. 1968. Potassium–argon dating of igneous and metamorphic rocks with applications to the Basin ranges of Arizona and Sonora. *In*: HAMILTON, E. I. & FARQUHAR, R. M. (eds) *Radiometric Dating for Geologists*. Intersciences, London, 1–71.

DASZINNIES, M. C., JACOBS, J., KOHN, B. & GRANTHAM, G. H. 2004. A combined apatite fission-track and (U-Th)/He study of the passive continental margin of northern Mozambique: preliminary results. *In*: SCIENTIFIC COMMITTEE (eds) *Joint Earth Sciences Meeting: Société Géologique de France – Geologische Vereinigung, Strasbourg*, 29.

DINGLE, R. V. & SCRUTTON, R. A. 1974. Continental breakup and the development of the Post-Paleozoic sedimentary basins around southern Africa. *Geological Society of America Bulletin*, **85**, 1467–1474.

DI VINCENZO, G., VITI, C. & ROCCHI, S. 2003. The effect of chlorite interlayering on ^{40}Ar–^{39}Ar biotite dating; an ^{40}Ar–^{39}Ar laser-probe and TEM investigations of variably chloritised biotites. *Contributions to Mineralogy and Petrology*, **145**, 643–658.

DODSON, M. H. 1973. Closure temperature in cooling geochronological and petrological systems. *Contributions to Mineralogy and Petrology*, **40**, 259–274.

EBY, G. N., RODEN TICE, M., KRUEGER, H. L., EWING, W., FAXON, E. H. & WOOLLEY, A. R. 1995. Geochronology and cooling history of the northern part of the Chilwa alkaline province, Malawi. *Journal of African Earth Sciences*, **20**, 275–288.

GALBRAITH, R. F. & LASLETT, G. M. 1993. Statistical models for mixed fission track ages. *Nuclear Tracks and Radiation Measurements*, **21**, 459–470.

GALLAGHER, K., BROWN, R. & JOHNSON, C. 1998. Fission track analysis and its applications to geological problems. *Annual Reviews in Earth and Planetary Sciences*, **26**, 519–572.

GALLAGHER, K., HAWKESWORTH, C. J. & MANTOVANI, M. J. M. 1994. The denudation history of the onshore continental margin of SE Brazil inferred from apatite fission-track data. *Journal of Geophysical Research*, **99**, 18 117–18 145.

GLEADOW, A. J. W. 1981. Fission track dating methods: what are the real alternatives? *Nuclear Tracks and Radiation Measurements*, **5**, 3–14.

GRANTHAM, G. H., KEIDAN, H. ET AL. 2005*a*. New U-Pb SHRIMP Zircon data from NE Mozambique: implications for Mozambique Belt evolution and Gondwana E–W or N–S? *In*: PANKHURST, R. J. & VEIGA, G. D. (eds) *Gondwana 12, Mendoza, Argentina*, 182.

GRANTHAM, G. H., MABOKO, M. & EGLINGTON, B. M. 2003. A review of the evolution of the Mozambique Belt and implications for the amalgamation and dispersal of Rodinia and Gondwana. *In*: YOSHIDA, M., WINDLEY, B. F. & DASGPTA, S. (eds) *Proterozoic East Gondwana: Supercontinent Assembly and Breakup*. Geological Society, London, Special Publications, **206**, 401–425.

GRANTHAM, G. H., MACEY, P., INGRAM, B. A., RADEMEYER, M., EGLINGTON, B., KEIDAN, H. & AZEVEDO, S. 2005*b*. The chemistry and age of Karoo-age andesitic lavas along the northern Mozambique coast. *In*: PANKHURST, R. J. & VEIGA, G. D. (eds) *Gondwana 12, Mendoza, Argentina*, 183.

GREEN, P. F. 1981. A new look at statistics in fission-track dating. *Nuclear Tracks and Radiation Measurements*, **5**, 77–86.

GROVE, M. & HARRISON, T. M. 1996. ^{40}Ar* diffusion in Fe-rich phlogopite. *American Mineralogist*, **81**, 940–951.

HABGOOD, F. 1963. *The Geology of the Country West the Shire River Between Chikwawa and Chiromo*. Bulletin of Geological Survey Nyasaland, **14**.

HAMES, W. E. & BOWRING, S. A. 1994. An empirical evaluation of the argon diffusion geometry in muscovite. *Earth and Planetary Science Letters*, **124**, 161–167.

HARRISON, T. M. 1981. Diffusion of (super 40) Ar in hornblende. *Contributions to Mineralogy and Petrology*, **78**, 324–331.

HARROWFIELD, M., HOLDGATE, G. R., WILSON, CH. J. L. & MCLOUGHLIN, S. 2005. Tectonic significance of the Lambert graben, East Antarctica: Reconstructing the Gondwanan rift. *Geology*, **33**, 197–200.

HOLLICK, L., BAUER, W. *ET AL.* 2006. Structural evolution in the N Moçambique-Milange, Mocuba and Insaca-Gurue sheets. *In*: SCIENTIFIC COMMITTEE (eds) *CAG 21 Abstract Volume, Maputo, Mozambique*, 68–70.

HURFORD, A. J. 1990. International Union of Geological Sciences Subcommission on Geochronology Recommendation for standardization of fission track dating calibration and data reporting. *Nuclear Tracks and Radiation Measurements*, **17**, 233–236.

JACOBS, J., FANNING, C. M., HENJES-KUNST, F., OLESCH, M. & PAECH, H.-J. 1998. Continuation of the Mozambique Belt into East Antarctica: grenville-age metamorphsim and polyphase Pan-African high-grade events in central Dronning Maud Land. *Journal of Geology*, **106**, 385–406.

JACOBS, J. & THOMAS, B. 2004. Himalayan-type indenter-escape tectonics model for the southern part of the late Neoproterozoic–early Palaeozoic East African Antarctic orogen. *Geology*, **32**, 721–724.

JARITZ, W., KREUZER, H., MUELLER, P. & HARRE, W. 1977. Die Vulkanitserien im Küstengebiet von Nordmoçambique. *Geologisches Jahrbuch. Reihe B: Regionale Geologie Ausland*, **26**, 147–165.

JONCKHEERE, R. C. & WAGNER, G. A. 2000. On the thermal stability of fossil and neutron-induced fission-tracks in natural titanite. *Nuclear Instruments and Methods – Physics Research B*, **168**, 78–87.

KOHN, B. P., WAGNER, M. E., LUTZ, T. M. & ORGANIST, G. 1993. Anomalous Mesozoic thermal regime, central Appalachian Piedmont; evidence from sphene and zircon fission-track dating. *Journal of Geology*, **101**, 779–794.

KRIEGSMAN, L. M. 1995. The Pan-African event in East Antarctica: a view from Sri Lanka and the Mozambique Belt. *Precambrian Research*, **75**, 263–277.

KRÖNER, A. 1991. African linkage of Precambrian Sri Lanka. *Geologische Rundschau*, **80**, 429–440.

KRÖNER, A., SACCHI, R., JAECKEL, P. & COSTA, M. 1997. Kibaran magmatism and Pan-African granulite metamorphism in northern Mozambique: single zircon ages and regional implications. *Journal of African Earth Science*, **25**, 467–484.

KRÖNER, A., WILLNER, A. P., HEGNER, E., JAECKEL, P. & NEMCHIN, A. 2001. Single zircon ages, PT evolution and Nd isotopic systematics of high-grade gneisses in southern Malawi and their bearing on the evolution of the Mozambique belt in southeastern Africa. *Precambrian Research*, **109**, 257–291.

KUIPER, Y. D. 2002. The interpretation of inverse isochron diagrams in $^{40}Ar/^{39}Ar$ geochronology. *Earth and Planetary Science Letters*, **203**, 499–506.

KUSKY, T. M., ABDELSALAM, R. D., TUCKER, R. D. & STERN, R. J. 2003. Evolution of the East African and related orogens, and the assembly of Gondwana. *Precambrian Research*, **123**, 81–85.

LAWYER, L. A., GAHAGAN, L. M. & DALZIEL, I. W. D. 1998. A tight fit early Mesozoic gondwana, a plate reconstruction perspective. *Memoir of the National Institute of Polar Research*, **53**, (Special Issue), 214–229.

LEE, J. K. W. 1993. The argon release mechanisms of hornblende in vacuo. *Chemical Geology*, **106**, 133–170.

LEE, J. K. W. 1995. Multipath diffusion in geochronology. *Contributions to Mineralogy and Petrology*, **120**, 60–82.

LEE, J. K. W., ONSTOTT, T. C., CASHMAN, K. V., CUMBEST, R. J. & JOHNSON, D. 1991. Incremental heating of hornblende in vacuo: implications for $^{40}Ar–^{39}Ar$ geochronology and the interpretation of thermal histories. *Geology*, **19**, 872–876.

LEEDER, M. R. 1991. Denudation, vertical crustal movements and sedimentary basin infill. *Geologische Rundschau*, **80**, 441–458.

LO, C. H., LEE, J. K. W. & ONSTOTT, T. C. 2000. Argon release mechanisms of biotite in vacuo and the role of short-circuit diffusion and recoil. *Chemical Geology*, **165**, 135–166.

LUDWIG, K. R. 2001. *Isoplot 2.49 – A Geochronological Toolkit for Microsoft Excel*. Berkeley Geochronology Center Special Publication 1a, **4**.

LULIN, J. M. 1985. *Un nouveau gîte à Nb, Ta, U, REE d'origin magmatique en Afrique orientale: le complex ealcalin tectonisé de Meponda, Précambrien, province de Niassa, Moçambique*. PhD thesis, University of Orléans.

MCDOUGALL, I. & HARRISON, T. M. 1999. *Geochronology and Thermochronology by the $^{40}Ar/^{39}Ar$ Method*. Oxford University Press, New York, 269.

MCINTYRE, G. A., BROOKS, C., COMPSTON, W. & TUREK, A. 1966. The statistical assessment of Rb–Sr isochrones. *Journal of Geophysical Research*, **71**, 5459.

MITCHELL, J. G. 1968. The argon-40/argon-39 method for potassium/argon age determination. *Geochimica et Cosmochimica Acta*, **32**, 781–790.

MÖLLER, A., MEZGER, K. & SCHENK, V. 2000. U–Pb dating of metamorphic minerals: Pan-African metamorphism and prolonged slow cooling of high pressure granulites in Tanzania, East Africa. *Precambrian Research*, **104**, 123–146.

MORLEY, C. K., HARANYA, W., PHOOSONGSEE, S., PONGWAPEE, S., KORNSAWAN, A. & WONGANAN, N. 2004. Activation of rift oblique and rift parallel pre-existing fabrics during extension and their effect on deformation style: examples from the rifts in Thailand. *Journal of Structural Geology*, **26**, 1803–1829.

NAESER, C. W. & MCKEE, E. H. 1970. Fission-track and K–Ar ages of Tertiary ash-flow tuffs, north-central, Nevada. *Geological Society of America Bulletin*, **81**, 3375–3384.

NORCONSULT CONSORTIUM. 2007. *Mineral resources management capacity building project, Republic of Mozambique, Geological Mapping Lot 1,* Sheet Explanation, 32 Sheets, Scale: 1/25000, Report No B6,

National Directorate of Geology, Republic of Mozambique.

ONSTOTT, T. C., PHILLIPS, D. & PRINGLE-GOODELL, L. 1991. Laser microprobe measurement of chlorine and argon zonation in biotite. *Chemical Geology*, **90**, 145–168.

PERITS, F., AHLBRANDT, T. *ET AL.* 2002. C, Ver. 2.0. U.S. Department of the Interior and U.S. Geological Survey. *Open File Report*, **97-470A**, 1–50.

PINNA, P. & MARTEAU, P. 1987. *Carta geological de Moçambique, 1:1000,000 with explanatory notes.* Instituto Naçional de Geological, Maputo.

PINNA, P., JOURDE, G., CALUEZ, J.-Y., MROZ, J. P. & MARQUES, J. M. 1993. The Mozambique Belt in northern Mozambique: neoproterozoic (1100–850 Ma) crustal growth and tectogenesis and superimposed Pan-African (800–550 Ma) tectonism. *Precambrian Research*, **62**, 1–59.

REEVES, C. V., SAHU, B. K. & DE WIT, M. 2002. A re-examination of the paleo-position of Africa's eastern neighbours in Gondwana. *Journal of African Earth Sciences*, **34**, 101–108.

REX, D. C. & GUISE, P. G. 1995. Evaluation of argon standards with special emphasis on time scale measurements. *In*: ODIN, G. S. (ed.) *Phanerozoic Time Scale. Bulletin of Liaisian Information IUGS Subcomission Geochronology*, **13**, 21–23.

ROBERTS, M. P., GRANTHAM, G. H. *ET AL.* 2005. Preliminary pressure temperature determinations on granulite-facies rocks of the Mugeba Klippe, north central Moçambique. *In*: SCIENTIFIC COMMITTEE (eds) *GEO 2005, Abstract Volume, Durban, South Africa, University of Kwazulu-Natal*, 193–194.

RODDICK, J. C. 1978. The application of isochron diagrams in $^{40}Ar/^{39}Ar$ dating: a discussion. *Earth and Planetary Science Letters*, **41**, 233–244.

ROGERS, J. J. W., UNRUG, R. & SULTAN, M. 1995. Tectonic assembly of Gondwana. *Journal of Geodynamics*, **19**, 1–34.

SACCHI, R., CADOPPI, P. & COSTA, M. 2000. Pan-African reactivation of the Lurio segment of the Kibaran Belt system: a reappraisal from recent age determinations in northern Mozambique. *Journal of African Earth Science*, **30**, 629–639.

SACCHI, R., MARQUES, J., COSTA, M. & CASATI, C. 1984. Kibaran events in the southernmost Mozambique Belt. *Precambrian Research*, **25**, 141–159.

SALMAN, G. & ABDULA, I. 1995. Development of the Mozambique and Ruvuma sedimentary basins, offshore Mozambique. *Sedimentary Geology*, **96**, 7–41.

SCHANDELMEIER, H., BREMER, F. & HOLL, H.-G. 2004. Kinenmatic evolution of the Morondava rift basin of SW Madagascar – from wrench tectonics to normal extension. *Journal of African Earth Science*, **38**, 321–330.

STEIGER, R. J. & JÄGER, E. 1977. Subcommission on geochronology: convention on the use of decay constants in geo- and cosmochronology. *Earth and Planetary Science Letters*, **36**, 359–362.

STÜWE, K., WHITE, L. & BROWN, R. 1994. The influence of eroding topography on steady-state isotherms. Application to fission-track analysis. *Earth and Planetary Science Letters*, **124**, 63–74.

THOMAS, R. J., BAUER, W. *ET AL.* 2006. Mozambique Belt in the Milange-Mocuba-Malema area, Moçambique. *In*: SCIENTIFIC COMMITTEE (eds) *CAG 21 Abstract Volume, Maputo, Mozambique*, 161–163.

VAN DER BEEK, P. 1995. *Tectonic Evolution of Continental Rifts: Inferences From Numerical Modelling and Fission Track Thermochronology*. PhD thesis, Amsterdam, Vrije University.

VAN DER BEEK, P., ANDRIESSEN, P. & CLOETHINGH, S. 1995. Morphotectonic evolution of rifted continental margins: inferences from a coupled tectonic-surface process model and fission track thermochronology. *Tectonics*, **14**, 406–421.

VAN DER BEEK, P., CLOETHINGH, S. & ANDRIESSEN, P. 1994. Mechanisms of extensional basin formation and vertical motions at rift flanks: constraints from tectonic modelling and fission track thermochronology. *Earth and Planetary Science Letters*, **121**, 417–433.

VAN DER BEEK, P., MBEDE, E., ANDRIESSEN, P. & DELVAUX, D. 1998. Denudation history of the Malawi and Rukwa Rift flanks (East African Rift System) from apatite fission track thermochronology. *Journal of African Earth Science*, **26**, 363–385.

VERNIERS, J., JOURDAN, P. P., PAULIS, R. V., FRASCA-SPADA, L. & BOCK, F. R. D. 1989. The Karroo Graben of Metangula Northern Mozambique. *Journal of African Earth Science*, **9**, 137–158.

VIOLA, G., HENDERSON, I., BINGEN, B., FEITIO, P., THOMAS, R., HOLLICK, L. & JACOBS, J. 2006. A new tectonic framework for northern Mozambique. *In*: SCIENTIFIC COMMITTEE (eds) *CAG 21 Abstract Volume, Maputo, Mozambique*, 168–169.

VISSER, J. N. J. & PRAEKELT, H. E. 1996, Subduction, mega-shear systems and Late Palaeozoic basin development of Gondwana. *Geologische Rundschau*, **86**, 632–646.

VISSER, J. N. J. & PRAEKELT, H. E. 1998. Late Palaeozoic crustal block rotations within the Gondwana sector of Pangea. *Tectonophysics*, **287**, 201–212.

WAGNER, G. A. & VAN DEN HAUTE, P. 1992. *Fission Track Dating*. Enke, Stuttgart.

WOPFNER, H. 1993. Structural development of Tanzania Karoo basins and the break-up of Gondwana. *In*: FINLAY, A. E. A. (ed.) *Gondwana Eight*. Balkema, Hobart, 531–539.

WOPFNER, H. 1994. The Malagasy, a chasm in the Tethyan margin of Gondwana. *Journal of Southeast Asian Earth Sciences*, **9**, 451–461.

WOPFNER, H. 2002. Tectonic and climatic events controlling deposition in Tanzanian Karoo basins. *Journal of African Earth Science*, **34**, 167–177.

WOPFNER, H. & KAAYA, C. Z. 1991. Stratigraphy and morphotectonics of Karoo deposits of the northern Selous Basin, Tanzania. *Geological Magazine*, **128**, 319–334.

ZIEGLER, P. A. & CLOETHING, S. 2004. Dynamic processes controlling the evolution of rifted basins. *Earth-Science Reviews*, **64**, 1–50.

Denudation along the Atlantic passive margin: new insights from apatite fission-track analysis on the western coast of South Africa

A. KOUNOV[1]*, G. VIOLA[2], M. DE WIT[3] & M. A. G. ANDREOLI[4]

[1]*Institute of Geology and Paleontology, Basel University, 4056 Basel, Switzerland*

[2]*NGU, Geological Survey of Norway, 7491 Trondheim, Norway*

[3]*AEON and Department of Geological Sciences, UCT, 7701 Rondebosch, South Africa*

[4]*South African Nuclear Energy Corporation, PO Box 582, 0001 Pretoria, South Africa, and School of Geosciences, University of the Witwatersrand, Private Bag 3, 2050 Wits, South Africa*

**Corresponding author (e-mail: a.kounov@unibas.ch)*

Abstract: Apatite fission-track (AFT) data from two traverses across the Great Escarpment of the western coast of South Africa are used to reconstruct the tectonic evolution and denudation history of this sector of the Atlantic passive margin. Fission-track ages range between 180 and 86 Ma. Modelling of this data identifies two distinct cooling events. The first event, between 160 and 138 Ma, is recorded only by the rocks above the escarpment in the Karoo area, and is tentatively linked to post-Karoo magmatism (*c.* 180 Ma) thermal relaxation. The second, between 115 and 90 Ma, results instead from a tectonically induced denudation episode responsible for the removal of up to 2.5 km of crust across the coastal zone in front of the escarpment and less than 1 km on the elevated interior plateau. Based on these results, it is suggested that the Cretaceous is the time when most of the elevated topography of Southern Africa was generated, with only a minor Cenozoic contribution.

Continental passive margins have been the target of geomorphological studies since the early 1920s and different models have been suggested for their morphotectonic development (e.g. du Toit 1926; King 1953; Ollier 1985; Partridge & Maud 1987; Gilchrist & Summerfield 1990). Passive margins with high elevation have been of particular interest due to their impressive geomorphological features, including a low-lying coastal plain separated from an elevated inland region by seaward-facing escarpments. In the early schemes of margin evolution, the formation of the escarpment was generally attributed to the downflexing of the lithosphere and the development of a broad monocline that was subsequently eroded by backwearing during successive denudational phases (e.g. King 1953; Ollier 1985; Ollier & Marker 1995; Partridge & Maud 1987; Seidl *et al.* 1996). This type of evolutionary model relied mostly on geomorphological observations and on the correlation of erosional surfaces over great distance. The main problem with these models, however, is the lack of reliable age constraints on the erosion surfaces and quantitative information about the rates of denudation, which in turn hamper reliable large-scale correlations.

Over the past two decades a number of low-temperature thermochronological studies, including apatite fission-track (AFT), (U–Th)/He and cosmogenic nuclide analysis, have provided new quantitative geochronological constraints on the denudational history of passive margins, thus allowing the testing of earlier evolutionary models (e.g. Moore *et al.* 1986; Brown *et al.* 1990, 2000; Gallagher *et al.* 1998; Gallagher & Brown 1999; Fleming *et al.* 1999; Cockburn *et al.* 2000; Persano *et al.* 2002; Kounov *et al.* 2007, 2008). AFT and (U–Th)/He thermochronology are of particular value when attempting to understand the morphotectonic history of continental margins because these methods provide quantitative estimates of the amount of rock removal in the upper *c.* 3–5 km of the continental crust over timescales of millions to hundreds of millions of years. This in turn allows the reconstruction of the early denudation stages, whereas later short-term and more site-specific erosional processes and their rates are better studied with measurements of *in situ*-produced cosmogenic nuclides.

Based on these recent geochronological advances, new conceptual models have been proposed that differ from the classical escarpment retreat (backwearing) schemes essentially in the character of the post-break-up tectonics, the initial position and subsequent migration of the drainage divide, and in the spatial and temporal pattern of related denudation (e.g. Gilchrist & Summerfield 1990; Gilchrist *et al.* 1994; Cockburn *et al.* 2000; Brown *et al.* 2002).

From: LISKER, F., VENTURA, B. & GLASMACHER, U. A. (eds) *Thermochronological Methods: From Palaeotemperature Constraints to Landscape Evolution Models*. Geological Society, London, Special Publications, **324**, 287–306.
DOI: 10.1144/SP324.19 0305-8719/09/$15.00

The regional patterns of post break-up denudation along the continental margin of Southern Africa, as reconstructed from several AFT and cosmogenic nuclides studies, are generally incompatible with a simple model of landscape development involving steady retreat of an escarpment initially formed at the coast at the time of break-up (e.g. Fleming *et al.* 1999; Brown *et al.* 2000, 2002; Cockburn *et al.* 2000; Van der Wateren & Dunai 2001; Kounov *et al.* 2007, 2008; Tinker *et al.* 2008). Reported denudation rates along the coastal plain since the break-up are an order of magnitude lower than the values necessary for the escarpment retreat model.

These recent geochronological data require, instead, more complex models whereby the escarpment that formed at the coast at the time of the continental break-up was rapidly destroyed by rivers flowing from an interior divide towards the rapidly changing base level (Cockburn *et al.* 2000; Brown *et al.* 2002). This interpretation is in agreement with numerical models of surface evolution that emphasize the importance of drainage divides in controlling the location and evolution of major escarpments (e.g. Gilchrist *et al.* 1994; van der Beek & Braun 1999). More recently, however, these numerical models were criticized by Moore & Blenkinsop (2006), based on field observations that apparently support a scarp retreat model for the evolution of the Drakensberg escarpment (cf. King 1953). Thus, the details of the evolution of the great southern African escarpment still remain elusive.

Detailed AFT analyses on outcrop and borehole samples from the South African margin suggest the presence of a first period of accelerated denudation in the early Cretaceous (140–120 Ma) followed by a second in the Mid-Cretaceous (100–80 Ma: Brown *et al.* 2002; Tinker *et al.* 2008). A Late Cretaceous episode of accelerated denudation (80–60 Ma) was reported from the Drakensberg Escarpment area (Brown *et al.* 2002), whereas Raab *et al.* (2002) suggest a discrete period of accelerated cooling, beginning at about 70 Ma, in northern Namibia related to the reactivation of earlier basement structures (e.g. the Waterberg thrust), possibly caused by changes in the spreading geometry in the South Atlantic and SW Indian Ocean (e.g. Nürnberg & Müller 1991).

These studies indicate in general that a total of approximately 5 km of rocks were eroded during the Cretaceous, whereas during the Cenozoic the amount of total denudation decreased dramatically to less than about 1 km, approaching the present-day denudation rates as determined from cosmogenic nuclides data (Fleming *et al.* 1999; Cockburn *et al.* 2000; Van der Wateren & Dunai 2001; Kounov *et al.* 2007).

Whereas the southern (Tinker *et al.* 2008) and the eastern (Brown *et al.* 2002) coast of South Africa were already investigated by detailed AFT studies with the goal of understanding the local evolution and denudation history, the Atlantic passive margin along the west coast remains poorly studied. Our AFT work contributes further to the understanding of the Atlantic African margin by providing a new, detailed reconstruction of the spatial and temporal patterns of denudation along two transects across the western passive margin of South Africa and its interior since the onset of continental rifting.

The western margin of South Africa

Summary geological framework

The western coast of South Africa is characterized by the occurrence of diverse lithologies, from Proterozoic metamorphic rocks to Neoproterozoic and Mesozoic sedimentary and igneous successions. The Mesoproterozoic Namaqualand Metamorphic Province comprises intensely deformed supracrustal sequences intruded by numerous pre-, syn- and post-tectonic granitoids (e.g. Johnson *et al.* 2006). Exhumation of the Namaqualand metamorphic rocks during the Early Neoproterozoic (between 1000 and 800 Ma) was followed by deposition of the Gariep Supergroup in a pull-apart basin (Gresse 1995). The sediments of the Neoproterozoic–Cambrian Vanrhynsdorp Group unconformably overlie the Gariep Supergroup (Johnson *et al.* 2006). The Vanrhynsdorp and Gariep Group rocks were deformed and metamorphosed during the Pan-African orogeny between 650 and 480 Ma at relatively low temperatures. Proterozoic and Neoproterozoic rocks are unconformably overlain by the Ordovician–Devonian thick siliciclastic sequences of the Cape Supergroup (Johnson *et al.* 2006). In the study area the Cape Supergroup is represented mainly by the quartzite-dominated rocks of the Table Mountain Group (Johnson *et al.* 2006) (Fig. 1). From the late Carboniferous to the Early Jurassic, the Karoo sedimentary succession (Karoo Supergroup), which consists of several kilometres of clastic sediments, was deposited in the Karoo Basin. This is considered as a foreland basin possibly formed in response to orogenic loading of the Cape Fold Belt to the south (Johnson *et al.* 2006). The Cape Fold Belt formed during crustal shortening related to the subduction and accretion of the paleo-Pacific plate beneath Gondwana (de Wit & Ransome 1992). The break-up of Gondwana (*c.* 170–150 Ma: Hawkesworth *et al.* 1999) started with the separation of west Gondwana (Africa and South America) from east Gondwana (Australia, Antarctica, India and New Zealand), post-dating the extrusion of the

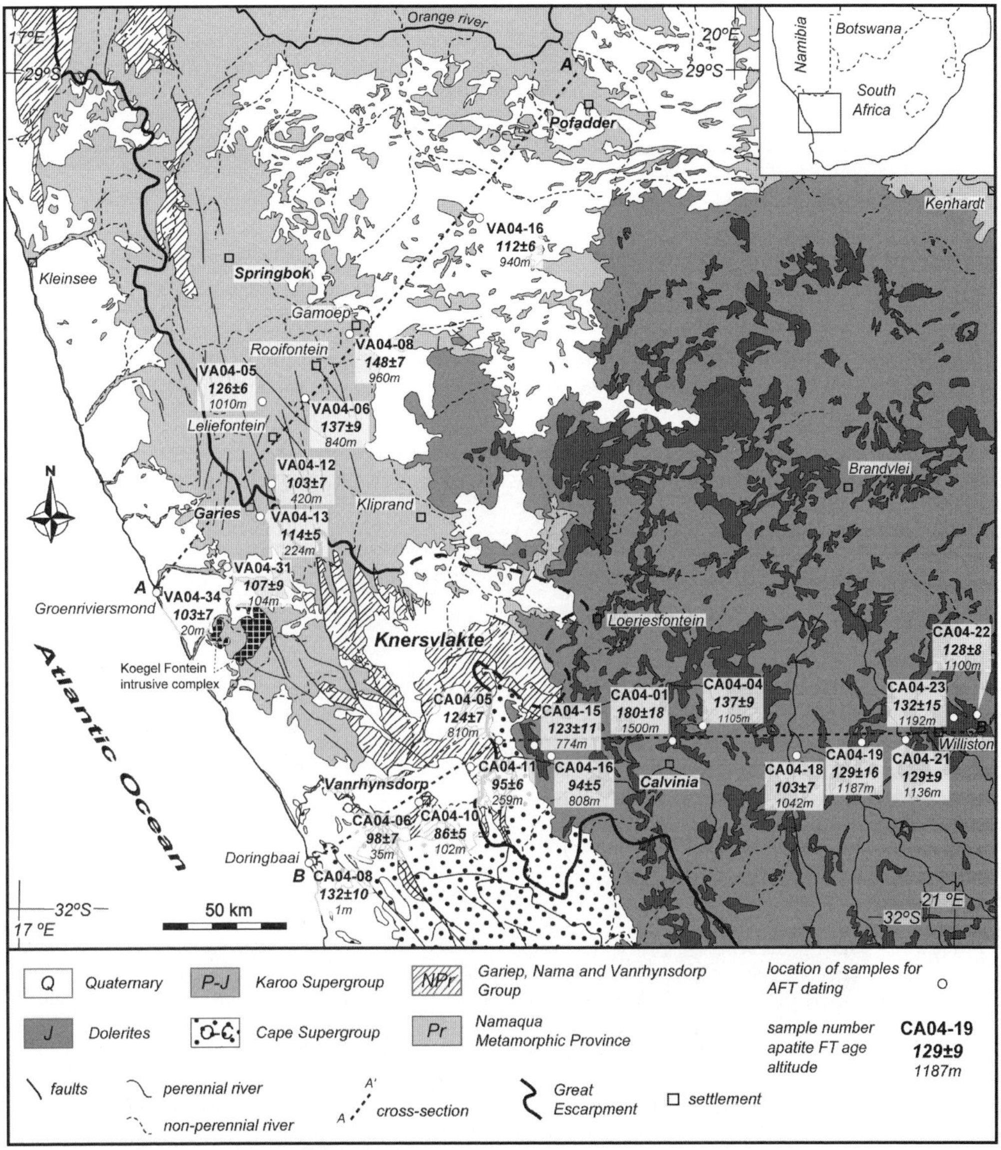

Fig. 1. Geological map of the study area (western South Africa) with the locations of the analysed samples and obtained fission-track ages. Lines A–A′ and B–B′ trace the sections of Figure 4.

voluminous and extensive continental flood basalts of the Drakensberg Group (184–174 Ma: Jourdan *et al.* 2007) exposed in the SW part of South Africa and with the emplacement of numerous dolerite sills (Karoo magmatics) preferentially intruded into the flat-lying beds of the Karoo Supergroup (Cox 1992; Duncan *et al.* 1997). Continental rifting between South America and Africa began during the Late Jurassic (*c.* 150 Ma: Nürnberg & Müller 1991; Stern & de Wit 2004; Trumbull *et al.* 2007). The rifting was accompanied by the intrusion of syenite and granite plutons, as well as dolerite dykes, between 137 and 125 Ma along the margin (e.g. Eales *et al.* 1984; Trumbull *et al.* 2007). One prominent intrusion is the Rietport granite, part of the Koegel Fontein intrusive complex (133.9 ± 1.3 Ma: De Beer *et al.* 2002), now cropping out to the NW of Vanrhynsdorp (Fig. 1).

The rifting–drifting transition was marked by the initiation of sea-floor spreading in the southern Atlantic at about 134 Ma (Rabinovich & LaBrecque 1979; Eagles 2007). The southern Atlantic opened by northwards propagation of the spreading centre over a period of around 40 Ma. During the drifting period, a number of mafic alkaline intrusions, including kimberlites and related rocks, intruded across southern Africa. Two distinct intrusion peaks have been reported at 145–115 and 95–80 Ma, corresponding to kimberlite Group II and I, respectively (e.g. Smith *et al.* 1985; Basson & Viola 2004).

In the study area a wide range of Cenozoic sedimentary deposits cover the low coastal plain (Fig. 1). Raised marine terraces occurring at various elevations along the coast have been linked to Miocene, Pliocene and Quaternary transgressions (Pether 2000).

Geomorphology

The SW African high-elevated passive margin comprises a gently inclined low coastal plain and an elevated inland plateau separated by a seaward-facing escarpment (the 'Great Escarpment' of King 1953) (Fig. 2).

In the southern part of the study area (Fig. 2), the low-lying coastal plain [from 0 to 200 m asl (above sea level) in altitude and up to 80 km wide], is

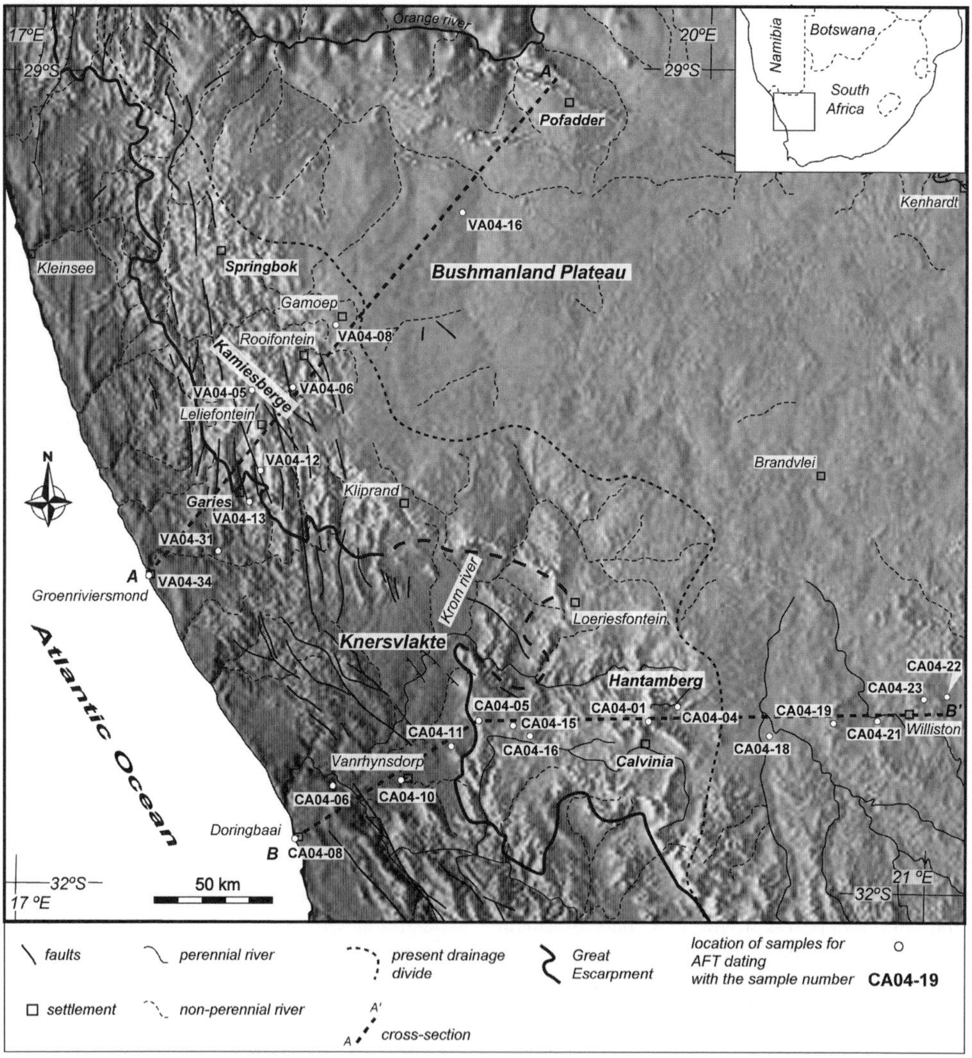

Fig. 2. Shaded relief map of the study area (western South Africa) with the locations of the analysed samples.

developed mainly across the Gariep and Vanrhynsdorp (Nama) Group phyllites and siltstones (Johnson *et al.* 2006) (Fig. 1). Close to the coast, isolated remnants of quartzite-dominated rocks of Cape Supergroup (Johnson *et al.* 2006) also crop out.

Inland, the coastal plain is flanked by a well-defined escarpment that, near Vanrhynsdorp, forms an approximately 600 m-high, subvertical cliff. The summit of the escarpment is capped by up to 50 m of subhorizontal Table Mountain Group quartzites and quartz-conglomerates. To the north, the top of the escarpment cliff gradually loses altitude, and in the area known as the Knersvlakte (Fig. 2) the quartzite cap disappears and the escarpment is morphologically much less pronounced. Here, in places, it is eroded by the Krom River and its tributaries (Kounov *et al.* 2008, Fig. 2).

East of the escarpment the inland plateau has a mean elevation of about 1000 m and generally low relief (Fig. 2). The top of the plateau is underlain by the Karoo Supergroup siliciclastic sediments (Johnson *et al.* 2006) intruded by numerous subhorizontal Mid-Jurassic Karoo dolerite sills and several subvertical dykes (Fig. 1). Small hills (kopies), formed by the relatively resistant dolerite sills, rise locally above the plateau floor. For example, in the area around Calvinia (*c.* 1000 m) a number of these high kopies, with densely spaced dolerites and steep scarps, rise up to 700 m above the general plateau (e.g. Hantamberg: Fig. 2).

The landscape of the northern part of the study area (Namaqualand) is defined by three main geomorphic units: an approximately 50 km-wide sandy coastal plain; an approximately 30-km wide escarpment zone; and an internal elevated plateau (the Bushmanland Plateau) at a mean altitude of 950 ± 50 m (Fig. 2). The escarpment here is much less defined from a geomorphological perspective than further to the south. Indeed, it lacks a proper scarp and is, instead, represented by a broad highland region, which includes the Kamiesberge Mountains, reaching up to 1770 m in altitude (Fig. 2). The Namaqualand area is underlain by the high-grade granitic gneisses and granulites of the Mesoproterozoic Namaqua Metamorphic Province (Johnson *et al.* 2006) (Fig. 1).

Structures

The large-scale Mesozoic–Cenozoic structural grain of the west coast of South Africa is characterized by complex sets of faults. In the Namaqualand area two major sets of faults strike N to NNW and NW, respectively. They can be easily identified on satellite and aerial images, especially along the escarpment area and in the elevated plateau, which is not covered by Quaternary deposits (Fig. 3). Most of these faults show top-to-the-west normal displacement, but small-scale thrusts and grabens that strike N to NNW have also been documented (Brandt *et al.* 2005; Viola *et al.* 2005). Several major north–south-trending normal faults juxtapose the Late Palaeozoic Nama Group sediments against the metamorphic basement rocks of the Namaqualand Metamorphic Province (Fig. 1). It is generally difficult to constrain precisely the age of these faults because of the common lack of marker horizons. Nevertheless, the fault structures above the escarpment in the Namaqualand area have been described broadly as late Mesozoic–Cenozoic based on the age of the rocks they dissect (Brandt *et al.* 2005; Viola *et al.* 2005). Significant lateral variation in the depth of denudation owing to post-Cretaceous normal faulting has been recognized in Namaqualand (Brown 1992). Neotectonic activity along these structures is indicated by a large number of seismic events recorded in the region (e.g. Viola *et al.* 2005).

Apatite fission-track analysis

Fission-track results

In this section we present AFT data from two traverses that run roughly east-west and, thus, perpendicular to the trend of the Great Escarpment in the region (location shown in Figs 1 & 2).

Analytical procedures for AFT analysis follow those outlined by Seward (1989). Etching of the apatite grains was carried out with 7% HNO_3 at 21 °C for 50 s. Irradiation was carried out at the ANSTO facility, Lucas Heights, Australia. Microscopic analysis was completed at University of Cape Town using an optical microscope with a computer-driven stage ('Autoscan' software from Autoscan Systems Pty Ltd, Melbourne, Australia). All ages were determined using the zeta approach (Hurford & Green 1983) with a zeta value of 341 ± 10 for CN5 (Table 1, analyst: A. Kounov). They are reported as central ages (Galbraith & Laslett 1993) with a 1σ error (Table 1). The magnification used was 1250×, at which horizontal confined track lengths and etch-pits diameters (D_{par}) were also measured. Between 5 and 10 etch pits were measured per dated grain, depending on the quality and the density of the track pits on the grain surface. The D_{par} values of the analysed samples are between 1.6 and 2.7 μm, with the average relative error of less than 5%.

Our northernmost traverse, between Groenriviersmond and Pofadder, is located within the gneissic basement of the Namaqua Metamorphic Province (Fig. 1) and is referred here to as the Namaqua Traverse. Our southern traverse, referred to as the Karoo Traverse, between Doringbaai and

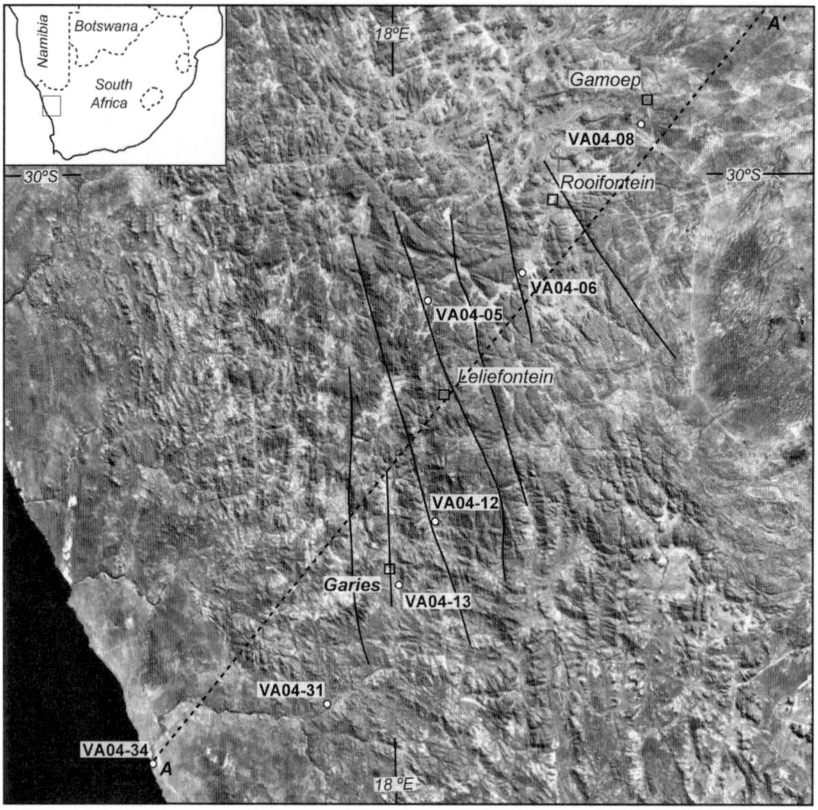

Fig. 3. Landsat image of the Namaqualand area with interpreted traces of the steep fault structures shown in Figure 4. Many other lineaments, probably corresponding to Mesozoic faults, are also easily visible. Line A–A′ trace the section of Figure 4.

Williston, crosses the Gariep and Nama Group sediments, as well as the Cape and Karoo Supergroup deposits and Karoo dolerites (Fig. 1). The locations of the samples are shown in Figure 1 and summarized in Table 1.

The stratigraphic age of the 22 analysed samples ranges from Precambrian (Namaqua metamorphics) through to early Jurassic (Karoo dolerites; Table 1). Samples yield AFT ages ranging between 180 and 86 Ma, with the older ages characteristic for the continental interior above the escarpment (Fig. 1). All samples passed the χ^2 test and have AFT ages significantly younger than their stratigraphic ages, indicating that they were affected by temperatures between 60 and 110 °C or higher after their deposition or formation (Green & Duddy 1989; Corrigan 1993). The samples have mean track lengths between 12.3 and 14.54 μm, with a standard deviation of 0.88–2.41 μm (Table 1). Samples with less than 35 measured track lengths were not taken into further consideration because of their low number of confined track measurements, insufficient for a robust modelling approach.

Samples with the youngest AFT ages (114–86 Ma) show mostly narrow, unimodal track-length distributions (standard deviation 0.88–1.48 μm) and lack short tracks (<*c.*11 μm: Fig. 4). This suggests that they experienced fast cooling from temperatures higher than about 110 °C along a rather simple cooling path at the time indicated by the AFT age (Laslett *et al.* 1987). Track-length distributions for the older samples are, instead, significantly different. Histograms show generally broader distributions (standard deviation 1.24–2.41 μm) with 'tails' of shorter tracks (<*c.* 11 μm: Fig. 4). This indicates that these samples experienced a more complex thermal history during which they spent a significant amount of time at temperatures between about 110 and 60 °C (apatite partial annealing zone) before final exhumation, which caused a shortening of their old tracks (Green & Duddy 1989; Corrigan 1993).

Table 1. *Apatite fission-track results*

Sample number	Latitude (S)/ Longitude (E)	Altitude (m)	Lithology	Stratigraphic division	Number of grains	$\rho_d(N_d)$ (10^6 cm^{-2})	$\rho_s(N_s)$ (10^6 cm^{-2})	$\rho_i(N_i)$ (10^6 cm^{-2})	$P(\chi^2)$ (%)	U (conc.) (ppm)	Central age ($\pm 1\sigma$) (Ma)	MTL ($\pm 1\sigma$) (μm)	SD (N) (μm)	D_{par} (μm)
Namaqua Traverse														
VA04/34	S30.85997/E17.57692	20	Bi gneiss	NMP	17	1.118 (4687)	2.703 (467)	4.949 (855)	98.8	61	103.3 ± 6.8	14.06 ± 0.12	1.12 (86)	2.2
VA04/31	S30.76619/E17.88967	104	Bi gneiss	NMP	23	1.079 (4146)	0.415 (250)	0.709 (427)	99.9	9	106.8 ± 9.2	14.29 ± 0.10	0.93 (86)	2.3
VA04/13	S30.59236/E18.00863	224	granitic gneiss	NMP	22	1.003 (4217)	2.771 (1926)	4.119 (2863)	81.7	51	114.0 ± 5.1	13.79 ± 0.31	1.14 (102)	2.1
VA04/12	S30.48478/E18.05954	420	granitic gneiss	NMP	17	0.968 (4146)	1.254 (410)	1.985 (649)	99.94	29	103.4 ± 7.4	14.54 ± 0.25	1.48 (36)	2.4
VA04/05	S30.19328/E18.04745	1010	charnockite	NMP	22	1.250 (4306)	2.646 (1309)	4.433 (2193)	90.92	44	125.9 ± 6.1	13.72 ± 0.12	1.24 (102)	1.7
VA04/06	S30.15173/E18.21009	840	granitic gneiss	NMP	13	1.092 (4306)	3.790 (621)	5.103 (836)	99.92	61	136.8 ± 8.5	13.42 ± 0.25	1.57 (39)	1.6
VA04/08	S29.93714/E18.40922	960	charnockite	NMP	21	1.052 (4306)	2.317 (1530)	2.771 (1830)	96.97	35	148.3 ± 7.1	12.32 ± 0.22	2.41 (123)	1.6
VA04/16	S29.52382/E18.93592	940	gneiss	NMP	22	1.017 (4687)	1.389 (902)	2.141 (1391)	99.93	27	111.5 ± 6.0	14.04 ± 0.12	1.35 (128)	2.3
Karoo Traverse														
CA04/08	S31.81319/E18.23548	1	sandstone	TMG	13	1.003 (4217)	1.259 (353)	1.619 (454)	88.54	22	131.6 ± 10.3	12.42 ± 0.61	2.10 (12)	2.3
CA04/06	S31.63409/E18.40824	35	sandstone	VG	13	0.997 (4292)	1.366 (411)	2.352 (707)	43.35	37	98.0 ± 7.3	12.59 ± 0.45	0.91 (4)	2.1
CA04/10	S31.60948/E18.70766	102	sandstone	VG	22	0.956 (4306)	1.029 (707)	1.939 (1332)	99.9	27	86.0 ± 4.9	13.87 ± 0.09	0.88 (102)	2.2
CA04/11	S31.48021/E18.92288	259	gravel	VG	22	12.141 (4292)	0.691 (492)	1.500 (1067)	79.77	18	94.7 ± 6.0	13.89 ± 0.13	1.27 (101)	2.2
CA04/05	S31.37293/E19.01697	825	conglomerate	TMG	26	10.505 (4217)	0.874 (729)	1.251 (1044)	88.77	18	123.9 ± 7.3	13.76 ± 0.25	1.65 (44)	2.5
CA04/15	S31.39204/E19.18904	774	dolerite	Karoo magm.	22	11.418 (4292)	0.347 (264)	0.545 (414)	100	6	122.9 ± 10.5	14.76 ± 0.57	1.8 (10)	2.7
CA04/16	S31.43291/E19.26326	808	conglomerate	Dwyka Gp	23	9.772 (4446)	1.368 (1092)	2.408 (1922)	42.74	30	94.2 ± 4.9	13.84 ± 0.12	1.21 (101)	2.2
CA04/01	S31.38068/E19.78422	1520	dolerite	Karoo magm.	31	11.697 (4217)	0.254 (220)	0.277 (240)	100	3	180.3 ± 17.9	14.36 ± 0.24	1.27 (28)	2.5
CA04/04	S31.32342/E19.91601	1080	dolerite	Karoo magm.	22	1.269 (4306)	0.757 (474)	1.186 (743)	76.16	12	136.6 ± 9.2	13.35 ± 0.26	1.98 (56)	2
CA04/18	S31.43123/E20.30881	1024	sandstone	Ecca Gp	20	1.068 (4446)	1.500 (492)	2.628 (861)	99.51	31	103.1 ± 6.7	13.28 ± 0.39	1.29 (11)	2.2
CA04/19	S31.38337/E20.59016	1187	dolerite	Karoo magm.	15	0.947 (4446)	0.319 (132)	0.396 (163)	98.54	5	128.7 ± 15.6	14.79 ± 0.23	0.46 (4)	3.1
CA04/21	S31.35553/E20.79923	1036	sandstone	Ecca Gp	15	0.974 (4687)	1.979 (419)	2.524 (534)	97.89	34	128.7 ± 9.4	13.58 ± 0.24	1.52 (42)	2.3
CA04/23	S31.28741/E20.99069	1192	dolerite	Karoo magm.	20	0.997 (4306)	0.308 (150)	0.392 (191)	99.87	5	131.8 ± 9.2	15.78 ± 0.34	0.758 (5)	2.7
CA04/22	S31.26801/E21.0839	1100	sandstone	Ecca Gp	21	1.214 (4146)	1.309 (552)	2.102 (886)	85.02	21	127.7 ± 8.1	13.38 ± 0.13	1.31 (95)	2.5

All ages are central ages (Galbraith 1981). $\lambda D = 1.55125 \times 10^{-10}$. A geometry factor of 0.5 was used. Zeta = 341 ± 10 for CN5 glass. Irradiations were performed at the ANSTRO facility, Lucas Heights, Australia. $P(\chi^2)$ is the probability of obtaining χ^2 values for ν degrees of freedom, where ν= number of crystals − 1. ρ_d, ρ_s and ρ_i represent the standard, sample spontaneous and induced track densities, respectively. MTL is the mean track length. SD is the standard deviation. D_{par} is the mean track pit length. NMP, Namaqua Metamorphic Province; TMG, Table Mountain Group; VG, Vanrhynsdorp Group; Dwyka Gp, late Carboniferous–early Permian (Karoo Supergroup); Ecca Gp, Permian (Karoo Supergroup); Karoo magm., Karoo magmatics (*c.* 183 Ma, Duncan *et al.* 1997).

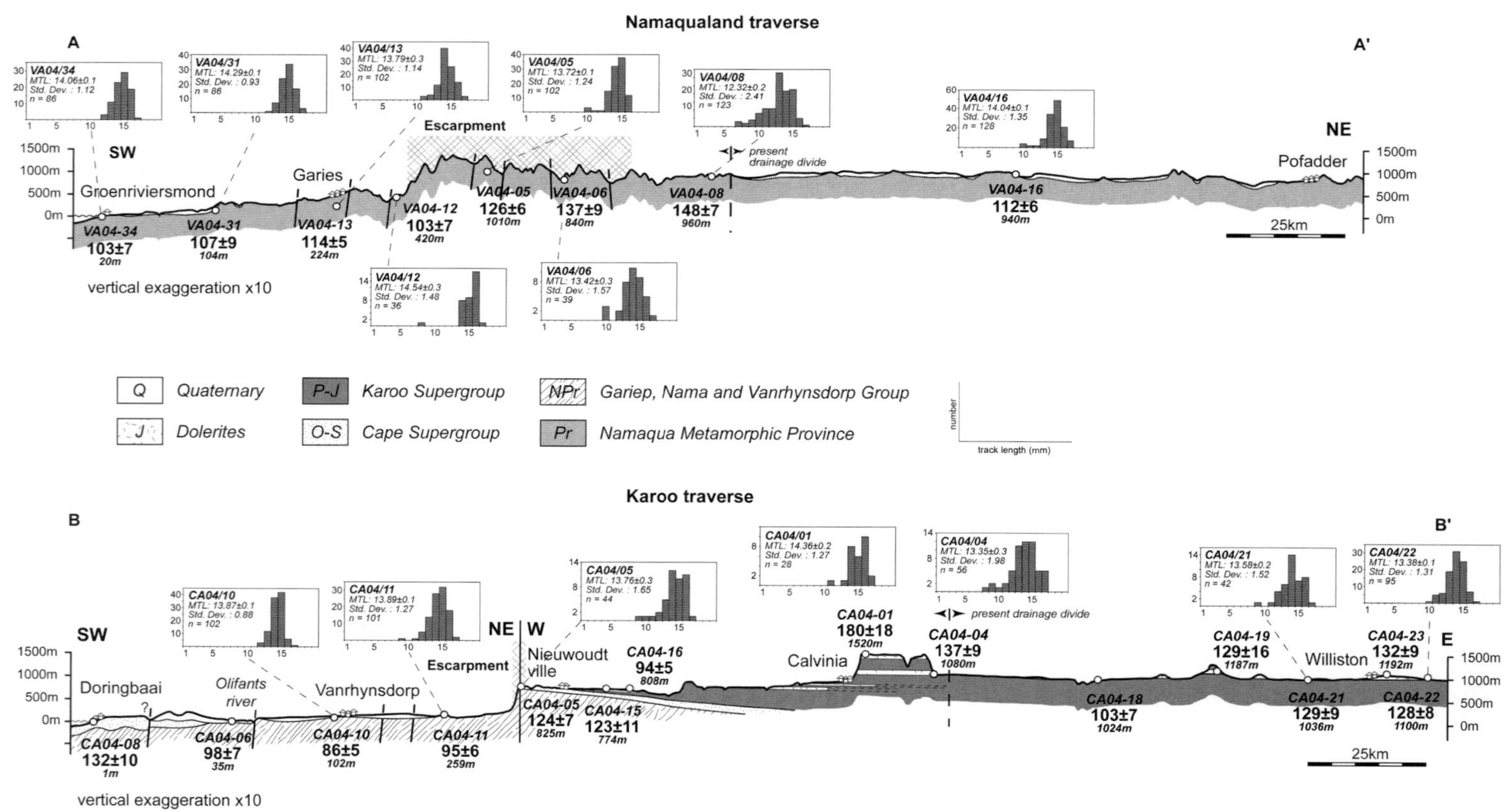

Fig. 4. Cross-sections A–A′ between Groenriviersmond and Pofadder (Namaqua Traverse) and B–B′ between Doringbaai and Williston (Karoo Traverse) with the locations of apatite fission-track samples, relative ages and track-length distribution histograms.

Namaqua Traverse. Along this traverse eight samples from crystalline rocks of the Namaqua Metamorphic Province were analysed. Apatite ages range from 148±7 to 103±7 Ma. Samples show a strong correlation between age and distance from the coast, with age increasing systematically towards the interior of the continent (Figs 4 & 5). Two minor exceptions are samples VA04/12 and VA04/16 (Figs 4 & 5). The age–altitude plot for this traverse shows some disturbance (Fig. 5).

Karoo Traverse. This traverse contains two samples from the Palaeozoic Table Mountain Group, three from the Neoproterozoic Vanrhynsdorp Group, four from the Karoo Group sediments (including the Dwyka and Ecca formations) and five from Early Jurassic Karoo dolerites (Table 1). Apatite ages range from 180±18 to 86±5 Ma. The youngest samples, characterized by longer mean-track lengths, are found on the coastal plain (Figs 4 & 5). Three exceptions are samples CA04/08, CA04/16 and CA04/18. The correlation plots between age and distance from the coast and altitude appear disturbed (Fig. 5). The topographically highest sample (a dolerite dyke, CA04/01) yields an AFT age of 180±18 Ma that corresponds closely to its presumed emplacement age (*c.* 183 Ma, $^{40}Ar/^{39}Ar$ age: Duncan *et al.* 1997) (Fig. 4).

Apatite fission-track thermal modelling

Fission tracks in apatites are formed continuously through time at an approximately uniform initial length. Upon heating, tracks are then gradually annealed or shortened to a length that is determined by the maximum temperature to which apatites are exposed and the time. For example, tracks are completely annealed at a temperature of 110–120 °C for a period of 10^5–10^6 years. These annealing characteristics allow the construction of time–temperature paths by inverse modelling.

We have modelled our AFT data to quantify the timing and amount of cooling at specific locations along the studied traverses. Modelling of the apatite age and track-length distribution data was carried out with the program Monte Trax by Gallagher (Gallagher 1995), using an initial track length of 16.3 μm. Age and track-length distribution parameters, as well as user-defined time (t)–temperature (T) boxes, are used as input data, allowing just one change of direction of the t–T path within each individual t–T box. To select t–T points and define a cooling path, the program uses a Genetic Algorithm probabilistic approach

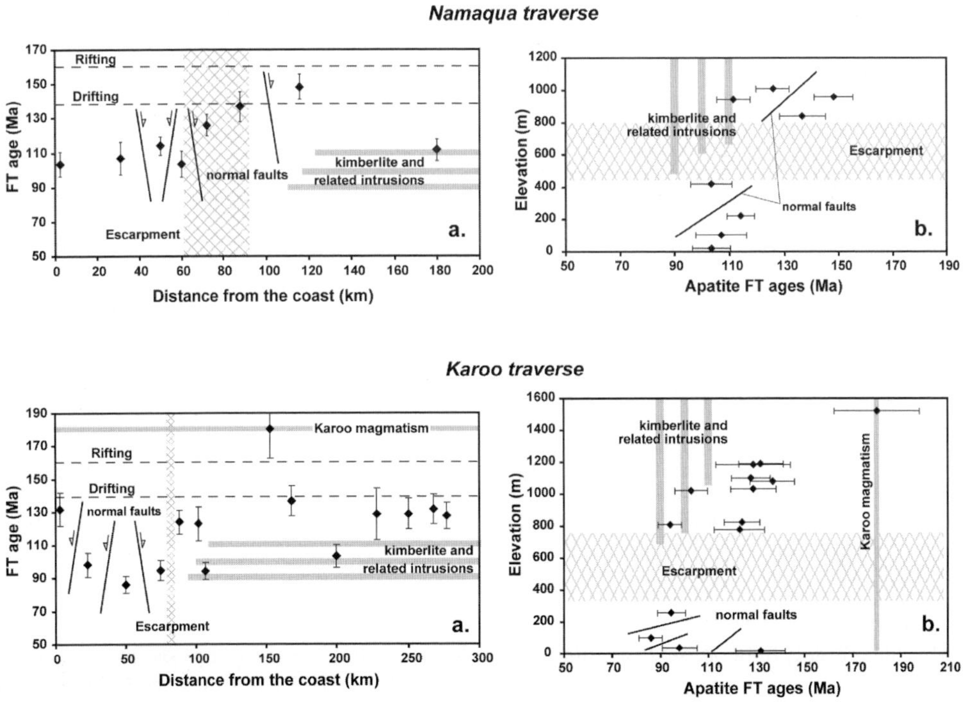

Fig. 5. Relationship between apatite fission-track ages from the Namaqualand and Karoo traverses v. (**a**) distance from the coast and (**b**) elevation.

(Gallagher & Sambridge 1994), which optimizes the stochastic production of successive generations of thermal history models. The predicted fission-track parameters are then quantitatively compared to the observed (measured) values, and the level of consistency between the two sets is used to choose the thermal history that is most consistent with the observed data. It is important to remember that the best thermal history obtained during this process is not necessarily the only possible. Other thermal histories may match the observed data similarly well and it is therefore imperative to consider as many other geological constraints as possible to determine the most likely path.

It has been shown that the annealing properties of apatites are controlled principally by their chlorine/fluorine ratio (e.g. Green *et al.* 1986; Ketcham *et al.* 1999; Barbarand *et al.* 2003). Given that a positive correlation has been demonstrated between Cl wt% and apatite D_{par} (Donelick 1993), we have systematically measured apatite track etch-pit diameters (Table 1). Most of the samples modelled in our study have D_{par} values between 2.0 and 2.5 μm (Table 1). This is close to the value that we have obtained for the Durango apatite standard (2.36 ± 0.02 μm; $n = 100$) etched under the same conditions used for the rest of our samples. The composition of Durango apatite was therefore used with the Laslett algorithm in our models (Laslett *et al.* 1987). Samples VA04/05, VA04/06 and VA04/08, which have D_{par} values lower than those measured for the Durango apatite (between 1.6 and 1.7 μm, Table 1), may have been annealed at temperatures lower than those indicated by the models. The net amount of denudation would in this case be overestimated by modelling, but the timing of individual cooling events would instead remain unaffected. These samples were therefore also used for the modelling.

In order to discard the possibility that variations in AFT ages may be due to variable apatite compositions and not to different thermal histories, we have plotted the measured D_{par} against the corresponding AFT ages. The lack of any positive correlation along the traverses suggests that the compositional influence is negligible (Fig. 6). Moreover, no particular positive correlation between the age and D_{par} is found when individual grain measurements are plotted. In eight samples (out of 22 samples analysed) a positive correlation is shown but with R^2 values (coefficient of determination) less than 0.1. Only four samples show R^2 values between 0.1 and 0.3.

Modelling results from our study are illustrated in Figure 7.

Namaqua Traverse. Modelling of the AFT data along this traverse has to take into account the

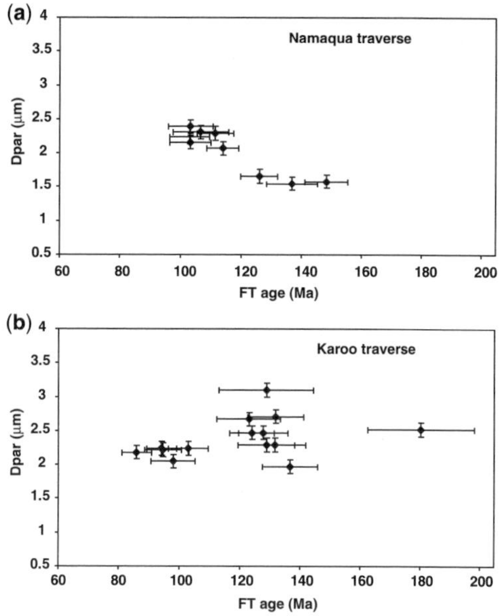

Fig. 6. Apatite fission-track ages plotted against D_{par} values for individual samples from (**a**) the Namaqua and (**b**) the Karoo traverses. No correlation is present.

uncertainty concerning the pre-rift thermal history of the samples and especially of those above the escarpment. The age (114–103 Ma) and track-length distribution (mean track lengths between 13.9 and 14.54 μm) of the samples collected along the coastal plain (Fig. 4) suggest that they cooled from temperatures higher than 110 °C at the time indicated be their AFT age, thus only after the opening of the South Atlantic.

The last known significant thermal event affecting this area was linked to the late Neoproterozoic Pan-African orogeny. The time of the post-orogenic cooling is not known. Most probably the Namaqualand area was covered by the Permo-Triassic Karoo deposits, but their areal extent and their local maximum thickness are not known. All these uncertainties in the pre-rift thermal history of the samples above the escarpment makes the modelling a serious challenge. We therefore allowed a large degree of freedom for the pre-rifting (pre-150 Ma) thermal conditions in our models and applied tighter time–temperature constraints only for the last 150 Ma time window. Input time–temperature boxes were defined for three main time intervals: 150–120 Ma, 120–80 Ma and 80–0 Ma (Fig. 7). These periods overlap with phases of accelerated cooling in Southern Africa reported by previous studies (Brown *et al.* 2002; Raab *et al.* 2002; Tinker *et al.* 2008).

The results for samples both above and below the escarpment show one distinct period of fast

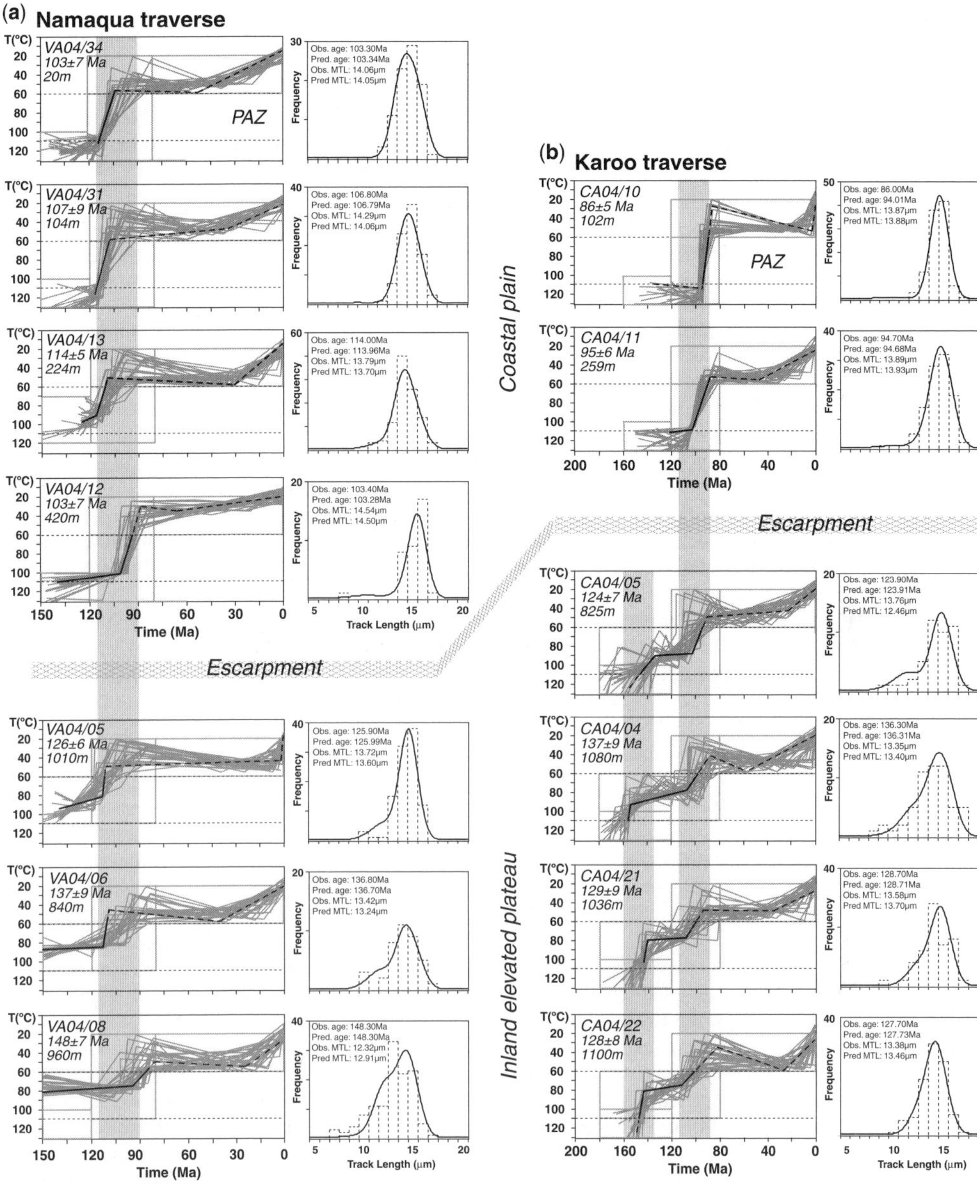

Fig. 7. Modelled thermal histories for samples, and comparison between the predicted fission-track parameters and the observed data from (**a**) the Namaqualand and (**b**) the Karoo traverses with their AFT age and altitude. The shaded vertical bands represent the cooling events at 160–138 Ma and 115–90 Ma. Horizontal dashed lines within individual models at 60–110 °C bracket the partial annealing zone (PAZ) for apatite within the temperature limits assigned by Laslett *et al.* (1987). The thick black lines represent the best-fit paths, and the grey lines the best 50 modelled paths. The dashed segments of the thermal histories at temperatures lower than 60 °C indicate only a possible continuation of the thermal history because the annealing model is not sufficiently sensitive below 60 °C.

cooling between 115 and 90 Ma. Before 115 Ma the samples above the escarpment underwent a very slow cooling between 90 and 80 °C, whereas the samples from the coastal plain were still at temperatures higher than 110 °C.

Karoo Traverse. AFT ages from this traverse are interpreted as being fully reset during the Karoo igneous event (*c.* 180 Ma: Duncan *et al.* 1997). Based on zircon fission-track analysis, Brown *et al.* (1994) suggested maximum palaeotemperatures of at least 250 ± 50 °C at about 190 ± 10 Ma for the Karoo sediments. Time–temperature constrains identical to those used in the Namaqua traverse were adopted as input to the model of the samples above the escarpment. This allows the program to search for possible changes of direction in the time–temperature path within three main intervals: 150–120 Ma, 120–80 Ma and 80–0 Ma (Fig. 7). Considering their age and track-length distribution (Fig. 4), the samples from the coastal plain were fully reset before 150 Ma.

Modelling from this traverse reveals two distinct cooling events (Fig. 7). The first one is between 160 and 138 Ma, and is recorded only by the samples above the escarpment. This first cooling event is followed by a period of quiescence during which the samples above the escarpment were at temperatures of between 80 and 90 °C for at least 30 Ma (Fig. 7).

Similar to the results from the Namaqua traverse, a period of rapid cooling between 115 and 90 Ma is indicated by these models (Fig. 7). This event is recorded by samples both above and below the escarpment. The fact that the samples from the coastal plain record just this event also suggests that they were at temperatures higher than 110 °C before 115 Ma.

Discussion

Denudation history

As shown above, our AFT analysis reveals two discrete phases of cooling, separated by a period of relative thermal stability.

160–138 Ma. This cooling event is constrained only by the thermal models of the samples situated inland from the present escarpment in the Karoo Traverse (Fig. 7). It is still not clear whether this event represents post-Karoo magmatism (*c.* 180 Ma) thermal relaxation or is, instead, the expression of tectonic processes related to rifting. The fact that the thermal models from the Namaqua Traverse above the escarpment show no evidence of significant cooling during this same time interval suggests that the thermal history of the samples from the Karoo Basin was, indeed, affected by the widespread Karoo magmatism (Fig. 7). We favour, therefore, a scenario whereby this cooling episode is linked to post-Karoo thermal relaxation. Most probably, all samples situated landward from the present escarpment were part of a stable continental interior during this period and, therefore, were not significantly affected by the tectonic and/or thermal processes related to the rifting itself.

It is not possible to reconstruct the thermal history of the samples along the present coastal plain during this period by AFT analysis because they were at temperatures higher than the AFT closure temperature at this time (Fig. 7).

138–115 Ma. Modelling of samples from the continental interior indicates this to be a period of quiescence, during which these rocks experienced cooling of less than 15 °C during *c.* 30 Ma (Fig. 7). This time coincides with the transition of the south Atlantic margin from an active, related to the rifting, to a passive mode (e.g. Nürnberg & Müller 1991).

The distinctly younger ages of samples situated above the escarpment (between 94 ± 6 and 112 ± 6 Ma: Fig. 4) are difficult to explain (e.g. CA04/16 and CA04/18 from the Karoo Traverse, and VA04/16 from the Namaqua Traverse). Although most of the previously published AFT ages from the elevated continental interior are older than 150 Ma, some scattered younger ages are also reported (Brown *et al.* 1990, 2000; Gallagher & Brown 1999). The lack of any particular patterns in their distribution and evidence for active post-rifting tectonics exclude the possibility that these data are related to major post-rifting denudation events. Nevertheless, Gallagher & Brown (1999) suggested that significant post-break-up denudation in the present-day continental interior could be possibly the result of 'structural reactivation' along unidentified tectonic lineaments.

It is also possible that these young ages are the result of the chemistry of the dated minerals, which allowed for much lower closure temperatures (hence the young ages), or that they are due to heating from local kimberlitic intrusions (Fig. 5). Two distinct peaks of several hundreds of intrusions have been reported from southern Africa at 145–115 and 95–80 Ma (e.g. Basson & Viola 2004; Trumbull *et al.* 2007).

The modelling of two samples from the coastal plain (VA04/13 and VA04/12) and one from the escarpment area (VA04/05) (Fig. 7) along the Namaqualand Traverse also suggests relatively slow cooling during this period. AFT analysis does not allow the presence of possible differences in the amount of denudation to be established between the coast and the continental interior.

115–90 Ma. All modelled samples show a distinct second episode of accelerated cooling between 115 and 90 Ma. In particular, most of the samples from the coastal plane cooled from temperatures above 110 °C (Fig. 7). If the geothermal gradient is known, it is possible to estimate the amount of denudation for each sample. The present-day geothermal gradient in the Karoo Basin is about 20 °C km^{-1} (Gough 1963). Brown *et al.* (2002) and Tinker *et al.* (2008) derived palaeo-geothermal gradients of 20–25 °C km^{-1} for the Middle Cretaceous in Southern Africa from AFT analysis on borehole samples. Given the difficulties in constraining palaeo-geothermal gradients, however, we limit ourselves here to a rough estimation of the patterns of denudation along the two traverses for the period 115–90 Ma for three possible different palaeo-geothermal gradients: 20, 25 and 30 °C km^{-1} (Fig. 8). Considering the lithological difference of the samples analysed along the two traverses, one could assume that high geothermal gradients are more probable for the Namaqua Traverse, underlined entirely by gneissic rock. Nevertheless, such a possible difference would not affect the differential denudation between the coastal plain and the continental interior or across some of the tectonic structures observed along the two traverses.

Figure 8 suggests that there is a substantial difference in the amount of denudation between the coastal plain and the elevated plateau in the period 115–90 Ma. The amount of denudation ranges from a maximum of 1.5–2.7 km along the coast to less than 1 km above the escarpment for each of the selected geothermal gradients. Moreover, along the Namaqualand traverse denudation changes significantly and abruptly across individual faults within the faulted area. This suggests a prolonged tectonic activity during or subsequent to this period that affected both the coastal plain and the present day escarpment area (Figs 2 & 3) and is confirmed by the disturbed age–altitude plot for the same traverse (Fig. 5).

Sample VA04/12 was collected from within the footwall of a large normal fault (Fig. 2), mapped on the 1 000 000 geological map of South Africa (Keyser 1998) and was easily visible on satellite images (Fig. 3). Thermal modelling for this sample predicts fast cooling between 100 and 90 Ma, thus slightly later than for the other samples. We suggest that this may reflect the time of faulting (Fig. 7). Active tectonics along the coastal area could be responsible for an elevated geothermal gradient. However, even for an elevated geothermal gradient (30 °C km^{-1}), the estimated denudations from the coast and the escarpment area are significantly higher than the maximum calculated denudation for the elevated plateau (Fig. 8).

It is not clear whether the differential denudation observed between the coastal plain and the plateau along the Karoo Traverse is in part also due to faulting, as shown for the case of the Namaqua Traverse. There are no observed or mapped faults in the escarpment area of the southern traverse, although several faults are mapped along the coastal plain and the age–altitude plot shows some disturbance (Fig. 5). Top-to-the-west normal faulting could explain the relatively old age of sample CA04/08 (132 ± 10 Ma) collected at the coast (Figs 1 & 4). Differential uplift related to flexural isostatic rebound along the coast due to the enhanced denudation should also be considered. Irrespective of whether accompanied by tectonic movements or not, the large difference in the amount of denudation between the coastal plain and the continental interior must be a direct consequence of differential erosion. High denudation along the coast was probably the result of significant erosion by high-energy river systems that flowed down to the low Atlantic base level from a drainage divide, which, already at this time, coincided with the present-day position of the escarpment (as previously suggested along the south African margin by Cockburn *et al.* 2000; Brown *et al.* 2002; Tinker *et al.* 2008). It must be pointed out that the erosion along the coastal plain was probably facilitated by the presence of the Gariep and Vanrhynsdorp (Nama) Group phyllites and siltstones. The continental interior was instead eroded slowly possibly by low-energy river systems, a difference that explains the substantial change in the amount of denudation across the escarpment (Fig. 8).

Other evidence for the regional character of the substantial increase in denudation and uplift during the Mid-Cretaceous is as follows. (1) the Early Cretaceous Rietport granite (133.9 ± 1.3 Ma: De Beer *et al.* 2002), intruded in the coastal area, was already exhumed close to the surface by the end of the Palaeocene, as suggested by the presence of high-level olivine melilitite plugs intruded into the granite at around 56 Ma (De Beer *et al.* 2002). (2) Increased Mid-Cretaceous denudation has already been reported by other low-temperature thermochronological studies on the western South African margin (e.g. Brown *et al.* 1990, 2000). (3) A significant amount of uplift was also observed offshore Namaqualand, where seismic profiles and boreholes show significant erosional horizons marking the Aptian regression (121–112 Ma; e.g. Gerrard & Smith 1983; Brown *et al.* 1995) (Fig. 9). From the Aptian until the end of the Turonian (*c.* 90 Ma), increasing amounts of offshore clastic deposits and other minor unconformities are reported (Gerrard & Smith 1983; Paton *et al.* 2007) (Fig. 9). (4) Hirsch (2008) recently reported a second rifting phase from 117 to 95 Ma deduced from subsidence analysis in the Orange Basin. This rifting phase was related to subsidence in the outer shelf contemporaneous with uplift in the inner shelf. This observation is consistent with the tectonically induced uplift of the continental margin during the same period suggested here. (5) In western Central Namibia, the regional-scale Waterberg thrust, which brought the Neoproterozoic Damara basement over Jurassic sandstones, may also indicate an important, Mid-Cretaceous

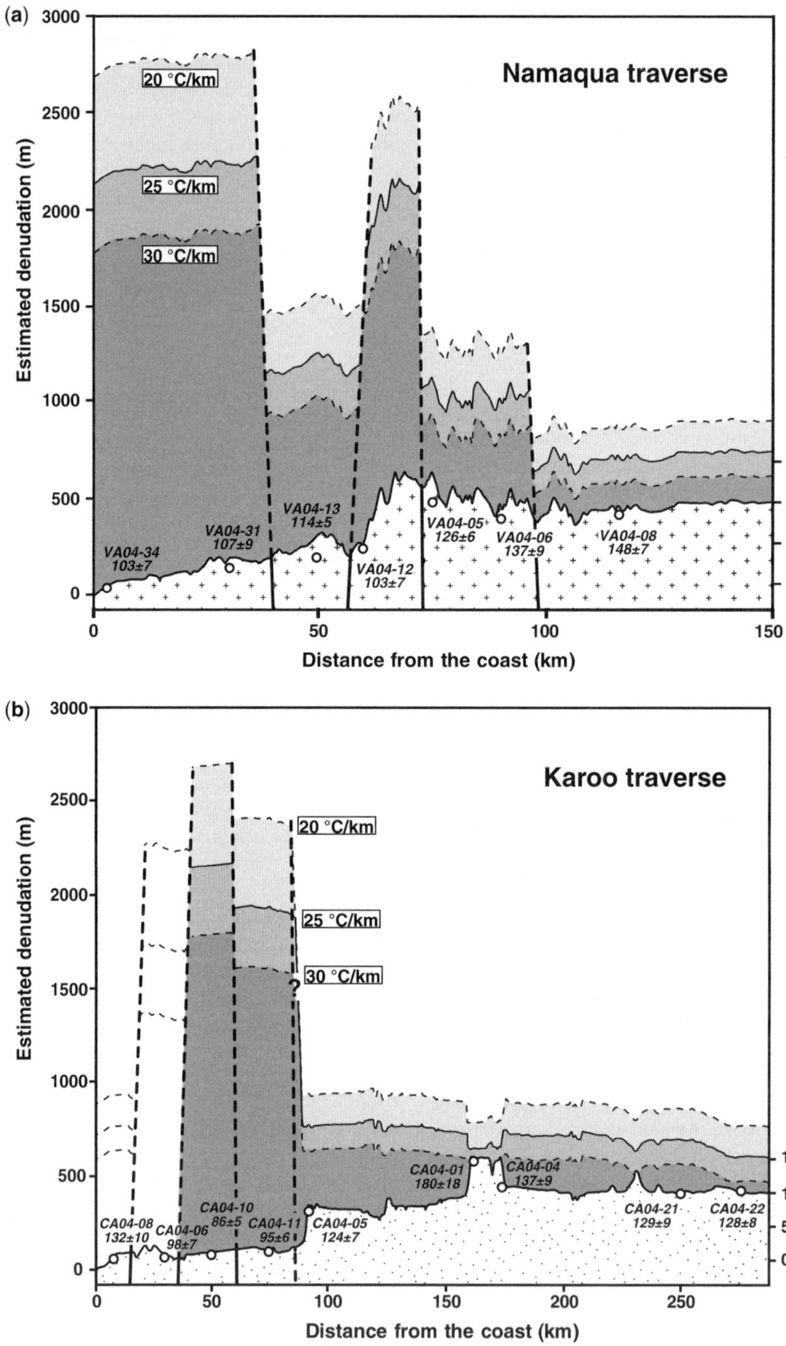

Fig. 8. Estimated denudation for the period between 115 and 90 Ma calculated from AFT data modelling and based on three possible palaeo-geothermal gradients of 20, 25 and 30 °C km^{-1} for: (**a**) the Namaqualand Traverse and (**b**) the Karoo Traverse. For samples CA04/06 and CA04/08 from the Karoo Traverse the presented dashed lines for the estimated denudation are only suggested. The samples were not modelled because of the insufficiency of the track length records.

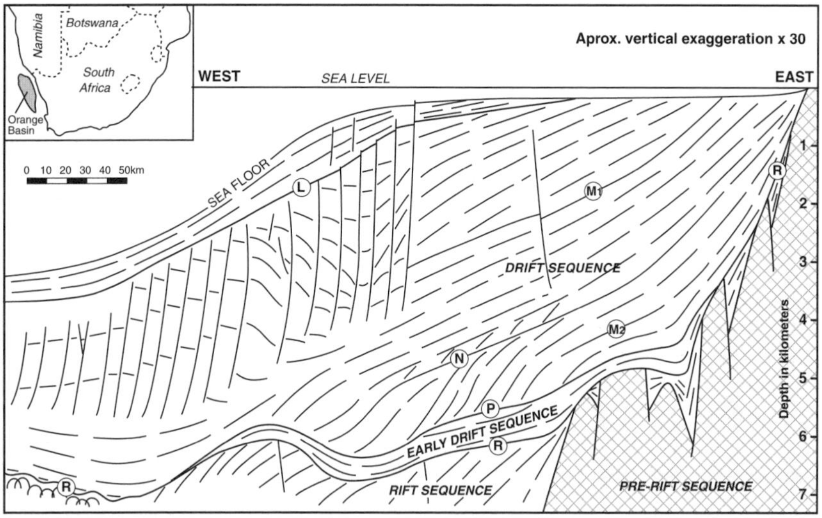

Fig. 9. Diagrammatic section across Orange Basin based on well and seismic data (from Gerrard & Smith 1983). Main unconformities: R, drift onset unconformity (Valanginian); P, Aptian unconformity; L, base Tertiary unconformity. Secondary unconformities: M_2, Albian; N, Cenomanian; M_1, Coniacian.

episode of crustal shortening and uplift (cf. Raab *et al.* 2002; Viola *et al.* 2005 and references therein).

Collectively, these data support a model of significant Mid-Cretaceous uplift and denudation along the western South African margin. The demonstrated significant changes in amount of denudation between the continental interior and its margin are here tentatively related to the fact that already at that time the drainage divide was located close to its present position, thus separating high-energy river systems rapidly eroding the coastal area from low-energy systems developed in the elevated plateau. These observations allow us to suggest that it is probably during this time that most of South African present-day topography was established (see also Doucouré & de Wit 2003; de Wit 2007).

Post-90 Ma. The post Mid-Cretaceous denudation of the study area cannot be constrained by AFT analysis because, by then, the sampled rocks were already exhumed to crustal levels corresponding to temperatures lower than 60 °C, thus beyond the resolution of the method. Nevertheless, some authors have reported the existence of a Late Cretaceous phase of accelerated denudation based on fission-track analysis from other areas, such as the Drakensberg Mountains (Brown *et al.* 2002) and northern Namibia (Raab *et al.* 2002, 2005).

Much lower rates for the present-day denudation in southern Africa are estimated from cosmogenic nuclides analysis (e.g. Fleming *et al.* 1999; Cockburn *et al.* 2000; Brown *et al.* 2002; Kounov *et al.* 2007). Whilst some extrapolate low denudation rates to the entire Cenozoic on the basis of the prevailing aridity of the climate and the lack of substantial uplift throughout that period (e.g. Cockburn *et al.* 2000), other authors consider the Cenozoic as the main period of uplift, topographic development and escarpment formation in Southern Africa (e.g. Partridge & Maud 1987; Partridge 1998; Burke 1996).

Whilst there is still no consensus on this, our study allows us to conclude with confidence that:

- Cenozoic denudation in the study area was less than 2–3 km because it did not bring to the surface rocks from deeper crustal levels (e.g. there are no samples with fission-track ages younger than Late Cretaceous);
- present-day erosional rates are at least an order of magnitude lower than during the Cretaceous (e.g. Kounov *et al.* 2007).

In addition, offshore seismic stratigraphy has revealed deposition of up to 2 km of sediments along the western African margin since the Mid-Cretaceous, suggesting substantial subsidence during this time as well as considerable onshore denudation. On the other hand, much thinner Tertiary succession occurs above a significant unconformity (Gerrard & Smith 1983; Brown *et al.* 1995; Aizawa *et al.* 2000; Paton *et al.* 2007) Tinker *et al.* 2008) (Fig. 9). This unconformity is described as a well-developed marine planation surface witnessing the uplift of the offshore margin (Aizawa *et al.* 2000). Nevertheless, it must be mentioned that the mean annual sediment input along the SW coast of Africa (as measured by sediment accumulation rates in the main Orange Basin depocentre) decreased progressively from the late Cretaceous to the Neogene (Dingle & Hendey 1984; Rust & Summerfield 1990). This decrease is confirmed by the observation that Late Cretaceous lake sediments can still be found as crater deposits infilling the craters of Cretaceous olivine melilites and kimberlites in the Namaqualand area above the escarpment (de Wit 1999).

We suggest that Cenozoic pulses of uplift and denudation have existed and played a role in the development of the main geomorphological features of the passive margin, although less important than those during the Cretaceous,

Evidence of Cenozoic, and even Quaternary, tectonics, including fault reactivation and sedimentation, have indeed been reported both offshore and onshore Namaqualand and Southern Namibia by a number of authors (e.g. Andreoli *et al.* 1996; Brandt *et al.* 2005; Viola *et al.* 2005).

Driving mechanism for the Mid-Cretaceous denudation

The AFT results presented in this study reveal the existence of a distinct period of active tectonic uplift accompanied by denudation between 115 and 90 Ma on the western coast of South Africa. Tectonically driven pulses of accelerated denudation have already been reported from the Southern Africa passive margin (Brown 1992; Brown *et al.* 2002; Raab *et al.* 2002). Although their existence is now well documented by low-temperature thermochronology analysis and by offshore seismic and borehole data, their driving forces remain still largely unknown (Gerrard & Smith 1983; Brown *et al.* 1995, 2002; Aizawa *et al.* 2000).

Gilchrist & Summerfield (1990) were the first to suggest that flexural isostatic rebound, resulting from the high denudation rate on the evolving coast flank of the rifted margin, may have played a significant role in the margin upwarp. They demonstrated that this process could generate 600 m of uplift on the Southern African margin, and estimated the total denudation along the coast to be between 2 and 3 km. Although this mechanism may have been important during margin development, it cannot explain the 5 km denudation across the Southern African margin since continental break-up that has been reported in previous AFT studies, nor the tectonically induced pulses of denudation (Brown *et al.* 1990; Raab *et al.* 2002, 2005).

Some authors claim that climate changes could induce uplift (e.g. Molnar & England 1990; Wobus *et al.* 2003), whereas others support exactly the opposite view, suggesting that tectonic processes could result in climate changes (e.g. Raymo & Ruddiman 1992; Burbank *et al.* 2003). During the Cretaceous Southern Africa was characterized by hot climatic conditions, whereas in the Middle Miocene the cold Benguela current led to a more arid climate (e.g. Uenzelmann-Neben *et al.* 2007 and references therein). Several studies suggest that this cold and arid climate prevailed for a significant part of the Cenozoic (e.g. Cockburn *et al.* 2000). It is, however, unlikely that the dramatic decrease in denudation rates (an order of magnitude) observed from the Cretaceous–Cenozoic boundary (e.g. Cockburn *et al.* 2000; Brown *et al.* 2002; Tinker *et al.* 2008) was controlled only by climate changes. More recent analogues are also against such a theory. For example, present-day denudation rates in the Drakensberg escarpment area (Fleming *et al.* 1999), which is characterized by humid and warm environment, are similar to those reported for the west coast of southern Africa, where climatic conditions are, instead, significantly dry (Cockburn *et al.* 2000; Van der Wateren & Dunai 2001). Also, Cenozoic denudation rates from the western coast of Southern Africa are similar to those reported for the same period in Madagascar and Sri Lanka, where warm and humid climate conditions have prevailed since the Cretaceous (Hewawasam *et al.* 2003; Seward *et al.* 2004; Vanacker *et al.* 2007).

In summary, all of these observations suggest that an important acceleration in denudation rate could have been caused only by a significant regional uplift, with climate-driven erosion acting only as a second-order factor.

Some authors relate post-break-up tectonic processes in Southern Africa to dynamic processes in the mantle (Doucouré & de Wit 2003; Burke 1996; de Wit 2007; Tinker *et al.* 2008). Burke (1996) suggested that the highlands of Southern Africa are mainly Cenozoic in age (possibly as young as 30 Ma), whereas other authors present evidence that much of the present topography was formed during the Cretaceous (Doucouré & de Wit 2003; de Wit 2007; Kounov *et al.* 2008). They all agree, however, that substantial uplift is probably associated with the tomographically imaged low-velocity zone in the lower mantle–core boundary, called the African Superswell (e.g. Lithgow-Bertelloni & Silver 1998). According to this dynamic topography model, present-day topography would be a dynamic feature formed in response to vertical stresses at the base of the Southern African lithosphere that generated positive buoyancy in the mid–lower mantle.

We believe that this scenario accounts efficiently for the results and observations of our paper and also of other recent AFT studies (Brown *et al.* 2002; Raab *et al.* 2002; Tinker *et al.* 2008).

Tinker *et al.* (2008) correlated periods of increased denudation with peaks of kimberlite emplacement, and the formation of the Parana and Agulhas igneous provinces, thus suggesting a causative link between lower-mantle upwelling processes and increased denudation. Certainly, peaks of kimberlitic activity were related to the emplacement of hot magma at the base of the lithosphere, which triggered diffuse crustal uplift. This large-scale uplift was accompanied by tectonic activity probably related to reactivation of fault structures along the continental margin.

Conclusions

AFT results across the western coast of South Africa and its interior reported in this paper are consistent with the existence of a discrete, tectonically induced, Mid-Cretaceous pulse of substantial denudation. Thermal modelling of the new fission-track data indicates up to 2.5 km of denudation in the coastal zone and less than 1 km on the elevated interior plateau during this phase of accelerated denudation. Greater tectonic activity occurred along the passive margin and its interior during this period than in the Cenozoic. This suggests that the Mid-Cretaceous was probably the time when most of the present-day Southern African high-elevation topography was formed.

The spatial patterns of denudation reported here suggest localized, post-rifting and fault-controlled uplift along the passive margin.

It is tentatively suggested that the substantial Mid-Cretaceous pulse of denudation is a direct consequence of significant uplift associated with the African Superswell, a tomographically imaged low-velocity zone at the lower-mantle–core boundary.

This work was supported by the Claude Leon Foundation post-doctoral fellowship (to Alexandre Kounov) and by Inkaba yeAfrica funds. Fission-track dating was carried out at the University of Cape Town (UCT). We would like to thank M. Tredoux for providing a Macintosh computer for the fission-track laboratory at UCT. Constructive reviews by A. Henk and an anonymous reviewer, and the editorial work of F. Lisker helped to improve the quality of the paper. This is AEON contribution 68, and Inkaba yeAfrica contribution number XX.

References

AIZAWA, M., BLUCK, B., CARTWRIGHT, J., MILNER, S., SWART, R. & WARD, J. 2000. Constrains on the geomorphological evolution of Namibia from the offshore stratigraphic record. *Communs of the Geological Survey of Namibia*, **12**, 337–346.

ANDREOLI, M. A. G., DOUCOURÉ, M., VAN BEVER DONKER, J., BRANDT, D & ANDERSEN, N. J. B. 1996. Neotectonics of Southern Africa – a review. *Africa Geoscience Review*, **3**, 1–16.

BARBARAND, J., CARTER, A., WOOD, I. & HURFORD, T. 2003. Compositional and structural control of fission-track annealing in apatite. *Chemical Geology*, **198**, 107–137.

BASSON, I. & VIOLA, G. 2004. Passive kimberlite intrusion into actively dilating dyke-fracture arrays: evidence from fibrous calcite veins and extensional fracture cleavage. *Lithos*, **76**, 283–297.

BRANDT, D., ANDREOLI, M. & MCCARTHY, T. S. 2005. The Late Mesozoic paleosoils and Cenozoic fluvial deposits at Vaalputs, Namaqualand, South Africa. Possible depositional mechanism and their bearing on the evolution of the continental margin. *South African Journal of Geology*, **108**, 267–280.

BROWN, L. F., BENSON, J. M. *ET AL.* 1995. *Sequence stratigraphy in Offshore South African divergent basins, an atlas on exploration for Cretaceous Lowstand Traps by SOEKOR (Pty) Ltd.* AAPG Studies in Geology, **41**.

BROWN, R. W. 1992. *A Fission Track Thermochronology Study of the Tectonic and Geomorphic Development of the Sub-aerial Continental Margins of Southern Africa.* PhD thesis, La Trobe University, Melbourne, Australia.

BROWN, R. W., GALLAGHER, K. & DUANE, M. 1994. A quantitative assessment of the effects of magmatism on the thermal history of the Karoo sedimentary sequence. *Journal of African Earth Sciences*, **18**, 227–243.

BROWN, R. W., GALLAGHER, K., GLEADOW, A. J. W. & SUMMERFIELD, M. A. 2000. Morphotectonic evolution of the South Atlantic margins of Africa and South America. *In*: SUMMERFIELD, M. A. (ed.) *Geomorphology and Global Tectonics.* Wiley, Chichester, 255–284.

BROWN, R. W., RUST, D. J., SUMMERFIELD, M. A., GLEADOW, A. J. W. & DE WIT, M. C. J. 1990. An Early Cretaceous phase of accelerated erosion on the south-western margin of Africa: Evidence from apatite fission track analysis and the offshore sedimentary record. *Nuclear Tracks and Radiation Measurement*, **17**, 339–350.

BROWN, R. W., SUMMERFIELD, M. A. & GLEADOW, A. J. W. 2002. Denudation history along a transect across the Drakensberg Escarpment of southern Africa derived from apatite fission track thermochronology. *Journal of Geophysical Research*, **107**, (B12), 2350, doi:10.1029/2001JB000745.

BURBANK, D. W., BLYTHE, A. E. *ET AL.* 2003. Decoupling of erosion and precipitation in the Himalayas. *Nature*, **426**, 652–655.

BURKE, K. 1996. The African Plate. *South African Journal of Geology*, **99**, 341–409.

COCKBURN, H. A. P., BROWN, R. W., SUMMERFIELD, M. A. & SEIDL, M. A. 2000. Quantifying passive margin denudation and landscape development using a combined fission-track thermochronometry and cosmogenic isotope analysis approach. *Earth and Planetary Science Letters*, **179**, 429–435.

CORRIGAN, J. D. 1993. Apatite fission-track analysis of Oligocene strata in South Texas, USA; testing annealing models. *Chemical Geology*, **104**, 227–249.

COX, K. G. 1992. Karoo igneous activity and the early stages of the break-up of Gondwanaland. *In*: STOREY, B. C., ALABASTER, T. & PANKHURST, R. J. (eds) *Magmatism and the Causes of Continental Break-up*. Geological Society, London, Special Publications, **68**, 137–148.

DE BEER, C. H., GRESSE, P. G., THERON, J. N. & ALMOND, J. E. 2002. *The Geology of the Calvinia Area. Explanation of 1:250 000-scale Sheet 3118 Calvinia*. Council for Geoscience, South Africa.

DE WIT, M. C. J. 1999. Post-Gondwana drainage and the development of diamond placers in Western South Africa. *Economic Geology*, **94**, 721–740.

DE WIT, M. J. 2007. The Kalahari Epeirogeny and climate change: differentiating cause and effect from core to space. *South African Journal of Geology*, **110**, 367–392.

DE WIT, M. J. & RANSOME, I. G. D. (eds) 1992. Regional inversion tectonics along the southern margin of Gondwana. *Inversion Tectonics of the Cape Fold Belt, Karoo and Cretaceous Basins of Southern Africa*. Balkema, Rotterdam, 15–21.

DINGLE, R. V. & HENDEY, B. Q. 1984. Late Mesozoic and Tertiary sediment supply to the eastern Cape basin (SE Atlantic) and paleo-drainage system in southwestern Africa. *Marine Geology*, **56**, 13–26.

DONELICK, R. A. 1993. Apatite etching characteristics versus chemical composition. *Nuclear Tracks and Radiation Measurements*, **21**, 604.

DOUCOURÉ, C. A. & DE WIT, M. J. 2003. Old inherited origin for the present near bimodal topography of Africa. *Journal of African Earth Sciences*, **36**, 371–388.

DUNCAN, R. A., HOOPER, P. R., REHACEK, J., MARSH, J. S. & DUNCAN, A. R. 1997. The timing and duration of the Karoo igneous event, southern Gondwana. *Journal of Geophysical Research*, **102**, 18 127–18 138.

DU TOIT, A. 1926. *Geology of South Africa*. Oliver & Boyd, Edinburgh.

EAGLES, G. 2007. New angles on South Atlantic opening. *Geophysical Journal International*, **166**, 353–361.

EALES, H. V., MARCH, J. S. & COX, K. G. 1984. The Karoo igneous province: an introduction. *In*: ERLANK, A. J. (ed.) *Petrogenesis of the Volcanic Rocks of the Karoo Province*. Geological Society of South Africa, Special Publications, **13**, 1–26.

FLEMING, A., SUMMERFIELD, M. A., STONE, J. O. H., FIFIELD, L. K. & CRESSWELL, R. G. 1999. Denudation rates for the southern Drakensberg escarpment, SE Africa, derived from in-situ-produced cosmogenic ^{36}Cl: initial results. *Journal of the Geological Society, London*, **156**, 209–212.

GALBRAITH, R. F. 1981. On statistical-models for fission track counts. *Journal of the International Association for Mathematical Geology*, **13**, 471–478.

GALBRAITH, R. F. & LASLETT, G. M. 1993. Statistical models for mixed fission-track ages. *Nuclear Tracks and Radiation Measurements*, **21**, 459–470.

GALLAGHER, K. 1995. Evolving temperature histories from apatite fission-track data. *Earth and Planetary Sciences Letters*, **136**, 421–435.

GALLAGHER, K. & SAMBRIDGE, M. 1994. Genetic algorithms: A powerful tool for large-scale nonlinear optimization problems. *Computational Geosciences*, **20**, 1229–1236.

GALLAGHER, K. & BROWN, R. 1999. The Mesozoic denudation history of the Atlantic margins of southern Africa and southeast Brazil and the relationship to offshore sedimentation. *In*: CAMERON, N. R., BATE, R. H. & CLURE, V. S. (eds) *The Oil and Gas Habitats of the South Atlantic*. Geological Society, London, Special Publications, **153**, 41–53.

GALLAGHER, K., BROWN, R. & JOHNSON, C. 1998. Fission track analysis and its applications to geological problems. *Annual Review of Earth and Planetary Science*, **26**, 519–572.

GERRARD, I. & SMITH, G. C. 1983. Post-Paleozoic succession and structure of the southwestern African continental margin. *In*: WATKINS, J. S. & DRAKE, C. L. (eds) Studies in Continental Margin Geology. *AAPG, Memoir*, **34**, 49–74.

GILCHRIST, A. R. & SUMMERFIELD, M. A. 1990. Differential denudation and flexural isostasy in formation of rifted-margin upwarps. *Nature*, **346**, 739–742.

GILCHRIST, A. R., HENK, K. & BEAUMONT, C. 1994. Post-Gondwana geomorphic evolution of southwestern Africa: Implications for the controls on landscape development from observations and numerical experiments. *Journal of Geophysical Research*, **99**, 12 211–12 228.

GOUGH, D. I. 1963. Heat flow in the Southern Karoo. *Proceedings of the Royal Society of London, Series A*, **272**, 207–230.

GREEN, P. F. & DUDDY, I. R. 1989. Some comments on paleotemperature estimation from apatite fission track analysis. *Journal of Petroleum Geology*, **12**, 111–114.

GREEN, P. F., DUDDY, I. R., GLEADOW, A. J. W., TINGATE, P. R. & LASLETT, G. M. 1986. Thermal annealing of fission tracks in apatite, 1. A qualitative description. *Chemical Geology*, **59**, 237–253.

GRESSE, P. G. 1995. The Late Pan-African Vanrhynsdorp foreland thrust-fold belt in southern Namaqualand, South Africa. *Journal of African Earth Sciences*, **21**, 91–105.

HAWKESWORTH, C., KELLEY, S., TURNER, S., LE ROEX, A. & STOREY, B. 1999. Mantle processes during Gondwana break-up and dispersal. *Journal of African Earth Sciences*, **28**, 239–261.

HEWAWASAM, T., VON BLANCKENBURG, F., SCHALLER, M. & KUBIK, P. 2003. Increase of human over natural erosion rates in tropical highlands constrained by cosmogenic nuclides. *Geology*, **31**, 597–600.

HIRSCH, K. 2008. Integrating structural and sedimentological observations with numerical lithospheric models to assess the 3D evolution of the South African continental passive margin. PhD Thesis, Free University, Berlin, Germany.

HURFORD, A. J. & GREEN, P. F. 1983. The zeta age calibration of fission-track dating. *Chemical Geology*, **41**, 285–317.

JOHNSON, M. R., ANHEAUSSER, C. R. & THOMAS, R. J. (eds) 2006. *The Geology of South Africa*. Geological Society of South Africa, Johannesburg.

JOURDAN, F., FÉRAUD, G., BERTRAND, H. & WATKEYS, M. K. 2007. From flood basalts to the inception of oceanization: Example from the 40Ar/39Ar high resolution picture of the Karoo large igneous province, *Geochemistry, Geophysics, Geosystems*, **8**, Q02002, doi:10.1029/2006GC001392.

KETCHAM, R. A., DONELICK, R. A. & CARLSON, W. D. 1999. Variability of apatite fission-track annealing kinetics III: Extrapolation to geological time scales. *American Mineralogist*, **84**, 1235–1255.

KEYSER, N. (compiler). 1998. *Geological Map of the Republic of South Africa, 1997* (released 1998). Council for Geoscience, Pretoria.

KING, L. C. 1953. Canons of landscape evolution. *Geological Society of America Bulletin*, **64**, 721–752.

KOUNOV, A., NIEDERMANN, S., DE WIT, M. J., VIOLA, G., ANDREOLI, M. & ERZINGER, J. 2007. Present denudation rates at selected sections of the South African escarpment and the elevated continental interior based on cosmogenic ^{3}He and ^{21}Ne. *South African Journal of Geology*, **110**, 235–248.

KOUNOV, A., VIOLA, G., DE WIT, M. J. & ANDREOLI, M. 2008. A Mid Cretaceous paleo-Karoo River Valley across the Knersvlakte plain (northwestern coast of South Africa): Evidence from apatite fission-track analysis. *South African Journal of Geology*, **111**, 409–420.

LASLETT, G. M., GREEN, P. F., DUDDY, I. R. & GLEADOW, A. J. W. 1987. Thermal annealing of fission tracks in apatite 2. A Quantitative Analysis. *Chemical Geology (Isotope Geoscience Section)*, **65**, 1–13.

LITHGOW-BERTELLONI, C. & SILVER, P. 1998. Dynamic topography, plate driving forces and the African superswell. *Nature*, **395**, 269–272.

MOLNAR, P. & ENGLAND, P. 1990. Late Cenozoic uplift of mountain ranges and global climate change: chicken or egg? *Nature*, **346**, 2934.

MOORE, A. & BLENKINSOP, T. 2006. Scarp retreat pinned drainage divide in the formation of the Drakensberg escarpment, southern Africa. *South African Journal of Geology*, **109**, 599–610.

MOORE, M. E., GLEADOW, A. J. W. & LOVERING, J. F. 1986. Thermal evolution of rifted continental margins: New evidence from fission tracks in basement apatites from southeastern Australia. *Earth and Planetary Science Letters*, **78**, 255–270.

NÜRNBERG, D. & MÜLLER, R. D. 1991. The tectonic evolution of the South Atlantic from late Jurassic to present. *Tectonophysics*, **191**, 27–53.

OLLIER, C. D. 1985. Morphotectonics of continental passive margins with great escarpment. *In*: MORISAWA, M. & HACK, J. T. (eds) *Tectonic Geomorphology*. Allen & Unwin, Boston, MA, 3–25.

OLLIER, C. D. & MARKER, M. E. 1995. The great escarpment of southern Africa. *Zeitschrift für Geomorphologie, Supplementband*, **54**, 37–56.

PARTRIDGE, T. C. 1998. Of diamonds, dinosaurs and diastrophism: 150 million years of landscape evolution in southern Africa. *South African Journal of Geology*, **101**, 167–185.

PARTRIDGE, T. C. & MAUD, R. R. 1987. Geomorphic evolution of southern Africa since the Mesozoic. *South African Journal of Geology*, **90**, 179–208.

PATON, D. A., DI PRIMO, R., KUHLMANN, G., VAN DER SPUY, D. & HORSFIELD, B. 2007. Insights into the petroleum system evolution of the southern Orange Basin, South Africa. *South African Journal of Geology*, **110**, 261–274.

PERSANO, C., STUART, F. M., BISHOP, P. & BARFOD, D. N. 2002. Apatite (U–Th)/He age constrains on the development of the Great Escarpment on the southeastern Australian passive margin. *Earth and Planetary Science Letters*, **200**, 79–90.

PETHER, J., ROBERTS, D. L. & WARD, J. D. 2000. Deposits of the West Coast. *In*: PARTRIDGE, T. C. & MAUD, R. R. (eds) *The Cenozoic of Southern Africa*. Oxford University Press, UK, 33–54.

RAAB, M. J., BROWN, R. W., GALLAGHER, K., CARTER, A. & WEBER, K. 2002. Late Cretaceous reactivation of major crustal shear zones in northern Namibia: constraints from apatite fission track analysis. *Tectonophysics*, **349**, 75–92.

RAAB, M. J., BROWN, R. W., GALLAGHER, K., WEBER, K. & GLEADOW, A. J. W. 2005. Denudational and thermal history of the Early Cretaceous Brandberg and Okenyenya igneous complexes on Namibia's Atlantic passive margin. *Tectonics*, **24**, TC3006, doi:10.1029/2004TC001688.

RABINOVICH, P. D. & LABRECQUE, J. 1979. The Mesozoic South Atlantic Ocean and evolution of its continental margin. *Journal of Geophysical Research*, **84**, 5973–6002.

RAYMO, M. E. & RUDDIMAN, W. F. 1992. Tectonic forcing of Late Cenozoic climate. *Nature*, **359**, 117–122.

RUST, D. J. & SUMMERFIELD, M. A. 1990. Isopach and borehole data as indicators of rifted margin evolution in southwestern Africa. *Marine and Petroleum Geology*, **7**, 277–287.

SEIDL, M. A., WEISSEL, J. K. & PRATSON, L. F. 1996. The kinematics and pattern of escarpment retreat across the rifted continental margin of SE Australia. *Basin Research*, **12**, 301–316.

SEWARD, D. 1989. Cenozoic basin histories determined by fission track dating of basement granites, South Island, New Zealand. *Chemical Geology*, **79**, 31–48.

SEWARD, D., GRUJIC, D. & SCHREURS, G. 2004. An insight into breakup of Gondwana: Identifying events through low-temperature thermochronology from basement rocks of Madagascar. *Tectonics*, **23**, TC3007, doi:10.1029/2003TC001556.

SMITH, C. B., ALLSOPP, H. L., KRAMERS, J. D., HUTCHINSON, G. & RODERICK, J. C. 1985. Emplacement ages of Jurassic-Cretaceous South African kimberlites by the Rb–Sr method on phlogopite and whole rock samples. *Transactions of the Geological Society of South Africa*, **88**, 249–266.

STERN, C. R. & DE WIT, M. J. 2004. Rocas Verdes ophiolite, southernmost South America: remnants of progressive stages of development of oceanic type crust in a continental back arc basin. *In*: DE DILEK, Y. & ROBINSON, P. T. (eds) *Ophiolites in Earth History*. The Geological Society, London, Special Publications, **218**, 665–683.

TINKER, J., DE WIT, M. & BROWN, R. 2008. Mesozoic exhumation of the southern Cape, South Africa, quantified using apatite fission track thermochronology, *Tectonophysics*, **455**, 77–93.

Trumbull, R. B., Reid, D. L., de Beer, C., van Acken, D. & Romer, R. L. 2007. Magmatism and continental breakup at the west margin of southern Africa: A geochemical comparison of dolerite dikes from northwestern Namibia and the Western Cape. *South African Journal of Geology*, **110**, 477–502.

Uenzelmann-Neben, G., Schlüter, P. & Weigelt, E. 2007. Cenozoic oceanic circulation within the South African gateway: indications from seismic stratigraphy. *South African Journal of Geology*, **110**, 275–294, doi:10.2113/gssajg.110.2/3.275.

Vanacker, V., von Blackenburg, F., Hewawasam, T. & Kubik, P. W. 2007. Constraining landscape development of the Sri Lankan escarpment with cosmogenic nuclides in river sediment. *Earth and Planetary Science Letters*, **253**, 402–414.

van der Beek, P. A. & Braun, J. 1999. Controls on post-mid-Cretaceous landscape evolution in the southeastern highlands of Australia: Insights from numerical surface process models. *Journal of Geophysical Research*, **104**, 4945–4966.

Van der Wateren, F. M. & Dunai, T. J. 2001. Late Neogene passive margin denudation history- cosmogenic isotope measurements from the central Namib desert. *Global and Planetary Change*, **30**, 271–307.

Viola, G., Andreoli, M., Ben-Avrahama, Z., Stengeld, I. & Reshef, M. 2005. Offshore mud volcanoes and on land faulting in southwestern Africa: neotectonic implications and constraints on the regional stress field. *Earth and Planetary Science Letters*, **231**, 147–160.

Wobus, C. W., Hodges, K. V. & Whipple, K. X. 2003. Has focused denudation sustained active thrusting at the Himalayan topographic front? *Geology*, **31**, 861–864.

Steady-state exhumation pattern in the Central Andes – SE Peru

G. M. H. RUIZ[1,2]*, V. CARLOTTO[3], P. V. VAN HEININGEN[1] & P. A. M. ANDRIESSEN[1]

[1]*Vrije Universiteit (Vu), Amsterdam, The Netherlands*

[2]*Present address: University of Neuchâtel, Neuchâtel, Switzerland*

[3]*INGEMMET, Lima, Peru*

**Corresponding author (e-mail: geoffrey.ruiz@unine.ch)*

Abstract: The Western Cordillera of SE Peru is part of the Central Andes and is situated to the west of the Eastern Andes from which it is separated by the northern termination of the Altiplano – the Inter-Andean Valley. It is a volcanic–volcano-detrital chain that developed in the Palaeogene, and is characterized by a 4000 m-high mean altitude whose origin is poorly constrained. We selected a vertical profile in the region of Abancay to trace the record the evolving uplift and erosion history of the Andean orogen. Fission-track data on both apatite and zircon crystals were completed on plutonic rocks of the Tertiary Andahuaylas–Yauri batholith. Ages ranged between 24 and 14 Ma, and 38 and 30 Ma, respectively. Thermal modelling was completed for the whole profile and does not, like age–altitude relationships, show evidence of any clear disruption of the exhumation paths since 38 Ma either by sedimentary burial and/or rapid exhumation. One of the noteworthy aspects of the data is that exhumation was steady at a rate of 0.17 km Ma^{-1} from the late Eocene until at least the middle Miocene (38–14 Ma). The uplift of the Western Cordillera was thus probably steady for this period with sedimentary deposition restricted to the present-day Altiplano and Inter-Andean Valley regions.

The uplift chronology of the Central Andes is contentious (Gregory-Wodzicki 2000; Lamb & Davis 2003; Ghosh *et al.* 2006; Garzione *et al.* 2006, 2008; Sempere *et al.* 2006; Hartley *et al.* 2007). It is currently considered that deformation and uplift began in the Western Cordillera (WC) in the Eocene, developed later and more slowly in the Eastern Cordillera (EC) (40–10 Ma: McQuarrie *et al.* 2008), whereas the Subandean Zone (SAZ) probably developed as early as the Middle Miocene (Barnes *et al.* 2008; McQuarrie *et al.* 2008). Phases of deformation and uplift are often constrained by lateral association with Bolivia further south, with the exception of two recent studies from the WC further south that demonstrated that it was affected in the Late Miocene by a phase of uplift (Schildgen *et al.* 2007; Thouret *et al.* 2007). Palaeomagnetic and $^{40}Ar/^{39}Ar$ studies (Gilder *et al.* 2003; Rousse *et al.* 2003, 2005) indicate that major mountain building developed 12–10 Ma ago in the EC coevally with ridge subduction and transpressional deformation in the Altiplano region. Additional palaeomagnetic data indicate that counter-clockwise rotation probably occurred earlier, i.e. before 15 Ma, in the fore-arc region between Bolivia and Peru (Roperch *et al.* 2006), and as early as the late Eocene–early Oligocene in the northern termination of the Altiplano in SE Peru (Roperch *et al.* 2008). Establishing long-term order rates and patterns of mountain exhumation, both among and within individual ranges, is therefore a high priority. Thermochronology is an indispensable tool for such studies and fission-track dating is the best of all methods because of its associated thermal modelling, but also because it allows an immediate and reliable estimation of exhumation through an internal pair method on apatite and zircon minerals.

We complement and add to the previous studies by presenting the first apatite (AFT) and zircon (ZFT) fission-track thermochronological ages from a vertical profile located in the Western Cordillera of southern Peru near the town of Abancay (Fig. 1). The apatite partial annealing zone (APAZ) and zircon (ZPAZ) in the fission-track system lies between 60–120 °C (Green *et al.* 1989) and 210–270 °C (Tagami *et al.* 1996). Hence, it is particularly useful for evaluating low-temperature thermal histories, i.e. those affecting the upper 8–10 km of the WC, depending on the geothermal gradient. We have chosen this part of the WC because it exhibits: (1) crystalline rocks potentially rich in apatite and zircon; and (2) high slope gradients. Eight bedrocks were sampled along a vertical profile, with an average incremental difference in altitude of 150 m. Results indicate that exhumation was slow and steady since the Late Eocene with no evidence of a recent change,

From: LISKER, F., VENTURA, B. & GLASMACHER, U. A. (eds) *Thermochronological Methods: From Palaeotemperature Constraints to Landscape Evolution Models*. Geological Society, London, Special Publications, **324**, 307–316.
DOI: 10.1144/SP324.20 0305-8719/09/$15.00

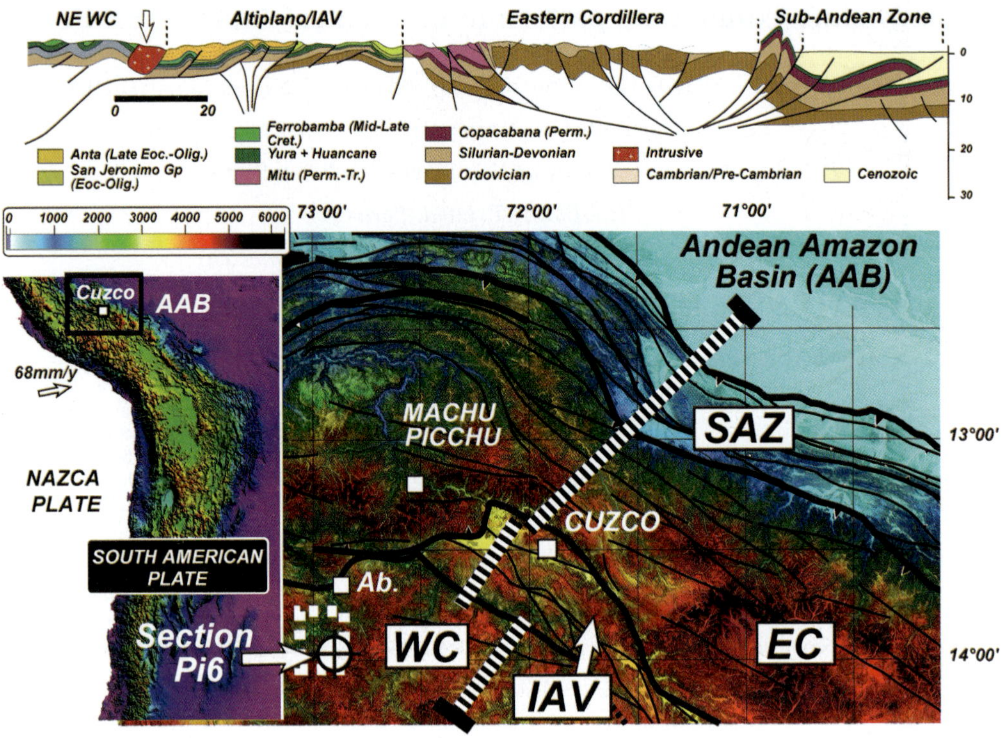

Fig. 1. Morphotectonic map of the Central Andes in southern Peru with Nazca–South American plate convergence (Norabuena *et al.* 1999). Location map in a DEM (Digital Elevation Model) from South America (Cornell Universtiy) showing the extent of the studied areas. The thick black lines correspond to boundaries between the different morphotectonic domains: the Andean Amazon Basin (AAB), the Sub-Andean Zone (SAZ), the Eastern Cordillera (EC), the Inter-Andean Valley (IAV) and, finally, the Western Cordillera (WC). Pi6, vertical profile sampled for fission-track analysis in the Andahuaylas–Yauri batholith. Thin black lines, faults. Thick dashed white and black lines, section illustrated to the top (Carlotto 2006). Ab., town of Abancay.

which is very different from earlier records from the Western Cordillera (Schildgen *et al.* 2007; Thouret *et al.* 2007).

Geodynamic setting

The tectonic framework is as uniform as can be expected over such a large area. The Andes are thrust southeastwards over the South American craton, whereas the subduction of the oceanic Nazca plate, with the buoyant Nazca ridge on top (Hampel 2002; Espurt *et al.* 2007; Clift & Ruiz 2008), beneath the South American plate occurs further west at a rate of 68 mm $year^{-1}$ and towards the ENE (Norabuena *et al.* 1999) (Fig. 1). The Andes in the region of Cuzco–Abancay mark a clear SW–NE offset in the strike of the Andes (Fig. 1) that corresponds to the NW termination of the Bolivian orocline (Fig. 1). The Andes in this area comprise of two different mountain ranges, the Eastern (EC) and Western (WC) Cordilleras, each having an elevation greates than 6000 m and separated by the 3800 m-high northern termination of the Altiplano, i.e. the Inter-Andean Valley (IAV; Fig. 1).

The WC is a volcanic–volcano-detrital chain that developed approximately 40–50 Ma ago (Jaillard *et al.* 2000) and is still characterized by 5000–6500 m-high active volcanoes that disappear north of about 14–15°S. Parts of the WC are composed of Tertiary plutons, such as the Andahuaylas–Yauri batholith in the region of Abancay (Fig. 2) (Noble *et al.* 1984; Perello *et al.* 2003). Earlier evolution, i.e. late Permian–middle Jurassic, corresponds to lithospheric thinning and a phase of rifting, whereas marine–alluvial sedimentation took place in the WC from middle Jurassic to middle Cretaceous (Sempere *et al.* 2002). Thickening and loading of the crust beneath the WC resulted in the development of a Palaeocene–Eocene foreland basin with continental deposits at the boundary zone between the WC and the Altiplano (Carlotto *et al.* 2005). Early Eocene sedimentation is not preserved in the WC, but exists in the IAV near Cuzco

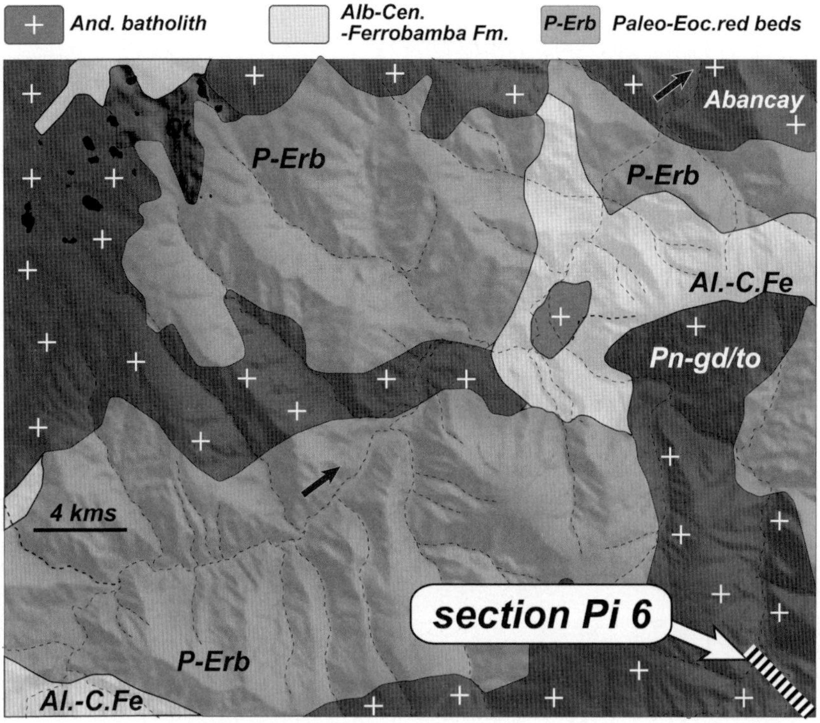

Fig. 2. Geological map (modified from INGEMMET) of the studied area with the sample site (bottom right) on top of a 90 m-resolution DEM (CGIAR-CSI). Dashed thin black lines, rivers; And., Andahuaylas–Yauri.

with a 5 km-thick pile of continental series deposits. The scope of this paper is not a complete structural and stratigraphic study, so for further details the reader should refer to Laubacher (1978), Carlotto (1998) and Sempere *et al.* (2002).

Results and interpretation

Zircon fission-track (ZFT) data

ZFT ages range between 36.1 ± 1.8 and 31.3 ± 1.6 Ma, whereas all data pass χ^2 test (Gailbraith 1981) (Table 1), suggesting non-dispersion in individual grain ages. ZFT ages are older towards the top of the section and younger towards the bottom (Fig. 3), suggesting that all samples belong to a thermally homogeneous block that cooled through the ZPAZ in the late Eocene–early Oligocene. Such a phase post-dated the Early–Middle Eocene age of intrusion for this batholith (Noble *et al.* 1984; Perello *et al.* 2003).

Apatite fission-track (AFT) data

AFT ages are younger than ZFT ages, and comprised ages between 22.5 ± 1.9 and 15.5 ± 1.5 Ma (Table 1 and Fig. 3). Between 21 and 46 track lengths were measured per sample, with mean values varying between 11.20 and 14.04 μm; whereas standard deviation (SD) and D_{par} (Ketcham *et al.* 1999) measurements range between 1.9 and 0.8 μm, and 2.37 and 2.98 μm, respectively (Table 1 and Fig. 4). Mean track lengths (MTL) and SD are plotted v. age (Fig. 4): no clear pattern of long track length and short SD is visible, thus indicating that this part of the WC did not undergo any rapid phase of cooling through the 120–60 °C temperature window. In addition, track-length distributions are roughly unimodal (Fig. 5), corroborating that these samples underwent a simple phase of slow cooling between 120 and 60 °C.

Temperature–time modelling

Inverse temperature–time modelling was performed for all samples from bottom to top of the profile using the AFTsolve program (Ketcham *et al.* 2000). Input parameters for modelling are AFT ages, track lengths and D_{par} distributions, as well as ZFT ages and geological constraints, such as the imposed presence near the surface at a given time. Two sets of thermal modelling were

Table 1. *Apatite fission-track (AFT) and zircon fission-track (ZFT) ages from the Andahuaylas–Yauri batholith in the Western Cordillera of southern Peru (Figs 1 & 3)*

ID	Altitude (m)	Lat.	Long.	N	$P\chi^2$ (%)	ρ_s (N_s) (×10^6)	ρ_i (N_i) (×10^6)	ρ_d (N_d) (×10^6)	Age (Ma)	±1σ (Ma)	U (ppm)	MTL (μm)	SD	D_{par} (μm)	SD
Apatite															
Pi6.1	3870	−13.9853	−72.8747	23	10.0	2.274 (456)	20.740 (4159)	11.015 (6155)	22.5	1.9	23.37	11.21 ± 0.34 (41)	1.9	2.68 ± 0.22 (40)	0.25
Pi6.2	3800	−13.9833	−72.8737	16	100.0	1.554 (168)	13.793 (1491)	10.907 (6155)	22.0	1.9	16.87	13.40 ± 0.23 (31)	1.3	2.78 ± 0.29 (20)	0.32
Pi6.3	3650	−13.9806	−72.8746	20	99.5	1.741 (323)	16.199 (3005)	10.798 (6155)	20.8	1.4	17.88	14.04 ± 0.16 (37)	1.0	2.87 ± 0.20 (44)	0.24
PI6.4	3450	−13.9790	−72.8772	20	98.0	0.685 (137)	6.210 (1242)	10.690 (6155)	21.1	2.0	7.07	12.76 ± 0.21 (46)	1.4	2.37 ± 0.21 (28)	0.23
Pi6.5	3250	−13.9759	−72.8815	19	100.0	0.787 (148)	9.048 (1701)	10.582 (6155)	16.5	1.5	10.44	13.13 ± 0.18 (34)	1.1	2.55 ± 0.23 (43)	0.35
Pi6.6	3100	−13.9735	−72.8810	20	87.0	1.036 (203)	12.010 (2354)	10.474 (6155)	16.2	1.3	13.96	13.36 ± 0.17 (21)	0.8	2.61 ± 0.27 (31)	0.29
Pi6.7	3000	−13.9727	−72.8853	20	93.0	1.092 (141)	12.680 (1637)	10.366 (6155)	16.0	1.5	16.25	12.89 ± 0.26 (35)	1.6	2.72 ± 0.22 (23)	0.29
Pi6.8	2850	−13.9680	−72.8941	20	93.0	0.862 (140)	10.228 (1662)	10.258 (6155)	15.5	1.5	12.48	12.45 ± 0.21 (30)	1.1	2.98 ± 0.31 (17)	0.30
Zircon															
Pi6.1	3870	−13.9853	−72.8747	19	47.0	7.771 (1088)	8.436 (1181)	6.485 (7200)	36.1	1.8	51.08	8.69 ± 0.32 (11)	1.1	–	–
Pi6.2	3800	−13.9833	−72.8737	20	71.0	10.811 (1287)	11.719 (1395)	6.379 (7200)	35.5	1.7	76.41	9.70 ± 0.48 (5)	1.0	–	–
Pi6.3	3650	−13.9806	−72.8746	19	96.0	13.627 (1146)	14.828 (1247)	6.274 (7200)	34.8	1.7	96.34	–	–	–	–
Pi6.4	3450	−13.9790	−72.8772	17	32.0	20.555 (3443)	22.322 (3739)	6.168 (7200)	34.93	1.2	163.15	–	–	–	–
Pi6.5	3250	−13.9759	−72.8815	20	55.0	14.403 (3356)	15.562 (3626)	6.063 (7200)	33.8	1.2	110.33	–	–	–	–
Pi6.6	3100	−13.9735	−72.8810	20	66.0	18.809 (4493)	20.857 (4982)	5.957 (7200)	32.4	1.1	143.63	–	–	–	–
Pi6.7	3000	−13.9727	−72.8853	20	97.0	10.108 (2153)	11.028 (2349)	5.757 (7200)	31.9	1.3	76.83	–	–	–	–
Pi6.8	2850	−13.9680	−72.8941	20	59.0	14.242 (2009)	15.816 (2231)	5.746 (7200)	31.3	1.6	109.58	9.79 ± 0.30 (11)	1.0	–	–

Apatite and zircon separates were counted by G.M.H. Ruiz using ζ'calibration factors of 359 ± 11 (CN5 standard glass, 1250× magnification) and 121 ± 3 (CN1 standard glass, 1600× magnification-oil), respectively, and irradiated at the Petten facility. Apatites were etched in 5.5 M HNO_3, at 21 °C for 20 s (Carlson *et al.* 1999), whereas zircons were etched in a eutectic melt of NaOH–KOH at 220 °C (Gleadow *et al.* 1976). Numbers in parentheses indicate the number of measured track lengths (MTL) and D_{par}. (Ketcham *et al.* 1999) values. N, number of dated grains per sample; SD, standard deviation. The time–temperature modelling of sample Pi6.6 was not completed because of the small number of track-length measurements (21).

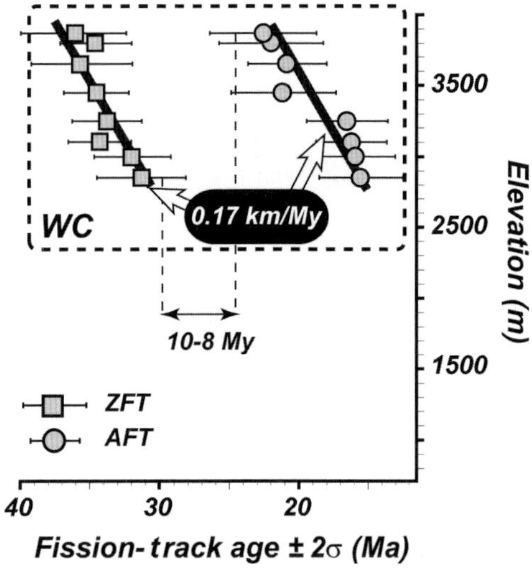

Fig. 3. AFT and ZFT age–altitude plots from the Andahuaylas–Yauri batholith in the Western Cordillera (WC). Fission–track ages for both apatite and zircon gets younger towards the bottom of the profiles (Table 1). Grey lines, best correlation fit paths within 2σ error bars.

performed, one without any geological constraints – modelling 'free' (Fig. 5, left) – and one with an imposed presence at/or near the surface in about the early Oligocene (Fig. 5, right) because of the presence of continental sediments (Anta Formation) on top of the batholith in the IAV to the south of Cuzco (Carlotto *et al.* 2005).

Temperature–time models for the first set indicate that the upper sample (Pi6.1) entered the apatite partial annealing zone (APAZ, 120–60 °C) at about 30 Ma, whereas the bottom sample (Pi6.8) entered approximately 20 Ma ago (Fig. 5, right). Similarly, the uppermost sample left the APAZ about 18–20 Ma ago, about the same time as the lowermost sample entered it, and the lowermost sample left the zone in the last 5 Ma.

The second set of models yielded good-quality Kolmogorov–Smirnov (K–S) tests, such as the first one giving results of between 0.64 and 0.93, and 0.72 and 0.98, respectively (Fig. 5). Thermal modelling for the six lowermost samples suggest a total resetting to a temperature greater than 120 °C in the late Oligocene–early Miocene, whereas a less protracted phase of heating is modelled for the two upper ones (Fig. 5, right). Such a phase was followed by a phase of cooling starting from about

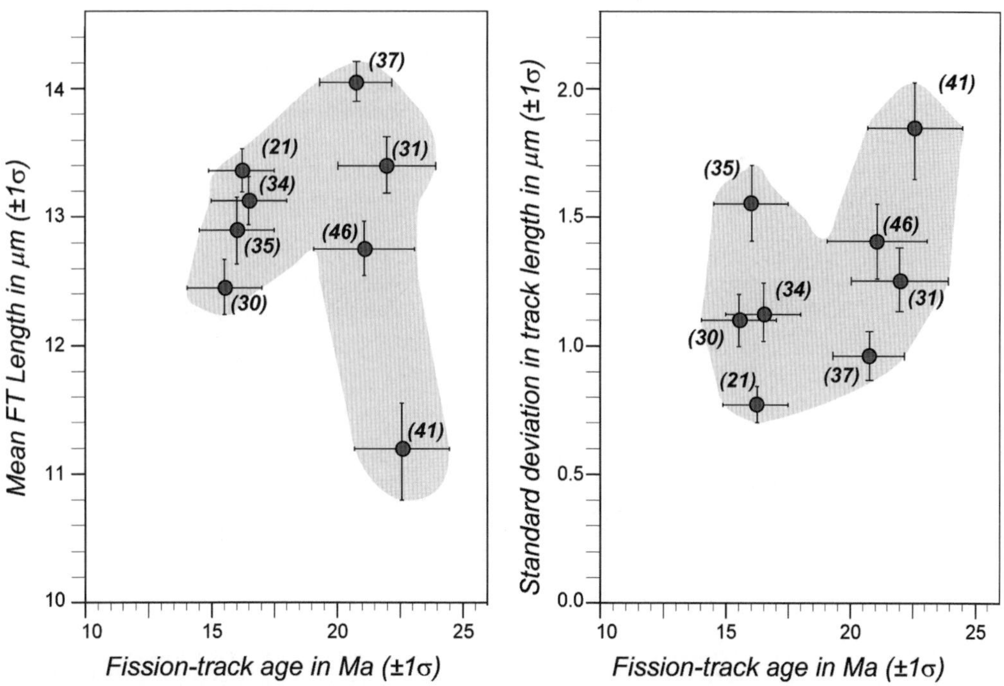

Fig. 4. Relationship between the AFT age and the mean track length (left), and standard deviation (right) – section Pi6. The relationship is shown for all samples that yielded adequate track length data (all errors are $\pm 1\sigma$). Shading shows the general trend of the data.

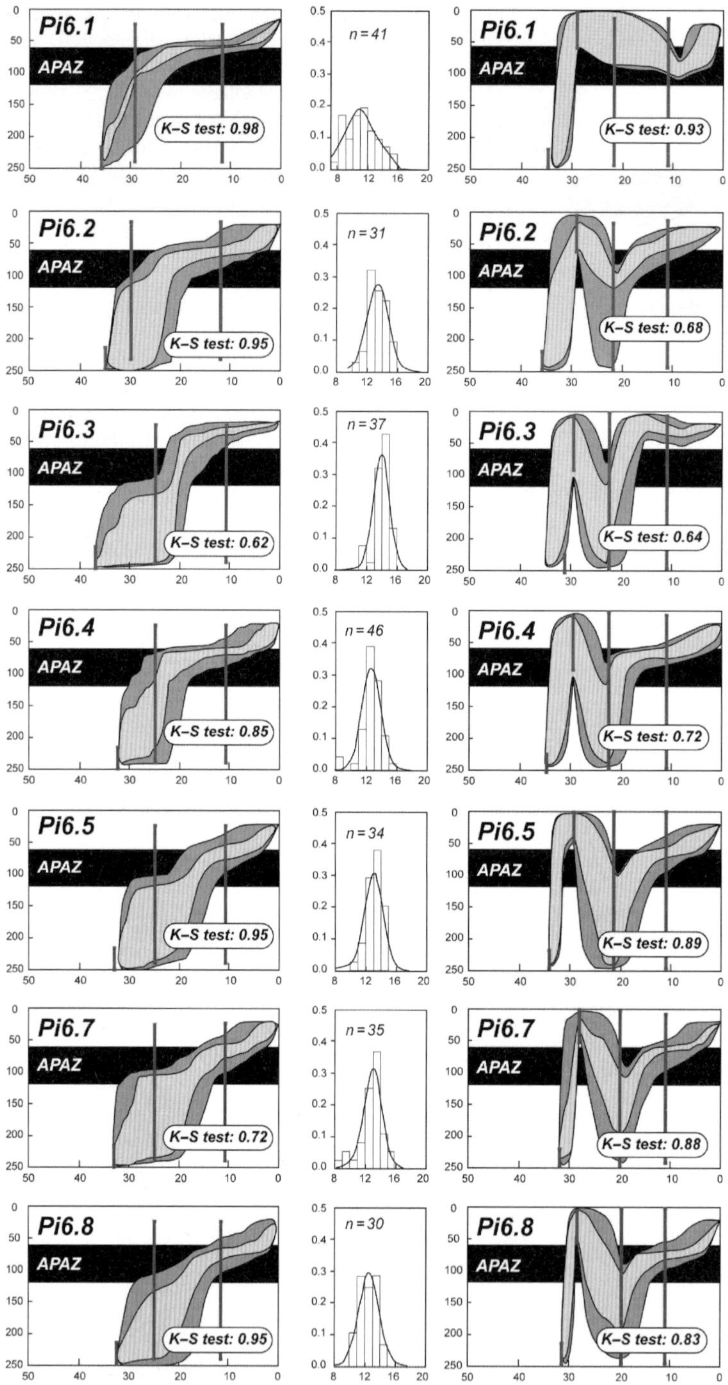

Fig. 5. Temperature–time (T–t) modelling of seven samples from the Andahuaylas–Yauri batholith (see Figs 1 & 2 for the locations, and Table 1 for details). T–t modelling was completed following the procedures described in Ketcham *et al.* (2000) with geological and temperature (ZFT ages) constraints (vertical grey bars). (Left) Modelling 'free' (see text), i.e. without any geological constraints. (Right) Models with an imposed near-surface presence in the early Oligocene (vertical grey bar). (Middle) Track-length histogram (X-axis, length in μm; Y-axis, relative representation; n, number of measured confined tracks). The best T–t paths fitting both the AFT age and the track-length distribution (right) and the Kolmogorov–Smirnov (K–S) test are shown: dark (65%) and light (95%) confidence. APAZ, apatite partial annealing zone.

20 Ma until 17–16 Ma. The most recent thermal history cannot be constrained because models are outside, or at the upper edge, of the 120–60 °C (APAZ) temperature window, with the exception of the bottom sample (Pi6.8; Fig. 5, right), which points towards a loosely constrained and ultimate phase of cooling starting at around 5–4 Ma.

Differences between the two sets of models are roughly restricted for most of the samples in the left-hand part of the models, i.e. pre-24–20 Ma (Fig. 5), and correspond to the the geological constraint we added. Models were forced to reach temperatures lower than 60 °C, i.e. the near surface at *c.* 31–28 Ma, but had time to reach temperatures of at least 120 °C, i.e. total resetting, before models of the first set enter the APAZ (Fig. 5). The two thermal models for the uppermost sample (Pi6.1) do not have any similarity because the early entrance into the APAZ for the first set is anterior to, or almost synchronous with, its near-surface position in the second one. Hence, these two models are much contrasted, with a clear phase of cooling starting at arount 10–9 Ma for the second set (Fig. 5, right). This late phase of cooling is absent from lower samples where no fault system exists between sample sites. All this suggests that the true time–temperature paths for this part of the Western Cordillera most probably correspond to the ones yielded by the first configuration, i.e. without: (1) any near-surface presence in the late Eocene–early Oligocene; and (2) rapid and important burial in the late Oligocene–early Miocene.

Age–elevation plots

AFT and ZFT ages are plotted v. altitude (Fig. 3). Slopes and age differences from bottom to top of the profile are almost identical for the two thermochronometers, pointing towards an apparent exhumation rate of about 0.17 km Ma^{-1} between 36 and 31 Ma (ZFT), and 23 and 15 Ma (AFT). This rate is low and in agreement with the periods of slow cooling we evidenced through the APAZ (120–60 °C) by thermal modelling (Fig. 5, left). Hence, we are confident that exhumation was slow and at a rate of 0.1–0.2 km Ma^{-1} in this region of the WC for these two periods.

Discussion and conclusions

Late Eocene–Oligocene

The late Eocene–Early Oligocene epoch was characterized by a period of slow exhumation in the WC near Abancay (Fig. 1), as evidenced by thermal modelling and ZFT age–altitude relationships (Figs 3 & 5). This period most probably postponed the emplacement of the Andahuaylas–Yauri batholith of Eocene age (Noble *et al.* 1984; Perello *et al.* 2003) until shallow crustal levels. Some K/Ar data exist from the Eastern Cordillera 150 km further to the NE (Farrar *et al.* 1988; Kontak *et al.* 1990*a*, *b*) that are interpreted as reflecting the development of a NW-oriented tectono-thermal zone in relation to an important phase of mountain building at approximately 38 Ma. Such as event would correspond to the so-called Incaic phase of Steinmann (1929), which was related in the 1990s to a change towards rapid convergence between the South American and Farallon plates (Mégard 1984; Pardo-Casas & Molnar 1987). This phase of deformation can probably be traced in the WC of southern Peru through shortening (Carlotto 2006), but was not associated with rapid (as evidenced by thermal modelling) nor important exhumation because late Eocene ZFT ages are still present today at the surface. As previously stated the subsection on 'Temperative–time modelling', a 5000–6000 m-thick pile of late Eocene–middle Oligocene syn-orogenic sediments has been reported in the IAV near Cuzco (Carlotto 1998; Carlotto *et al.* 2005). Such a pile is absent today in the WC, suggesting either non-deposition or post-depositional erosion in the WC. If deposited where our profile was selected, this pile would have been heated by sedimentary burial of the Andahuaylas–Yauri batholith to the extent of 5.5 km × 25 °C km^{-1}, i.e. *c.* 137.5 °C, as illustrated on the right-hand side of Figure 5. However, we reject the post-depositional erosion solution because time–temperature modellings are not consistent from the bottom to the top of the profile (see earlier).

All combined data suggest that the IAV was most probably limited to the west in the Eocene–Oligocene by the slowly exhuming WC, and possibly (because quantitative constraints are still poor) to the east by a developing Eastern Cordillera.

Miocene

Exhumation occurred at a similar rate of 0.17 km Ma^{-1} for the early Miocene (22–15 Ma) and late Eocene–Early Oligocene in the WC, whereas a gap of 9 Ma prevails between the two periods (Fig. 3). We interpret this time lag to correspond to the temperature gap between the partial annealing zones of the zircon and apatite systems because a total resetting to temperatures of more than 120 °C at a time between the two periods is unlikely (see the previous subsection). An identical rate for these two epochs is thus not a coincidence but rather characterized a sole and continuous phase of exhumation from about 36 to 15 Ma.

Different basins of Eocene–Miocene age existed in the IAV (Fig. 1). Clasts indicate that source rocks were mainly composed of quartzites, carbonates, volcanics and plutonic rocks (see the compilation in Carlotto *et al.* 2005). Carbonate and quartzite clasts were probably sourced from sediments that constituted Mesozoic basins in the present-day EC and WC (Sempere *et al.* 2002), whereas volcanic activity prevailed in the region from at least 40 Ma (see the next subsection). Clasts of plutonic origin are only encountered in adjacent basins to the NE of the Andahuaylas–Yauri batholith (e.g. Descanso Basin: Carlotto *et al.* 2005) from the Middle to Late Miocene, suggesting that the upper levels of the batholith reached the surface not before the Middle Miocene. This additional evidence is in total agreement with the slow phase of cooling we modelled for the middle–late Miocene (Fig. 5, left). Since deposition in the Miocene, shortening from 45% to 25% occurred in the IAV, which was constrained by surface data (Carlotto 2006).

Palaeo-geothermal gradient

Our pair fission-track ages indicate that exhumation was continuous and at a rate of 0.17 km Ma^{-1} from *c.* 35 until 15 Ma. A palaeo-geothermal gradient can be deduced from these analyses using the following formula:

$$G = \frac{\Delta T}{\Delta l} = \frac{T_c(\mathrm{ZFT}) - T_c(\mathrm{AFT})}{\Delta l}$$

where T_c(ZFT) and T_c(AFT) are the temperatures of closure of the AFT and ZFT systems that can be approximated to 230 and 110 °C, respectively (Reiners & Brandon 2006). Δl is an artificial depth that we calculate using the exhumation rate of 0.17 km Ma^{-1}, and the almost identical (within error bars) age difference, Δt, of *c.* 14–15 Ma we produced using the AFT and ZFT systems from a single sample, e.g. the lowermost (Pi6.8) or the uppermost one (Pi6.1: Fig. 3 and Table 1):

$$\Delta l = \frac{\Delta l}{\Delta t} \times \frac{\Delta t}{1} = 0.17 \times 15 = 2.55 \text{ km}.$$

In turn, the geothermal gradient becomes:

$$G = \frac{\Delta T}{\Delta l} = \frac{230 - 110}{2.55} = 47\ ^\circ\mathrm{C\ km}^{-1}.$$

A value of 47 °C km^{-1} is high but in the perfect range of geothermal gradients for the upper 12 of the crust in the Andes, and more generally in volcanic arcs (Henry & Pollack 1988). The geothermal gradient we calculated most probably prevailed from 36 Ma, i.e. the ZFT age of the uppermost sample, until about 15 Ma, i.e. the AFT age of the lowermost dated level, suggesting continuous volcanic activity in this region until at least 15 Ma, which is corroborated by the presence of volcanic levels in most of the neighbouring Eocene–Late Miocene basins (Carlotto *et al.* 2005).

Uplift of the Western Cordillera of southern Peru

The depth of closure for the ZFT system depends on the geothermal gradient. Using the geothermal gradient we calculated for the Late Eocene–Middle Miocene of approximately 47 °C km^{-1}, the depth of closure corresponds to 230/47, i.e. *c.* 5 km. Bedrocks from the Andahuaylas–Yauri batholith today yield late Eocene ZFT ages suggesting that denudation has not exceeded 5 km since. Interestingly, the WC is today characterized by a mean altitude of 3500–4000 m to the south of Abancay (Fig. 1). This prompts the question: when did the WC acquire its topography? Our results suggest: (1) a continuous exhumation from the late Eocene until the middle Miocene, independent of the high contemporaneous geothermal gradient that prevailed in this region of the Central Andes; and (2) there has been no phase of heating/burial since the late Eocene for the Andahuaylas–Yauri batholith near Abancay. We thus favour a model involving continuous uplift from at least the late Eocene with sediment deposition in basins located today in the IAV, which was later tightened up in the middle to late Miocene (Carlotto 1998, 2005, 2006; Rousse *et al.* 2003, 2005). The locus of deformation has since been transferred along the foothills of the Andes in SE Peru (Hermoza 2004), which yield very young AFT ages (Fig. 3). The possible causes are many and possibly coupled, these are: (a) the westward drift of the South American plate (Russo & Silver 1996); (b) the thickening of the crust (Isacks 1988); (c) the sediment starvation along the trench because of aridity in the fore-arc (Hartley 2003); and (d) the ridge subduction below the South American plate (Hampel 2002; Rousse *et al.* 2003; Espurt *et al.* 2007, 2008; Clift & Ruiz 2008); the latter two increasing friction between the subducting Nazca and South American plates (Lamb & Davis 2003), and transferring shortening within the interior of the South American plate (Espurt *et al.* 2008).

However, whatever processes are considered to have shaped the geology of SE Peru, none generated rapid exhumation in the Western Cordillera near Abancay from the Late Eocene until the Middle Miocene. The exhumation pattern is rather slow and steady for this period, at a rate of about 0.17 km Ma^{-1}, whereas the post-middle Miocene record would probably be traceable using a thermochronometer with a lower temperature of closure, i.e. U–Th/He on apatite (70 °C: Farley 2000), similar to what was undertaken further south by Schildgen *et al.* (2007) to ascribe a late Miocene age for the uplift of the western margin of the Andean plateau.

We wish to thank A. Hartley, B. Ventura, D. Seward and F. Lisker for constructive reviews; J. Jacay, T. Sempere, E. Jaillard and P. Roperch for discussions on the regional geology; and J. Salichon and M. Ruiz for field-work assistance.

References

BARNES, J. B., EHLERS, T. A., MCQUARRIE, N., O'SULLIVAN, P. B. & TAWACKOLI, S. 2008. Thermochronometer record of central Andean plateau growth, Bolivia (19.5S). *Tectonics*, **27**, TC3003, doi:10.1029/2007TC002174.

CARLOTTO, V. 1998. *Evolution Andine au niveau de Cuzco (Pérou)*. PhD thesis, University of Grenoble, France.

CARLOTTO, V. 2006. El acortamiento y la deformación Andina en el Sur del Perú: Cusco–Abancay–Sicuani. *Boletín Sociedad Geológica del Perú*, **101**, 91–119.

CARLOTTO, V., JAILLARD, E. *ET AL.* 2005. Las Cuencas Terciarias sinorogénicas en el Altiplano y la Cordillera Occidental del Sur del Perú. *Boletín Especial Alberto Giesecke Matto, Sociedad Geológica del Perú*, **6**, 103–126.

CARLSON, W. D., DONELICK, R. A. & KETCHAM, R. A. 1999. Variability of apatite fission-track annealing kinetics: I. Experimental results. *American Mineralogist*, **84**, 1213–1223.

CLIFT, P. D. & RUIZ, G. M. H. 2008. How does the Nazca Ridge subduction influence the modern Amazonian foreland basin? Comment. *Geology*, **36**, doi: 10.1130/G24355C.1.

ESPURT, N., BABY, P. *ET AL.* 2007. How does the Nazca Ridge subduction influence the modern Amazonian foreland basin? *Geology*, **35**, 515–518.

ESPURT, N., FUNICIELLO, F., MARTINOD, J., GUILLAUME, B., REGARD, V., FACENNA, C. & BRUSSET, S. 2008. Flat subduction dynamics and deformation of the South American plate: Insights from analog modeling. *Tectonics*, **27**, TC3011, doi:10.1029/2007TC002175.

FARLEY, K. A. 2000. Helium diffusion from apatite: general behaviour as illustrated by Durango fluorapatite. *Journal of Geophysical Research*, **105**(B2), 2903–2914.

FARRAR, E., CLARK, A. H., KONTAK, D. J. & ARCHIBALD, D. A. 1988. The Zongo–San Gaban zone: eocene foreland boundary of the central Andean orogen, northwest Bolivia and southeast Peru. *Geology*, **16**, 55–58.

GAILBRAITH, R. F. 1981. On statistical models for fission tracks counts. *Mathematical Geology*, **13**, 471–478.

GARZIONE, C. N., HOKE, G. D. *ET AL.* 2008. Rise of the Andes. *Science*, **320**(5881), 1304–1307.

GARZIONE, C. N., MOLNAR, P., LIBARKIN, J. C. & MACFADDEN, B. J. 2006. Rapid late Miocene rise of the Bolivian Altiplano: evidence for removal of mantle lithosphere. *Earth and Planetary Science Letters*, **241**(3–4), 543–556.

GHOSH, P., GARZIONE, C. N. & EILER, J. M. 2006. Rapid uplift of the Altiplano revealed through $^{13}C-^{18}O$ bonds in paleosol carbonates. *Science*, **311**, 511–514.

GILDER, S., ROUSSE, S., FARBER, D., MCNULTY, B., SEMPERE, T., TORRES, V. & PALACIOS, O. 2003. Post-Middle Oligocene origin of paleomagnetic rotations in the Upper Permian to Lower Jurassic rocks from northern and southern Peru. *Earth and Planetary Sciences Letters*, **210**, 233–248.

GLEADOW, A. J. W., HURFORD, A. J. & QUAIFE, R. D. 1976. Fission-track dating of zircon: improved etching techniques. *Earth and Planetary Science Letters*, **33**, 273–276.

GREEN, P. F., DUDDY, I. R., LASLETT, G. M., HEGARTY, K. A., GLEADOW, A. J. W. & LOVERING, J. F. 1989. Thermal annealing of fission tracks in apatite – Qualitative modelling techniques and extensions to geological timescales. *Chemical Geology*, **79**, 155–182.

GREGORY-WODZICKI, K. M. 2000. Uplift history of the Central & Northern Andes: a review. *Geological Society of America Bulletin*, **112**, 1091–1105.

HAMPEL, A. 2002. The migration history of the Nazca Ridge along the Peruvian active margin: a re-evaluation. *Earth and Planetary Sciences Letters*, **203**, 665–679.

HARTLEY, A. J. 2003. Andean uplift and climate change. *Journal of the Geological Society, London*, **160**, 7–10.

HARTLEY, A., SEMPERE, T. & WORNER, G. 2007. A comment on 'Rapid late Miocene rise of the Bolivian Altiplano: Evidence for removal of mantle lithosphere' by C. N. Garzione *et al.* [Earth and Planetary Sciences Letters, **241**(2006), 543–556]. *Earth and Planetary Sciences Letters*, **259**, 625–629.

HENRY, S. G. & POLLACK, H. N. 1988. Terrestrial heat flow above the Andean subduction zone, Bolivia and Peru. *Journal of Geophysical Research*, **93**, 15,153–15,162.

HERMOZA, W. 2004. *Dynamique de tectono-sédimentaire et restauration séquentielle du rétro-bassin d'avant-pays des Andes centrales*. PhD thesis, Université Paul Sabatier-Toulouse III.

ISACKS, B. L. 1988. Uplift of the Central Andean plateau and bending of the Bolivian Orocline. *Journal of Geophysical Research*, **93**(B4), 3211–3231.

JAILLARD, E., HÉRAIL, G., MONFRET, T., DIAZ-MARTINEZ, BABY, P., LAVENU, A. & DUMONT, J.-F. 2000. Tectonic evolution of the Andes of Ecuador, Peru, Bolivia and northernmost Chile. *In*: CORDANI, U., MILANO, E. J., THOMAS FILHO, A. & CAMPOS, D. A. (eds) *Tectonic Evolution of South America. 31st International Geological Congress*, Rio de Janeiro, 481–559.

KETCHAM, R. A., DONELICK, R. A. & CARLSON, W. D. 1999. Variability of apatite fission track annealing kinetics III: Extrapolation to geological time scales. *American Mineralogist*, **84**, 1235–1255.

KETCHAM, R. A., DONELICK, R. A. & DONELICK, M. B. 2000. AFTSolve: A program for multi-kinetic modelling of apatite fission-track data. *Geological Materials Research*, **2**(1), 32.

KONTAK, D. J., FARRAR, E. A., CLARK, A. H. & ARCHIBALD, D. A. 1990*a*. Eocene tectono-thermal rejuvenation of an upper Paleozoic–lower Mesozoic terrane in the Cordillera de Carabaya, Puno, south-eastern Peru, revealed by K–Ar and 40Ar/39Ar dating. *Journal of South American Earth Sciences*, **3**(4), 231–246.

KONTAK, D. J., FARRAR, E. A., CLARK, A. H., ARCHIBALD, D. A. & BAADSGAARD, H. 1990*b*. Late Paleozoic–early Mesozoic magmatism in the Cordillera de Carabaya, Puno, south-eastern Peru, geochronology and petrochemistry. *Journal of South American Earth Sciences*, **3**(4), 213–230.

LAMB, S. & DAVIS, P. 2003. Cenozoic climate change as a possible cause for the rise of the Andes. *Nature*, **425**, 792–797.

LAUBACHER, G. 1978. *Géologie de la Cordillière orientale et de l'Altiplano au nord et nord-ouest du Lac Titicaca (Pérou)*. Travaux et documents de l'ORSTOM, Paris, **95**.

MCQUARRIE, N., BARNES, J. B. & EHLERS, T. A. 2008. Geometric, kinematic, and erosional history of the central Andean Plateau, Bolivia (15–17S). *Tectonics*, **27**, TC3007, doi:10.1029/2006TC002054.

MÉGARD, F. 1984. The Andean orogenic period and its major structure in central and northern Peru. *Journal of the Geological Society, London*, **141**, 893–900.

NOBLE, D. C., MCKEE, E. H., EYZAGUIRRE, V. R. & MAROCCO, R. 1984. Age and regional tectonic and metallogenetic implications of igneous activity and mineralization in the Andahuaylas–Yauri belt of Southern Peru. *Economic Geology*, **79**, 172–176.

NORABUENA, E. O., DIXON, T. H., STEIN, S. & HARRISON, C. G. A. 1999. Decelerating Nazca–South America and Nazca–Pacific plate motions. *Geophysical Research Letters*, **26**, 3405–3408.

PARDO-CASAS, F. & MOLNAR, P. 1987. Relative motion of the Nazca (Farallon) and South American plate since Late Cretaceous times. *Tectonics*, **6**, 223–248.

PERELLO, J., CARLOTTO, V. *ET AL*. 2003. Porphyry-style alteration and mineralization of the Middle Eocene to Early Oligocene Andahuaylas–Yauri Belt, Cuzco region, Peru. *Economic Geology*, **98**, 1575–1605.

REINERS, P. W. & BRANDON, M. T. 2006. Using thermochronology to understand orogenic erosion. *Annual Reviews in Earth and Planetary Sciences*, **34**, 419–466.

ROPERCH, P., CARLOTTO, V. & CHAUVIN, A. 2008. Block rotations within the northern Peruvian Altiplano. *In*: *7th International Symposium on Andean Geodynamics (ISAG 2008, Nice), Extended Abstracts*, University of Nice, France, 469–472.

ROPERCH, P., SEMPERE, T. *ET AL*. 2006. Counterclockwise rotation of late Eocene-Oligocene fore-arc deposits in southern Peru and its significance for oroclinal bending in the central Andes. *Tectonics*, **25**, TC3010, doi:10.1029/2005TC001882.

ROUSSE, S., GILDER, S., FARBER, D., MCNULTY, B., PATRIAT, P., TORRES, V. & SEMPERE, T. 2003. Paleomagnetic tracking of mountain building in the Peruvian Andes since 10 Ma. *Tectonics*, **22**(5), 1048, doi:10.1029/2003TC001508.

ROUSSE, S., GILDER, S., FORNARI, M. & SEMPERE, T. 2005. Insight into the Neogene tectonic history of the northern Bolivian Orocline from new paleomagnetic and geochronologic data. *Tectonics*, **24**, TC6007, doi:10.1029/2004TC001760.

RUSSO, R. M. & SILVER, P. G. 1996. Cordillera formation, mantle dynamics, and the Wilson cycle. *Geology*, **24**(6), 511–514.

SCHILDGEN, T. F., HODGES, K. V., WHIPPLE, K. X., REINERS, P. W. & PRINGLE, M. S. 2007. Uplift of the western margin of the Andean plateau revealed from canyon incision history, southern Peru. *Geology*, **35**(6), 523–526.

SEMPERE, T., CARLIER, G. *ET AL*. 2002. Late Permian–Middle Jurassic lithospheric thinning in Peru and Bolivia, and its bearing on Andean-age tectonics. *Tectonophysics*, **345**, 153–181.

SEMPERE, T., HARTLEY, A. & ROPERCH, P. 2006. Comment on 'Rapid uplift of the altiplano revealed through C-13–O-18 bonds in paleosol carbonates'. *Science*, **314**(5800), 760.

STEINMANN, G. 1929. *Geologie von Peru*. Carl Winters Universitätsbuchhandlung, Heidelberg.

TAGAMI, T., CARTER, A. & HURFORD, A. J. 1996. Natural long-term annealing of the zircon fission-track system in Vienna Basin deep borehole samples: constraints upon the partial annealing zone and closure temperature. *Chemical Geology*, **130**, 147–157.

THOURET, J.-C., WÖRNER, G., GUNNEL, Y., SINGER, B., ZHANG, X. & SOURIOT, T. 2007. Geochronologic and stratigraphic constraints on canyon incision and Miocene uplift of the Central Andes in Peru. *Earth and Planetary Science Letters*, **263**, 151–166.

Combined titanite and apatite fission-track data from Gjelsvikfjella, East Antarctica – another piece of a concealed intracontinental Permo-Triassic Gondwana rift basin?

B. EMMEL[1]*, J. JACOBS[1,2] & M. C. DASZINNIES[1]

[1]*Fachbereich Geowissenschaften, Universität Bremen, PF 330440, 28334 Bremen, Germany*

[2]*Department of Earth Science, University of Bergen, Allegaten 41, 5007 Bergen, Norway*

**Corresponding author (e-mail: emmel@uni-bremen.de)*

Abstract: Titanite and apatite fission-track ages from Gjelsvikfjella and the eastern Mühlig–Hofmann Mountains, East Antarctica, range between 516 ± 50 and 323 ± 30 Ma and 366 ± 16 and 186 ± 9 Ma, respectively. The thermochronological data set indicates differential cooling of two tectonic blocks (Hochlinfjellet–Festninga and Risemedet). Inverse modelled time–temperature paths suggest that the Hochlinfjellet–Festninga block cooled at first below 60 °C during the mid-Palaeozoic, whereas the Risemedet block cooled during the earliest Triassic. Differential cooling is most probably related to physical separation along active faults, which is associated with Gondwana-wide intracontinental rifting. This tectonic activity shaped the landscape in the study area along structures running perpendicular to the continental margin of Dronning Maud Land. A rift locus was possibly located along the Penck–Jutul graben west of the study area. In contrast to other parts of Dronning Maud Land, Jurassic magmatism and initial break-up between East Africa and East Antarctica did not influence the apatite FT data. Modelled apatite fission-track data indicate the onset of final cooling since the Early Cretaceous, suggesting post-Cretaceous unroofing of the palaeosurfaces in eastern Dronning Maud Land.

The Gjelsvikfjella (GF) and Mühlig–Hofmann Mountains (MHm) belong to western Dronning Maud Land (DML), and are located between about 3° and 4°E at 72°S in East Antarctica (Fig. 1). DML is part of an east–west-trending mountain range with elevations rising up to 3000 m following approximately 200 km inland the contour of the continental margin (Fig. 1). DML is one example of an escarpment separating a high-elevated interior plateau (*c.* 3000 m) from a low-lying plain (±sea level) along the continental margin. The Norwegian station Troll (Fig. 1) is located in GF and, thus, a relatively good infrastructure for fieldwork is given. Hence, the area has been the subject of intensive research concerning, in particular, the Precambrian metamorphic and the Cenozoic landscape evolution (e.g. Näslund 2001; Jacobs *et al.* 2003*a*, *b*; Paulsson & Austrheim 2003). However, the mountain range comprises a variety of metamorphic rock types, but lacks sufficient brittle structures and sedimentary rocks to constrain the timing and magnitude of its morpho-tectonic development. In such a setting, low-temperature thermochronological techniques such as fission-track (FT) and (U–Th)/He dating offer the unique possibility of reconstructing the upper-crustal tectono-thermal history. In western DML (Fig. 1) Permo-Carboniferous sedimentary rocks (Larsson *et al.* 1990) overlay an undated palaeosurface and suggest a surface development prior to their deposition (Näslund 2001). Vitrinite reflectance data from Carboniferous coal seams (Bauer *et al.* 1997) and apatite FT data (Jacobs & Lisker 1999) imply that this surface was buried kilometres deep after its initial pre-Carboniferous unroofing. In our study area valuable bases for the reconstruction of the post-Carboniferous tectono-thermal evolution are palaeosurface remnants that crop out at Jutulsessen, Hoggestabben and in the western MHm (see the photographs in Fig. 1). Combining these time–temperature constraints with titanite and apatite FT data, it is possible to determine the chronology and type of rock cooling from approximately 300 °C to below 60 °C and, finally, to conditions with exposed bedrock. The inferred cooling paths can contribute to a better understanding of the escarpment-forming processes in DML and the possible relationship to ancient proximate parts within the Gondwana supercontinent.

Geological setting

The GF and the eastern MHm expose polyphase metamorphosed and deformed supracrustal basement rocks that are intruded by various plutonic rocks, which were also partly deformed, metamorphosed and migmatized (Jacobs *et al.* 2003*a*). The basement rocks comprise banded gneisses,

From: LISKER, F., VENTURA, B. & GLASMACHER, U. A. (eds) *Thermochronological Methods: From Palaeotemperature Constraints to Landscape Evolution Models*. Geological Society, London, Special Publications, **324**, 317–330.
DOI: 10.1144/SP324.21 0305-8719/09/$15.00

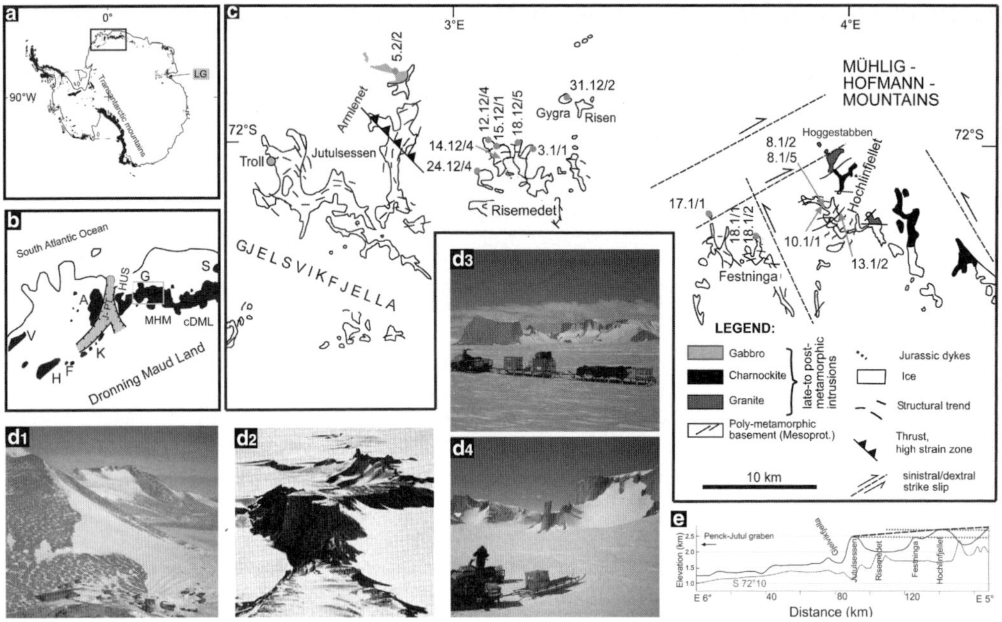

Fig. 1. (**a**) Map of the Antarctic continent. Black areas indicate outcrops. The box shows the location of panel b. (**b**) Map of the study area and adjacent areas. (**c**) Simplified geological map of Gjelsvikfjella and western Mühlig–Hofmann Mountains, with structural pattern and sample locations indicated (from Jacobs *et al.* 2003*a*; Owada *et al.* 2003). (**d**) Photographs (d_1–d_4) showing remnants of a palaeosurface near the Troll station (taken from http://www.polarfoundation.org) at the Jutulsessen nunatak (oblique aerial photograph of the eastern part of the Jutulsessen nunatak is taken from Näslund 2001), the Hoggestabben and within the Mühlig–Hoffmann Mountains. (**e**) East–west profile across the study area parallel to the continental margin at S72°00 (black line) and S72°10 (grey line), derived from the GTOPO 30 DEM dataset. The dashed line indicates the average trend of the palaeosurface. The dotted lines show another possible interpretation of two horizontal surfaces. Abbreviations: A, Ahlmannryggen; cDML, Central Dronning Maud Land; F, Fossilryggen; GF, Gjelsvikfjella; H, Heimefrontfjella; HUS, H.U. Sverdrupfjella; J-PT, Penck–Jutul graben; K, Kirwanveggen; LG, Lambert Graben; MHm, Mühlig–Hofmann Mountains; S, Schirmacher Oases; V, Vestfjella.

migmatites, charnockites, syenites and granites that are penetrated by several generations of mafic dykes (mostly not dated).

Chronology of the basement development

U–Pb zircon sensitive high-resolution ion microprobe (SHRIMP) and secondary ion mass spectrometer (SIMS) ages of zoned zircons give the best age constraint for the pre- and syn-metamorphic development. Thereby, U–Pb ages from zircon cores indicate late Middle Proterozoic crystallization of the protoliths (Jacobs *et al.* 2003*a*, *b*). The U–Pb distribution within the rims of the zoned zircons suggests metamorphism of the protoliths at about 550–500 Ma (Jacobs *et al.* 2003*a*, *b*; Paulsson & Austrheim 2003). In DML metamorphism took place under medium-pressure granulite-facies conditions, reaching temperatures of about 830 ± 20 °C (Markl & Piazolo 1999). This tectono-thermal event is interpreted as the collisional stage between East and West Gondwana (Markl & Piazolo 1999; Bauer *et al.* 2003). Late tectonic (*c.* 500 Ma) syenites, charnockites and granites occur at Stabben (Paulsson & Austrheim 2003), Hochlinfjellet (Jacobs & Bauer 2001) and in the eastern MHm (Ohta *et al.* 1990), respectively. K–Ar and Ar/Ar biotite and muscovite ages ranging between *c.* 510 and 410 Ma (Jacobs *et al.* 1995, 1996, 1999) document the initial post-collisional cooling history of DML.

No Beacon Group but Early Permian surfaces?

Along the Transantarctic Mountains (Fig. 1), a Devonian–Triassic siliciclastic succession known as the Beacon Supergroup (Barrett 1991) overlie the basement rocks. The proximal sedimentary rocks from the study area are the Permo-Carboniferous rocks from western DML (Fig. 1) (e.g. Larsson *et al.* 1990). Their maximum thickness reaches up to 140 m (Bauer *et al.* 1997). The sedimentary sequence starts with glacial deposits that discordantly overlay the crystalline basement. Palynologic analyses of the basal diamictites indicate an uppermost Carboniferous–Early Permian age (Larsson *et al.* 1990). Cross-bedded sandstones with intersected shales (with thin coal seams) overlay the glacial deposits. In GF and the western MHm Permo-Carboniferous sedimentary rocks have not been described. However, Näslund (2001) suggests that flat high-elevated plateau surfaces at Jutulsessen (Fig. 1d_2) are relicts of a regional Early Permian palaeosurface. The plain surfaces at the Hoggestabben and in the MHm (Fig. 1d_3 and d_4) are similar to the surfaces observed at the Jutulsessen. These palaeosurfaces were possibly part of a larger Early Permian surface extending over 700 km from Heimefrontfjella in western DML to the MHm (Näslund 2001).

Post-Permian structures and Jurassic volcanism

In East Antarctica denudation started at least during the Permo-Triassic, when intracontinental rift systems developed within the Gondwana supercontinent (e.g. Catuneanu *et al.* 2005; Harrowfield *et al.* 2005). The present crustal thickness has been estimated using a gravity based crustal model and indicates thinning from more than 40 km in the hinterland of the GF to about 20 km along the continental margin (Llubes *et al.* 2003, p. 112, fig. 7). Adjacent to the study area, the Penck–Jutul graben (Fig. 1b) is thought to be a Palaeozoic–Mesozoic rift system (Grantham & Hunter 1991). Within GF, mafic dykes (not yet dated) of probably Jurassic age penetrate the basement rocks and are associated with intense synchronous jointing (Jacobs & Bauer 2001). Dykes were mapped at Risemedet and Festninga (Fig. 1c). The occurences of volcanic rocks related to the Jurassic magmatism in DML indicates a decreasing thermal influence from west to east. At Vestfjella (Fig. 1b), continental flood basalts crop out as lava piles of up to about 900 m thickness overlaying basement rocks penetrated by associated dyke swarms (Luttinen *et al.* 1998). The basalts erupted mainly during a short period between 184 and 175 Ma, dated by $^{40}Ar/^{39}Ar$ feldspar and/or whole-rock ages (Duncan *et al.* 1997; Zhang *et al.* 2003), and magmatism lasted until *c.* 160 Ma as indicated by a $^{40}Ar/^{39}Ar$ phlogopite age (Luttinen *et al.* 2002). The base of the Vestfjella lava is unexposed, but basalt-hosted fragments suggest the occurrence of Permian sedimentary rocks beneath the volcanic rocks (Luttinen & Furnes 2000). Riley *et al.* (2005) dated mafic dykes from Ahlmannryggen (central DML) by $^{40}Ar/^{39}Ar$ analysis, and revealed two stages of magma emplacement at *c.* 190 and *c.* 178 Ma. In central DML, following these igneous activities, a conjugated fault system yielding offsets of some tens to hundreds of metres developed. At Hoggestabben right-lateral faults overprint Jurassic magmatic dykes (Owada *et al.* 2003, figs 5 and 6, pp. 114–115).

Post-Jurassic offshore sedimentary record

The post-Jurassic evolution of the East Antarctic continent is derived exclusively from offshore data. A seismic study by Hinz & Krause (1982) indicates that seaward-dipping reflectors (Early Jurassic volcanic rocks) are unconformably overlain by sedimentary strata that are considered to be of Cenozoic age. Between 25°W and 9°E their thickness ranges from approximately 0.7 to 2.9 km. Ocean Drilling Project (ODP) Leg 113, sites 689, 692 and 693 from the Maud Rise to the NE of the study area record sediments from the latest Cretaceous to the Cenozoic (Kennett & Barker 1990; O'Connell 1990). At site 693 terrigenous sediments dominate the Late Cretaceous (O'Connell 1990). ODP data offshore of DML suggest a temperate–subtropical Late Cretaceous continental margin (Kennett & Barker 1990). During the mid-Tertiary the climate changed dramatically, forming the Antarctic ice sheet, which has reached its continental proportions since the latest Eocene–Early Oligocene (summarized in Ingólfsson 2004). The ice-sheet growth is documented by the stepwise increase in $\delta^{18}O$ during the Miocene into the Pliocene (Miller *et al.* 1991). Shallow drilling at site KK9601 (1400 m) at the continental margin of western DML penetrated coarse Quaternary sediments (Kristoffersen *et al.* 2000). The pebble lithologies includes predominantly Jurassic continental flood basalts (dated using K–Ar to 168 ± 5 Ma: Kristoffersen *et al.* 2000) and minor contributions from metamorphic basement rocks.

Fission-track analyses

Titanite and apatite mineral concentrates from samples collected in the GF and eastern MHm were separated using conventional crushing, sieving, magnetic and heavy liquid techniques. Populations of several hundred grains were embedded in epoxy, and thereafter ground and polished to expose internal mineral surfaces. Titanites were etched individually for 17–22 min in a mixture of concentrated HF, HNO_3, HCl and H_2O (1:2:3:6), and apatites were etched for *c.* 50 s in 5% HNO_3 at room temperature to reveal spontaneous tracks (e.g. Wagner & Van den Haute 1992). Samples were irradiated in channels 7 and 8 of the Thetis reactor at the Institute for Nuclear Sciences, Ghent (Belgium) using total thermal neutron fluxes of 1×10^{16} n cm^{-2} (apatite) and

0.6×10^{16} n cm^{-2} (titanite). Mica detectors were etched for 15 min in 40% HF (room temperature) to reveal induced tracks. Titanite and apatite fission tracks were counted using an optical microscope with a total magnification of 1250×. Apatite FT confined lengths and etch-pit diameters (D_{par}) were measured with a magnification of 2000×. Fission-track ages were calculated using the Zeta approach (Hurford & Green 1983) and are given as central ages with their one sigma (1σ) errors.

Combined titanite and apatite FT data allow the cooling of samples to be constrained from *c.* 300 to 60 °C. The closing temperature of the titanite FT system is estimated to be 275 ± 25 °C (Kohn *et al.* 1993). *In situ* observations at the deep continental drill hole 'Kontinentale Tiefbohrung' in Germany give an estimate for the partial annealing zone of the titanite fission-track system of between 310 and 265 ± 10 °C (Coyle & Wagner 1998). The partial annealing zone of the apatite FT system is constrained to a temperature range of between 110 ± 10 and 60 °C (Green *et al.* 1986*a*, *b*; Wagner & van den Haute 1992 and references therein).

Previous fission-track studies in Dronning Maud Land

A reconnaissance FT study of mainly basement rocks from central DML (70°–72°S, 8°–14°E) gave titanite, zircon and apatite FT ages ranging from 551 ± 51 to 293 ± 28 Ma, 364 ± 47 to 237 ± 31 Ma and 315 ± 18 to 83 ± 3 Ma, respectively (Meier 1999). Meier (1999) suggests that the titanite and zircon FT ages reflect the continuous cooling of central DML after the Pan-African metamorphism. During the Late Palaeozoic–Early Mesozoic, separation of basement into discrete blocks occurred. This was followed by an Early Jurassic phase of enhanced exhumation related to the Bouvet–Karoo mantle plume activity. Flexural isostatic upward movement of the lithosphere controlled the central DML passive margin evolution during the mid-Cretaceous (Meier 1999).

Apatite FT ages from Heimefrontfjella in western DML (Fig. 1b) are reported by Jacobs & Lisker (1999). They range between 172 ± 17 and 81 ± 8 Ma, with a mean track length (MTL) varying between *c.* 13.5 and 12.0 μm. Permo-Carboniferous sedimentary rocks yield apatite FT ages of between *c.* 140 and 115 Ma indicating a tectono-thermal overprint after deposition. Based on modelled apatite FT data, the authors suggested that total annealing of fission tracks occurred at *c.* 180 Ma caused by the burial of the basement underneath up to about 1500–2000 m of Jurassic lava. Thereafter, samples cooled to approximately 80 °C during the Jurassic and a second cooling step occurred in the Early Cretaceous (Jacobs & Lisker 1999).

Results and interpretation of fission-track analyses

The titanite FT ages of four samples are 516 ± 50 Ma (sample 18.1/2), 367 ± 23 Ma (Sample 24.12/2), 342 ± 30 Ma (Sample 14.12/4) and 323 ± 30 Ma (Sample 5.1/2) (Table 1). All samples are from similar elevations (*c.* 1500 m), indicating that a general age increase from west (Jutulsessen, 323 ± 30 Ma) to east (Festninga, 516 ± 50 Ma) is related to a different thermo-tectonic development.

Apatite FT ages of 13 samples range between 366 ± 16 and 186 ± 9 Ma. The MTL of eight samples vary between 12.92 ± 0.18 and 11.71 ± 0.22 μm, with standard deviations ranging between 2.27 and 1.77 μm (Table 1). Three samples failed the χ^2-test, indicating mixed single-grain age populations with possibly different apatite chemistries and complex cooling histories.

In general, the apatite FT ages increase from the west (Jutulsessen) to the east (Hochlinfjellet), concordant with increasing topography, sample elevations and titanite FT ages. The samples from Jutulsessen and Risemedet were collected at elevations of approximately 1500 m, and are characterized by apatite FT ages ranging between *c.* 215 and 185 Ma. Only sample 14.12/4 (Fig. 1c and Table 1) is significantly older (277 ± 20 Ma) than the others, whereby only nine single-grain ages could be dated (Table 1). In addition, the sample is characterized by short tracks (MTL: 11.77 ± 0.29 μm), with a high standard deviation value of 2.26 μm (Table 1). However, the mean D_{par} value does not indicate substantial different apatite annealing kinetics. The only obvious difference is the lithology; the sample was taken from a Precambrian mafic dyke.

The samples from Festninga and Hochlinfjellet were collected at elevations ranging between 1500 and 2060 m, and give apatite FT ages of between 366 and 273 Ma. The oldest (18.1/1: 366 ± 16 Ma) and the youngest sample (17.1/1: 273 ± 22 Ma) are from the Festninga nunatak, both separated from the samples located in the Hochlinfjellet and from each other by transform faults (Fig. 1). Samples from Hochlinfjellet yielded within their 1σ errors overlapping apatite FT ages, from 334 ± 20 to 300 ± 17 Ma. In addition, the narrow range of MTL spanning from 12.5 to 11.7 μm indicates cooling of an undisturbed crustal block.

Without sample 18.1/1, which is separated by faulting, the apatite FT ages from Festninga–Hochlinfjellet show an undisturbed age–elevation

Table 1. *Fission-track results from basement rocks of the Gjelsvikfjella and the western Mühlig–Hofmann Mountains*

Sample Location	Elev. (m) Mineral	No of grains Rock type	(N_s) ρ_s ($\times 10^5$ cm^{-2})	(N_i) ρ_i ($\times 10^5$ cm^{-2})	(N_d) ρ_d ($\times 10^5$ cm^{-2})	$P(\chi^2)$ (%)	Fission-track age $\pm 1\sigma$ (Ma)	Mean track length $\pm 1\sigma$ (µm) SD (µm) (No.)	Mean D_{par} (µm) $\pm$ SD (No.)
8.1/2	2060	17	806	592	6614	46	320 ± 21	–	1.39 ± 0.21
Hochlinfjellet	Apatite	Intermed. dyke	16.43	12.07	14.35				(85)
8.1/5	1900	20	3556	2616	6614	18	316 ± 14	11.91 ± 0.22	1.60 ± 0.23
Hochlinfjellet	Apatite	Intermed. dyke	47.80	35.16	14.34			2.18 (98)	(396)
10.1/1	2060	17	1527	1075	6614	4	334 ± 20	12.49 ± 0.19	1.68 ± 0.41
Hochlinfjellet	Apatite	Mafic dyke	25.27	17.79	14.34			1.94 (100)	(385)
13.1/2	2000		935	726	6614	95	300 ± 17	11.96 ± 0.28	1.58 ± 0.21
Hochlinfjellet	Apatite	Migmatic gneiss	8.29	6.44	14.34			2.06 (54)	(96)
17.1/1	1800	13	793	680	6614	18	273 ± 22	11.71 ± 0.29	1.54 ± 0.20
Festninga	Apatite	Migmatite	15.01	12.87	14.34			2.27 (60)	(65)
18.1/1	1500	18	2448	1518	6614	36	366 ± 16	11.98 ± 0.19	1.61 ± 0.20
Festninga	Apatite	Leucosome	35.59	22.07	14.34			1.90 (100)	(390)
3.1/1	1520	20	1123	1222	6614	53	216 ± 11	–	1.52 ± 0.26
Risemedet	Apatite	Augen gneiss	13.83	15.05	14.35				(99)
14.12/4	1500	9	558	472	6614	30	277 ± 20	11.77 ± 0.29	1.64 ± 0.25
Risemedet	Apatite	Mafic dyke	30.57	25.86	14.35			2.26 (63)	(47)
15.12/1	1500	14	811	913	6614	2	215 ± 16	12.92 ± 0.18	1.80 ± 0.25
Risemedet	Apatite	Augen gneiss	25.89	29.15	14.36			1.77 (100)	(370)
18.12/5	1500	16	1864	2356	6614	20	186 ± 9	–	1.75 ± 0.22
Risemedet	Apatite	Intermed. gneiss	31.17	39.39	14.36				(86)
24.12/4	1500	16	1314	1475	6614	37	208 ± 11	12.67 ± 0.2	1.61 ± 0.20
Risemedet	Apatite	Augen gneiss	36.79	23.67	14.35			1.82 (85)	(335)
31.12/2	1570	18	1104	1373	6614	0	195 ± 14	–	1.57 ± 0.23
Gygra	Apatite	Granite dyke	17.50	21.77	14.35				(100)
5.1/2	1455	20	599	659	6614	18	216 ± 16	–	1.59 ± 0.23
Jutulsessen	Apatite	Gabbro	8.26	9.09	14.35				(95)
18.1/2	1500	12	1148	271	15914	3	516 ± 50	–	–
Festninga	Titanite	?	40.52	9.56	21.16				
14.12/4	1500	14	1044	380	15914	10	342 ± 30	–	–
Riesemedet	Titanite	Mafic dyke	38.01	13.84	21.11				
24.12/2	1500	13	1269	439	15914	32	367 ± 23	–	–
Risemedet	Titanite	Augen gneiss	112.93	39.07	21.05				
5.1/2	1455	18	767	301	15914	40	323 ± 25	–	–
Jutulsessen	Titanite	Gabbro	13.54	5.31	21.22				

Apatite (titanite) fission-track ages were calculated using an IRMM-540 (CN2) dosimeter glasses with a zeta value for B. Emmel of $\zeta_{IRMM\text{-}540} = 333 \pm 9$ ($\zeta_{CN2} = 124 \pm 3.25$). Fission-track ages are reported as central ages.

Elev., elevation; ρ_s, ρ_i, ρ_d, N_s, N_i, N_d, density and number of counted spontaneous, induced and dosimeter glass tracks; $P(\chi^2)$, chi-square probability.

relationship, with increasing ages with elevation (Fig. 2). The oldest sample from the highest elevation revealed the longest measured track lengths of this sample set. The apatite FT ages plotted v. the MTL revealed a half 'boomerang' which might be indicative of the transition from a fast cooling episode to slowly cooling one at *c.* >330 Ma. This trend is confirmed by the apatite FT age v. SD relationship, whereby the slowly cooled samples revealed the highest values (Fig. 2).

Modelling strategy

Apatite FT data, including single-grain apatite FT ages, track lengths, etch-pit diameters and standard deviations, were used to model specific hypothetical time–temperature (t–T) paths. An inverse modelling procedure applying the program 'HeFTy' (Ketcham 2005) was carried out using the annealing algorithm of Ketcham *et al.* (1999). As a kinetic parameter we used the D_{par} values. Because our etching protocol is different to the etching conditions proposed by Ketcham *et al.* (1999), we calibrated the measured etch pits against the Ketcham model. Therefore, we calculated a ratio by measuring D_{par} values for the common apatite FT age standards etched with our conditions (Durango, 1.58 μm; Fish Canyon, 2.16 μm) compared to published D_{par} values (Durango, 1.83 μm; Fish Canyon, 2.43 μm) from Donelick *et al.* (1999). All D_{par} measurements used for modelling were calibrated with the determined calibration factor of 1.142. Only samples with more than 80 measured track length's were chosen for modelling.

The apatite FT modelling procedure was constrained by external t–T parameters.

- The titanite FT ages give the starting conditions (the samples from the Hochlinfjellet–Festninga were at 265–310 °C at 516 ± 50 Ma; the samples from the Risemedet were at 265–310 °C at 345 ± 25 Ma).
- Palaeosurface remnants from Jutulsessen and MHm, from which the samples were collected, have been inferred to have reached surface conditions prior to the deposition of Permo-Carboniferous sediments in the Heimefrontfjella (Näslund 2001).
- Vitrinite reflectance (Bauer *et al.* 1997) and apatite FT data (Jacobs & Lisker 1999) from Permo-Carboniferous sedimentary rocks in western DML indicate post-depositional reheating to 80–100 °C. In addition, (U–Th)/He ages from the MHM and central DML span between *c.* 230 and 140 Ma (Emmel *et al.* 2006), suggesting a similar reheating scenario.
- A present surface temperature of approximately −20 °C is the final modelling constraint. Offshore ODP records indicate the development of glacial conditions within Antarctica since the Late Eocene (e.g. Ehrmann *et al.* 1992) and thus we allowed the model to cool below 0 °C since about 45 Ma.

For five samples, modelling was performed until 20 t–T paths show a goodness of fit (goodness-of-fit probability [GOF] ≥0.5) were found.

Results of inverse modelling

In the following the results of inverse modelling procedures are discussed using the best-fit modelled time–temperature path (Fig. 3). Samples from Hochlinfjellet (8.1/5 and 10.1/1) indicate similar cooling histories, with a phase of rapid cooling from about 260 to below 60 °C between 400 and 360 Ma. Thereafter, samples were reheated to maximum temperatures of *c.* 85 °C before final cooling to below 60 °C at about 120–110 Ma. The modelled apatite FT data of sample 18.1/1 from Festninga indicate an earlier first cooling step at *c.* 450 Ma. Subsequently, the sample was reheated to a maximum temperature of around 80 °C before final cooling to surface conditions started synchronous with the samples from Hochlinfjellet at about 120 Ma.

The best-fit modelled paths of samples from Risemedet (24.12/4 and 15.12/1) indicate a first phase of accelerated cooling from approximately 260 to below 60 °C at *c.* 240 Ma. Samples were reheated to maximum temperatures of *c.* 90–75 °C before final cooling to surface condition synchronous with samples from the Hochlinfjellet and Festninga at approximately 130–120 Ma.

Discussion

The thermochronological results presented earlier indicate a protracted tectono-thermal history of Gjelsvikfjella and western MHM including: (i) fast cooling after the latest metamorphism (titanite FT data); (ii) differential cooling during the Late Carboniferous–Triassic intracontinental Gondwana rift phase (titanite FT and apatite FT data); (iii) possible reheating after the Carboniferous and Triassic; and (iv) final crustal cooling to below 60 °C since the Late Cretaceous (modelled apatite FT data).

The data set thus allows the cooling history to be divided into four distinctive thermo-tectonic periods, discussed in the following subsections.

Post Pan-African cooling

The last major tectonic and metamorphic event affecting the basement rocks of DML occurred during the final amalgamation of Gondwana along the East African–Antarctic Orogen (Jacobs &

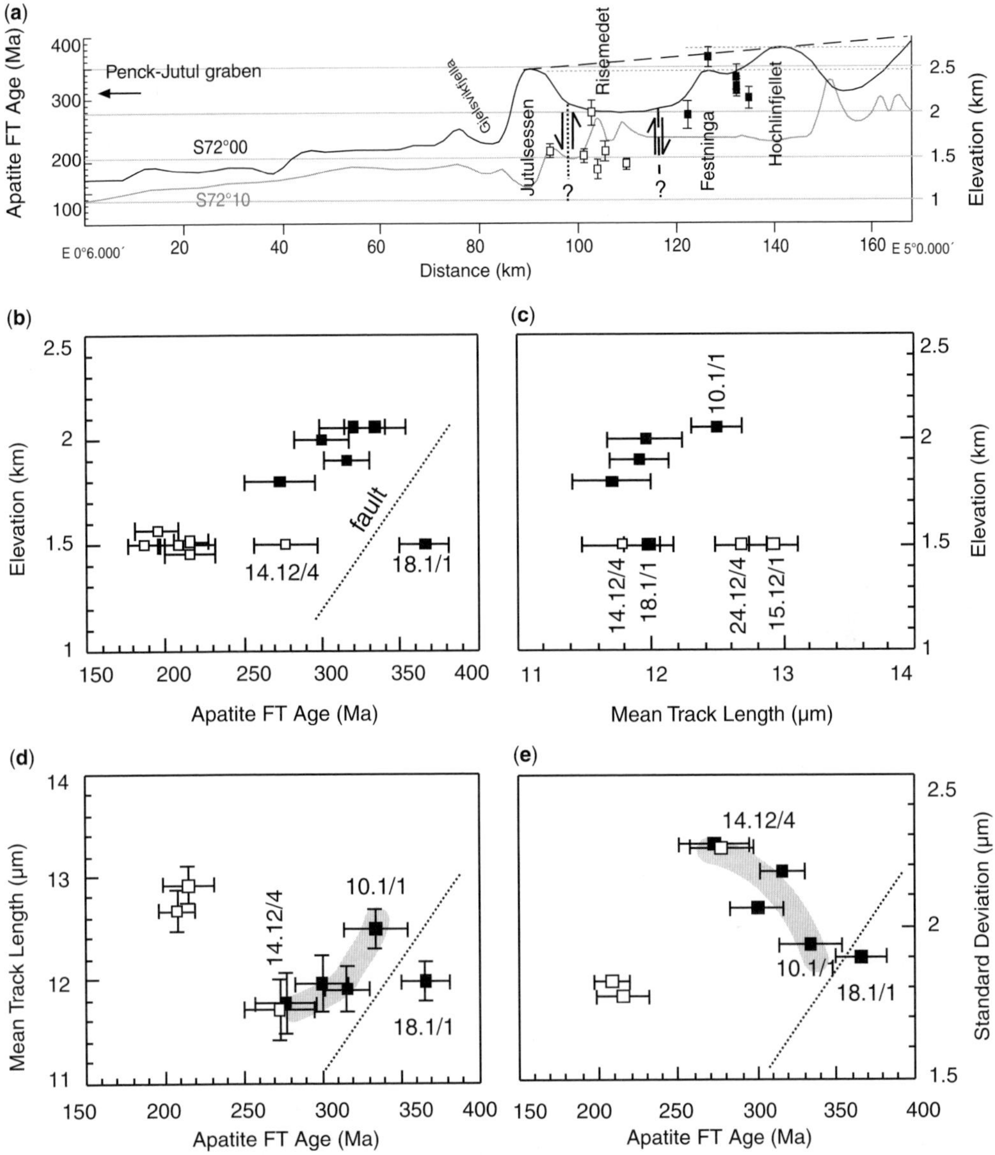

Fig. 2. Fission-track results from Gjelsvikfjella and the eastern Mühlig–Hofmann Mountains. Samples from Hochlinfjellet–Festninga are displayed as black boxes, and samples from Risemedet and Jutulsessen as white boxes. (**a**) Margin-parallel east–west profile across the working area at S72°00 (black line) and S72°10 (grey line) with apatite FT ages. (**b**) Apatite FT age–elevation relationship. Samples from the Risemedet are all from similar elevations, giving similar apatite FT ages. Samples from the Hochlinfjellet–Festninga show the expected age increase with elevation. Sample 18.1/1 does not follow the trend apparently caused by block separation strike-slip structures. (**c**) Apatite mean track lengths plotted against elevation. The sample set from the Hochlinfjellet–Festninga shows an increase in track length with elevation. (**d**) Apatite FT age v. mean track-length diagram showing a half 'boomerang' type distribution of the pre-Jurassic samples, indicating an episode of relatively fast cooling prior to 330 Ma. This half boomerang is supported by (**e**) the apatite FT age v. standard deviation relationship.

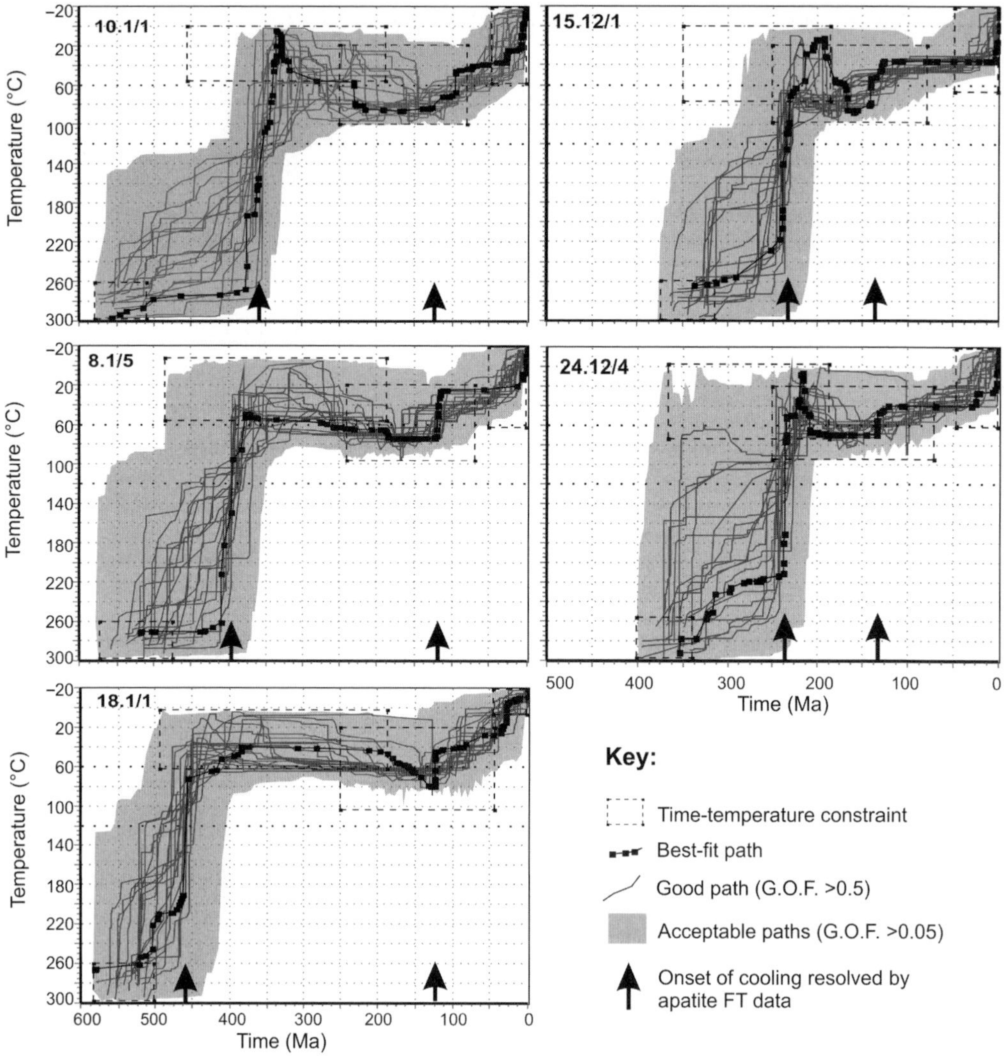

Fig. 3. Modelled cooling paths of representative samples from the Hochlinfjellet, Festninga and Risemedet nunataks. Onset of first cooling below *c.* 110 °C occurred at Hochlinfjellet during the 400–450 Ma episode. In contrast, onset of the first cooling at Risemedet occurred at *c.* 240 Ma. During this time both crustal segments were most probably separated along structures perpendicular to the continental margin. GOF, goodness of fit.

Thomas 2004) between about 550 and 500 Ma, as constrained by zircon U–Pb SHRIMP and SIMS ages (Jacobs *et al.* 2003*a*, *b*; Paulsson & Austrheim 2003). Our titanite FT data document the post-tectonic cooling to below *c.* 300 °C. Sample 18.1/2 from the Festninga nunatak has a titanite FT age of 516 ± 50 Ma. The titanite FT age overlaps within its 1σ error with the rim zircon SHRIMP age of 510 ± 14 Ma obtained from sample 18.1/2 at the same locality (Jacobs *et al.* 2003*a*). These ages suggest fast cooling from approximately 800 to *c.* 300 °C soon after the latest metamorphism. K–Ar and Ar/Ar biotite and muscovite ages from DML, ranging between *c.* 510 and 410 Ma (Jacobs *et al.* 1995, 1996, 1999), support this interpretation. In contrast, the samples from the Risemedet and Jutulsessen nunataks, with titanite FT ages between *c.* 370 and 320 Ma, indicate a slower post-metamorphic cooling (<500 Ma) or, alternatively, an Early Carboniferous tectonic–thermal event that did not affect the sample from the Festninga nunatak.

Intracontinental rifting – relationship to a central Gondwana rift system

All apatite FT ages (*c*. 320–180 Ma), except sample 18.1/1 (366 ± 16 Ma) and two titanite FT ages from Risemedet, match the timing of the intracontinental rifting and initial break-up of Gondwana at *c*. 300–180 Ma (e.g. Cantenau *et al.* 2005; Harrowfield *et al.* 2005). The oldest apatite FT ages from Hochlinfjellet overlap within their 1σ errors, with the youngest titanite FT ages from the Risemedet indicating differentially exhumed crustal levels. In addition, inverse modelled t–T paths argue for a differential first cooling step at Hochlinfjellet (*c*. 450–350 Ma) and Risemedet (*c*. 240 Ma) (Fig. 3). Both constraints display differential cooling of the Risemedet and the Hochlinfjellet–Festninga nunataks during Palaeozoic–Mesozoic times (Figs 1c & 3). All inverse modelled t–T paths show a second cooling step related to reheating (Fig. 3). In DML, no indications exist for a regional Permo-Triassic volcanic and/or metamorphic event. Since the end of the Palaeozoic, the main geological process affecting the working area was continental rifting related to the fragmentation of the Gondwana supercontinent. Onshore sediments associated with rifting are known from Heimefrontfjella, and corresponding vitrinite reflectance data argue for reheating to temperatures of *c*. 80 °C after Late Carboniferous–Early Permian deposition (Bauer *et al.* 1997). The modelled cooling related to Mesozoic reheating (Fig. 3) can best be explained by burial heating due to a kilometre thick sedimentary cover above the dated basement rocks. The maximum palaeotemperatures obtained by inverse modelling of the apatite FT data vary between about 70 and 90 °C before final cooling started in the Late Cretaceous (Fig. 3). Based on apatite FT data, Lisker *et al.* (2003) modelled Cretaceous palaeo-geothermal gradients for the basement along the Lambert rift, an intra-continental Gondwana rift system situated in East Antarctica. The gradients range between 29 ± 7 °C km^{-1} (adjacent to the rift) and 19 ± 4 °C km^{-1} (remote to the rift). If we assume a generic Mesozoic geothermal gradient ranging between 30 and 20 °C km^{-1}, reheating of up to *c*. 70–90 °C would be due to burial by a sedimentary cover with a thickness of 2.3–4.5 km. During the Permo-Triassic, Antarctica was part of the Gondwana supercontinent (Fig. 4) and since the Carboniferous large intracontinental Gondwana rift basins have developed (Veevers 2004). Within these basins, kilometre-scale thick sedimentary

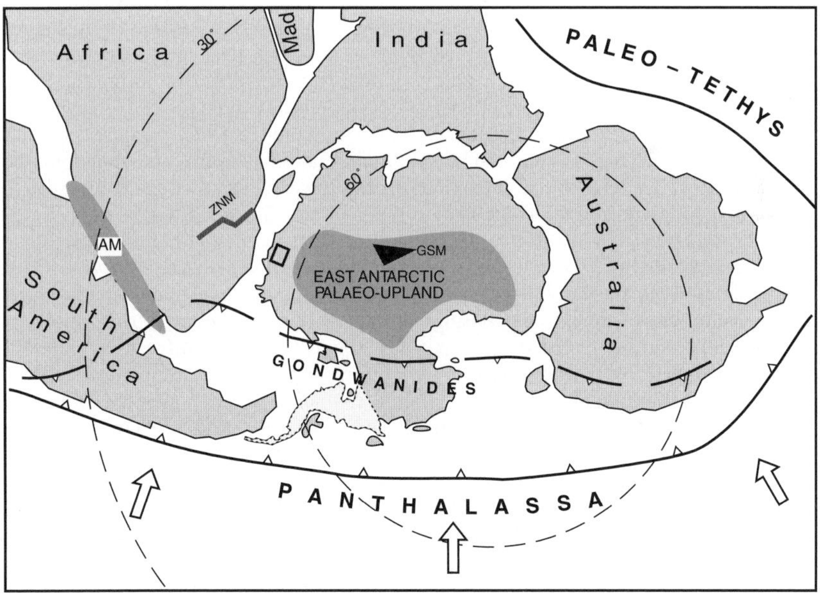

Fig. 4. Reconstruction of the Early Permian southern Pangea continent configuration (redrawn from Veevers *et al.* 1994). The black box shows the location of the study area adjacent to the Permo-Triassic upland in central Antarctica (dark grey colour). The Penck–Jutul graben is located west of the box. The dashed line indicates an evolving rift shoulder from the Zambezi rift system to the northern Mozambican basement (ZNM). The arrows indicate subduction along the panthalassan margin of Gondwana. Abbreviations: AM, Atlantic Mountains; Mad, Madagascar; GSM, Gamburtsev Mountains.

successions were deposited (e.g. Catuneanu *et al.* 2005). The centre of Antarctica was probably a palaeo-highland and served as a source for the intracontinental rift basins on the margins of the upland (Veevers *et al.* 1994). Another possible source for sediments is the evolving rift system along the line from the Zambezi rift system to the north Mozambican basement (Fig. 4). Titanite FT ages from northern Mozambique, ranging between *c.* 285 and 220 Ma, suggest the cooling of a denuding east–west-trending rift flank during the Permian (Daszinnies *et al.* 2009). Within the Gondwana supercontinent, DML had a proximal position to the palaeo-highland in Antarctica and the evolving rift shoulder in northern Mozambique (Fig. 4), and, thus, DML was possibly subjected to permanent sediment input during the Permo-Triassic. Further indications for rift-related tectonics are given by sub-ice depressions mapped NW of the Heimefrontfjella and between the H. U. Sverdrupfjella and the Kirwanveggen (e.g. Grantham & Hunter 1991). These are interpreted to be part of a Palaeozoic–Mesozoic rift system with its locus at the Penck–Jutul graben (e.g. Grantham & Hunter 1991). A recent study of the sub-glacial topography of DML visualizes a second smaller-scaled NW–SE-trending depression between GF and the MHm parallel to the Penck–Jutul graben (Steinhage *et al.* 1999, p. 270, fig. 5). However, proximal Permo-Carboniferous sediments crop out at the Heimefrontfjella and at the Fossilryggen approximately 500 km SW of the study area. The preserved thicknesses of these sequences are relatively thin (up to 140 m), but vitrinite reflectance data (Bauer *et al.* 1997) suggest palaeotemperatures similar to those from the modelled apatite FT data of the present study. Therefore, we propose a scenario with fast differential cooling during the Palaeozoic and Mesozoic followed by reheating due to sedimentary overburden. Whilst the Hochlinfjellet–Festninga cooled quickly during the 450–350 Ma episode, the Risemedet was exhumed later during the Triassic. Differential cooling of the Risemedet and the Hochlinfjellet–Festninga was guided by brittle structures, and exhumation was probably related to the evolution of an intracontinental rift. These rift-related tectonics shaped the landscape parallel to the continental margin and are reflected in the present topography along the escarpment (Fig. 1e). This suggested a long-term influence on the present landscape, which is supported by a comparative study from Indian, Australian and East Antarctic rift systems by Harrowfield *et al.* (2005). The authors proposed that the more than 300 km-wide Antarctic continental shelf is primarily a Permian feature and was only a little modified by the Mesozoic break-up of Gondwana.

In the study area, a large-scale palaeosurface at the Jutulsessen, possibly stretching to the Hochlinfjellet (see the photographs in Fig. 1), is suggested to be an erosional surface formed prior to the Early Permian (Näslund 2001). The FT data indicate that this surface must have been developed before a modelled reheating event, i.e. prior to *c.* 300 Ma. During the Triassic, segmentation of the palaeosurface occurred due to differential cooling of the basement blocks (Fig. 3) along brittle structures.

Jurassic magmatism and Gondwana break-up

Two samples from Risemedet (18.12/5 and 31.12/2) yielded Jurassic apatite FT ages (*c.* 195–178 Ma), and basaltic dykes (most probably Jurassic) were mapped at the Festninga and Risemedet nunataks (Fig. 1) suggesting that igneous activities affected the area. Jurassic magmatic activities are well known and described in DML (see earlier in this paper). Basalts were erupted during a short period between *c.* 184 and 175 Ma (Duncan *et al.* 1997; Zhang *et al.* 2003), and are associated with mantle plume activities (Storey 1995). Volcanism clearly predates the development of the first oceanic crust between Africa and Antarctica at approximately 155 Ma (Jokat *et al.* 2003). This igneous event is interpreted to have reset all apatite FT ages in the Heimefrontfjella, western DML (Jacobs & Lisker 1999). Moreover, Jacobs & Lisker (1999) suggested that a 1.5–2 km-thick lava pile, associated with mantle-plume-related volcanism during the Early Jurassic, covered the basement rocks. However, as the titanite FT ages from the study area are unaffected by this Jurassic thermal event, an exceptional heat source in the lower crust is considered to be less likely. Titanite and zircon FT ages from central DML are all far older than 180 Ma and, therefore, also argue against a plume beneath this area (Meier 1999). Furthermore, the inverse modelled t–T paths do not indicate a Jurassic thermal overprint (Fig. 3). Thus, we suggest that the 'Jurassic' apatite FT ages are not associated with cooling after the total resetting due to mantle-plume-related volcanism.

Gjelsvikfjella and western MHm are situated more than 500 km away from Heimefrontfjella. Obviously, only marginal igneous activities related to a mantle plume affected our working area. Instead, our data probably give evidence for the eastern thermal limit of the Bouvet–Karoo mantle plume.

Cretaceous development of the continental margin

Modelled apatite FT data (Fig. 3) revealed a distinct cooling step during the Early Cretaceous (*c.* 140–120 Ma) and cooling to below 60 °C at least since the Late Cretaceous (*c.* 90 Ma). Within DML apatite FT ages related to Cretaceous cooling are known from the Heimefrontfjella (Jacobs & Lisker 1999) and the Schirmacher Oases (Meier 1999). At the latter location (elevation *c.* 50 m) Meier (1999) yielded apatite FT ages ranging between approximately 117 and 83 Ma, suggesting that these samples were at temperatures of about 110 °C during the Cretaceous. Our modelled t–T paths yielded maximum palaeotemperatures of approxiamtely 75–90 °C during the Cretaceous, suggesting varying palaeo-temperatures of around 20–35 °C for samples from the Schirmacher Oases and GF at this time. The samples presented here are from elevations of about 1500–2000 m (Table 1), giving an elevation difference between Schirmacher Oasis and GF of 1450–1950 m. Using Cretaceous palaeogeothermal gradients of approximately 20–30 °C km^{-1} (Lisker *et al.* 2003), the palaeotemperature differences would equal a vertical elevation difference of about 650–1750 m. This suggests that parts of the present relief are inherited from the Cretaceous when erosion along the margin (Schirmacher Oases) was much higher than in the hinterland (Gjelsvikfjella). Interestingly, sedimentary records in OPD Leg 113, site 689 from the Maud Rise NE of the study area start with Late Cretaceous sediments, and sedimentation continues until the Cenozoic (Kennett & Barker 1990), and at site 693 the Late Cretaceous is dominated by terrigenous sediments.

The triggers of the Cretaceous cooling episode were most probably the opening of the south Atlantic, and the onset of rifting between Antarctica and the India–Sri Lanka block. The separation of Africa and South America started in the Early Cretaceous (e.g. Hay *et al.* 1999), contemporaneous with rifting between Sri Lanka and Antarctica (Lawyer *et al.* 1991). In addition, a temperate–subtropical Late Cretaceous climate along the margin of East Antarctica is indicated by ODP data (Kennett & Barker 1990). Both factors may have contributed to the margin development, whereby rifting caused mechanical instabilities (strike-slip structures in Fig. 1) and the climate led to enhanced denudation due to chemical weathering.

Cenozoic unroofing and glacial erosion

Because already minor inclined surfaces are unstable over geological periods (e.g. Bonov *et al.* 2006 and references therein), the occurrence of pre-Permian palaeosurfaces in our working area (Näslund 2001) suggests a geological recent unroofing to surface conditions. The modelled t–T paths suggest that a sedimentary overburden preserved the palaeosurfaces at least until the Late Cretaceous (Fig. 3). Based on geomorphological interpretations from Jutulsessen, Näslund (2001) proposed erosion by wet-based glaciers during the Middle Eocene–Late Eocene. Since the middle Miocene, the development of relatively stable cold-based ice sheets preserved the gross morphology of this region. Furthermore, Näslund (2001) suggests that since the Oligocene the main erosion was by ice streams that only deepened the tectonic troughs separating the isolated nunataks (Fig. 1). Modelled time–temperature paths indicate cooling below the sensitivity of apatite fission-track analysis (<60 °C) since about 100 Ma. Cooling below *c.* 60 °C can not be resolved with apatite FT data. However, still about 80 °C (with a mean surface temperature of −20 °C) are left for rock cooling until the samples reach surface conditions potentially containing information for a distinct, but undetected, cooling step. ODP data indicate post-Cretaceous phases of basement erosion, e.g. sediments from Leg 113, Site 689 from the Maud Rise NE of the working area show a mica layer identified at depths related to the *c.* 42–44 Ma episode. This is interpreted to be caused by an increase in erosion in response to the development of glacial conditions in East Antarctica (Diester-Haass 1995).

Conclusions

Combined titanite and apatite FT ages suggest differential cooling of basement blocks during Palaeozoic–Mesozoic intracontinental Gondwana rift evolution. During this time the characteristics of the present topography along the evolving continental margin developed. The data support the assumptions of a major Permian architecture of the east Antarctic shelf (Harrowfield *et al.* 2005). The main cooling event affecting all samples occurred prior to the Jurassic passive margin evolution. In contrast to western DML, a significant Jurassic tectono-thermal event is not recorded at GF and the MHm. A second cooling step from about 90 to below 60 °C occurred during the Late Cretaceous where parts of the present morphology parallel to the margin developed. Differential erosion shaped the landscape relief between the Late Cretaceous and the Oligocene. Fission-track data indicate a post-Cretaceous unroofing of palaeosurfaces in central DML.

This project was financially supported in part by Deutsche Forschungsgemeinschaft grant JA 617/24. We thank A. Läufer for providing samples from the Mühlig Hofmann Mountains. The quality of the paper benefited from reviews by J.-O. Näslund and M. Zattin.

References

BARRETT, P. J. 1991. The Devonian to Triassic Beacon Supergroup of the Transantarctic Mountains and correlatives in other parts of Antarctica. *In*: TINGEY, R. J. (ed.) *The Geology of Antarctica*. Oxford Monographs on Geology and Geophysics, **17**, 120–152.

BAUER, W., HAGEMANN, H. W., POSCHER, R. F., SACHSENHOFER, R. F. & SPAETH, G. 1997. Permian coals from Western Dronning Maud Land – composition, environment, and the influence of Jurassic magmatism on their maturity. *In*: RICCI, C. A. (ed.) *The Antarctic Region: Geological Evolution and Processes*. Siena, 945–951.

BAUER, W., FIELITZ, W., JACOBS, J., FANNING, C. M. & SPAETH, G. 2003. Mafic dykes from Heimefrontfjella and implications for the post-Grenvillian to pre-Pan-African geological evolution of western Dronning Maud Land, Antarctica. *Antarctic Science*, **15**, 379–391.

BONOV, J. M., LIDMAR-BERGSTRÖM, K. & JAPSEN, P. 2006. Palaeosurfaces in central West Greenland as reference for identification of tectonic movements and estimation of erosion. *Global and Planetary Change*, **50**, 161–183.

CATUNEANU, O., WOPFNER, H., ERIKSSON, P. G., CAIRNCROSS, B., RUBIGE, B. S., SMITH, R. M. H. & HANCOX, P. J. 2005. The Karoo basins of south-central Africa. *Journal of African Earth Sciences*, **43**, 211–253.

COYLE, D. A. & WAGNER, G. A. 1998. Positioning the titanite fission-track partial annealing zone. *Chemical Geology*, **149**, 117–125.

DASZINNIES, M. C., JACOBS, J., WARTHO, J.-A. & GRANTHAM, G. H. 2009. Post Pan-African thermo-tectonic evolution of the north Mozambican basement and its implication for the Gondwana rifting. Inferences from $^{40}Ar/^{39}Ar$ hornblende, biotite and titanite fission-track dating. *In*: LISKER, F., VENTURA, B. & GLASMACHER, U. A. (eds) *Thermochronological Methods: From Palaeotemperature Constraints to Landscape Evolution Models*. Geological Society, London, Special Publications, **324**, 25–50.

DISTER-HAASS, L. 1995. Middle Eocence to early Oligocene paleoceanography of the Antarctic Ocean (Maud Rise, ODP Leg 113, Site 689): change from a low to a high productivity ocean. *Palaeogeograhy, Palaeoclimatology, Palaeoecology*, **113**, 311–334.

DONELICK, R. A., KETCHAM, R. A. & CARLSON, W. D. 1999. Variability of apatite fission-track annealing kinetics; II, Crystallographic orientation effects. *American Mineralogist*, **84**, 1224–1234.

DUNCAN, R. A., HOOPER, P. R., REHACEK, J., MARSH, J. S. & DUNCAN, A. R. 1997. The timing and duration of the Karoo igneous event, southern Gondwana. *Journal of Geophysical Research*, **102**, 18,127–18,138.

EHRMANN, W. U., MELLES, M., KUHN, G. & GROBE, H. 1992. Significance of clay mineral assemblages in the Antarctic Ocean. *Marine Geology*, **107**, 249–273.

EMMEL, B., JACOBS, J., DASSZINIES, M. C. & CROWHURST, P. 2006. Combined apatite fission-track and single grain apatite (U–Th)/He ages – a possibility to identify thermal overprinted crustal segments in Dronning Maud Land (East Antarctica). *In*: *European Conference on Thermochronology*. Schriftenreihe der Deutschen Geologischen Gesellschaft, **49**, 40.

GRANTHAM, G. H. & HUNTER, D. R. 1991. The timing and nature of faulting and jointing adjacent to the Pencksökket, western Dronning Maud Land, Antarctica. In: THOMSON, M. R. A., CRAME, J. A. & THOMSON, J. W. (eds) *Geological Evolution of Antarctica*. Cambridge University Press, Cambridge, 47–51.

GREEN, P. F., DUDDY, I. R., GLEADOW, A. J. W., TINGATE, P. R. & LASLETT, G. M. 1986*a*. Fission-track annealing in apatite; track length measurements and the form of the Arrhenius plot. *Nuclear Tracks and Radiation Measurements*, **10**, 323–328.

GREEN, P. F., DUDDY, I. R., GLEADOW, A. J. W., TINGATE, P. R. & LASLETT, G. M. 1986*b*. Thermal annealing of fission tracks in apatite 1. A Qualitative Description. *Chemical Geology*, **59**, 237–253.

HARROWFIELD, M., HOLDGATE, G. R., WILSON, C. J. L. & MCLOUGHLIN, S. 2005. Tectonic significance of the Lambert graben, East Antarctica: reconstructing the Gondwanan rift. *Geology*, **33**, 197–200.

HAY, W. W., DECONTO, C. N. *ET AL*. 1999. An alternative global Cretaceous palaeogeography. *In*: BARRERA, E. & JOHNSON, C. (eds) *The Evolution of Cretaceous Ocean/Climate Systems*. Geological Society of America, Special Publications, **332**, 1–48.

HINZ, K. & KRAUSE, W. 1982. The continental margin of Queen Maud Land, Antarctica: Seismic sequences, structural, elements and geological development. *Geologisches Jahrbuch*, **E23**, 17–41.

HURFORD, A. J. & GREEN, P. F. 1983. The zeta age calibration of fission-track dating. *Isotope Geoscience*, **1**, 285–317.

INGÓLFSSON, Ó. 2004. The Quaternary glacial and climate history of Antarctica. *In*: EHLERS, J. & GIBBARD, P. L. (eds) *Quaternary Glaciations of the World, Part III*. Kluwer, Dordrecht, 3–43.

JACOBS, J. & BAUER, W. 2001. Gjelsvikfjella and Mühlig–Hofmann-Gebirge (E-Antarctica): another piece of the East Antarctic Orogen? *Zeitschrift für Geologische Wissenschaften*, **152**, 249–259.

JACOBS, J. & LISKER, F. 1999. Post permian tectono-thermal evolution of western Dronning Maud Land, East Antarctica: an apatite fission-track approach. *Antarctic Science*, **11**, 451–460.

JACOBS, J. & THOMAS, R. J. 2004. Himalayan-type indenter-escape tectonics model for the southern part of late Neoproterozoic–early Paleozoic East African–Antarctic orogen. *Geology*, **32**, 721–724.

JACOBS, J., AHRENDT, H., KREUTZER, H. & WEBER, K. 1995. K–Ar, ^{40}Ar-^{39}Ar and apatite fission track evidence for Neoproterozoic and Mesozoic basement rejuvenation events in the Heimefrontfjella and Mannefallknausane (East Antarctica). *Precambrian Research*, **75**, 251–263.

JACOBS, J., BAUER, W. & FANNING, C. M. 2003*b*. New age constraints for Grenville-age metamorphism in western central Dronning Maud Land (East Antarctica), and implications for the palaeogeography of Kalahari in Rodinia. *International Journal of Earth Sciences*, **92**, 301–315.

JACOBS, J., BAUER, W., SPAETH, G., THOMAS, R. J. & WEBER, K. 1996. Lithology and structure of the Grenville-aged (~1.1 Ga) basement of Heimefrontfjella (East Antarctica). *Geologische Rundschau*, **85**, 800–821.

JACOBS, J., FANNING, C. M. & BAUER, W. 2003*a*. Timing of Grenville-age vs. Pan-African medium- to high grade metamorphism in western Dronning Maud Land (East Antarctica) and significance for correlations in Rodinia and Gondwana. *Precambrian Research*, **125**, 1–20.

JACOBS, J., THOMAS, R. J., ARMSTRONG, R. A. & HENJES KUNST, F. 1999. Age and thermal evolution of the Mesoproterozoic Cape Meredith Complex, West Falkland. *Journal of the Geological Society, London*, **156**, 917–928.

JOKAT, W., BOEBEL, T., KÖNIG, M. & MEYER, U. 2003. Timing and geometry of early Gondwana breakup. *Journal of Geophysical Research*, **108**(B9), 2428.

KENNETT, J. P. & BARKER, P. F. 1990. Latest Cretaceous to Cenozoic climate and oceanographic developments in the Weddell Sea, Antarctica: an Ocean Drilling perpective. *In*: BARKER, P. F., KENNETT, J. P. *ET AL.* (eds) *Proceedings of the Ocean Drilling Program, Scientific Results*, **113**, 937–960.

KETCHAM, R. A. 2005. Forward and Inverse modeling of low-temperature thermochronometry data. *Reviews in Mineralogy and Geochemistry*, **58**, 275–314.

KETCHAM, R. A., DONELICK, R. A. & CARLSON, W. A. 1999. Variability of apatite fission-track annealing kinetics: III. Extrapolation to geological time scales. *American Mineralogist*, **84**, 1235–1255.

KRISTOFFERSEN, Y., STRAND, K., VORREN, T., HARWOOD, D. & WEBB, P. 2000. Pilot shallow drilling on the continental shelf, Dronning Maud Land, Antarctica. *Antarctic Science*, **12**, 463–470.

KOHN, B. P., WAGNER, M. E., LUTZ, T. M. & ORGANIST, G. 1993. Anomalous Mesozoic thermal regime, central Appalachian Piedmont; evidence from sphene and zircon fission track dating. *Journal of Geology*, **101**, 779–794.

LARSSON, K., LINDSTROM, S. & GUY-OHLSON, D. 1990. An early Permian palynoflora from Milorgfjella, Dronning Maud Land, Antarctica. *Antarctic Science*, **2**, 331–344.

LAWVER, L. A., ROYER, J.-Y., SANDWELL, D. T. & SCOTESE, C. R. 1991. Crustal development: Gondwana break-up – evolution of the Antarctic continental margins. *In*: THOMPSON, M. R. A., CRAME, J. A. & THOMPSON, J. (eds) *Geological Evolution of Antarctica*. Cambridge University Press, Cambridge, 533–540.

LISKER, F., BROWN, R. & FABEL, D. 2003. Denudational and thermal history along a transect across the Lambert Graben, northern Prince Charles Mountains, Antarctica, derived from apatite fission track thermochronology. *Tectonics*, **22**, 1055.

LLUBES, M., FLORSCH, N., LEGRESY, B., LEMOINE, J.-M., LOYER, S., CROSSLEY, D. & RÉMY, F. 2003. Crustal thickness in Antarctica from CHAMP gravimetry. *Earth and Planetary Science Letters*, **212**, 103–117.

LUTTINEN, A. V. & FURNES, H. 2000. Flood basalts of Vestfjella: jurassic magmatism across an Achaean–Proterozoic lithospheric boundary in Dronning Maud Land, Antarctica. *Journal of Petrology*, **41**, 1271–1305.

LUTTINEN, A. V., RAMO, O. T. & HUHMA, H. 1998. Neodymium and strontium isotopic and trace element composition of a Mesozoic CFB suite from Dronning Maud Land, Antarctica: implications for lithosphere and asthenosphere contributions to Karoo magmatism. *Geochimica et Cosmochimica Acta*, **62**, 2701–2714.

LUTTINEN, A. V., ZHANG, X. & FOLAND, K. A. 2002. 159 Ma Kjakebeinet lamproites (Dronning Maud Land, Antarctica) and their implications for Gondwana breakup processes. *Geological Magazine*, **139**, 525–539.

MARKL, G. & PIAZOLO, S. 1999. Stability of high-Al titanite from low-pressure calcsilicates in light of fluid and host-rock composition. *American Mineralogist*, **84**, 37–47.

MEIER, S. 1999. Paleozoic and Mesozoic tectono-thermal history of central Dronning Maud Land, East Antarctica – evidence from fission-track thermochronology. Berichte zur Polarforschung, Alfred-Wegner-Institut für Polar- und Meeresforschung, Bremerhaven.

MILLER, K. G., WRIGHT, J. D. & FAIRBANKS, R. G. 1991. Unlocking the ice house: Oligocene–Miocene oxygen isotopes eustasy and margin erosion. *Journal of Geophysical Research*, **96**, 6829–6848.

NÄSLUND, J.-O. 2001. Landscape development in western and central Dronning Maud Land, East Antarctica. *Antarctic Science*, **13**, 302–311.

O'CONNELL, S. B. 1990. Sedimentary facies and depositional environment of the Lower Cretaceous East Antarctic margin: sites 692 and 693. *Proceedings of the Ocean Drilling Program, Scientific Results*, **113**, 71–88.

OHTA, Y., TORUDBAKKEN, B. O. & SHIRAISHI, K. 1990. Geology of Gjelsvikfjella and western Muhlig-Hofmannfjella, Dronning Maud Land, east Antarctica. *Polar Research*, **8**, 99–126.

OWADA, M., BABA, S., LAÜFER, A., ELEVOLD, S., SHIRAISHI, K. & JACOBS, J. 2003. Geology of eastern Mühlig–Hofmannfjella and Filchnerfjella in Dronning Maud Land, East Antarctica: a preliminary report on a Japan–Norway–Germany joint geological investigation. *Polar Geoscience*, **16**, 108–136.

PAULSSON, O. & AUSTRHEIM, H. 2003. A geochronological and geochemical study of rocks from Gjelsvikfjella, Dronning Maud Land, Antarctica – Implications for Mesoproterozoic correlations and assembly of Gondwana. *Precambrian Research*, **125**, 113–138.

RILEY, T. R., LEAT, P. T., CURTIS, M. L., MILLAR, I. L., DUNCAN, A. R. & FAZEL, A. 2005. Early–middle Jurassic dolerite dykes from Western Dronning Maud Land (Antarctica): identifying mantle source in the Karoo large igneous province. *Journal of Petrology*, **46**, 1489–1524.

STEINHAGE, D., NIXDORF, U., MEYER, U. & MILLER, H. 1999. New maps of the ice thickness and subglacial topography in Dronning Maud Land, Antarctica, determined by means of airborne radio-echo sounding. *Annals of Glaciology*, **29**, 267–272.

STOREY, B. C. 1995. The role of mantle plumes in continental break-up: case histories from Gondwanaland. *Nature*, **377**, 301–308.

VEEVERS, J. J. 2004. Gondwanaland from 650–500 Ma assembly through 320 Mamergerin Pangea to 185–100 Ma breakup: supercontinental tectonics via stratigraphy and radiometric dating. *Earth Science Reviews*, **68**, 1–132.

VEEVERS, J. J., POWELL, C. M., COLLINSON, J. W. & LOPEZ-GAMUNDI, M. 1994. Synthesis. *In*: VEEVERS, J. J. & POWELL, C. MCA. (eds) *Permian–Triassic Pangean Basins and Foldbelts Along the Panthalassan Margin of Gondwanaland.* Geological Society of America, Memoirs, **184**, 331–352.

WAGNER, G. A. & VAN DEN HAUTE, P. 1992. *Fission-Track Dating*. Ferdinand Enke Verlag, Stuttgart, 285 pp.

ZHANG, X., LUTTINEN, A. V., ELLIOT, D. H., LARSSON, K. & FOLAND, K. A. 2003. Early stages of Gondwana breakup: The $^{40}Ar/^{39}Ar$ geochronology of Jurassic basaltic rocks from western Dronning Maud Land, Antarctica, and implications for the timing of magmatic and hydrothermal events. *Journal of Geophysical Research B:-Solid-Earth*, **108**, ECV 7-1–7-19.

Fission-track analysis of the Atotsugawa Fault (Hida Metamorphic Belt, central Japan): fault-related thermal anomaly and activation history

RYUJI YAMADA[1]*, HASBAATOR ONGIRAD[2], TATSUO MATSUDA[1], KENTARO OMURA[1], AKIRA TAKEUCHI[3] & HIDEKI IWANO[4]

[1]*National Research Institute for Earth Science and Disaster Prevention, Tennodai 3-1, Tsukuba 305-0006, Japan*

[2]*Public Works Research Institute, Minamihara 1–6, Tsukuba 305-8516, Japan*

[3]*Department of Earth Science, University of Toyama, Gofuku 3190, Toyama 930-8555, Japan*

[4]*Kyoto Fission-Track Co. Ltd, Minamitajiri-cho 44-4, Kyoto 603-8832, Japan*

**Corresponding author (e-mail: ryamada@bosai.go.jp)*

Abstract: Fission-track (FT) thermochronology was applied to the Atotsugawa Fault in the Hida Metamorphic Belt, central Japan, to detect any ancient thermal anomaly associated with fault displacements. Apatite and zircon grains from gouges (*c.* 2 cm wide) and fractured rocks (*c.* 10 cm) at six fracture zones within a 15 m-wide fault zone were dated. Most of the zircon (120–150 Ma) and apatite (44–60 Ma) ages agree well with emplacement ages for the granites that intrude the Hida Belt. The discordance in zircon and apatite FT ages is interpreted to reflect cooling due to regional uplift and associated erosion. A thermal anomaly was identified at one of the gouge zones that showed an exceptionally young apatite age (*c.* 32 Ma) with an unimodal FT length distribution. It presumably indicates secondary heating induced by frictional slip during an earthquake, possibly giving a younger limit of the initiation of the activity in the Hida Belt.

In order to investigate a long-term landscape evolution of a certain region, it is important to reveal the activity history of the faults that exist in the region. Abundant frictional heat during a fault displacement is implied by the occurrence of pseudotachylyte, a melt rock in the fault zone (e.g. Sibson 1975). The thermal footprint within rocks in fault zones can possibly be detected by low-temperature thermochronometry (e.g. Scholz *et al.* 1979). Fission-track (FT) thermochronology has recently been applied to detect the thermal anomaly in fault zones, such as the Nojima Fault in central Japan (Murakami *et al.* 2002; Yamada *et al.* 2007*a*), and the San Gabriel fault zone in southern California (d'Alessio *et al.* 2003). The FT method is suitable for detecting thermal anomaly associated with fault activity because: (1) the minerals commonly used for analysis (apatite, zircon) are relatively resistant to weathering so that they are likely to survive under the hydrothermal conditions in some fault zones; (2) closure temperatures for FT methods are relatively low (*c.* 110 and 240 °C for apatite and zircon, respectively; e.g. Gallagher *et al.* 1998), so they can serve as sensitive indicators for thermal events in the upper crust; and (3) well-documented annealing kinetics (e.g. Laslett & Galbraith 1996; Yamada *et al.* 2007*b*) allow quantitative modelling of a rock's thermal history (e.g. Ketcham *et al.* 1999). From the view point of seismology, the amount of heat produced along a fault is essential information when investigating fault strength, which controls the earthquake generation process. The significance of this kind of approach is to assess directly the parameters related to fault strength with natural samples that are marked by ancient earthquakes, independent from laboratory friction experiments.

In this study, we applied FT thermochronology to reveal an activity history of the Atotsugawa Fault, central Japan, by detecting palaeo-thermal anomalies induced by fault activities. This fault, 60 km in length, lies within the Hida Belt, which cuts through Palaeozoic metamorphics and Mesozoic granites. The most marked characteristic of this fault is the coexistence of creeping and locking sections along the same fault system (Geographical Survey Institute, Japan 1997). It is important to know when and how the activity started in order to understand the mechanism of earthquake occurrence, as well as to reveal the regional tectonic history. We sampled gouges and fractured rocks at six fracture zones within a 15 m-wide fault zone in

From: LISKER, F., VENTURA, B. & GLASMACHER, U. A. (eds) *Thermochronological Methods: From Palaeotemperature Constraints to Landscape Evolution Models*. Geological Society, London, Special Publications, **324**, 331–337.
DOI: 10.1144/SP324.22 0305-8719/09/$15.00

the eastern segment of the fault zone, where possible creeping has been reported (Geographical Survey Institute, Japan 1997). The merit of using samples from small fracture zones is that the dimensions and alignment of samples can directly be measured for characterizing the thermal anomaly in detail. To identify an ancient thermal anomaly related to the fault, we compared FT ages of the fracture zone with those of the granitic rocks that intrude the metamorphic country rock (Matsuda *et al.* 1997). The present analysis seems to have identified an ancient thermal anomaly in a sample of the fault gouge. Also, the regional uplift and erosion history in the Hida Belt will be discussed based on the discordance in apatite and zircon FT ages found in this analysis.

Outline of the Atotsugawa Fault

The Atotsugawa Fault is a strike-slip fault with a right-lateral displacement along 60 km surface trace (Matsuda 1966). The strike of the fault trace is approximately N60°E, and the fault plane is almost vertical (90° ± 10°) near the surface. There have been large historical earthquakes along the fault; one of the largest events was the 1858 Ansei Hietsu earthquake with *M* 7.0 (Mikumo *et al.* 1988). A spatial heterogeneity of the seismicity along the Atotsugawa Fault is suggested by the distribution of epicentres beneath the trace of the fault (e.g. Mikumo *et al.* 1988). The seismicity in the central segment of the fault is rather low compared with the eastern and western segments. This segment seems to correspond to a possible creep zone implied by the geodetic observation of the baseline length across the fault trace. The Geographical Survey Institute, Japan (1997) found evidence of a creep-like movement, with a rate of approximately 1.5 mm year^{-1} in the central segment, and no observable baseline change in the western segment.

A geological map of the Hida Belt around the Atotsugawa Fault is shown in Figure 1a. The basement consists of the Hida metamorphic rocks of Triassic or earlier age, and two types of granites, namely Funatsu and Shimonomoto types, ranging in age from Triassic to Jurassic. Marine and non-marine sedimentary rocks of the Tetori group, ranging in age from Jurassic to Cretaceous, unconformably overlie the metamorphic and granitic units. Several geochronological studies have been undertaken on the metamorphic and granitic units in the Hida Belt. The reported ages cluster around three periods of approximately 170–190, 210–230 and 330 Ma (Ota & Itaya 1989). Sohma & Kunugiza (1993) argued that these periods reflect the main metamorphism of granulite facies at 330 Ma, regional metamorphism of amphibolite facies at 250–200 Ma and contact metamorphism resulting from the intrusion of Funatsu-type granites at 170–200 Ma, respectively. In order to reveal the cooling history and the effect of regional uplift, Matsuda *et al.* (1997) measured the ages of the Shimonomoto-type granites with various dating methods including FT data (Fig. 2b). They concluded that: (1) Shimonomoto-type granites intruded at 200–180 Ma and cooled to about 300 °C at 150 Ma; and (2) since 150 Ma the granites have slowly cooled to the surface due to uplift and erosion at a uniform rate. The initiation of fault activity is presumed to be during the late Tertiary based on geological and geomorphological studies (Matsuda 1966).

Samples and experiments

Six fracture zones are within an outcrop 15 m long near the portal of the Kamioka Mine prospect tunnel, located on the right bank 1.5 km upstream from the confluence of the Atotsugawa and the Takahara rivers (*c.* 350 m altitude: Fig. 1b). Deformation and alteration of the Hida metamorphic rocks in the fracture zones were more advanced than those in the surrounding area. Each fracture zone consists of a gouge zone 1–3 cm wide, with fractured rocks in a surrounding zone 10–15 cm wide on both sides of the fault gouge. FT samples were collected from both the gouge and fractured rocks (200–400 g each) about 10 cm apart in each of the six fracture zones. These were called samples ATG1G-6G (gouges) and ATG1F-6F (fractured rocks) (Table 1). The conditions of these samples were not good in general because many grains had cracks on grain surfaces probably due to deformation in the fault zone. To investigate the effect of the fracture zones, two reference samples were also sampled where no fractures were observed, one from the vicinity of the fault outcrop (ATGR1) and another from the Miyagawa area, that is presumed to be a locked segment of the fault (ATGR2) located about 15 km west of the outcrop. Consequently, 14 samples were analysed in total. The locations of the samples are identified by the distance to the tentative reference point, shown in Figure 1b.

FT analysis was carried out using the external detector method (ED1: Gleadow 1981). Ages were calculated following the zeta approach (Hurford 1990). A detailed description of the experimental procedure followed and the system calibration are documented in Danhara *et al.* (2003). Thirty grains with good shapes and homogeneous spontaneous track distributions were randomly chosen from each sample, except for the apatites from ATG1C and 3C. Sufficient numbers of grains were not obtained from these two samples because of the poor condition of the grains.

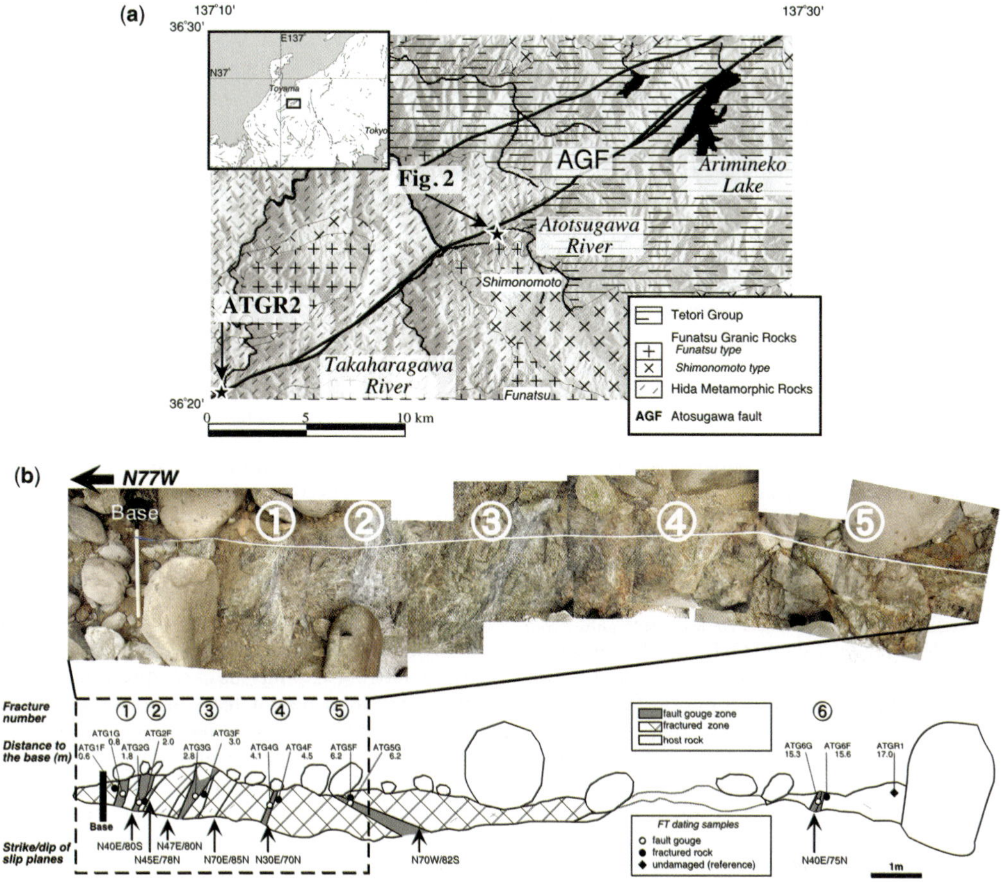

Fig. 1. (**a**) A simplified geological map around the Atotsugawa Fault (modified after Yamada *et al.* 1989; Kano *et al.* 1999). The location of ATGR2 sample is also shown. (**b**) Sketches of the fault outcrop and sampling locations. Distances were measured to the tentative reference point, shown as 'Base' in the pictures.

Results and interpretation

Analytical results are listed in Table 1. For some samples that did not pass the χ^2 test (Galbraith 1981) at the 5% significance level, the mean ρ_s/ρ_i ratios were quoted (Wagner & Van den Haute 1992). Lower $P(\chi^2)$ values for zircon samples with the ED1 method can be explained by non-Poissonian variation that is attributed to the difference in uranium contents above and below the observed internal surfaces due to the zoned distribution of uranium (Danhara *et al.* 1991). Lower $P(\chi^2)$ values for apatites indicate a broad range of single-grain ages within each sample, probably due to partial resetting or uneven thermal effects by secondary heating events.

FT ages are plotted against sample distance to the tentative reference point (Fig. 2a). Open and solid circles (as shown in Fig. 1b) indicate fault gouge and fractured rocks, respectively. For comparison, the age distribution for the apatite and zircon samples of the Shimonomoto-type granitic rocks (Fig. 2b; modified after Matsuda *et al.* 1997) are also shown as shaded bands behind the plot. Zircon FT ages are concordant among samples within 2σ errors, and their distribution also agrees with that of the granitic rocks. The effect of fracture zones is negligible with respect to the zircon ages, as indicated by the insignificant difference in mean age of samples from fracture zones (131 ± 6 Ma, 1σ) and that of reference samples (142 ± 6 Ma), where the mean ages are weighted by the square of the error of the ages. Apatite FT ages are concordant among samples within 2σ error, including those with lower $P(\chi^2)$ values, and their distribution also agrees with that of the granitic rocks; except for ATG1G, a gouge sample from a fracture zone. Compared with zircon FT ages, apatite FT ages are

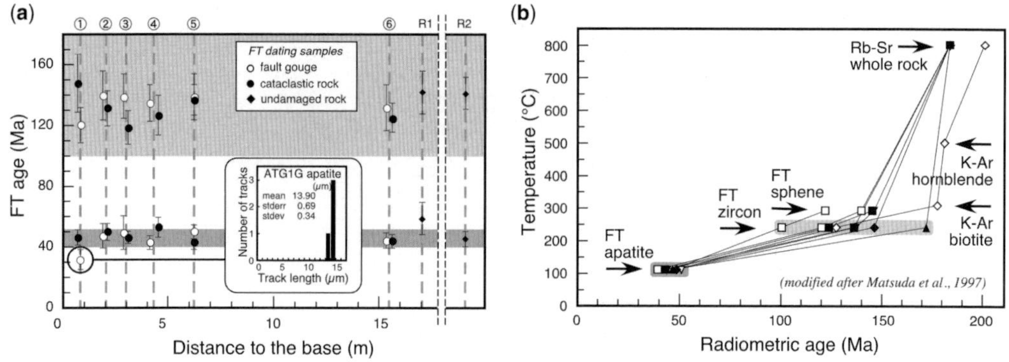

Fig. 2. (**a**) Distribution of apatite and zircon FT ages in the fault outcrop samples. Open and solid circles (as shown in Fig. 1b) indicate fault gouge and fractured rocks, respectively. Length distribution of apatite FTs for sample ATG1G is also shown. Shaded bands behind the plot indicate the apatite and zircon FT age distributions of the Shimonomoto-type granites (Matsuda *et al.* 1997). (**b**) Cooling histories of the Shimonomoto-type granites revealed by various radiometric dating methods, modified after Matsuda *et al.* (1997). Different symbols indicate samples from different rock bodies. Error bars are 2σ.

exceedingly young for each rock samples analysed. With the exclusion of ATG1G apatite, the difference in mean age of samples from fracture zones (47.2 ± 2.6 Ma, 1σ) and that of reference samples (48.7 ± 3.1 Ma) is negligible.

The age of ATG1G apatite was 32.1 ± 3.2 Ma, a significantly younger age at the 2σ level, compared with other apatites samples within a fault outcrop 15 m long. In order to characterize the thermal history of this sample, the distribution of track lengths was analysed, as shown in Figure 2a. A typical narrow-length distribution of confined tracks (Gallagher *et al.* 1998) indicates that this sample cooled rapidly after it underwent the secondary heating up to the temperature range over the partial annealing zone. Because many of the grains in this sample have cracks on their surfaces, there were not enough confined tracks to enable numerical modelling of the thermal evolution of this sample.

Discussion

Regional uplift and erosion of the Hida Belt

The discordance in apatite and zircon FT ages, which have different closure temperatures, is considered to reflect thermotectonic evolution due to slow cooling of the host rocks. Matsuda *et al.* (1997) revealed the slow cooling process of the Shimonomoto-type granitic rocks by means of thermochronological analyses including the FT method, and suggested that the slow cooling process may be attributed to regional uplift and erosion of the Hida Belt. Both apatite and zircon FT ages of the Hida metamorphic rocks obtained in this study are mostly concordant with those of the Shimonomoto-type granitic rocks (Matsuda *et al.* 1997), except the apatite age from one of the gouge samples. This indicates that both the Hida metamorphic rock and the Shimonomoto-type granites in the Hida Belt were uplifted concurrently, at least to a level corresponding to temperatures cooler than the closure temperature of the zircon FT system (*c.* 240 °C). Because samples were collected only from a small faulted outcrop, it is not adequate to use them for detailed landscape evolution. The ranges of zircon and apatite FT ages were, however, concordant with those of granites that intrude this region, and the sampling area was confirmed not to have heterogeneous exhumation history.

Skarn deposits in the Kamioka Mine, distributed widely in the middle of the Hida Belt, were formed by chemical metasomatism of limestone in the Hida metamorphic rocks. The time of the formation of these deposits could be marked by thermal impact related to hydrothermal metamorphism. The formation of the deposits has been dated as Late Cretaceous–Early Paleogene (hornblende K–Ar ages of 63.8–67.5 Ma: Sato and Uchiumi 1990). These reported ages are, however, considerably younger than zircon FT ages of the metamorphic rocks obtained in this study, although the zircon FT system is more thermally sensitive than the hornblende K–Ar system. This might be because the deposits were formed by very local hydrothermal activities in each metallogenetic provinces and so did not overprint the zircon FT ages in the studied area, reflecting the regional uplift history.

Table 1. *Results of FT analysis*

Sample	Distance (m)	No. of crystals	Spontaneous		Induced		$P(\chi^2)$ %	$\overline{\rho_s/\rho_i} \pm \sigma$	Dosimeter		Age ± 1σ (Ma)
			ρ_s	N_s	ρ_i	N_i			ρ_d	N_d	
ZIRCON											
ATG1G	0.8	30	12.99	(5027)	1.406	(544)	54		0.06940	(3553)	121 ± 6
ATG1F	0.6	30	15.37	(9988)	1.418	(922)	<5	11.35 ± 0.72	0.06943	(3555)	148 ± 9
ATG2G	1.8	30	12.94	(7093)	1.285	(704)	<5	10.77 ± 0.62	0.06919	(3543)	140 ± 8
ATG2F	2.0	30	14.28	(7282)	1.412	(720)	94		0.06922	(3544)	132 ± 6
ATG3G	2.8	30	15.65	(5197)	1.458	(484)	90		0.06913	(3539)	140 ± 7
ATG3F	3.0	30	20.70	(5961)	2.264	(652)	13		0.06916	(3541)	119 ± 5
ATG4G	4.1	30	17.86	(6626)	1.728	(641)	9		0.06946	(3556)	135 ± 6
ATG4F	4.5	30	16.77	(5064)	1.725	(521)	22		0.06949	(3558)	127 ± 6
ATG5G	6.2	30	10.03	(8782)	1.022	(895)	<5	10.69 ± 0.57	0.06931	(3549)	139 ± 7
ATG5F	6.2	30	18.11	(11481)	1.726	(1094)	7		0.06934	(3550)	137 ± 5
ATG6G	15.3	30	14.22	(6273)	1.508	(665)	<5	10.17 ± 0.57	0.06925	(3546)	132 ± 7
ATG6F	15.6	30	15.21	(9232)	1.582	(960)	51		0.06928	(3547)	125 ± 5
ATGR1	17.0	30	18.02	(5963)	1.650	(546)	26		0.06937	(3552)	142 ± 7
ATGR2	(∼15 000)	30	11.72	(12196)	1.075	(1119)	80		0.06910	(3538)	142 ± 5
APATITE											
ATG1G		12	0.27	(124)	1.321	(601)	72		0.9321	(3728)	32.1 ± 3.2
ATG1F		30	0.70	(817)	2.403	(2804)	<5	0.303 ± 0.016	0.9317	(3727)	47.1 ± 2.5
ATG2G		30	0.46	(423)	1.621	(1480)	<5	0.318 ± 0.027	0.9060	(3624)	48.1 ± 4.1
ATG2F		30	0.72	(605)	2.152	(1803)	37		0.9063	(3625)	50.7 ± 2.6
ATG3G		23	0.28	(122)	0.836	(366)	41		0.9055	(3622)	50.4 ± 5.4
ATG3F		30	0.64	(817)	2.076	(2632)	27		0.9058	(3623)	46.9 ± 2.2
ATG4G		30	0.42	(620)	1.493	(2189)	34		0.9313	(3725)	44.0 ± 2.2
ATG4F		30	0.52	(787)	1.547	(2338)	<5	0.346 ± 0.020	0.9309	(3724)	53.7 ± 3.1
ATG5G		30	0.61	(939)	1.807	(2801)	90		0.9072	(3629)	50.7 ± 2.2
ATG5F		30	0.28	(408)	0.972	(1403)	12		0.9074	(3630)	44.0 ± 2.7
ATG6G		30	0.68	(453)	2.279	(1520)	9		0.9066	(3626)	45.1 ± 2.6
ATG6F		30	0.51	(630)	1.719	(2137)	7		0.9069	(3628)	44.6 ± 2.3
ATGR1		30	0.64	(461)	1.881	(1364)	<5	0.383 ± 0.033	0.9324	(3730)	59.6 ± 5.1
ATGR2		30	0.23	(494)	0.750	(1626)	9		0.9049	(3620)	45.9 ± 2.6

Distances are measured to the tentative reference point, shown in Figure 1b. Track densities (ρ) are in 10^6 tracks cm^{-2}, while the numbers of tracks (N) are given in parentheses. Samples were analysed with the external detector method, using the ζ-calibration with NIST-SRM612 glass ($\zeta = 330 \pm 5$ for apatite and 380 ± 3 for zircon; Danhara *et al.* 2003). $P(\chi^2)$ is the probability of obtaining a χ^2 value for ν degrees of freedom, where ν = (number of crystals − 1) (Galbraith 1981). Some samples did not pass the χ^2 test, as can be derived from the $P(\chi^2)$ value. For these samples, the mean $\overline{\rho_s/\rho_i}$ ratios and the errors are quoted.

Thermal anomaly in a fracture zone

Secondary heating is presumed for ATG1G because the discordance in apatite ages was found at a very narrow range in a fracture zone. This indicates the existence of a thermal anomaly in the gouge in the fracture zone, and a relationship between the secondary heating and local geological phenomena at a gouge. Local change in apatite FT age around a fault plane was also found for the San Gabriel Fault (d'Alessio *et al.* 2003) and the Nojima Fault (Murakami & Tagami 2004; Yamada *et al.* 2007*a*).

Possible sources for the secondary heating are: (1) frictional heat during a fault displacement; and (2) heat transfer or dispersion via geothermal fluids at the interior of the crust (e.g. Murakami *et al.* 2002). The circulation of hot fluids within the fracture zone is a possible heat source associated with faults, because a fracture zone can be a channel of high-temperature fluid from the deep crust when an earthquake occurs. This is, however, implausible in this case because in the investigated area only a small, approximately 2 cm-wide, gouge zone is present, and if the fluid conveys the heat in such a narrow channel it must be higher than 500 °C according to the numerical assessment of heat diffusion (Mizoguchi *et al.* 2009), which may also have affected the apatite ages of fractured rocks beside ATG1A. Therefore, frictional heat generated during a fault displacement that formed the gouge in the fracture zone may have caused the thermal anomaly. Although this fault gouge zone could have been activated several times, it may safely be assumed that the secondary heating was caused by a single displacement at the first fracturing of the gouge zone because of following reasons: (1) it is expected that the fault strength may be maximum and the amount of the frictional heat is the largest at the first fracturing of a particular gouge zone; (2) the effect of accumulated heat is not significant because frictional heat generated by each activation should diffuse into the vicinity of rocks with normal thermal diffusivity during the recurrence interval of this fault system (*c.* 3000 years: Takeuchi *et al.* 2003); and (3) FT annealing is more sensitive to a change in temperature than to duration, as indicated by the decrease in temperatures in the partial annealing zone of approximately 40 °C (apatite) and 20 °C (zircon) for a magnitude of longer annealing duration, estimated by using the published kinetics (e.g. Laslett & Galbraith 1996; Yamada *et al.* 2007*b*).

If the apatite FT age of sample ATG1G was affected by the frictional heat during a fault displacement at the gouge zone, it can be presumed that this age may approximate a younger limit of the initiation of the activity of the Atotsugawa Fault, to which this gouge zone belongs. This is concordant with Matsuda (1966), who suggested that the activity of the Atotsugawa Fault started in the late Tertiary, based on the displacement of landforms and the slip rate of the recent activity of the Atotsugawa Fault.

The numerical modelling of the heat generation by frictional slip and the subsequent heat conduction with respect to the geometry and alignment of the samples can provide the constraints on estimates of fault strength. If a single displacement on the fracture zone is assumed to cause the secondary heating of the ATG1G sample, the amount of the heat can be constrained by the fact that the zircon FT age was not reset to the apatite age.

Conclusions

Both apatite and zircon FT ages of the Hida metamorphic rocks obtained from a fault outcrop are mostly concordant with those of the Shimonomoto-type granitic rocks that intrude the Hida Belt (Matsuda *et al.* 1997). The discordance in these ages suggests that both the metamorphics and the granites in the Hida Belt were uplifted concurrently, at least to a level corresponding to cooler temperatures than the closure temperatures of the zircon FT system (*c.* 240 °C).

An apatite FT age from one of the gouge zones was significantly younger than other samples at the 2σ level with an unimodal FT length distribution, indicating the secondary heating up to the temperature range over the partial annealing zone. This thermal anomaly in a small, approximately 2 cm-wide gouge zone may have been caused by frictional heat during a fault displacement at the first fracturing that formed the gouge in the fracture zone. Therefore, the apatite FT age of sample ATG1G (32.1 ± 3.2 Ma) may approximate a younger limit of the activity of the Atotsugawa Fault. This is consistent with a previous estimate of the initiation during the late Tertiary, based on geological and geomorphological studies (Matsuda 1966).

We thank Drs E. Fukuyama, A. Kubo and F. Yamashita for their help with sample collection, and Dr K. Mizoguchi for his discussion on thermal history analyses. We appreciate detailed reviews by Prof. P. Kamp and Dr F. Lisker.

References

D'ALESSIO, M. A., BLYTHE, A. E. & BÜRGMANN, R. 2003. No frictional heat along the San Gabriel fault, California: evidence from fission-track thermochronology. *Geology*, **31**, 541–544.

DANHARA, T., IWANO, H., YOSHIOKA, T. & TSURUTA, T. 2003. Zeta calibration values for fission track dating with a diallyl phthalate detector. *Journal of the Geological Society of Japan*, **109**, 665–668.

DANHARA, T., KASUYA, M., IWANO, H. & YAMASHITA, T. 1991. Fission-track age calibration using internal and external surfaces of zircon. *Journal of the Geological Society of Japan*, **97**, 977–985.

GALBRAITH, R. F. 1981. On statistical models for fission track counts. *Mathematical Geology*, **13**, 471–488.

GALLAGHER, K., BROWN, R. & JOHNSON, C. 1998. Fission track analysis and its applications to geological problems. *Annual Review of Earth and Planetary Science*, **26**, 519–572.

GEOGRAPHICAL SURVEY INSTITUTE, JAPAN. 1997. Crustal movements in the Chubu and the Hokuriku Districts. *Report of the Coordinating Committee for Earthquake Prediction*, **57**, 520–524.

GLEADOW, A. J. W. 1981. Fission-track dating method: what are the real alternatives? *Nuclear Track Detection*, **2**, 105–117.

HURFORD, A. J. 1990. Standardization of fission-track dating calibration: recommendation by the Fission Track Working Group of the I.U.G.S. Subcommission of Geochronology. *Chemical Geology*, **80**, 171–178.

KANO, K., HARAYAMA, T. *ET AL*. 1999. *Geological Map of Japan 1:200 000, Kanazawa*. Geological Survey of Japan, Tsukuba.

KETCHAM, R. A., DONELICK, R. A. & CARLSON, W. D. 1999. Variability of apatite fission-track annealing kinetics: III. Extrapolation to geological time scales. *American Mineralogist*, **84**, 1235–1255.

LASLETT, G. M. & GALBRAITH, R. 1996. Statistical modelling of thermal annealing of fission tracks in apatite. *Geochimica et Cosmochimica Acta*, **60**, 5117–5131.

MATSUDA, T. 1966. Strike-slip faulting along the Atotsugawa Fault. *Bulletin of the Earthquake Research Institute*, **44**, 1179–1212 (in Japanese with English abstract).

MATSUDA, T., GOTO, A., MORINAGA, H. & KANO, T. 1997. Thermal history of Jurassic 'Shimonomoto-type granitoids' in the Hida belt, Southwest Japan. *Fission Track News Letter*, **10**, 43–44 (in Japanese).

MIKUMO, T., WADA, H. & KOIZUMI, M. 1988. Seismotectonics of the Hida region, central Honshu, Japan. *Tectonophysics*, **147**, 95–119.

MIZOGUCHI, K., YAMADA, R. & FUKUYAMA, E. 2009. Fault strength inferred from frictional heat detected by fission-track thermochronology: an example of the Atotsugawa fault, Japan. *Tectonophysics* (submitted).

MURAKAMI, M., TAGAMI, T. & HASEBE, N. 2002. Ancient thermal anomaly of an active fault system: zircon fission-track evidence from Nojima GSJ 750 m borehole samples. *Geophysical Research Letters*, **29**, 2123, doi: 10.1029/2002GL015679.

MURAKAMI, M. & TAGAMI, T. 2004. Dating pseudotachylyte of the Nojima fault using the zircon fission-track method. *Geophysical Research Letters*, **31**, L12604, doi: 10.1029/2004GL020211.

OTA, K. & ITAYA, T. 1989. Radiometric ages of granitic and metamorphic rocks in the Hida metamorphic belt, central Japan. *The Bulletin of the Hiruzen Research Institute, Okayama University of Science*, **15**, 1–25.

SATO, K. & UCHIUMI, S. 1990. K–Ar ages and mineralization of the Kamioka Pb–Zn skarn deposit in the Hida terrain, Japan. *Mining Geology*, **40**, 389–396 (in Japanese with English abstract).

SCHOLZ, C. H., BEAVAN, J. & HANKS, T. C. 1979. Frictional metamorphism, argon depletion, and tectonic stress on the Alpine fault, New Zealand. *Journal of Geophysical Research*, **84**, 6770–6782.

SIBSON, R. H. 1975. Generation of pseudotachylite by ancient seismic faulting. *Geophysical Journal of the Royal Astronomical Society*, **43**, 775–794.

SOHMA, T. & KUNUGIZA, K. 1993. The formation of the Hida nappe and the tectonics of Mesozoic sediments: the tectonic evolution of the Hida region, Central Japan. *Memoirs of the Geological Society of Japan*, **42**, 1–20 (in Japanese with English abstract).

TAKEUCHI, A., ONGIRAD, H. & AKIMITSU, T. 2003. Recurrence interval of big earthquakes along the Atotsugawa fault system, central Japan: results of seismogeological survey. *Geophysical Research Letters*, **30**, 8011, doi: 10.1029/2002GL014957.

WAGNER, G. & VAN DEN HAUTE, P. 1992. *Fission-track Dating*. Kluwer, Dordrect.

YAMADA, N., NOZAWA, T., HARAYAMA, S., TAKIZAWA, F. & KATO, H. 1989. *Geological Map of Japan 1:200 000, Takayama*. Geological Survey of Japan.

YAMADA, R., MATSUDA, T. & OMURA, K. 2007*a*. Apatite and zircon fission-track dating from the Hirabayashi-NIED borehole, Nojima Fault, Japan: evidence for anomalous heating in fracture zones. *Tectonophysics*, **443**, 153–160, doi: 10.1016/j.tecto.2007.01.014.

YAMADA, R., MURAKAMI, M. & TAGAMI, T. 2007*b*. Statistical modelling of annealing kinetics of fission tracks in zircon; Reassessment of laboratory experiments. *Chemical Geology*, **236**, 75–91.

Index

Note: Page numbers in *italics* denote figures, those depicted in **bold** denote tables.

absolute calibration 42
absolute dating 10
accelerator mass spectrometry 222
advective heat transfer 121, 122
    isotherms 120–121
AFT *see* apatite fission track
age, predicted and observed 106–107, 121–122
age standards 37, 42, 49, **50**, *51*, 159, 296, 322
    zircon U–Th/He 49
age-elevation relationship 6–8, 95, *96*, 105, *106*, 108, 111, 123, 313, 320–322
    East Antarctica *323*
    Gotthard transect, Switzerland *116*
    TFT 276
AHe *see*, apatite U–Th–Sm/He
Alpine belt 141–152
    bedrock apatite FT ages 141–143, *142*
    erosion pattern 148–149
    Gotthard transect, Switzerland *122*
    tectonic setting 141–143, *142*
    see also Po Basin; Po Delta; Po Plain
Altai Mountains *238*
    basin map *242*
    digital terrain model *238*
    tectonic setting 237–239
Andaman Flysch, origin 65–69
Andaman Islands 65–70
    apatite grain ages *68*
    apatite Nd ratios *68*, *69*
Andean orogen 307
annealing 1, 89, 295
    algorithm 73, 322
    and apatite chemistry 213, 296
    kinetics 331
    model 73–74, 89, *189*
    properties 296
    time dependence 3
    time-temperature dependence 3, 5, 336
    zircon 47
    see also multi-kinetic annealing model; radiation enhanced annealing
apatite 1, 213
    ages *127*, **130**
    annealing properties 173, 296
    chemical composition 73–74, 156, 162, 184, 203, 248, 252, 296
    chlorine content 73, 193, 200, 203, 212–213, *212*, 296
    chlorine/fluorine ratio 296
    F-apatite annealing equation 173
    fluorine content 73, 296
    neodymium isotopic composition 60
    partial annealing zone 170, 177, 229, 248, *250–251*, 252, *297*, 307, 311, *312*, 313, 320
    provenance 57–72
    uranium content *41*
    see also bedrock apatite
apatite fission track 1, 118, 143, 307
    ages 64–65
    analysis 320–322
    analytical methods 164, 168–170, 194–200, 223, 291
    applications of 10–11
    effect of sedimentary processes on 144
    and erosion rates 228–230
    length distribution *334*
    modelling 5, 322
    and neodymium data 69, 70
    provenance studies *9*, 69
    resetting 69
apatite fission track ages 131–132, 218, *294*
    distribution 133, 134–135, *134*
    and elevation *295*, *297*
    and etch pit diameter *296*
    interpretation 203
    and lithology 200
    and mean track length 322
    radial plots 170, *172*, *188*
    spatial distribution *295*
    and uranium content 207–209, *212*, 213
    vs. chlorine content *201*
apatite fission track data **77**, *78–79*, 156, 217, 237
    Ardennes 167–179, *169*, 170–173, **171**, *172*
    Australia *211*
    calibration 74
    Canada *210*, *211*
    comparison **80**, *81*
    convective heat transfer 94–96
    Corsica **231**, *232*
    Finland 202–213, *209*, *211*, *212*
    Gjelsvikfjella 317–330
    Gotthard transect, Switzerland *114*, **115**
    Himalayas *63*, *64*, **66**, *67*
    horizontal confined track length 202, *203*
    interpretation 3, 94–96
    isotherm shape 113–116
    radial plots *159*, *161*
    Sierra de Cameros **157–158**, 162, 164
    South Africa 287–306
    Teletskoye graben 244–245, 248–254, **249**
apatite neodymium data *59*, 69, 70
    Andaman Islands *68*, *69*
    Bhutan *61*
    database for Himalayas 62
    detrital sediments 61, *62*
apatite (U–Th–Sm)/He 1
    analytical methods 194–200
    analytical uncertainties 200
    cratonic lithosphere 194
    multi grain analysis 200–202
    single grain analysis 200–202
apatite (U–Th–Sm)/He age 193
    anomalous 194
    and effective uranium concentration 206, *206*
    and grain size 204–206
apatite (U–Th–Sm)/He data
    Finland **197–199**, 203–207, *205*
    partial retention zone 204
    reliability 203

apatite (U–Th)/He age *107*, 153
  Sierra de Cameros 156, *161*, **162**, 162, *163*
Ar/Ar dating 1, *54*, 307, 324
  age spectra 237
  analysis methods 265–266, 283
  feldspar ages 319
  hornblende 261–286
  phlogopite ages 319
  Teletskoye graben 244, 246, *247*
  see also biotite Ar/Ar dating; hornblende Ar/Ar data; muscovite Ar/Ar ages
Ardennes
  AFT data 167–179, *169*, *172*, *174*
  AFT interpretation 170–173
  cooling episodes 177
  crustal structure *176*
  differential exhumation *176*, 177
  exhumation events 167, 173, 177
  late-post Variscan evolution 167–179
  regional geology 167–168
  sampling methods 168–170
  structural controls 175–177
  structural units *169*
  tectono-thermal evolution 167
  thermal history modelling 173–175
  thermal regimes 175–177
  time-temperature histories 173, *174*, 175, 177
Argon ratio *see* Ar/Ar; K/Ar dating
Atlantic passive margin, denudation 287–306
Atotsugawa fault 332
  fission track ages and sample distance 333, *334*
  fission track analysis 331–337
  geological map *333*
  samples 331–332, *333*
Australia, AFT data *211*
Australian passive margin 47
  AHe & AFT 49
  landscape evolution 49–54
  thermal history *52*
  zircon data **51**
automatic counting
  background correction 31, *32*
  benefits 35–36
  manual comparison *35*
  problems 26
  procedure 31–35
  threshold and segmentation 31–32, *33*
  track overlaps 32–34

basin analysis 11–12
Be *see* beryllium
bedrock apatite
  fission track ages Alpine belt 141–143, *142*
  fission track ages Po Delta 147–148
  Nd/Nd ratios 61–62, *62*
Belgium 167–179
Bengal Delta 64, *65*, **66**, *67*
beryllium
  analysis methods 222–223
  concentration 217–235, **229**
  exposure ages 218
beryllium weathering rates 222, **229**
  Corsica 227–228
Bhutan *67*, 70
Bhutan river sands 61, *61*, *62*
biotite Ar/Ar ages 318, 324
biotite Ar/Ar data, Mozambican basement 266–270, **267–270**, **271**, *273*, 275–276
biotite Ar/Ar dating 244, *247*, 248
  Mozambican basement 261–286
biotite Rb–Sr cooling ages 277, *277*
Bohemian Massif 181–192, 182
  geotectonic map *183*
bulk earth values 58–59
Buluk Member Tuff 42

Canada, AFT data *210*, *211*
Central Andes
  AFT data 309, **310**
  deformation phases 307
  digital elevation map *308*, *309*
  erosion history 307
  exhumation pattern 307–316
  geodynamic setting 308–309
  geological map *309*
  morphotectonic map *308*
  tectonic framework 308
  uplift history 307
  ZFT data 309, **310**
chemical etching 1, 37–46
  effect on LA-ICP-MS 37–46, **39**
chemical weathering 220, 327
chondritic uniform reservoir (CHUR) 59, *59*
chronometry, multi method 237–259
climate, and denudation rate 218–220
closure temperature 4, 100, 118, 314, 331
  cross over 11, 194
  depth 105
  neodymium 59
  titanite 276, 320
closure temperature isotherms 95–96, *96*
  model assumptions 87
coincidence mapping 25–36, 32, *33*
  aluminium coating 29–31, *30*
  method 27, *28*
  operator review 34–35, *34*
  results 35
  sample preparation 27–28
convective heat transfer 87–98, 121, 122
Corsica 217–235
  AFT data **231**, *232*
  beryllium weathering rates 227–228
  cross sections *230*
  geomorphic evidence 230, 233
  glacier extension *221*
  lateral moraines 223, *223*, *226*, *227*
  lithologies *219*
  palaeosurfaces 220, *233*
  precipitation *219*, *230*, 231–233
  regional setting 218–220, *218*
  rock castle *233*
  sampling locations *219*, *221*
  weathering evidence 220–222
cosmogenic nuclide analysis 217
cratons 11
crossover ages 11, 194
crustal processes, shallow 47–56
crustal thickness, gravity based crustal model 319

Dabie Shan, China 100, *101*
denudation 5–6, *7*, 10–11, 181, 261, 298–299, 299–301, *300*, 327
  Atlantic passive margin 287–306
  Eastern Alps 128, 131
  mantle upwelling 303
  mechanisms 302–303
  Mozambican basement 282–283
  tectonically induced 303
denudation induced cooling 47
denudation rate 92, *93*, 94, 217, 302
  and climate 218–220
deposition *9*
detrital apatites *65*, 141–152, 153
  Nd isotopic composition 61, *62*
  Po Delta **147**
detrital fission track ages *65*, 144–147
  dating method 144–145
detrital fission track data, advantages 143
detrital geochronology *126*
  techniques 141
detrital sediments, apatite Nd isotopic composition 61, *62*
detrital thermochronology 8–9, 57–72, 128, 135
digital image acquisition 26–27
double dissolution 204
Dpar *see* etch pit diameter
Dronning Maud Land
  differential cooling 326
  fission track studies 320–322
  Jurassic magmatism 319
  metamorphism 318
  palaeosurfaces 317
  rift related tectonics 326
  sample elevations 320
  tectonic activity 317
  topography 320
Durango apatite 49, 159, 296, 322

East African margin 261, 265
East Antarctica
  AFT ages 325–326
  age-elevation *323*
  basement development 318
  Cenozoic unroofing 327
  continental margin development 326–327
  cooling history 322, *324*
  denudation 319
  differential cooling 327
  fission track results **321**, *323*
  geological setting 317–319
  Gjelsvikfjella 317–330
  glacial erosion 327
  intracontinental rifting 325
  Jurassic magmatism 326
  offshore sedimentary record 319, 327
  outcrop map *318*
  palaeosurfaces 319, 322
  palaeotectonic reconstruction *325*
  palaeotemperatures 327
  post Pan-African cooling 322–324
  sedimentary sequence 318–319, 325–326
  tectono-thermal history 322–327
  TFT ages 322, 325
  U–Pb ages 318
  vitrinite reflectance 322
Eastern Alps
  Austroalpine unit 126–127, *127*, 128, 134
  denudation history 128, 131
  fission track age distribution 128
  fission track ages **130**
  fission track litho-terrain 133–135
  geology 126–127, *129*
  palaeogeography 131
  palaeogeology 135, *136*
  pebble population dating 125–140, **130**
  Penninic units 126–127, *127*, 128, 134
  provenance studies 125–140
erosion 65, 141–152
  climate driven 302
  human activity 224–225
  patterns 141, 148–149, 149
erosion rates 87, **229**
  long term 228–231, 233
  and precipitation 231–233
  short term 148, 230–231, 233
  and tectonic activity 217, 231
erosion-transport interval 9
etch pit diameter 73, 75, *76*, 80–83, *82*, 113, 156, *161*, 164, 184, 203, 296, *296*, 309, 320, 322
etching 319
  bulk etch rate 84
  conditions 322
  degrees of 80–83, *83*, 84
  non-quantitative 74, 76–84, *76*, **80**
  quantitative 74, 76–84, *78–79*, **80**
  time 80, 83–84
  track density 84
  see also chemical etching
etching techniques 73–85, 74–76, **75**, 168
  AFT 73–85, **77**, *78–79*
  calibration 73–85
  comparison of 73–85, **75**, **80**, *81*, *82*
  data variation 76–79
  nitric acid 74
exhumation 5–6, *9*, 10, 65
  cooling 175, 177
  Hida metamorphic belt 334
  partial annealing zone *6*
  passive margins 49–53
  steady state 307–316
exhumation rate 8, 57, 99, 108, 111, 112, 116, 121–122, 122, *122*, 123, 143, 149, 313
  Gotthard transect, Switzerland *122*
  Peru 314
  rapid 282
  topography reconstruction 105–107, *105*
exposure age 222
  glacial boulders *223*, 224–227, **224**, *226*, *227*
extensional setting 94
external detector method 164, 193, 209, *209*, *210*, 211, *211*, *212*, 223, 332
external detectors 80–83, *82*, 168, 320

Fast Fourier Transform 102
fault displacement, frictional heat 331, 336
fault plane, local AFT age 336
fault strength 331, 336

fault systems 175–177
activation history 331–337
thermal anomalies 331–337
fault zone 87–98, 91, 120–121
conductivity 92, *93*, 94
displacement *120*, 121, 331, 336
thermal anomaly 336
fault-thrust system 181
faults 299
differential cooling 317
isotherm perturbation 121
South Africa 299
Fennoscandia Shield
AFT thermochronology 193–216
denudation history 193
finite difference model *101*
Finland 193–216
AFT age and AHe age *202*
AFT age and chlorine content *201*
AFT ages *196*, 213
AFT data 76, **195**, 200–203, *203*, *209*, *211*, *212*
analytical methods 194–200
apatite U–Th–Sm/He data **197–199**, 203–207, *205*, 213
data interpretation 202–213
sample locations *196*
samples 194–195
study area *196*
thermal history modelling 203, 207
time-temperature inversion models *208*
Fish Canyon Tuff 42, 49, **50**, *51*, 322
fission decay constant 2
fission track 1, 37
age distribution 128, 143
age types 5
analysis error 25
analysis methods 1–2, 137, 319–320
automated counting 25–36
density 1
discrimination 25
evolution of method 2–3
grain age 147–148, **147**
grain age distribution 2, 141, 143, 144, *145*, *146*
isotherm perturbation 111–124
litho-terrain 126
modelling 89–90
multi-kinetic annealing model 5, 84, 207, *208*
radiation enhanced annealing 194
source terrain 125
spontaneous/induced track ratio 2, 211, *212*
surface parallel confined tracks 2
unimodal length distribution 331
fission track age standards 322
fission track ages 141–152
palaeogeology 135, *136*
fission track data
bedrock vs. detrital 143–144
convective heat transfer 87–98
interpretation 87–98
fission track length 1, 79, 228, *232*
distribution *68*, 89, *160*, 184, *294*, 334
histograms *188*
see also mean track length
fluid flow 87, 336
two-dimensional 88
fluorapatite, PAZ 3
fluorine vs. AFT ages *161*
Fourier series 99
France 167–179

geomorphic evidence 217
geothermal fluids 336
geothermal gradient 99, 100, 103–104, *104*, 108, 112, 117, 298–299, 307
rift margins 281
steady state 87
topography reconstruction 105, *105*
Germany 181–192
glacial boulders
dating 223–227, *223*
exposure age *223*, 224–227, **224**, *226*, *227*
weathering girdles *225*, *226*, *228*
glacial erosion 327
glaciation *221*
Gondwana
break up 325, 326
rift basin 317–330
rift evolution 327
rifting onset 261–286
Gotthard transect, Switzerland 111–124
AFT data *114*, **115**
age-elevation relationship *116*
exhumation rates *122*
geology 112, *113*, *114*, 117, **118**
isotherm modelling 116–120, *119*
modelling methodology 112–113
samples *114*
granite
cooling history *334*
emplacement age 331

He *see* helium
heat flow 203, 213, 229, 281
heat flux 93, *93*, 94
heat production 116, 117–118, 118, **118**, 120, 121, 203, 213
rate 122, 123
heat transfer 117, 336
time dependent 88
see also advective heat transfer; convective heat transfer; heat flow; heat flux; thermal conductivity
helium 3
loss and recoil correction 49
partial retention zone 206
production-diffusion model 4
retentivity 194, 204, 206
see also apatite U–Th–Sm/He; U–Th/He dating; zircon U–Th/He
Hida metamorphic belt 331–337
exhumation history 334
fission track analysis **335**
geology 332
regional uplift and erosion 334
Himalayas 57
AFT data *63*, *64*, **66**, *67*
ancient sediments 65–69

exhumation history 65
geological map *58*
neodymium data *60*, *61*, 62, *63*, *64*, *67*
recent sediment record 64–65
regional geology 60–61
whole rock data *60*, *61*, *69*
see also Bengal Delta; Bhutan river sands
hornblende Ar/Ar data 266, **267–270**, 270–275, **271**, *272*, 275–276
hydraulic head difference 92–93, *93*, 94
hydrothermal fluid flow 87

Iberian Range 153–166
ICP-MS *see* inductively coupled plasma mass spectroscopy
image acquisition system 28–31
hardware 28
image capture sequence 28–29, *29*
image analysis technique 25, 26
inductively coupled plasma mass spectroscopy (ICP-MS) 193, 209, *209*, *210*, 211, *211*
see also laser ablation inductively coupled plasma mass spectrometry (LA-ICP-MS)
internal pair method 307
intracontinental rift 325
basin 265, 317–330
evolution 155, 326
isostatic uplift 280
isotherm
advective heat transfer 120–121
amplitude 103, *103*
calculated *104*, *106*, 108–110
factors affecting 103–104
see also palaeoisotherm
isotherm modelling *117*
AFT data 113–116
errors 101, *102*, 108–110
Gotthard transect, Switzerland 116–120, *119*
shape 113–116, 116–120
three dimensional *119*
two dimensional *120*
isotherm perturbation 111–124
boundary conditions 112
correction 112
faults 121
magnitude 112
topography 111

Japan 331–337

K/Ar dating 1, 313
ages 318, 324

landscape evolution 49–54, 331, 334
landscape evolution modelling 8–9
landscape evolution studies 10–11
laser ablation inductively coupled plasma mass spectrometry (LA-ICP-MS) 3, 26, 36, 37–46, 57, 60, 204, *205*
apatite results 40–41
isotope concentrations **44**
methods 38–40
operating conditions *40*
see also inductively coupled plasma mass spectroscopy (ICP-MS)
laser pit size *40*
lead *see* U/Pb ages, zircon SHRIMP U–Pb dating
Lusatian Block
AFT data 181–192, 184, **185–186**, *187*, *188*
AFT interpretation 184–187
annealing model *189*
denudation history 181–192
geological evolution 181–182
geological setting 182–184
late Cretaceous denudation 187–190
late Palaeogene denudation 190
mean track length 184, *187*, *188*
sample depth 184, *187*
study area *182*
tectonic activity 181, 183–184, 190–191
thermal history 181–192, *189*
thermal history modelling 184–187, 187–190, *189*
track length distribution 184, *188*

mass spectrometry *see* accelerator mass spectrometry; inductively coupled plasma mass spectroscopy (ICP-MS); laser ablation inductively coupled plasma mass spectrometry (LA-ICP-MS); secondary ion mass spectrometer (SIMS)
mean track length 156, 184, *188*, 193, *209*, *210*, *211*, 248, **249**, *250–251*, 320, 322
and age 309, *311*, *323*
and elevation *323*
and uranium content 207–209, 213
metamorphism 318, 334
mica detectors 320
Mozambican basement
Ar/Ar interpretation 270–275, *277*
Axial Granulite Complex 264, 278
biotite Ar/Ar data 266–270, **267–270**, **271**, *273*, 275–276
cooling rates 275–276
denudation history 279–282, 282–283
early Palaeozoic cooling 278–279
geological setting 264–265
hornblende Ar/Ar data 266, **267–270**, 270–275, **271**, *272*, 275–276
igneous activity 278
inferred rift geometry *281*
late Palaeozoic cooling 279–282
litho-tectonic units *263*
Lurio Belt foreland 264–265
metamorphism 278, 282
multistage cooling history 282–283
pressure-temperature estimates 282
rift related cooling 279–280
sample locations *263*, *276*
shear zone 282
spatial TFT age pattern *281*
TFT ages *276*
TFT analysis 270, **274**, *275*, 276–278
thermal history 278–282
thermo-tectonic evolution 261–286
time-temperature paths *277*
topographic map *276*
Mozambican rift structure *262*
multi-kinetic annealing model 5, 84, 207, *208*
muscovite Ar/Ar ages 244, 246, *247*, 318, 324

Nd *see* neodymium
neodymium
  bulk rock database for Himalayas 62
  closure temperature 59
  measurement 57–72
  whole rock ratios 59–60, *59*, *60*, *61*
neodymium data 57
  and AFT 69, 70
  applications 61–69
  bedrock apatite ratios 61–62
  bulk sediment 59
  Himalayas *63*, *64*, *67*
  interpretation of 59
  methodology 60
  whole rock distortion 69
neutron activation technique 37

optical luminescence dating (OSL) 63, *65*, **66**
orogenic belts
  analysis 10
  research 125
orogenic source area 141

palaeodrainage reconstruction 135
palaeogeodynamics 137
palaeogeography, Eastern Alps 131
palaeogeological reconstruction 135–137
palaeogeothermal gradient 7, 11, 153, 314, 325, 327
  Sierra de Cameros 162–164
  South Africa *300*
palaeoisotherm 99
  depth 100
  uncertainties 100–103
palaeomagnetic studies 307
palaeosurface 220–222, *220*, 317, *318*, 326
  Corsica *233*
  East Antarctica 319, 322
  tectono-thermal evolution 317
  weathering 220–222
palaeotopography
  evolution 120
  reconstruction 99–110
Pan-African metamorphism 261–286
Pangaea, reconstruction *325*
partial annealing, degree of 163, 203
partial annealing zone 3, 5, 91, 143, *160*, 167, 184, 193, 203, 334
  apatite 170, 177, 229, 248, *250–251*, 252, *297*, 307, 311, *312*, 313, 320
  exhumed *6*
  residence in 213
  titanite 3, 278, 320
  zircon 3, 307, 313
passive margins 47, 49–54, **51**, *52*, 217, 287–306
  analysis 10–11
  evolution 287
  exhumation 49–53
  geomorphology 287
  post break up denudation 288
PAZ *see* partial annealing zone
pebble population dating 125–140, *126*
  exhumation history 125–126
  geology 128–129
  gneiss 128–133
  interpretation 133–135
  provenance 135
  results 129–131
  samples 129, 133, *134*
  sampling strategy 137
  source interpretation 132–133, 135
permeability 94
Peru *see* Central Andes; SE Peru
plutonic systems, cooling of 53–54
Po Basin 141
Po Delta 141, *142*
  bedrock fission track ages 147–148
  dating method 144–145
  detrital apatites **147**
  fission track results 145–147
  grain age distributions *146*, 147–148, **147**
  probability density plots *146*
  sampling 144–145
  sediment source *144*
  single grain ages *145*
Po Plain, source areas 147–148, *147*
potassium *see* K/Ar dating
precipitation
  Corsica *219*, *230*, 231–233
  and denudation rates 217
  and erosion rates 231–233
precision of topography reconstruction (PTR) 100–103, 101–103, *103*, 105
probability density plots 2
  Po Delta *146*
  sediment contribution 148
provenance studies 8–10, 58, 69, 125–140, 141, 149
  indicators 9
  source interpretation 132–133
pseudotachylyte 331

radiation damage trails *see* fission track
radiation enhanced annealing 193–216, 206–207, 207–213
radiometric dating *334*
random effects model 2
Rb–Sr biotite cooling ages 277, *277*
reflected light images 25, 26, 27, *30*, *33*
  interference bands *30*, 31
  problems 29–31
regolith downwearing 220–222, 223–227, **228**
  rates 230
rift basins 265
rift margins 282, 283
  evolution models 279
  geothermal gradients 281
  uplift model 279–282
rifting, time-temperature paths *53*
rock properties 112
root mean square 106, *106*
rubidium *see* Rb–Sr biotite cooling ages

samarium *see* Sm–Nd geochemistry, U–Th–Sm/He dating, apatite U–Th–Sm/He
samples
  density 103–104
  horizontal profiles 112
  preparation methods and modelling 84

size determination 104, 108
vertical profiles 6–7, 95, 105, *106*, 307
Scotland
Hebridean Igneous Province **51**, 53, 54, *54*
mafic plutons 47
Muck Tuff 49, **51**
zircon data **51**
SE Peru
AFT age vs altitude *311*, 313
AFT data **310**
cooling phases 314
deformation phases 313, 314
Eocene-Oligocene 313
exhumation pattern 307–316, 314–315
exhumation rate 313, 314
Miocene 313–314
thermal history 311–313
thermal modelling 307
time-temperature modelling *312*, 313
uplift 314–315
Western Cordillera 314–315
ZFT age vs altitude *311*, 313
ZFT data **310**
secondary ion mass spectrometer (SIMS) 318, 324
sediment provenance 61
see also provenance studies
sediment reworking *144*
sediment storage *144*
sedimentary basin analysis 11, *12*
sedimentary overburden 326
sensitivity analysis 99
SHRIMP sensitive high resolution ion microprobe *see* zircon SHRIMP U–Pb dating
Siberian Altai Mountains, thermo-tectonic evolution 237–259
Sierra de Cameros
AFT age populations *163*
AFT data **157–158**, *161*, 162
apatite U–Th/He ages 156, *161*, **162**, 162, *163*
exhumation 153–166
fission track interpretation 159, 162
fission track length distribution *160*
fission track results 156
geological framework 155–156
geological map *154*
palaeogeothermal gradient 159, 162
sample locations *159*, *161*
tectonic models 153–155, *155*, 162
thermal events *163*
thermal modelling 156–159, *160*
ZFT data **157–158**, 159, 162
Sierra Nevada, California
palaeotopography 107–108, *107*
U–Th/He data 99
SIMS *see* secondary ion mass spectrometer (SIMS)
single grain fission track age distributions 3, *3*, *131*
Sm–Nd geochemistry 58–60
software 5
AFTSolve 90, 159, 170, 207, *250–251*, 309
BINOMFIT 164, 170
FEFLOW 87, 88
HeFTy 322
TERRA 117, *118*
soil downwearing *see* regolith downwearing
soil etching 226–227
South Africa
AFT analysis 287–306
AFT results 291–295, **293**, *294*, *295*, 296–298
Cenozoic denudation 301–302
cooling phases 287, 298
denudation history 287, 298–302
denudation mechanisms 299–301, *300*, 302–303
fault structures 291, *292*, 299
geological framework 288–290
geological map *289*
geomorphology 290–291
mid Cretaceous denudation 299, 302–303
palaeogeothermal gradients *300*
post magmatism thermal relaxation 298
relief map *290*
sample locations *289*, *290*, 292, *294*
seismic data *301*
tectonic evolution 287, 299
thermal modelling 295–298, *297*
unconformities 301, *301*
uplift 299–301, *300*
Spain 153–166
AFT data 76
strontium *see* Rb–Sr biotite cooling ages
Switzerland 111–124

tectonic activity
and erosion rates 217, 231
interpretation 95, *95*
processes 12
South Africa 299
uplift 280
tectonic evolution 237
palaeosurfaces 317
tectonic models 153
Sierra de Cameros 162
Teletskoye graben 237–259
AFT data 244–245, 248–254, **249**
AFT interpretation 252–254
Ar/Ar dating 244, 246, *247*
basement structure 239–242
Cenozoic reactivation 243, 252
features *239*
formation of 243
geodynamics 239–242
geological map *240*
lithology **241**
Mesozoic reactivation 242, 252
samples **241**, 244
tectonic evolution 254–255, *255*
thermal history 254–255, *255*
thermal history model 248, *250–251*
time-temperature paths *250–251*
zircon SHRIMP U–Pb dating 244, 245–246, *245*
temperature-time paths *see* time-temperature paths
TFT *see* titanite fission track
Th *see* thorium
thermal annealing behaviour 89
thermal anomalies
fault activity 331
fluid circulation 87
thermal conductivity 116, 117, 118, **118**, 119, 121, 122, 203

thermal field
    perturbation 100
    steady state 100
thermal gradient *see* geothermal gradient
thermal history 10, 53, 173–175, 181, 187–190, 203, 207, 237, 331
thermal modelling 73, *90*, 112, 153, 181, 296–298
    Ardennes 173–175
    boundary conditions 88
    comparisons *118*
    constraints 173
    convective heat transfer *89*
    Finland 203, 207
    fluid flow 88–90
    FT ages and distance *93*
    Gotthard transect 111–124
    limitations 94
    Lusatian Block 187–190
    model parameters *91*, 92–93
    parameters 100, 118–120, **119**
    quantitative 331
    reference model 90–91, *91*, *92*, *95*
    SE Peru 307
    Sierra de Cameros 156–159, *160*
    South Africa 295–298
    technique 295–296
thermal structure, upper crust 8, 87
thermochronology 1–23, *12*
    internal pair method 307
    low temperature 4–5, 9–12, 99–110
    modelling 4–5
    multi-method approaches 10, 11
    and topography 8
thorium 4
    concentration 202, 204, *205*
    microinclusions 204
    zonation 204
    see also apatite U–Th–Sm/He; U–Th/He thermochronology; U/Th ratios; zircon U–Th/He
time-temperature histories 73, 173
    Ardennes 173, *174*, 175, 177
time-temperature paths 87, 89, *91*, *93*, 228, 295, 322, *324*, 327
    envelopes 3, *160*, 184–187
    Finland *208*
    inverse modelling *208*, 317, 322, 325
    model input parameters 309–311
    modelling *53*, 167, 203, 229, *232*, 309–313, *312*
    modelling constraints 309–311, *312*, 322
    Mozambican basement *277*
    SE Peru 313
    Teletskoye graben *250–251*
titanite 1
    closure temperature 320
    partial annealing zone 3, 278, 320
titanite fission track
    age-elevation *275*, 276
    age-latitude *275*, *279*, 280
    ages 324
    analysis 320–322
    closure temperature 276
    Gjelsvikfjella 317–330
    methods 266
    Mozambican basement 261–286, 270, **274**, *275*, 276–278
topography 99, 116, 117, 118, 123
    amplitude 103, *103*, 112
    apatite thermochronology 8
    isotherm perturbation 112
    three dimensional 122
    wavelength 103–104, *104*, 112
topography reconstruction *101*, 107, *107*, 108
    exhumation rate 105–107, *105*
    geothermal gradient 105, *105*
    see also precision of topography reconstruction (PTR)
track length *see* fission track length
transmitted light images 25, 26, 27, *33*
trapping model 194, 206

U–Th/He thermochronology 1, 3–4, 118, 287, 317, 322
    Sierra Nevada, California 99
U/Pb ages *54*, 237
    East Antarctica 318
U/Th ratios 42–45, **44**
unroofing 317
uplift 5–6, *7*, 65, 299–301
    isostatic 280
    regional 302
uranium 2, 3
    micro-inclusions 204
    ratios 42–45, **44**
    zonation 204
    see also apatite U–Th–Sm/He; U–Th/He thermochronology; U/Pb ages; U/Th ratios; zircon U–Pb SHRIMP; zircon U–Th/He
uranium concentration 37–46, 204, *205*
    analysis variation *40*
    effective 194, 206, *206*
    measurement method 38–40
    sequential measurement *41*
uranium content 193, 202, 207–209, 209–210, *209*, *210*, *211*
    and AFT age *212*, 213
    apatite *41*
    LA-ICP-MS **39**
    and mean track length 213
    zircon *42*

vitrinite reflectance 11, *12*, 317, 325
    East Antarctica 322

weathering 217–235
    see also chemical weathering
weathering girdles 224–225, *225*, 226, *226*, *228*
weathering rates 217
    beryllium analysis 222
    and climate effect 227–228
    and lithology 227–228, 230
    measurement 222
whole rock ages 319

zeta calibration approach 2, 223, 320, 332
ZHe *see* zircon U–Th/He

zircon 1
  closure temperature 47
  euhedral shape 132, *132*
  LA-ICP-MS, results 41–42
  partial annealing zone 3, 307, 313
  uranium content *42*
zircon age standards **43**, 49, **51**
zircon ages *127*, **130**, 131–132
zircon fission track 47, 143–144, 307
  analytical methods 164
zircon fission track data 156
  probability density plots 133
  radial plots *159*, *161*
  Sierra de Cameros **157–158**, 159, 162
zircon SHRIMP U–Pb dating 237, 282, 318, 324
  Teletskoye graben 244
zircon U–Th/He 47–56, *51*, *54*, 153
  age standards 49, *51*
  analytical techniques 48–49
  applications 47, 49–54